Food Engineering

Students entering the food processing stream need to acquire knowledge of concepts and analytical skills together with knowledge of their applications. ***Food Engineering: Principles and Practices*** explains the different unit operations in food processing with an emphasis on the principles of food engineering as well as the different types of equipment used for the purpose.

An approach in which propounding concepts and theory is immediately followed by numerical examples makes this book unique among food engineering textbooks. The examples, which are thoroughly explicated, have been taken, in general, from different competitive examinations and have been selected with practical applications for a better appreciation and understanding by the students. In the case of equipment, the constructional and operational features are discussed along with the specialty features of these types of equipment for better understanding of their applications.

Key Features:

- Merges a presentation of food engineering fundamentals with a discussion of unit operations and food processing equipment;
- Reviews concepts comprehensively with suitable illustrations and problems;
- Provides an adequate number of examples with different levels of difficulty to give ample practice to students;
- Explains equipment units in three broad subheadings: construction and operation, salient features, and applications.

This book is written as a textbook for students of food processing and food technology. Therefore, the book is meant for undergraduate and graduate students pursuing food processing and food technology courses. It also serves as a reference book for shop floor professionals and food processing consultants.

Food Engineering

Principles and Practices

Sanjaya K. Dash
Pitam Chandra
Abhijit Kar

CRC Press
Taylor & Francis Group
Boca Raton London New York

CRC Press is an imprint of the
Taylor & Francis Group, an **informa** business

Designed cover image: ©Shutterstock

First edition published 2024
by CRC Press
6000 Broken Sound Parkway NW, Suite 300, Boca Raton, FL 33487-2742

and by CRC Press
4 Park Square, Milton Park, Abingdon, Oxon, OX14 4RN
CRC Press is an imprint of Taylor & Francis Group, LLC

Library of Congress Cataloging-in-Publication Data

Names: Dash, Sanjaya K., author. | Chandra, Pitam, author. | Kar, Abhijit, author.
Title: Food engineering : principles and practices / by Sanjaya K. Dash, Pitam Chandra, Abhijit Kar.
Description: First edition. | Boca Raton : CRC Press, [2023] | Includes bibliographical references and index.
Identifiers: LCCN 2022032316 (print) | LCCN 2022032317 (ebook) | ISBN 9781032258003 (hbk) | ISBN 9781032231853 (pbk) | ISBN 9781003285076 (ebk)
Subjects: LCSH: Food science. | Food industry and trade.
Classification: LCC TP370 .D375 2023 (print) | LCC TP370 (ebook) | DDC 664—dc23/eng/20220901
LC record available at https://lccn.loc.gov/2022032316
LC ebook record available at https://lccn.loc.gov/2022032317

ISBN: 9781032258003 (hbk)
ISBN: 9781032231853 (pbk)
ISBN: 9781003285076 (ebk)

DOI: 10.1201/9781003285076

Typeset in Times
by Deanta Global Publishing Services, Chennai, India

Contents

Section B   Unit Operations

Section B1   Primary Processing, Separation, Size Reduction and Mixing

Section B2   Processing Involving Application of Heat

Section B3   Processing Involving Removal of Heat or Moisture

Section B4   Processing at/ near Ambient Temperature

Section B5   Food Packaging and Material Handling

Section C Food Quality, Safety and Waste Disposal

Preface

Food is a multi-dimensional topic requiring considerations related to the sources of raw materials, methods and extent of processing, adequacy for meeting the nutritional needs of a wide range of consumers, the health and wellness of consumers, and sustainability of mankind and ecology. Food is not only the fuel needed to drive human activity, but also the nourishment for the human mind, maintaining a happy balance among all living as well as non-living entities on this planet, our mother Earth. People with varied backgrounds and education join the food domain, seeking to understand and contribute to improvements in production, processing, trade, policy frameworks, and fulfilment of individual and societal needs. The stakeholders crisscross the food domain in their bids to succeed, and they need exposure to topics as varied as science, engineering, nutrition, health, wellness, environment, ecology, and spirituality.

This process of learning has become more complicated due to the paradigms created by the disruptive nature of pandemics. There is going to be increased emphasis on self-learning. Since education in the physical mode cannot be taken for granted, study materials, including textbooks, tutorials, and training modules, need to be appropriate for self-learning without requiring too much cross-referencing.

We conceived this book as a consequence of our teaching experience, namely, that students found it convenient to learn from a book if it provided the concepts and the applications side by side in an inclusive manner. In addition, such an approach will also benefit those stakeholders from allied subjects who cannot be expected to be proficient in the basics of engineering. Thus, the book aims to constitute a blend of engineering principles and practices as applied to food in a comprehensive manner. Although the book is intended to serve as a textbook of food engineering for an undergraduate program in food processing/technology, it can also very well serve as a helpful guide for other food disciplines, such as food science, food trade, food cold chain, food nutrition, etc.

The book is divided into two parts: the first part presents engineering principles as they apply to the food industry and the second part builds applications on the first part. Once a student has gained proficiency in the first part, the second part serves as a logical extension of the first. The topics that constitute the first part include dimensions and units, material and energy balances, reaction kinetics, psychrometry, fluid properties and flows, heat and mass transfer, refrigeration and air conditioning, and water activity and thermo-microbiology. We are convinced that the topics presented in the first part serve as a strong foundation for a learner to imbibe the knowledge presented in the second part on unit operations.

Each chapter in the first part begins with the development of theory and equations, relevant material properties, and applications. A chapter in the second part begins with relating to the relevant section(s) in the first part, augmentation with additional theory if needed, and presentation in focus of the details of the unit operation. The presentation for a specific unit operation will generally include the description of the equipment related to the unit operation. We have, as far as possible, tried to guide the reader at the end of each chapter in recapitulating the essence of the chapter and the main points of learning. The second part has been expanded to include today's concerns and developments in food biotechnology, food quality and safety, shelf-life determination, and waste treatment.

We acknowledge the authors, publishers, and manufacturers for their direct or indirect contributions through their knowledge, research, and development of materials used in this book. We acknowledge the many papers, books, and conference proceedings, which have been consulted in the preparation of this book.

The authors wish to express their gratitude to M/s Arvinda Blenders, Ahmedabad; M/s DP Pulverizer Industries, Mumbai; M/s Shiva Engineers, Pune; and M/s Bajaj Processpack Limited, Noida, India for their generosity to permit us to use the pictures of their products for value addition to the material presented in the book.

Gratitude is also due to M/s Taylor and Francis, M/s Elsevier, M/s Academic Press, M/s Copyright Clearance Center, M/s F&B News, M/s Kitab Mahal, New Delhi, M/s *ARCC* Journals, M/s Chemical World, M/s MDPI, M/s Springer Nature, and the American Chemical Society.

We are thankful to M/s Taylor and Francis and its very helpful officers Stephen Zollo, Laura Piedrahita, Bharath Selvamani, and Marsha Hecht for the pleasant experience of book publishing.

The authors are deeply indebted to their families for creating conducive ecosystems leading to the preparation of the book. We thank our students who took our courses for their feedback and the institutions for encouraging us to undertake this mission.

We hope that the book, as intended, will meet the expectations of students and teachers in food engineering/technology. In addition, we would like this book to assist stakeholders from allied disciples, including food science, nutrition, trade, and environment, in meeting their professional goals. Post the COVID-19 pandemic, we expect this book to be a potent self-study aid. It goes without saying that the concerted efforts of the authors over months entailed the loss of precious family time. However, this hardship will have been worthwhile if the book delivers on the teaching and learning front.

Sanjaya K. Dash
Pitam Chandra
Abhijit Kar

About the Authors

Sanjaya Kumar Dash is currently Dean, College of Agricultural Engineering and Technology, Odisha University of Agriculture and Technology, Bhubaneswar. He earned his doctorate from ICAR – Indian Agricultural Research Institute, New Delhi in 1999 and his Master of Engineering from Rajasthan Agricultural University, Udaipur in 1989. He is a 1987 graduate in Agricultural Engineering from Odisha University of Agriculture and Technology, Bhubaneswar. Dr Dash has completed special programs overseas at Hebrew University, Jerusalem; University of Göttingen, Germany; Ohio State University; Michigan State University; University of Saskatchewan; and Wageningen University. The Samanta Chandra Sekhar Award of the Government of Odisha and the Best Teacher Award from Odisha University of Agriculture and Technology are among the several awards and recognitions that he has received.

Pitam Chandra retired as Director, ICAR-Central Institute of Agricultural Engineering, Bhopal in 2014. He was Assistant Director General (Proc. Engg) at the Indian Council of Agricultural Research, New Delhi during 2003–2009. Pitam received his doctorate degree from Cornell University, USA, in 1979 and MS degree from University of Manitoba, Canada, in 1976. His 1974 Bachelor of Technology (Hon) in Agricultural Engineering is from GB Pant University of Agriculture and Technology, Pantnagar. Pitam worked as a post-doctoral Research Associate at North Carolina State University, Raleigh, USA, during 1979–1980 before accepting the position of Associate Professor at Junagadh Agricultural University, Junagadh (erstwhile Gujarat Agricultural University). Pitam

has been successful in introducing appropriate greenhouse technology in India, leading to intensification of horticultural production in the country and export of flowers from India. It was for his efforts in greenhouse technology that he received the Rafi Ahmed Kidwai Award from the Indian Council of Agricultural Research, New Delhi. Pitam continues to teach after retirement, first at the National Institute of Food Technology Entrepreneurship and Management (2014–2017) and now at Sharda University.

Abhijit Kar is currently Director of the ICAR – National Institute of Secondary Agriculture, Ranchi, Jharkhand, India. He earlier served as Principal Scientist in Division of Food Science and Postharvest Technology at the Indian Agricultural Research Institute, New Delhi. He received Ph.D. from ICAR in the year 2002. His Master of Technology degree is from G.B. Pant University of Agriculture and Technology, Pantnagar. He graduated from Odisha University of Agriculture and Technology, Bhubaneswar. He pursued Post-Doctoral Research on Non-destructive Quality Evaluation using Imaging Techniques at Purdue University, Indiana, USA, in 2010. Abhijit is a certified auditor/lead auditor of Food Safety Management Systems in addition to being a certified GLOBALGAP and intellectual property professional. Abhijit is a recipient of several prestigious awards and recognitions, such as the Lal Bahadur Shastri Outstanding Young Scientist Award, the Jawaharlal Nehru Award for outstanding post-graduate agricultural research, and the J.C. Anand Memorial Gold Medal for outstanding contribution in the field of Post-Harvest Technology.

1

Introduction

1.1 Importance of Food Processing

Food is the basic need for the existence of humanity. Therefore, food has been discussed, researched, and professed intensively over thousands of years. Although mankind has ensured its survival through growing and storage of food, the progress of human civilization required undertaking such non-food activities as housing, transport, clothing, safety, defense, health, etc. History presents ample evidence that while agriculture had the major share in the economy of a society until a couple of hundred years ago, science and engineering-based developments occurred faster and the relative share of agriculture in the society's income kept decreasing. Globally, agriculture contributes only 4% to gross domestic product (GDP). This is not to say that agriculture is losing its importance. Agriculture is the source of raw materials for food, the essential requirement for humanity's survival. It indicates only that non-agricultural activities are being undertaken more intensively for all-around societal development, possibly at the cost of urgent matters pertaining to agriculture and food.

There are some serious concerns related to agriculture and the food sector. There were 770 million people, about one-tenth of the global population, on the Earth who did not get enough food in the year 2020. At the same time, the world lost and wasted about one-third of the food that was produced by using the world's resources. It is safe to assume that more than enough food is being produced globally to meet the gross nutritional requirements of the population; the food supply chain is inefficient, leading to colossal quantitative losses and wastages (about 1.3 billion tons); as well as access to food is insufficient. It is evident that the issues of food loss and waste are essential for improving global food and nutritional security. Minimization of the losses and wastes will also contribute to meeting climate goals as well as to reducing stress on the environment. Contaminated and adulterated food affects the health of about one in ten people globally and, consequently, about 0.4 million people lose their lives each year. Food for more than two billion people remains deficient in essential vitamins and minerals. Another two billion people in the world are obese, even in developing countries, and the major cause is wrong dietary regimes.

There is a preponderance of rural populations moving to urban areas in search of a better life. It is estimated that about two-thirds of the world's population will be living in cities by 2050. Keeping in view the characteristics of urban and rural areas, present food production and processing activities will need a paradigm shift. Unlike the present, certain high moisture foods will need to be grown and processed in urban areas to make the whole affair more efficient and cost-effective. Greater emphasis will need to be placed on

a circular agricultural economy to create jobs, better food, higher income, and a cleaner environment.

Today, societal developments have necessitated that the food be processed commercially and made available on demand to meet the requirements of consumers from different sections of society. The need for food varies with the age and physiology of the consumer, location, season, culture, ethics, profession, social status, religion, and many other factors. Therefore, the challenges for food processing today are in terms of customizing the needs of a large spectrum of consumers to ensure their wellness while ensuring sustainability.

More recent progress in the food processing sector relates to retaining nutritive value, flavor, and aesthetics of food after ensuring its safety aspects. Customization of individual diets for nutrition, health, and wellness is becoming increasingly important for the sustainability of individuals, societies, and the environment.

The practices of cleaning, sun drying, storing, and improving palatability of foods have been carried out for thousands of years. Experience taught man to develop the techniques of food preservation and cooking that have been improvised and scaled up. Some food-processing activities whose applications have a long history include drying, fermentation, pickling, smoking, and salting. Recent history documents the evolution of the practices of pasteurization and canning, which have become ubiquitous. Food processing has seen its fastest growth during the twentieth century, and it continues today, albeit with greater challenges. The two world wars had their flip sides in terms of their catalytic role in promoting mass manufacturing of canned foods and ready-to-eat package meals with longer shelf life. Space programs have brought food safety into sharper focus. Technologies such as evaporation, spray drying, freeze drying, and the use of preservatives have made it possible to increase the shelf lives of foods with additional interventions of packaging and appropriate storage environment. Ingredients such as artificial sweeteners, flavors, and colors help to create differentiated products appealing to different consumer groups. Microwave ovens, blenders, and other appliances in kitchens have helped to take the drudgery of cooking meals away. Innovation-led, large-scale food processing plants make it possible to quickly produce and package foods. Therefore, it is no wonder that ready-to-eat, ready-to-cook, baked, and many other similar food products meet the expectations of varying consumer profiles.

1.2 Food Engineering Applications

Engineering is the application of science for the benefit of mankind. While science lets us understand and deduce the

DOI: 10.1201/9781003285076-1

principles of nature, engineering utilizes these principles to develop gadgets and processes to be utilized by mankind. Depending on understanding the surroundings, man has strived to engineer solutions to his problems over millennia. Tools for growing food and for hunting, utensils and tools for cooking food, the practice of setting curd, and the practices of drying for preservation of food are some of the early engineering interventions in the food sector to make human life comfortable.

Food engineering is the application of engineering principles and processes for handling, preservation, and processing of food with a view to derive the maximum benefits sustainably. Although food science theoretically precedes food engineering, science and engineering work together simultaneously to find specific solutions for food-related problems. The physical manifestation of food science and food engineering is in the form of food technology. While science and engineering work behind the scenes, it is food technology in the form of a gadget, a process, or a product that benefits the user.

Food processing must not be seen in isolation. It is well understood that the quality of raw materials dictates the quality of processed products. Therefore, it is very necessary to have functional linkages with the producers of raw materials for food processing. The raw food materials must be produced using good agricultural practices responsibly. Obviously, food safety and environmental sustainability are at the core of good agricultural practices. Then, the processing needs to be carried out using appropriate food safety protocols and with due consideration given to workers' health and environmental sustainability. The job of a food processor does not end here. Instead, processed food must fully meet the standards of subsequent handling and consumption. Food engineers need to identify and undertake all such avenues where engineering could bridge the gap between food availability and its consumption. For example, why do consumers in developed countries end up wasting a significant portion of their groceries? These wastages lead to all-around losses of money, natural resources, and environment.

Commercialized food processing in the recent past has come in for criticism for reasons that include the health of the consumers and inefficient processing technologies. Fingers are pointed at processed foods for several present-day ailments. Also, the current food systems are seen to threaten people's health and the environment through generation of excessive wastes and pollutants. The input resources in food processing are not optimally utilized, thus leading to unsustainability. Today, when society is moving toward elimination of home cooking, commercially processed food must effectively substitute for home-cooked food, whereby, the food needs to be safe, satiating, nutritious, fulfilling religious and cultural expectations, affordable, and environmentally sustainable. Thus, food engineering today needs to assimilate all these needs of the consumer while developing food-processing solutions. Fortunately, knowledge in this domain is already abundant, existing in terms of recipes and customs of food from different regions globally. This knowledge will need to be integrated with contemporary science to arrive at the processes, devices, and products for today's consumers.

The food engineering domain includes the application of basics of food, physics, biochemistry, microbiology, and mathematics to such operations as heating, cooling, drying, cooking, size reduction, conveyance, pasteurization, sterilization, filtration, mixing, and extraction for appropriate process and equipment design, monitoring, and control. With the advancement of science and other branches of engineering, the domain of food engineering continues to grow. Moreover, food has both backward linkages to address input issues and forward linkages to deal with consumer satisfaction, health and wellness, by-products management, and environmental sustainability.

Food is perceived as a very important component of human wellness, which is far more than just nutrition and health. Food has a spiritual dimension as well. When a person is satisfied with his/her physical and mental wellness, the person endeavors to strive for the greater good of humanity. Thus, food is differentiated in terms of body requirements in relation to work, mental needs, and nourishment of soul. Food engineers need to work with all concerned to ensure that customized foods are made available to everyone using a correct mix of ingredients and processing technology.

1.3 About this Book

In view of the above, while focusing on the core contents of an engineering course, to be successful, a food engineer needs to internalize a comprehensive view of the food sector and of human health. In fact, an appreciation for the biological nature of food, its raw materials, and human nutrition imparts relevance to the profession of food engineering. Food engineers need to shoulder the challenges of increasing the efficiency of the food supply chain and enabling people to have adequate access to adequate food.

The food processing sector is growing fast, faster than agriculture. As a comparison, global agricultural production in 2019 was valued at about USD 4.0 trillion whereas the value of processed food was USD 5.9 trillion in the same year, and it is estimated that it will grow up to USD 7.7 trillion by 2026. There is still a large potential to tap as far as food processing is concerned, and, therefore, food processing will continue to outpace agriculture. More and more universities and institutions have begun offering educational programs in food processing and related disciplines. The course Unit Operations of Food Processing is taught as an integral part of most of these educational programs. It is hoped that the present effort will not only make the topics interesting and understandable, with adequate numerical examples and more emphasis on the details of the equipment, but also that students will look forward to the unfolding of wider food processing issues in the near future. This book explains concepts, equipment, and applications in such a manner so as to strengthen the capability of the student as an engineer in and a consultant to the food industry. The equipment units are explained in three broad subheadings, namely, construction and operation, salient features, and applications, which help to give a better understanding of the working principles and to impart a better comprehension of the differences between different types of equipment used for the same activity.

This effort has been made in view of the teaching experience of the authors, where it was realized that students entering the food-processing stream needed knowledge both of the concepts and computational and analytical skills and of their applications. Consequently, the propounding of concepts and theory is immediately followed by numerical examples. The approach of presenting the principles and numerical examples side by side is essentially to handhold the students in developing confidence to understand not only the theory, but also its application for problem solving. There have been several advances recently in terms of new techniques and equipment for solving food quality and safety-related concerns globally. The latest technological developments have been included in the book, thus updating its contents and enhancing its usefulness in empowering students to compete globally.

Section A

Basic Concepts and Principles

2

Dimensions and Units

The comparison of any physical quantity in terms of its numerical value is called *measurement*. The standard in which the quantities of similar nature are measured is called *unit*. Understanding the units and measurements is very important as it deals with the tool of comparison and in establishing the validity of any theory. For understanding the basic food processing operations and machineries, for comparing the different methods and equipment as well as for analyzing the economics of the systems, it is important to have a proper knowledge of the units and measurements.

A *dimension* is a measure of a physical variable (without numerical values), while a unit is a way to assign a number or measurement to that dimension. For example, length is a dimension, but it is measured in units of meters (m), centimeters or in feet or inches. The dimension of a physical quantity is more fundamental than some scale unit that is used to express the amount of that quantity. The study of the relationship between physical quantities with dimensions and units of measurement is called *dimensional analysis*. The knowledge of dimensions is important for the conversion of units as well as for understanding the relation between the units. It also helps in checking the accuracy of the equations.

A concise treatment of dimensions, units, measurement, and analysis relevant to food engineering is presented below.

2.1 Physical Quantity

Food engineering, like any other branch of engineering, deals with physical parameters that are measurable and quantifiable. A physical quantity has to have its magnitude and unit. Take for example, the length of an object that is measured as 10 meter. It means that the object's length is ten times the length unit that is 1 meter. For, convenience in writing, "meter" is abbreviated as "m" while writing as unit. Therefore, the value of the physical quantity is written as 10 m. Thus, a unit is first chosen for expressing a physical quantity and then the number indicates the multiples of that unit expressing the physical quantity. A physical quantity could either be a ratio (defined dimensionless magnitude), a scalar (defined magnitude no direction), or a vector (defined magnitude and direction).

A few examples of *dimensionless quantities* are friction factor, relative humidity, refractive index, and relative permeability. Complex products of simpler quantities, such as Reynolds number, defined as the $\dfrac{\rho.v.L}{\mu}$, where ρ is mass density, v is speed, L is length, and μ is dynamic viscosity, are also dimensionless. Number of molecules, number of thermally accessible states in statistical thermodynamics, and number of energy levels represent another class of dimensionless quantities. All of these counting quantities are taken to have the SI unit 'one.'

Mass of an object is a *scalar*; it has a defined magnitude and the associated unit. Since velocity has a defined direction and a magnitude, it is a *vector*. There is a fourth category of physical quantities, referred as *tensor*, which is a multi-dimensional vector (magnitude, direction, and the plane that the component acts on).

A physical quantity may be either a *fundamental* or a *derived quantity*. Fundamental quantities are then used to express derived quantities. There are seven fundamental quantities as given in Table 2.1.

All other physical quantities are called as derived quantities and expressed in terms of the fundamental quantities.

2.2 Dimensions

Properties of physical entities are classified in different qualitative terms. Duration of a process, such as evaporation, is expressed in terms of time and not in terms of mass or distance. These qualitative terms are better known as dimensions. As indicated above, there are seven fundamental quantities. These are the minimum number of quantities that have permitted the representation of all other quantities in their terms. The square brackets added to the letters show that the quantity used is a dimensional quantity. The dimensions of any other quantity will involve one or more of these basic dimensions. Dimensions of any given physical quantity are, thus, the product of the fundamental quantities raised to different powers. The four dimensions more frequently used in food processing are mass, length, time, and temperature.

Example 2.1 Find the dimensions of pressure.

SOLUTION:

Pressure is defined as force per unit area, i.e., $P = F/A$, where F is the force exerted on an object and A is the projected area perpendicular to the force F. Area A is the multiplication of two linear dimensions, say length and breadth ($L \times L = L^2$). Force acting on an object, F, is obtained by multiplying the acceleration term with the mass of the object. Rate of change of velocity of the object ($v/T = L.T^{-2}$) with time is termed as acceleration. Therefore, dimensionally,

$$P = F / A$$

$$= \left(MLT^{-2}L^{-2} \right)$$

$$= \left(ML^{-1}T^{-2} \right)$$

DOI: 10.1201/9781003285076-3

TABLE 2.1

Fundamental quantities

S. No.	Fundamental quantity	Fundamental dimension	SI Unit	Abbreviation
1	Time	T	second	s
2	Length	L	meter	m
3	Mass	M	kilogram	kg
4	Temperature	θ	kelvin	K
5	Luminous intensity	cd	candela	cd
6	Amount of substance	mol	mole	mol
7	Electrical current	A	ampere	A

Clearly, pressure, which is a derived physical quantity, has been expressed in terms of fundamental dimensions of mass, length, and time. You can derive the dimensions of any physical quantity as explained above. Dimensions for a few physical quantities have been shown to be derived from fundamental dimensions in Table 2.2.

Example 2.2 Find the dimensions of Planck's constant.

SOLUTION:

According to Planck, $E = h.v$, or $h = E/v$

where E = energy of photon, h = Planck's constant, and v = frequency of photon
Substituting the dimensions of all physical quantities

$$[h] = [M\ L^2\ T^{-2}] / [T^{-1}] = [M\ L^2\ T^{-1}]$$

Example 2.3 If an equation of force is given as Force = X / Density, find the dimensions of X.

SOLUTION:

$$[X] = [\text{Density}] \times [\text{Force}] = [ML^{-3}] \times [MLT^{-2}]$$
$$= [M^2 L^{-2} T^{-2}]$$

TABLE 2.2

Physical quantities derived from fundamental quantities and their dimensions

Quantity	Formula	Dimensions
Area	Length × Length	L^2
Volume	Length × Length × Length	L^3
Velocity	Length/ Time	$L^1\ T^{-1}$
Momentum	Mass × Velocity	$M^1 L^1 T^{-1}$
Acceleration	Velocity/ Time	$L^1\ T^{-2}$
Force	Mass × Acceleration	$M^1\ L^1\ T^{-2}$
Pressure	Force/ Area	$M^1\ L^{-1}\ T^{-2}$
Energy	Force × Length	$M^1\ L^2\ T^{-2}$
Power	Energy/ Time	$M^1\ L^2\ T^{-3}$

2.3 Units

2.3.1 Fundamental Units

As mentioned earlier, the fundamental quantities of length, mass, time, thermodynamic temperature, amount of substance, luminous intensity, and electrical current have their units defined as the fundamental units. Therefore, any system of units for all physical quantities would consist of both fundamental and derived units. The three common systems in vogue are as follows.

1. Fundamental units in CGS system are centimeter (cm), gram (g) and second (s). The CGS system is also called Gaussian system.
2. Fundamental units in MKS system, also known as Giorgi system, are meter, kilogram, and second.
3. Fundamental units in FPS system, also known as British system, are foot, pound, and second.

With a view to compare the values of these fundamental units globally, there is now an international agreement on the use of these systems of units, called SI (Système International d'Unités) system. SI system is a comprehensive, coherent, and rationalized system. The following are the precise standards that regulate the measurement of the fundamental units in SI system.

Length: When light travels in vacuum for $(299,792,458)^{-1}$ s, the distance covered is one meter.

Mass: Mass of one kilogram is expressed in terms of Planck's constant, h, taken as $6.62607015 \times 10^{-34}$ J.s (equal to kg.m.s^{-2}) where length and time are defined using the speed of light and the hyperfine transition frequency of Cesium[133], respectively. Also, one kilogram is the mass of 5.0188×10^{25} numbers of $_6C^{12}$ carbon atoms.

Time: One second is the duration of 9192631770 periods of the radiation of the ground state of the caesium[133] atom at rest at a temperature of 0 K corresponding to the transition between the two hyperfine levels.

Electrical Current: Amount of current that would produce a force equal to 2×10^{-7} newton per meter of length in two straight parallel conductors of negligible circular cross-section, placed 1 m apart, in vacuum and having infinite length is one ampere.

Thermodynamic Temperature: It is an absolute temperature scale, expressed in kelvin (K), where the triple point of water is defined as 273.16 K and 0.01 °C. Temperature difference of 1 K is equal to 1 °C.

Amount of Substance: A mole is the amount of substance that has the same number of elementary entities as the number of carbon (C^{12}) atoms in 0.012 kilogram.

Luminous Intensity: Unit of luminous intensity of a source in a given direction is candela (cd). One candela is the radiant intensity of 1/683 watt per steradian of monochromatic radiation emitted by the source at the frequency of 540×10^{12} hertz.

In addition to the fundamental units, there are *supplementary units* as well as derived units. Examples of supplementary units are steradian (sr) for solid angle and radian (rad) for plane angle. An arc of the circle equivalent to its radius subtends an angle of one radian at the center of the circle (1 radian = 180/π = 57° 17' 45''). One steradian is the solid angle subtended by one square meter surface area of a sphere at its center when the radius of the sphere is one meter. Value of solid angle of a sphere is 4 π sr.

2.3.2 SI Derived Units

Quantities neither fundamental nor supplementary are known as derived. It is possible to define any quantity in terms of the seven fundamental quantities with the help of quantity equations. For obtaining the *SI derived units* for these derived quantities, the seven SI fundamental units are used. Table 2.3 has a few examples of such SI derived units.

Twenty-two SI derived units have been given special names and symbols for ease of understanding and convenience.

Annexure I details the conversion factors for different unit systems for different quantities.

2.4 Extensive and Intensive Properties

Property of a given system is any of its characteristic. Mass, pressure, temperature, thermal conductivity, volume, and viscosity are a few examples of properties. Properties of materials and systems can be *extensive* or *intensive*. An extensive property depends upon the size or the contents of the system. For example, total volume, total mass, and energy content depend on the size of the material under consideration. Intensive properties are independent of the mass of the system. Temperature, pressure, density, specific heat, etc. are examples of intensive properties.

2.5 Prefixes

While dealing with magnitudes of physical quantities, the values may differ considerably, e.g., the mass of the sun is 2×10^{30} kg while that of an electron is 9.1×10^{-31} kg. In such cases, the representation of extremely small or extremely large numbers is facilitated by using appropriate prefixes. Table 2.4 gives the prefixes for representing fractions and multiples of SI units.

2.6 Significant Figures

If a micrometer could measure a dimension up to only three significant figures, the measured value could not be expressed with more than three significant figures. Measurement systems have

TABLE 2.3

SI derived units that have special symbols and names

S. No.	Quantity	SI Units	Symbol	Fundamental unit
1.	Force	newton	N	kg. m. s^{-2}
2.	Pressure, stress	pascal (N.m^{-2})	Pa	kg. m^{-1}. s^{-2}
3.	Power	Watt, J.s^{-1}	W	kg. m^2. s^{-3}
4.	Energy, work	joule, N. m, W. s	J	kg. m^2. s^{-2}
5.	Catalytic activity	katal	kat	s^{-1}·mol
6.	Celsius temperature	degree Celsius	°C	K
7.	Radionuclide activity	becquerel	Bq	s^{-1}
8.	Dose equivalent	sievert, J.kg^{-1}	Sv	m^2.s^{-2}
9.	Absorbed dose, specific energy	gray, J.kg^{-1}	Gy	m^2.s^{-2}
10.	Frequency	hertz	Hz	s^{-1}
11.	Inductance	henry, W$_b$. A^{-1}	H	kg.m^2.s^{-2}.A^{-2}
12.	Magnetic flux density	tesla, W$_b$. m^{-2}	T	kg. s^{-2}.A^{-1}
13.	Magnetic flux	weber, V.s	W$_b$	kg.m^2.s^{-2}.A^{-1}
14.	Electric capacitance	farad	F	kg^{-1}.m^{-2}. A^2.s^4
15.	Electric conductance	siemen, mho	S	A^2.s^3.kg^{-1}.m^{-2}
16.	Electric resistance	ohm, V.A^{-1}	Ω	kg-m^2. A^{-2}.s^{-3}
17.	Electrical potential difference, emf	volt	V	kg-m^2.A^{-1}. s^{-3}
18.	Electrical charge	coulomb	C	s·A
19.	luminance	Lux, lm. m^{-2}	lx	cd.m^{-2}
20.	Illuminance	Lux, lm. m^{-2}	lx	cd.m^{-2}
21.	Solid angle	steradian	sr	m^2·m^{-2}
22.	Plane angle	radian	rad	m. m^{-1}

TABLE 2.4

SI prefixes

deka, da	hecto, h	kilo, k	mega, M	giga, G	tera, T	peta, P	exa, E	zeta, Z	yotta, Y
10^1	10^2	10^3	10^6	10^9	10^{12}	10^{15}	10^{18}	10^{21}	10^{24}

deci, d	centi, c	milli, m	micro, μ	nano, n	pico, p	femto, f	atto, a	zepto, z	yocto (y)
10^{-1}	10^{-2}	10^{-3}	10^{-6}	10^{-9}	10^{-12}	10^{-15}	10^{-18}	10^{-21}	10^{-24}

limited accuracy and precision. Therefore, measured value of a physical quantity could have only as many digits as we have confidence in. Significant figures in a numerical value are the number of digits that are trustworthy. Therefore, the measured values could have only so many significant figures as are commensurate with the specifications of the measuring system. As far as a given measured quantity is concerned, the rules governing the number of significant figures in the quantity are as follows.

(1) In case of a measured value, non-zero digits only are taken as significant. Numbers of significant digits in 42.3, 243.4, and 24.123 are three, four, and five, respectively.

(2) When a zero appears between non-zero digits, it is taken as a significant figure. For example, zeros in 5.03, 5.604, and 4.004 constitute significant figures.

(3) However, zeros are insignificant if they appear to the left of a number. There are three significant figures in 0.543, only two in 0.045, and only one in 0.009.

(4) Zeros may exist to the right of a number. Such zeros, also called trailing zeros, are significant. Zeroes in 4.330, 433.00, and 343.000 are significant figures.

(5) Number of significant figures in exponential notation is given by its numerical portion. Thus, 1.54×10^{-2} and 1.32×10^4 both have three significant figures.

2.6.1 Rounding Off

The practice of rounding off the numerical values in a measurement or a calculation follows the following rules.

(1) In cases of rounding off, the value of the preceding digit remains the same if the digit needed to be dropped is less than 5. For example, 7.92 becomes 7.9 after rounding off. Likewise, 3.64 will be rounded off to 3.6.

(2) The value of the preceding digit is increased by one if the digit needed to be dropped is more than 5. Thus, 6.47 will be rounded off to 6.5 and 22.88 to 22.9.

(3) Preceding digit remains the same, if it is even, when the digit needed to be dropped is 5 or 5 followed by zeros. Therefore, 3.250 becomes 3.2 and 12.650 becomes 12.6 on rounding off.

(4) In case the digit to be dropped is 5 or 5 followed by zeros and if it is odd, the preceding digit is raised by one. For example, 9.550 is rounded off to 9.6 and 22.150 to 22.2.

2.6.2 Significant Figures in Calculations

Measured values in an experiment are often needed to undergo arithmetic operations to arrive at final values. It needs to be realized that precision in measurement may not be the same for all observations. In such cases, the final value cannot be more precise than the least precise measured value. Number of significant digits in such calculations is governed by the following two rules.

(1) When numbers in an addition or subtraction have different precisions, the resultant value cannot have more number of decimal places than the number of decimal places that exist in the least precise number.

For example, the three values 33.3, 3.11, and 0.313 have different number of digits after decimal point. As per the rule given above, the final value will have only one digit after decimal place, i.e., 36.7.

(2) The consideration for rounding off the result of a multiplication or division is to retain only the number of significant figures that exists in the least precise term in the calculation. For example, 158.43×0.35 (two significant figures) = 55.4505. As the answer should have two significant figures, it will be 55.45.

2.7 Order of Magnitude

Physical quantities are often expressed as $M \times 10^n$ units where n is an integer and M varies between 1 and 10. When magnitude of a quantity is expressed in terms of the power of 10 then the number representing the power is termed as the order of the magnitude. It is necessary to first round off the given value such that M is equal to 1 before determining the power of 10, i.e., the value of n. While rounding off a number, the last digit is ignored if it is less than 5. However, increase the preceding digit by one when the last digit is 5 or more than five.

Take, for example, the case of speed of light. Its value in vacuum is 3×10^8 m.s^{-1} $\approx 10^8$ m.s^{-1} (ignore 3 being less than 5). In case of an electron's mass, the value is 9.1×10^{-31} kg and it becomes 10^{-30} kg after rounding off since $9.1 > 5$.

Check Your Understanding

1. A box contains two glass balls of 2.15 g and 12.39 g. Mass of the box itself is 2.3 kg. Determine the resultant mass of the box with due consideration of the correct number of significant figures.

2. A rectangular sheet has length of 1.5 cm and width of 1.203 cm. Determine the rectangular sheet's face area to the correct number of significant figures.

    Answer: 2314 g

3. The force resisting the motion of a body with a velocity v through a fluid at rest is represented by $F = C_d v^2 A \rho$, where C_d represents drag coefficient and A is area of cross-section perpendicular to the direction of motion. Determine the dimensions of C_d.

4. Determine the dimensions of Stefan–Boltzmann constant from the equation of blackbody thermal radiation.

5. Derive the dimensions of dynamic viscosity from Newton's law of viscosity.

6. Convert the units of thermal resistance, electrical conductivity, and pressure from FPS system of units to SI system.

7. The gas constant occurs in the ideal gas law as PV = nRT. The value of R in SI system is 8.3145 kg.m^2 mol^{-1}.K^{-1}.s^{-2}. Determine the value of R in FPS system of units.

8. A high-pressure processing system operates at a pressure of 800 atmospheres. Calculate the pressure value in SI units.

    Answer: 81060 kPa

9. If velocity (v), force (F), and energy (E) are considered as the fundamental units, what will be the dimension of mass?

    Answer: $[M] = [Ev^{-2}F^0]$

Objective Questions

1. What is the difference between dimension and unit of measurement?

2. Name some dimensionless quantities and explain how they are helpful in analysis of different practical situations.

3. What are the different fundamental quantities? What are their dimensions and SI units?

4. Find the dimensions of (a) force, (b) work, (c) viscosity, (d) heat energy, (e) momentum, (f) inductance, (g) electrical resistance, and (h) electrical capacitance.

5. Differentiate between the extrinsic property and intrinsic property with suitable examples.

6. What is order of magnitude in expressing the magnitude of a quantity?

BIBLIOGRAPHY

Acharya, B.P. and Das, R.N. 2000. *A Course on Numerical Analysis*. Kalyani Publishers, New Delhi.

Balagurusamy, E. 2000. *Numerical Methods*. McGraw-Hill, Noida, India

Brennan, J.G., Butters, J.R., Cowell, N.D. and Lilly, A.E.V. 1976. *Food Engineering Operations*, 2nd ed. Elsevier Applied Science, London.

Dash, S.K. and Sahoo, N.R. 2012. *Concepts of Food Process Engineering*. Kalyani Publishers, New Delhi.

Huntley, H.E. 1974. *Dimensional Analysis*. Dover Publ., New York.

Langhaar, H.L.1951. *Dimensional Analysis and Theory of Models*. John Wiley & Sons Inc., New York.

NCERT. 2006. *Physics, Part I*. National Council of Educational Research and Training, New Delhi.

Rao, D.G. 2009. *Fundamentals of Food Engineering*. PHI learning Pvt. Ltd., New Delhi.

Sahay, K.M. and Singh, K.K. 1994. *Unit Operations in Agricultural Processing*. Vikas Publishing House, New Delhi.

Toledo, R.T. 2007. *Fundamentals of Food Process Engineering*, 3rd edition. Springer, New York.

3

Material and Energy Balance

Material and energy balances form the bedrock of food processing operations and, therefore, are essential for those connected to food processing and industry. The application of material balance is important in calculating the yield from a process or to know the quantity of raw materials required for a particular output and for control of processing operations. For any commercial manufacturing process, first the mass balance is carried out during the product development stage, then in pilot scale operation so that the masses of different streams can be calculated for the commercial process. Finally, the mass balance is carried out for the commercial process to check if there is any variation from the earlier predicted values and corrective measures are taken. When there is any change in the process, material balances are needed to be reassessed. This is applicable for both batch and continuous processes. In food industries, the mass balance often decides the amount of final product from the raw material(s) and thus the costing of the final product is carried out considering other cost inputs such as energy, manpower, etc.

Similarly, the energy balance is very important. It can give us an idea of the energy input and output in a process so that the losses or inefficiencies of the system can be determined. The heat exchangers and other processing equipment can be designed only after knowing the energy requirements and, thus, the energy balance. As the cost of energy is a major component for the cost of the final product, new and improved processes and equipment are being developed for using energy more efficiently in food processing systems.

3.1 Basic Principles

The material and energy balances are based on the laws of conservation of mass and energy. The *law of mass conservation* states that in a steady state condition, the amount of material entering a system must equal the amount leaving it. Under steady state, there is no accumulation of mass or energy in the system although the system is in continuous operation. Actually, in most of the continuous operations there is a phase of unsteady state at the start of the operation. The steady state is attained after a period of adjustment. For example, when we connect a hosepipe with a tap and allow the water to flow through the pipe, initially there is a lag and the mass entering the hosepipe may not be the same as the mass going out. After some time, however, we would observe the masses to be the same. The *law of energy conservation*, also known as first law of thermodynamics, states that energy can be neither created nor destroyed.

The application of laws of mass and energy conservation for analyzing the processes and products is known as the *material and energy balance*.

For the mass and energy balance, first, there is selection of a system and, second, the basis on which the balances are to be done. Here, system refers to anything that can be defined by boundaries. For example, if we need to know the amount of dry materials obtained from a kg of wet material going into the dryer in a fruit processing plant, we take the dryer as a system and consider all the streams going into and out of the dryer. However, if we need to know how much raw material is required for producing 1000 kg dried fruits in a food processing plant, then we will take the whole plant (including the receiving section, washing section, drying section, packaging section, etc.) and consider all the streams going into and out of the plant. In the second case the dryer will be a component in the whole system and the mass flowing into and out of the dryer may not be at all important for the calculations. Thus, the boundary line should be located such that the dryer (or in the second case the complete plant) is completely enclosed within the boundary line. The same principle will also be applicable for the heat flow into and out of an equipment or system.

In case of multi-component systems, it is a good practice to draw arbitrary boundary lines and then to equate all streams entering the system with all streams leaving the system. Everything outside the boundary is considered outside the system. The boundary lines should be carefully drawn so that all streams that enter or leave the system would cross the boundary line; no stream should be missed in writing the mass balance equations.

A basis is the parameter on which the further calculations are done and all results are calculated with respect to it. It is usually expressed in terms of weight of some material or component entering or leaving the system or in terms of time. For example, in a milk dryer the basis can be 100 kg of fluid milk entering into the dryer or the amount of milk going in per unit time. It is always convenient to select a component that enters the system in the minimum number of streams and leaves in the minimum number of streams as the basis. For continuous systems the flow rate per unit time is taken.

It is often convenient to carry out component balances before analyzing the total mass balance in a multi-component system. For example, we may take the dry solids instead of total milk as the basis in the above example. There can also be balance for components such as carbohydrates, proteins, salts, etc.

As many equations as possible are written with the available information and then the unknown quantities are found out by solving the above equations.

Before taking up the task of carrying out the material and mass balances, it is useful to review the features of properties of materials, dimensions, and units that facilitate the computations.

DOI: 10.1201/9781003285076-4

3.2 Mass Balance

In food processing, a simple form of mass balance equation can be written as

$$\text{Mass in} = \text{Mass out} + \text{Mass stored} + \text{Losses} \quad (3.1)$$

3.2.1 Applications in Different Unit Operations

The following are some illustrative examples of the application of the concepts presented above.

Case I. Mass Balance in Evaporation Process

Let us consider the case (Figure 3.1(a)) where a dilute juice is evaporated continuously to produce a concentrated product.

Let F, V, and P be the masses of feed (juice or liquor) entering the evaporator, vapor and concentrated product going out of the evaporator, and X (with subscripts) are the fractions of solids in different streams. We can redraw Figure 3.1(a) showing the different masses and the fractions of solids (within brackets) as in Figure 3.1(b). Check that the dotted line represents the boundary line for the system. In this case two streams are entering and three streams are leaving the system.

As per mass balance,

$$F = V + P \quad (3.2)$$

Here F, V, and P may have the units of kg (total quantity in a batch system) or kg/hour (or per minute or second), i.e. the rate of flow for a continuous system.

In the figure, W_s is the amount of steam used for evaporation and W_{sc} is the amount of condensate after evaporation. W_s and W_{sc} will also have the units of kg or kg.h^{-1}, as mentioned above. The total amount of steam fed into the system goes out as condensate (without mixing with the solution), hence the terms are ignored while writing equation 3.2.

The solid balance equation can then be written by equating the solid entering the system (with the feed) with the solid going out of the system (with the product and vapor).

$$F.X_f = P.X_p + V.X_v \quad (3.3)$$

If the solid escaping with the vapor is considered negligible, then the above equation can be further modified as

$$F.X_f = P.X_p \quad (3.4)$$

For the present case, two simplified equations, 3.2 and 3.4, are available. The parameters that are related to the process are F, V, P, X_f, and X_p. Thus, if three of these parameters are known, the above two equations can be solved to find out the two unknown quantities. For example, if the amount of feed and the solid concentration in the feed and desirable solid concentration in the product are known, then the amount of water to be evaporated from the juice and the product that will be available after the evaporation process can be easily calculated even before the start of the process.

Case II. Mass Balance in Mixing Process

Suppose we have a juice with 12% solids. Some amount of sugar solution (say having 50% sugar) is to be mixed with the juice to bring it to 30% solids concentration. Now we can calculate the amount of sugar needed by mass balance. The schematic diagram is shown in Figure 3.2.

In this case we get the following equations:

$$F_1 + F_2 = P \quad (3.5)$$

$$\text{and, } F_1 \times X_{f1} + F_2 \times X_{f2} = P \times X_p \quad (3.6)$$

Suppose instead of sugar solution we had dry sugar with us, then $X_{f2} = 1$ and the above equation can be modified as

$$F_1 \times X_{f1} + F_2 = P \times X_p \quad (3.7)$$

The above principles may be extended to any other problem associated with mass balance.

Case III. Mass Balance in Separation Process

Suppose we have milk with 5% fat content. In a cream separator, we remove some amount of cream that has 45% fat. With the help of mass balance, it is possible to find out the amount of fat in skim milk.

The schematic diagram is drawn as Figure 3.3.

In this case we get the following equations:

$$F = P_1 + P_2 \quad (3.8)$$

$$\text{and, } F \times X_f = P_1 \times X_{p1} + P_2 \times X_{p2} \quad (3.9)$$

In this case P_1 corresponds to skim milk and P_2 to cream and X is the fraction of fat in the components. Suppose instead

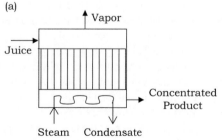

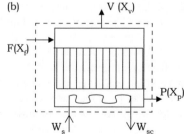

FIGURE 3.1 (a) Different streams entering and going out of an evaporation system, (b) Representation of different mass flow components in the system

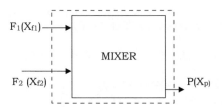

FIGURE 3.2 Mass flow components in a mixing system

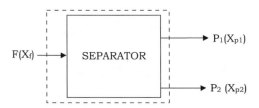

FIGURE 3.3 Mass flow components for a separator (one stream in and two streams out)

of cream, we are removing only water (i.e. P_2 denotes water), and the fraction of fat in P_2 is zero, then the equation becomes similar to the mass balance of evaporation process.

The above principles may be extended to any other problem associated with mass balance.

The following examples explain material balance in mixing operations.

> **Example 3.1 One hundred kg of cow milk (fat content 5%) is available. (a) Find the final fat percentage in the milk if 20 kg of water is added to it. (b) How much water should be added to it so that it has the standard of toned milk (3% fat)? (c) What will be the amount of water to be added to make it double toned (1.5% fat)?**

SOLUTION:

Since the fat is 5%, the amount of fat in 100 kg milk is 5 kg.

(a) Now, 20 kg of water is added to it, but the amount of fat remains constant (i.e., 5 kg)
So, the final fat content = (5/120) × 100 = 4.17%

(b) $100 \times (0.05) = (100 + x) \times (0.03)$
or, 5/0.03 = 100 + x
or, x = 166.67 − 100 = 66.67 kg
Hence, water to be added to 100 kg of milk to make it toned milk is 66.67 kg.

(c) For this case, the fat will be 1.5%

Thus, $100 \times (0.05) = (100 + x) \times (0.015)$

or, 5/0.015 = 100 + x

or, x = 333.33 − 100 = 233.33 kg

Hence, water to be added to 100 kg of milk is 233.33 kg to make it double toned milk.
(In the above problem, we have ignored the solids not fat (SNF) in the milk, but in actual practice, skim milk powder is also added along with water to standardize the SNF content.)

Example 3.2 The milk has 2% fat. What needs to be done if 3% fat is required in the milk?

SOLUTION:

In 100 kg of milk there is 2 kg fat.

If M kg water is added to 100 kg of milk, then the percent fat will be $\dfrac{2}{100 + M} \times 100$

Thus, $\dfrac{2}{100 + M} \times 100 = 3$

or, 100 + M = 200/3 = 66.67 kg

So, M = 66.67 − 100 = −33.33 kg

Hence, the need is to remove 33.33 kg of water to bring the fat content up to 3% level.

If we use specific mass balance equation,

$$F = 100 \text{ kg}, X_f = 0.02, X_p = 0.03$$

$$\text{or, } 100 \times 0.02 = P \times 0.03$$

$$\text{Thus, } P = (100 \times 0.02) \div 0.03 = 66.67 \text{ kg.}$$

i.e., 33.33 kg of water needs to be removed to raise the fat content to 3%.

We can also increase the percent of fat by adding cream to the milk. Say, if the cream contains 40% fat, then the amount of cream that is to be added can also be found out by the specific mass balance as follows.

$$\text{Here, } F_1 \times X_{f1} + F_2 \times X_{f2} = (F_1 + F_2) \times X_p$$

$$\text{or, } 100 \times (0.02) + C \times (0.4) = (100 + C) \times (0.03)$$

$$\text{or, } C(0.4-0.03) = 100(0.03-0.02)$$

$$\text{or, } C = 1/0.37 = 2.703 \text{ kg}$$

Thus, 2.703 kg of cream (40% fat content) may be added to make the final fat to 3%.

Example 3.3 A sample 'A' of milk with 1.5% fat is mixed with another sample 'B' having 6.5% fat to get the final fat as 3.16%. What should be the ratio of two milks?

SOLUTION:

In this situation

$$F_1 \times X_{f1} + F_2 \times X_{f2} = (F_1 + F_2) \times X_f$$

$$\text{or, } F_1 \times (0.015) + F_2 \times (0.065) = (F_1 + F_2) \times 0.0316$$

$$\text{or, } F_1 \times (0.015) + F_2 \times (0.065) = F_1 \times 0.0316 + F_2 \times 0.0316$$

or, $F_1 (0.0166) = F_2 (0.0334)$

or, $F_1 / F_2 = 2.01$

Hence, the ratio of the two milks A and B will be 2.01:1.

Example 3.4 One thousand kg of milk having 3% fat & 8.5% SNF has been evaporated so that the final fat concentration is 6.0%. After that 200 kg of sugar is added to the milk. What will be the final TSS (total soluble solid) of milk?

SOLUTION:

For the first situation, as discussed above, the weight of milk may be taken as 1000 kg.

$$1000 \times (0.03) = P \times (0.06)$$

$$P = 500 \text{ kg}$$

Then 200 kg of sugar was added.

So final weight of the milk = 500 + 200 = 700 kg

Again going for the mass balance for the total solids,

$$1000 \times (0.115) + 200 = 700 \times (X)$$

$$X = 315/700 = 0.45 = 45 \%$$

Thus, the final TSS of the milk is 45%.

Example 3.5 It is required to prepare skim milk after removing fat from whole milk containing 4.7% fat. The skim milk should have 90.2% water, 4% protein, 4.9% carbohydrate, 0.2% fat, and 0.8% ash. Calculate the composition of whole milk assuming that only fat is removed and there is no other loss during processing.

SOLUTION:

Let the basis be 100 kg of skim milk. It contains 0.2% fat. The fat removed from the whole milk is x kg.
   Thus, total original fat = x + 0.2 kg
   Total original mass = (100 + x) kg
   The original fat content was 4.7%,

thus $\dfrac{x + 0.2}{x + 100} = 0.047$

$$x + 0.2 = 4.7 + 0.047 \, x, \, x = 4.72 \text{ kg}$$

Alternatively, the fat can also be calculated with the mass balance equations.
   Fat in whole milk = Fat in skim milk + Fat removed

$$F(X_f) = P_1(X_{p1}) + P_2$$

Here F is the whole milk, P_1 is the skim milk, and P_2 is the fat removed. Thus $F = P_1 + P_2$

$$(100 + P_2) (0.047) = 100 (0.002) + P_2$$

or $4.7 + 0.047P_2 = 0.2 + P_2$
i.e., $0.953 \, P_2 = 4.5$, i.e. $P_2 = 4.72$ kg
   The other compositions can be calculated as follows.

water $= \dfrac{90.2}{104.72} = 86.13\%$, protein $= \dfrac{4}{104.72} = 3.82\%$,

carbohydrate $= \dfrac{4.9}{104.72} = 4.68\%$, ash $= \dfrac{0.8}{104.72} = 0.76\%$

Example 3.6 One thousand kg of wheat is dried from an initial moisture content of 20% (wet basis) to a final moisture content of 10% (wet basis). How much water is taken out from the crop?

SOLUTION:

We can write the equations as

Mass of moist wheat = Mass of dry matter in wheat
        + mass of moisture (evaporated)

Mass of dry matter in moist wheat = Mass of dry
            matter in dry wheat

(as there is no dry matter in evaporated moisture)
In equation form:

$$F = P + V$$

i.e., $1000 = P + V$

$$F \times X_f = P \times X_p$$

Thus, $(1000) \times (0.8) = (P) \times (0.9)$ (check that the solid fraction = 1−moisture fraction)

$$P = (1000) \times (0.8) / (0.9) = 888.889 \text{ kg}$$

$$V = 1000\text{-}888.889 = 111.111 \text{ kg}$$

Hence, 111.111 kg of water is evaporated from the wheat.

Example 3.7 Cauliflower is dried from 84% total moisture content to 6% moisture content. What will be the product yield, if 10% weight of the material is lost during sorting and removing the outer leaves?

SOLUTION:

Let us take the basis as 100 kg cauliflower. Taking out 10% losses, mass dried is 90 kg.

During drying the solid content remains constant, only the moisture is taken out.

Initial solid (fraction) = (100-84)/100 = 0.16

Final solid (fraction) = (100-6)/100 = 0.94

If F and P are the weights of wet and dried products and X is the respective solid fractions,

$$F(X_f) = P(X_p)$$

$$90(0.16) = P(0.94)$$

$$P = 15.32 \text{ kg}$$

Thus, the yield ratio = 15.32/100 = 15.32%.

Example 3.8 One thousand kg of groundnut seeds, having 48% oil, are fed to a mechanical screw type oil expeller. The other components of the seeds are 25% protein, 16% carbohydrates, 6% moisture, and 5% other constituents. It was observed that the meal coming out of the expeller contains 7% oil. To recover this oil, the meal is fed to a solvent extraction system. After solvent extraction, the residual oil in the meal is 0.5%. The meal is then dried to 5% moisture content before storage. Thus, the whole process utilizes three pieces of equipment, namely, an oil expeller, a solvent extractor, and a dryer. Calculate the different mass streams of the individual equipment.

SOLUTION:

The basis is 1000 kg of groundnut seeds. Thus, it has 480 kg oil, 250 kg protein, 160 kg carbohydrates, 60 kg moisture, and 50 kg other constituents. The non-oil components are 520 kg.

As given, there are three pieces of equipment, namely, the expeller, the solvent extractor, and the dryer. Thus, the mass flow diagram can be as follows.

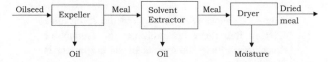

The problem will be solved by taking each piece of equipment as a system and obtaining the mass balance for that system.

I. Mass balance in expeller:

The different streams (along with their notations) going in and out of the expeller are shown in the figure.

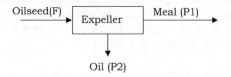

Here the oil is only extracted and hence the non-oil contents in F and P1 remain the same. If X represents the mass fraction of the non-oil matters in the different streams, then, $X_F = 0.52$ and $X_{P1} = 0.93$.

$$F(X_F) = P(X_{P1}), \text{ i.e., } 1000(0.52) = P1(0.93)$$

$$P1 = 1000(0.52)/0.93 = 559.14 \text{ kg}$$

Oil expressed = 1000-559.14 = 440.86 kg

II. Mass balance in solvent extraction:

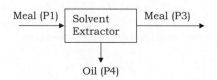

Here also, as oil is extracted the non-oil contents of P1 and P3 remain the same. The non-oil fraction in P3 is 99.5%, i.e. $X_{P3} = 0.995$.

$$\text{Thus, } 559.14 \times (0.93) = P3 \times (0.995)$$

$$P3 = 522.613 \text{ kg}$$

Oil extracted = 559.14 – 522.613 = 36.526 kg

Oil in meal (P3) = 522.613 x 0.005 = 2.613 kg

III. Mass balance during drying:

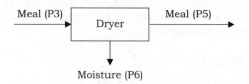

Here only moisture is removed. Thus, the solids (non-moisture constituents) remain the same in the meal (P3) and meal (P5). The non-moisture materials constitute the oil, carbohydrates, proteins, as well as other constituents

Thus, the non-moisture materials in the dried product
= 2.613 + 160 + 250 + 50 = 462.613 kg

Thus, P5 × (0.95) = 462.613 kg

P5 = 486.96 kg

Example 3.9 Thirty kg of sugar is mixed thoroughly with the help of a stirrer with 120 kg of apple juice (10% concentration). Then the apple juice has to be evaporated to bring it to 60% solids concentration. Find out the amount of water evaporated.

SOLUTION:

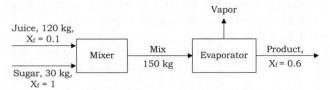

For the first unit operation (mixing)

$$F_1 \times X_{f1} + F_2 \times X_{f2} = (F_1 + F_2) \times X_P$$

$$120 \times (0.1) + 30 \times (1) = 150 \times (X_P)$$

$$\text{or, } X_P = 42/150 = 0.28 \text{ or } 28\%$$

For the second unit operation (evaporation)

$$F = P + V$$

$$F \times X_F = P \times X_P$$

$$\text{Thus, } 150 \times (0.28) = P \times 0.6$$

$$\text{or, } P = 70 \text{ kg}$$

$$V = 150 - 70 = 80 \text{ kg}$$

Thus, 80 kg of water is to be evaporated from the apple juice + sugar mix solution.

Example 3.10 Milk and sugar are mixed in a ratio of 60:40 (w/w) and then the mixture is evaporated until the solids are 75% to obtain sweetened condensed milk. What will be the yield of sweetened condensed milk from 1000 kg milk having 12% solids? Also find out the amount of water removed in the evaporator.

SOLUTION:

The amount of milk is 1000 kg.
   Thus, the amount of sugar added is
$1000 \times (40/60) = 666.67$ kg
   The mass balance equation for mixing is

$$1000 \ (0.12) + 666.67 \ (1.00) = (1000 + 666.67) \times X_{Mix}$$

$$\text{or, } X_{Mix} = (120 + 666.67)/1666.67 = 0.472$$

This is the concentration of the product after mixing from which the water is evaporated to make it to 75% solids. If C is the amount of final condensed milk

$$\text{or, } (1666.67) \ (0.472) = C \ (0.75)$$

$$C = 1048.89 \text{ kg}$$

$$\text{or, } V = 1666.67 - 1048.89 = 617.77 \text{ kg}$$

Thus, the amount of water evaporated is 617.77 kg

Example 3.11 In a method of preparation of orange juice, the process involves three stages. In the first stage, the fresh extract containing 12% solids is passed through a strainer from which two streams, namely, the strained juice (stream A) and pulpy juice (stream B) are obtained in the ratio of 3:1. The strained juice then goes to the evaporator in the 2nd stage for removal of moisture. The evaporated product has 60% solids. In the 3rd stage, the evaporated juice goes to a blender in which it is mixed with the pulpy juice obtained from the strainer (stream B). The final concentration of the blended product is 40% solids. If the system is fed with 100 kg juice, what is the amount of concentrated juice coming out of the evaporator?

SOLUTION:

The flow chart for the process can be shown as in the figure.

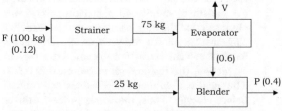

The overall mass balance equations can be written as

$$F = P + V$$

$$F \times X_F = P \times X_P$$

If we take the basis F as 100 kg

$$100 \ (0.12) = P \ (0.4)$$

$$P = 30 \text{ kg}$$

$$V = 100 - 30 = 70 \text{ kg}$$

This 70 kg vapor is removed from the 75 kg juice entering into the evaporator, i.e. the amount of concentrated juice coming out of the evaporator is 75 – 70 = 5 kg.

Example 3.12 In the above example, find out the concentration of the 25 kg pulp and 75 kg strained juice coming out of the strainer.

SOLUTION:

In this case, 25 kg of fresh juice with solid concentration X is mixed with 5 kg of concentrated juice (solid concentration 0.6) to give the final juice of 30 kg with solid fraction 0.4.

$$\text{Thus, } 25 \ (X) + 5 \ (0.6) = 30 \ (0.4)$$

or, X = (12–3)/25 = 0.36

The concentration of the 75 kg juice can be found out as

$$100 (0.12) = 75 (X_1) + 25 (0.36)$$

$$X_1 = (12–9)/75 = 0.04$$

3.2.2 Pearson Square Method

This is an alternate method for finding out the resultant concentration of a mixture of two streams with varying concentrations.

Consider, we have to find out the ratio of 1.5% fat (doubled toned) milk and 80% cream to obtain the final 3% fat milk.

The steps are as follows.

1. A square as shown below is drawn.
2. Write the fat percentage of the feed materials on the left top and bottom corners.
3. Write the final fat percentage at the center of the square.
4. Write the differences between the final fat percentages and the feed fat percentages on the diagonally opposite corners. In case the value is negative, write the absolute value.
5. The values obtained on the opposite corners are the ratios of the materials to be finally combined.

For the present case, the values will be as shown in Figure 3.4.

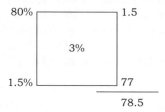

FIGURE 3.4 Mass balance using Pearson's square method

Thus, the ratio of the milk and the cream will be 77:1.5 or 51.33:1.

In this case, 77 kg of milk containing 1.5% fat will be added to 1.5 kg of cream containing 80 % fat to make 78.5 kg of 3% fat milk.

Some more examples of the Pearson's square method are given below.

> **Example 3.13 What should be the ratio of milk with 3% fat and cream with 65% fat to make the product with 5% fat?**
>
> **SOLUTION:**
>
> This problem can be solved by "Pearson's square" method.

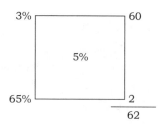

Using the procedure indicated above, all the values are written in the square. As observed, 60 kg milk of 3% fat needs to be added to 2 kg cream of 65% fat to give 62 kg milk product of 5% fat.

3.2.3 Calculation of Moisture Loss in Agricultural Commodities

The moisture content of a substance is expressed in percent by weight, either on wet basis (w.b.) or on dry basis (d.b.).

$$\text{Moisture content}\left(\text{wet basis}\right),\ M_w(\%) = \frac{W_w}{W_t} \times 100 \tag{3.10}$$

$$\text{Moisture content}\left(\text{dry basis}\right),\ M_d(\%) = \frac{W_w}{W_d} \times 100 \tag{3.11}$$

Here W_t is the total weight of the material, W_d and W_w are the weights of dry matter and moisture, respectively.

$$W_t = W_d + W_w \tag{3.12}$$

From the above equations, the relationship between M_d and M_w can be obtained as

$$M_d = \frac{M_w}{100 - M_w} \times 100,\ \text{and}\ M_w = \frac{M_d}{100 + M_d} \times 100 \tag{3.13}$$

During a hydration or dehydration process, the denominator in case of moisture content (dry basis) does not change. Therefore, calculations of moisture gain or loss are simpler when working with moisture content (dry basis).

> **Example 3.14 If wheat has a moisture content of 35% dry basis, what is its moisture content on a wet basis?**
>
> **SOLUTION:**
>
> Moisture content on dry basis
>
> $$= \frac{\text{weight of moisture}}{\text{weight of dry matter}} = 35\% = \left(\frac{35}{100}\right)$$

Moisture content on wet basis

$$= \frac{\text{weight of moisture}}{\text{weight of dry matter} + \text{weight of moisture}}$$

$$= \left(\frac{35}{100 + 35}\right)$$

$$= 0.259 = 0.259 \times 100\% = 25.9\%$$

Moisture content can also be expressed in fractions as kg of water (kg total weight)$^{-1}$, as we have discussed in equations 3.3 and 3.4 and the subsequent examples. Some more examples are stated below.

Example 3.15 One thousand kg of wheat is dried from 20% (wet basis) moisture content to 10% (wet basis) moisture content. How much water is taken out from the crop?

SOLUTION:

The given data are arranged as follows. Check that even though the moisture content and total weights are changing, the solids content remains constant. So, if we take the solids content as the basis, we can straightway find out the final weight and the water to be removed.

	Amount of grain	Moisture content (w.b.)	Dry matter content, %	Dry matter in kg
Initial condition	1000 kg	20%	80%	800 kg
Final condition	P	10%	90%	800 kg

From the above table, if P is the final weight of the wheat grain,

$$80\% \text{ of } 1000 \text{ kg} = 90\% \text{ of } P$$

$$\text{or, } 800 = (90/100) \times P$$

$$P = 888.89 \text{ kg}$$

Thus, the amount of water taken out = 1000 − 888.89 = 111.11 kg

Example 3.16 Five hundred kg of corn with initial moisture content of 25% (dry basis) is dried to 10% moisture content (dry basis). Find the weight reduction.

SOLUTION:

The given data is written as follows.

	Amount of corn	Moisture content (d.b.)	Dry matter content, %	Dry matter in kg
Initial condition	500 kg	25%	?	?
Final condition	P	10%	?	?

In this question, the moisture content is given to be on dry basis. But the values of 500 kg or the P that we would like to find out are the total weights (wet weights) of the materials and cannot be directly related to the dry basis moisture content. So, we have to convert the moisture content dry basis to that on wet basis to find out the answer. So, we modify the table as follows.

	Amount of corn	Moisture content (d.b.)	Moisture content (w.b.)	Dry matter content	Dry matter in kg
Initial condition	500 kg	25%	(25/125)×100 = 20%	80%	400 kg
Final condition	P	10%	(10/110)×100 = 9.09%	90.909%	400 kg

$$80\% \text{ of } 500 \text{ kg} = 90.909\% \text{ of } P$$

$$\text{or, } 400 = (90.909/100) \times P$$

$$P = 440 \text{ kg}$$

Thus, the amount of water taken out = 500 − 440 = 60 kg.

If a product loses moisture during a drying process, its total weight continually reduces. However, there is no change of the dry matter content. Hence, the weight of the dry matter is taken as the basis (or the moisture content is mentioned on dry basis) for scientific purposes such as the determination of the rate of moisture loss, plotting the drying curves, etc. Interpretation becomes easy and convenient to understand when the moisture content is stated on dry basis as it is directly proportional to the amount of moisture present in the commodity at any time. Check that the moisture content stated on wet basis is not directly proportional to the amount of moisture at any time.

The moisture content is also expressed as a fraction of total moisture present to the total weight of the substance and is denoted as X. We have solved examples 3.8 and 3.9 on this basis.

3.3 Different Forms of Concentrations

Depending on the components used in calculating the mass balance, the following forms can be used for expressing the concentrations of a solute in a solution.

Concentration (W/W) is the weight of the particular material (solute dissolved) divided by the total weight of the solution, kg.kg^{-1}.

Concentration (W/V) is the quantity of solute by weight per unit volume of solution, kg.m^{-3}.

Molar Concentration (M) is the number of molecular weights of the solute (moles) in a solution. It is also expressed as molarity and the unit is mol.m^{-3} or mol. liter^{1}.

Mole Fraction is the ratio of the number of moles of the solute/solvent to the total number of moles of all components present in the solution.

Molarity is the moles of solute per liter of solution.

Molality is the quantity of solute in moles per kg of the solvent. Molality is also known as molal concentration.

Brix is the soluble solid concentration (W/W) expressed as percentage.

In most food processing calculations, percentage signifies percentage by weight (w/w) unless otherwise specified. We considered the concentrations as percentage by weight (w/w) in the earlier examples. Similarly, the term "mole" used in this book would mean a mass of the material equal to its molecular weight in kilograms (kg-mole).

Gas concentrations are primarily measured in gas weight per unit volume or as partial pressures. The gases are assumed to follow the perfect gas law, $pV = nRT$, where p is the pressure, V the volume, R the gas constant, T the absolute temperature, and n the number of moles. R is equal to 0.08206 m^3.atm. $mole^{-1}$. K^{-1}.

The molar concentration of a gas is then obtained from $n/V = p/RT$

and, the weight concentration is nM/V where M is the molecular weight of the gas.

Example 3.17 A salt solution is prepared by adding 12 kg common salt (NaCl) to 100 kg water. The density of the liquid is 1188 kg.m⁻³. Find out the concentration of salt in this solution as a (a) weight/weight fraction, (b) weight/volume fraction, (c) mole fraction, (d) molar concentration. Take the molecular weight of salt as 58.5 kg.mole⁻¹.

SOLUTION:

There is 12 kg salt in 112 kg solution.

Thus, the weight/weight fraction = 12/(100 + 12)
= 0.107 or 10.7%.

The density of solution is 1188 kg.m⁻³,
112 kg of solution contains 12 kg salt
1188 kg solution would contain
$\frac{12}{112} \times 1188 = 127.286$ kg salt,
Thus, the fraction (weight/volume)
= 127.286/1000 = 0.127 or 12.7%
Now, in 112 kg solution, moles of water
= 100/18 = 5.56, and moles of salt = 12/58.5 = 0.2051

Thus, mole fraction of salt $= \frac{0.205}{5.56 + 0.205} = 0.0356$

Molar concentration (M) = 127.286/58.5
= 2.17 moles.m⁻³

For dilute solutions, the mole fraction can be approximated as the moles of salt/mole of water, i.e. in this case it is 0.205/5.56. As the solution becomes more dilute, this approximation improves and becomes close to the actual.

Example 3.18 Assuming that air has 79% nitrogen and 21% oxygen by weight at 1.3 atmosphere pressure and 25°C temperature, calculate:

  (a) **mean molecular weight of air,**
  (b) **mole fraction of oxygen,**
  (c) **oxygen concentration in mole.m⁻³ and kg.m⁻³**

Take the molecular weights of nitrogen and oxygen as 28 and 32.

SOLUTION:

Let the basis be 100 kg of air.

Thus, moles of nitrogen = 79/28 = 2.82 and moles of oxygen = 21/32 = 0.656

Total moles = 2.82 + 0.656 = 3.476

Mean molecular weight = 100/3.476 = 28.77

Mole fraction of oxygen in air
$= \frac{0.656}{2.82 + 0.656} = 0.1887$

Substituting the values in the gas equation, pV= nRT

$$1.3 \times 1 = n \times (0.08206 \times 303)$$

$$n = 0.05228$$

Weight of air in 1 m³ = n × mean molecular weight = 0.05228 × 28.77 = 1.504 kg

21% oxygen weighs 0.21 × 1.504 = 0.316 kg

In moles, it is 0.316/32 = 0.0099 moles.m⁻³

The mole fraction of a dissolved gas in a liquid is determined as the number of moles of gas using the gas laws, taking the volume of the liquid as the volume, and then directly calculating the number of moles of liquid.

In some cases, a reaction takes place, and the material balances have to be adjusted accordingly.

Example 3.19 A soft drink is prepared with 3 parts (volumes) of CO₂ with 1 part of water at atmospheric pressure and 0°C temperature. If there is no other component in the drink, find out the mass and mole fractions of the CO₂ in the drink.

SOLUTION:

Let us take the basis as 1000 kg (1 m³) of soft drink, Thus the volume of CO₂ will be 3 m³.

Substituting the values in the equation, pV=nRT

$$1 \times 3 = n \times (0.08206 \times 273.15)$$

n = 0.134 moles

Thus, weight of added $CO_2 = 0.134 \times 44 = 5.889$ kg

Mass fraction of CO_2 in the drink $= 5.889/(1000 + 5.889)$
$$= 5.854 \times 10^{-3}$$

Mole fraction of CO_2 in the drink

$$= \text{Moles of } CO_2 / \left(\text{Moles of } CO_2 + \text{Moles of water} \right)$$

$$= \frac{0.134}{(1000 / 18) + 0.134} = 2.40 \times 10^{-3}$$

3.4 Energy/Heat Balance

As per the *law of conservation of energy*, in a steady state condition, the energy input to a unit operation should be equal to the output energy and energy stored. Energy may appear in many forms as heat/enthalpy, work, internal energy, mechanical energy, electrical energy, etc.

The energy balance equation, in general, can be written as:

$$\text{Energy in} = \text{Energy out} + \text{Energy stored} + \text{Energy loss}$$
$$\text{from the system} \qquad (3.14)$$

Thus, after deciding the system and boundary line that should cross all the streams entering and going out of the system, the heat and all other forms of energy leaving or entering the system have to be equated.

In food processing, it is important to recognize that food is a major source of energy by itself, and, therefore, needs to be considered in the overall energy balance. It is also important to note that the system may transform energy from one form to another during the process. In most food processing applications, it is assumed that the amount of heat converted to other forms of energy, such as work, is negligible.

As in the case of mass balance, many equations are possible to be written on the basis of available information. It is possible to write enthalpy balances around various items of equipment or stages of process or the complete processing plant. Thereafter, from available energy balances, unknown quantities could be found out.

Some reactions in food processing are exothermic and some are endothermic. However, the heat quantities are small as compared to other forms of energy and, therefore, may be ignored. The heat used in calculations can be either sensible heat (heat involved in change in temperature of the material) or latent heat (heat required to change the phase).

In the MKS system the unit of heat is calorie. There is at times a confusion between the dietary Calorie and the physical calorie. While both are energy units, the dietary Calorie (denoted as upper case "C") equals 1000 thermal calories. One calorie in thermal exchanges is the energy needed to raise (or lower) the temperature of one gram of liquid water by one degree centigrade. Since one calorie of energy is too small to describe the energy content of food, in dietary terminology the thermal kilocalorie is used and is called simply a Calorie. Thus, a food Calorie, equal to 1000 cal, is the energy needed to change the temperature of one kg of liquid water by one degree centigrade. Remember that the food Calorie is often spelled with an upper case "C" to distinguish it from the thermal calorie ("c").

Example 3.20 Eight thousand kg of apples are to be frozen from an initial temperature of 26°C to the final temperature of -18°C in 6 hours. Find the rate of heat removal if the maximum demand is twice the average demand. Apple has 86% water. Assume the freezing temperature is -2°C. Take the specific heat of food solids as 1.25 kJ. kg⁻¹.°C⁻¹, that of water as 4.2 kJ.kg⁻¹.°C⁻¹ and the specific heat of ice as 2.1 kJ.kg⁻¹.°C⁻¹. Assume the latent heat of vaporization as 288.1 kJ.kg⁻¹.

SOLUTION:

The specific heat of food solids is given as 1.25 kJ .kg⁻¹.°C⁻¹and that of water is 4.2 kJ.kg⁻¹.°C⁻¹. The apple has 86% water. Thus, in this case:

specific heat of apples above freezing point

$$= 4.2 \times 0.86 + 1.25 \times 0.14 = 3.612 + 0.175$$

$$= 3.787 \text{ kJ.kg}^{-1}.°C^{-1},$$

specific heat of apples below freezing point

$$= 2.1 \times 0.86 + 1.25 \times 0.14 = 1.806 + 0.175$$

$$= 1.981 \text{ kJ.kg}^{-1}.°C^{-1},$$

Total enthalpy change per kg commodity

$$= \text{sensible heat above freezing} + \text{latent heat}$$

$$+ \text{sensible heat below freezing}$$

$$= \left[26 - (-2) \right](3.787) + 0.86 \times 335 + \left[-2 - (-18) \right](1.981)$$

$$= 106.036 + 288.1 + 31.696 = 425.832 \text{ kJ.kg}^{-1}.$$

Total enthalpy change for 8000 kg apples $= 8000 \times 425.832 = 3406656$ kJ

Total time is 8 h, the heat removal rate
$$= 3406656/(6 \times 3600) = 157.715 \text{ kW}$$

And, if the maximum rate of heat removal is twice the average,

the required rate of heat removal
$$= 2 \times 157.715 = 315.43 \text{ kW}.$$

Example 3.21 One thousand cans of vegetable soup (each can weighs 40 g and contains 0.3 kg of soup) is being autoclaved to 100°C (average can temperature). After this, cooling water at 4°C is used to cool the cans up to 25°C. What is the quantity of cooling water required if the water leaves at 20°C?

Assume the soup and the can metal have the specific heats of 4.0 kJ.kg⁻¹.°C⁻¹ and 0.36 kJ. kg⁻¹.°C⁻¹, respectively. Ignore the heat losses.

SOLUTION:

Let us take a datum temperature of 0°C for the calculations (as the final temperature of the cans is 30°C). Let w be the quantity of cooling water required;

Heat Entering

Heat of the cans at 100°C

= weight of cans × temperature above datum × specific heat

$$= (1000 \times 0.04) \times (100 - 0) \times 0.36 \text{ kJ} = 1440 \text{ kJ}.$$

Heat of the can contents (soup)

= weight of soup × temperature above datum × specific heat

$$= (1000 \times 0.3) \times (100 - 0) \times 4.0 = 120000 \text{ kJ}$$

Heat of the water

= weight of water × temperature above datum × specific heat

$$= w \times (4 - 0) \times 4.186 = 16.744 \text{ w kJ}.$$

Heat leaving

Heat in cans = 1000 × 0.04 × (25-0) × 0.36 = 360 kJ

Heat in can contents = 1000 × 0.3 × (25 –0) × 4.0
= 30000 kJ

Heat in water = w × (20 - 0) × (4.186) = 83.72 w kJ

Total heat entering = total heat leaving

Thus, 1440 + 120000 + 16.744 w
= 360 + 30000 + 83.72 w

or 66.976 w = 91080

or w = 1359.89 kg

The above calculations can also be performed by taking the datum temperature of 25°C (as the final temperature of the cans is 25°C) and the same result is obtained.

Example 3.22 Determine the quantity of ice to be mixed at 0°C adiabatically with 1 kg water at 50°C such that the temperature of the resulting water becomes 0°C? Take the specific heat of water as 4.18 kJ.kg⁻¹.°C⁻¹ and the latent heat of melting of ice at 0°C as 335 kJ.kg⁻¹.

SOLUTION:

Heat lost by 1 kg water = m.C_p.Δt = (1)(4.18)(50) kJ

Let m be the amount of ice required and λ be the latent heat of fusion, then heat gained by ice = m. λ

Thus, (m)(335) = (1)(4.18)(50) kJ

or, m = 0.623 kg

Example 3.23 A 1-mm diameter wire with 15 cm length is immersed in water and the water is made to boil by passing electric current through it. If the water is at atmospheric pressure (it boils at 100°C) and h = 5600 W.m⁻².K⁻¹, what is the value of electrical power that must be supplied to maintain the wire surface at 110°C in equilibrium state?

SOLUTION:

This is a condition of heat balance, where the heat supplied by the wire is taken by the water. To maintain constant temperature, the heat taken by the water should be same as heat supplied by the wire.

Heat taken by water = q = h A (t₁-t₂)

In this case, A = π d L = π (0.001) (0.15) = 4.71 × 10⁻⁴ m²

q = (5600 W.m⁻².K⁻¹) (4.71 × 10⁻⁴ m²) (110-100) K
= 26.376 W

Thus, 26.376 W of electrical power must be applied.

Example 3.24 A hot surface of 40 cm × 40 cm is held at 120°C, on which air flows at 25°C. The plate is 6 cm thick and the thermal conductivity of the material is 120 W.m⁻¹.K⁻¹. If the convective heat transfer coefficient on the plate surface exposed to air is 45 W.m⁻².K⁻¹, how much heat is transferred to air? If it is observed that 160 W heat is lost from the plate by radiation, calculate the inside plate temperature.

SOLUTION:

Heat lost by convection to air,

$$q = h A (t_1 - t_2)$$

$$= (45)(40 \times 40)(120 - 25) = 684 \text{ W}$$

Heat lost by radiation = 160 W

Heat transfer by conduction through the plate = heat lost by convection + heat lost by radiation

or, k. A (Δt/L) = 684 + 160 = 844 W

$$\Delta t = \frac{(844)(0.06)}{(0.4)(0.4)(120)} = 2.637°C$$

The inside plate temperature is 120 + 2.637 = 122.637°C

More problems on heat and mass balance will be discussed in different chapters as drying, evaporation, mixing, etc.

Check Your Understanding

1. Suppose milk is available with 4.5 % fat. The milk is desired to have 3% fat only. Find the ratio in which the milk and water need to be mixed?

    Answer: $F_1: F_2 = 2: 1$

2. What should be the ratio in which cream with 80% fat is mixed with double-toned milk so as to get 3% fat milk?

    Answer: $F_1: F_2 = 0.77:0.015 = 51.33: 1$

3. Twenty-five kg of sugar are added to 100 kg of orange juice having 10% solids concentration. Find out the final solid concentration of the juice.

    Answer: 28%

4. Standard toned milk is evaporated in such a way that the total solids in the final product are 35%. What is the amount of evaporation? What is the amount of concentrated milk?

    Answer: Per 100 kg of milk, evaporation = 67.1 kg, final product will be 32.9 kg.

5. The jam is prepared by addition of fruit juice with sugar and then evaporating the mixture to the desired concentration level. In one such process, 40 parts of fruit juice having 12% dissolved solids is mixed with 60 parts sugar. Then the mix is evaporated to a final solid concentration of 65%. What yield of jam is expected from the fruit?

    Answer: 2.49 kg of the jam per kg of fruit

6. Determine the ratio of milk with 3% fat and cream with 88% fat to make a milk product of 3.5% fat? Use Pearson's square method.

    Answer: 84.5 kg of 3% fat milk are required to be added to 0.5 kg of 88% cream to obtain 85 kg of 3.5% fat milk.

7. One thousand kg of sunflower seeds (51% oil, 21% protein, 20% carbohydrates, 7.2% moisture and 0.8% other constituents) are fed to a mechanical oil expeller. After removal of the oil, the meal contained 8% oil. Thus the meal was again fed to a solvent extraction system to extract more oil. After solvent extraction, the meal contained 0.5% oil. The meal was then dried to 5% moisture content before storage. Find out the oil extracted in the expeller and the solvent extractor and the moisture removed in drying.

    Answer: Oil expressed in mechanical expeller = 467.39 kg; Oil extracted in solvent extraction = 40.15 kg; Moisture removed in drying = 49.87 kg

8. In the manufacture of sweetened condensed milk, the milk is mixed with sugar to give a mixture of 65 parts milk to 45 parts of sugar. The mixture is then evaporated until the solids are 70%. What yield of sweetened condensed milk can be expected from 500 kg milk that has 14% solids? Find out the amount of water evaporated from milk and sugar added to the product.

    Answer: 594.72 kg final product and 251.43 kg water evaporation

9. A continuous fermentation system is used for growing Baker's yeast. Flow residence time of 12 h is maintained in the fermenter of 16 m^3 volume. A 2% inoculum is put in the growth medium and then into the fermenter. The inoculum contains 1.4% of yeast cells. Doubling time of yeasts in the fermenter is 3 h. The fermenter effluent containing the broth is fed to a continuous centrifuge and the result is a yeast cream that has 95% of the total yeast in the broth. The yeast concentration in the yeast cream is 8%. Determine the flow of the yeast cream and that of the residual broth from the centrifuge. Assume that the broth has the same density as that of water, i.e. 1000 kg m^{-3}.

    Answer: Flow rate of yeast rich stream 70.91 kg h^{-1}, flow rate of residual broth 1262.08 kg h^{-1}

10. A soft drink is carbonated with 3.3 parts (volumes) of CO_2 with 1 part of the water at atmospheric pressure and 0°C temperature. Calculate the mass and mole fractions of the CO_2 in the drink. Neglect other components in the drink.

    Answer: Mass fraction of $CO_2 = 6.44 \times 10^{-3}$; mole fraction of $CO_2 = 2.63 \times 10^{-3}$

11. Eight thousand kg of apples are to be frozen from an initial temperature of 30°C to the final temperature of -18°C. The total freezing time is to be 8 h. Find the freezing demand if the maximum demand is twice the average demand. Apples have 86% water. Assume the freezing temperature is -2°C. Take any other assumption as necessary.

    Answer: 245 kW

12. Eight hundred cans of corn soup have been autoclaved to 100°C average can temperature. The cans are, subsequently, cooled to 30°C by cooling water. What is the quantity of cooling water required if it enters at 8°C and leaves at 25°C?

    Assume the soup and the can metal have the specific heats of 4.0 kJ.kg^{-1}.°C^{-1} and 0.50 kJ.kg^{-1o}C^{-1}, respectively. Each can weighs 40 g and contains 0.3 kg of soup. Assume that the heat content is 1.6×10^4 kJ above 30°C. Ignore the heat loss through the walls.

    Answer: 960.1 kg

Objective Questions

1. Explain the laws of conservation of mass and energy.

2. With schematic diagrams, explain the different mass and energy balance equations in case of (a) continuous evaporator, (b) continuous tunnel dryer, (c) batch dryer.

3. Explain Pearson's square method of mass balance.

4. Differentiate between
   (a) Molar concentration and mole fraction
   (b) Molarity and molality
   (c) Concentration (w/v) and concentration (w/w)

BIBLIOGRAPHY

Ahmad, T. 1997. *Dairy Plant Engineering and Management*, 4th ed. Kitab Mahal, New Delhi.

Brennan, J.G., Butters, J.R., Cowell, N.D. and Lilly, A.E.V. 1976. *Food Engineering Operations*, 2nd ed. Elsevier Applied Science, London.

Charm, S.E. 1978. *Fundamentals of Food Engineering*, 3rd ed. AVI Publishing Co., Westport, Connecticut.

Dash, S.K. and Sahoo, N.R. 2012. *Concepts of Food Process Engineering.* Kalyani Publishers, New Delhi.

Datta, A.K. 2001. *Transport Phenomena in Food Process Engineering.* Himalaya Publishing House, Mumbai.

Earle, R.L. and Earle, M.D. 2004. *Unit Operations in Food Processing*, Web Edition. The New Zealand Institute of Food Science & Technology, Inc., Auckland. https://www.nzifst.org.nz/ resources/unitoperations/index.htm

Fellows, P.J. 2000. *Food Processing Technology.* Woodhead Publishing, Cambridge, UK.

Heldman, D.R. (Ed.) 1981. *Food Process Engineering.* AVI Publishing Co., Westport, Connecticut.

Kessler, H.G. 2002. *Food and Bio Process Engineering*, 5th ed. Verlag A Kessler Publishing House, Munchen, Germany.

Rao, D.G. 2009. *Fundamentals of Food Engineering.* PHI learning Pvt. Ltd., New Delhi.

Sahay, K.M. and Singh, K.K. 1994. *Unit Operations in Agricultural Processing.* Vikas Publishing House, New Delhi.

Singh, R.P. and Heldman, D.R. 2014. *Introduction to Food Engineering.* Academic Press, San Diego, CA.

Toledo, R.T. 2007. *Fundamentals of Food Process Engineering*, 3rd ed. Springer, New York.

Watson, E.L., and Harper, J.C. 1988. *Elements of Food Engineering*, 2nd ed. Van Nostrand Reinhold, New York.

4

Reaction Kinetics

The concept of the rate of reaction or reaction speed is important in food processing operations. Most of the reactions occurring in food involve conversion of one component to another. A common example is conversion of sucrose to glucose and fructose. Many forms of spoilage of food and degradation of nutrients occur due to chemical reactions within the food. Hence, a knowledge of the rate of reaction is required to decide the processing parameters or to determine the potential shelf life of any food.

4.1 Chemical Kinetics and Order of Reaction

The process of chemical reactions occurring in food can be explained like any other chemical reaction. *Chemical kinetics* is the component of physical chemistry that studies reaction rates. The energy level of the reactants (reacting substances) has to be first increased to a reasonable level for the reaction to start. This energy, which must be supplied to the reactants before the reaction can start, is called the *activation energy.* Activation energy is the threshold value acquired by the reactant(s) before reaching the transition state. The reaction then begins and continues either in a forward or in a backward direction.

The reaction rate or speed of reaction can be slow, as in the browning of a cut apple or the rusting of a metal, or fast, as oxygen reacts with magnesium to form magnesium oxide (Mg + O → MgO) or the *Maillard reaction,* which is a non-enzymatic reaction occurring between sugars and proteins, which lead to browning of some foods as they are heated. The rate of a reaction can be simply expressed as the rate of change in concentration of a reactant. It may also be given as the rate of generation of a new substance. Suppose component A is converted to component B in a particular reaction. The *reaction rate* can be given as the rate of change in concentration of the component A or B. The concentration change can, thus, be given in $mol.dm^{-3}. s^{-1}$ (or $mol. l^{-1}.s^{-1}$), i.e., "moles per cubic decimeter per second" or "moles per liter per second." If B is a gas, then the rate during the reaction can be expressed as the volume given off per unit time at any particular time (say in $cm^3 s^{-1}$).

In any chemical reaction the change in concentration of the reacting substances affects the rate of reaction; these are related by a mathematical expression that is defined as the *rate equation*. The rate equations for different situations are explained below.

Suppose in an experiment it is observed that the reaction rate and the concentration of one of the reactants (say A) are directly proportional to each other, it can be written as

$$\text{rate} \,\alpha\, [A] \qquad (4.1)$$

In this equation, [A] denotes the concentration of A in $mol. l^{-1}$ and the rate has a unit of $mol.l^{-1}.s^{-1}$.

The above equation can be expressed as

$$\text{rate} = k\,[A] \qquad (4.2)$$

Here k is the *rate constant* or *reaction rate coefficient.*

Similarly, in another situation where the reaction rate is proportional to $[A]^2$, the equation would be written as

$$\text{rate} = k\,[A]^2 \qquad (4.3)$$

Thus, the generalized equation will be

$$\text{rate} = k\,[A]^a \qquad (4.4)$$

Here a is the order of reaction.

For experiments involving a reaction between two components A and B, a generalized relationship can be represented as follows:

$$\text{rate} = k\,[A]^a[B]^b \qquad (4.5)$$

The superscripts "a" and "b" in equation 4.5 denote the *order of reaction* with respect to A and B, respectively.

In the rate equations, the rate constant is constant for a given reaction only if the concentration of the reactants is changed. However, as we will discuss later, k varies with change in temperature or in the presence of a catalyst.

It is not possible to deduce the order of a reaction from the reaction's chemical equation. The orders of reactions may be different from the actual number of molecules colliding (*Stoichiometry molecularity*). It is only in elementary reactions that the Stoichiometry molecularity and order of reaction are the same, that is, in one-step reactions.

The order of a reaction need not necessarily be a whole number, it can also have zero and fractional values. In very simple forms the orders of reaction are 0, 1, or 2.

4.1.1 Zeroth Order Reaction

If the reaction rate is constant for a particular component, then the superscript of the concentration in equation 4.4 will have a zero value, and hence it is a zeroth order reaction.

$$\text{rate} = k = k\,[A]^0 \qquad (4.6)$$

The equation expresses that the reaction rate is independent of the reactant's concentration.

DOI: 10.1201/9781003285076-5

To represent the change in concentration (C) of a particular component with time (θ), the above equation can be written as

$$dC/d\theta = -k \qquad (4.7)$$

Integrating within the time limits, 0 to θ (at time 0, the concentration is C_0 and at time θ, the concentration is C),

$$C_0 - C = k\theta \qquad (4.8)$$

Zeroth order reactions are less common and are generally observed when the reactants saturate a material, such as a surface or a catalyst, used in the reactions. The chemical reaction $2HI_{(g)} \rightarrow H_{2(g)} + I_{2(g)}$ is an example of a zeroth order reaction. Similarly, when gaseous ammonia decomposes on a hot platinum surface at high pressure, the rate of reaction remains constant.

$$2NH_{3(gas)} \rightarrow N_{2(gas)} + 3H_{2(gas)}$$

Half-life of a chemical reaction is the time elapsed for the depletion of the reactant to fifty percent of its initial value. If C_0 is the initial concentration of the reactant for a zeroth order reaction then $C_0/2k$ is the half-life. If C is given in $mol.l^{-1}$ and the time is given in s, the rate constant will have a unit of $mol.l^{-1}.s^{-1}$.

4.1.2 First Order Reaction

A first order reaction is written as follows:

$$\text{reaction rate} = k\,[A] \qquad (4.9)$$

In this situation, also called a unimolecular reaction, concentration of only one reactant determines the rate of the reaction. Other reactants, if present, will be of zeroth order.

Molecularity of a reaction represents the number of reacting species (which include atoms, ions, or molecules) that participate in the reaction (reactions taking place in one step) and collide at the same time for the chemical reaction to take place.

As we will be discussing in the chapter on thermal processing, a very common example of this order reaction is the bacterial death with time under thermal processing.

The hydrolysis of sucrose forming glucose and fructose is another example of first-order reaction.

$$C_{12}H_{22}O_{11} + H_2O \rightarrow C_6H_{12}O_6 + C_6H_{12}O_6$$

Here only one component is used for the reaction. Though there is the presence of water it does not affect the reaction.

The equation 4.9 can be modified to represent the change in concentration (C) of a particular component with time (θ) as follows:

$$dC/d\theta = -k\,C \qquad (4.10)$$

Integrating within the time limits, 0 to θ (at time 0, the concentration is C_0 and at time θ, the concentration is C),

$$\int_{C_0}^{C} \frac{dC}{C} = -k \int_0^{\theta} d\theta$$

$$\text{or, } \ln\left(\frac{C}{C_0}\right) = -k\theta$$

$$\text{or } \ln\left(\frac{C_0}{C}\right) = k\theta \qquad (4.11)$$

$$\text{i.e. } \ln C = \ln C_0 - k\theta \qquad (4.12)$$

The above equation 4.11 can also be written as

$$\log\left(\frac{C_0}{C}\right) = \frac{k\theta}{2.303} \qquad (4.13)$$

$$\text{i.e. } \log C = \log C_0 - \frac{k\theta}{2.303} \qquad (4.14)$$

The above equation is of the form, $y = C + mx$ (Here C is a constant)

A straight line is obtained on a plotting of log C versus time. Here the ordinate is $\log_{10}C$, the abscissa is the time, θ, the y intercept is $\log_{10}C_0$, and the line slope $b = -k$. The derivative dy/dx gives the slope. The dimension of rate constant will be $time^{-1}$.

4.1.3 Second Order Reaction

The reactions follow the equation,

$$\text{rate} = k\,[A]^2 \qquad (4.15)$$

As discussed above, for a second order rate reaction,

$$\frac{dC}{d\theta} = -kC^2 \qquad (4.16)$$

Integration of this equation within limits yields

$$\frac{C - C_0}{CC_0} = -k\theta \quad \text{or} \quad \frac{C_0 - C}{CC_0} = k\theta \qquad (4.17)$$

In this case the rate constant is obtained by plotting the ratio $(C_0 - C)/CC_0$ with respect to time on a plain graph paper. The slope of this line will be k. It will have the unit of $l.mol^{-1}.s^{-1}$.

Most of the gas phase reactions are of second order.

It is unlikely to observe elementary reactions of third order.

4.1.4 Alternate Modes of Expressing the First and Second Order Reactions

We can also derive the rate constants in a slightly different approach as follows.

For a first order reaction, if the initial concentration value is "a" and the concentration after time θ is (a-x), then

$$\frac{dx}{d\theta} = -k(a - x) \qquad (4.18)$$

x is a variable here.

Thus,

$$\frac{dx}{a-x} = -k.d\theta$$

or $\left[\ln(a-x)\right]_{x=0}^{x=x} = -k\left[\theta\right]_0^\theta$

The final solution will be

$$\ln\frac{a}{a-x} = k\theta \qquad (4.19)$$

and $\log(a-x) = \log a - \frac{k\theta}{2.303} \qquad (4.20)$

Check that the equations 4.19 and 4.20 are of the same forms as equations 4.11 and 4.14.

When in a reaction, two components, A and B, are involved,

$$A + B \rightarrow P \qquad (4.21)$$

The concentration changes as follows:

$$a + b \rightarrow (a-x) + (b-x) \qquad (4.22)$$

Thus, $\frac{dx}{d\theta} = -k(a-x)(b-x) \qquad (4.23)$

or, $\frac{dx}{(a-x)(b-x)} = -kd\theta \qquad (4.24)$

i.e. $\log\frac{b(a-x)}{a(b-x)} = -\frac{k\theta(a-b)}{2.303} \qquad (4.25)$

If the initial concentrations are equal (a=b) and the change in concentration is denoted as x, then the equation 4.23 can be written as

$$\frac{dx}{d\theta} = -k(a-x)^2 \qquad (4.26)$$

We get by integration

$$-\frac{1}{(a-x)} = -k\theta + M$$

At $\theta = 0$, $x = 0$, and thus, $M = -1/a$

Hence, $\frac{x}{a(a-x)} = -k\theta \qquad (4.27)$

For equation 4.25, k is determined by plotting $\log\frac{b(a-x)}{a(b-x)}$-vs-

time, or when a = b, the graph is plotted as $\left(\frac{x}{a-x}\right)$-vs-time.

A more general presentation of the order of reaction can be as follows:

$$\frac{dC}{d\theta} = -k(C)^m \qquad (4.28a)$$

$$\int_a^{a-x}\frac{dC}{C^m} = -k\int_0^\theta d\theta \qquad (4.28b)$$

$$\frac{a^{1-m} - (a-x)^{1-m}}{1-m} = k\theta \qquad (4.29)$$

4.1.5 Overall Order of the Reaction

In case of a reaction that involves many reactants, the concentration of each reactant is raised to some power, and, in that situation, the *overall order of the reaction* is represented by the summation of all individual orders, whereas the powers of the individual reactants are the orders of individual reaction. For example, in equation 4.5, the overall order is a+b. If the values of a and b are 1 each, then this is an overall second order reaction.

For the reaction $CO + NO_2 \rightarrow CO_2 + NO$, the reaction equation can be written as

$$rate = k[CO][NO_2] \qquad (4.30)$$

The reactions with respect to CO and NO_2 are individually first order reactions, both a and b are 1 and the reaction is of second order (a + b = 2).

In the chemical reaction $O_2 + 2NO \rightarrow 2NO_2$, the reaction equation can be written as

$$rate = k[O_2]^1[NO]^2 \qquad (4.31)$$

It is an overall third order reaction, a + b = 3. As mentioned, third order reactions are rare.

4.1.6 Other Orders of Reactions

As in a bimolecular reaction between adsorbed molecules, with respect to a reactant, reactions can have an undefined order.

The concentration of that reactant remains constant, which is in great excess as compared to other reactants; however, a pseudo constant is obtained, and its concentration is included in the rate constant. If the reactant having constant concentration is B, then

$$r = k[A]^a[B] = k'[A]^a \qquad (4.32)$$

The situation is also true if there is a catalyst in the reaction as the concentration of catalyst does not change in the reaction. Here, a *pseudo first order* rate equation is obtained for the second order rate equation. This is why it will be more appropriate to take the conversion of sucrose into glucose and fructose as a pseudo first order reaction.

The hydrolysis of esters by dilute mineral acids is an example of a pseudo first order reaction as excess amount of water is involved in the reaction.

$$CH_3COOCH_3 + H_2O \rightarrow CH_3COOH + CH_3OH$$

TABLE 4.1

Units of rate constants

Order of reaction	Unit of rate constant
Zero	$mol.l^{-1}.s^{-1}$
One	s^{-1}
Two	$l.mol^{-1}.s^{-1}$
n	$l^{n-1}.mol^{1-n}.s^{-1}$

The approximation of pseudo first order reaction has another practical utility. For example, the measurement of a second order reaction rate is difficult since the concentrations of the two simultaneously changing reactants need to be monitored. In this situation the pseudo first order approximation can be taken.

As mentioned before, the orders of reactions can be non-integers also. For example, the order of the reaction of the decomposition of ethanol to methane and carbon monoxide is 1.5 with respect to ethanol. Such reactions are known as *broken-order reactions*.

A *mixed-order reaction* is the one where the order of a reaction changes during the process due to the change in operating variables as pH.

Conversion of ozone (order 2) to oxygen (order 1) is one of the rare cases of *negative-order reactions*.

As it is observed from the different rate equations, different orders of reactions call for different units of rate coefficients (Table 4.1).

The order of a reaction is generally found out experimentally because the coefficients of a balanced equation do not permit its determination.

Example 4.1 If we plot bacterial population with time (min) on a semi log chart, we obtain a straight line indicating that it is a first order reaction. If the slope of the curve has been obtained as 0.12, what will be the value of reaction rate constant?

SOLUTION:

The straight line slope will be k/2.303.
   Thus, $k = 0.12 \times 2.303 = 0.276$ min^{-1}.
Check that the reaction rate constant has the unit of min^{-1} being a first order reaction.

Example 4.2 In the reaction, $2N_2O_5 \rightarrow 4NO_2 + O_2$ the concentration (mol. l^{-1}) of N_2O_5 changed from 0.88 to 0.24 in 5 min at 320 K. Determine the rate of reaction if it is a reaction of first order.

SOLUTION:

The reaction rate is obtained by plotting the concentrations vs time on a semi log paper.
   The slope of the line = k/2.303
   $= (\log(C_1) - \log(C_2))/\theta$
   Thus, $k/2.303 = \log(0.88/0.24)/5$
   $= 0.5643/5 = 0.1128$ min^{-1}
The rate of reaction is $k = 0.2599$ min^{-1}

Example 4.3 Initial concentration of N_2O_5 for the above reaction occurring at the same condition is kept at 0.240 mol. l^{-1}. What will be the concentration after 2.0 min?

SOLUTION:

We have already found out that the rate of reaction is $k = 0.2599$ min^{-1}
   Thus, for the second situation

$$\log(0.24/C) = (0.2599)(2)/2.303 = 0.2257$$

$$\text{or, } 0.24/C = 1.6815$$

$$\text{or, } C = 0.1427 \text{ mol. l}^{-1}$$

Example 4.4 In the above situation, what will be the time required to change the concentration (mol. l^{-1}) from 0.24 to 0.08?

SOLUTION:

$$\ln(C_0/C) = k\theta$$

$$\ln(0.24/0.08) = (0.2599)(\text{min}^{-1})\theta(\text{min})$$

$$\theta = 1.0986/0.2599 = 4.22 \text{ min}$$

Example 4.5 In the above situation, what will be the time required to decompose half of the sample?

SOLUTION:

$$\ln(C_0/C) = k\theta$$

$$\ln(2) = (0.2599)(\text{min}^{-1})\theta(\text{min})$$

$$0.693 = (0.2599)(\text{min}^{-1})\theta(\text{min})$$

$$\theta = 2.66 \text{ min}$$

Check that it will be always ln (2) for decomposition of half of the sample on the left-hand side. Thus, the initial concentration is not important.

The above implies that, the time required for a reactant to decompose to one-half of its initial value is fixed independent of concentration.

The *half-life* for a first order reaction is given as

$$\theta_{1/2} = 0.693/k$$

k is the first order reaction rate constant. In other words, half-life and the rate constant k are inversely correlated.

Half-life is short if k is large whereas a relatively long half-life is there for a slow reaction, i.e., a small k.

TABLE 4.2

Comparison of different orders of reactions

Order	Rate equation expression	Relationship between concentration and time	Half-life	Linear plot
2	rate = kC^2	$\dfrac{1}{C} - \dfrac{1}{C_0} = k\theta$	$\dfrac{1}{C_0 k}$	$(1/C)$-vs-θ
1	rate = kC	$\log\left(\dfrac{C_0}{C}\right) = \dfrac{k\theta}{2.303}$	$\dfrac{0.693}{k}$	$\log C$-vs-θ
0	rate = k	$C_0 - C = k\theta$	$\dfrac{C_0}{2k}$	C-vs-θ

Example 4.6 Determine the half-life of a second order reaction.

SOLUTION:

For a reaction of second order, $\dfrac{C_0 - C}{CC_0} = k\theta$

In this case $C = C_0/2$

Thus, the above equation can be modified as

$\dfrac{C_0/2}{C_0^2/2} = k\theta$

i.e., $\theta = \dfrac{1}{C_0 k}$

Table 4.2 gives a comparison of different orders of reactions.

With the above understanding, the reaction rate constant and orders are found experimentally as follows. For example, during the decomposition of $N_2O_5(g)$, the reaction takes place as follows:

$$2N_2O_5(g) \rightarrow 4NO_2(g) + O_2(g)$$

The form of the rate law for decomposition of $N_2O_5(g)$ will be, Rate = $k[N_2O_5]^a$.

To determine "a," N_2O_5 concentration is initially prepared in a flask and the rate of reduction in concentration is measured. Then, the initial concentration is changed and the new rate of decomposition of N_2O_5 is measured. We can find the order of the decomposition reaction by comparing these rates. The *method of initial rates* is the approach to find reaction order in this manner. The following example further explains the whole process.

Example 4.7 In the reaction, $2N_2O_5(g) \rightarrow 4NO_2(g) + O_2(g)$, it is observed that the rate of reaction changed from 0.0136 M. l^{-1}. s^{-1} to 0.0272 M. l^{-1}. s^{-1}, when the initial concentration of N_2O_5 was changed from 7.2×10^{-4} M.l^{-1} to 1.44×10^{-3}M. l^{-1}. Find out the order of reaction.

SOLUTION:

$$\frac{\text{rate1}}{\text{rate2}} = \frac{k\left[N_2O_5\right]_1^a}{k\left[N_2O_5\right]_2^a}$$

i.e., $\dfrac{(0.0272\text{M.l}^{-1}.\text{s}^{-1})}{(0.0136\text{M.l}^{-1}.\text{s}^{-1})} = \dfrac{(1.44\times 10^{-3}\text{M.l}^{-1})^a}{(7.2\times 10^{-4}\text{M.l}^{-1})^a}$

Or, $\dfrac{2}{1} = (2)^a$, i.e., a = 1,

i.e., it is a first order reaction.

If there are two reactants, as in $H_2(g) + I_2(g) \rightarrow 2HI(g)$, it is expected that the rate of reaction will be affected by the concentration of both reactants.

$$\text{Rate} = k[H_2]^a[I_2]^b$$

Thus, we conduct a set of two experiments maintaining the initial concentration of one component constant and varying that of the second reactant and another set of two experiments keeping the initial concentration of second reactant constant. The following example illustrates the method.

Example 4.8 In the reaction of $H_2(g) + I_2(g) \rightarrow 2HI(g)$, the initial concentrations for hydrogen gas and iodine gas and the reaction rates (M.s^{-1}) at 750K are as follows. Find out the orders of reactions.

Experiment number	$[H_2]_0$ (M.l^{-1})	$[I_2]_0$ (M,l^{-1})	Rate (M.l^{-1}. s^{-1})
1	0.14	0.14	4.2×10^{-4}
2	0.28	0.14	8.4×10^{-4}
3	0.28	0.28	16.8×10^{-3}

SOLUTION:

By comparing the rates,

$$\frac{\text{rate 1}}{\text{rate 2}} = \frac{k\left[H_2\right]_1^a\left[I_2\right]_1^b}{k\left[H_2\right]_2^a\left[I_2\right]_2^b}$$

i.e., $\dfrac{4.2\text{x}10^{-4}\text{M.l}^{-1}\text{s}^{-1}}{8.4\text{x}10^{-4}\text{M.l}^{-1}\text{s}^{-1}} = \dfrac{k(0.14\text{M.l}^{-1})^a(0.14\text{M.l}^{-1})^b}{k(0.28\text{M.l}^{-1})^a(0.14\text{M.l}^{-1})^b}$

$$0.5 = (0.5)^a (1.0)^b$$

Thus, a = 1.

By proceeding as above, we find that b =1. Thus, this is an overall second order reaction with first order for each reactant.

Once the orders of reaction are determined, substituting the values in the rate equation gives the rate constants.

Example 4.9 Determine the reaction rate constant in the above example of overall second order reaction.

SOLUTION:

$$k(0.14\text{M.l}^{-1})^a(0.14\text{M.l}^{-1})^b = 4.2\times 10^{-4}\text{M.l}^{-1}.\text{s}^{-1}$$

$$k(0.14\text{M.l}^{-1})^2 = 4.2\times 10^{-4}\text{M.l}^{-1}.\text{s}^{-1}$$

$k = (4.2 \times 10^{-4})/ (0.14 \times 0.14) = 2.14 \times 10^{-2}$ l. (M.s)$^{-1}$

Example 4.10 During the reaction of $2N_2O_5 \rightarrow 4NO_2 + O_2$, the concentration of N_2O_5 (mol.l^{-1}) was plotted with respect to time (min) on a semi-log paper and the plot was observed to be a straight line with a slope of -0.12min^{-1}. Determine the concentration of N_2O_5 after 5 min if the initial concentration is 0.18 mol. l^{-1}.

SOLUTION:

The data are plotted on a semi-log paper. Thus, the slope of the line is k/2.303.

In this case, $k/2.303 = 0.12$min^{-1}

Now, $\log(C_0/C) = (k/2.303)\theta$

$\log (0.18/C) = (0.12) (5) = 0.6$

$0.18/C = 3.98$

$C = 0.045$ mol. l^{-1}

In this case we can also write as $k = 0.276$ min^{-1} and use the equation $\ln(C_0/C) = k\theta$.

4.2 Factors Influencing Reaction Rate

The reaction rate is affected by the following factors.

Nature of the reaction. Nature of reaction, physical state of the reacting species (solid, gas, or liquid), their number, the complexity of the reaction, etc.

Concentration of reactants. In general, an increase in the concentrations of reactants leads to an increase in the frequency of collisions and, therefore, an increase in the reaction rate.

Pressure. The concentrations of reactants change with a change in pressure, which affects the rate of the reaction. However, for the reactants in condensed-phase (solids or liquids), the rate constant is almost independent of pressure in the range normally encountered in industry. Therefore, the pressure effect is neglected in practice.

Temperature. Molecules move more quickly and collide more vigorously with an increase in temperature, which enhances the probability of bond cleavages and rearrangements. Besides, molecular collisions are more successful since the colliding particles have the necessary activation energy. All these factors increase the reaction rate. As compared to the food stored in a refrigerator, food spoils more quickly at room temperature. Milk becomes sour quickly when stored at room temperature – these are examples of food reactions being temperature dependent.

The temperature dependence of a reaction rate is generally explained by the Arrhenius equation. As a rule of thumb for many reactions, every 10°C increase in temperature leads to a doubling of the reaction rate.

In some rare situations, the reaction rate is not affected by the temperature (called non-Arrhenius behavior) or there may be a decrease in the reaction rate with an increase in temperature (anti-Arrhenius). The rate constant decreases with a temperature increase for those reactions that have no activation barrier (e.g., some radical reactions).

Surface area. The reaction rate increases with an increase in the surface area as more solid particles are exposed for the reaction. Stirring strongly influences the rates for heterogeneous reactions.

Order of reaction. It controls the reaction rate by managing the reactant concentration (or pressure).

Solvent. The properties of the solvent in a solution and the ionic strength affect the reaction rate.

Intensity of electromagnetic radiation. Electromagnetic radiation imparts energy to the reactants and may accelerate the reactions. An increase in the radiation intensity will increase the reaction rate.

Presence of catalyst. The rate of reaction is accelerated by the presence of a catalyst in both forward and reverse directions.

Besides, reaction rates for the same molecule could be different because of the kinetic isotope effect if it has different isotopes.

Out of all the factors mentioned above, the temperature is the most important one affecting the rate of reaction. Also, as we discussed, the rate equation consists of three parameters, viz., concentration; the rate constant, k; and the reaction order. Thus, all the above factors, except concentration and order, are considered in the rate constant, k.

4.3 Influence of Temperature on Rate of Reaction

The *Arrhenius equation* (proposed by Svante Arrhenius, 1889) is used to show the temperature dependence of reaction rates. The variations of diffusion coefficients, the death rate of microorganisms in thermal processing, creep rates, etc. with respect to temperature are modelled using this equation. The changes in the rates of most reactions in food with temperature are also described reasonably by the Arrhenius equation.

The following is the Arrhenius equation.

$$k = Ae^{-\frac{E_a}{RT}} \qquad (4.33)$$

where the rate constant is k, E_a is the *activation energy*, and T is the absolute temperature in Kelvin. R is the universal gas constant (8.31434 J.(g-mol)$^{-1}$. K^{-1}) and A is the *pre-exponential factor* (or simply the *pre-factor*).

The equation indicates that there is exponential decay of the rate constant in accordance with the exponent $-E_a/RT$. The factor RT is essentially the average kinetic energy, and, thus, the exponent is the activation energy divided by the average kinetic energy. The negative sign indicates that the rate will be smaller for higher E_a/RT. In other words, the reaction rate is inversely proportional to the activation energy and directly proportional to the temperature.

The activation energy can be lowered by the use of catalysts. In fact, the activation energy of a reaction without any catalyst is higher in comparison to the activation energy of the catalyzed reaction. Thus, the above equation also implies that the effect of temperature is more for an un-catalyzed reaction than the corresponding catalyzed reaction. The exponential change indicates that the effects of these factors are quite significant.

The equation may also be expressed as

$$k = Ae^{-\frac{E_a}{k_B T}} \qquad (4.34)$$

In this equation, the universal gas constant has been replaced by the Boltzmann constant, k_B. Accordingly, E_a in this equation will have different units from those in equation 4.33. In the equation 4.33, E_a has the unit common in chemistry, i.e., energy per mole, while equation 4.34 uses the unit common in physics, i.e., energy per molecule.

The units of A, the pre-exponential factor, is the total number of collisions per second that leads to the occurrence of reaction or not (Check that k is the number of collisions per second in a continuing reaction). The likelihood of any given collision resulting in a reaction is given by $e^{-E_a/RT}$.

The units of A are different according to the order of the reaction and are the same as that of the rate constant. The unit is s^{-1} for a first order reaction, and, hence, it is known as the reaction's *attempt frequency* or *frequency factor*.

Since the temperature range encountered in kinetic studies is small, the activation energy can be assumed to be temperature independent. Similarly, the effect of temperature on the pre-exponential factor can be neglected under a wide range of practical conditions. However, in "barrier-less" diffusion-limited reactions, A is dominant as compared to the $e^{-\frac{E_a}{RT}}$ factor, and hence should not be ignored.

Equation 4.33 yields after taking the natural logarithm

$$\ln(k) = -\frac{E_a}{RT} + \ln(A) \qquad (4.35)$$

The equation is of the form $y = mx + C$, where $x = T^{-1}$.

Therefore, the result is a straight line when ln (k) is plotted against T^{-1}. Here, (-R) times the slope of the straight line is the activation energy. The intercept gives the value of A.

As governed by the Maxwell-Boltzmann law, the exponential part of the Arrhenius equation indicates the component of reactant molecules that possesses enough kinetic energy for the reaction. This fraction can vary between zero to one (depending on the values of E_a and

T). If the fraction is zero, then the equation 4.35 reduces to k = A. In other words, if either the kinetic energy of all molecules exceeded E_a (which is not common) or the activation energy were zero, the fraction of molecules that would react is represented by A.

The factor specifically relating to molecular collisions is the pre-exponential factor A. It relates to the frequency of molecules with enough energy colliding in the correct orientation to initiate a reaction. The value of A is determined experimentally because it depends on temperature and varies with different reactions. It is not advisable to extrapolate the pre-exponential factor in view of the linearity of $ln(k)$ over only a narrow temperature range.

The relative molecular orientation is important at the point of collision in some reactions; therefore, a *steric factor* (denoted by ρ) is introduced and A is defined as

A=Zρ, where Z is the *frequency factor*.

The value of ρ is estimated by comparing the observed rate constant with the one in which *A* is assumed to be the same as Z in some cases because it is otherwise difficult to assess.

In comparing two rate constants, we can write that

$$\ln(k_2) - \ln(k_1) = \ln(A) - \frac{E_a}{RT_2} - \ln(A) + \frac{E_a}{RT_1} \qquad (4.36)$$

$$\ln\frac{k_2}{k_1} = \frac{E_a}{R}\left(\frac{1}{T_1} - \frac{1}{T_2}\right) \qquad (4.37)$$

The activation energy can thus be found out from experiments at only two temperatures.

Example 4.11 What will be the reaction's activation energy, which doubles its rate between 20°C to 30°C ?

SOLUTION:

$$\ln\frac{k_2}{k_1} = \frac{E_a}{R}\left(\frac{1}{T_1} - \frac{1}{T_2}\right)$$

$$E_a = \frac{(8.314)\ln(2/1)}{\dfrac{1}{(273.15+20)} - \dfrac{1}{(273.15+30)}} = \frac{(8.314)(0.693)}{(.0034 - 0.00329)}$$

$$= 52378.2 \text{ J.mol}^{-1} = 52.378 \text{ kJ.mol}^{-1}$$

Example 4.12 In an experiment conducted at sea level, a particular food took 6 minutes for denaturation of proteins in it. If the same food were cooked at a higher altitude (where boiling point of water was 98°C), the time of denaturation was observed to be 8 minutes. If the reaction

(denaturation of proteins) is considered to be of first order, estimate the activation energy for the protein denaturation reaction.

SOLUTION:

Considering the reaction to be of first order,

$\ln(C_0 / C) = k\theta$

If the initial and final concentrations are the same, then k is inversely proportional to θ.

Thus, $k_2 / k_1 = \theta_1 / \theta_2$ (let 1 be the sea level conditions and 2 be at higher altitude.)

In this situation, $k_2 / k_1 = \theta_1 / \theta_2 = 6 / 8 = 0.75$, and the temperatures are 373.15°C and 371.15°C.

$$\ln(0.75) = \frac{E_a}{R}\left(\frac{1}{T_1} - \frac{1}{T_2}\right)$$

$$E_a = \frac{(8.314)\ln(0.75)}{\dfrac{1}{373.15} - \dfrac{1}{371.15}} = \frac{-2.3919}{-1.444 \times 10^{-5}}$$

$$= 165632.64 \text{ J.mol}^{-1}$$

Another modification of the equation 4.33 is

$$\frac{d(\ln k)}{dT} = \frac{E_a}{RT^2} \tag{4.38}$$

Other modified forms of the Arrhenius equation are also used for special situations. Two of such equations are given below.

$$k = A\left(\frac{T}{T_0}\right)^n e^{-E_a/RT} \tag{4.39}$$

where n is a unit less power and T_0 is a reference temperature. The value of n usually lies between −1 and 1.

$$k = Ae^{\left(-E_a/RT\right)\beta} \tag{4.40}$$

Here β is a dimensionless number. The main purpose of β is to make the model fit the data. However, this dimensionless number may sometimes have theoretical relevance, for example, relating to a range of activation energies.

When side products or reaction intermediates are formed, the term used for the products is "rate of appearance" and that for the reactants is "rate of disappearance."

Another equation, known as the Eyring equation (1935), also known as the Eyring–Polanyi equation, describes the chemical reaction rate and energy relationship.

It is written as

$$k = \left(\frac{k_B T}{h}\right)\exp\left(\frac{\Delta S}{R}\right)\exp\left(-\frac{\Delta H}{RT}\right) \tag{4.41}$$

or, in a linear form as follows

$$\ln\frac{k}{T} = -\frac{\Delta H}{R}\cdot\frac{1}{T} + \ln\left(\frac{k_B}{h}\right) + \left(\frac{\Delta S}{R}\right) \tag{4.42}$$

In these equations, ΔH is the enthalpy of activation, R is the gas constant, k_B = Boltzmann constant, h = Planck's constant,

and ΔS is the entropy of activation. The plot of $\ln(k/T)$ with respect to $(1/T)$ gives a straight line, which has the slope of $-\Delta H/R$ and thus ΔH is obtained. The intercept is $\ln\left(\dfrac{k_B}{h}\right) + \left(\dfrac{\Delta S}{R}\right)$. It gives the value of ΔS.

Check Your Understanding

1. For a reaction, if the rate constants in $M^{-1}.s^{-1}$ are 2.8 at 350 K and 16.4 at 500 K, what is the reaction's activation energy (in $kJ.mol^{-1}$)?

     Answer: 17.147 $kJ.mol^{-1}$

2. Find the rate constant for a reaction at 100°C when 4.5 $M^{-1}.s^{-1}$ is the pre-exponential factor and 260 kJ.mol^{-1} is the activation energy.

     Answer: 1.8×10^{-36} $M^{-1}.s^{-1}$.

3. Find the rate constant at 200°C if the rate constant and activation energy at 100°C are 1.8×10^{-36} $M^{-1}.s^{-1}$ and 260 $kJ.mol^{-1}$.

     Answer: $8.86 \times 10^{-29} M^{-1}.s^{-1}$

4. What will be the activation energy at 25°C, if 10 $M^{-1}s^{-1}$ is the pre-exponential factor and the rate constant is 7.6 $M^{-1}.s^{-1}$?

     Answer: 680.28 $J.mol^{-1}$.

5. For a reaction if the activation energy is 280 $kJ.mol^{-1}$ and the rate constant is 10 $M^{-1}.s^{-1}$ at 25°C, find out the temperature at which the rate constant will be 15 $M^{-1}.s^{-1}$.

     Answer: 26.07°C

Objective Questions

1. What are the governing equations of zero first, and second order reactions? Also mention the units of rate constants for these orders of reactions.

2. What is a pseudo first order reaction?

3. What is the molecularity of a reaction?

4. What is the half-life of a reaction? What is the half-life of a first order reaction in terms of the rate constant and initial concentration?

5. Name the factors that influence the rate of a reaction.

6. Explain the effect of temperature on the rate of reaction.

7. Differentiate between

    (a)   order of reaction, reaction rate, and rate constant;

    (b)   activation energy and pre-exponential factor;

    (c)   zeroth order reaction and first order reaction (with examples).

BIBLIOGRAPHY

Evans, M.G. and Polanyi M. 1935. Some applications of the transition state method to the calculation of reaction velocities, especially in solution. *Trans. Faraday Soc.* 31: 875–894. doi:10.1039/tf9353100875.

Eyring, H. 1935. The activated complex in chemical reactions. *J. Chem. Phys.* 3 (2): 107–115.

Ghosal, S.K., Sanyal, S.K. and Datta, S. 1993. *Introduction to Chemical Engineering.* Tata McGraw Hill, New Delhi.

Kessler, H.G. 2002. *Food and Bio Process Engineering,* 5th ed. Verlag A Kessler Publishing House, Munchen, Germany.

Perry, R.H., Chilton, C.H. and Kirkpatrick, S.D. 1963. *Chemical Engineers Handbook,* 4th ed. McGraw-Hill Book Co., New York.

Peters, M.S. 1954. *Elementary Chemical Engineering.* McGraw Hill Book Co., New York.

Rao, D.G. 2009. *Fundamentals of Food Engineering.* PHI learning Pvt. Ltd., New Delhi.

Skinner, G.B. 1974. *Introduction to Chemical Kinetics.* Academic Press, New York.

Toledo, R.T. 2007. *Fundamentals of Food Process Engineering,* 3rd ed. Springer, New York.

5

Psychrometry

The calculation of air required during convective air drying needs knowing the physical and thermal conditions of the air. These conditions can be expressed in terms of temperature, humidity ratio, relative humidity, enthalpy, etc. and explain the capability of the drying air to absorb moisture and the extent to which the commodity could be dried.

The study of moist air is known as *psychrometry*. When water vapor is mixed with dry air, the mixture is termed *moist air*. A knowledge of moist air properties is necessary for studying the drying of materials and designing of drying systems using natural or heated air. Though the water vapor quantity in moist air, used for drying operation, is small (from about 0.002 kg to 0.1 kg water vapor per kg of dry air), it significantly influences the drying operation. The psychrometric calculations are usually based on the standard atmospheric conditions at mean sea level, i.e., air temperature of 15°C and atmospheric pressure of 101.325 kPa.

5.1 Psychrometric Properties of Air-Water Vapor Mixture

The different psychrometric properties of air-water vapor mixture are as follows.

Dry bulb temperature (t or t_{db}): Temperature indicated by a thermometer with its sensor exposed to the air-water vapor mixture is its dry bulb temperature.

Water vapor partial pressure at saturation (p_{ws}): When moisture is continuously added to air, it becomes saturated with water vapor and its relative humidity approaches 100% at a given temperature and pressure. Any additional water added to the air will become liquid water; it may be either mist or droplets. Water vapor partial pressure at saturation is the value of the partial pressure at which the moist air is in saturation condition, i.e., the moist air is in a state of equilibrium with free water on a flat surface. The temperatures of the air and water in the moist air at saturation partial vapor pressure are equal.

An increase in moist air temperature leads to an increase in the saturation vapor pressure of the moist air and the air at higher temperature can hold more moisture before saturation. An empirical relationship of saturation vapor pressure (p_{ws}) with temperature for moist air is given as follows:

$$p_{ws} = 6.11 \times 10^{S} \qquad (5.1)$$

where p_{ws} is in millibars (hPa or hecto Pascals), and $S = \dfrac{7.5\,t}{237.7 + t}$, t is in °C.

Example 5.1 Determine the saturation vapor pressure of moist air at 25°C dry-bulb temperature.

SOLUTION:

Equation 5.1 can be used for finding the solution.

$$p_{ws} = 6.11 \times 10^{S}$$

Here $S = \dfrac{7.5\,t}{237.7 + t}$

i.e., $S = 0.71374$

$$p_{ws} = 6.11 \times 10^{0.71374} = 31.606 \text{ mb}$$

$$= 31.61 \text{ hPa}$$

Albright (1990) suggested the following relationship for determining the saturation vapor pressure of moist air at different temperatures.

$$\ln(p_{ws}) = \frac{A_1}{T} + A_2 + A_3 T + A_4 T^2 + A_5 T^3 + A_6 T^4 + A_7 \ln(T)$$

$$(5.2)$$

In the above equation, p_{ws} is the saturation vapor pressure in Pa and T is the absolute temperature, K.

The constants of the above equation for the temperatures between 0 to 200°C are as follows:

A_1	A_2	A_3	A_4	A_5	A_6	A_7
-5.8002 $\times 10^3$	1.3915	-48.6402 $\times 10^{-3}$	41.7648 $\times 10^{-6}$	-14.452 $\times 10^{-9}$	0	6.546

The constants for -100 to 0°C temperature range are also available (Albright 1990).

Example 5.2 The moist air temperature is 30°C and RH is 40%. Find its saturation vapor pressure at normal atmospheric pressure.

SOLUTION:

Using the equation 5.2,

DOI: 10.1201/9781003285076-6

TABLE 5.1

Saturation vapor pressure at different temperatures

Temperature, °C	p_{ws}, kPa	Temperature, °C	p_{ws}, kPa
0	0.611	50	12.333
10	1.228	60	19.92
20	2.338	70	31.16
25	3.168	80	47.34
30	4.242	90	70.10
40	7.375	100	101.325

$$\ln(p_{ws}) = \frac{-5.8002 \times 10^3}{303.15} + 1.3915 + (-)48.6402 \times 10^{-3}(303.15)$$

$$+ 41.7648 \times 10^{-6}(303.15)^2 + (-)14.452 \times 10^{-9}(303.15)^3$$

$$+ 6.546 \ln(303.15) = 8.3593$$

Thus, p_{ws} = exp (8.3593) = 4269.7678 Pa
The value of p_{ws} can also be directly obtained from
a steam table. Table 5.1 shows the saturation vapor
pressure values for some selected temperatures.

Relative humidity (RH or φ): Relative humidity represents the extent to which the moist air is saturated by
water vapor and is given as the actual partial water
vapor pressure of moist air divided by the partial
water vapor pressure at saturation at identical temperature and atmospheric conditions.

Relative humidity is also given by actual mole
fraction of water vapor in moist air sample divided
by the mole fraction of water vapor if the moist
air were saturated at the same temperature and
pressure.

$$\varphi = p_w/p_{ws} = x_w/x_{ws} \qquad (5.3)$$

Humidity ratio (W): It is also known as absolute
humidity, and it is given by the amount of water
vapor evaporated into a unit amount of dry air. Its
unit is kg of moisture per kg of dry air.

Molecular weights of water vapor and air have a ratio of
0.62198 (18.01534/28.9645).

$$\text{Thus, } W = 0.622 \frac{x_w}{x_a} \qquad (5.4)$$

Assuming water vapor and air act as perfect gases,

$$p_aV = n_aRT, \ p_wV = n_wRT, \text{ and } pV = nRT \qquad (5.5)$$

where p is the atmospheric pressure ($p=p_a+p_w$) and n for
moist air is the total number of moles of air and water ($n=n_a+n_w$)

$$\text{Thus, } p_a/p=x_a; \text{ and } p_w/p = x_w$$

$$\text{and, } W = 0.622 \frac{p_w}{p_a} \text{ or } W = 0.622 \frac{p_w}{p - p_w} \qquad (5.6)$$

$$\text{Similarly, } W_s = 0.622 \frac{p_{ws}}{p - p_{ws}} \qquad (5.7)$$

If the dry bulb temperature, pressure, and relative humidity are
known, then W can be calculated (p_{ws} and p_w can be obtained
from t and φ, respectively). The equations can also be used to
calculate relative humidity if t, p, and W are known.

**Example 5.3 What will be the partial water vapor
pressure for moist air if the humidity ratio is 0.04
kg.kg^{-1}? Take the total pressure of the moist air
as 101.325 kPa.**

SOLUTION:

Air and water have their molecular weights as
28.9645 and 18.015, respectively.

Therefore, the mole

$$\text{fraction of water} = \frac{\dfrac{0.04}{18.015}}{\dfrac{0.04}{18.015} + \dfrac{1}{28.96}}$$

$$= 0.0022 / (0.0345 + 0.0022) = 0.0599$$

Therefore, the water vapor partial pressure (p_w) =
(101.325) (0.0599) = 6.07 kPa

**Example 5.4 Moist air at standard atmospheric
pressure has 60°C dry bulb temperature and
40% RH. Saturation partial water vapor pressure at 60°C is 19.92 kPa. Determine the humidity ratio of the moist air.**

SOLUTION:

Saturation water vapor pressure at t= 60°C is, p_{ws} =
19.92 kPa = 19920 Pa
The actual partial water vapor pressure will,
therefore, be

$$p_w = \varphi \cdot p_{ws} = (0.4) (19920) = 7968 \text{ Pa, and}$$

$$W = 0.62198 \frac{p_w}{(p - p_w)} = 0.62198 \frac{7968}{(101325 - 7968)}$$

$$= 0.053 \text{ kg.} (\text{kg dry air})^{-1}$$

W corresponding to 60°C t_{db} and 40% RH can be
read from psychrometric chart as 0.053 kg per kg
of dry air.

Degree of saturation (μ): Another way of expressing
the amount of moisture present in moist air is the
degree of saturation (μ), which is given as follows:

$$\mu = W / W_s \qquad (5.8)$$

Another expression, *percent humidity* is defined as

$$\text{Percent humidity} = \mu \times 100 \qquad (5.9)$$

The relationship between degree of saturation and relative humidity is given as

$$\phi = \frac{\mu}{1 - (1 - \mu)\dfrac{p_{ws}}{p}} \qquad (5.10)$$

Clearly, the degree of saturation can be seen to be related to relative humidity non-linearly. The μ and φ are equivalent for dry air and saturated air, but their values are different at intermediate values of humidity.

Example 5.5 In the problem 5.4, what is the degree of saturation of air? Also find the percent humidity of the moist air.

SOLUTION:

$$W_s = 0.62198 \frac{p_{ws}}{(p - p_{ws})}$$

$$= 0.62198 \frac{19920}{(101325 - 19920)} = 0.152 \text{ kg.}(\text{kg dry air})^{-1}$$

Degree of saturation = $W/W_s = 0.3486$

Per cent humidity = Degree of saturation $\times 100 = 34.86\%$

Note that percent humidity is not the same as relative humidity.

Specific volume (v): Volume of moist air per unit mass of dry air is known as the *specific volume* or *humid volume* of the moist air.

$$v = V_a / M_a = V / 28.9645 \, n_a \qquad (5.11)$$

From the perfect gas relationship, $p = p_a + p_w$

i.e., $v = \dfrac{R_a T}{p - p_w}$

If this relationship is combined with the definition of W, we obtain

$$v_{dry\ air} = \frac{1}{p} R_a T (1 + 1.6078 W) \qquad (5.12)$$

The specific volume, v, of the air-water vapor mixture is then given as

$$v = \frac{1}{p} R_a T \frac{(1 + 1.6078W)}{1 + W}, \text{ where p in Pa} \qquad (5.13)$$

Another equation commonly used for humid volume is

$$v = \frac{22.4}{29}\left(\frac{T}{273}\right) + \frac{22.4}{18} W \left(\frac{T}{273}\right)$$

$$= \frac{22.4}{273}(T)\left(\frac{1}{29} + \frac{W}{18}\right) \qquad (5.14)$$

$$= (0.00283 + 0.00456\ W)T \text{ m}^3.\text{kg}^{-1}$$

T is the absolute temperature.

Volume of one gram-mole of any gas at 273 K and 1 atm is 22.4 liter.

Dew point temperature (t_{dp}): When ice is put in a glass of water, the moisture condenses on the outer surfaces of the glass. This happens because when the surrounding moist air, coming in contact with the glass surface, is cooled below a specific temperature, the water vapor in the cooled air condenses. This specific temperature is known as the dew point temperature (t_{dp}). At saturation, the dry bulb temperature (t) and dew point temperature (t_{dp}) are the same. Alternatively, when moist air is sensibly cooled to such an extent that it becomes saturated, then the air temperature attained is known as the *dew point temperature*.

The dew point temperature depends on p_w and can be estimated as follows:

For temperatures between –60°C and 0°C

$$t_{dp} = -60.45 + 7.0322 \ln(p_w) + 0.37 \left[\ln(p_w)\right]^2 \quad (5.15)$$

For temperatures between 0° and 70°C,

$$t_{dp} = -35.957 - 1.8726 \ln(p_w) + 1.1689 \left[\ln(p_w)\right]^2$$

$$(5.16)$$

In the above equations t_{dp} is in °C and p_w in pa.

Humid heat (S): It is the heat needed to increase the temperature by one degree of unit mass of dry air along with its water vapor. The moist air's humid heat is written as follows.

$$S = 1.006 + 1.805W \text{ kJ.kg}^{-1}. \text{ K}^{-1} \qquad (5.17)$$

In the above equation, 1.006 and 1.805 (both in kJ.kg^{-1}. K^{-1}) are the specific heats of dry air and water vapor, respectively. In most of the food drying operations, the value of W is very small, hence the value of S can be assumed as 1.006 kJ.kg^{-1}. K^{-1}.

Note that heat capacity is an extensive property, i.e., the value is affected by the system's size. It is possible to convert the quantity into an intensive property if it is taken per unit quantity of substance, volume, or mass. *Molar heat capacity*, for example, is obtained by dividing the heat capacity by the number of moles of the pure substance. Similarly, *specific heat capacity* is the heat capacity per unit mass of the material and is commonly known as specific heat. In some cases, the term *specific heat* refers to the ratio of the specific heat capacity of a given substance to the specific heat capacity of the reference material at the same temperature. In specific situations, volumetric heat capacity is also used.

Enthalpy (h): Moist air's enthalpy is obtained when water vapor's enthalpy is added to that of dry air. It is an extensive property.

$$h = h_a + h_w \qquad (5.18)$$

Considering 0°C as the reference temperature and 1 atm as reference pressure, the enthalpy at t (°C) is

$$h = 1.006\, t + W\left(h_{fg} + 1.805\, t\right)$$
$$= \left(1.006 + 1.805\, W\right) t + W.h_{fg} \qquad (5.19)$$

Heat of vaporization (h_{fg}) of water at 0°C is 2501 kJ. kg^{-1}. It is a function of temperature and can be described by $h_{fg} = 2501 - 2.42 \times t$ for temperatures between 0° and 65°C. However, as the value of $2.42 \times t$ is very small as compared to 2501, for most practical applications h_{fg} is taken as 2501 $kJ.kg^{-1}$.

Example 5.6 For the conditions of moist air given in example 5.4, find the specific volume, dew point temperature, and enthalpy.

SOLUTION:

Specific volume of moist air is found as

$$v = \frac{1}{p} R_a T \left(\frac{1 + 1.6078\, W}{1 + W} \right), \text{ p in Pa}$$

$$= \frac{1}{101325}(287.055)(333.15)\left(\frac{1 + 1.6078 \times 0.053}{1 + 0.053} \right) m^3.kg^{-1}$$

$$= 0.973\, m^3.kg^{-1}$$

Dew point temperature,

$$t_d = -35.957 - 1.8726 \ln\left(p_w\right) + 1.1689\left(\ln\left(p_w\right)\right)^2$$

$$= -35.957 - 1.8726 \ln\left(7968\right) + 1.1689\left(\ln\left(7968\right)\right)^2$$

$$= -35.957 - 16.82 + 94.327 = 41.55°C$$

The enthalpy of moist air is, therefore, obtained as

$$h = 1.006\, t + W\left(2501 + 1.805\, t\right)$$

$$= 1.006(60) + 0.053(2501 + 1.805 \times 60)$$

$$= 198.65\, kJ.kg^{-1}$$

Example 5.7 Moist air has a humidity ratio 0.053 kg per kg dry air at 60°C and 40% RH. Determine the humid heat of the moist air.

SOLUTION:

The humid heat (specific heat) of moist air is given by

$$S = 1.006 + 1.805\, W\ kJ.kg^{-1}.°C^{-1}$$

$$= 1.006 + 1.805(0.053) = 1.102\ kJ.kg^{-1}.°C^{-1}$$

Example 5.8 Temperature and pressure of moist air in a room are 26.7°C and 101.325 kPa, respectively. Partial pressure of water vapor in the moist air is 2.76 kPa. Determine (a) humidity ratio, (b) saturation humidity, (c), relative humidity, and (d) degree of saturation. Take 3.5 kPa as the water's saturation vapor pressure (pws) at 26.7°C.

SOLUTION:

The following equation gives the humidity ratio.

$$W = 0.622\left(\frac{p_w}{p - p_w} \right)$$

$$W = (0.622)\frac{2.76}{101.325 - 2.76} = 0.0174\ kg.\left(kg\ dry\ air\right)^{-1}$$

At 26.7°C, the saturation vapor pressure (p_{ws}) is given as 3.5 kPa.
  Saturation humidity,

$$W_s = 0.622\frac{p_{ws}}{p - p_{ws}}$$

or

$$W_s = (0.622)\frac{3.5}{101.325 - 3.5} = 0.0222\ kg.\left(kg\ dry\ air\right)^{-1}$$

Relative humidity,

$$RH = p_w/p_{ws} = 2.76/3.5 = 0.788 = 78.8\%$$

Degree of saturation = W/W_s = 0.0174/0.0222 = 0.784.

Example 5.9 If the air, at atmospheric pressure, is at 50°C temperature and 15% RH, what will be its absolute humidity and specific volume? The saturation vapor pressure of water is 12.35 kPa at 50 °C.

SOLUTION:

$$\text{As RH} = p_w/p_{ws}$$

$$p_w = \text{RH} \times p_{ws} = 0.15 \times 12.35 = 1.8525 \text{ kPa}$$

The absolute humidity,

$$W = 0.622 \frac{p_w}{p - p_w}$$

$$= (0.622) \frac{1.8525}{101.325 - 1.8525}$$

$$= 0.01158 \text{ kg water.} (\text{kg dry air})^{-1}$$

Specific volume is obtained as

$$v = (50 + 273.15) \times (0.00456 \times 0.01158 + 0.00283) \text{m}^3.\text{kg}^{-1}$$

$$= 0.93157 \text{ m}^3.(\text{kg dry air})^{-1}$$

Example 5.10 Assume that 2500 kJ is the thermal energy needed to convert 1 kg of water to water-vapor at 50°C. Also assume that the specific heats of air, water, and water-vapor remain constant at 1.0, 4.18, and 1.88 kJ.kg⁻¹.°C⁻¹, respectively, between 0°C and 50°C. If atmospheric air contains 0.02 kg water-vapor per kg dry air, estimate the enthalpy (kJ per kg of dry air) at 50°C.

SOLUTION:

The enthalpy can be obtained as

$$h = (c_a + c_w W) t + \lambda W$$

$$= (1.0 + 1.88(0.02))(50) + (2500)(0.02)$$

$$= 101.88 \text{ kJ.}(\text{kg dry air})^{-1}$$

Example 5.11 The saturation and absolute humidities of air in a room at total pressure of one atmosphere are 0.04 and 0.022 kg water per kg dry air, respectively. Determine the relative humidity, partial water vapor pressure, humid heat, saturation water vapor pressure, and humid volume.

SOLUTION:

Atmospheric humidity, $W = 0.022$ kg. (kg dry air)⁻¹;
$$p = 101.325 \text{ kPa}$$

$$p_w = p \left(\frac{1.611W}{1 + 1.611W} \right)$$

Thus,

$$p_w = 101.325 \left(\frac{1.611 \times 0.022}{1 + 1.611 \times 0.022} \right) = 3.46 \text{ kPa}$$

Similarly, saturation humidity, $W_s = 0.04$
Thus,

$$p_{ws} = 101.325 \left(\frac{1.611 \times 0.04}{1 + 1.611 \times 0.04} \right) = 6.122 \text{ kPa}$$

Relative humidity, $\phi = \dfrac{p_w}{p_{ws}} = \dfrac{3.46}{6.122} = 0.565 = 56.5\%$

Humid heat, $C_s = 1.006 + 1.88 \, W \text{ kJ.kg}^{-1}°\text{C}^{-1}$

$$= 1.006 + 1.88 \times 0.022 = 1.047 \text{ kJ.kg}^{-1}.°\text{C}^{-1}$$

Humid volume, $V = (t + 273)(0.00283 + 0.00456 \, W)$

$$= (25 + 273.15)(0.00283 + 0.00456 \times 0.022)$$

$$= 0.873 \text{ m}^3.(\text{kg dry air})^{-1}$$

Wet bulb temperature (t_{wb}). This is another indicator of the air moisture. There are two types of wet bulb temperature, i.e., *thermodynamic* t_{wb} *and psychrometric* t_{wb}.

The *psychrometric* t_{wb} is indicated by a thermometer whose bulb is covered with a wick and wetted with distilled water. The air flow around the wick should be approximately 3 m.s⁻¹. In this situation, temperature of the thermometer bulb along with that of the wick decreases and goes below the dry-bulb temperature of the surrounding air. This temperature will continue to fall till the rate of heat transfer from the air to the wick balances the heat transfer required for the evaporation of water from the wick into the air stream.

For saturated air, t_{wb} will be equal to the t_{db}. If RH is lower than 100%, the two temperatures will be different. However, the t_{wb} will never be more than the corresponding t_{db}. The difference between t_{db} of given moist air and its t_{wb} is known as *wet bulb depression*; a clear indication that the t_{wb} is lower than its corresponding t_{db} for non-zero RH. The depression keeps getting reduced with increasing relative humidity and, ultimately, the depression is zero for saturated air.

Let us consider a situation when moist air comes in contact with water on a flat surface. If the air is not saturated, there would be water vapor exchange. As the surface is flat, the surface tension effects can be ignored. Considering that the thermal energy needed to evaporate water will come only from the air (i.e., the process takes place under an adiabatic condition), the temperature of air will decrease and the humidity ratio of air will increase. The process will continue until the air is fully saturated and the limiting temperature at which the air will become saturated is the thermodynamic t_{wb}. The adiabatic condition means that the thermal energy needed to evaporate water will come only from the sensible heat of air and that the enthalpy is conserved in the evaporation process.

Making a heat balance of the wick, the amount of heat lost by vaporization can be written as

$$q = M_v.N_v.\lambda_v.A \tag{5.20}$$

In this equation q is in kW, M_v is the molecular weight of water vapor, N_v is the kg mol water evaporating per m².s, A (m²) is the surface area, λ_v (kJ.(kg water)$^{-1}$) is the latent heat of evaporation at the wet bulb temperature, t_{wb}.

While writing this equation, we have ignored the minor sensible heat change of vaporized liquid as well as the heat transfer due to radiation.

The value of N_v can be given as

$$N_v = \frac{k_v}{x_M}(y_v - y)$$

where k_y (kg mol.s^{-1}.m^{-2}.(mol frac)$^{-1}$) is the mass transfer coefficient, y_v is the mole fraction of water vapor in the air (gas) at the surface and y is the mole fraction in the gas. x_M is the log mean inert mole fraction of the air.

For a dilute mixture, x_M is taken as 1.

$$\text{Thus, } q = M_v k_v (y_v - y)\lambda_v A \qquad (5.21)$$

Further, if M_a is the molecular weights of air (gas), the mole fraction of the water vapor in terms of humidity ratio (W) can be written as

$$y = \frac{W/M_v}{W/M_v + 1/M_a} \qquad (5.22)$$

For small values of W (as encountered in drying applications), the above equation can be written as

$$y = W.(M_a/M_v)$$

Using these, the equation 5.21 can also be modified as

$$q = M_a.k_v.\lambda_v.A.(W_v - W)$$

The convective heat transfer from the air (gas) at t to the wick at t_{wb} is

$$q = h_c(t - t_{wb})\,A$$

Here h_c is the heat-transfer coefficient between air (gas) and surface of liquid.

$$\text{Thus, } M_a.k_v.\lambda_v.A.(W_v - W) = h_c(t - t_{wb})A$$

$$(W - W_v)/(t - t_{wb}) = -h_c/\,M_a.k_v.\lambda_v$$

$$\text{i.e., } \frac{W - W_v}{t - t_{wb}} = -\frac{h_c/M_a.k_v}{\lambda_v} \qquad (5.23)$$

The equation 5.23 states that if (W– W_v) is plotted against (t–t_{wb}) on a temperature/ humidity chart, the result is a straight line. The slope of the line will be $h_c/\,M_a.k_v.\lambda_v$. This line is known as *psychrometric line*.

In this equation $h_c/\,M_a.k_v$ is known as the *psychrometric ratio*. This term is often used to correlate the heat and mass transfer of water vapor air mixture.

Even though we assume adiabatic condition, the air gains a little enthalpy, and the water loses a like amount. The difference is the enthalpy content of the evaporated water prior to evaporation. Hence the constant wet bulb temperature lines do not exactly correspond to the constant enthalpy lines. A close examination of the psychrometric chart also shows that the constant wet-bulb temperature lines deviate slightly from being parallel. However, for most practical situations that we encounter, we can assume them to be parallel.

When both a psychrometric line and an adiabatic saturation line are plotted for the same point, the relation between the lines depend on the relative magnitudes of humid heat (S) and the psychrometric ratio, i.e. $h_c/M_a.k_v$. The value of the psychrometric ratio for air-water vapor mixture is approximately 0.96 to 1.005. As the value of humid heat of moist air is also around this value, the adiabatic saturation lines can also be taken as the wet bulb lines. In other words, the two temperatures are approximately equal for moist air. But for other gas-vapor mixtures, it may not be the same.

As discussed above, for ordinary conditions for air-water vapor mixture,

$$h_c/M_a.k_v \cong c_s \qquad (5.24)$$

c_s represents saturation condition. This relation is known as *Lewis relation*.

There are iterative procedures to find out wet bulb temperature of moist air. It can also be found out quickly from the psychrometric chart. The psychrometric chart is discussed in Section 5.2.

Example 5.12 Find the mass transfer coefficient of water vapor if the psychrometric ratio of water vapor air mixture is about 1 kJ.kg^{-1}.K^{-1} and the heat transfer coefficient is 10 W.m^{-2}.K^{-1}.

SOLUTION:

$$\text{Psychrometric ratio} = \frac{h}{M_B k_y} = 1000\ \text{J.kg}^{-1}\text{K}^{-1}$$

$$\text{Thus, } k_y = h/(M_B \times 1000) = 10/(18 \times 1000)$$
$$= 5.55 \times 10^{-4}\ \text{kg mole.m}^{-2}.\text{s}^{-1}$$

5.2 Psychrometric Chart and Its Applications

When physical and thermal properties of moist air are represented graphically, the result is a psychrometric chart (Figure 5.1 and Annexure II). While W is the ordinate in the chart, t is the abscissa. The other parameters shown on the psychrometric chart are t_{wb}, t_d, h, v, and φ. With the knowledge of any two moist air properties, the remaining moist air properties can be found.

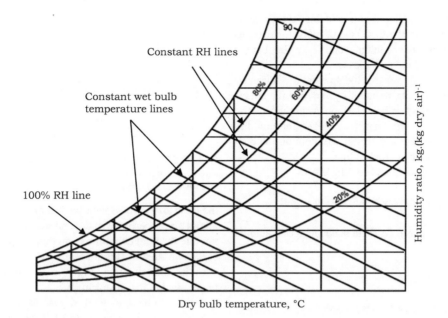

FIGURE 5.1   Psychrometric chart

The psychrometric chart plays a vital role in the design, analysis, and optimization of various food engineering systems and processing equipment, specifically where there is heat and moisture transfer between the food and the surrounding air. Drying, storage, and chilling are some common operations where psychrometric chart has applications.

Consider a drying process. Suppose the ambient air at a temperature of 30°C and 85% RH is heated to 60°C before entering the dryer. The psychrometric chart can be used to find out the RH of the air after the heating process. The change in humidity ratio and the water holding capacity of the air can also be determined. Alternatively, on the basis of desired removal of moisture, the amount of dry air that has to be supplied for drying can be determined. If the drying air needs to be recycled, the amount of fresh air at the specific temperature and RH to be mixed with the dryer exhaust air can be quantified.

Similarly, in an air conditioning process, a knowledge of psychrometrics can help in finding the conditions of final air after cooling through the air-cooling device and also after mixing of two air streams. The required number of air changes can also be considered. The psychrometric chart has also been very helpful in the management of the environment in livestock buildings and greenhouses.

More applications of the psychrometric chart have been given under the section on psychrometric processes.

Another form of graph for finding moist air properties, known as the *Mollier Chart*, has W and enthalpy as the independent variables.

5.3 Psychrometric Processes

A psychrometric process can be defined as a change in the state of moist air caused by adding or removing, individually or in combination, thermal energy and water vapor. Some such changes represented on a psychrometric chart are discussed below.

In Figure 5.2 (a), the processes 1-2 and 1-3 represent sensible heating and cooling, respectively. This occurs when a heat exchanger is employed for heating or cooling the given moist air, without physical contact with the heating or cooling medium, i.e., neither any moisture is removed from or added to the air. In the process, t, t_{wb}, h, v, and φ change, whereas the t_{dp}, W, and p_w do not change.

If moist air is heated by direct injection of steam, there will be heating and humidification (as in Figure 5.2 b). In this process, t, t_{wb}, t_{dp}, h, W, p_w, and v increase, whereas φ may decrease or increase depending on the extent of humidification.

During drying or evaporative cooling, the wet bulb temperature and enthalpy of the moist air remain constant and the process is shown as Figure 5.2 (c). Evaporative cooling takes place when moist air comes in contact with a wet medium for a specific length of time to permit a process of adiabatic saturation. This will continue until the air attains complete saturation. During drying, the air loses sensible heat for the evaporation of moisture from the commodity and, at the same time, the

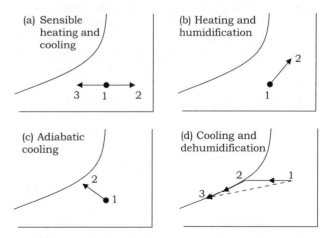

FIGURE 5.2   Changes in conditions of moist air shown on a psychrometric chart

evaporated moisture is added to the air, and thus, the air gains latent heat. In the process, the total heat of moist air remains unchanged. There is no change in the h and t_{wb} values, but the t decreases and t_{dp}, W, φ, and p_w increase.

In the convective drying process, the ambient air is usually first heated to reduce its relative humidity and then the hot air is entered into the drying chamber. Thus, on the psychrometric chart, the condition of air will change on a constant humidity line during sensible heating and then it will cool on the constant enthalpy (wet bulb temperature) line.

During the sensible heating and cooling process

$$m_{a1}h_1 + q = m_{a2}h_2, \text{ (q may be positive or negative)}$$

Here m_a is the mass of air. As there is no change in amount of dry air, $m_{a1} = m_{a2}$

$$\text{Thus, } q = m_a (h_2 - h_1) \qquad (5.25)$$

For steady flow conditions, q causes a temperature change.

$$\Delta t = \frac{q}{m_a c_{pa}} \text{ or } \Delta t = \frac{q}{\Delta h}, \ c_{pa} \text{ is the specific heat at constant}$$

pressure.

During the heating and humidifying,

$$m_{a1}h_1 + m_w h_w + q = m_{a2}h_2$$

$$m_{a1}W_1 + m_w = m_{a2}W_2$$

$$\text{Thus, } m_w = m_a(W_1 - W_2)$$

$$q = m_a (h_2 - h_1) - m_w h_w \qquad (5.26)$$

If the temperature of the air is reduced to less than the dew point temperature, then there will be loss of moisture due to condensation. First, when air cools, there will be no change of W until it reaches the saturation line at its dew point. After that, the process will be along the saturation line and water will be removed by condensation to ensure that the value of the saturation humidity is not exceeded. This is a process of cooling and dehumidification (Figure 5.2 d). The amount of water removed is the difference between the initial and the final humidity ratios per unit mass of dry air. The enthalpy change can also be separated into sensible heat change and latent heat change. In the figure the sensible heat change takes place at 1-2 and the latent heat change at 2-3. The values of t, t_{wb}, h, W, and v decrease and φ increases to 100%.

During the process of cooling and dehumidification,

$$m_{a1}h_1 = m_{a2}h_2 + q + m_c h_c \text{ and } m_{a1}W_1 = m_{a2}W_2 + m_c$$

$$\text{Thus, } m_c = m_a(W_1 - W_2)$$

$$q = m_a (h_1 - h_2) - m_c h_c \qquad (5.27)$$

Another condition arises, when two air streams are mixed, such as during drying, when the exhaust air is recirculated and mixed with fresh heated air, or during air conditioning when

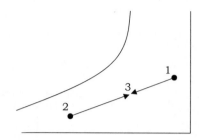

FIGURE 5.3 Mixing of two streams of moist air presented on a psychrometric chart (1 and 2 are the conditions of two streams before mixing; 3 is the condition after mixing)

the fresh air is mixed with the indoor air. In this situation the mixed air properties will depend on the individual air properties and the final condition is on the line joining the condition of the two streams (Figure 5.3).

Under this situation, $m_{a1} + m_{a2} = m_{a3}$, $m_{a1}W_1 + m_{a2}W_2 = m_{a3}W_3$, and $m_{a1}h_1 + m_{a2}h_2 = m_{a3}h_3$

Here m_a corresponds to the dry air mass flow rate, and the subscripts 1, 2, and 3 are for streams 1, 2, and the resultant stream.

Further simplifying the above equations, we obtain

$$W_3 = \frac{m_{a1}W_1 + m_{a2}W_2}{m_{a1} + m_{a2}} \qquad (5.28)$$

$$\text{and, } h_3 = \frac{m_{a1}h_1 + m_{a2}h_2}{m_{a1} + m_{a2}} \qquad (5.29)$$

Example 5.13 The temperature of moist air, initially at 70% RH and 30°C t_{db}, is raised to 70°C. The hot air is allowed to enter into a dryer for the drying process, and the air temperature after drying has been recorded as 45°C. Using a psychrometric chart, obtain the different psychrometric properties of the air just before and after drying.

SOLUTION:

The psychrometric properties of the air before drying and after drying are as follows

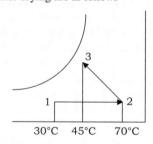

Psychrometric properties	Before drying	After drying
Dry bulb temperature	70°C	45°C
RH	10%	48%
Wet bulb temperature	34°C	34°C
Humidity ratio, kg.kg⁻¹	0.019	0.030
Enthalpy	29 kcal.kg⁻¹ (121.22 kJ.kg⁻¹)	29 kcal.kg⁻¹ (121.22 kJ.kg⁻¹)
Humid volume, m³.kg⁻¹	1	0.0945

Example 5.14 In a dryer, air at 35°C t_{db} and 25°C t_{wb} and flow rate of 60 m³.min⁻¹ is being mixed with the other air stream at 65°C t_{db} and 50°C t_{wb} and flow rate of 40 m³.min⁻¹. What are the dry bulb and wet bulb temperatures of the mixed air?

SOLUTION:

The given conditions are

Psychrometric properties	Stream 1	Stream 2
t_{wb}	25°C	50°C
t_{db}	35°C	65°C
Rate of air flow, m³	60 m³.min⁻¹	40 m³.min⁻¹

By using a psychrometric chart (Annexure II), the following properties are obtained for both the streams:

Psychrometric properties	Stream 1	Stream 2
Enthalpy (h), kJ.(kg dry air)⁻¹	104.5	273.79
Humid volume, m³.(kg dry air)⁻¹	0.91	1.08

From the humid volume values, we can find out the mass of dry air in two streams as follows.

$$m_1 = 60/0.91 = 65.93 \text{ kg.min}^{-1}$$

$$m_2 = 40/1.08 = 37.03 \text{ kg.min}^{-1}$$

Now, the enthalpy of the mixture, h_3 can be given as:

$$h_3 = \frac{m_1 h_1 + m_2 h_2}{m_1 + m_2} = 165.38 \text{ kJ.}\left(\text{kg dry air}\right)^{-1}$$

Thus, the final condition of the moist air will be on the straight line connecting the conditions of streams 1 and 2, and it will have the enthalpy value of 165.38 kJ. (kg dry air)⁻¹.

From the psychrometric chart, the t_{wb} and t_{db} of the mixture are 40°C and 46 °C, respectively.

(Alternatively, we can find the humidity ratios of the two streams of air as W_1 and W_2 and find out the final humidity ratio as $W_3 = \dfrac{m_1 W_1 + m_2 W_2}{m_1 + m_2}$ and then the other properties from the psychrometric chart corresponding to W_3 on the straight line connecting the conditions of streams 1 and 2. Check that we will get W_3 as 0.046 kg. (kg dry air)⁻¹.

Example 5.15 Air at 22°C and 80% RH is required to be heated to 65°C. Using the psychrometric chart, find the psychrometric properties of the heated air. Also, find the amount of heat per kg of dry air that needs to be added to heat it.

SOLUTION:

The conditions of the heated air as obtained from the psychrometric chart are as follows

Psychrometric properties	Values
Dry bulb temperature	65°C
RH	8%
Wet bulb temperature	30.5°C
Humidity ratio, kg.(kg dry air)⁻¹	0.013
Enthalpy, kcal. (kg dry air)⁻¹	24.5
Humid volume, m³. (kg dry air)⁻¹	0.98

The enthalpy of the air before drying (from psychrometric chart) = 13 kcal. (kg dry air)⁻¹

The quantity of heat added can be given as:

$$h_1 - h_2 = 24.5 - 13.0 = 11.5 \text{ kcal. (kg dry air)}^{-1}$$

Example 5.16 Moist air at a temperature of 46°C was passed over a cold surface so that the air was cooled gradually. Traces of moisture began to appear on the surface when the air temperature reached 22°C. Determine the moist air's relative humidity.

SOLUTION:

The saturation temperature is 22°C. Moving on a constant humidity line, the point of intersection at the 46°C t_{db} line indicate that the moist air has the humidity ratio of 0.017 kg per kg and the relative humidity as 30%.

Example 5.17 In a batch dryer, the rate of air supply is 400 m³.min⁻¹. Determine the heat requirement per hour for the drying process if the atmospheric air, initially at 30°C and 60% RH, is required to be raised to 70°C.

SOLUTION:

Using the psychrometric chart (Annexure II):

enthalpy of atmospheric air (h_1) = 17 kcal.(kg dry air)⁻¹

enthalpy of hot air (h_2) = 27.5 kcal.(kg dry air)⁻¹

Thus, the amount of heat added = $h_1 - h_2$ = 27.5-17.0
= 10.5 kcal.(kg dry air)⁻¹

The air supply is given in m³. min⁻¹

From the psychrometric chart, the moist air's specific volume is 0.88 m³.(kg dry air)⁻¹ at 30°C and 60% RH.

Thus, air flow rate = 400/0.88 = 454.545 kg.min⁻¹

Heat added = (454.545) (10.5) = 4772 kcal.min⁻¹
= 19950 kJ.min⁻¹

Example 5.18 Air at 40°C and 50% RH is cycled past the cooling coils to a final temperature of 20°C. Using a psychrometric chart, determine

the psychrometric properties of air after it is cooled. Also find the sensible and latent heats removed, and the water vapor condensed per kg of dry air.

SOLUTION:

On the psychrometric chart, the air temperature will reduce along a constant humidity ratio line until saturation and then further cool along the 100% RH line (as in Figure 5.2 d).

The initial and final points will be as given on the table.

Psychrometric properties	Point 1	Point 2	Point 3
t_{db}, °C	40		20
t_{dp}, °C	27.5		
t_{wb}, °C	30.5		20
RH, per cent	50	100	100
Humidity ratio, kg. (kg dry air)$^{-1}$	0.0235	0.0235	0.0145
Humid volume, m³. (kg dry air)$^{-1}$	0.92		0.85
Enthalpy, kJ.(kg dry air)$^{-1}$	102.41	87.78	58.52

The enthalpies at point 1 and 2 are given above as 102.41 and 87.78 kJ. (kg dry air)$^{-1}$

The enthalpy at point 3 is 58.52 kJ. (kg dry air)$^{-1}$

Sensible heat removed,

$$h_s = (1 \text{ kg}) (102.41 - 87.78) \text{ kJ.kg}^{-1} = 14.63 \text{ kJ. (kg dry air)}^{-1}$$

Latent heat removed,

$$h_l = (1 \text{ kg}) (87.78 - 58.52) \text{ kJ.kg}^{-1} = 29.26 \text{ kJ. (kg dry air)}^{-1}$$

The quantity of water vapor condensed,

$$W_1 - W_2 = 0.0235 - 0.0145 = 0.009 \text{ kg. (kg dry air)}^{-1}$$

Alternatively, the amount of condensed water can be found from the enthalpy changes in latent heat.

Heat of vaporization at 20°C

$$h_{fg} = 2501 - 2.42 \text{ t}$$

$$= 2501 - 2.42(20) = 2452.6 \text{ kJ.} (\text{kg dry air})^{-1}$$

Therefore, when per kg of dry air 29.26 kJ is removed, the amount of water condensed is

$$29.26 / 2452.6 = 0.011 \text{ kg. (kg dry air)}^{-1}$$

The minor difference is due to errors in the approximation of values from the psychrometric chart and the calculated h_{fg} value at 20°C.

Example 5.19 Determine the final state of moist air if 10 kJ is removed from 1.5 kg of dry air originally at 30°C and 50% RH. Take p_{ws} at 30°C as 4.242 kPa. Also find the humidity ratio of air.

SOLUTION:

If the specific heat of moist air is taken as 1.006 kJ. kg^{-1}.°C^{-1}, then the temperature change (using equation $q = m_a c_{pa} \Delta t$) is

$$\Delta t = -10 \text{ kJ}/(1.5)(1.006) = -6.63°C$$

Hence, the final temperature will be
30 - 6.63 = 23.37°C

$$\text{Given } p_{ws} = 4.242 \text{ kPa}$$

$$p_w = (0.5)(p_{ws}) = 2.121 \text{ kPa}$$

$$W = 0.622 \frac{p_w}{p - p_w}$$

$$= (0.622)\frac{2.121}{101.325 - 2.121} = 0.0133 \text{ kg water.} \left(\text{kg dry air}\right)^{-1}$$

As it is a sensible cooling process, the humidity ratio remains constant at 0.0133 kg (kg dry air)$^{-1}$. Thus, the final temperature is 23.37°C and W=0.0133 kg. (kg dry air)$^{-1}$.

The other properties can be determined as before either from a psychrometric chart or by using equations.

The final relative humidity is 72% (from psychrometric chart).

Alternatively, the change in enthalpy can be calculated as
$\Delta h = q/m_a = -10/1.5 = -6.67$ kJ. (kg dry air)$^{-1}$ and then the other conditions can be obtained from the psychrometric chart.

Example 5.20 Consider an air-water vapor mixture where one kilogram of air at 90% relative humidity and 30°C temperature is mixed with 2 kg of air at 30°C and an absolute humidity of 0.0103 kg.(kg dry air)$^{-1}$. The air mixture is then heated to 60°C. The water vapor pressures are 4.11 and 19.93 kPa at 30°C and 60°C, respectively. Total pressure of the system is 101.325 kPa. Assume average molecular weight of air as 29 kg (kg mole)$^{-1}$ at 30°C. Find the resultant relative humidity of the air- water vapor mixture without using a psychrometric chart.

SOLUTION:

Psychrometric properties of the 1st air stream are

$$m_a = 1 \text{ kg, } t_a = 30°C, \text{ RH}_a = 90\%, p_{wsa} = 4.11 \text{ kPa}$$

Psychrometric properties of the 2nd air stream are

$$m_b = 2 \text{ kg, } t_b = 30°C, W_b = 0.0103 \text{ kg. (kg dry air)}^{-1}$$

For water vapor of the first stream, the partial pressure, $p_{w1} = (5.11)(0.9) = 3.699$ kPa

The humidity ratio for the 1st stream,

$$W_a = 0.622 \frac{p_w}{p - p_w}$$

$$= (0.622)\frac{3.699}{101.325 - 3.699} = 0.0236 \text{ kg.} \left(\text{kg dry air}\right)^{-1}$$

For the mixed air stream, the humidity ratio (W_c) is obtained as follows

$$W_c = \frac{m_a W_a + m_b W_b}{m_a + m_b} = \frac{(1)(0.0236) + (2)(0.0103)}{(1+2)}$$

$$= 0.0147 \text{ kg.}\left(\text{kg dry air}\right)^{-1}$$

As a result, for the mixed air the partial water vapor pressure is obtained as follows.

$$W = 0.622 \frac{p_w}{p - p_w}$$

or, $$1.611 W = \frac{p_w}{p - p_w}$$

or, $$p_w = p\left(\frac{1.611W}{1 + 1.611W}\right)$$

Thus,

$$p_w = 101.325\left(\frac{1.611 \times 0.0147}{1 + 1.611 \times 0.0147}\right) = 2.339 \text{ kPa}$$

The mixed air RH is $(2.339/19.93) \times 100 = 11.75\%$

Example 5.21 Air at 10% relative humidity and 60°C temperature passes through a continuous dryer. The exit air is at 45°C. How much water is removed per kg air? Also, what will be the drying air volume to remove 15 kg water per hour?

SOLUTION:

Inlet air conditions read from the psychrometric chart are as follows.

    a. humidity ratio (W_1) of the drying air = 0.012 kg kg⁻¹

    b. specific volume = 0.964 m³. kg⁻¹

The drying air follows the wet bulb line, therefore,

    humidity ratio (W_2) = 0.0184 kg. kg⁻¹

Moisture removed = 0.0184 - 0.012 = 0.0064 kg. (kg)⁻¹

One kg, i.e., 0.964 m³, of drying air can remove 0.0064 kg water.

Thus, air volume required to remove 15 kg.h⁻¹

$$= \left(15 / 0.0064\right) \times 0.964$$

$$= 2259.4 \text{ m}^3\text{h}^{-1}$$

Example 5.22 The ambient air is at 24°C temperature and 80% RH. It is heated to 70°C for drying of a commodity kept in a tray dryer. The air leaving the dryer is at 80% RH. This air is heated again to 70°C and flows over another set of trays. The exit air from this dryer is at 80% RH. If 1200 m³.h⁻¹ is the air flow rate, obtain (a) water removed per hour, and (b) the energy needed to heat the air.

SOLUTION:

Air humidity at 24°C and 80% RH is read from the psychrometric chart as 0.015 kg.kg⁻¹. Besides, the specific volume and enthalpy values read were 0.862 m³ kg⁻¹, and 60.7 kJ kg⁻¹. The process is depicted on a psychrometric chart as follows

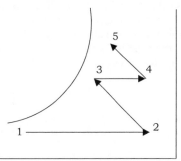

The heating process is along a constant humidity line; thus, the enthalpy is found to be 112.86 kJ. (kg dry air)⁻¹ by following the constant humidity line to a temperature of 70°C.

The drying process proceeds along the constant wet-bulb line and, thus, approaching the RH of 80% results in an air temperature of 35.5°C and a humidity ratio of 0.030 kg. (kg dry air)⁻¹.

Heating the air to 70°C again requires moving along the constant humidity line and the enthalpy read from a psychrometric chart is 150.48 kJ. (kg dry air)⁻¹.

Then, the humidity ratio obtained is 0.042 kg.kg⁻¹ when moving along the wet-bulb line to 80% RH.

Thus, the total energy input = enthalpy change of the moist air,

$$= 150.48 - 60.7$$

$$= 89.78 \text{ kJ.}\left(\text{kg dry air}\right)^{-1}$$

Total water removed = change in humidity ratio of air

$$= 0.042 - 0.015 = 0.027 \text{ kg.}\left(\text{kg dry air}\right)^{-1}$$

The air supply is 1200 m³.h⁻¹ = 1200/0.862 kg h⁻¹ = 1392.11 kg h⁻¹ = 0.387 kg.s⁻¹

Energy increase for the air = 89.78×0.387 kJ. s⁻¹

$$= 34.74 \text{ kW}$$

Moisture removed from dryer = $0.027 \times 0.387 = 0.010449$ kg s⁻¹ = 37.616 kg.h⁻¹

Example 5.23 The ambient air is at 45°C and 80% RH. A room of 1400 m³ volume has to be conditioned to 30°C and 60% RH. The process followed is the initial cooling of the air to

condense out a sufficient amount of moisture and then reheating to the desired condition. Find the amount of water to be removed from the air, the requirement of reheating of the air, and the air temperature after cooling. Assume that there are four air changes per hour.

SOLUTION:

The change of air conditions in a psychrometric chart will be as shown in the figure.

At 45°C and 80% RH (condition 1), the humidity ratio and the enthalpy read from the psychrometric chart are 0.051 kg.kg⁻¹ and 175.56 kJ.kg⁻¹, respectively.

The humidity ratio and the enthalpy at 30°C and 60% RH (condition 4) are 0.016 kg.kg⁻¹ and 71.896 kJ.kg⁻¹, respectively. At this condition, 0.88 m³.kg⁻¹ is the specific volume.

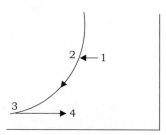

There are four air changes per hour, thus

Mass of air to be conditioned = $(1400 \times 4)/0.88$
$$= 6363.63 \text{ kg.h}^{-1}$$

Water removed per kg of dry air = $0.051 - 0.016$
$$= 0.035 \text{ kg.kg}^{-1}$$

Mass of water removed per hour = 6363.63×0.035
$$= 222.7 \text{ kg.h}^{-1}$$

Reheating required (from condition 3 to condition 4)
$$= (71.896 - 62.7) = 9.196 \text{ kJ.kg}^{-1}$$

Total reheat power required = 6363.63×9.196
$$= 58520 \text{ kJ.h}^{-1} = 16.25 \text{ kW}$$

Psychrometrics and the psychrometric chart are very important tools in understanding and analyzing the interrelationship of the important properties of air. In food processing operations, the chart is very helpful in applications where there is heat and moisture transfer. It helps to decide the various processes to be followed so as to achieve the required condition of air. Besides, as we have discussed, this is also very helpful in environment management applications in buildings, greenhouses, and livestock buildings.

Check Your Understanding

1. A heater is placed in front of a continuous countercurrent dryer. Air at 40°C and 70% RH is fed into the heater from which the air exits at 65°C. If saturation vapor pressure at 40°C and 65°C are 0.074 bar and 0.250 bar, respectively, what will be the relative humidity of the air coming out of the heater and entering the dryer?

   Answer: 21%

2. Air at 40°C temperature has a partial vapor pressure of 2.4 kPa. If the universal gas constant is 8.314 kJ.(kg mole)⁻¹.K⁻¹ and total pressure is 101.325 kPa, what is the humid volume of air?

   Answer: 0.907.m³.(kg dry air)⁻¹

3. Air at 70°C and 0.015 humidity ratio is cooled adiabatically by spraying water. The final temperature of the air is 55°C. Specific heat capacities of dry air and water vapor are 1.005 and 1.88 kJ.kg⁻¹.K⁻¹, respectively, and the latent heat of vaporization of water at 0°C is 2501.7 kJ.kg⁻¹. What is the absolute humidity of the outlet air?

   Answer: 0.021 kg water per kg dry air

4. Two kg mass of air at 40°C with 0.023 kg water vapor per kg dry air is mixed with 3 kg mass of air at 20°C with 0.008 kg water vapor per kg dry air to produce 5 kg mass of air at 60% relative humidity at 28°C. Assume all the streams are at normal atmospheric pressure (101.325 kPa). What is the saturation vapor pressure of water at 28°C?

   Answer: 3.7166 kPa

5. The partial pressure of water vapor in a moist air sample of relative humidity 70% is 1.6 kPa, the total pressure being 101.325 kPa. Moist air may be treated as an ideal gas mixture of water vapor and dry air. The relation between saturation temperature (T_s in K) and saturation pressure (p_s in kPa) for water is given as $\ln (p_s/p_0) = 14.317 - 5304/T_s$, where $p_0 = 101.325$ kPa. Find the dry bulb temperature of the moist air.

   Answer: 19.89°C

6. Air at a temperature of 20°C and 750 mm Hg pressure has a relative humidity of 80%. If the vapor pressure of water at 20°C is 17.5 mm Hg, determine the percentage humidity of the air.

   Answer: 79.62%.

Objective Questions

1. Draw a psychrometric chart and show the moist air properties that can be read from the chart.

2. On a psychrometric chart show the lines of (a) sensible heating, (b) evaporative cooling, (c) drying, (d) mixing of two streams of air, and (e) cooling and dehumidifying.

3. Differentiate between the following (for moist air)

   (a) Relative humidity and humidity ratio;

   (b) Wet bulb temperature and dew point temperature;

   (c) Specific volume and density;

(d) Relative humidity and degree of saturation;

(e) Humid heat and enthalpy.

BIBLIOGRAPHY

Albright, L.D. 1990. *Environment Control for Animals and Plants.* ASAE, St. Joseph, MI.

Brooker, D.B., Bakker Arkema, F.W. and Hall, C.W. 1974. *Drying Cereal Grains.* The AVI Publ. Co., Westport, Connecticut.

Chakraborty, A. 1995. *Post-Harvest Technology of Cereals, Pulses and Oilseeds.* IBH Publishing Co., Oxford.

Henderson, S.M. and Perry, R.L. 1980. *Agricultural Process Engineering.* The AVI Publ. Co., Westport, Connecticut.

Rao, D.G. 2009. *Fundamentals of Food Engineering.* PHI Learning Pvt. Ltd., New Delhi.

Sahay, K.M. and Singh, K.K. 1994. *Unit Operations in Agricultural Processing.* Vikas Publishing House, New Delhi.

Singh, R.P. and Heldman, D.R. 2014. *Introduction to Food Engineering.* Academic Press, San Diego, CA.

6

Fluid Properties and Flow Behaviors

Fluid flow and heat and mass transfer are two important processes involved in food processing. Heat and mass transfers occur in evaporation, drying, freezing, pasteurization, sterilization, etc. Similarly, fluid flow occurs in all heat exchangers as well as in many of the above processes. There are several liquid foods. There are applications of steam and compressed air, both gases, in food processing. Therefore, basic concepts of fluid flow are important in understanding different food processing operations. In this section we will discuss some properties of fluids and characteristics of fluid flow important in food processing applications; heat transfer will be discussed separately.

Fluids, even under pressure, flow without disintegration. Fluids include gases, liquids, and certain solids. Behavior of liquids and gases, collectively known as fluids, at rest and in motion constitutes the subject matter of *fluid mechanics*. Statics, dynamics, and kinematics are the three major segments of fluid mechanics. While fluids at rest are studied under *fluid statics*, fluid motion under the influence of external forces is the subject matter of *fluid dynamics*, and fluid motion without external forces forms the subject matter of *fluid kinematics*.

6.1 Density and Specific Gravity

$$\text{Density (or mass density)} = \rho = \text{mass / volume} \quad (6.1)$$

It is given as kg.m⁻³ under SI system.

Densities of liquids, under normal temperature and pressure ranges that we encounter, may be constant. However, variations in pressure and temperature have a considerable influence on gas densities.

$$\text{Specific weight or weight density} = w = \text{weight / volume} \quad (6.2)$$

Specific volume (υ, expressed in m³.kg⁻¹) is the volume that a unit mass of fluid would occupy.

$$\upsilon = 1/\rho \quad (6.3)$$

When weight density of a gas is divided by the weight density of air taken at room temperature (20°C), the ratio is commonly known as *specific gravity* or *relative density* of the gas. Specific gravity or relative density of a gas is also equivalent to the ratio of density of the gas to the density of air. In case of a liquid, the ratio of the liquid's weight density to the weight density of water measured at 4°C (water is the densest at 4°C) is called specific gravity of the liquid, which is also equivalent to the ratio of density of the liquid to the density of water. Density of water is 1×10^3 kg.m⁻³ at 4°C.

6.2 Viscosity and Viscoelasticity

In a fluid, molecules are constantly in random motion. However, in the absence of any force on the fluid, the net velocity is zero in a particular direction. Fluid molecules begin to slip past one another under pressure and the flow is said to occur in a particular direction in any given plane. It also implies that adjacent molecules experience a velocity gradient.

Viscosity is that property of a fluid that resists the movement of one layer of fluid over the adjacent layer, i.e., it is a measure of resistance to a fluid's flow (expressed as the *coefficient of viscosity*). The term in the food industry often used to describe it is the *consistency* of the product. As we will discuss later, fluid foods in general, except for a narrow range, exhibit non-Newtonian behavior and, hence, using the term *viscosity* for consistency may lead to erroneous interpretations.

To understand more on the physical representation of viscosity, we consider the fluid flow between two plates as shown in Figure 6.1. Let there be two surfaces at a distance dy and the surfaces move at velocities v and v+dv in the same direction. Let an ideal viscous fluid be present between the surfaces. The response of the fluid between the two surfaces can be explained with the shear stress (τ) and shear strain (υ) relationship.

> *Ideal fluid* is that hypothetical fluid that has zero viscosity and is incompressible. In other words, the frictional effects during fluid flow are zero. Ideal fluid has no property other than density. All fluids are real, i.e., the fluid viscosities are non-zero.

With reference to the Figure 6.1, the rate of change of velocity v with distance dy from the wall (dv/dy) is proportional to shear force per unit area, i.e., the shear stress.

$$\tau \, \alpha \frac{dv}{dy} \quad (6.4)$$

$$\text{or} \quad \tau = \mu \frac{dv}{dy} \text{ or } \tau = \mu(v) \quad (6.5)$$

Proportionality constant, μ, in the above equation is the *coefficient of dynamic viscosity*. The velocity gradient (dv/dy) created between the two surfaces is the shear rate. The momentum flows from a region of high velocity to low velocity, and, hence, often the equation 6.5 has a negative sign on the right-hand side, i.e., the equation is written as

DOI: 10.1201/9781003285076-7

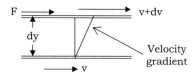

FIGURE 6.1　Velocity gradient in a fluid across two surfaces moving at differential velocities

$$\tau = \mu\left(-\frac{dv}{dy}\right) \tag{6.5b}$$

Thus, the shear stress necessary to produce unit rate of shear strain is called as *dynamic viscosity*, or just viscosity.

An ideal material may behave as *elastic*, *plastic*, or *viscous*. A material is said to be an ideal elastic body when its stress and strain are related as follows:

$$\tau = Ev \tag{6.6}$$

The above expression represents *Hooke's law*; here E is known as *Young's modulus* or *elasticity modulus*. When a body is stressed in either tensile or compression mode, the above equation is valid. A similar relationship is applicable when either shear stress or hydrostatic pressure acts on the body; *shear modulus* is the name of the proportionality coefficient in case of shear stress, and it is known as *bulk modulus* in case of hydrostatic pressure.

Ideal plastic behavior can be seen when force on a block of material kept on a flat surface is gradually increased to move the block. It is observed that the block would not move until a minimum amount of stress, known as *yield stress*, is applied. Once the yield stress is exceeded, the block movement continues indefinitely.

The third is ideal viscous behavior and a category of liquid foods exhibits this behavior.

The above also indicates that elasticity, plasticity, and viscosity are the three fundamental rheological parameters. Generally, no food product is purely elastic, plastic, or viscous and, hence, more complex analyses using these basic concepts are often used to represent the rheology of real food products.

A combination of elastic and viscous behavior is a better representation of many food products. This gives rise to *visco-elastic behavior*. A *visco-elastic fluid* is a fluid that partially returns to its original form when the applied stress is released. The behavior is both solid and liquid like. Rate of strain could greatly influence the resultant stress-strain relationship for the material. A material is said to exhibit a *linear visco-elastic* response if its shear stress-shear strain relationship is only a function of time. Whereas the behavior is termed as *non-linear visco-elastic* if stress as well as time affect the stress-strain relationship. Most of the food products exhibit this behavior. Theory of non-linear viscoelasticity is still not fully understood. Therefore, the theory of

linear visco-elasticity is mostly used in characterization of parameters for food products.

6.2.1　Newtonian and Non-Newtonian Fluids

Equation 6.5 is known as *Newton's law of viscosity*, and Newtonian fluids follow this law. Fluid viscosity is constant in *Newtonian fluids*; shear stress depends on shear strain. The examples of general Newtonian fluids include water, custard, and blood plasma. The plot between shear stress and shear rate for Newtonian fluids is a straight line passing through the origin of the coordinate system. However, the shear stress-shear strain relationship is not a straight line for many fluids as shown in Figure 6.2. *Non-Newtonian fluids* are those that do not conform to Newton's viscosity law. In other words, a fluid is *non-Newtonian* if its viscosity changes with a change in shear strain. Corn starch in water is a good example of a non-Newtonian fluid. It flows like a thick liquid, such as honey, but it feels like a solid.

There are situations where the change in viscosity is independent of time. In situations where the viscosity is changing, we will use the term *apparent viscosity*. When the apparent viscosity increases with increase in stress, it is known as shear thickening behavior and the fluid is known as *dilatant fluid*. Dilatant fluid behavior corresponds to one or more of (a) swelling or changing shape of the dispersed phase, (b) cross linking of molecules with each other, and (c) trapping molecules of the dispersion medium, when subjecting it to a shearing action. Dilatant behavior is observed when flocculation begins in a colloidal suspension. As mentioned above, corn starch in water is a good example of dilatant fluid and, hence, it is used as a common thickening agent in cooking. Another example of a dilatant material is sand completely soaked in water; a dry patch is created under your foot while walking on wet sand.

When the apparent viscosity decreases with increasing stress it is known as shear thinning behavior and the fluid is known as *pseudoplastic fluid*. Emulsions and suspensions exhibit

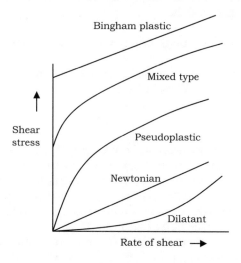

FIGURE 6.2　Variations in shear stress–shear rate relationship for different fluids

pseudoplastic behavior because the dispersed phase tends to coalesce in the flow stream. In such a situation, the particles align themselves to reduce the resistance. Some examples of pseudoplastic fluids are whipped cream, ketchup, molasses, syrups, ice, and blood. The most common non-Newtonian fluids are pseudoplastic fluids.

The time dependent viscosity behaviors are shown by *rheopectic fluids* in which apparent viscosity increases with duration of stress and the *thixotropic fluids* in which the apparent viscosity has a limited decrease with duration of stress. The examples of rheopectic fluids are printer ink and gypsum paste; examples of thixotropic fluids include pectin gel, some clays, semen, yoghurt, xanthan gum solutions, hydrogenated castor oil, etc.

The *Bingham plastic behavior* requires a threshold yield stress (τ_y) before a viscous behavior is exhibited. The following equation describes such a response.

$$\tau = m\left(-\frac{dv}{dy}\right) + \tau_y \tag{6.7}$$

The parameter (m) is *plastic viscosity* in the above equation.

Both pseudoplastic and dilatant fluids can be described by the following two parameter model. One of the parameters is *consistency coefficient* (m) and the other is *flow behavior index* (n).

$$\tau = m\left(-\frac{dv}{dy}\right)^n \tag{6.8}$$

As shown in Figure 6.2, the curve for the dilatant fluid is concave upward (n is more than 1) and that for the pseudoplastic fluids is concave downward (having n less than 1). The above shear equation is also known as *power-law model*.

In the above equation 6.8,

if n = 1, then the fluid is Newtonian;

if n > 1, then the fluid is dilatant;

if n < 1, then the fluid is pseudoplastic.

A general equation known as the *Herschel-Buckley model* has been proposed as follows, which also includes the Bingham plastic and quasi-plastic behaviors.

$$\tau = m\left(-\frac{dv}{dy}\right)^n + \tau_y \tag{6.9}$$

$$\text{Or,} \quad \tau = m(v)^n + \tau_y \tag{6.10}$$

In this equation, if n=1 and τ_y has some constant value, then the fluid is *Bingham plastic*.

Certain fluids behave as either dilatant or pseudoplastic fluids after reaching their initial yield points. In such cases, flow characteristics of the fluid are better represented by an equation similar to equation 6.9 requiring three parameters. Such equations are generally known as *quasi-plastic* or *mixed* types.

In most experiments, even though we do not know whether the fluid is Newtonian or Non-Newtonian, an apparent viscosity is measured, which would exert the same resistance to flow as that by a Newtonian fluid at the specified rate of shear. Apparent viscosity in a pseudoplastic fluid decreases with increasing shear rate whereas it increases with increasing shear rate in dilatant fluids.

After rearranging the equations 6.8 and 6.9 we get the results as follows

$$\tau = \left[m(v)^{n-1}\right]v \tag{6.11}$$

$$\tau - \tau_0 = \left[m(v)^{n-1}\right]v \tag{6.12}$$

Equations 6.11 and 6.12 resemble equation 6.5, and the multiplying factor of v in the above equation is termed the *apparent viscosity*, μ_{app}.

$$\mu_{app} = m(v)^{n-1} \tag{6.13}$$

For the fluids having their characteristics represented by equations 6.11 and 6.12, μ_{app} can also be expressed as

$$\mu_{app} = \frac{\tau}{v} \tag{6.14}$$

$$\mu_{app} = \frac{\tau - \tau_0}{v} \tag{6.15}$$

Although the units for apparent viscosity are the same units as those of dynamic viscosity, the magnitude varies with the shear rate. Therefore, the value of apparent viscosity needs to be specified corresponding to the shear rate at which it is measured.

6.2.2 Dynamic and Kinematic Viscosity

Dynamic or *absolute viscosity* has the dimensions of $ML^{-1}T^{-1}$ and is expressed in terms of SI units of $N.s.m^{-2}$ or Pa.s or poise. One poise is defined as the viscosity of the fluid if a tangential force of $1\ N.m^{-2}$ on the surface is required to maintain a relative velocity of $1\ m.s^{-1}$ between two layers of a liquid 1 m apart.

$$1\ Pa.s = 10\ poise = 1000\ cp$$

Dynamic viscosity of water at 20°C is 1 centipoise (cP) or 0.01 poise.

When dynamic viscosity of a fluid is divided by its density, the resultant value is termed as *kinematic viscosity*.

$$v = \mu / \rho \tag{6.16}$$

The kinematic viscosity has the unit of $m^2.s^{-1}$ or $cm^2.s^{-1}$ or Stoke.

$$1\ Stoke = 1\ cm^2.s^{-1} = 10^{-4}\ m^2.s^{-1}$$

The viscosities of some common food materials are given in Annexure III.

6.2.3 Effect of Temperature on Viscosity

Viscosities of liquids and gases respond differently to temperature changes. Viscosity of a gas increases as its temperature increases. Viscosity of a liquid, on the other hand, decreases with increase in liquid temperature. The relationships between the viscosities at different temperatures for different types of fluids are given as follows:

$$\text{For liquids,} \quad \mu = \mu_0 \left(\frac{1}{1 + \alpha t + \beta t^2} \right) \qquad (6.17)$$

where t = temperature, °C and α, β are constants.

For water, $\mu_0 = 1.79 \times 10^{-3}$ Pa.s, $\alpha = 0.03368$, $\beta = 0.000221$

Check that 1 Pa.s = 1000 cP, water has a viscosity of about 1 cP.

$$\text{For a gas,} \quad \mu = \mu_0 + \alpha t - \beta t^2 \qquad (6.18)$$

In this equation, the μ and μ_0 have the units of Pa.s.

For air $\mu_0 = 0.000017$ Pa.s, $\alpha = 0.56 \times 10^{-7}$, $\beta = 0.1189 \times 10^{-9}$
$$(\mu_0 = 0.02 \text{ cP})$$

The dynamic and kinematic viscosities of air at two temperatures are given in Table 6.1.

In contrast the viscosity of water decreases with increase in temperature asymptotically. The dynamic and kinematic viscosities of water at different temperatures between 0° and 100°C are given in Table 6.2.

Another relationship that expresses the relationship between viscosity of water and temperature, accurate to within 2.5% between 0° and 370°C, is given as

$$\mu = A \times 10^{\frac{B}{T-C}} \qquad (6.19)$$

Here μ is in Pa.s and water temperature, T, is in K.

The values of B and C are 247.8 and 140, respectively, in the unit of K. The value of A is 2.414×10^{-5} Pa.s,

The Arrhenius equation can also be utilized to find the viscosity at different temperatures, if the value at one temperature is known.

TABLE 6.1

Dynamic and kinematic viscosities of air at two temperatures

Temperature	Dynamic viscosity	Kinematic viscosity
15°C	1.81×10^{-5} kg.m^{-1}s^{-1} (18.1 µPa.s)	1.48×10^{-5} m^{2}s^{-1} (14.8 cS)
25°C	1.86×10^{-5} kg.m^{-1}s^{-1} (18.6 µPa.s)	1.57×10^{-5} m^{2}s^{-1} (15.7 cS)

TABLE 6.2

Temperature dependence of dynamic and kinematic viscosities of water

Temperature, °C	Dynamic viscosity, mPa.s (= $\times 10^{-3}$ Pa.s)	Kinematic viscosity, m^2.s^{-1}
0	1.787	1.787
10	1.519	1.307
20	1.002	1.004
30	0.798	0.801
40	0.653	0.658
50	0.547	0.553
60	0.467	0.475
70	0.404	0.413
80	0.355	0.365
90	0.315	0.326
100	0.282	0.290

6.2.4 Measurement of Viscosity

Dynamic and kinematic viscosities can be measured in many different ways. A few common methods are given below.

Viscosity cups help in determining a fluid's kinematic viscosity. The method requires filling a bowl with the fluid of unknown viscosity. There is a small opening at the bottom of the bowl. The time taken by the fluid to flow out of the bowl is correlated to viscosity.

In case of a vibrational viscometer, an electromechanical resonator measures the drag offered by the fluid while oscillating in the test fluid, when immersed. Power input for keeping the resonator vibrating at a constant amplitude is measured and correlated to the fluid viscosity.

A rotational viscometer requires the measurement of the torque that an object (usually, a disk or a bob) requires in the test fluid to maintain a predetermined speed. The shear stress and shear rate can also be recorded. As the shear force is applied externally, it is the dynamic viscosity that the rotational viscometer measures.

A capillary viscometer calls for the measurement of time taken by a defined volume of the fluid of unknown viscosity to flow through a capillary tube of known length and diameter. There is generally an upper and a lower mark on the tube and the time that the fluid takes to flow past these two marks is proportional to the fluid's kinematic viscosity.

A falling sphere viscometer, as the name implies, measures the time required by a sphere of known density while falling under gravity through a graduated tube filled with the fluid of known density and unknown viscosity. It is assumed that the sphere attains terminal velocity in the fluid before the time-distance measurement begins and that the tube diameter is at least 10 times the sphere diameter. The fluid should, obviously, be of such transparency that the sphere movement could be observed.

A consistometer is another device in which the fluid flow is directed through a metal trough by opening a gate at the inlet end with a small cross-section. Using a stopwatch, the distance travelled by the fluid in a given time is noted with the help of graduations on the trough. This method is convenient and, therefore, more popular for viscosity measurement of such products as ketchup and mayonnaise.

6.3 Laminar and Turbulent Flow

There are different types of fluid flow that are encountered in practice. Fluid flow can be either steady or unsteady, uniform or non-uniform, rotational or irrotational, laminar or non-laminar, compressible or incompressible flows, and one dimensional or multi-dimensional.

A flow is said to be *rotational* if the fluid particles rotate about their own mass center. Fluid particle rotation at a point in the flow field requires the determination of angular velocities of two differential linear elements of fluid particles, originally perpendicular to each other, and taking the average. The flow is said to be *irrotational* when the net rotation is zero.

In *laminar flow*, the fluid particles move along the parallel and straight stream-lines. The fluid particles in laminar flow appear to move in laminas (layers), sliding smoothly over the adjacent layers. Laminar flow occurs at very small velocity and/or high viscosity and is also termed *stream-line flow* or *viscous flow*.

Fluid particles under higher fluid velocity move in a zigzag fashion and there is intermixing of laminas. This is known as *turbulent flow*. Due to the turbulence, eddies are formed, which are responsible for high energy loss. Turbulence is caused by the contact between the flowing stream with solid boundaries or two fluid layers moving at substantially different velocities; the former is termed as wall turbulence and the latter as free turbulence.

Reynolds number (Re) of the flow, a dimensionless number, can indicate if the flow is turbulent or laminar.

$$\text{Re} = \frac{\rho D v}{\mu} \tag{6.20}$$

Here v indicates the flow velocity, ρ is the fluid density, μ is the fluid viscosity, and D is the diameter of pipe through which the fluid is flowing. The Re is also expressed as the ratio of inertia force to viscous force.

$$\text{Re is also written as} \quad \text{Re} = \frac{GD}{\mu} \tag{6.21}$$

where G is the mass flow rate of the fluid per unit area, $kg.m^{-2}.s^{-1}$.

In case of a pipe flow, the Reynolds number below 2100 indicates laminar flow. If the Re is more than 4000, the flow is a turbulent flow. A Reynolds numbers between 2100 and 4000 is termed as a transition region where it is not possible to know if the flow is laminar or turbulent.

Example 6.1 Internal diameter of pipe in a heat exchanger is 50 mm. Oil with a kinematic viscosity of 21.4 Stokes gives a discharge of 50 l.s⁻¹ through this pipe. Is the flow laminar or turbulent?

SOLUTION:

The radius of the pipe = 25 mm = 25×10^{-3} m

The area of the pipe is

$$A = \pi \times (25 \times 10^{-3})^2 = 1.96 \times 10^{-3} \text{ m}^2$$

Therefore, velocity

$$v = q/A = 50 \times 10^{-3} \text{ (m}^3.s^{-1}) / 1.96 \times 10^{-3} \text{ (m}^2) = 25.46 \text{ m. s}^{-1}$$

Kinematic viscosity, $v = 21.4$ Stokes $= 21.4 \times 10^{-4}$ m².s⁻¹

$$\text{Thus,} \quad \text{Re} = \frac{\rho D v}{\mu} = \frac{Dv}{v} = \frac{(0.05)(25.46)}{21.4 \times 10^{-4}} = 595$$

Since the Reynolds number is less than 2100, the flow is laminar.

6.4 Boundary Layer

As discussed earlier, the velocity in a pipe vanishes at the pipe wall and is the maximum at the center. Pipe wall or the solid boundary has an effect on the flow characteristics. This is true for all situations; the effect being confined to a thin fluid layer adhering to the solid boundary. The layer is called *boundary layer* and shear and shear forces are confined to this region of the fluid. However, the effect is very limited for fluids moving at low velocities or possessing high viscosities. The flow remains streamlined in the boundary layer. The heat transfer rate, as compared to that in the bulk flow, is lower in the boundary layer. The turbulent flow produces a thinner boundary layer and, hence, the heat transfer is improved.

The flow beyond the boundary layer is known as *potential flow*. In the potential flow the fluid behavior may approach an ideal fluid, i.e., incompressible with zero viscosity. Potential flow is also called irrotational flow as neither circulations nor eddies could form within the stream. Since there is almost no friction, mechanical energy is not dissipated into heat.

6.5 Relationship between Volume of Flow, Pressure Difference, and Viscosity of Fluid Flowing in a Tube

6.5.1 For Newtonian Fluid

For obtaining a relationship between the volume of flow, pressure difference, and viscosity of a Newtonian fluid flowing in a tube, it is assumed that the fluid is incompressible and the fluid properties are time independent. The flow is steady and laminar and there is no tangential or radial component of the fluid velocity and there is no-slip condition at the wall. It is also assumed that the fluid viscosity is unaffected by pressure. The measurements are made at constant temperature.

Figure 6.3 shows the velocity profile of a fluid flowing in a tube. Let the radius of the tube through which the fluid is flowing be R. Let Δp i.e. ($p_1 - p_2$) be the applied pressure causing the fluid to move through the tube. An annular ring of infinitesimal thickness dr is considered at a distance r from the

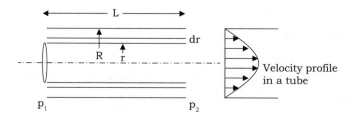

FIGURE 6.3   Fluid velocity profile in a tube

center. The shear stress that results at r is τ and the velocity at this point is v.

The force required to move liquid in radius r
$$= \text{pressure} \times \text{area} = \Delta p \times \pi r^2 \qquad (6.22)$$

The force that opposes the flow $= 2 \pi r L \tau \qquad (6.23)$

where τ = shear stress
Now balancing the forces at radius r,

$$\Delta p \pi r^2 = 2 \pi r L \tau$$

$$\text{Or,} \quad \Delta p . \pi r^2 = 2 \pi r L \mu \left(-\frac{dv}{dr} \right)$$

$$\text{i.e.} \quad -dv = \frac{\Delta p . r}{2\mu L} . dr$$

Integrating within limits, i.e., at radius r, v = v and at r = R, v = 0

$$-\int_0^v dv = \int_R^r \frac{\Delta p . r}{2\mu L} . dr = \frac{\Delta p}{2\mu L} \left[\frac{r^2}{2} \right]_R^r$$

The solution of above gives, $v = \dfrac{\Delta p . R^2}{4\mu L} \left[1 - \left(\dfrac{r}{R} \right)^2 \right]$

$$(6.24)$$

This equation is of the form of a parabola showing that the velocity distribution in a capillary tube is parabolic. We can modify this equation to find the maximum and mean velocities.

The velocity at the center (at r = 0) of the tube will be the maximum.

$$\text{Thus,} \quad v_{max} = \frac{\Delta p . R^2}{4\mu L} \qquad (6.25)$$

The mean velocity will be obtained from the equation $\bar{v} = \dfrac{q}{\pi R^2}$

where q, the volumetric flow rate, can be expressed as

$$q = \int_0^R 2\pi r . dr . v \qquad (6.26)$$

The above expression can be extended as

$$q = \int_0^R 2\pi r . dr . v = 2\pi \int_0^R r . dr . v$$

$$= 2\pi \int_0^R r . dr \frac{\Delta p . R^2}{4\mu L} \left[1 - \left(\frac{r}{R} \right)^2 \right] = \frac{2\pi \Delta p . R^2}{4\mu L} \left[\int_0^R r . dr - \int_0^R \frac{r^3 dr}{R^2} \right]$$

$$= \frac{2\pi \Delta p . R^2}{4\mu L} \left[\frac{r^2}{2} - \frac{r^4}{4R^2} \right]_0^R = \frac{\pi \Delta p . R^2}{4\mu L} \left[r^2 - \frac{r^4}{2R^2} \right]_0^R$$

$$= \frac{\pi \Delta p . R^2}{4\mu L} \left[R^2 - \frac{R^4}{2R^2} \right] = \frac{\pi \Delta p . R^2}{4\mu L} \left[\frac{R^2}{2} \right]$$

$$\text{or,} \quad q = \frac{\pi \Delta p . R^4}{8\mu L} \qquad (6.27)$$

The equation 6.27 is the *Poiseuille's equation*.

$$\bar{v} = \frac{q}{\pi R^2} = \frac{\pi \Delta p . R^4}{8\mu L} \bigg/ \pi R^2 = \frac{\Delta p . R^2}{8\mu L} = \frac{v_{max}}{2} \qquad (6.28)$$

Thus, the mean velocity in a capillary tube is determined to be half of the maximum velocity.

The category of capillary tube rheometers includes a variety of instruments working on the above principle where the fluid under consideration is forced to move through a tube of known configuration. The measured pressure gradient and the fluid's volumetric flow rate through the tube influence the shear stress-shear strain relationship. As viscosity is greatly affected by temperature, the measurement is done under isothermal conditions.

Example 6.2 A liquid food at the rate of 200 l.min^{-1} is being pumped through a 50 m long pipe of 3.5 cm diameter. The pressure at the pipe end is atmospheric. What would be the pressure at the pump discharge end if the viscosity and density of the food are 0.1 Pa.s and 1020 kg.m^{-3}, respectively?

SOLUTION:

The Poiseuille's equation for flow rate in a pipe is given by,

$$q = \frac{\pi \Delta p R^4}{8\mu L}$$

Here $q = 200 \times 10^{-3}$ m³. min⁻¹ = $(200 \times 10^{-3}/60)$ m³.s⁻¹

Now, $\Delta p = \dfrac{8\mu Lq}{\pi R^4} = \dfrac{8 \times 0.1 \times 50 \times 200 \times 10^{-3}}{(\pi) \times (1.75 \times 10^{-2})^4 \times 60} = 452518.84$ Pa

Total pressure at the pipe end = 101.325 kPa + 452.52 kPa
$= 553.843$ kPa

Alternatively,
Cross-sectional area of pipe,

$$A = \frac{\pi}{4}D^2 = \frac{\pi}{4}0.035^2 = 9.62 \times 10^{-4}\,\text{m}^2$$

Velocity through the pipe,

$$v = q/A = \frac{(0.2/60)}{9.62 \times 10^{-4}} = 3.46\ \text{m.s}^{-1}$$

$$\mathrm{Re} = \frac{\rho D v}{\mu} = \frac{1020 \times 0.035 \times 3.46}{0.1} = 1235.22$$

Thus, the flow is laminar.

The pressure drop (Δp) across two ends of the pipe for laminar flow, is given by

$$\Delta p = \frac{32\mu vL}{D^2} = \frac{32 \times 0.1 \times 3.46 \times 50}{(0.035)^2} = 452.52\ \text{kPa}$$

Note that $\dfrac{32\mu vL}{D^2} = \dfrac{8\mu Lq}{\pi R^4}$

Thus, $(p_{atm} + \Delta p)$ is the pressure generated at the discharge of the pump

$$= 101.325 + 452.52 = 553.843\ \text{kPa}$$

Example 6.3 A Newtonian liquid flows through a tube having 4 mm diameter and 50 cm length in laminar regime under a gauge pressure of value A. What will be the required pressure if the diameter of the tube is increased to 5 mm and the length to 100 cm for the same volumetric flow rate? The liquid is discharged at atmospheric pressure in both the cases.

SOLUTION:

Given $L_1 = 0.5$ m, $D_1 = 0.004$ m,
thus $R_1 = 0.002$ m

$L_2 = 1.0$ m, $D_1 = 0.005$ m, thus $R_2 = 0.0025$ m

If the flow rates are the same, then as per the Poiseuille's equation,

$$\frac{\pi \Delta p_1 R_1{}^4}{8\mu L_1} = \frac{\pi \Delta p_2 R_2{}^4}{8\mu L_2}$$

or $\Delta p_2 = \Delta p_1 \dfrac{L_2}{L_1}\left(\dfrac{R_1}{R_2}\right)^4$

or $\Delta p_2 = \Delta p_1 \left(\dfrac{1}{0.5}\right)\left(\dfrac{0.002}{0.0025}\right)^4 = 0.8192\ \Delta p_1$

$= 0.8192\ \text{A}\,(\text{in this case})$

Shear Rate at the Wall of a Tube for Newtonian Fluid

For a Newtonian fluid flow through a tube, the velocity profile versus average velocity is given as follows:

$$v = 2\bar{v}\left[1 - \left(\frac{r}{R}\right)^2\right] \qquad (6.29)$$

Differentiating the above equation

$$\frac{dv}{dr} = 2\bar{v}\left[\frac{2r}{R^2}\right] \qquad (6.30)$$

Thus, for a Newtonian fluid, shear rate at the wall $(r = R)$ is

$$\left.\frac{dv}{dr}\right|_w = \frac{4\bar{v}}{R} \qquad (6.31)$$

6.5.2 For Non-Newtonian Fluid

$$\text{Shear stress } \tau = m\left(-\frac{dv}{dr}\right)^n$$

Now, as in the previous case, we proceed by balancing the forces at radius r

$$\Delta p.\pi r^2 = 2\,\pi r L \tau$$

Now, $\Delta p.\pi r^2 = 2\,\pi r L m\left(-\dfrac{dv}{dr}\right)^n$

Proceeding as above and integrating within limits of R and r, we obtain

$$v = \left(\frac{\Delta p}{2mL}\right)^{1/n} \frac{n}{n+1}\left(R^{\frac{n+1}{n}} - r^{\frac{n+1}{n}}\right) \qquad (6.32)$$

The maximum velocity is at $r = 0$,

Thus, $v_{max} = \left(\dfrac{\Delta p}{2mL}\right)^{1/n} \dfrac{n}{n+1}R^{\frac{n+1}{n}}$ (6.33)

Similarly, the average velocity,

$$\bar{v} = \frac{q}{\pi R^2}$$

Now, $q = \int_0^R 2\pi r.dr.v$

$= \int_0^R 2\pi r.dr\left(\frac{\Delta p}{2mL}\right)^{1/n}\frac{n}{n+1}\left(R^{\frac{n+1}{n}}-r^{\frac{n+1}{n}}\right)$

Integrating the above and noting that $\bar{v}=\frac{q}{\pi R^2}$, the following equation is obtained

$$\bar{v}=\left(\frac{\Delta p}{2mL}\right)^{1/n}\left(\frac{n}{3n+1}\right)R^{\frac{n+1}{n}} \qquad (6.34)$$

6.6 Laminar Flow between Two Parallel Plates

The maximum velocity of a falling film along an inclined flat surface can be given as

$$v_{max}=\frac{\rho g\delta^2}{2\mu}\cos\alpha \qquad (6.35)$$

where g is the gravitational acceleration, δ is the thickness of falling film, and α is the angle of inclination with the horizontal.

The average velocity can be written as

$$\bar{v}=\frac{\rho g\delta^2}{3\mu}\cos\alpha \qquad (6.36)$$

If the surfaces are vertical, the angle α is 0. Thus, the above equations reduce to

$$v_{max}=\frac{\rho g\delta^2}{2\mu} \qquad (6.37)$$

$$\bar{v}=\frac{\rho g\delta^2}{3\mu} \qquad (6.38)$$

The volumetric flow rate q in a channel of width w is

$$q=A\bar{v}=\delta\times w\times\frac{\rho g\delta^2}{3\mu}\cos\alpha=\frac{\rho g w\delta^3\cos\alpha}{3\mu} \qquad (6.39)$$

Here A is the cross-sectional area. In case of vertical wall $\cos\alpha = 1$,

Thus, $q=\frac{\rho g w\delta^3}{3\mu}$ (6.40)

Example 6.4 A sample of oil is made to flow on a vertical wall as a film. What would be the thickness of the film if the density and viscosity of the oil are 940 kg.m^{-3} and 0.32 Pa.s, respectively, and the mean velocity of the film is 0.04 m.s^{-1}?

SOLUTION:

$v=\frac{\rho g\delta^2}{3\mu}$, or $\delta=\sqrt{\frac{3\mu v}{\rho g}}=\sqrt{\frac{3\times0.32\times0.04}{940\times9.81}}=2.04\times10^{-3}$ m

$= 2.04$ mm

6.7 Viscous Force on a Body Moving Through a Fluid

When a spherical ball is dropped in a fluid, the viscous force acting on the ball is dependent on the ball's velocity through the fluid (v), fluid viscosity (μ), and the ball's radius (a).

Thus, $F\propto \mu.a.v$ (6.41)

The proportionality constant has been found to be 6π. The equation, thus, modifies to

$$F=-6\pi.\mu.a.v \qquad (6.42)$$

Equation 6.42 is known as Stokes' law (after Sir George G. Stokes, a British scientist). Since retarding force is opposite to the direction of the object's motion, the negative sign is introduced in Stokes' law. The law is applicable in the processes of settling and sedimentation for separation of solids from fluids.

6.8 Terminal Velocity

Terminal velocity is the constant velocity (which is also the maximum velocity) attained by a body when falling through a viscous fluid. Upon acquiring terminal velocity, the force balance can be written as follows.

upward viscous force + upward buoyant force = weight of the body

If v_t is denoted as the terminal velocity, then the equation can be written as

$$6\pi.\mu.a.v_t=(4/3)\pi.a^3(\rho_p-\rho_f)g \qquad (6.43)$$

where, ρ_f and ρ_p are the densities of the fluid and the falling body.

i.e. $v_t=\frac{2}{9}.\frac{a^2}{\mu}(\rho_p-\rho_f)g$ (6.44)

The above expression indicates that the terminal velocity is

a) directly proportional to the difference in the particle and fluid densities and the square of the particle radius, and

b) inversely proportional to the viscosity of the medium.

6.9 Pressure at a Point in a Static Fluid

Consider a static fluid mass in a reservoir and a point in the static fluid. Pascal's law states that the pressure or intensity of pressure at the selected point is the same in all directions.

Hydrostatic law governs the magnitude of pressure at any point in a fluid at rest. The law states that pressure in a vertical direction increases or decreases in proportion to the height of the fluid column at that point.

$$\text{i.e., } \delta p / \delta z = \rho.g$$

$$\text{or } p = \rho.g.z, \text{ i.e. } z = \rho.g/p \qquad (6.45)$$

Here, z is called the pressure head.

The fluid pressure at a point is given in the units of $N.m^{-2}$ or Pa in SI system.

$$1 \ N.m^{-2} = 1 \ Pa = 10^{-5} \ bar$$

One atmosphere pressure equals 1.01325 bar.

> ***Gauge pressure and vacuum.*** In many food processing applications, we have to maintain a pressure that is either higher or lower than the atmospheric and the pressures are usually expressed as *gauge pressure* and *vacuum*. When absolute pressure is more than atmospheric pressure, *gauge pressure* is absolute pressure minus atmospheric pressure. However, when absolute pressure is lower than the atmospheric pressure, *vacuum* is atmospheric pressure minus absolute pressure.

6.10 Equations of Motion

Newton's second law of motion states that the net force acting on a fluid element in x direction, F_x, is the mass of the fluid element, m, multiplied by the acceleration in x direction, a_x.

$$F_x = a_x.m \qquad (6.46)$$

In a general case of fluid flow, the forces present are gravity (F_g), pressure (body) (F_p), viscous (F_v), turbulence (F_t), and compression (F_c).

Thus, we can write the equation as

$$F_x = (F_g)_x + (F_v)_x + (F_p)_x + (F_t)_x + (F_c)_x \qquad (6.47)$$

If F_c is negligible, as in case of water, the resulting net force

$$F_x = (F_g)_x + (F_v)_x + (F_p)_x + (F_t)_x \qquad (6.48)$$

This is *Reynold's equation of motion*.

If F_t is negligible, the resulting equation is called *Navier-Stokes equation*.

$$F_x = (F_g)_x + (F_v)_x + (F_p)_x \qquad (6.49)$$

If the flow is assumed to be ideal then viscous force (F_v) is zero, and the equation is called *Euler's equation*.

Euler's equation of motion is otherwise given by,

$$\delta p / \rho + gdz + vdv = 0 \qquad (6.50)$$

6.11 Bernoulli's Theorem

The resultant equation after integrating Euler's equation of motion is *Bernoulli's equation*.

$$\int \delta p / \rho + \int vdv + \int gdz = \text{constant} \qquad (6.51)$$

Density (ρ) for an incompressible flow is constant and the resultant equation is as follows:

$$p / \rho + gz + v^2/2 = \text{constant} \qquad (6.52)$$

$$\text{or, } p / \rho g + v^2/2g + z = \text{constant} \qquad (6.53)$$

$$\text{i.e., } p/w + v^2/2g + z = \text{constant} \qquad (6.54)$$

Taking the terms in the above equation on a per unit fluid weight basis, z = specific potential head or specific potential energy, $p/\rho g$ = specific pressure head or specific pressure energy, and $v^2/2g$ = specific kinetic head or specific kinetic energy.

Bernoulli's theorem for a steady state, frictionless, and incompressible fluid flow states that total specific energy of the flowing fluid remains constant. The *total specific energy* includes the specific potential energy (z), specific pressure energy ($p/\rho g$), and specific kinetic energy ($v^2/2g$).

In other words, in a steady flow of inviscid and incompressible fluid, the sum of pressure, elevation, and velocity heads remains constant along a streamline if no energy is added or taken out by an external source. In incompressible fluid flow, there is no appreciable change in temperature, hence, internal energy is ignored.

Bernoulli's equation has been derived by making the assumptions that the fluid is ideal and incompressible and that the flow is steady and irrotational. However, because of simplicity and ease of calculations, this equation is widely used for pressure and velocity calculations in a flowing fluid.

> **Example 6.5 A liquid, having the density of 1280 $kg.m^{-3}$ and flowing at 26 $m^3.h^{-1}$ through a horizontal pipe (with 5 cm internal diameter), exerts 46 kPa pressure. What will be the pressure if the pipe diameter is reduced to 4 cm?**
>
> **SOLUTION:**
> **For case I:**
>
> Flow rate = 26/3600 $m^3.s^{-1}$, i.e., $7.22 \times 10^{-3} \ m^3.s^{-1}$
>
> Area of pipe = $(\pi/4) \times D^2 = (\pi/4) \times (0.05)^2$
> $= 1.963 \times 10^{-3} \ m^2$

Velocity of flow = Flow rate / area
$= 7.22 \times 10^{-3} / (1.963 \times 10^{-3}) = 3.677$ m.s^{-1}

For case II:

Area of pipe $= (\pi/4) \times D^2 =$
$(\pi/4) \times (0.04)^2 = 1.256 \times 10^{-3}$ m^2

Velocity of flow = Flow rate / area =
$7.22 \times 10^{-3} / (1.26 \times 10^{-3}) = 5.73$ m.s^{-1}

Using Bernoulli's equation,

$$\frac{p_1}{\rho_1} + \frac{v_1^2}{2} + z_1 g = \frac{p_2}{\rho_2} + \frac{v_2^2}{2} + z_2 g$$

$$\frac{46 \times 10^3}{1280} + \frac{3.677^2}{2} + 0 = \frac{p_2}{1280} + \frac{5.73^2}{2} + 0$$

or, $35.9375 + 6.76 = p_2/1280 + 16.417$

i.e. $p_2/1280 = 26.2805$

Hence, $p_2 = 33639$ Pa $= 33.64$ kPa

Example 6.6 A U tube manometer is placed below the pipeline through which water flows. If the manometer reading is 6 cm and the density of the manometric fluid is 14000 kg.m^{-3}, what is the pressure drop along the two points of the pipe?

SOLUTION:

Given, $h = 0.06$ m and $p_m = 14000$ kg.m^{-3}

If the two points are denoted as A and B, and the corresponding total pressures as p_A and p_B,

$$p_A + (x + h)1000 = p_B + 1000\, x + 14000\, h$$

or, $p_A + (x + 0.06)\, 1000 = p_B + 1000\, x + 14000 \times 0.06$

$$p_A + 1000\, x + 60 = p_B + 1000\, x + 14000 \times 0.06$$

i.e., $p_A - p_B = 840 - 60 = 780$ kg.m^{-2}

6.12 Continuity Principle

The continuity principle in fluid dynamics is essentially the statement of conservation of mass; fluid mass is conserved in a flow field. The continuity principle is useful in a plethora of situations involving changes in fluid properties, flow geometries, and fluid dynamics. The continuity principle in its simplest form for incompressible flow in one direction is as follows:

$$\rho_1 v_1 A_1 = \rho_2 v_2 A_2 + A \frac{\partial}{\partial t}(\rho v) \qquad (6.55)$$

The above equation can be used when the fluid flow is unidirectional. It would need three-dimensional mass balance if the fluid flow were multi-directional. The time derivative term at steady state simply vanishes. In case of an incompressible fluid, the density does not change, and, hence, the continuity equation reduces to

$$A_1 v_1 = A_2 v_2 \qquad (6.56)$$

6.13 Momentum Balance

The approach to characterize a fluid flow in any situation requires consideration of conservation of mass, momentum, and energy.

Momentum (kg.m.s^{-1}) = mass (kg) $\times$ velocity (m.s^{-1})

Rate of momentum flow (kg.m.s^{-2}) = mass rate of flow (kg.s^{-1}) $\times$ velocity (m.s^{-1})

Momentum transfer occurs across streamlines when velocity gradient exists in a flowing fluid. The momentum transfer rate per unit area (kg.m^{-1}.s^{-2}), d(mV/A)/dθ, is termed *momentum flux*.

Momentum flux has the same unit as that of stress or pressure. Pressure drop in fluid flow through a pipe is caused by momentum flux between streamlines in a direction perpendicular to the direction of flow. The following equation represents the momentum balance.

ΣF + incoming momentum flux per unit time
= outgoing momentum flux per unit time
+ accumulation   (6.57)

All the external forces, e.g., stress, atmospheric pressure, restraints, etc., on the flow regime are added together in ΣF.

6.14 Pressure Drop and Friction Coefficient for a Fluid Flow in a Pipe

The equation for pressure drop due to friction in a pipe under laminar flow is as follows:

$$p_1 - p_2 = \frac{4fL\rho\bar{v}^2}{2D}$$

where v is the average velocity and L and D are the length and diameter of the pipe; f is known as the *Fanning friction factor*.
The friction head loss is also given by

$$h_f = \frac{4fLv^2}{2gD} \qquad (6.58)$$

The pressure drop for a pipe of given length can be given by

$$\frac{p_1 - p_2}{\rho g} = h_f = \frac{32\mu v L}{\rho g D^2}, \qquad (6.59)$$

where pipe length is L and the mean velocity is v. The equation is known as *Hagen-Poiseuille formula*.

Thus, head loss, h_f, for viscous fluid flow in a pipe from the above equation is

$$h_f = \frac{32\mu vL}{\rho gD^2} \qquad (6.60)$$

where f = coefficient of friction between the pipe and fluid.

Using the above two equations

$$\frac{32\mu vL}{\rho gD^2} = \frac{4fLv^2}{2gD}$$

$$f = \frac{32\mu vL2gD}{4Lv^2\rho gD^2} = \frac{16\mu}{v\rho D}$$

$$f = \frac{16}{Re} \qquad (6.61)$$

For Reynold's number between 4000 to 10^6, $f = \dfrac{0.079}{Re^{1/4}}$

$$(6.62)$$

Example 6.7 The Fanning friction factor (f) for a flow through a smooth pipe in the turbulent flow regime is given by $f = mRe^{-0.2}$, m being a constant. If the velocity of water in a section of the pipe is 1.8 m.s^{-1}, the frictional pressure drop has been observed as 12 kPa. What will be the pressure drop across this section (in kPa) when the velocity of water is doubled?

SOLUTION:

We know that $\Delta p = \dfrac{4fl\rho v^2}{2D}$

As $f = mRe^{-0.2}$, as per the above equation, $\Delta p \alpha v^{1.8}$

i.e., $\dfrac{\Delta p_1}{\Delta p_2} = \left(\dfrac{v_1}{v_2}\right)^{1.8}$

For the present case,

$$\Delta p_2 = \Delta p_1 \left(\frac{v_2}{v_1}\right)^{1.8} = (12)(2)^{1.8} = 41.78 \text{ kPa}$$

6.15 Measurement of Fluid Flow

The flow measurement can be done with the help of mechanical flowmeters, pressure-based meters, or variable area flowmeters. Newer techniques, such as optical flow meters and thermal mass flow meters, are also available. A brief discussion follows.

Flow rate in mechanical flowmeters is determined by recording the time needed to fill a specific volume and then dividing the volume by the recorded time. Continuous flow rate measurements require a mechanism for filling and emptying buckets on a continuous basis that assists in dividing the flow without letting it out of the tube. The mechanism can be in the form of reciprocating pistons in cylinders, gear teeth coupled with the inner wall of a meter, or a progressive opening created by rotating oval gears or a helical screw. Thus, these devices are labeled piston meter, rotary piston, or gear meters.

In case of a *turbine flow meter*, the rotation of turbine blades caused due to a flow of fluid is measured to determine the flow rate. In the case of a *Woltman meter*, also called a *helix meter*, a rotor with helical blades is placed axially in the flow stream. Larger size Woltman meters are more popular. A single jet meter has radial vanes on its impeller and a jet impinges on the impeller. In case of a multiple jet or multi-jet meter, there is a vertical shaft on which an impeller is mounted to rotate horizontally. The single jet and multi-jet meters have a similar working principle except that there are several ports directing the jets at the impeller in a multi-jet system whereas it is only one jet in a single jet system. A multi-jet meter is better since it minimizes uneven wear on its shaft and impeller.

In a *paddle wheel meter*, there are magnets mounted on the pedals perpendicular to the direction of flow. When exposed to flow, the pedals begin to rotate. There is a sensor to monitor the movement of magnets. The sensor, in turn, generates voltage and frequency signals in proportion to the flow rate.

The *Pelton wheel* or a *radial turbine* also translates the flow around an axis to flow rates.

The *current meter* involves propeller type blades which convert the signal to an electronic output. Such a device is often used for testing of hydroelectric turbines.

The *pressure-based meters* work on Bernoulli's principle. Fluid pressure loss is measured by using either a venturi tube, laminar plates, a nozzle, or an orifice. Change in pressure before and within the constriction is measured with the help of pressure sensors. Alternatively, dynamic pressure can be derived from the measurement of static and stagnation pressures.

The *Dall tube* is a venturi meter, but shorter, such that it experiences a lower pressure drop as compared to an orifice plate. Therefore, these meters are quite common for the measurement of flow rates in large pipe works.

While using the *Pitot tube* for flow measurement, one end of the tube is inserted into the flow to measure the stagnation pressure. The fluid velocity is determined from the difference between the stagnation pressure and the static pressure at the flow boundary.

The *impact probes* (also known as *averaging pitot tubes*) permit the measurement of multi-dimensional velocity. There are three or more holes on a specific pattern on the measuring tip. The velocity vector is measured in two dimensions if there are three holes. The three-dimensional velocity vector can be measured using the five holes arranged in the form of a "plus" sign.

Cone meters are based on the basic principle that governs the operation of venturi and orifice type meters. However, the upstream and downstream piping is the same. In addition to serving as a differential pressure producer, the cone also acts as a conditioning device.

A *Linear resistance meter*, also called laminar flowmeter, is suitable to measure very slow flows where differential pressure is linearly proportional to the fluid flow and fluid viscosity. Such fluid flows are also known as viscous drag flow or laminar flow. It may be noted that orifice plates, venturi, and other meters measure turbulent flow rates. The flow element of a laminar flow meter can be a long porous plug, a single long capillary tube, or a bundle of such tubes.

A *variable area meter*, for example a rotameter, has a variable cross-sectional area in which there is a float. The end of the smallest area side is connected to the flow pipe. The float takes different positions as per the volumetric flow rate. The floats are usually in the shape of spheres and spherical ellipses. In an alternate device, a spring-loaded tapered plunger, fitted in an orifice, is made to deflect by connecting the flow to the orifice.

Optical flowmeters use laser techniques to measure the actual flow of particles. The time required by some particles to move in between two laser beams is recorded by photodetectors. Such methods can measure within a wide range of flow and can be used in varied environmental conditions.

Flow rates in open channels can be determined by making a measurement of either the height of the flow or the flow velocity. *Dye testing* and *acoustic methods* can also be used for the flow rate determinations. The dye testing method involves the release of a known amount of dye in the flow stream per unit time. The dye concentration is measured after complete mixing at some downstream location.

Acoustic Doppler velocimetry makes use of the Doppler shift effect to measure the instantaneous velocity of particles. The ultrasonic Doppler shift is caused due to the ultrasonic beam by the flow of particles in the flowing fluid. This Doppler shift then is correlated to the fluid flow rate.

Thermal mass flowmeters measure mass flow rates by using heated elements and temperature sensors. The sensors essentially measure the rate of heat absorbed by the fluid and correlate it to the flow rate. This category includes mass air flow sensors fitted into automobile engines to measure the air intake rate.

Vortex flowmeters involve placing a shredder bar in the path of the fluid. The fluid flow rate is found to be proportional to the frequency at which these vortices alternate sides.

In addition, some non-contact type electromagnetic flow meters, magnetic flow meters are also available.

Fluid flow is closely associated with most food processing operations as most of the raw materials and utilities are fluids. The fluid foods or raw materials need to be transported from one section to another. Fluid also flows in the processing equipment, such as the evaporators, heat exchangers, and mixers, and the efficacy of any processing operation is greatly decided by the fluid flow properties. Fluid properties such as density and viscosity change during processing, which also need to be considered during the design of the food processing equipment. The different batches of fluid food products must be consistent and, hence, the product's viscosity is monitored. Rheological properties help characterization of the products. Thus, understanding a food processing system is never complete without a proper understanding of the fluid mechanics and flow behavior.

Check Your Understanding

1. A salt solution (1250 kg.m^{-3}) flowing at the rate of 30 m^3.h^{-1} through a horizontal 6 cm (internal) diameter pipe exerts 40 kPa pressure. What will be the pressure if the pipe diameter is reduced to 4 cm?

    Answer: 17.95 kPa

2. The pressure drop between two points of a pipe is measured using a U tube manometer. The manometer is placed below the pipeline through which the water flows. Determine the pressure drop along the two points of the pipe if the manometer reading is 8 cm and the density of manometric fluid is 13570 kg.m^{-3}.

    Answer: 1277 kg.m^{-2}

3. Water flows at the rate of 0.4 m^3.min^{-1} in a 7.5 cm diameter pipe at a pressure of 70 kPa. If the pipe reduces to 4cm diameter, then calculate the new pressure in the pipe. The density of water is 1000 kg.m^{-3}.

    Answer: 57.1 kPa

4. Whole milk having a density of 1030 kg.m^{-3} and kinematics viscosity of 2.06×10^{-6} m^2.s^{-1} flows with an average velocity of 20 mm.s^{-1} through a circular tube of length 10 mm and having an inside diameter of 25.4 mm. Calculate the pressure drop along the tube length.

    Answer: 0.021 Pa

5. A fluid is moving in a pipe of inner diameter D. This pipe is connected to another pipe of diameter 2D for conveying the fluid to a longer distance. What will be the ratio of the Reynolds numbers in both the pipes?

    Answer: $Re_1 = 2 \times Re_2$

6. A sugar syrup (density=1040 kg.m^{-3} and viscosity 1600×10^{-6} Pa.s) is required to be pumped into a tank (1.5 m diameter and 3 m height) by a 3 cm inside diameter pipe. If the liquid is required to flow under laminar conditions, what will be the minimum time to fill the tank?

    Answer: 19.36 h

7. A chocolate mix at 100°C is flowing through a 2 cm diameter and 4 m long stainless-steel tube at 13.2 kg .min^{-1}. The density of the mix is 1750 kg.m^{-3} and its viscosity at 100°C is 2 Pa s. What will be the pressure drop for this flow?

    Answer: 255998.4 Pa

8. Water is flowing at a rate of 0.5 m^3.s^{-1} in a horizontal pipeline of inside diameter 0.5m. The density and kinetic viscosity of the water is 1000 kg.m^{-3} and 10^{-6} m^2.s^{-1}, respectively. Assume the friction value to be 0.0093 and acceleration due to gravity as 9.81 m.s^{-2}. What will be the required power per unit length of the pipeline (in W.m^{-1}) to maintain a constant flow rate?

    Answer: 30.14

Objective Questions

1. Explain Newton's law of viscosity and find out the dimension of viscosity.

2. Explain the effect of temperature on the viscosity of liquids and gases.

3. Derive a relationship between the flow rate and pressure difference in a pipe with inner radius R and length L.

4. Explain (a) Poiseuille's law, (b) Stokes' law, (c) surface tension of a liquid, (d) Pascal's law, (e) Bernoulli's theorem, and (f) friction coefficient for fluid flow in a pipe.

5. Differentiate between the following:

   (a) Viscosity and viscoelasticity;

   (b) Dynamic viscosity and kinematic viscosity;

   (c) Newtonian fluid and non-Newtonian fluid;

   (d) Linear visco-elasticity and non-linear viscoelasticity;

   (e) Pseudoplastic fluid and dilatant fluid;

   (f) Thixotropic fluid and rheopectic fluid;

   (g) Bingham plastic behavior and pseudoplastic behavior;

   (h) Laminar flow and turbulent flow.

BIBLIOGRAPHY

Ahmad, T. 1997. *Dairy Plant Engineering and Management*, 4th ed. Kitab Mahal, New Delhi.

Bansal, R.K. 1998. *A Text Book of Fluid Mechanics*. Laxmi Publications, New Delhi.

Charm, S.E. 1978. *Fundamentals of Food Engineering*, 3rd ed. AVI Publ., Westport, Connecticut.

Chaudhry, M.H. 1993. *Open Channel Flow*. Prentice Hall, Hoboken.

Chow, V.T. 1959. *Open Channel Hydraulics*. McGraw-Hill, New York.

Datta, A.K. 2001. *Transport Phenomena in Food Process Engineering*. Himalaya Publishing House, Mumbai.

Earle, R.L. and Earle, M.D. 2004. *Unit Operations in Food Processing*, Web Edn. The New Zealand Institute of Food Science & Technology, Inc., Auckland https://www.nzifst.org.nz/ resources/unitoperations/index.htm

Heldman, D.R. (Ed.) 1981. *Food Process Engineering*. AVI Publ. Co., Westport, Connecticut.

Henderson, S.M. and Perry, R.L. 1976. *Agricultural Process Engineering*. AVI Publ Co., Westport, Connecticut.

Holman, P. 1996. *Experimental Methods for Engineers*. McGraw-Hill, New York.

Kokini, J.L., and Plutchok, G.J. 1987. Viscoelastic properties of semisolid foods and their bio-polymeric, components, *Food Technol.* 41(3):89.

Kreiger, I.M. and Maron, S.H. 1952. Direct determination of flow curves of non-Newtonian fluids. *J. Appl. Phys.* 33:147.

McCabe, W.L., Smith, J.C. and Harriot, P. 1993. *Unit Operations of Chemical Engineering*, 5th ed. McGraw-Hill, Inc., New York.

Modi, P.N. and Seth, S.M. 2000. *Hydraulics & Fluid Mechanics*. Standard Book House, New Delhi.

Ramanuthan, S. *Hydraulics, Fluid Mechanics & Hydraulic Machines*. Dhanpat Rai & Sons, Delhi.

Rao, M.A., Shallenberger, R.S., and Cooley, H.J. 1987. Effect of temperature on viscosity of fluid foods with high sugar content. In: M.Le-Maguer and P. Jelen (eds.) *Food Engineering and Process Applications*, Vol. I,. Elsevier Applied Science, New York.

Singh, R.P. and Heldman, D.R. 2014. *Introduction to Food Engineering*. Academic Press, San Diego, CA.

Toledo, R.T. 2007. *Fundamentals of Food Process Engineering*, 3rd ed. Springer, New York.

Watson, E.L., and Harper, J.C. 1989. *Elements of Food Engineering*, 2nd ed. Van Nostrand Reinhold, New York.

7

Heat and Mass Transfer

The movement of energy due to temperature difference from one region to another within a body or between two bodies is called heat transfer. It is common in most food processing operations, such as drying, freezing, canning, blanching, sterilization, etc. Heat transfer differs from thermodynamics in that thermodynamics is conceptually based on equilibrium or infinitesimal deviations from equilibrium while heat transfer occurs from non-equilibrium conditions, specifically, finite temperature differences.

7.1 Modes of Heat Transfer

Heat or thermal energy transfer in a given situation can occur due to either one or more of the three modes, i.e., conduction, radiation, and convection. *Conduction* is the transfer of thermal energy in a material through vibrating contiguous molecules of the material. The molecules do not move from their positions during conduction. When the energy transfer rate is large enough to make the molecules move around and transfer their energy contents to other molecules, the mode of heat energy transfer is known as *convection*. The molecular movement may be caused either by density difference or by agitation. When thermal energy transfer is caused by electromagnetic waves emanating from the material, the mode of heat transfer is known as *radiation*. Although more than one mode of heat transfer may actually be present in a given situation, the effect of any particular mode may be negligible and, thus, ignored.

The heat transfer rate may or may not vary with time. When the temperatures and material properties do not change with time, thermal fluxes remain unchanged, the heat transfer is called *steady state*. An example is the heat transfer that occurs in a metallic plate when the temperatures of two sides of the plate are constant. Thermal energy transfer through the wall of a cold store may also be considered to be in a steady state if the storage and the ambient temperatures are held constant.

Unsteady state heat transfer is more common in food processing operations as the temperatures of a heating or cooling medium and/or food keep changing. These situations are more complex than steady state conditions. But by use of some available charts and by making suitable assumptions, they can be analyzed in simpler ways.

Steady periodic heat transfer is a special case of unsteady state heat transfer where conditions do change with time but in a cyclic fashion (as in response to daily cycles of ambient conditions).

7.2 Conduction

Thermal conduction is the diffusion of thermal energy through a continuous medium. It is the only means of heat transfer through solid, opaque objects. In this case, there is an interchange of kinetic energy (or internal energy) between contiguous molecules without an appreciable displacement of molecules. Atoms at a higher temperature possess more kinetic energy than those at lower temperatures. Thus, in a solid matrix the energy is transferred to less active neighbors by oscillations of atoms and transport of free electrons. Energy in conduction can also be transferred by free electrons as happens in metallic solids.

Conduction can also occur in fluids through elastic collisions among the molecules. But it is only in laminar flow and in the laminar sub-layer of the turbulent boundary layer adjacent to solid objects that conduction dominates overall heat transfer.

During heat transfer through any medium, the properties of the substance or medium, such as thermal diffusivity, thermal conductivity, and specific heat, considerably influence the rate of heat transfer.

7.2.1 Effect of Thermal Conductivity

A property of the materials involved in the exchange of thermal energy, *thermal conductivity*, is the rate of thermal exchange through a unit thickness of material for unit temperature difference per unit area normal to the direction of heat flow. Thermal conductivity values indicate whether thermal energy is conducted through the material with ease or not. Comparing different materials, the one with the highest value of thermal conductivity is said to be the best heat conductor; the one with the lowest value is the poorest conductor. Thermal conductivity is an *intensive property*, i.e., it is independent of the size of the system.

Fourier's first law states that the amount of heat transferred per unit area per unit time under steady state is proportional to the temperature gradient. The law is mathematically expressed as follows:

$$\frac{q}{A} = -k \frac{dt}{dx} \qquad (7.1)$$

$$\text{i.e. } q = -k.A.\frac{dt}{dx} \qquad (7.2)$$

Here q (joule.s^{-1} or watt in SI unit) is heat flow per unit time and A (m^2) represents the area perpendicular to the heat flow direction. The term (q/A) on the LHS is known as heat flux. Infinitesimal temperature difference between two points, separated by distance dx, is represented by dt. Thus, dt/dx is termed as temperature gradient. Since thermal energy transfer takes place from higher temperature to lower temperature, the negative sign is introduced to maintain consistency in spatial

DOI: 10.1201/9781003285076-8

and flow directions. The proportionality constant, k, is termed as thermal conductivity with SI units as $W.m^{-1}.K^{-1}$. Remember that temperature difference, Δt, has the same magnitude when expressed on either the Celsius or the Kelvin scale.

Heat or energy is conducted in solids by two mechanisms. First, it is conducted by free electrons, which move through the metal lattice. Second, thermal energy is transferred through vibrations among contiguous molecules. The thermal conductivity values of solids vary quite widely. The values for some materials used in food processing equipment and in agricultural structures are given in Annexure IV.

In case of gases, molecules move in continuous random motion. During this movement, the collision of molecules with one another causes the exchange of energy and momentum. The molecules while moving from a high temperature region to a low temperature region carry kinetic energy with them and this energy is transferred to lower energy molecules when they collide. The gases as hydrogen have smaller molecules, can move faster, and exhibit higher thermal conductivities.

In the case of liquids also, the molecules in contact with a high energy surface, gain energy, and they collide with lower energy molecules. But since the molecules have higher packing density as compared to gases, the molecular force fields influence the energy exchange among them considerably. Liquids' thermal conductivity values vary with temperature and a linear equation of the form $k = a + bT$ adequately expresses the variation; a and b are empirical constants. Thermal conductivities of most liquids, being incompressible, are not affected by the pressure.

Thermal Conductivity of Foods

Composition of a material and, in some cases, the physical orientation of molecules influence its thermal conductivity. The following factors affect the thermal conductivity of foods.

1. Type and nature of the food (cell structure, the amount of air trapped between the cells);
2. Moisture content;
3. Surrounding temperature;
4. Pressure of surroundings.

When foods have high moisture contents, their thermal conductivities tend to be closer to that of water. Ice has a thermal conductivity of about $2.24~W.m^{-1}.K^{-1}$, which is nearly four times that of water ($k = 0.56~W.m^{-1}.K^{-1}$). Therefore, it is not surprizing that frozen foods have thermal conductivity values that are about 3 to 4 times more than those of unfrozen ones.

Some empirical equations have been proposed for obtaining thermal conductivity of different food materials. Most of these expressions for thermal conductivity consider that the product possesses two or more phases. Thus, the thermal conductivity is calculated by taking into consideration the contributions of thermal conductivities of these constituents.

The thermal conductivity of wheat, within a moisture range of 10–20% (dry basis), can be expressed as

$$k~(W.m^{-1}.K^{-1}) = 0.07 + 0.002 \times M \qquad (7.3)$$

where M is the moisture content in percent dry basis.

The thermal conductivity of single grain ranges from 0.35 to $0.70~W.m^{-1}.K^{-1}$ and bulk grain varies between 0.116 and 0.174 $W.m^{-1}.K^{-1}$. Thermal conductivity of air at atmospheric pressure varies in the range of 0.02–$0.026~W.m^{-1}.K^{-1}$.

For fruits and vegetables with more than 60% water, k can be approximated as

$$k~(W.m^{-1}.K^{-1}) = 0.148 + 0.00493 \times M \qquad (7.4)$$

For a two-component anisotropic system, the thermal conductivity is dependent on the direction of heat flow,

$$k~(W.m^{-1}.K^{-1}) = k_s.X_s + k_w.X_w \qquad (7.5)$$

where $k_s = 0.15~W.m^{-1}.K^{-1}$ and $k_w = 0.6~W.m^{-1}.K^{-1}$

The values of thermal conductivity for pure water at different temperatures have been proposed by Choi and Okos (1986) as

$$k_w = 5.7109 \times 10^{-1} + 1.7625 \times 10^{-3}~t - 6.7036 \times 10^{-6}~t^2 \qquad (7.6)$$

In this equation t is in °C and is valid from −40 to 150°C.

Rahman et al. (1997) have proposed a general power law correlation for thermal conductivity of selected fruits considering the thermal conductivity as a function of water content, porosity, and temperature.

$$\frac{\alpha}{1 - \varepsilon_a + \dfrac{k_a}{(k_w)_r}} = 0.996 \left(\frac{T}{T_r} \right)^{0.713} X_w^{0.285} \qquad (7.7)$$

Here, ε is the volume fraction, subscripts a and w denote air and water, subscript r refers to reference temperature, X is the mass fraction, and T is in K.

The Maxwell equation, also used for modeling the thermal conductivity of foods, is given as

$$k = k_c \left[\frac{1 - \left(1 - \dfrac{ak_d}{k_c} \right) b}{1 + (a-1)b} \right], \qquad (7.8)$$

where $a = \dfrac{3k_c}{2k_c + k_d}$, and $b = \dfrac{V_d}{V_c + V_d}$

In this equation, k = conductivity of mixture, k_c = conductivity of continuous phase, k_d = conductivity of dispersed phase, V_d = volume of dispersed phase, and V_c = volume of continuous phase.

Several other equations and models are available for determination of thermal conductivity of different foods. Carson (2006) has made a review of the thermal conductivity models for foods. Equations for predicting thermal conductivity of foods have been mostly developed as functions of temperature. Orientation of fibers, specifically in meats, influences their ability to conduct heat; when measured along the fibers, thermal conductivity is 15–30% higher in comparison to that measured across the fibers. Some standard values for thermal conductivities of different foods are given in Annexure V.

As discussed above, the thermal conductivity of any material (including foods) is affected significantly by its moisture content. A reduction in moisture content during the drying process can cause substantial reduction of thermal conductivity of foods. In addition, the other factors, such as temperature, pressure, etc., also affect the thermal conductivity. During freeze drying, the surroundings are at a much lower pressure than the atmosphere, therefore, the food's thermal conductivity is lower.

7.2.2 Effect of Specific Heat

Specific heat of any material is the amount of heat that is required to be added to or removed from a substance of unit mass at 15.5°C to change its temperature by 1 unit. It has the SI unit as $J.kg^{-1}. K^{-1}$. This gives an idea of the heat involved in raising and lowering the temperature of a mass of substance.

$$c_p = \frac{q}{m.\Delta t} \qquad (7.9)$$

Specific heat of any product is influenced by its constituents, temperature, pressure, and moisture content. Food commodities will exhibit significant changes in their specific heat values as they are dried or wetted since water has a much higher specific heat in comparison to those of solids.

Specific heat for a material at constant volume is lower than that at constant pressure. In fact, pressure in most of the food processing applications is kept constant and, therefore, it is the specific heat at constant pressure that is used.

The specific heat is measured by a *calorimeter*, generally a simple thermos vacuum bottle. The differential scanning calorimeter (DSC) is also used for measuring specific heat.

Several empirical expressions have been proposed for estimating the specific heats of food materials based on the relative proportions of their constituents.

A simple way of expressing the specific heat of any food can be as follows:

$$c_p (J.kg^{-1}.K^{-1}) = C_d.(1-X_w) + C_w.(X_w) \qquad (7.10)$$

where c_d and c_w are the specific heats of dry grain and that of moisture (in $J.kg^{-1}.K^{-1}$) and X_w is the moisture content in fraction.

The specific heat of bone dried grain varies from 1.46 to $1.88 \times 10^3 \ J.kg^{-1}.K^{-1}$ ($0.35–0.45 \times 10^3 \ cal.kg^{-1}.K^{-1}$). The above linear relationship between c_p and X_w exists above 8% moisture level only.

Chen (1985) suggested the following simple expression for specific heat of unfrozen food.

$$c_p = 4.19 - 2.30X_s - 0.628X_s^3 \qquad (7.11)$$

Here c_p is in $kJ.kg^{-1}.K^{-1}$, and Xs is the mass fraction of solids in food.

In case of freezing, there is a phase change, and, thus, the latent heat involved during the phase change must be considered. As the latent heat is not released at a constant temperature, but over a range of temperatures, a new term, *apparent*

specific heat, is used to account for both the sensible and the latent heat effects. Schwartzberg (1976) has proposed the following equation for apparent specific heat:

$$c_a = c_u + (X_b - X_{wo})\Delta c + EX_s \left(\frac{RT_0^2}{M_w T^2} - 0.8\Delta c \right) \qquad (7.12)$$

where c_a is the apparent specific heat, c_u is the specific heat of food above initial freezing point, X_b and X_{wo} are the mass fractions of bound water and water above initial freezing point, Δc is the difference between specific heats of water and ice, E is the ratio of relative molecular masses of water M_w and food solids M_s. R is the universal gas constant = 8.314 kJ.(kg mol·K)$^{-1}$, T_s freezing point of water = 273.2 K.

Chen (1985) developed the following model as a modified form of a model developed by Siebel (1892) for specific heat.

$$c_a = 1.55 + 1.26X_s + \frac{X_s RT_0^2}{M_s T^2} \qquad (7.13)$$

Here c_a is the apparent specific heat, $kJ.(kg.K)^{-1}$, X_s is mass fraction of food solid, T_0 is the freezing point of water = 273.2 K, M_s is the relative molecular mass of soluble solids in food. T is the temperature of food, °C.

If the relative molecular mass of the soluble solids is unknown, the following equation can be used

$$c_a = 1.55 + 1.26X_s + \frac{(X_{wo} - X_b)L_0 t_f}{t^2} \qquad (7.14)$$

L_0 is the latent heat of fusion of water.

The specific heat of water, salts, carbohydrates, proteins, and lipids can be taken as 4.187, 0.837, 1.423, 1.549, and 1.674, respectively (all in $kJ.kg^{-1}.K^{-1}$). Thus, a simple equation can be written to find the specific heat of a food material as follows:

$$c_p \left(kJ.kg^{-1}.K^{-1} \right) = \left(0.837 \times X_{salt} + 1.423 \times X_{carbohydrate} + 1.549 \right.$$
$$\left. \times X_{protein} + 1.674 \times X_{lipid} + 4.187 \times X_{water} \right) \qquad (7.15)$$

The above equation can be modified for a sugar solution as

$$c_p (kJ.kg^{-1}.K^{-1}) = (1.423 \times X_{carbohydrate} + 4.187 \times X_{water}) \qquad (7.16)$$

Specific heat of ice is 2.1 $kJ.kg^{-1}.K^{-1}$, i.e., the specific heat of a product being frozen would continue to decrease during the process of freezing. The specific heats of selected foods and materials used in food engineering applications are given in Annexures VI and VII.

7.2.3 Effect of Thermal Diffusivity

When thermal conductivity (k) of a material is divided by the product of its density (ρ) and specific heat (c_p), the resulting quantity is termed *thermal diffusivity*, α. Thermal diffusivity of a material indicates the speed with which thermal energy gets dispersed throughout the material. Higher diffusivity means a higher dispersal rate and vice versa.

$$\alpha = \frac{k}{\rho c_p} \qquad (7.17)$$

It can be checked that the thermal diffusivity has the unit of $m^2.s^{-1}$.

The thermal diffusivity values of selected foods are given in Annexure VIII.

7.2.4 Difference between Thermal Conductivity and Thermal Diffusivity

Thermal diffusivity indicates the rate at which a body having a non-uniform temperature distribution attains thermal equilibrium. Consequently, there will be no further heat transfer within the body once the temperature distribution in it is uniform. It relates to unsteady state heat transfer. However, thermal conductivity is derived from Fourier's heat conduction law, which is relevant for heat transfer under steady state.

While thermal conductivity decides the amount of heat that will flow in a material, thermal diffusivity determines how fast the heat will disperse within it or flow out of it. Thermal conductivity relates to the thermal conduction for unit length; thermal diffusivity relates to temperature distribution over the entire length or surface or volume.

Both thermal conductivity and thermal diffusivity are important parameters that affect the transfer of thermal energy through a material. Bodies with higher thermal diffusivity can quickly adjust their temperatures to those of their surroundings as they are able to conduct heat more quickly as compared to their volumetric heat capacities ($\rho.c_p$). If a body has high heat capacity and density, the thermal diffusivity may be low even if the thermal conductivity is relatively high.

As discussed earlier, thermal conductivity of ice increases to almost four times that of liquid water. The specific heat of ice reduces to almost half that of liquid water. It is, therefore, not surprising that frozen foods have about nine to ten times higher thermal diffusivity in comparison to those of unfrozen ones.

7.2.5 Steady-State Thermal Conduction

Thermal energy that enters the system must leave in a steady-state process. Therefore, in this condition, no heat storage is possible, and the temperatures and heat fluxes remain unchanged with time. Under steady state, the rate of heat transfer by conduction with a temperature difference of Δt between both ends ($\Delta t = t_1 - t_2$) in a slab of length L is given by

$$q = kA \frac{t_1 - t_2}{L} \qquad (7.18)$$

This is Fourier's law of heat conduction.

Considering a homogenous wall of thickness L, having the temperatures at two ends (boundaries) as t_1 and t_2, if the temperature at any distance x is t, then

$$q = kA \frac{t_1 - t}{x}$$

Using this equation and the equation 7.18

$$\frac{t_1 - t}{x} = \frac{t_1 - t_2}{L}$$

$$\text{or,} \quad t = t_1 + \frac{x}{L}(t_2 - t_1) \qquad (7.19)$$

This equation indicates that for a steady state heat transfer situation in one dimension in a homogenous solid (i.e., with uniform and isotropic properties) without any internal heat source, the temperature field will be linear with distance of heat flow. Other coordinate systems do not necessarily yield temperature profiles linear in distance.

Example 7.1 If t_1 is the temperature on one face of the slab and t_2 on the other face, such that $t_1 > t_2$, and the conduction heat transfer follows Fourier's law, find an expression of temperature variation in the slab as a function of distance x from the face having temperature t_1.

SOLUTION:

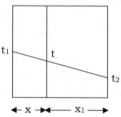

Let us assume that t is the temperature at a distance x from the face that is maintained at t_1.

Steady state heat transfer within the slab is the same at any cross section. Therefore, using Fourier's law, the rate of heat conduction is represented as follows:

$$q = kA \frac{t_1 - t}{x} = kA \frac{t - t_2}{x_1}$$

$$\text{or} \quad \frac{t_1 - t}{x} = \frac{t - t_2}{x_1}$$

$$\text{or } t_1.x_1 - t.x_1 = t.x - t_2.x$$

$$\text{or } t(x + x_1) = (t_1 x_1 + t_2 x)$$

$$\text{or} \quad t = \frac{t_1 x_1 + t_2 x}{x + x_1}$$

Example 7.2 A concrete wall of 20 m^2 area and 34 cm thickness has 40°C and 6°C temperatures at the opposite faces of the wall. Assuming thermal conductivity of the concrete wall to be 0.5 $W.m^{-1}.K^{-1}$, (a) represent the spatial temperature variation within the wall, and (b) determine the rate of conduction through the wall.

SOLUTION:

Temperature field for steady-state conduction in one dimension can be written as follows:

$$t = t_1 + (t_2 - t_1)\frac{x}{L}$$

Here, $t_1 = 40°C$ $t_2 = 6°C$

and $x = 0$ when $t_1 = 40°C$

$$t = 40 + (6 - 40) \times (x / 0.34)$$

$$= 40 - (34x / 0.34) = 40 - 100x$$

Therefore, the temperature field is expressed as $t = 40 - 100x$

The conduction heat transfer rate, q, is given as
$q = k.A. (\Delta t/L) = (0.5) (20) (40-6)/0.34 = 1000.0$ W

Thermal Conductance and Thermal Resistance

The equation 7.18 can also be written as $q = U.A. \Delta t$, where $U = k/L$

Here U ($=k/L$) is known as *unit area thermal conductance*. Thermal conductance is an extensive property, whereas thermal conductivity is intensive. The inverse of unit area thermal conductance, U, is termed *thermal resistance*, R.

$$q = A\frac{\Delta t}{R}, R = L / k \qquad (7.20)$$

The equation is analogous to the flow of electricity in a circuit. In the case of electricity, the flow of current (I) is expressed as the ratio of potential difference (V) and the total resistance (R), i.e. I = V/R, and thus, R = V/I.

Heat transfer in composite sections can be easily represented with the help of the concepts of thermal conductance or thermal resistance.

Conduction Heat Transfer in Series

Heat transfer in a composite wall having different materials can be found out by considering it as heat transfer in series in individual sections. This is similar to having electrical resistances in series. In electricity flow, the total circuit resistance in series equals the individual resistances added together. Heat transfer also follows the same relationship, i.e.,

$$R = R_1 + R_2 + ... + R_n \qquad (7.21)$$

In the wall of a cold store, different layers for water proofing, insulation, etc. are provided in the basic concrete construction. In this situation, heat has to pass in series through materials with different thermal properties, as shown in Figure 7.1.

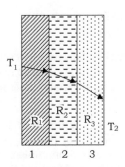

FIGURE 7.1 Conduction heat transfer in a composite wall made up of three materials

This electrical analogy helps in simplifying the heat transfer problems. As an example, for the above wall having three sections, the rate of thermal conduction is represented as follows:

$$\frac{q}{A} = \frac{\Delta t}{R_1 + R_2 + R_3} \qquad (7.22)$$

If the areas of the three media are equal, the equation is written as follows:

$$q = \frac{A.\Delta t}{\dfrac{x_1}{k_1} + \dfrac{x_2}{k_2} + \dfrac{x_3}{k_3}} \qquad (7.23)$$

If the three sections have different areas, the modified equation will be as follows:

$$q = \frac{\Delta t}{\dfrac{x_1}{A_1 k_1} + \dfrac{x_2}{A_2 k_2} + \dfrac{x_3}{A_3 k_3}} \qquad (7.24)$$

In the above equations, the subscripts 1, 2, and 3 represent the three media in series through which the heat is being transferred.

In the above equations, the numerator Δt is known as the *thermal potential* and the denominator is known as the *resistance to thermal energy flow*.

Conduction Heat Transfer in Parallel

In a cold storage room, heat transfer takes place through the roof, walls, doors, windows, and other sections that are considered to be parallel paths of thermal energy flow. Here, areas of heat transfer of different paths are different, and the resistance used in the equation must be the total resistance and not the unit area resistances.

The equivalent resistances in parallel heat transfer can be given as

$$\frac{A_{total}}{R'} = \sum \left(A_{individual} \Big/ R_{individual} \right) \qquad (7.25)$$

A_{total} is the sum of all areas through which thermal energy is transferred and $R_{individual}$ is the unit area thermal resistance of each path. In the above equation, R' is the unit area thermal resistance averaged over all heat transfer paths.

Conduction Heat Transfer Both in Series and in Parallel

In this situation, each heat flow path series has to be examined first and its unit area thermal resistance has to be calculated. Then the calculations for all parallel paths have to be done.

Example 7.3 A wall (8 m × 3 m) of a cold storage room is made up of three layers: the inner layer of wood board (5 mm thick), middle layer of cork board (25 mm thick) and the outer brick layer (10 cm thick). What will be the rate of heat transfer through the wall if the outside and inside temperatures of the wall are 42°C and 5°C. What are the interfacial temperatures? Thermal conductivities of the materials may be taken as 0.15 $W.m^{-1}.K^{-1}$ for wood, 0.043 $W.m^{-1}.K^{-1}$ for cork board, and 0.69 $W.m^{-1}.K^{-1}$ for brick.

SOLUTION:

Given, A= area of heat transfer = 8 × 3 = 24 m²

$$t_1 = 5°C \text{ and } t_4 = 42°C$$

The unit area thermal resistance of each layer are calculated as follows.

$$R_1 = L_1/k_1 = 0.005/0.15 = 0.0333 \text{ m}^2.K.W^{-1}$$

$$R_2 = L_2/k_2 = 0.025/0.043 = 0.5814 \text{ m}^2.K.W^{-1}$$

$$R_3 = L_3/k_3 = 0.1/0.69 = 0.145 \text{ m}^2.K.W^{-1}$$

Total unit area thermal resistance is $R = R_1 + R_2 + R_3 = 0.7597 \text{ m}^2.K.W^{-1}$

Heat flux is, $q = \dfrac{A\Delta t_1}{R} = \dfrac{24 \times (42-5)}{0.7597} = 1168.88 \text{ W}$

The interfacial temperatures can be determined as follows:

Heat flux, $q = \dfrac{A\Delta t_1}{R_1}$ or, $\Delta t_1 = \dfrac{qR_1}{A}$

$$= 1168.88 \times (0.0333/24) = 1.622°C$$

Similarly, $\Delta t_3 = \dfrac{qR_3}{A} = 1168.88 \times (0.145/24) = 7.06°C$

Now, $t_2 = t_1 + \Delta t_1 = 6.62°C$

$$t_3 = t_4 - \Delta t_3 = 42-7.06 = 34.94°C$$

Hence, the interfacial temperatures are 6.62°C and 34.94°C, respectively.

Example 7.4 A composite of 5 mm wood-board inner layer, a corkboard middle layer, and a 10 cm brick outer layer is used for constructing a cold store wall. The maximum temperature of the outside air is expected to be 50°C and the cold store temperature must not exceed 5°C. The coefficient of convective thermal energy exchange with respect to the inside air is 100 $W.m^{-2}.K^{-1}$ and that with respect to the outside air is 10 $W.m^{-2}.K^{-1}$. Find the corkboard thickness to ensure that the heat loss does not exceed 10 $W.m^{-2}$. Thermal conductivities of different components of the wall may be assumed to be the same as given in example 7.3.

SOLUTION:

Given, $k_w = 0.15 \text{ W.m}^{-1}.K^{-1}$, $\Delta x_1 = 0.005$ m

$$k_c = 0.043 \text{ W.m}^{-1}.K^{-1}$$

$$k_b = 0.69 \text{ W.m}^{-1}.K^{-1}, \Delta x_3 = 0.1 \text{ m}$$

$$h_i = 100 \text{ W.m}^{-2}.K^{-1}$$

$$h_o = 10 \text{ W.m}^{-2}.K^{-1}$$

$$q = 10 \text{ W.m}^{-2}$$

$$q = \dfrac{t_o - t_i}{\dfrac{1}{h_i} + \dfrac{\Delta x_1}{k_w} + \dfrac{\Delta x_2}{k_c} + \dfrac{\Delta x_3}{k_b} + \dfrac{1}{h_o}},$$

or, $10 = \dfrac{50-5}{\dfrac{1}{100} + \dfrac{0.005}{0.15} + \dfrac{\Delta x_2}{0.043} + \dfrac{0.1}{0.69} + \dfrac{1}{10}}$

or, $10 = \dfrac{45}{0.288 + \dfrac{\Delta x_2}{0.043}}$

$$0.288 + \Delta x_2 / 0.043 = 4.5$$

or, $\Delta x_2 = 0.181$ m

The required thickness of corkboard is 0.181 m.

Example 7.5 There is a wall of 4 m height and 20 m length, in which the doors and windows occupy 20% of the area. The unit area thermal resistance of the doors and windows have been found to be 3.2 $m^2.K.W^{-1}$. The remaining portion of the wall has a unit area thermal resistance of 4.8 $m^2.K.W^{-1}$. What is the wall's average thermal resistance on unit area basis? Find the heat transfer rate when the temperatures of two sides of the wall differ by 30°C.

SOLUTION:

Please note that we have to find the heat transfer rate. Thus the answer would be in $J.s^{-1}$ or W.

We have to find the total unit area thermal resistance as parallel heat transfer takes place.

The total wall area is 80 m², the doors and windows occupy 20%, i.e., 16 m² area.

Now,

$$\frac{A}{R} = \frac{A_1}{R_1} + \frac{A_2}{R_2}, \text{or} \ \frac{80}{R} = \frac{16}{3.2} + \frac{64}{4.8} = 18.333$$

or, R = 80/18.333 = 4.3636 m².K.W^{-1}
Heat flow through the wall q = A.Δt/R =
(80)(25)/4.3636 = 458.33 W

Heat Conduction in Cylindrical Coordinate System

Heat transfer through a pipe, as happens in a tubular heat exchanger or a cooling duct, is an example of heat conduction in a cylindrical coordinate system. In a cylindrical coordinate system, total thermal resistance must be used rather than unit area thermal resistance.

Heat conduction through a cylinder or pipe walls with r_1 and r_2 as the inner and outer radii (with corresponding temperatures t_1 and t_2) and length L is given as follows:

$$q = 2\pi L k \frac{t_1 - t_2}{\ln\left(\frac{r_2}{r_1}\right)} \qquad (7.26)$$

Another way of expressing the above equation is as follows:

$$q = kA_m \frac{t_1 - t_2}{r_2 - r_1} \qquad (7.27)$$

where A_m = log mean area = $\dfrac{A_2 - A_1}{\ln\left(A_2/A_1\right)} = \dfrac{2\pi L(r_2 - r_1)}{\ln\left(r_2/r_1\right)}$

The equation 7.26 can also be expressed as

$$q = \frac{t_1 - t_2}{\left[\ln\left(r_2/r_1\right)/2\pi Lk\right]} \qquad (7.28)$$

Thus, the thermal resistance is R = $\dfrac{\ln\left(r_2/r_1\right)}{2\pi kL}$ (7.29)

The heat transfer in composite walls in a cylindrical coordinate system can be presented like the conduction in a composite slab by taking the individual heat resistances in series. If a cylindrical wall is made up of three composite sections, then heat conduction can be given as follows:

$$q = \frac{t_i - t_o}{\dfrac{\ln\left(r_1/r_i\right)}{2\pi Lk_1} + \dfrac{\ln\left(r_2/r_1\right)}{2\pi Lk_2} + \dfrac{\ln\left(r_o/r_2\right)}{2\pi Lk_3}} \qquad (7.30)$$

While the subscripts i and o denote the cylinder's inner and outer sides, the subscripts 1 and 2 denote the interfaces from the inner to the outer side. Thermal conductivities of the three materials are k_1, k_2, and k_3 from the inner to the outer side.

Example 7.6 A 20 cm internal diameter metal duct has a 2 mm thick wall. It is covered with a 4 cm insulation layer. The temperatures at the outer surface of the duct and of the outer surface of the insulation are 4°C and 60°C, respectively. Thermal conductivity of the metallic duct is 68 W.m^{-1}.K^{-1} and of the insulation material is 0.04 W.m^{-1}.K^{-1}. Determine the heat conduction rate through the insulation per meter length of duct. Assume steady state heat flow.

SOLUTION:

Thermal resistance to conduction in cylindrical coordinates is written as follows:

$$R = \frac{\ln\left(r_o/r_i\right)}{2\pi kL}$$

Here, L = 1 m (we have to find the heat transfer rate per meter length)

For the insulation layer, r_i = 102 mm and r_o = 142 mm

Hence, R = $\dfrac{\ln\left(\dfrac{142}{102}\right)}{(2\pi)(0.04)(1)} = 1.3164 \ \text{K.W}^{-1}$

Thermal energy exchange from the surface of the insulated pipe per unit length

$$\Delta t/R = (60\text{-}4)/1.3164 = 42.539 \ \text{W}$$

General Heat Transfer Equations and Temperature Field within a Homogenous Wall

The Laplacian $\nabla^2 t$ for one-dimensional heat transfer is

$$\nabla^2 t = \frac{1}{n^m} \cdot \frac{d}{dn}\left(n^m \frac{dt}{dn}\right) \qquad (7.31)$$

m = 0, 1, and 2 for Cartesian, cylindrical, and spherical coordinate systems. The "n" is x for Cartesian coordinates and r for cylindrical/ spherical coordinates.

For example, the equation in Cartesian coordinates is written as follows:

$$\nabla^2 t = \frac{d}{dx}\left(\frac{dt}{dx}\right) = \frac{d^2 t}{dx^2} \qquad (7.32)$$

The equation in cylindrical coordinates is written as follows:

$$\nabla^2 t = \frac{1}{r} \cdot \frac{d}{dr}\left(r \frac{dt}{dr}\right) = \frac{d^2 t}{dr^2} + \frac{1}{r}\frac{dt}{dr} \qquad (7.33)$$

Fourier's conduction equation with no internal heat generation can be written as follows:

$$\nabla^2 t = \alpha^{-1} \frac{\partial t}{\partial \theta} \qquad (7.34)$$

This equation applies to time dependent heat transfer problems.

For steady-state heat conduction and a uniformly distributed heat source in the body, the *Poisson equation* is obtained.

$$\nabla^2 t + \frac{q_{gen}}{k} = 0 \tag{7.35}$$

This situation can be approximated when solar energy is absorbed by a brick wall or a plastic glazing.

If the heat conduction is under a steady state, the equation 7.34 reduces to the *Laplace equation*.

$$\nabla^2 t = 0 \tag{7.36}$$

This equation can be applied for heat transfer in metals (with good thermal conductivity) or even in thin walls where the response time is so much less that steady-state conditions can be assumed.

The temperature field in a homogenous wall can also be obtained with the Laplace equation.

$$\text{Here, } \frac{d^2 t}{dx^2} = 0 \tag{7.37}$$

which can be integrated twice to yield

$$t = C_1 x + C_2 \tag{7.38}$$

At $x = 0$, $t = t_1$, hence $C_2 = t_1$
and at $x = L$, $t = t_2$, thus, $t_2 = C_1 L + C_2$
or $C_1 = (t_2 - t_1)/L$
i.e., for $0 < x < L$, the temperature field is

$$t = t_1 + \left(t_2 - t_1\right)\frac{x}{L} \tag{7.39}$$

Temperature Field within a Homogenous Cylindrical Shell

Assume a homogenous cylindrical shell of inner radius r_i and outer radius r_o. Let the inner and outer surfaces are at t_i and t_o, respectively.

The relevant equation for the cylindrical coordinate system from equation 7.31 is

$$\frac{d^2 t}{dr^2} + \frac{1}{r}\frac{dt}{dr} = 0 \tag{i}$$

Substituting $S = dt/dr$, the above equation is simplified as

$$\frac{dS}{dr} + \frac{S}{r} = 0 \tag{ii}$$

If equation (ii) is multiplied by $r.dr$, it becomes

$$rdS + Sdr = 0 \tag{iii}$$

$$\text{i.e., } d(rS) = 0 \tag{iv}$$

Equation (iii) can be integrated once to

$$rS = C_1 \tag{v}$$

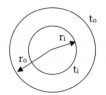

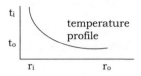

FIGURE 7.2   Temperature profile in a homogenous cylindrical shell

Substituting dt/dr for S, we get $r.\dfrac{dt}{dr} = C_1$

i.e., $dt = C_1 \dfrac{dr}{r}$

Integrating this equation

$$t = C_2 + C_1 \ln r \tag{vi}$$

Applying the boundary conditions, yields

$$t_i = C_2 + C_1 \ln r_i \tag{vii}$$

$$\text{and } t_o = C_2 + C_1 \ln r_o \tag{viii}$$

From these two equations, the constants C_1 and C_2 are obtained as

$$C_1 = \frac{t_o - t_i}{\ln(r_o / r_i)} \text{ and } C_2 = \frac{t_i \ln r_o - t_0 \ln r_i}{\ln(r_o / r_i)}$$

and the temperature field is expressed as follows:

$$t = \frac{t_o - t_i}{\ln(r_o / r_i)} \ln r + \frac{t_i \ln r_o - t_0 \ln r_i}{\ln(r_o / r_i)} \tag{7.40}$$

The equation shows a logarithmic increase or decrease in the temperature of a homogenous cylindrical shell under steady-state heat conduction (depending on the boundary conditions) and the cylindrical shell radius controls the thermal gradient. The temperature profile will be as shown in Figure 7.2.

7.3 Convection

Convection is generally used to denote thermal energy transport through fluids by large-scale eddying motions. The term *convective heat transfer*, however, refers to the transfer of thermal energy between a fluid and a solid. Thermal gradient within a fluid causes density gradient and this ultimately causes bulk molecular motion within the fluid. This motion helps energy transfer. For example, when milk is heated in a kettle, heat is transferred from the metallic pan to the milk and there is a convection current set within the milk for bulk heat transfer everywhere in the milk. Fluid motion leads to faster heat transfer in comparison to conduction.

Convective thermal energy exchange between a solid object and a fluid involves a boundary layer, a thin and relatively stagnant fluid layer attached to the solid. The boundary layer resists thermal energy exchange between the bulk fluid and the solid. Fluid velocity in the boundary layer is much smaller than that in the free stream. Besides, material properties in the boundary layer are different from those of the bulk fluid. There is an inverse proportionality between the value of the coefficient of thermal energy exchange and the boundary layer thickness. If turbulence is induced in the fluid, the result is a thinner boundary layer with accelerated thermal exchange.

If thermal conductivity in the boundary layer is k and the boundary layer thickness is δ, the boundary layer heat transfer can be represented as follows:

$$q = k.A \ (t_{solid} - t_{liquid})/\delta \qquad (7.41)$$

However, boundary layer thickness is difficult to determine because of its thinness. Therefore, for simplified calculations the ratio k/δ is taken as h (or h_c), the *convective heat transfer coefficient*. The modified equation for convective thermal energy exchange rate is as follows:

$$q = h \ A \ (t_{solid} - t_{liquid}) = h \ A \ \Delta t \qquad (7.42)$$

where, h $(W.m^{-2}.K^{-1})$ is the convective thermal energy exchange coefficient and A (m^2) is the interface area between the fluid and the solid through which the thermal exchange occurs. Proportionality of the rate of thermal exchange with a temperature difference between the two bodies is also postulated in Newton's law of cooling.

Factors influencing the coefficient of convective thermal energy exchange between the participating fluid and the solid surface include fluid properties (e.g., specific heat, density, viscosity), temperature difference, gravity force, and characteristic dimension. Calculation of h is very complicated, cumbersome, and, hence, four dimensionless numbers are universally used to correlate the above properties to get a value of h. One is the Reynolds number. The other three numbers are as follows:

$$\text{Nusselt number, Nu} = h.L/k \qquad (7.43)$$

$$\text{Grashof number, Gr} = \frac{L^3 \Delta t.\rho^2 g\beta}{\mu^2} \qquad (7.44)$$

$$\text{Prandtl number, Pr} = \mu.c_p/k \qquad (7.45)$$

where h $(W.m^{-2}.K^{-1})$ is the coefficient of convective thermal energy exchange at the solid-liquid interface, k is the thermal conductivity, L (m) is the characteristic dimension (diameter or length), c_p is the specific heat at constant pressure, g is the acceleration due to gravity, and β $(m.m^{-1}.K^{-1})$ is the thermal expansion coefficient. The characteristic dimension should represent the boundary layer's growth (or thickness). For a cylinder with cross flow perpendicular to the cylinder axis, the

outer diameter of the cylinder is taken as the characteristic dimension. For natural convection, the characteristic dimension can be the height of a vertical plate or the diameter of a sphere. For complex shapes, either the equivalent diameter or the fluid volume per unit area of the solid surface is taken as the characteristic dimension.

The Nusselt number (Nu) is the ratio of the magnitude of convective thermal exchange and the magnitude of conduction thermal exchange. Thus, Nu indicates the intensity of the convective heat transfer process as compared to heat conduction.

The Prandtl number (Pr) compares the momentum diffusion with the diffusion of thermal energy from the solid surface to the boundary layer.

The Grashof number (Gr) compares the magnitude of buoyancy forces with that of the viscous drag force within the liquid.

The ratio of inertial to viscous forces is termed the Reynolds number. It expresses the level of turbulence. As discussed in the previous chapter, in a pipe flow of finite length, the flow having Re less than 2100 is considered laminar and more than 4000 is considered to be turbulent. The flows having Re values between 2100 and 4000 are said to be in the transient zone. Convection thermal energy exchange occurs due to either natural or forced convection.

7.3.1 Natural Convection

Convective thermal energy exchange is termed *natural* or *free convection* if the fluid motion is induced by density differences caused only by temperature gradients. Convective currents generated in the fluid are governed only by the difference between the temperatures of the fluid and the solid surface.

The Nusselt number in natural convection, Nu, is a function of the Rayleigh number $(Ra = Gr_f \times Pr_f)$. The subscript f indicates that the dimensionless numbers need to be determined by using properties corresponding to film temperature, t_f.

$$t_f = \frac{t_w + t_\infty}{2}$$

In general,

$$Nu = M \ (Gr_f \times Pr_f)^n \qquad (7.46)$$

In the case of turbulent flow in a horizontal cylinder, the values of M and n are 0.13 and 0.33. The values are 0.53 and 0.25 if the flow is laminar. The M and n values for a few general cases are given in Annexure IX.

The temperature range under standard environment conditions is quite narrow and some simple equations for natural convection are given in Annexure X for dry air conditions at standard atmospheric pressure and 20°C. For example, a simple equation for natural convection heat transfer for horizontal cylinder in laminar and turbulent ranges are as follows:

$$\text{For laminar flow } h = 1.32(\Delta t/L)^{0.25} \qquad (7.47)$$

$$\text{For turbulent flow } h = 1.24(\Delta t)^{0.33} \qquad (7.48)$$

The flow type, whether laminar or turbulent, is decided by the numerical value of the product of Gr and Pr. If (Gr×Pr) is between 10^4 and 10^8, then the flow is laminar, and if it is between 10^8 and 10^{12}, the flow is turbulent.

For standard air, (Gr×Pr) can be taken as

$$(Gr.Pr) = 10^8 L^3 \Delta t \qquad (7.49)$$

where, L is in m and Δt in K.

Some other empirical relationships for Nu for natural convection conditions are given below.

For a vertical isothermal surface, the Nu is given as (Churchill and Chu, 1975b)

$$Nu_x^{1/2} = 0.825 + \frac{0.387 \times Ra_x^{1/6}}{\left[1 + (0.492 / Pr)^{9/16}\right]^{8/27}}, \text{ for } 10^{-1} < Ra_x < 10^{12}$$

$$(7.50)$$

The following equation is also appropriate for a constant heat flux condition (Churchill and Chu, 1975b).

$$Nu_x = 0.68 + \frac{0.670 \times Ra_x^{1/4}}{\left[1 + (0.492 / Pr)^{9/16}\right]^{4/9}}, \text{ for } Ra_x < 10^9$$

$$(7.51)$$

The properties are evaluated at the film temperature.

For a vertical surface and the condition of constant heat flux

$$Nu = M (Gr_f \times Pr_f)^n \qquad (7.52)$$

where M = 0.60, n =1/5 for $10^5 < Gr < 10^{11}$

M = 0.17, n =1/4 for $2 \times 10^{13} < Gr < 10^{16}$

For isothermal horizontal cylinders, the heat transfer by convection is given as (Churchill and Chu, 1975a)

$$Nu_d^{1/2} = 0.60 + 0.387 \left(\frac{Gr \times Pr}{\left[1 + (0.559 / Pr)^{9/16}\right]^{16/9}} \right)^{1/6},$$

$$\text{for } 10^{-5} < Gr.Pr < 10^{12} \qquad (7.53)$$

An equation used for the laminar range is as follows (Churchill and Chu, 1975a):

$$Nu_d = 0.36 + \frac{0.518(Gr \times Pr)^{1/4}}{\left[1 + (0.559 / Pr)^{9/16}\right]^{4/9}} \text{ for } 10^{-6} < Gr.Pr < 10^9$$

$$(7.54)$$

For horizontal surfaces with constant heat flux, the equations are as follows

For an upward facing heated surface,

$$Nu_x = 0.13(Gr_x \times Pr)^{1/3}, \text{ for } Gr_x.Pr < 2 \times 10^8 \qquad (7.55)$$

$$\text{and } Nu_x = 0.16(Gr_x \times Pr)^{1/3}, \text{ for } 2 \times 10^8 < Gr_x.Pr < 10^{11}$$

$$(7.56)$$

For a downward facing heated surface,

$$Nu_x = 0.58(Gr_x \times Pr)^{1/5}, \text{ for } 10^6 < Gr_x.Pr < 10^{11} \qquad (7.57)$$

All properties, except β, in the above equations need to be evaluated at an equivalent temperature, t_e, defined as follows:

$$t_e = t_w - 0.25(t_w - t_\infty) \qquad (7.58)$$

where t_w represents the average temperature of the wall.

The following empirical equations can be used for spherical bodies involved in heat transfer due to natural convection.

$$Nu = 2 + 0.43(Gr \times Pr)^{1/4}, \ 1 < Gr.Pr < 10^5 \qquad (7.59)$$

Amato and Tien's equation for spheres experiencing heat transfer due to natural convection in water is as follows (Amato and Tien, 1960):

$$Nu = 2 + 0.5(Gr \times Pr)^{1/4}, \ 3 \times 10^5 < Gr_d.Pr < 8 \times 10^8$$

$$(7.60)$$

Churchill's equation for natural convective heat transfer for spheres is as follows (Churchill, 1983):

$$Nu = 2 + \frac{0.589(Gr \times Pr)^{1/4}}{\left[1 + (0.469 / Pr)^{9/16}\right]^{4/9}}, \ Pr > 0.5 \text{ and } Gr.Pr < 10^{11}$$

$$(7.61)$$

7.3.2 Forced Convection

Devices such as paddles or pumps are used to agitate the fluid involved in thermal energy exchange. Fluid velocity, object geometry, and the thermo-physical properties have a strong bearing on the magnitudes of the convective heat transfer coefficients in such situations.

For forced convection, Nu is a function of Re and Pr.

$$Nu = a.Re^b Pr^c \qquad (7.62)$$

Here a, b, and c are constants and are obtained from experimental data.

Different modifications of the above form of equation are also used, which have been developed through experimental validations. The following equation is used for forced convection in a streamline pipe flow.

$$Nu = 1.62\left(Re.Pr.\frac{d}{L}\right)^{0.33} \tag{7.63}$$

where, pipe length is L (m) and pipe diameter is d (m). The equation is valid for Re.Pr.(d/L) >120 and when all physical properties are measured at the mean bulk fluid temperature.

An empirical equation for thermal energy exchange due to forced convection is as follows:

$$Nu = a.Re^{b} Pr^{c}\left(\frac{\mu_b}{\mu_w}\right)^{d}\left(\frac{L}{d}\right)^{e} \tag{7.64}$$

where a, b, c, d, and e are constants, μ_b and μ_w are fluid viscosities evaluated at bulk and wall temperatures, respectively. In addition, the characteristic dimension of the surface is L and the pipe diameter is d. The values of the coefficients are different depending on whether the flow is laminar or turbulent. Table 7.1 gives the values of the constants (Holman, 1997).

Similar constants have also been empirically obtained for liquid/ gas flows past cylinders and spheres. For example, for gas flowing past a sphere (325≤ Re ≤ 70000), the constants a, b, and c are 0.4, 0.6, and 0.33. For liquid flow normal to a cylinder with $0.1 \leq Re \leq 300$, the values of a, b, and c are 0.5, 0.6, and 0.3. The values of d and e are zero for both these cases.

Forced Convection for Streamline Flow in a Pipe

In case of laminar fluid flow in horizontal tubes, the following *Sieder-Tete* empirical relationship can be used to represent convective heat transfer.

$$Nu = 1.86\times(Re_d Pr)^{0.33}\left(\frac{d}{L}\right)^{0.33}\left(\frac{\mu}{\mu_s}\right)^{0.14} \tag{7.65}$$

Check that the constants are as given in Table 7.1. For using the above equation, μ_w corresponds to the wall temperature and all other properties are evaluated at the mean bulk temperature of the fluid. Arithmetic average of the inlet and outlet temperature differences is used for finding the average coefficient of convective thermal energy exchange. The equation is valid for Re_d Pr (d/L) > 10 and not appropriate for very long tubes as the heat transfer coefficient value would be zero.

Forced Convection for Turbulent Flow in a Pipe

The following relationship is applicable for turbulent fluid flow in pipes.

$$Nu = 0.027\times Re_d^{0.8} Pr^{0.33}\left(\frac{\mu}{\mu_s}\right)^{0.14} \tag{7.66}$$

TABLE 7.1

Constants to be used in equation 7.64 for forced convection heat transfer in a pipe

Reynolds no.	a	b	c	d	e
Less than 2100	1.86	0.33	0.33	0.14	0.33
More than 10^4	0.027	0.8	0.33	0.14	0

Here μ is fluid viscosity corresponding to the bulk temperature of the fluid whereas μ_s corresponds to the temperature of the heat transfer surface. The validity of this equation is generally assured for $0.6 \leq Pr \leq 16000$, $Re_d \geq 6000$, and $L/d \geq 60$. The expected accuracy of the results is high because the equation accounts for the change in the viscosities, μ and μ_s. An iterative process is used for obtaining the solution from the correlation since the viscosity factor changes as the Nusselt number changes.

The *Dittus-Boelter equation* (for turbulent flow) can be used for calculating the Nusselt number for small temperature differences across the fluid, specifically applicable for smooth tubes. The equation is given as follows:

$$Nu = 0.023 Re_d^{0.8} Pr^{n} \tag{7.67}$$

Here the internal diameter of the circular duct is d. The constant, n, is 0.4 for the fluid getting heated and 0.3 for the fluid getting cooled. The equation is valid for $0.6 \leq Pr \leq 160$, $Re_d \geq 10000$, $L/D \geq 10$ and the wall temperature being not too different from that of the fluid. While viscosity is determined at the mean temperature of the film, all other physical properties are evaluated at the mean bulk temperature of the fluid.

The *Gnielinski correlation* for turbulent pipe flow is given below.

$$Nu_d = \frac{(f/8)(Re_d - 1000)Pr}{1 + 12.7(f/8)^{1/2}(Pr^{2/3} - 1)} \tag{7.68}$$

Here a Moody chart can be used to obtain f, the Darcy factor. For smooth tubes, the following correlation developed by Petukhov serves the purpose.

$$f = (0.79\ln(Re_d) - 1.64)^{-2} \tag{7.69}$$

The range of validity for the above correlation is $0.5 \leq Pr \leq 2000$, and $3000 \leq Re_d \leq 5\times10^6$.

Gnielinski has also suggested the following two equations for two different situations.

For $0.5 \leq Pr \leq 1.5$, and $10^4 \leq Re \leq 5\times10^6$,

$$Nu = 0.0214(Re_d^{0.8} - 100)Pr^{0.4} \tag{7.70}$$

and for, $1.5 \leq Pr \leq 500$, and $3000 < Re < 10^6$,

$$Nu = 0.012(Re_d^{0.87} - 280)Pr^{0.4} \tag{7.71}$$

The above equation is appropriate for fully developed tube flows that are turbulent. The following equation proposed by Nusselt is appropriate for such conditions as flows not fully developed prevailing at a tube entrance region.

$$Nu_d = 0.036 Re_d^{0.8} Pr^{1/3}\left(\frac{d}{L}\right)^{0.055} \quad \text{for } 10 < L/d < 400 \tag{7.72}$$

All these empirical equations (7.64–7.72) can have up to 25% errors. The Petukov equation gives more accurate results for fully developed turbulent flow in smooth tubes.

$$Nu_d = \frac{(f/8)(Re_d \times Pr)}{1.07 + 12.7(f/8)^{1/2}(Pr^{2/3}-1)}\left(\frac{\mu_b}{\mu_w}\right) \quad (7.73)$$

All properties in this equation, except the viscosities, correspond to $t_f = (t_w+t_b)/2$. In this equation, the value of n is 0 for situations of constant heat flux or for gases. The value of n is 0.11 for $t_w > t_b$, and 0.25 for $t_w < t_b$.

The equation for friction factor is as follows:

$$f = (1.82 \log_{10} Re_d - 1.64)^{-2} \quad (7.74)$$

Equation 7.74 is applicable for $0.5 < Pr < 2000$, $10^4 < Re_d < 5 \times 10^6$ and $0.8 < \mu_b/\mu_w < 40$.

For tubes at constant wall temperature with fully developed laminar flow, the *Hausen equation*, as given below, is used.

$$Nu_d = 3.66 + \frac{0.0668(d/L)(Re_d \times Pr)}{1 + 0.04\left[(d/L)(Re_d \times Pr)\right]^{2/3}} \quad (7.75)$$

The above equation gives the coefficient of convective thermal energy exchange averaged over the entire length of the tube. The equation also suggests that for sufficiently long tubes the value of the Nusselt number becomes constant at 3.66.

The coefficient of convective thermal energy exchange for cross flow across cylinders is obtained as follows:

$$Nu = C\left(\frac{v_\infty d}{v_f}\right)^n Pr_f^{1/3} \quad (7.76)$$

Table 7.2 lists the constants n and C (Knudsen and Katz, 1958; Holman, 1997).

The following equation represents the coefficient of convective thermal energy exchange from liquids to cylinders in cross flow.

$$Nu = (0.35 + 0.56 Re_f^{0.52}) Pr_f^{0.3} \quad (7.77)$$

The equation 7.77 is valid for $10^{-1} < Re_f < 10^5$.

Following *Eckert and Drake equations* are suitable for tube heat transfer in cross flow.

$$Nu = (0.43 + 0.50 Re^{0.5}) Pr^{0.38}\left(\frac{Pr_f}{Pr_w}\right)^{0.25} \text{ for } 1 < Re < 10^3$$
$$(7.78)$$

$$Nu = 0.25 Re^{0.6} Pr^{0.38}\left(\frac{Pr_f}{Pr_w}\right)^{0.25} \text{ for } 10^3 < Re < 2 \times 10^5$$
$$(7.79)$$

TABLE 7.2

Values of C and n for use in equation 7.76

Re_d	n	C
40000–4×10⁵	0.805	0.0266
4000–40000	0.618	0.193
40–4000	0.466	0.683
4–40	0.385	0.911
0.4–4	0.330	0.989

During the use of these equations for gases, the fluid properties are determined at the film temperature and the Prandtl number may be dropped. For liquids, the fluid properties are estimated at free stream temperature and the Prandtl number is retained.

Another relationship, after Whitaker (1972), is as follows.

$$Nu = (0.4 Re^{0.5} + 0.06 Re^{2/3}) Pr^{0.4}\left(\frac{\mu_\infty}{\mu_w}\right)^{0.25} \quad (7.80)$$

This is applicable for $0.25 < \mu_\infty/\mu_w < 5.2$, and $40 < Re < 10^5$, $0.65 < Pr < 300$. The value of μ_w corresponds to the temperature of film whereas all other properties correspond to the free stream temperature.

The local Nusselt number for a flat plate under laminar flow at a distance x from the plate edge is given as follows:

$$Nu_x = 0.332 \times Re_x^{1/2} Pr^{1/3}, \quad Pr > 0.6 \quad (7.81)$$

The average Nusselt number for this flow condition is as follows:

$$Nu_x = 2 \times 0.332 \times Re_x^{1/2} = 0.664 \times Re_x^{1/2} Pr^{1/3}, \quad Pr > 0.6$$
$$(7.82)$$

For airflow in a pipe at standard atmospheric pressure, a simplified equation can be used for a turbulent situation as follows:

$$h = B\frac{G^{0.8}}{d^{0.2}} \quad (7.83)$$

Here, B is a coefficient computed from thermal properties of air and is dependent on temperature. G is the mass flow of air in the duct ($=\rho.V$) and d is the diameter of the pipe (Use hydraulic diameter in case of non-circular cross-section, d= 4A/P; A = area and P = perimeter). The value of B for a temperature of 40°C can be taken as 3.24.

Example 7.7 Water, flowing in a steel pipe (diameter 0.02 m) is to be cooled from 40°C to 30°C. The velocity of water in the steel pipe is 1.8 m.s⁻¹. The inside surface temperature of the steel pipe is maintained at 25°C. The physical properties of water at mean bulk temperature of the fluid area are specific heat = 4.16 kJ.kg⁻¹, density = 990 kg.m⁻³, thermal conductivity = 0.62 W.m⁻¹.K⁻¹, viscosity = 7.6×10⁻⁴ Pa.s. Find out the convective heat transfer coefficient of water.

SOLUTION:

From the given values,

$$Re = \frac{\rho.d.v}{\mu} = (990 \times 0.02 \times 1.8)/7.6 \times 10^{-4} = 46894$$

$$Pr = \frac{\mu.c_p}{k} = \frac{4.16 \times 1000 \times 7.6 \times 10^{-4}}{0.62} = 5.1$$

As per the Re value, the flow is turbulent, hence using the Dittus–Boelter equation,

$$Nu = 0.023\,Re^{0.8}\,Pr^{1/3}$$

$$Nu = 0.023 \times 46894^{0.8} 5.1^{1/3} = 216.012$$

$$\frac{hD}{k} = 216.012$$

$$h = 216.012 \times 0.62 \times \frac{1}{0.02} = 6696.372\ W.m^{-2}K^{-1}$$

7.4 Conduction and Convection in Series

The equation 7.42, the expression for the convective heat transfer between a solid and a fluid, can be written as

$$q = \frac{\Delta t}{1/hA} \tag{7.84}$$

In the above equation, Δt is the potential of heat transfer and $1/hA$ is the resistance to heat transfer.

If there is a wall or surface that is heated by a hot fluid on one side, there will be heat transfer by convection from the hot fluid to the wall and conduction through the wall. Once the wall is heated, it will also transfer heat to the fluid on the other side, if there is a temperature gradient. This happens when heat is transferred through the wall of a cold store or through a plate heat exchanger. Similarly, in a tubular heat exchanger, there is heat transfer between fluids through a metallic pipe or cylinder. These are examples of conduction and convection in series.

For conduction and convection in series through a wall/surface (Figure 7.3 a), the equation 7.22 can be modified and written as

$$q = \frac{\Delta t}{\dfrac{1}{h_i A_i} + \dfrac{x_1}{k_1 A_1} + \dfrac{1}{h_o A_o}} \tag{7.85}$$

The energy exchange coefficients, h_i and h_o, correspond to the inside and outside surfaces, respectively. The parameters x and

k correspond to the heat exchanging wall. If the area of heat transfer is the same, the above equation can be simplified as

$$\frac{q}{A} = \frac{\Delta t}{\dfrac{1}{h_i} + \dfrac{x_1}{k_1} + \dfrac{1}{h_o}} \tag{7.86}$$

The above equation is similar to the equation 7.23.

In this case, the overall resistance is given by

$$R = \frac{1}{U} = \frac{1}{h_i} + \frac{x_1}{k_1} + \frac{1}{h_o} \tag{7.87}$$

Overall heat transfer coefficient (OHTC) is represented by U and $q = U.A.\Delta t$.

For heat transfer through a metallic pipe or cylinder, if the inner and outer radii for a hollow cylinder are r_1 and r_2, respectively (Figure 7.3 b), the thermal energy exchange for a hollow cylinder is represented as follows:

$$q = \frac{\Delta t}{\dfrac{1}{h_i A_i} + \dfrac{\ln(r_2/r_1)}{2\pi Lk} + \dfrac{1}{h_o A_o}} \tag{7.88}$$

The overall resistance for thermal energy exchange in this case is given as follows:

$$R = \frac{1}{UA} = \frac{1}{h_i A_i} + \frac{\ln(r_2/r_1)}{2\pi Lk} + \frac{1}{h_o A_o} \tag{7.89}$$

Example 7.8 A steel pipe with its inner and outer diameters as 25 mm and 30 mm, respectively, is carrying steam at 121°C. The coefficient of convective thermal energy exchange for the interior surface of the steel pipe is 5000 W.m⁻².K⁻¹. The steel pipe on its outside surface has a 10 mm glass wool insulation. The thermal energy exchange coefficient for the outside still air is 10 W.m⁻². K⁻¹ at 30°C. Thermal conductivities for steel and glass wool are 43 and 0.031 W.m⁻¹.K⁻¹, respectively. What would be the overall resistance of heat transfer per unit interior area of the pipe? Determine thermal energy loss from the steam per unit pipe length.

SOLUTION:

Given, ID = 25 mm, h_i = 5000 W.m⁻².K⁻¹

OD = 30 mm, h_o = 10 W.m⁻².K⁻¹

t_i = 121°C, t_o = 30°C

k_s = 43 W.m⁻¹.K⁻¹, k_i = 0.031 W.m⁻¹.K⁻¹

For using in equation for overall resistance, the radii are calculated as

r_1 = 0.0125 m, r_2 = 0.015 m, r_3 = 0.025 m

Overall heat transfer resistance (Considering L as unity)

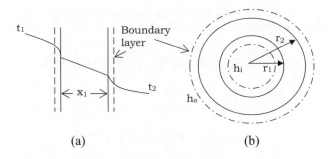

t_1 | Boundary layer | r_2 | h_i | r_1 | t_2 | h_o | x_1

(a)   (b)

FIGURE 7.3 Transfer of heat between two fluids (a) through a flat surface/ wall, and (b) through a metallic pipe/cylinder

$$R = \frac{1}{h_i A_i} + \frac{\ln\left(r_2/r_1\right)}{2\pi k_s} + \frac{\ln\left(r_3/r_2\right)}{2\pi k_{gw}} + \frac{1}{h_o A_o}$$

i.e., $R = \dfrac{1}{(5000)(\pi)(0.025)(1)} + \dfrac{\ln\left(0.015/0.0125\right)}{2\pi(43)}$

$\qquad + \dfrac{\ln\left(0.025/0.015\right)}{2\pi(0.031)} + \dfrac{1}{10(\pi)(0.05)(1)}$

$\qquad = 2.54N65 \times 10^{-3} + 6.748 \times 10^{-3} + 2.6226 + 0.6366$

$\qquad = 3.2624 \text{ K.W}^{-1}$

Hence, $q = (t_i - t_o)/R = (121-30)/3.2624 = 27.89$ W

Example 7.9 A heating duct of 0.4 m×0.5 m cross-section is being used to convey 40°C air at a velocity of 3 m.s⁻¹. Find the coefficient of convective thermal energy exchange between the air and the duct wall. Assume the air density to be 1.3 kg.m⁻³.

SOLUTION:

For air flow inside a duct at standard atmospheric pressure and turbulent conditions, the equation 7.83 applies.

$$h = B\frac{G^{0.8}}{d^{0.2}}$$

The mass flow rate,

$$G = (1.3 \text{ kg.m}^{-3})(3 \text{ m.s}^{-1}) = 3.9 \text{ kg.m}^{-2}.\text{s}^{-1}$$

Hydraulic diameter is

$$d = 4(0.2 \text{ m}^2)/(1.8 \text{ m}) = 0.44 \text{ m}$$

Coefficient of convective heat transfer is obtained as follows:

$$h = 3.24 \times \frac{(3.9)^{0.8}}{(0.44)^{0.2}} = 11.34 \text{ W.m}^{-2}.\text{K}^{-1}$$

Example 7.10 In a double-pipe (tubular) heat exchanger, a liquid flowing in an inner pipe is heated by steam condensing on the outside. The temperature of the wall remains constant and the heat transfer coefficient has been observed as 1400 W.m⁻².K⁻¹ for the interior of the inner pipe. Find the heat transfer coefficient if the rate of liquid flow is doubled.

SOLUTION:

For flow in a pipe under turbulent conditions,

$$Nu = hD/k = 0.027 \,(Re)^{0.8}\,(Pr)^{0.33}\,(\mu_b/\mu_w)^{0.14}$$

Other parameters remaining constant,

$$h \propto Re^{0.8}$$

or, $h \propto (G\,D/\mu)^{0.8}$

or, $h \propto (G)^{0.8}$

Hence, $h_1/h_2 = (G_1/G_2)^{0.8}$

or, $1400/h_2 = (1/2)^{0.8} = 0.574$

or, $h_2 = 2437.54 \text{ W.m}^{-2}.\text{K}^{-1}$

Alternatively,

For turbulent flow (Re > 10000, 0.7 < Pr < 700, L/d < 60) and d << L, the following equations could be used to obtain the same result.

$$h \propto (Dv)^{0.8}$$

or, $h_1 = \left[(2D)\left(\dfrac{v}{4}\right)\right]^{0.8} = \left(\dfrac{D.v}{2}\right)^{0.8}$

where, v is the fluid velocity in m.s⁻¹.
This will also lead to the above answer.

Example 7.11 A horizontal pipe with outside diameter 80 mm and surface temperature 100°C is used for convective heating of a room. The pipe surface is maintained at 100°C by the warm water flowing through it. Determine the coefficient of convective thermal exchange and the rate of convective thermal energy loss from the pipe if room air is maintained at 30°C. Also, determine the thermal energy lost per unit pipe length.

SOLUTION:

A pipe can be considered as a horizontal cylinder.
Calculate Rayleigh number, Gr.Pr, to determine if the fluid flow is turbulent or laminar.

$$Gr.Pr = 10^8 L^3(\Delta t) = 10^8 (0.08)^3 (70) = 3584000$$

$$= 0.35 \times 10^7$$

Since Gr.Pr is within the range of 10^4 and 10^8, the flow is laminar.

For laminar flow, coefficient of convective energy exchange is determined as follows:

$$h = 1.32\left(\Delta t/L\right)^{0.25} = 1.32\left(70/0.08\right)^{0.25} = 7.18 \text{ W.m}^{-2}.\text{K}^{-1}$$

The heat flux, (q/A), also denoted as q″, is calculated as follows:

$$q'' = h.\Delta t = 7.18\,(70) = 502.54 \text{ W.m}^{-2}$$

The pipe area per unit length for thermal energy exchange is now obtained as follows:

$$A = \pi\, d\, L = \pi\, (0.08)\, (1.0) = 0.251\ \mathrm{m^2.m^{-1}}$$

$$\text{or, } q = (502.54)\, (0.251) = 126.3\ \mathrm{W.m^{-1}}.$$

Thus, the pipe heat loss is 502.54 $\mathrm{W.m^{-2}}$ and the heat loss per meter length is 126.3 $\mathrm{W.m^{-1}}$.

Example 7.12 What would be the coefficient of surface convective heat transfer in the above problem if the outside diameter of the duct is 0.8 m, all other factors remaining constant? Also, find the heat loss rate from the duct due to convection.

SOLUTION:

Proceeding as above,

$$\mathrm{Gr.Pr} = 10^8 L^3\, (\Delta t) = 10^8\, (0.8)^3\, (80) = 35.84 \times 10^8$$

The flow is, therefore, turbulent.

The coefficient of convective thermal exchange is represented as follows:

$$h = 1.24\, (70)^{0.33} = 5.038\ \mathrm{W.m^{-2}.K^{-1}}$$

$$q'' = h.\Delta t = 5.038\, (70) = 352.69\ \mathrm{W.m^{-2}}$$

The thermal exchange surface per unit pipe length can now be determined as follows:

$$A = \pi\, D\, L = \pi\, (0.8)(1.0) = 2.51\ \mathrm{m^2.m^{-1}}$$

Thus $q = (352.69)\, (2.51) = 885.27\ \mathrm{W.m^{-1}}$

7.5 Unsteady State Heat Transfer

When during the heating or cooling process, the temperature changes with time, the heat transfer process is said to be in *unsteady state*. Thus, both time and location affect the temperature (recall that temperature varies with location only under steady state).

In unsteady state, there may be heat generation or heat loss (negative heat generation) within the object. Parameters for temperature determination at a given point within a food during processing include time taken for heating or cooling and spatial coordinates of point of interest in the food besides material properties. When a metal slab is heated, initially there is storage of heat within the body and considerable time is taken for the transfer of heat from one end to the other end, which is also a good example of unsteady state heat transfer.

7.5.1 Heat Conduction through a Body with Internal Heat Generation

Let us consider a rectangular object having dimensions dx×dy×dz as shown in Figure 7.4 with G ($\mathrm{W.m^{-3}}$) as internal heat generation.

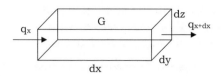

FIGURE 7.4 Heat conduction through a rectangular object with heat generation

In this case

$$\text{Heat in } q_x = -kA\frac{dt}{dx} = -k\left(dy.dz\right)\frac{dt}{dx} \quad (7.90)$$

$$\text{Heat out } q_{x+dx} = q_x + \frac{\partial}{\partial x}\left(q_x\right)dx \quad (7.91)$$

Under unsteady state condition,

heat in + heat generated = heat out + heat accumulation

Thus,

$$q_x + G\left(dx.dy.dz\right) = q_{x+dx} + \rho V c_p \frac{dt}{d\theta}$$
$$= q_x + \frac{\partial}{\partial x}\left(q_x\right)dx + \rho V c_p \frac{dt}{d\theta} \quad (7.92)$$

$$\text{or, } G\left(dx.dy.dz\right) = \frac{\partial}{\partial x}\left(-k.dy.dz.\frac{dt}{dx}\right)dx + \rho V c_p \frac{dt}{d\theta}$$

$$\text{or, } G = \frac{\partial}{\partial x}\left(-k.\frac{dt}{dx}\right) + \rho c_p \frac{dt}{d\theta}$$

$$= -k.\frac{\partial^2 t}{\partial x^2} + \rho c_p \frac{dt}{d\theta}$$

$$\text{If } G = 0, \text{ then } \rho c_p \frac{dt}{d\theta} = k.\frac{\partial^2 t}{\partial x^2}$$

$$\text{or, } \frac{\partial t}{\partial \theta} = \frac{k}{\rho c_p}.\frac{\partial^2 t}{\partial x^2} = \alpha \frac{\partial^2 t}{\partial x^2} \quad (7.93)$$

Equation 7.93 presents time-dependent heat transfer in x direction. The above equation indicates that the heat transfer rate is affected by thermal diffusivity, i.e., temperature of the food material undergoes a change depending on its properties and the temperature of the heating medium. The analytical solution of this governing equation is attainable for only a few simplified shapes, e.g., infinite slab, sphere, and infinite cylinder.

Most of the process duration in many food operations passes through an unsteady state. For example, during the boiling of tomato juice in an open kettle or sterilizing milk in cans the unsteady state process is of much longer duration than the steady state and there is significant temperature variation with time during most of the process.

7.5.2 Unsteady State Conduction and Convection Heat Transfer with Negligible Internal Resistance

First, we will discuss the internal and external resistances to heat transfer when combined conduction and convection take place. If we heat peas in hot (or boiling) water, there will be convective heat transfer from the water to the peas and then there will be conduction heat transfer within the peas. The solid object poses resistance to conduction heat transfer due to its internal structure; there is also resistance to convection heat transfer posed by the fluid layer surrounding the solid. Similarly, when a food slab is frozen, there will be convective heat transfer from the surface of the apple to the surrounding air and the heat will be conducted from the inner parts of the apple to the surface layer. Thus, there will be internal and external resistances to heat transfer depending upon the heating or cooling conditions.

A dimensionless number, *Biot number* or Bi, is used to represent the relative strength of these two resistances.

$$Bi = \frac{\text{Internal resistance to heat transfer}}{\text{external resistance to heat transfer}} \quad (7.94)$$

Representing the internal resistance as L/k (L is the characteristic dimension) and the external resistance as 1/h, the equation for Biot number can be written as

$$Bi = \frac{L/k}{1/h}$$

$$Bi = \frac{hL}{k} \quad (7.95)$$

In an object, the characteristic dimension, L, represents the shortest distance between its surface and center such that L =V/A.

For a cylinder of length H and radius r, $\quad L = \frac{\pi r^2 H}{2\pi r H} = \frac{r}{2}$

For a sphere with radius r, $\quad L = \frac{(4/3)\pi r^3}{4\pi r^2} = \frac{r}{3}$

For a square rod with sides × and length H, $\quad l = \frac{x.x.H}{4.x.H} = \frac{x}{4}$

A Biot number more than 40 means that the numerator value is much higher in comparison to the value in the denominator and, therefore, the heat transfer surface resistance may be neglected. Conversely, a Biot number value less than 0.1 indicates that the solid's internal resistance to heat transfer is negligible. Between these two values both internal and external resistances have finite values. When the thermal conductivity of a solid is very high, as in the case of a metal, there will be negligible internal resistance, but for foods, as the thermal conductivity is relatively low, the internal resistance cannot be ignored.

Negligible internal resistance also means that there will not be much temperature variation within the body and heat is transferred instantaneously within the body. Such a situation also develops when a liquid food is stirred while being heated.

A heat transfer process of this type can be described mathematically as follows:

Let a food be heated in a steam-jacketed kettle. Steam at a temperature t′ is used to heat the food from t_0. If the fluid temperature at any time θ is t,

$$(\rho V)c_p (dt/d\theta) = -UA(t' - t)$$

The product of fluid density and volume (ρ.V) has been used in lieu of mass flow rate, ṁ, to take the changes in density and volume of the fluid into consideration during the heating process. Here, U is the heat transfer coefficient and A is the heating surface area. c_p is the specific heat of the food.

Thus, $\quad -\int_{t_0}^{t} \frac{dt}{t'-t} = \int_{0}^{\theta} \frac{U.A}{\rho V c_p} d\theta$

or, $\quad \log_e \frac{t'-t}{t'-t_0} = -\frac{U.A}{\rho V c_p} \theta$

or, $\quad \frac{t'-t}{t'-t_0} = e^{-\frac{UA}{\rho V c_p}\theta} \quad (7.96)$

Here in the above equation $\frac{\rho V c_p}{UA}$ is called the *time constant* (ζ). If the fluid temperature is plotted against time, the graph will be as shown in the Figure 7.5 and the fluid temperature will vary as follows.

$$t = t' - (t' - t_0)e^{-\frac{UA}{\rho V c_p}\theta}. \quad (7.97)$$

For the situation when the heated medium temperature becomes equal to that of the heating medium (i.e. t = t′),

$$(t' - t_0)e^{-\theta/\zeta} = 0, \text{ or } \theta \text{ becomes equal to } \infty.$$

It indicates that the heated medium would take a very long (~infinite) time to reach the heating medium temperature.

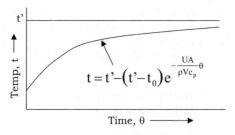

FIGURE 7.5 Temperature change of a fluid during unsteady state heating

However, the time constant can be reduced to increase the heat gain by the body by increasing either U or A.

An unsteady state thermal process approaches steady state if it is continued for a long time.

The *time constant* (ζ) is also mentioned as the time taken for temperature difference between the heating and the heated media to reach 63.2% of the initial temperature difference.

$$\text{At that time } \frac{t'-t}{t'-t_0} = 0.368 - e^{-1}$$

To attain the temperature up to 99% of the initial temperature difference it will take 4.6ζ time.

Equation 7.96 can be further modified as follows.

$$\frac{t'-t}{t'-t_0} = e^{-\frac{hA}{\rho V c_p}\theta}$$

By multiplying k and (V/A) in both numerator and denominator and noting that the characteristic dimension L = volume (V)/area (A),

$$\frac{t'-t}{t'-t_0} = \exp-\left[\frac{h(V/A)}{k}\cdot\frac{k}{\rho.c_p}\cdot\frac{\theta}{(V/A)^2}\right]$$

$$= \exp-\left[\frac{hL}{k}.\alpha.\frac{\theta}{L^2}\right] = \exp-\left[\frac{hL}{k}.\frac{\alpha\theta}{L^2}\right]$$

$$\frac{t'-t}{t'-t_0} = e^{-(Bi.Fo)} \qquad (7.98)$$

Here Fo ($=\alpha\theta/L^2$) is the dimensionless *Fourier number*. As observed from this expression, Fo relates the thermal diffusivity with heating/cooling time and the size of the material. Bi is the Biot number as discussed earlier (equation 7.95).

Fo can be rearranged as

$$Fo = \frac{\alpha\theta}{L^2} = \frac{k}{\rho c_p}\cdot\frac{\theta}{L^2} = \frac{k(1/L)L^2}{\rho c_p L^3/\theta}$$

$$= \frac{\text{rate of heat conduction across L in volume } L^3 \left(W.K^{-1}\right)}{\text{rate of heat storage in volume } L^3 (W.K^{-1})}$$

$$(7.99)$$

Thus, the Fourier number (Fo) for an element of given volume is an expression of the heat conduction rate across characteristic dimension L per unit rate of heat storage. Therefore, a higher Fourier number in a given period of time indicates deeper heat penetration into the solid.

In some cases, a slightly modified form of the equation 7.96 is used to take care of the ignored parameters in developing the above equation by incorporating another constant A as follows:

$$\frac{t'-t}{t'-t_0} = Ae^{-\frac{hA}{\rho V c_p}\theta} \qquad (7.100)$$

Example 7.13 Forty liters of tomato soup will be heated from 30°C to a temperature of 90°C in a steam-jacketed kettle. Steam at 100°C is used in the jacket and the heat transfer coefficient is 1200 W.m⁻².K⁻¹. The heat transfer area of the kettle is 0.4 m². The density and specific heat of tomato soup can be taken as 630 kg.m⁻³ and 3.67 kJ.kg⁻¹. K⁻¹, respectively. Find out the time required for the heating if there is no other heat loss and gain for the soup.

The unsteady state heat transfer equation will be used.

$$\frac{t'-t}{t'-t_0} = e^{\left(\frac{-UA}{\rho.V.c_p}\right)\theta}$$

In this case, $\frac{t'-t}{t'-t_0} = \frac{100-90}{100-30} = \frac{10}{70} = 0.143$

The other values to be used in the above equation are as follows:

U = 1200 W.m⁻².K⁻¹, A = 0.4 m², V = 40 liters = 0.04 m³,

c_p = 3670 J.kg⁻¹.K⁻¹ and ρ = 630 kg.m⁻³

Thus, $\ln(0.143) = -\frac{(1200)(0.4)}{(630)(0.04)(3670)}\theta$

$$\theta = -\frac{\ln(0.143)(630)(0.04)(3670)}{(1200)(0.4)} = 374 \text{ s}$$

The time required will be 374 seconds or 6.23 min.

Example 7.14 For a can containing semisolid food, 5 minutes is required to increase the temperature from 80°C to 90°C and 7 minutes is required to raise from 80°C to 100°C (temperatures monitored at the center of the can). The saturated steam in the retort is at 125°C. The heating process stipulates achieving 120°C at the center of the can. How much time will be needed for the thermal process assuming the food's surface heat transfer resistance to be zero?

SOLUTION:

In this case, as two conditions are given, we will use equation 7.100.

$$\frac{t'-t}{t'-t_0} = Ae^{-\frac{hA}{\rho V c_p}\theta}$$

With the given values and using the symbol of the time constant, the above equation can be written for the two situations as follows:

For the first situation, $\frac{125-90}{125-80} = Ae^{-(1/\zeta)(5\times60)}$

$$(A)$$

For the second situation, $\dfrac{125-100}{125-80} = Ae^{-(1/\zeta)(7\times60)}$

(B)

There are two unknowns A and ζ, and two equations. The solutions yield,

$$\zeta = 356.717 \text{ and } A = 1.80$$

Substituting the value of ζ in equation (A)

$$0.7778 = Ae^{-300/356.717}$$

Now, with the same principle

$$\frac{125-120}{125-80} = Ae^{-(1/\zeta)(\theta)}$$

Or, $\theta = 994.646$ s $= 16.57$ min

Example 7.15 A spherical fish ball with a radius of 25.4 mm is at a uniform temperature of 15°C. It is suddenly brought to a cold chamber whose temperature is kept constant at -30°C. The convective heat transfer coefficient of 10 W.m^{-2}.°C^{-1}. The average thermo physical properties of fish are: k = 0.5 W.(m.°C)$^{-1}$, ρ = 970 kg.m^{-3} and c$_p$ = 2.45 kJ.kg^{-1}.°C^{-1}. What will be the temperature of the fresh fish ball after keeping it for one hour in the cold chamber?

SOLUTION:

The fish ball is taken as a sphere, for which
V/A = r/3 = 25.4/3 = 8.46 mm, θ = 1 h = 3600 s

$$\frac{t'-t}{t'-t_0} = Ae^{-\frac{hA}{\rho Vc_p}\theta}$$

$$\frac{t-(-30)}{15-(-30)} = e^{\left(\frac{-10}{970\times2.45\times1000\times0.00846}\right)\times3600}$$

$$t = 14.58°C$$

7.5.3 Heat Transfer with Finite Thermal Surface and Internal Resistances

Solutions for regular shapes, such as infinite slab, sphere, and infinite cylinder, have been obtained using the basic transient heat transfer equation (equation 7.96), and temperature-time charts have been plotted. These charts (Figure 7.6) are helpful in the heat transfer solutions for simple shaped foods. The charts relate the temperature ratio, the Fourier number, and the inverse of the Biot number. While the temperature ratio uses a log scale on vertical axis, the Fourier number goes on a linear horizontal axis.

The temperature ratio (or temperature factor) is $\dfrac{t_a - t}{t_a - t_i}$,

(7.101)

The surrounding medium's temperature is t_a and t_i is the initial temperature. The temperature at time θ during the process is t. The temperature ratio during the process represents the temperature change, at a given time, remaining to be accomplished before the process completion.

At the beginning of the processing, the temperature ratio is unity and it keeps on decreasing as the process continues.

In this chart, δ is the characteristic half dimension (it is the radius for a sphere or cylinder, and for a slab it is half of the thickness).

The graphs can be used for even those surface resistances that are close to zero. The lines for k/hδ = 0 specify that the surface heat transfer resistances are insignificant.

The steps for the analysis are as follows:

- First with the available information, k/hδ or the reciprocal of the Biot number is obtained.
- With the knowledge of temperatures of the surrounding medium, initial and final (desired) temperature of the commodity/can, the temperature ratio is obtained.
- In the third step, from the k/hδ value and temperature ratio, the Fourier number is obtained from the chart on the X-axis.
- The Fourier number ($\alpha\,\theta/\delta^2$) is further used to find the time.

Example 7.16 Carrot cubes measuring 1 cm × 1 cm × 1 cm need to be blanched before drying. Hot water at 98°C is used for blanching. The initial temperature of carrots is 25°C and it is desired that the temperature at the thermal center of the carrots should be 90°C. The density, specific heat, and thermal conductivity of carrots can be taken as 1040 kg.m^{-3}, 3.86 kJ.kg^{-1}.K^{-1} and 0.6 W.m^{-1}.K^{-1}. Find the time required for blanching if the heat transfer coefficient is 1400 W.m^{-2}.K^{-1}.

SOLUTION:

In this case, δ = 0.5 cm = 5 × 10^{-3} m

Thus, $\dfrac{k}{h.\delta} = \dfrac{0.6}{(1400)(5\times10^{-3})} = 0.086$

The temperature ratio, $\dfrac{t_a - t}{t_a - t_i} = \dfrac{98-90}{98-25} = 0.1096$

From the chart, corresponding to a slab and with the above values, the Fourier number is obtained as 1.16.

Thus, $\dfrac{k}{\rho.c_p}\cdot\dfrac{\theta}{\delta^2} = 1.16$

θ = (1.16)(5×10^{-3})^2(1040)(3.86×10^3)/(0.6) = 194 s

The time required for blanching is 194 seconds.

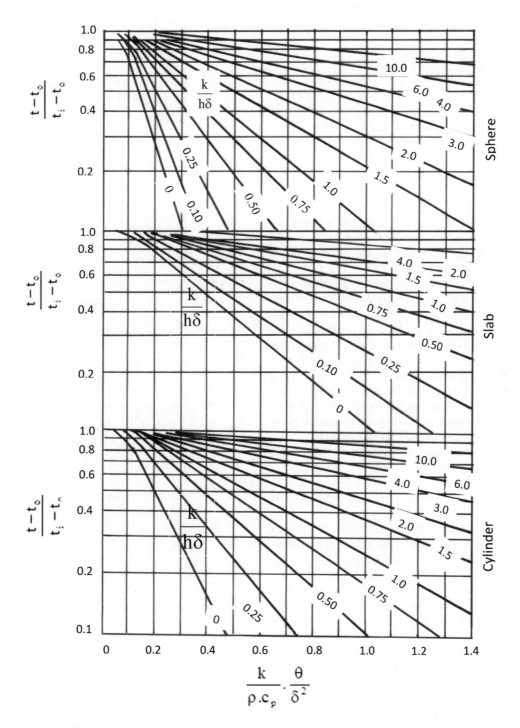

FIGURE 7.6 Chart for solution of unsteady state heat transfer (a) for a sphere, (b) for a slab, and (c) for a cylinder (After Henderson and Perry, 1976)

7.6 Combined Conduction and Convection Heat Transfer in Finite Objects

Heat transfer to a finite object, such as a cylindrical can, of finite cylindrical shape can be obtained by using the time-temperature plots for both infinite slab and infinite cylinder. First, considering an infinite cylinder, the Fourier and Biot numbers are calculated using the cylinder's radius as the characteristic dimension. The temperature ratio is determined using the charts as discussed in the above paragraph. The Fourier and

Biot numbers are then recalculated considering an infinite slab with characteristic dimension as the cylinder's half length (height). The temperature ratio is calculated for the infinite slab. The two temperature ratios, thus determined, are multiplied to yield the solution for the cylindrical can.

For the finite cylinder, temperature ratio can be obtained as follows:

$$\left(\frac{t_a - t}{t_a - t_i}\right)_{finite\ cylinder} = \left(\frac{t_a - t}{t_a - t_i}\right)_{infinite\ cylinder} \left(\frac{t_a - t}{t_a - t_i}\right)_{infinite\ slab}$$

(7.102)

And, similarly for a brick-shaped material,

$$\left(\frac{t_a - t}{t_a - t_i}\right)_{\text{finite brick shape}} = \left(\frac{t_a - t}{t_a - t_i}\right)_{\text{infinite slab width}}$$

$$\left(\frac{t_a - t}{t_a - t_i}\right)_{\text{infinite slab depth}} \left(\frac{t_a - t}{t_a - t_i}\right)_{\text{infinite slab height}} \quad (7.103)$$

7.7 Radiation Heat Transfer

Any object above an absolute zero temperature (0 K or -273.15°C) emits *thermal radiation*. It is the electromagnetic radiation emitted by the object and the spectral distribution determined by its temperature.

An object is made up of atoms that have sub-atomic particles, such as electrons. At 0 K there is no movement of these sub-atomic particles, and the energy is considered to be zero. At temperatures more than 0 K, the sub-atomic particles begin to vibrate and move around. This movement of sub-atomic particles results in the energy being given off as radiation. Since this radiation is due to the temperature of the object, it is termed *thermal radiation*. Thus, the electromagnetic radiation emitted due to the motion of thermally energized particles is termed the object's thermal radiation. The amount of radiant energy emitted by an object is dependent on its temperature and radiation properties.

All forms of radiation travel at the speed of light, 3×10^8 m.s^{-1}. It is also important that

$$c = \lambda.\nu \quad (7.104)$$

where c = speed of light, ν = frequency, and λ = wavelength

The equation indicates that radiation with a longer wavelength would have a smaller frequency and vice versa.

Figure 7.7 shows the frequencies and wavelengths of different electromagnetic radiations. The radiation traditionally considered to be involved in heat transfer is in the infrared band. However, visible light and UV radiation from the sun are examples of electromagnetic radiations, which, when absorbed, are also converted to thermal energy. Thus, thermal radiation can be considered to have wavelengths ranging between 0.1 and 100 μm (the wavelength of visible light is within 0.35 and 0.75 μm).

7.7.1 Transmission of Thermal Radiation and Emissive Power

Thermal radiation is transmitted as discrete quanta with each quantum having some energy.

$$E = h. \nu \quad (7.105)$$

Here the value of h, the *Planck's constant*, is 6.625×10^{-34} J.s.

As the energy can also be expressed as, $E = mc^2$ (this is Einstein's equation, which established the equivalence between energy and mass), we obtain

$$m = h. \nu / c^2$$

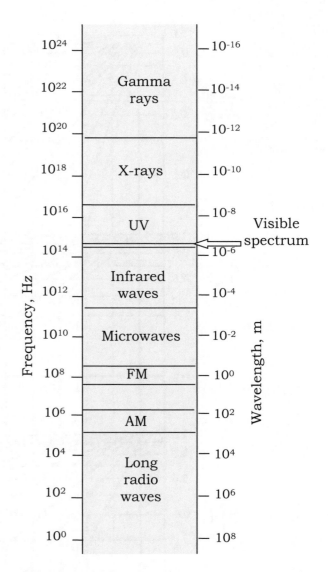

FIGURE 7.7   Radiation spectrum

$$\text{and, momentum} = c (h. \nu / c^2) = h. \nu / c \quad (7.106)$$

By using the principles of quantum-statistical thermodynamics and considering the radiation as a gas, the following equation is obtained for the radiation's energy density per unit wavelength and per unit volume.

$$u_\lambda = \frac{8\pi hc\lambda^{-5}}{e^{hc/\lambda kT} - 1} \quad (7.107)$$

where k is *Boltzmann constant* and is equal to 1.380649×10^{-23} J.K^{-1}.

Total thermal radiation emitted by a black body is obtained by integration of the radiation's energy density per unit wavelength, given above, over all wavelengths.

$$E_b = \sigma T^4 \quad (7.108)$$

This equation is known as the *Stefan-Boltzmann law* in which the energy emitted per unit area per unit time by a black body, E_b, has the unit of W.m^{-2} or J.s^{-1}.m^{-2}. The subscript b denotes black body, which is used to represent an ideal radiator. T is the absolute temperature and σ is the *Stefan-Boltzmann constant*.

$$\sigma = 5.669 \times 10^{-8} \text{ W.m}^{-2}.\text{K}^{-4}$$

E_b is also known as a black body's emissive power.

Conventionally, it is called a black body radiation as the materials appear black to the human eye. A lamp black is an example, which absorbs all radiation coming on to it. However, some materials that are not black in appearance may behave as a black body for specific radiations. As an example, snow, ice, and many white paints absorb all long wave thermal radiation and, hence, they are considered black for long-wave thermal radiations.

A *black body* is defined as an object that has emissivity of 1, i.e., it emits all the radiation that is produced due to its temperature. A black body is the perfect emitter of thermal radiation. For any non-black body, the equation 7.108 modifies as follows:

$$E = \varepsilon \, \sigma T^4 \qquad (7.109)$$

where the emissivity of the surface emitting the radiation is ε. It can be shown that a perfect emitter is also the perfect absorber for the same spectrum of radiation.

When electromagnetic radiation emitted by one surface is received and subsequently absorbed by another surface, it raises the energy level of the absorbing surface. As seen from equations 7.108 and 7.109, objects at high temperatures emit more electromagnetic radiation than those at low temperatures. Thus, the net transfer of thermal energy occurs from the higher temperature surface to the one at lower temperature.

As the thermal radiation leaves an object, the temperature of the body is lowered because of the reduction in energy content of the object. The converse occurs when an object absorbs thermal radiation, i.e., the temperature increases. Bodies in a closed system continue exchanging net radiation until their temperatures equalize.

7.7.2 Emissivity, Absorptivity, Reflectivity, and Transmissivity

When radiation falls on the surface of an object, a fraction of it is absorbed by the object, a fraction is reflected back, and the remaining radiation is transmitted through the object. Accordingly, these phenomena are known as *absorptance*, *reflectance*, and *transmittance* and the fractions are denoted as absorptivity (α), reflectivity (ρ), and transmissivity (ι). The magnitudes of these fractions are functions of the object's surface properties. The fraction absorbed by the surface is converted to thermal energy.

Two more terms, irradiance and radiosity, are also important for radiation heat exchange calculations; *irradiance* is the total radiation incident on a surface per unit time per unit area, and *radiosity* is the total radiation leaving a surface consisting of the component that is emitted due to its own temperature, the component that is reflected, and the component that is transmitted by the radiating body (Figure 7.8).

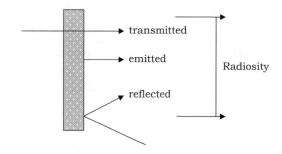

FIGURE 7.8 Difference between emitted, reflected, and transmitted radiation

As per the above discussion,

$$\alpha + \rho + \iota = 1 \qquad (7.110)$$

In case the body does not transmit thermal radiation, as happens in most opaque solid bodies, then the above equation reduces to

$$\rho + \alpha = 1 \qquad (7.111)$$

Emissivity (ε) is the ability of a surface to absorb incident radiation. Absorptance and emittance are opposites, physically, but numerically equal for a given surface and radiation at the same wavelength. For instance, a surface having emittance of 0.7 for the wavelength (at a specific surface temperature) will also have an absorptance of 0.7 for radiation having the same wavelength.

If there is a small body within an enclosure in thermal equilibrium, the absorbed energy must be equal to the energy given off by the body. This can be written as

$$E.A = q_i.A.\alpha \qquad (7.112)$$

If it is a black body, then $\alpha = 1$ and the equation becomes as follows:

$$E_b.A = q_i.A.1 \qquad (7.113)$$

From equations 7.112 and 7.113, $E/E_b = \alpha$

This implies that, at a given temperature, a body's absorptivity can be expressed as the emissive power of the body divided by the emissive power of a black body. With the similar reasoning, it is possible to show the following.

$$E/E_b = \varepsilon$$

$$\text{Therefore, } \varepsilon = \alpha \qquad (7.114)$$

Equation 7.114 represents the *Kirchhoff's identity*.

In addition to the surface characteristics, the emissivity of a material depends on the temperature and wavelength of radiation. Most surfaces do not absorb all incident radiation and reflect and/or transmit some part (s) of it. Such bodies are called *grey bodies* and have emissivity values less than 1. Table 7.3 gives the approximate emissivity values of some common materials.

TABLE 7.3

Approximate emissivity values for some materials

Material	Emissivity
Polished metal	< 0.05
Unpolished metal	0.7–0.25
White paper	0.91–0.95
Painted metal or wood	0.9
Brick	0.93
Concrete	0.94
Water	0.96
Ice	0.98
Burnt toast	1.00

As the radiation properties of an object are determined by the surface to a depth of only several wavelengths of the radiation involved, the thermal radiation properties of a surface can be changed by simply painting over it.

Emissive power of a body can be expressed in terms of *monochromatic emissivity*. Monochromatic emissivity of a body at a given temperature and wavelength is expressed as monochromatic emissive power of the body divided by that of a black body.

$$\varepsilon_\lambda = \frac{E_\lambda}{E_{b\lambda}} \tag{7.115}$$

A *grey body* absorbs incident radiation equally at all wavelengths, i.e., the monochromatic emissivity (ε_λ) for a grey body does not depend on wavelength and remains constant. A body is said to be a *diffuse body* if its properties do not change with direction. Therefore, a grey diffuse body has its properties independent of direction and wavelength.

The monochromatic black-body emissive power, $E_{b\lambda}$ is related to the energy density as

$$E_{b\lambda} = \frac{u_\lambda c}{4} \tag{7.116}$$

$$E_{b\lambda} = \frac{C_1 \lambda^{-5}}{e^{C_2/\lambda T} - 1} \tag{7.117}$$

where λ is the wavelength (m), $C_1 = 3.743 \times 10^8$ W.µm^4.m^{-2}, $C_2 = 1.4387 \times 10^4$ µm.K.

The *Wein's displacement law*, as given below, gives the values of maximum wavelength (i.e., the point of inflection) in the radiation curves.

$$\lambda_{max} T = 2897.6 \text{ µm.K} \tag{7.118}$$

When only two bodies are involved in thermal radiation exchange, the rate of heat transfer from one body to the other equals the net absorbed energy by the body remaining after the energy leaving the body is subtracted.

$$q_{1-2} = C\sigma A_1 (T_1^4 - T_2^4) \tag{7.119}$$

where $1/C = (1/\varepsilon_1 + 1/\varepsilon_2 - 1)$. The two surfaces have their emissivities as ε_1 and ε_2.

When a small body with emissivity ε_1 at a uniform temperature T_1 is exchanging thermal radiation with its surroundings at temperature T_2, the net radiation exchange from the small body to the infinite surroundings is written as follows:

$$q = \varepsilon_1 \sigma A_1 (T_1^4 - T_2^4) \tag{7.120}$$

The above equation can be used for calculating the heat exchange for a cake being baked in an oven or many such applications.

Example 7.17 A heated pipe surface, having the surface emissivity (ε) of 0.12, is at 150°C. Determine the radiation heat leaving the surface. What will be the heat flux if the surface is painted so that its emissivity changes to 0.9?

SOLUTION:

The pipe surface temperature is 423.15 K. Therefore, the thermal radiation flux is as follows:

$$q'' = (0.12)(5.6697 \times 10^{-8})(423.15)^4$$

$$= 218.126 \text{ W}$$

By application of paint the surface emittance is changed to 0.90.

Then, $q'' = (0.90)(5.6697 \times 10^{-8})(423.15)^4$

$$= 1635.99 \text{ W}$$

Example 7.18 A hot body (80°C), kept in a room has the emissivity of 0.85. The body is in thermal radiation equilibrium with all objects in the room being at 30°C. Determine the radiation heat loss from the hot body if the surface area of the body for radiation heat exchange is 0.6 m².

SOLUTION:

$$q = A_1 \varepsilon_1 \sigma \left(T_1^4 - T_2^4\right)$$

Here $T_1 = 353.15$ K and $T_2 = 303.15$ K

Thus, $q = (0.6)(0.85)(5.6697 \times 10^{-8})(353.15^4 - 303.15^4)$

$$= 205.537 \text{ W}$$

Example 7.19 A loaf of bread is being baked in an oven at 180°C temperature. Assuming that the bread temperature is 100°C and the surface emissivity is 0.86, what will be the rate of heat gain during the baking process? The surface area of the loaf for the radiation heat exchange is 0.06 m².

SOLUTION:

$$q = A_1 \varepsilon_1 \sigma \left(T_1^4 - T_2^4 \right)$$

$$= 0.06 \times 0.86 \times 5.6697 \times 10^{-8} \left(453.15^4 - 373.15^4 \right)$$

$$= 66.64 \text{ J s}^{-1}.$$

7.7.3 Radiation Shape Factor

Suppose there are two surfaces A_1 and A_2, for which the radiation heat exchange needs to be determined. All the radiation energy that leaves A_1 is not intercepted by A_2 and all the energy leaving A_2 is not intercepted by A_1. Thus, the heat exchange will have to be calculated considering the fraction of total energy leaving the first object that reaches the second one and vice versa. These fractions, known as radiation shape factors, are useful in analysis of radiation exchanges.

The *radiation shape factor* between two surfaces m and n is defined as the fraction of energy leaving surface m that reaches surface n and is denoted as F_{m-n}.

Thus, F_{1-2} = fraction of radiation that leaves surface 1 and reaches surface 2

F_{2-1} = fraction of radiation that leaves surface 2 and reaches surface 1

The radiation shape factor is a geometrical parameter and it is also known as the *view factor* or *configuration factor*.

Effect of Shape Factor in Radiation Heat Exchange

Let us consider two surfaces A_1 and A_2 exchanging radiation (Figure 7.9). The energy arriving at surface 2 after leaving surface 1 is $E_{b1} A_1 F_{12}$, and the energy arriving at surface 1 after leaving surface 2 is $E_{b2} A_2 F_{21}$.

Assuming both the surfaces to be black (so that the radiation incident on a surface is fully absorbed), the net radiation exchange is represented as follows:

$$q_{1-2} = E_{b1} A_1 F_{12} - E_{b2} A_2 F_{21} \tag{7.121}$$

If both surfaces are at the same temperature, then $T_1 = T_2$, $E_{b1} = E_{b2}$, and $q_{1-2} = 0$.

$$\text{Thus, } A_1 F_{12} = A_2 F_{21} \tag{7.122}$$

The net thermal radiation exchange can be written as follows:

$$q_{1-2} = A_2 F_{21}(E_{b1} - E_{b2}) = A_1 F_{12} (E_{b1} - E_{b2}) \tag{7.123}$$

The equality expressed by equation 7.122 is known as the *reciprocity relation*. General form of reciprocity relation can be written as follows:

$$A_m F_{m-n} = A_n F_{n-m} \tag{7.124}$$

This relationship has been developed for black surfaces. However, as long as the radiation is diffuse, this equation is applicable for non-black surfaces as well.

Next, we will develop a general relation for F_{12} (or F_{21}). Taking two elements with areas dA_1 and dA_2, let r be the line joining these elements. Let the angle between the surface normal to dA_1 and the line r be φ_1 and the angle between the surface normal to dA_2 and the line r be φ_2.

The projection of dA_1 on the line between the centers is

$$dA_1 \cos \varphi_1$$

The energy leaving dA_1 in the direction of angle φ_1 can be represented as follows:

$$I_b dA_1 \cos \varphi_1$$

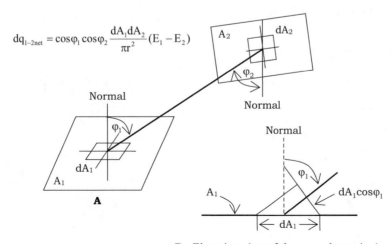

$$dq_{1-2net} = \cos \varphi_1 \cos \varphi_2 \frac{dA_1 dA_2}{\pi r^2} (E_1 - E_2)$$

B. Elevation view of the area shown in A

FIGURE 7.9 Radiation shape factor between two bodies

where, I_b represents the blackbody intensity. This equation will hold good provided it is assumed that the radiation's intensity is the same in all directions, i.e., the surface gives off diffuse radiation.

Similarly, for an area dA_n located at distance r from A_1, the radiation incident on dA_n is represented as follows:

$$dA_n = I_b dA_1 \cos\varphi_1 \frac{dA_n}{r^2} \qquad (7.125)$$

The area dA_n subtends a solid angle of dA_n/r^2. The area dA_n can also be written as follows:

$$dA_n = dA_2 \cos\varphi_2$$

Thermal radiation leaving dA_1 and arriving at dA_2 is

$$dq_{12} = E_{b1} \cos\varphi_1 \cos\varphi_2 \frac{dA_1 dA_2}{\pi r^2}$$

and energy arriving at dA_1 after leaving dA_2 is

$$dq_{21} = E_{b2} \cos\varphi_1 \cos\varphi_2 \frac{dA_1 dA_2}{\pi r^2}$$

Thus, the net energy transfer is given as

$$dq_{net21} = (E_{b1} - E_{b2}) \cos\varphi_1 \cos\varphi_2 \frac{dA_1 dA_2}{\pi r^2} \qquad (7.126)$$

The integral, according to equation 7.126, is either $A_1 F_{12}$ or $A_2 F_{21}$. The integral evaluation requires knowledge of A_1 and A_2.

Let us consider a case in which there is radiation exchange from infinitesimal area dA_1 to a finite flat disk A_2 (Figure 7.10). A ring of radius x with area dA_2 is taken as the element in A_2 for analysis.

$$dA_2 = 2\pi x.dx.$$

Noting that $\varphi_1 = \varphi_2$, and integrating over the area A_2 by applying equation 7.126,

$$dA_1 F_{dA_1-A_2} = dA_1 \int_{A_2} \cos^2\varphi_1 \frac{2\pi x.dx}{\pi r^2}$$

Now, $r = (R^2 + x^2)^{1/2}$, and $\cos\varphi_1 = \dfrac{R}{(R^2 + x^2)^{1/2}}$

Substituting the values, we have

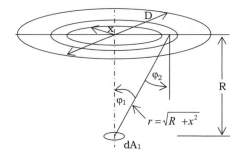

FIGURE 7.10 Radiation emitted by a small area to a disk

$$dA_1 F_{dA_1-A_2} = dA_1 \int_0^{D/2} \frac{2R^2 x.dx}{(R^2 + x^2)^2}$$

Thus, $dA_1 F_{dA_1-A_2} = -dA_1 \left[\dfrac{R^2}{(R^2 + x^2)}\right]_0^{D/2} = dA_1 \dfrac{D^2}{4R^2 + D^2}$

i.e., $F_{dA_1-A_2} = \dfrac{D^2}{4R^2 + D^2} \qquad (7.127)$

The radiation shape factors for other geometries have been calculated and are available in many heat transfer publications. Charts are also available for common shape factors, for example, for radiation exchange between two parallel plates or two concentric cylinders with finite length or two perpendicular rectangles with a common edge (Holman, 1997). After the shape factors are obtained from these charts or equations, the following equation can be used for the calculation of heat exchange by radiation.

$$q_{12} = A_1 F_{12}(E_{b1} - E_{b2}) = \sigma A_1 F_{12}\left(T_1^4 - T_2^4\right)$$

The above shape factor relations have been derived considering ideal surfaces (i.e., assuming that the surfaces perfectly diffuse the radiations). In reality, neither of the surfaces are perfectly diffuse nor is the radiation leaving any surface uniform in all directions. In addition, for real surfaces, thermal radiation properties are affected by the surface characteristics and the surroundings. The direction of the incident radiation as well as the wavelength also influence these thermal radiation properties. However, the dependence of the incident radiation's wavelength and the intensity could usually constitute very complicated functions of the characteristics of all surfaces in the surroundings and the surface temperatures.

Relationship between Shape Factors

Let us consider the Figure 7.11 (a) in which there is radiation exchange between three bodies. Considering that the total

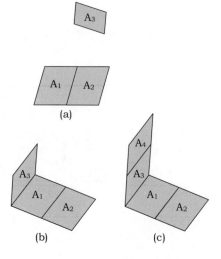

FIGURE 7.11 Some examples of surfaces exchanging heat by radiation

shape factor is the sum of the shape factors of its parts, we can write

$$F_{3-(1,2)} = F_{3-1} + F_{3-2} \qquad (7.128)$$

The above equation can be modified and represented as follows:

$$A_3.F_{3-(1,2)} = A_3.F_{3-1} + A_3.F_{3-2} \qquad (7.129)$$

Using the reciprocity theorem

$$A_3.F_{3-(1,2)} = A_{(1,2)}. F_{(1,2)-3}$$

$$A_3.F_{3-1} = A_1.F_{1-3}$$

$$A_3.F_{3-2} = A_2.F_{2-3}$$

Thus, $A_{(1,2)}. F_{(1,2)-3} = A_1.F_{1-3} + A_2.F_{2-3}$

In other words, total radiation arriving at surface is the sum of radiations from surfaces 1 and 2.

Similarly, for figure 7.11 (b), $F_{1-(2,3)} = F_{1-2} + F_{1-3}$.

For the situation as depicted in Figure 7.11 (c),

$$A_{(1,2)}. F_{(1,2)-(3,4)} = A_1.F_{1-(3,4)} + A_2.F_{2-(3,4)} \qquad (7.130)$$

Both $F_{(1,2)-(3,4)}$ and $F_{2-(3,4)}$ can be obtained from Figure 7.11 (b) and $F_{1-(3,4)}$ can be expressed as

$$A_1.F_{1-(3,4)} = A_1.F_{1-3} + A_1.F_{1-4} \qquad (7.131)$$

also, $A_{1-2}.F_{(1,2)-3} = A_1.F_{1-3} + A_2.F_{2-3} \qquad (7.132)$

Solving for $A_1.F_{1-3}$ from equation 7.132, inserting this in equation 7.131 and then substituting the values in equation 7.130, we obtain,

$$A_{(1,2)}.F_{(1,2)-(3,4)} = A_{(1,2)}.F_{(1,2)-3} - A_2.F_{2-3} + A_1.F_{1-4} + A_2.F_{2-(3,4)}$$

Except F_{1-4}, all shape factors can be evaluated from Figure 7.11.

Substituting the values in the above equation,

$$F_{1-4} = \frac{1}{A_1}(A_{1-2}F_{1,2-3,4} + A_2F_{2-3} - A_{1-2}F_{1,2-3} - A_2F_{2-3,4})$$

$$(7.133)$$

As a general relationship, $\sum_{j=1}^{n} F_{ij} = 1$, i.e., the total energy that arrives at all surfaces j from surface i is 1.

i.e., $F_{11} + F_{12} + F_{13} = 1$

It should also be remembered that the surfaces involved in the radiation exchange do not see themselves,

thus, $F_{11} = F_{22} = F_{nn} = 0$.

7.7.4 Radiation Heat Exchange for Typical Situations

We can find relationships for radiation heat exchange between grey bodies. In the case of radiation exchange between black bodies, the radiation energy that reaches a surface is absorbed completely. But for grey bodies (which are non-black for the specific waves), some part of the energy striking the surface may be reflected partly or entirely and may be reflected back to another surface. The analysis is more complicated when the energy is reflected multiple times between the surfaces.

Energy Balance on an Opaque Material

For the present analysis, we assume uniform distribution of G (irradiation) and J (radiosity) over each surface (Figure 7.12). Although the assumption is not realistic for grey diffuse surfaces, it is made for simplicity.

Now, if no energy is transmitted, then J is the sum of emitted and reflected energy.

i.e., $J = \varepsilon E_b + \rho G \qquad (7.134)$

As transmissivity in this case is zero, $\rho = 1 - \alpha = 1 - \varepsilon$.

Thus, $J = \varepsilon E_b + (1-\varepsilon) G \qquad (7.135)$

Net energy that leaves the surface is given as

$$\frac{q}{A} = J - G = \varepsilon E_b + (1-\varepsilon)G - G \qquad (7.136)$$

or, $q = \frac{\varepsilon A}{1-\varepsilon}(E_b - J) \qquad (7.137)$

The above equation can be written as follows:

$$q = \frac{E_b - J}{(1-\varepsilon)/\varepsilon A} \qquad (7.138)$$

In the above equation, the potential of the heat flow is the numerator and the heat flow resistance is the denominator.

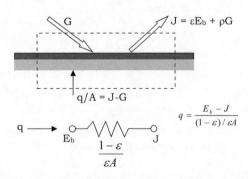

FIGURE 7.12 Energy balance on the surface of an opaque material

Energy Exchange between Two Grey Surfaces

Considering two surfaces exchanging radiant energy as shown in Figure 7.13, we analyze the heat transfer as follows:

Total radiation reaching surface 2 from surface 1 is $J_1A_1F_{12}$. Total radiation reaching surface 1 from surface 2 is $J_2A_2F_{21}$. The net energy exchange between the two surfaces is

$$q_{1-2} = J_1A_1F_{12} - J_2A_2F_{21}$$

But, $A_1F_{12} = A_2F_{21}$
Thus, $q_{1-2} = (J_1 - J_2)A_1F_{12} = (J_1 - J_2)A_2F_{21}$

$$\text{or, } q_{1-2} = \frac{J_1 - J_2}{1/A_1F_{12}} \tag{7.139}$$

As previously discussed, in this case also, we can develop a network element as shown in Figure 7.14(a). To develop a network, each surface has to be considered with the surface resistance, $(1-\varepsilon)/\varepsilon A$ and the space in between will be considered with resistance $1/A_mF_{mn}$. The figure represents thermal radiation exchange between the two surfaces only.

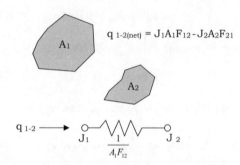

FIGURE 7.13   Energy exchange by radiation between two surfaces

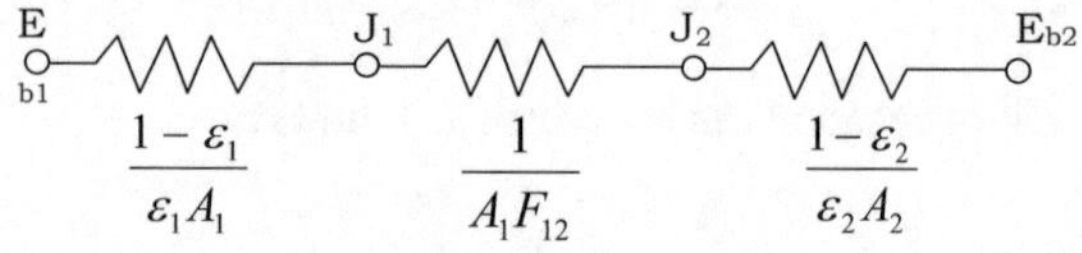

(a)   Network for radiation exchange between two surfaces seeing each other only

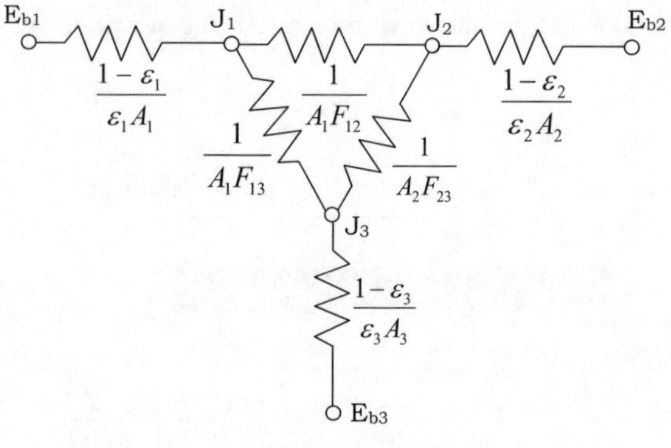

(b)   Network for radiation exchange among three surfaces seeing each other only

FIGURE 7.14   Radiation network for two situations with two and three surfaces

$$q_{net} = \frac{E_{b1} - E_{b2}}{\dfrac{(1-\varepsilon_1)}{\varepsilon_1 A_1} + \dfrac{1}{A_1F_{12}} + \dfrac{(1-\varepsilon_2)}{\varepsilon_2 A_2}} = \frac{\sigma\left(T_1^4 - T_2^4\right)}{\dfrac{(1-\varepsilon_1)}{\varepsilon_1 A_1} + \dfrac{1}{A_1F_{12}} + \dfrac{(1-\varepsilon_2)}{\varepsilon_2 A_2}} \tag{7.140}$$

Figure 7.14(b) shows the radiation network for three surfaces that exchange heat only among themselves. Here, the heat exchange between bodies 1 and 2, and that between bodies 1 and 3, are given as,

$$q_{1-2} = \frac{J_1 - J_2}{1/A_1F_{12}} \quad q_{1-3} = \frac{J_1 - J_3}{1/A_1F_{13}}$$

The derivations for equations for radiation heat exchange for other typical situations, such as surfaces with large areas and insulated surfaces, infinitely long parallel planes, two concentric cylindrical surfaces, concentric spheres, a small object situated in a large enclosure, etc., are available in standard textbooks on heat transfer. The final heat transfer equations for some such situations are given in Table 7.4.

Example 7.20 The view factor of a large cylinder (20 cm diameter × 60 cm length) from a coaxial smaller cylinder (10 cm diameter x 60 cm length) is 0.34. The view factor of the larger cylinder of itself (concave inner surface) is 0.25. What is the view factor of the larger cylinder with respect to either annular end?

SOLUTION:

Denoting the subscripts 1, 2, 3, and 4 for the surface of the inner cylinder, outer cylinder, and annular ends, respectively.

$$F_{12} = 0.34, \ F_{22} = 0.25, \text{ and } A_3 = A_4 = \pi\left(10^2 - 5^2\right)$$

$$= 235.5 \text{cm}^2$$

$$A_1F_{12} = A_2F_{21}$$

Considering $A = \pi d.L$, for the same length of the pipe, $10 \times 0.34 = 20 \times F_{12}$
i.e., $F_{12} = 0.17$

$$F_{21} + F_{22} + F_{23} + F_{24} = 1$$

As, $F_{23} = F_{24}$,

$$F_{23} = (1 - F_{21} - F_{22})/2 = \frac{(1 - 0.17 - 0.25)}{2}$$

$$= 0.58/2 = 0.29$$

Example 7.21 The heating surface of an oven has an emissivity of 0.7 with 0.1 m² surface area and is maintained at 280°C. The view factor of this surface with respect to a piece of bread of 0.01 m² surface area is 0.05. If the bread has emissivity of 0.3 and receives 10 W of energy through radiation from the heating surface, what will be the

TABLE 7.4

Equations for radiation heat exchange

Condition	Net radiation heat transfer	
Surfaces with large areas and insulated surfaces	$q_{net} = \dfrac{\sigma A_1(T_1^4 - T_2^4)}{\dfrac{A_1 + A_2 - 2A_1F_{12}}{A_2 - A_1(F_{12})^2} + \left(\dfrac{1}{\varepsilon_1} - 1\right) + \dfrac{A_1}{A_2}\left(\dfrac{1}{\varepsilon_2} - 1\right)}$	(7.141)
Infinitely long parallel planes	$\dfrac{q}{A} = \dfrac{\sigma(T_1^4 - T_2^4)}{\dfrac{1}{\varepsilon_1} + \dfrac{1}{\varepsilon_2} - 1}, A_1 = A_2$	(7.142)
Two concentric cylinders	$q_{net} = \dfrac{\sigma A_1(T_1^4 - T_2^4)}{\dfrac{1}{\varepsilon_1} + \dfrac{A_1}{A_2}\left(\dfrac{1}{\varepsilon_2} - 1\right)}$	(7.143)
Two concentric infinite cylinders	$q_{net} = \dfrac{\sigma A_1(T_1^4 - T_2^4)}{\dfrac{1}{\varepsilon_1} + \left(\dfrac{1}{\varepsilon_2} - 1\right)\left(\dfrac{r_1}{r_2}\right)}$ with $\dfrac{A_1}{A_2} = \left(\dfrac{r_1}{r_2}\right); \dfrac{r_1}{L} \to 0$	(7.144)
Small object situated in a large enclosure	$q_{net} = \sigma A_1 \varepsilon_1(T_1^4 - T_2^4)$, for $\dfrac{A_1}{A_2} \to 0$	(7.145)
Concentric spheres	$q_{net} = \dfrac{\sigma A_1(T_1^4 - T_2^4)}{\dfrac{1}{\varepsilon_1} + \left(\dfrac{1}{\varepsilon_2} - 1\right)\left(\dfrac{r_1}{r_2}\right)^2}$, with $\dfrac{A_1}{A_2} = \left(\dfrac{r_1}{r_2}\right)^2$	(7.146)

For concentric cylinders and spheres, the subscript 1 denotes the inner one.

steady-state bread surface temperature? Take the Stephan-Boltzmann constant as 5.67×10^{-8} W.m^{-2}.K^{-4}.

SOLUTION:

We use the equation 7.140

$$q_{net} = \dfrac{\sigma\left(T_1^4 - T_2^4\right)}{\dfrac{(1-\varepsilon_1)}{\varepsilon_1 A_1} + \dfrac{1}{A_1 F_{12}} + \dfrac{(1-\varepsilon_2)}{\varepsilon_2 A_2}} = \dfrac{A_1 \sigma\left(T_1^4 - T_2^4\right)}{\dfrac{(1-\varepsilon_1)}{\varepsilon_1} + \dfrac{1}{F_{12}} + \dfrac{(1-\varepsilon_2)}{\varepsilon_2}\cdot\dfrac{A_1}{A_2}}$$

(A)

Given, emissivity of oven surface, $\varepsilon_1 = 0.7$, emissivity of bread, $\varepsilon_2 = 0.3$; view factor, $F_{12} = 0.05$.

The denominator of the right-hand side of the above equation (A) is

$$\dfrac{(1-\varepsilon_1)}{\varepsilon_1} + \dfrac{1}{F_{12}} + \dfrac{(1-\varepsilon_2)}{\varepsilon_2}\cdot\dfrac{A_1}{A_2} = \dfrac{(1-0.7)}{0.7} + \dfrac{1}{0.05} + \dfrac{(1-0.3)}{0.3}\cdot\dfrac{0.1}{0.01}$$

$$= 43.758$$

Substituting the value in equation (A),

$$10 = 0.1 \times 5.67 \times 10^{-8} \times \left(553.15^4 - T_2^4\right) / 43.758$$

Or, $T_2 = 357.15$ K $= 84.51°$C

7.7.5 Radiation Heat Transfer Coefficient

Since radiation and convection energy transfers often occur simultaneously, it is convenient to put the radiation equation in a form similar to that of convection.

The convective heat transfer can be written as

$$q_{convection} = h_{convection}\cdot A\,(T_w - T_\infty) \qquad (7.147)$$

In similar lines, the radiation heat transfer can be written as

$$q_{radiation} = h_{radiation}\cdot A\,(T_1 - T_2) \qquad (7.148)$$

The sum of the above two will give the total heat transfer, and, hence, considering that the radiation exchanging surface and the fluid temperatures are the same, the following equation can be written for the total heat transfer.

$$q = (h_c + h_r)\,A\,(T_w - T_\infty)$$

It is convenient to use this equation when evaluating combined thermal exchange by natural convection from a hot body and radiation in an enclosure (e.g., a steam pipe in a room). Handbooks often tabulate the combined heat transfer coefficients. It is easy to see that while the coefficient of radiation heat transfer is strongly dependent on the temperature, convection coefficient is not.

Using the equation 7.143,

$$\frac{q}{A_1} = \frac{\sigma(T_1^4 - T_2^4)}{\frac{1}{\varepsilon_1} + \frac{A_1}{A_2}\left(\frac{1}{\varepsilon_2} - 1\right)} = h_r(T_1 - T_2) \quad (7.149)$$

Thus, $h_r = \dfrac{\sigma(T_1^2 + T_2^2)(T_1 + T_2)}{\dfrac{1}{\varepsilon_1} + \dfrac{A_1}{A_2}\left(\dfrac{1}{\varepsilon_2} - 1\right)}$ (7.150)

In the same way the resistance to radiation will be

$$R_{rad} = 1/h_r \quad (7.151)$$

7.8 Mass Transfer

Mass transfer is very important in food processing. All heat transfer processes within a fluid or between different fluids that mix with each other involve mass transfer. Some common examples of mass transfer in food processing are removal of moisture in a drying and evaporation process, osmotic dehydration, evaporative cooling process, dispersion of solids in liquids or mixing of two liquids, liquid-liquid or liquid-solid extraction, distillation, gas absorption, dehumidification, etc. In the case of drying or dehydration, the moisture of the commodity is removed usually by the application of heat. However, as we discuss in subsequent chapters, moisture can be removed from a hygroscopic commodity by changing the vapor pressure gradient without the application of heat. The solid-liquid extraction uses a solvent for extracting a liquid from the solid. The common examples are solvent extraction of oil from oil-bearing materials and extraction of essential oils and herbal products from barks and leaves and caffeine from coffee. Examples of liquid-liquid extraction are separation of aromatics, such as toluene, benzene, and xylene, from gasoline reformate or recovery of penicillin from fermentation broth. Distillation is the process used for fractionation of crude oil and separation of air. The process of gas absorption and stripping involves the use of a solvent for absorbing gases like CO_2 or H_2S from natural gas, stripping of volatile organic gases from water by steam, deodorization of oil in the refining process. The common process of absorption is removal of moisture from air by using silica gel during dehumidification. The ion exchange method is used for demineralization of water and separation of salts. The advanced techniques of membrane separation and super critical fluid separation involve mass transfer processes.

In the following paragraphs, we will discuss some general laws and theories of mass transfer. The theories of separation of the specific mass transfer operations and the equipment used for those will be discussed in subsequent chapters.

7.8.1 Theory of Mass Transfer

In a system, the mass transfer occurs in the direction of reducing concentration level. Like heat transfer, the rate of mass transfer also depends on the driving potential and the resistance. The mass transfer stops when the mass concentration gradient reduces to zero.

Both the mass diffusion on a molecular scale and the bulk mass transport resulting from the convection process are included in mass transfer.

Mass transfer can happen through mass diffusion on a molecular level, which is also known as the *molecular mass transfer*, and through convection, which is also called *convective mass transfer*. These processes are considered and analyzed in analogy to conduction and convection heat transfer.

Fick's law is used to represent the rate of mass diffusion, stating that a component's mass flux is proportional to the concentration gradient.

$$\frac{m_b}{A} = -D\frac{\partial C_b}{\partial x} \quad (7.152)$$

In the above equation, m_b/A (kg.h^{-1}.m^{-2}) is the mass flux, A (m²) is the area through which mass is flowing, D (m².h^{-1}) is the coefficient of mass diffusion, and C_b (kg.m^{-3}) is component B's mass concentration. Equation 7.152 is similar in form to Fourier's heat conduction law and the law of viscosity for Newtonian fluids.

The convective mass transfer is the passage of constituent material(s) between two immiscible fluids through a boundary surface. Here, the components in the mixture move with a considerable velocity. It can be further classified as natural convective or forced convective mass transfer in analogy with heat transfer.

The following equation gives convective mass transfer.

$$\frac{m_b}{A} = h_m\left(\Delta C_b\right) \quad (7.153)$$

Here, m_b/A is the mass flux and h_m is the *mass transfer coefficient* of component "b." This equation is also similar to the equation for convective heat transfer. The mass transfer coefficient depends on the properties of the fluid, flow characteristics of the fluid, the geometry of the transfer system, and the concentration gradient.

If there is a mass transfer between a surface and a fluid, there is the development of a concentration boundary layer. Mass transfer is analogous to momentum and energy transfers. The non-dimensional numbers corresponding to Nusselt and Prandtl numbers of convective heat transfer are the Sherwood and Schmidt numbers, respectively, in the case of mass transfer.

Sherwood (Sh) and *Schmidt* (Sc) numbers are defined as follows.

$$Sh = h_m L / D \quad (7.154)$$

$$Sc = \nu / D \quad (7.155)$$

where L is the characteristic dimension, D is the diffusivity, h_m is the convective mass transfer coefficient, and ν is the kinematic viscosity.

7.9 Heat, Mass, and Momentum Transfer Analogies

7.9.1 Reynolds Analogy

The boundary layer relationships for momentum transfer in case of a flat plate are the same as those for energy transfer for similar boundary conditions if the Prandtl number (Pr) = 1, if the pressure gradient is zero, and if the viscous dissipation can be considered negligible. In addition, there should be no heat source.

For such a case,

$$\text{Stanton number, } St = \frac{Nu}{Re.Pr} = \frac{h_c}{\rho.V.c_p} = \frac{f}{2} \quad (7.156)$$

where, f is the friction factor.

This equation relating heat and momentum transfers is known as Reynold's analogy.

7.9.2 Colburn Analogy

The Colburn analogy is obtained by modifying the above equation and is given below for Prandtl number range of 0.6 to 50.

$$Pr.St^{2/3} = f/2 \quad (7.157)$$

Note that diffusivity, D, plays the same role in the mass transfer equation as thermal diffusivity in the energy equation. Therefore, the analogy for a flat plate between mass and momentum transfer yields:

$$\frac{Sh}{Sc.Re} = \frac{h_m L}{D}.\frac{D}{\nu}.\frac{\nu}{VL} = \frac{h_m}{V} = \frac{f}{2} \quad (7.158)$$

Schmidt number can have values other than one. In that case,

$$\frac{Sh}{Sc.Re}Sc^{2/3} = \frac{f}{2} \quad (7.159)$$

It can now be shown that,

$$\frac{h_c}{h_m.\rho.c_p} = \left(\frac{\alpha}{D}\right)^{2/3} \quad (7.160)$$

The dimensionless number (α/D) is commonly encountered in the field of air conditioning and (α / D)$^{2/3}$, which is also dimensionless, is termed the *Lewis number.*

The above-mentioned analogies permit us to compute heat transfer coefficients if we know the friction factor. Then, the mass transfer coefficient can be obtained from the heat transfer coefficient.

Check Your Understanding

1. A particular wall (12 m × 8 m) of a cold storage room is made up of wood board (5 mm thick), cork board (25 mm layer), and an outer brick layer (10 cm). What will be the interfacial temperatures if the outside and inside temperatures of the wall are held at 40°C and 5°C. The thermal conductivities of brick, wood, and cork board are 0.69 W.m^{-1}. K^{-1}, 0.15 W.m^{-1}. K^{-1}, and 0.043 W.m^{-1}. K^{-1}.

   Answer: 6.536°C and 33.32°C

2. In a particular wall of 5 m height and 18 m length, the doors and windows occupy 30% of the wall area with 2.8 m².K.W^{-1} unit area thermal resistance. The remaining portion is well insulated and has 4.8 m².K.W^{-1} unit area thermal resistance. Find the heat transfer rate when the temperatures of two sides of the wall differ by 30°C.

   Answer: 683.04 W

3. A 6 cm insulation layer covers a 20 cm diameter metallic duct having 2 mm wall thickness. The temperature at the duct's exterior surface is 2°C and that at the insulated outer surface is 40°C. The thermal conductivity of the insulation material is 0.04 W.m^{-1}. K^{-1} and that for the metallic duct is 65 W.m^{-1}.K^{-1}. Assuming steady state heat flow, find the rate of heat conduction through the insulation per meter length of duct.

   Answer: 20.64 W

4. Liquid flow is turbulent when flowing through a double-pipe thermal energy exchanger. Steam condenses on the outside of the inner pipe such that the wall temperature remains constant. The coefficient of thermal energy exchange is 1200 W.m^{-2}.K^{-1} for the interior of the inner pipe. Determine the coefficient of convective thermal energy exchange for the condition when the rate of liquid flow through the pipe is doubled.

   Answer: 2090.59 W.m^{-2}.K^{-1}

5. If thermal conductivity, mass diffusivity, equimolar mass transfer coefficient based on concentration gradient, density, and specific heat capacity of air are 0.03 W.m^{-1}.K^{-1}, 2.4×10^5 m².s^{-1}, 0.3 m.s^{-1}, 1.0 kg.m^{-3} and 1.0 kJ.kg^{-1}.K^{-1}, respectively, find the convective heat transfer coefficient of air.

   Answer: 348.12 W.m^{-2}.K^{-1}

6. Peas having an average diameter of 6 mm are blanched to give a temperature of 85°C at the center. The initial temperature of peas is 15°C and the temperature of the hot water blancher is 95°C. The thermal conductivity, specific heat, and density of peas are 0.35 W.m^{-1}.K^{-1}, 3.3 kJ.kg^{-1}.K^{-1}, and 980 kg.m^{-3}, respectively. The heat transfer coefficient is 1200 W.m^{-2}.K^{-1}. If the value of the Fourier number (Fo) is 0.32, what will be the time of blanching?

   Answer: 26.6 s

7. A metal wire of 0.01 m diameter and thermal conductivity 200 W.m^{-1}.K^{-1} is exposed to a fluid stream with a convective heat transfer coefficient of 100 W.m^{-2} K^{-1}. What is the value of the Biot number?

   Answer: 0.0125

8. For a laminar flow of fluid in a circular tube, the convective heat transfer coefficient is h_1 at a velocity v_1. If the fluid properties remain constant and the velocity is reduced by half, what will be the value of the convective heat transfer coefficient?

   Answer: $0.794\ h_1$

9. In the case of laminar forced convection over a flat plate, at a specific stream velocity, v, the heat transfer coefficient has been found to be h_1. If the free stream velocity becomes $2 \times v$, what will be the value of the heat transfer coefficient (h_2)?

   Answer: $h_2 = \sqrt{2} \times h_1$

10. A metallic ball ($\rho = 2700$ kg.m^{-3} and $c_p = 0.9$ kJ. kg^{-1}°C^{-1}) of diameter 7.5 cm is allowed to cool in air at 25°C. When the temperature of the ball is 125°C, it is found to cool at a rate of 4°C per minute. Find the heat transfer coefficient. Ignore the thermal gradients inside the ball.

    Answer: 20.3 W.m^{-2}.K^{-1}

11. A container having a volume 282.7 cm^3 and a total surface area 245 cm^2 is completely filled with milk whose initial temperature is 25°C. The continually stirred milk container is suddenly exposed to a steam bath at 100°C. The overall heat transfer coefficient between steam and milk is 1136 W.m^{-2}.K^{-1}. The properties of milk are specific heat capacity = 3.9 kJ. kg^{-1}.K^{-1}, thermal conductivity = 0.54 W.m^{-1}.K^{-1}, and density = 1030 kg.m^{-3}. What is the time required to heat the milk up to a temperature of 85°C? Neglect the thermal resistance and heat capacity of the container walls.

    Answer: 65.7 seconds

12. Two small parallel plane square surfaces, each measuring 4mm $\times$ 4 mm, are placed 0.5 m apart (center to center) with a 30° angle between the radial distance and both the surface normals. What is the view factor between the two surfaces?

    Answer: 1.53×10^{-5}

13. There is a radiation heat exchange inside an annulus between two very long concentric cylinders. The radius of the outer cylinder is R_0 and that of the inner cylinder is R_i. What is the radiation view factor of the outer cylinder onto itself?

    Answer: $1 - (R_i / R_o)$

14. A solid sphere of radius $r_1 = 20$ mm is placed concentrically inside a hollow sphere of radius $r_2 = 30$ mm. What will be the view factor F_{21}?

    Answer: $F_{21} = 4/9$

Objective Questions

1. Explain why
   (a) a metal ball cools faster when kept in chilled water as compared to when kept at normal room temperature.
   (b) a green colored body heats faster under sun than a white colored body.
   (c) heat transfer in moist grain is more than that in dry grain.

2. Differentiate between
   (a) Steady state and unsteady state heat transfer;
   (b) Thermal conductivity and thermal diffusivity;
   (c) Black body and grey body;
   (d) Schmidt number and Sherwood number;
   (e) Emissivity and absorptivity.

3. Write short notes on
   (a) Fourier's law of heat conduction;
   (b) Unit area thermal conductance;
   (c) Biot number;
   (d) Time constant;
   (e) Stefan-Boltzmann law;
   (f) Radiation shape factor;
   (g) Fick's law of diffusion;
   (h) Reynolds analogy for heat, mass and momentum transfer.

4. Derive the expression for
   (a) Thermal resistance for conduction and convention in series;
   (b) Temperature field in a homogenous wall;
   (c) Heat transfer from one fluid to other through a pipe;
   (d) Unsteady state heat conduction with internal heat generation;
   (e) Time required for heating of a fluid from t_1 to t_2 in a jacketed kettle by using steam at 100°C;
   (f) Net heat transfer by radiation in between two infinitely long parallel planes.

BIBLIOGRAPHY

Albright, L.D. 1990. *Environment Control for Animals and Plants*. ASAE, St. Joseph, MI.

Amato, W.S. and Tien, C. 1960. Free convection heat transfer from isothermal spheres in water. *International Journal of Heat and Mass Transfer* 15: 327–339.

Brennan, J.G. (Ed.) 2006. *Food Processing Handbook*. 3rd ed. WILEY-VCH Verlag GmbH & Co. KGaA, Weinheim.

Carslaw, H.S. and Jaeger, J.C. 1959. *Conduction of Heat in Solids*, 2nd ed. Oxford University Press, Oxford. .

Carson, J.K. 2006. Review of effective thermal conductivity models for foods. *International Journals of Refrigeration* 29(6): 958–967.

Charm, S.E. 1978. *Fundamentals of Food Engineering*, 3rd Edn. AVI Publ. Co., Westport, CT.

Chen, C.S. 1985a. Thermodynamic Analysis of the Freezing and Thawing of Foods: Enthalpy and Apparent Specific Heat. *Journal of Food Science* 50(4):1158–1162.

Choi, Y., Okos, M.R. 1986. Effects of temperature and composition on the thermal properties of foods. In Le M. Maguer, &

P. Jelen (Eds.). *Food Engineering and Process Applications, Vol. 1: Transport Phenomena*, Elsevier, New York, 93–101.

Churchill, S.W. 1983. Free convection around immersed bodies. In: G.F. Hewitt (ed.) *Heat Exchanger Design Handbook*. Hemisphere Publishing Corp., Washington, 2.5.7-24.

Churchill, S.W. and Chu, H.S. 1975a. Correlating equations for laminar and turbulent free convection from a horizontal cylinder. *International Journal of Heat and Mass Transfer* 18: 1049–1053.

Churchill, S.W. and Chu, H.H. 1975b. Correlating equations for laminar and turbulent free convection from a vertical plate. *International Journal of Heat and Mass Transfer 18*, 1323–1329.

Earle, R.L. and Earle, M.D. 2004. *Unit Operations in Food Processing*. Web Edn. The New Zealand Institute of Food Science & Technology, Inc., Auckland. https://www.nzifst.org.nz/ resources/unitoperations/index.htm

Fellows, P.J. 2000. *Food Processing Technology*. Woodhead Publishing, Cambridge, UK.

Fricke, B.A. and Becker, B.R. 2001. Evaluation of Thermophysical Property Models for Foods. *HVAC&R Research* 7(4): 311–330.

Geankoplis, C.J. 2004. *Transport Processes and Separation Principles*. 4th ed. Prentice-Hall of India, New Delhi.

Gupta, C.P. and Prakash, R. 1994. *Engineering Heat Transfer*. Nem Chand and Bros., Roorkee.

Hallstrom, B., Skjoldebrand, C. and Tragardh, C. 1988. *Heat Transfer and Food Products*. Elsevier Applied Science, London.

Heldman, D.R. and Singh, R.P. 1981. *Food Process Engineering*. AVI Publishing Co., Westport, CT.

Henderson, S.M. and Perry R.L. 1955. *Agricultural Process Engineering*. John Wiley & Sons, New York.

Holman, J.P. 1997. *Heat Transfer*. 8th ed. Mc-Graw Hill, New York. .

https://www.researchgate.net/publication/233303640_Evaluation _of_Thermophysical_ Property_Models_for_Foods [accessed May 28 2022].

https://www.researchgate.net/publication/327690409_Review _of_the_Thermo-physical_properties_models_of_foods [accessed May 28 2022].

Jaffer, A. 2023. Natural Convection heat transfer from an isothermal plate. *Thermo* 3(1): 148–175.

Kessler, H.G. 2002. *Food and Bio Process Engineering*. 5th ed. Verlag A Kessler Publishing House, Munchen, Germany.

Khurmi, R.S. 2006. *Refrigeration and Air Conditioning*. S Chand and Co. Ltd., New Delhi.

Knudsen, J.D. and Katz, D.L. 1958. *Fluid Dynamics and Heat Transfer*. McGraw-Hill, New York.

Kumar, D.S. 2008. *Engineering Thermodynamics*. S.K. Kataria & Sons, Delhi.

Rahman, M.S., Chen, X.D. and Perera, C.O. 1997. An Improved Thermal Conductivity Prediction Model for Fruits and Vegetables as a Function of Temperature, Water Content and Porosity. *Journal of Food Engineering* 31:163–170.

Rao, D.G. 2009. *Fundamentals of Food Engineering*. PHI Learning, New Delhi.

Sahay, K.M. and Singh, K.K. 1994. *Unit Operations in Agricultural Processing*. Vikas Publishing House, New Delhi.

Schwartzberg, H.G. 1976. Effective Heat Capacities for the Freezing and Thawing of Food. *Journal of Food Science* 41(1):152–156.

Siebel, J.E. 1892. Specific Heat of Various Products. *Ice and Refrigeration* 2: 256–257.

Singh, R.P. and Heldman, D.R. 2014. *Introduction to Food Engineering*. Academic Press, San Diego, CA.

Toledo, R.T. 2007. *Fundamentals of Food Process Engineering*. 3rd Edn. Springer, New York.

Watson, E.L. and Harper, J.C. 1989. *Elements of Food Engineering*. 2nd ed.. Van Nostrand Reinhold, New York.

Whitaker, S. 1972. Forced convection heat transfer correlations for flow in pipes, past flat plates, single cylinders, single spheres, and for flow in packed beds and tube bundles. *AIChE Journal* 18(2): 361–371.

8

Refrigeration and Air Conditioning

The ability to cool and bring down the temperature of an object or an enclosure to a desired extent have been revolutionary developments as far as human comfort and food processing are concerned. The ability to lower the temperatures of food products has made it possible to store them for longer periods, transport them over longer distances, and maintain their quality and safety until their consumption. There are several temperature reduction technologies. We will discuss the basic principles and equipment in this chapter.

8.1 Basic Principles

Refrigeration is the process of temperature reduction and its maintenance for an object or enclosure to a level that is lower than the surroundings. It takes place by picking up the heat from the system being refrigerated and moving it into a system that exists at a higher temperature. Since heat flows by itself from a higher temperature to a lower temperature, the refrigeration process is opposite to that of the heating process. The second law of thermodynamics suggests that work needs to be done to move the heat from the reservoir at a lower temperature to the one at a higher temperature. Similar to the case of pumping power for raising a fluid as a function of depth, the amount of work done will increase with an increasing temperature difference between the two reservoirs.

> The *zeroth law of thermodynamics* states that if two bodies are each in thermal equilibrium with some third body, then they are also in equilibrium with each other. The *first law of thermodynamics* is a version of the principle of conservation of energy and states that heat, being a form of energy, cannot be created or destroyed. The *second law of thermodynamics* states that for a spontaneous process, the entropy of an isolated system increases. The second law helps in calculating the achievable thermal efficiency of a heat engine.

Air conditioning is defined as the simultaneous modification and control of temperature, moisture content, cleanliness, odor, and circulation of air in an enclosure. The control of pressure is also now considered a component of air conditioning. Depending on the set of desired conditions, air may require either heating or cooling for temperature control and either humidification or dehumidification for the control of moisture. Similarly, odor and circulation also need to be controlled to achieve the desired conditions. Since temperature is the most important parameter to be controlled in air conditioning, it is often referred to as refrigeration. Obviously, refrigeration and

air conditioning are related, but they are not similar. Air conditioning may be intended for either human comfort or safe and effective storage of food materials and products.

As the term refrigeration defines it, heat needs to be moved from the object being refrigerated to an object at higher temperature, and mechanical work is required to be done for this heat transfer. Removal of heat from an object lowers its temperature. A mechanical system carries out the movement of heat and uses a fluid, called *refrigerant*, to pick the heat up from the object being refrigerated and discharge it into a practically infinite sink at a higher temperature in a continuous manner repeating a cycle of steps. This cycle is called the *refrigeration cycle*. The heat of the refrigerated object is designated as the source and the place where the heat is discharged is called the sink. The sink of heat is assumed to be almost infinite times larger than the finite source. The refrigerant in the refrigeration cycle soaks up the heat from the source. Here the form of the refrigerant is liquid. The refrigerant boils upon picking the heat up from the source known as the evaporator, i.e., the space where the material is being refrigerated, and changes the state from liquid to vapor.

There are two ways in which the refrigeration cycle can function. One way is *vapor compression refrigeration cycle* and the other is *vapor absorption refrigeration cycle*. The two refrigeration cycles are now explained.

8.2 Refrigeration Systems Based on Vapor Compression Cycle

The low temperature refrigerant vapor at low pressure, after picking up the heat from the evaporator, goes to a compressor where the vapor is compressed, turning the gas into liquid at high temperature and pressure. Mechanical work is done by the compressor. This high temperature refrigerant at high pressure now passes through a condenser, where it loses heat to the sink. Heat transferred to the sink is the sum of heat picked up from the evaporator and the amount of work done by the compressor. The refrigerant goes through the expansion valve to become the refrigerant vapor with reduced temperature and pressure and ready to pick the heat up from the evaporator. Figure 8.1 is a schematic representation of the different components of a vapor compression refrigeration cycle.

8.2.1 Refrigeration Load and Coefficient of Performance

While undergoing a thermodynamic cycle, the first law of thermodynamics requires that net heat supplied to a system by its surroundings needs to be equal to the net work done by the system on the surroundings, i.e., the energy balance on

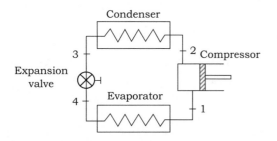

FIGURE 8.1 Schematic representation of a vapor compression refrigeration system

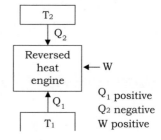

FIGURE 8.2 Energy flow to a system

the system under a steady-state condition (Figure 8.2) can be written as

$$Q_1 + Q_2 + W = 0 \qquad (8.1)$$

$$\text{or, } Q_1 + W = -Q_2$$

Obviously, Q_1 is the heat extracted from the object being refrigerated and the work done to extract this heat is W. The ratio of the refrigeration effect created (Q_1) and the energy expended to achieve the refrigeration (W) is known as the refrigeration cycle's *coefficient of performance* (COP).

$$\text{Coefficient of performance (COP)} = \frac{\text{Total refrigeration effect } (Q_1)}{\text{Energy expended (W)}}$$

$$(8.2)$$

Refrigeration effect (Q_{rf}) is defined as the heat extracted per unit quantity of refrigerant in the evaporator and is expressed in J.kg^{-1} or kJ.kg^{-1}.

$$Q_{rf} = (h_{out} - h_{in}) \qquad (8.3)$$

where h_{out} is the refrigerant's enthalpy at the evaporator exit, and h_{in} is the enthalpy at the entrance of the evaporator.

The *refrigerating load* (Q_{rl},) then is the refrigeration effect per unit time. It is represented as follows:

$$Q_{rl} = \dot{m}_r(h_{out} - h_{in}) \qquad (8.3a)$$

where $\dot{m}_r$ is the refrigerant's mass flow rate (kg.s^{-1}) through the evaporator. It is now convenient to define refrigerating capacity (J.s^{-1} or W) as the actual rate of heat extraction in the evaporator by the refrigerant.

Refrigeration capacity of a system is expressed as *ton of refrigeration* (TR). One ton of refrigeration, denoted as 1.0 TR, is the cooling effect provided by melting of one short ton (2000 lb) of ice in 24 hours. This definition of refrigeration capacity translates to different units as follows. Latent heat of melting is taken as 144 BTU.lb^{-1}.

$$1.0 \text{ TR} = (2000 \times 144) / (24 \times 60) = 200 \text{ BTU.min}^{-1}$$

$$\text{Since 1 BTU} = 1055 \text{ J, 1 TR} = (200 \text{ BTU} \times 1055)/60$$
$$= 3.5167 \text{ kW}$$

A refrigeration system's coefficient of performance (COP) is a dimensionless number. Therefore, both the refrigeration effect and the work input must be in the same units.

It has been proven that the highest value of COP for the refrigeration cycle operating between the temperatures of source and sink, T_1, and T_2, respectively, can only be for the Carnot refrigeration cycle.

The Carnot cycle is completely reversible and is, therefore, the ideal cycle. A real refrigeration cycle will, however, not be as efficient as the Carnot refrigeration cycle. The ideal refrigeration cycle consists of a compressor, a condenser, an expansion valve, and an evaporator. The Carnot cycle's coefficient of performance can now be written as follows:

$$COP_{carnot} = \frac{\text{Refrigeration effect}}{\text{Work done}} = \frac{T_1}{T_2 - T_1} \qquad (8.4)$$

Note that the Carnot cycle's COP is dependent on the temperatures of the source and the sink only. Moreover, COP increases with the increase in the source temperature and with the reduction in the sink temperature. Another interpretation is that the COP increases when the difference in the temperatures of the source and the sink decreases. A real refrigeration cycle's COP will always be lower than that of the Carnot cycle due to more work required to be done by the compressor and no work recovered from the expansion valve, therefore increasing the denominator for the same numerator value.

8.2.2 Subcooling and Superheating

The temperature of the condensed refrigerant can be reduced to a level below the saturation temperature for the liquid refrigerant's condensing pressure. This practice, known as *subcooling*, results in an increase in the refrigeration effect. The extent of subcooling depends upon the coolant employed during condensation, i.e., surface water, well water, atmospheric air, etc. It is also affected by the condenser's capacity and construction.

The subcooled refrigerant's enthalpy is represented as follows:

$$h_{sc} = h_{s,con} - c_{pr} (T_{s,con} - T_{sc}) \qquad (8.5)$$

where h_{sc} is the enthalpy of the subcooled liquid refrigerant, $h_{s,con}$ is the enthalpy of the saturated liquid refrigerant at the condensing temperature, c_{pr} is the specific heat of the refrigerant at constant pressure, $T_{s,con}$ is the saturation temperature of the liquid refrigerant at the condensing pressure, and T_{sc} is the temperature of the subcooled refrigerant. Enthalpy h_{sc} is almost equal to the enthalpy of the saturated liquid refrigerant at the subcooled temperature.

Superheating of the refrigerant helps in preventing slugging damage to the compressor. The degree of superheating depends on such factors as the construction of evaporator, the type of compressor, and the refrigerant properties.

Enthalpy of a refrigerant at various points in the refrigeration system could be estimated using polynomials. Enthalpy of a refrigerant at saturated liquid and saturated vapor state depends upon the saturation temperature or pressure, i.e., the saturated temperature and the saturated pressure for the refrigerant are interdependent. Therefore, it is expedient to estimate the refrigerant enthalpies from a relation of the following form.

$$h = f(T_s) \tag{8.6}$$

The enthalpy differences along the constant entropy lines can be estimated in close temperature ranges by a linear function as follows:

$$h_2 - h_1 = f(T_{s2} - T_{s1}) \tag{8.7}$$

The following polynomial can be used for estimating enthalpies of saturated liquid R-22 refrigerant in the saturated temperature range of -7°C to 50°C.

$$h_{lr} = 10.409 + 0.268 \times T_{sl} + 1.48 \times 10^{-4} \times T_{sl}^2 + 5.343 \times 10^{-7} \times T_{sl}^3 \tag{8.8}$$

For saturated vapor refrigerants (R-22) for its saturated temperatures in the same range, the similar polynomial can be as follows:

$$h_{vr} = 104.465 + 0.98445 \times T_{sv} - 1.226 \times 10^{-4} \times T_{sv}^2 \\ -9.861 \times 10^{-7} \times T_{sv}^3 \tag{8.9}$$

The coefficients in the above equations have been derived from the data available in ASHRAE tables and charts for refrigerants. Enthalpy changes for isentropic compression process of refrigerant between the initial and final states along the constant entropy line can also be estimated using a similar polynomial.

$$h_2 - h_1 = c_1 + c_2(T_{s2} - T_{s1}) + c_3(T_{s2} - T_{s1})^2 + c_4(T_{s2} - T_{s1})^3 \tag{8.10}$$

where subscripts s, 1, and 2 refer to saturated condition, compressor inlet, and compressor outlets, respectively. The values of c_1, c_2, c_3, and c_4 have been found to be -0.18165, 0.21502, -1.24×10⁻³, and 8.198 ×10⁻⁶ for R-22. In all the above equations the constants are specific to the specific refrigerant.

8.2.3 Multistage Refrigeration System

A multistage refrigeration system may have

1. Several compressors connected in series;
2. A combination of a high-stage and a low-stage compressor;
3. A combination of two separate refrigeration systems;
4. Single motor or prime mover-driven two or more impellers connected internally in a series.

Multi-stage refrigeration systems as compared to single-stage systems are more expensive and complicated, but there are advantages in terms of enhanced refrigeration effect and greater flexibility to ameliorate load fluctuations.

8.3 Refrigeration Systems Based on Vapor Absorption Cycle

A *vapor absorption cycle* (VAC), like a vapor compression cycle (VCC), has evaporation, condensing, compression, and expansion components. The two cycles differ in the way the vapor from the evaporator is processed. In the case of VAC, vapor first goes to an absorber, where it is absorbed in a liquid, and the solution of the vapor and the liquid then goes to a regenerator before going to the condenser. Ammonia, water, and lithium bromide have served as the common refrigerants.

Let ammonia be the refrigerant. The liquid ammonia enters the evaporator coils, picks up the heat there, and exits the evaporator as low-pressure gas. This gas now enters the absorber where water is present as the absorbing liquid as a weak ammonia solution. Now, the ammonia coming from the evaporator gets absorbed in the weak water-ammonia mixture, turning into a strong solution. In the process, the solution temperature increases, subsequently reducing the ammonia absorbing capacity of the solution. Slurry temperature is maintained in optimum range by adding cold water. The strong ammonia-water solution is now pumped to the regenerator where the water and ammonia turn into vapors by heating the solution with external sources such as steam, hot water, etc. On the top of the generator is an analyzer that picks up the ammonia vapors, leaving water vapors behind. The condensed water vapors are sent back to the generator. Any water vapor entrapped in the ammonia vapor has the potential to reduce the efficiency of refrigeration and even damage the system. The high temperature ammonia vapors at high pressure coming from the generator through the analyzer now enter the water-cooled or air-cooled condenser where the ammonia gas loses its heat and turns liquid. The high-pressure liquid ammonia then passes through the expansion valve, further cooling the ammonia, reducing its pressure, and getting converted to gaseous form. Thus, the refrigeration cycle is complete. It may be noted in the case of VAC that there is no mechanical compressor requiring power. An external source supplies heat to the generator, resulting in high pressure and high temperature.

The entire vapor absorption refrigeration cycle can now be recapitulated and its components can be listed as evaporator,

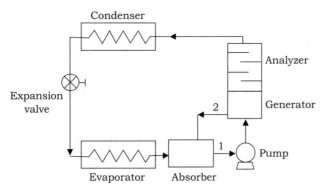

1: Strong solution, 2: weak solution

FIGURE 8.3 Schematic diagram of refrigeration system based on vapor absorption cycle

absorber, pump, generator, analyzer, condenser, and expansion valve. The vapor absorber cycle for refrigeration is shown in Figure. 8.3

The performance of a refrigerator working on the principle of vapor absorption cycle can be expressed as follows:

$$COP_{vars} = \frac{Q_e}{Q_g + W_p} \approx \frac{Q_e}{Q_g} \tag{8.11}$$

In this equation, Q_e is the heat extracted in the evaporator, Q_g is the heat supplied to the generator, and W_p is the work done by the pump.

If the condenser and the absorber reject the heat at temperature T_o, the generator receives heat at temperature T_g, and the evaporator temperature is T_e, then the maximum COP for the vapor absorption refrigeration system is represented as follows:

$$COP_{ideal.vars} = \frac{Q_e}{Q_g} = \frac{T_e}{T_0 - T_e} \times \frac{T_g}{T_g - T_0} \tag{8.12}$$

$$\text{i.e. } COP_{ideal.vars} = COP_{Carnot} \times \eta_{Carnot} \tag{8.13}$$

8.4 Air Conditioning

Air conditioning is the process of treating the air in an enclosure such that it is comfortable to the intended user of the enclosure, whether people, animals, or food. Often the most significant change is the change in temperature through either refrigeration or heating. In addition, the humidity regime in the enclosure also affects the degree of comfort. Air circulation and odor become critical in some cases. The space to be air conditioned is like the evaporator in a refrigeration system, the cooling load for air conditioning of the enclosure becomes the refrigeration load of the refrigeration system meant for the air conditioning.

The air conditioning load for the enclosure consists of external and internal loads. External loads are those that enter into the space being air conditioned through the enclosure envelop, i.e., conduction through walls, floor, and roof; solar and sky radiations; and infiltration-exfiltration. Internal loads include the rest, i.e., loads on account of human activity, product, lights, and other electro-mechanical appliances. Estimation of air conditioning load requires information about the location, site, and weather data. Operational information such as activity in the air-conditioned space, conditioned air attributes, frequency of door opening/closing, number of lights and their wattages, and number and specifications of any machinery.

The outdoor climatic conditions chosen for the load estimation represent long-term conditions representative of the location. These conditions are found in ASHRAE handbooks and many other publications. Solar loads chosen represent the clear sky solar radiation values for the selected month. The internal space occupancy considered for design is full occupancy. Both sensible and latent heat loads form the total load. Peak air conditioning load is used to design the refrigeration system.

8.5 Cooling Loads

The cooling loads for a refrigeration or air conditioning system need to be calculated to determine the amount of heat removal and, thus, the size of the equipment. In addition to the heat to be removed from the product (which is also known as the product load), there are also external loads due to heat transfer through roof and walls, floor, load due to electrical appliances, solar radiation, infiltration/exfiltration, load due to persons working in the area, etc. A brief discussion follows.

Heat transfer through roof and walls. It is essentially a conductive load represented as

$$q = U A \Delta T \tag{8.14}$$

ΔT is the difference in the temperatures on the outside and the inside of the air-conditioned space. The U and A correspond to the total roof area or wall area under consideration.

Heat transfer through floor. The floor of the space being conditioned exchanges heat with the deep ground and the ambient air. The temperature deeper in the ground remains almost constant year-round. The perimeter portion of the floor exchanges heat with its ambient environment. Therefore, the cooling load estimation with respect to the floor needs to consider these conditions. The cooling load is calculated as follows:

$$q = (U A \Delta T)_{deep\ ground} + (f P \Delta T)_{perimeter} \tag{8.15}$$

The deep ground ΔT is the difference between the air-conditioned space and the deep ground. The perimeter ΔT is between the air-conditioned space and the ambient air. P is the perimeter of the air-conditioned space and f is the perimeter heat transfer coefficient ($W.m^{-1}. K^{-1}$).

Solar load. The solar energy incident on the exterior surfaces of the roof and opaque walls is partly reflected from these surfaces and partly absorbed. The effect of solar energy absorption is taken into account through the concept of sol-air temperature. The sol-air temperature then represents the design outdoor air temperature. Sol-air temperatures ($T_{sol-air}$)

can be found either tabulated in heating, ventilation, and air conditioning (HVAC) handbooks or calculated as follows:

$$T_{sol-air} = T_{amb} + (\alpha.I_s - \varepsilon\,\Delta T) / h_{amb} \qquad (8.16)$$

where T_{amb} is the temperature of the outdoor air, α is the solar absorptivity of the surface, I_s is the incident solar radiation, ε is thermal emissivity of the surface, and ΔT is the temperature difference corresponding to thermal radiation exchange with the surroundings. The two terms within the brackets represent the net radiation, i.e., solar radiation input minus the radiation loss to the surroundings.

If the enclosure being air conditioned has some non-opaque surface capable of transmitting solar radiation directly into the enclosure, there is an additional term representing this direct solar gain.

$$q_{solar} = A \times Sc \times I_s + [U\,A\,\Delta T]_{solar} \qquad (8.17)$$

The first term on RHS is the direct solar energy input through the non-opaque surface and the second term is the conduction component from that area. Sc is the shading coefficient and I_s is the solar radiation intensity perpendicular to the surface.

Cooling load due to human activity. A person generates about 100 W to 600 W heat (both sensible and latent) depending upon the type of activity, age, and size of the person. The lower value corresponds to a sitting man and the higher value corresponds to a person doing some industrial activity.

Human cooling load, sensible (q_{hsen}) and latent (q_{hlat}), is estimated as follows:

$$q_{hsen} = N\,(h_s)(clf) \qquad (8.18)$$

$$q_{hlat} = N\,(h_l)(clf) \qquad (8.19)$$

where N is the number of people in the space, h_s is the sensible heat generated by a person, h_l is the latent heat generated by a person, and clf is the cooling load factor indicating the fraction of 24 hours that the person is present in the enclosure.

Heat generated by lights. The rate of heat generated by lights (q_{light}) is given as follows:

$$q_{light} = W.F_u.F_{bf}.clf \qquad (8.20)$$

where W is wattage of the lights, F_u is the utilization factor of the lights, F_{bf} is the ballast factor, and clf is the cooling load factor as explained under human activity head. It is a small component of the cooling load, especially, with more efficient LED-based luminaires.

Load due to electro-mechanical appliances. Each appliance needs to be considered for estimating its contributions to the cooling load. The method of estimation of these loads is as follows:

$$q_{ema} = (P \times F_{um} \times F_{lm} \times clf\,)/\eta \qquad (8.21)$$

where P is the power of the appliance, F_{um} is the utilization factor, F_{lm} is the load factor, η is the appliance efficiency, and clf is the cooling load factor.

Heat load by exfiltration-infiltration and ventilation. There may be leakages through the enclosure structure, doors, and windows that contribute to the cooling load. The following equations can be used to calculate the sensible, latent, and total components of the cooling load.

$$q_{i-e-s} = c_{pa} \times \rho \times IER \times \Delta T \qquad (8.22)$$

$$q_{i-e-l} = \lambda \times \rho \times IER \times \Delta W \qquad (8.23)$$

$$q_{i-e-t} = \rho \times IER \times \Delta h \qquad (8.24)$$

where q_{i-e-s}, q_{i-e-l}, and q_{i-e-t} are the sensible, latent, and total cooling load components, respectively. ΔT, ΔW, Δh are differences between outside and inside temperature, humidity ratio, and enthalpies, respectively. c_{pa} is the air specific heat, IER is the infiltration-exfiltration rate, λ is the latent heat of water vapor, and ρ is the air density. For calculating the cooling load due to ventilation, the same equations can be used after replacing IER by ventilation rate (VR).

Product load. If the air-conditioned space is used for cold storage, the product load needs to be calculated and added to the components calculated above to determine the cooling load of the cold storage. The product load has two components, one is the sensible heat load and the other is the load due to the product respiration.

$$q_{ps} = m \times c_{pp} \times (T_{pe} - T_s) \qquad (8.25)$$

$$q_{pres} = m \times r_{res} \qquad (8.26)$$

where T_{pe} is the product temperature entering the cold store, T_s is the cold store temperature, m is the product mass, c_{pp} is the product specific heat, and r_{res} is the product's heat of respiration.

$$
\begin{aligned}
\text{Total cooling load} = \text{Contributions from} \big(&\text{roof} + \text{walls} + \text{floor}\\
&+\text{solar} + \text{human activity} + \text{light} + \text{electro}\\
&-\text{mechanical appliances} + \text{Infiltration}\\
&-\text{exfiltration} + \text{ventilation} + \text{product}\big)
\end{aligned}
$$

8.6 Evaporative Cooling

One of the best examples of evaporative cooling is perspiration cooling of our body. During warm conditions, our body perspires and a thin layer of this perspiration appears on the skin ready to be evaporated. Even a mild breeze takes this perspiration away in vapor form, lowering the skin temperature because the heat for evaporation came from the skin. The principle of evaporative cooling, cooling through evaporation, is to lower the temperature of an object or a space by converting a part of its sensible heat into latent heat. As compared to a vapor compression refrigeration system, evaporative cooling does not involve a lot of external energy to achieve cooling, and water is the refrigerant in evaporative cooling. Since the cooling effect is achieved by converting a part of the system's sensible heat to latent heat, there is no change in the total enthalpy of the system. Therefore, *evaporative cooling* is also known as *adiabatic cooling*.

Evaporative cooling can be either a single-stage or a two-stage process. Consider an air stream, unsaturated with moisture passing through moist pads. The pads are kept moist by recirculation of water over the pads and the water also cools down due to the recirculation to the air's wet bulb temperature. Since the air is unsaturated with moisture, it has the potential to absorb moisture from the wet pads. The moisture picked up from the pads and subsequent vaporization reduces the air's dry bulb temperature in proportion to the rise in its humidity. Obviously, this cooling can continue only to the point where the air steam is saturated and no more potential for evaporation exists. The cooled air can now be used for space cooling or cooling of some objects. The air stream's wet-bulb temperature is the lowest temperature that can be achieved through the process of evaporative cooling. However, since the process of saturation is seldom complete, the efficiency of evaporative cooling (η_{ec}) is less than 100%.

$$\eta_{ec} = \frac{T_{db} - T_{ec}}{T_{db} - T_{wb}} \quad (8.27)$$

where T_{db} is the air stream's dry bulb temperature, T_{ec} is the actual temperature achieved through evaporative cooling, and T_{wb} is the wet bulb temperature. As discussed in Chapter 5, the denominator ($T_{db} - T_{wb}$) is called wet-bulb depression and represents the maximum available cooling potential. Under warm and humid conditions in places such as Kolkata and Chennai, wet-bulb depression is small because of high humidity conditions, and evaporative cooling, as a consequence, is very limited, say 5–10°C. Whereas, wet-bulb depression is quite large in places like Bikaner and Jodhpur because of a high dry-bulb temperature and low relative humidity; therefore, temperature in such places could be reduced by 20–25°C. As discussed in Chapter 5, the evaporative cooling process can be represented on a psychrometric chart on constant enthalpy lines. A total of 100% efficiency is achieved if the cooling process continues up to the 100% relative humidity line.

The above process of evaporative cooling is called a direct single-stage cooling process since the cooled air stream directly enters the application zone. Sometimes the conditioned air is too humid for the application. In such cases, a sensible heat exchanger is required to cool the air by the evaporatively cooled humid air. The cooling process then is called an *indirect cooling* process. The resultant efficiency of the indirect cooling (η_{eci}) is expressed as follows:

$$\eta_{eci} = \frac{T_{db} - T_{ec}}{T_{db} - T_{wbi}} \quad (8.28)$$

where T_{wbi} is the indirectly cooled air's wet-bulb temperature.

It is possible to carry out the evaporative cooling process based on two-stages instead of just one to attain temperatures lower than those attainable through single-stage evaporative

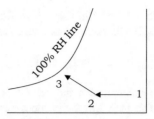

1-2: Indirect evaporative cooling
2-3: Direct evaporative cooling

FIGURE 8.4 Two-stage evaporative cooling process on a psychrometric chart

cooling. The two-stage cooling makes use of both direct and indirect single-stage evaporative processes. In the first stage, the air that needs to be cooled is indirectly cooled by the evaporatively cooled air stream through the heat exchanger. The process continues on a constant humidity ratio line and, as a result, the wet and dry-bulb temperatures of the targeted air come down. The second stage is the direct single-stage process where the previously cooled air now goes through the process of evaporative cooling (Figure 8.4).

The cooling efficiency of the two-stage evaporative cooling process is more than 100% and the air is less humid than that obtained in the single-stage process. The evaporative cooling efficiency of the two-stage process (η_{tsec}) is given as follows:

$$\eta_{tsec} = \frac{T_{db} - T_{ca}}{T_{db} - T_{wb}} \quad (8.29)$$

where T_{ca} is the temperature of the conditioned air.

8.7 Cryogenics

Cryogenics is the science that deals with the production of very low temperatures and their effect on matter. In practical terms, it considers temperatures below 120 K (-153°C). For food applications, temperatures below 223 K (-50°C) or lower are considered cryogenic. *Cryostat* is the system for a low temperature application.

There are two ways of reaching low temperatures; through the use of cryogens or by mechanical coolers. Liquid cryogen-based cryostats use the same principle as the traditional technique of using blocks of ice to keep food cold during storage and transport. With a liquid cryogen, evaporation of the liquid maintains the temperature of the cryostat at the boiling point of the cryogen. An alternative to using liquid cryogens to maintain a cryostat is to use a mechanical cooler. A mechanical cooler here consists of a room temperature compressor, which may be water cooled, and an expansion chamber within the cryostat. The governing principle is that helium gas is compressed at room temperature in the compressor, then sending it into the cryostat where it is expanded to produce the required cooling. The mechanical cooling systems confront issues of high initial and repair costs and vibrations in comparison to the use of cryogens. Common cryogenic gases widely used are hydrogen, nitrogen, oxygen, helium, fluorine, argon, and methane. However, the two common cryogens used in food

TABLE 8.1

Boiling points and expansion ratios of common cryogens

Gas	Boiling point, K	Expansion ratio
Carbon dioxide	194.7	1 to 789
Helium	3.3	1 to 757
Hydrogen	20.5	1 to 851
Methane	109.2	1 to 643
Neon	27.3	1 to 1438
Nitrogen	77.4	1 to 696
Oxygen	100.2	1 to 860

processing are nitrogen and carbon dioxide. Most cryogenic liquids are odorless and colorless when vaporized to gas.

The boiling points and expansion ratios of common cryogens are listed in Table 8.1.

8.8 Common Refrigerants

The fluids for use as refrigerants can be either primary or secondary refrigerants. The working fluids used directly for refrigeration systems based on vapor compression and vapor absorption cycles are known as *primary refrigerants*. As discussed earlier, primary refrigerants undergo phase change in the evaporator to provide refrigeration. The fluids that carry the thermal energy from one location to other without undergoing any phase change are termed *secondary refrigerants*.

Common secondary refrigerants include solutions of water and ethylene glycol, water and propylene glycol, and water and calcium chloride. These solutions are generally called brines or antifreezes, and they can be used when sub-zero temperatures are required. The freezing point of brine is lower than that of water because of the dissolved solutes and the concentration of the dissolved solutes affect the freezing point. *Eutectic point* is the concentration of the brine, which gives the lowest temperature without solidification. Pure water is also used as a secondary refrigerant if the operating temperature is above 0°C.

For a particular application, thermodynamic and thermophysical properties, environment, safety, as well as economics are responsible for the selection of a refrigerant.

Some desirable properties of the refrigerant are as follows:

- The refrigerant's mass flow rate per unit cooling capacity should not be high. Therefore, the selected refrigerant should have high latent heat of vaporization and low boiling point.
- The refrigerant in liquid form should have low specific heat to have a higher degree of sub-cooling and, consequently, a smaller amount of flash gas at the evaporator inlet.
- At the same time for less superheating, the vapor specific heat should be more.
- To reduce the size of compressor, the refrigerant should be in dense vapor form.
- The refrigerant's thermal conductivity in both liquid and vapor phases should be high.

- The refrigerant should have low miscibility with compressor oil.
- Achieving smaller frictional pressure drops requires the viscosities in both liquid and vapor phases to be small.
- It should have low toxicity and should be non-flammable.
- It should be safe for the environment in view of ozone depletion, global warming, etc.

8.8.1 Classification and Nomenclature of Refrigerants

The classification of fluids used as refrigerants is as follows:

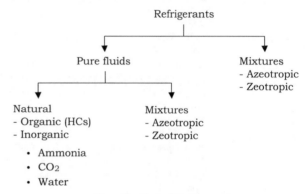

Classification of Refrigerants

A mixture of liquids having similar composition of its constituents in liquid and vapor phases is known as a *azeotropic mixture*, which boils at a constant temperature. The liquids forming azeotropic mixture themselves have different boiling points. As a consequence, the individual components neither evaporate nor condense at the same boiling point. The mixture is said to be in a temperature glide. The phase changes of the liquid components in a *zeotropic mixture* take place at a series of temperatures instead of the same temperature. The main difference between azeotropic and zeotropic mixtures is that while the dew point of an azeotropic mixture intersects the bubble point, the dew point of a zeotropic mixture is quite distinguishable from its bubble point.

A mixture of 95.63% ethanol and 4.37% water (by mass) is an example of an azeotrope; it boils at 78.2 °C. Although ethanol boils at 78.4 °C and water at 100 °C, the mixture boils at 78.2 °C. Note that the azeotropic mixture boils at a temperature lower than those of either of its constituents. The mixture of ethane, methane, nitrogen, propane, and isobutene is a zeotropic mixture.

A refrigerant is commonly designated by a numbering system, where a unique number depending upon the type of refrigerant, its chemical composition, molecular weight, etc. follows the letter R, an abbreviation for refrigerant.

Derivatives of alkanes ($C_n H_{2n+2}$), e.g., CH_4 (methane) and C_2H_6 (ethane), form the category of *fully saturated, halogenated compounds*. These refrigerants are named as R XYZ, where the first digit X represents one atom less than the number of carbon atoms actually present. The middle digit Y represents one more than the number of hydrogen atoms. The last digit Z represents the number of fluorine atoms. The digit X is taken as zero if there are only two digits in the unique number. The balance indicates the number of chlorine atoms.

In this form of nomenclature, R22 means X=0, Y=2, and Z=2, i.e., there is one C, one H, and two Fluorine. The number of (H+F) atoms is 3 (1+2). Therefore, the balance is 4 –3

= 1 and the number of chlorine atoms is one. Thus, R 22 is $CHClF_2$.

Similarly, R12 means X=0, Y=1, and Z=2, i.e., there is one C, no H, and two Fluorine atoms. The balance, 4 – 2= 2, there are two Cl. Thus, R12 is CCl_2F_2.

Using the pattern of the unique number explained above, R134a can be seen to represent $C_2H_2F_4$ (derivative of ethane).

TABLE 8.2

Properties of some common refrigerants

Name of the refrigerant	Properties	Applications	Remarks
R 11 (CFC)	Boiling point = 23.7°C h_{fg} at boiling point =182.5 kJ.kg^{-1} $t_{critical}$ = 197.98°C c_p/c_v = 1.13 ODP = 1.0 GWP = 3500	Earlier used in industrial heat pumps, large air conditioning systems	The use has gradually been phased out.
R12 (CFC)	Boiling point = -29.8°C h_{fg} at boiling point =165.8 kJ.kg^{-1} $t_{critical}$ =112.04°C c_p/c_v = 1.126 ODP = 1.0 GWP = 7300	Earlier used in small cold storages, small air conditioners, domestic refrigerators, and water coolers	The use has gradually been phased out; some old equipment may be using R12.
R22 (HCFC)	Boiling point = -40.8°C h_{fg} at boiling point =233.2 kJ.kg^{-1} $t_{critical}$ =96.02°C c_p/c_v = 1.166 ODP = 0.05 GWP = 1500	Earlier used in air conditioning systems, cold storages	The use is gradually being phased out.
R134a (HFC)	Boiling point = -26.15°C h_{fg} at boiling point =222.5 kJ.kg^{-1} $t_{critical}$ =101.06°C c_p/c_v = 1.102 ODP = 0.00 GWP = 1200	Home refrigerators, automobile A/Cs, water coolers, etc.	Highly hygroscopic. Used commonly in household refrigerators.
R 717 (Ammonia)	Boiling point = -33.35°C h_{fg} at boiling point = 1368.9 kJ.kg^{-1} $t_{critical}$ =133.0°C c_p/c_v = 1.31 ODP = 0.00 GWP = 0.0	Preferred in food processing, cold storages, ice plants, and frozen food cabinets	Highly efficient; excellent heat transfer properties; toxic; flammable; inexpensive; easily available; and causes corrosion in copper pipes
R 744 (CO$_2$)	Boiling point = -78.4°C h_{fg} at 40°C= 321.3 kJ.kg^{-1} $t_{critical}$ =31.1°C c_p/c_v = 1.3 ODP = 0.00 GWP = 1.0	Cold storages, air conditioning systems	Non-flammable and non-toxic; requires considerably higher operating pressures compared to ammonia; safer for use in refrigerated ships; very low critical temperature; inexpensive, eco-friendly; and available
R718 (H$_2$O)	Normal boiling point = 100°C h_{fg} at boiling point = 2257.9 kJ.kg^{-1} $t_{critical}$ =374.15°C c_p/c_v = 1.33 ODP = 0.00 GWP = 1.0	Absorption systems, steam jet systems	Large specific volume; eco-friendly; inexpensive and available
R600a (iso-butane)	Normal boiling point = -11.73°C h_{fg} at boiling point = 367.7 kJ.kg^{-1} $t_{critical}$ =135.0°C c_p/c_v = 1.086 ODP = 0.00 GWP = 3.0	Useful for water coolers as well as home refrigerators	Eco-friendly but inflammable. R-600a has a lower global warming potential compared to the R-134a and, hence, is gaining popularity for use in home refrigerators.

(ODP: Ozone depletion potential; GWP: Global warming potential)

HFC: Hydrofluorocarbon; HCFC: Hydrochlorofluorocarbons; hfg: latent heat of evaporation

The letter "a" denotes an isomer, i.e., molecules having the same chemical composition but a different atomic structure.

Inorganic refrigerants. Ammonia, carbon dioxide, and water are inorganic refrigerants. Any refrigerant of this category is designated by the digit 7 followed by its molecular weight rounded-off to a whole number. For example, ammonia (mol. weight 17), the refrigerant, bears the identification R 717. Water as a refrigerant is R718 and carbon dioxide is R744.

Mixtures. The azeotropic mixtures and non-azeotropic mixtures (also called zeotropic refrigerants) are designated by 500 and 400 series, respectively.

Azeotropic mixtures:

R 500: Mixture of R 12 (73.8 %) and R 152a (26.2%)
R 502: Mixture of R 22 (48.8 %) and R 115 (51.2%)
R503: Mixture of R 23 (40.1 %) and R 13 (59.9%)
R507A: Mixture of R 125 (50%) and R 143a (50%)

Zeotropic mixtures:

R404A: Mixture of R 125 (44%), R 143a (52%) and R 134a (4%)

R407A: Mixture of R 32 (20%), R 125 (40%) and R 134a (40%)

R407B: Mixture of R 32 (10%), R 125 (70%) and R 134a (20%)

R410A: Mixture of R 32 (50%) and R 125 (50%)

Hydrocarbons. Some hydrocarbons are also used as refrigerants, e.g., C_4H_{10} (n-butane) as R 600, C_3H_8 (propane) as R 290, C_4H_{10} (iso-butane) as R 600a, C_3H_6 as R1270, and unsaturated hydrocarbon C_2H_4 as R1150.

8.8.2 Properties of Common Refrigerants

The properties of some common refrigerants are given in Table 8.2.

The refrigerants such as Trichlorofluoromethane (also called Freon-11, CFC-11, or *R-11*), Dichlorodifluoromethane (R-12), and Chlorodifluoromethane or difluoromonochloromethane (also known as HCFC-22, or R-22), which were used earlier have been phased out because of their non-zero ozone-depleting potential (ODP). According to new policy guidelines, the use of chlorine and bromine containing refrigerants is being disallowed because these refrigerants directly affect the ozone depletion potential (ODP).

The CFC and HCFC families of refrigerants are now being replaced by the HFC family of refrigerants. R134a is from this family, which is also known as Tetrafluoroethane (CF_3CH_2F) or 1,1,1,2-Tetrafluoroethane, norflurane (INN), Freon 134a, Forane 134a, Genetron 134a, Florasol 134a, Suva 134a, or HFC-134a.

The R600a has higher latent heat of vaporization than R134a. The temperature of R600a can be easily reduced to a desired level with a small change in pressure. Specific heat (at constant pressure) of R600a is much higher than R134a, though the density is a bit low for R600a. These properties make R 600a a better refrigerant as compared to R134a. However, R600a is highly inflammable. Thus, when using R600a, the refrigerant circuit pipes must be fully secure.

Check Your Understanding

1. Differentiate between the vapor compression refrigeration cycle, and vapor absorption refrigeration cycle.
2. Define (a) COP, (b) ton of refrigeration, (c) Subcooling
3. Name the different heat loads in a cold store for fruits and vegetables. Explain how to calculate the solar load and the load due to electrical appliances.
4. Explain the principle of evaporative cooling of agricultural structures.
5. What are the common cryogens used in food industries? Which one of these is most commonly used and why?
6. Explain the classification and nomenclature of different refrigerants. State the chemical formulae/ commercial names of R 12, R22, R 134a, R 500, R 717, R 718, R 744, and R 600a.
7. What are the common refrigerants used in household refrigerators? State their relative advantages and disadvantages.

BIBLIOGRAPHY

Ananthanarayan, P.N. 2013. *Basic Refrigeration and Air Conditioning* 4th ed. McGraw Hill, New Delhi, p. 720.

Arora, C.P. 2000. *Refrigeration and Air Conditioning.* Tata McGraw Hill, New Delhi.

Arora, R.C. 2010. *Refrigeration and Air Conditioning.* PHI Learning, New Delhi .

Hundy, G.F., Trott, A.R. and Welch, T.C. 2008. *Refrigeration and Air-Conditioning.* 4th ed. Butterworth-Heinemann, Oxford.

Khurmi, R.S. and Gupta, J.K. 1987. *Textbook of Refrigeration and Air Conditioning.* S Chand Publishing, New Delhi.

Prasad, M. 2003. *Refrigeration and Air-Conditioning.* 2nd Edition. New Age International Publishers, New Delhi.

9

Water Activity and Thermobacteriology

It is very important that both raw food and processed food be kept protected from microorganisms. Among the different parameters that control the growth of microorganisms in food, temperature, pH, and water activity are very important. The term *water activity* will be used several times in subsequent chapters, namely, drying, thermal processing and evaporation. In fact, when we reduce the moisture content of a commodity for safe storage, basically we try to reduce the water activity. A lower water activity gives a longer shelf life. During food processing operations, the process engineers consider water activity to decide whether the food is shelf-stable or not. Often, water activity is taken as a critical control point in monitoring food safety. In addition, it is very important to understand the growth and death behavior of microorganisms during heat processing. It helps in designing the thermal processing parameters for a specific commodity. We will discuss how the death rate of microorganisms depends on the processing parameters, i.e., the temperature and time of processing. We will also discuss how the type and population of bacteria decide the processing parameters. This chapter is devoted to a discussion of water activity and thermo-bacteriology.

9.1 Water Activity

9.1.1 Importance of Water Activity in Storage of Food

The shelf life of any agricultural commodity is greatly influenced by its moisture content. However, all the moisture present in the commodity may not support the different biochemical, microbiological, or enzymatic activities that are responsible for food degradation. For example, even though pure ice crystals and pure liquid water have the same moisture content (100%), the ice crystals do not help these reactions at the same pace as liquid water does. Thus, it is actually not the total amount of moisture, but the amount of free moisture that contributes to water activity and that decides the shelf life of food. Therefore, the term *water activity* is used as an indicator of stability of any product under a given situation along with its moisture content.

Denoted as a_w, *water activity* is the chemical potential of the water present in a food, i.e., its ability to participate in chemical reactions.

The water activity of a food is affected by different parameters, including temperature, pH, oxygen, and CO_2 in the system. Since the water activity for pure water at any temperature above its freezing point is 1.0, the value of water activity can vary within 0 to 1. The a_w of high moisture foods (such as fruits

and vegetables and meat, etc.) is often more than 0.95. The highly perishable fresh foods can have a_w value of 1.00. For intermediate foods, such as jams, condensed milk, margarine, fruit cake, etc., a_w is between 0.6 to 0.9.

The dried foods can have water activities between 0–0.6. Some examples are dried vegetables (5% moisture) and whole milk powder (less than 4% moisture), which have a_w value of 0.20. Dried spices and noodles have water activity values of about 0.5.

It has been observed that for inhibiting the activity of most fungi, yeasts, and bacteria, the water activity of the food should be below 0.7, 0.8, and 0.9, respectively. Most of the microorganisms are commonly inhibited at water activity less than 0.6. For eliminating other deteriorative reactions, the safe water activity is less than 0.3.

Table 9.1 indicates the importance of water activity in foods.

As mentioned earlier in this section, the water activity reduces if the food temperature is reduced to freezing temperature. The ice at 0, -10, -20, and -50°C has the water activities of 1.00, 0.91, 0.82, and 0.62, respectively. Therefore, as compared to products at higher temperatures, frozen products have a longer safe storage period. Similarly, the water available for chemical reactions can be reduced by increasing the solids concentration in liquid foods. This is the basis for increasing the shelf life of foods by the addition of salt, sugar, etc. or through evaporation. The addition of chemical preservatives, acids, etc. also can reduce water activity.

9.1.2 Mathematical Expressions for Water Activity

Water activity is expressed in equation form as the ratio of the vapor pressure of water in a solution (p_s) to the vapor pressure of pure water (p_{ws}) at the same temperature.

$$a_w = \frac{p_s}{p_{ws}} \qquad (9.1)$$

The numerator can also be stated as the partial pressure in the head space of the material. The above equation also implies that the water activity of a commodity will be the same as the equilibrium relative humidity of the surrounding atmosphere at the equilibrium water content. In other words, if a food has a water activity value of 0.86, the food will neither gain nor loose moisture if stored in a place maintained at 86% RH. The presence of salt, sugar, and other dissolved solutes affects the vapor pressure in the head space and, hence, the water activity is also affected.

Some empirical equations have been proposed for determination of water activity. Most of them have been found to be suitable for specified ranges of water activities.

TABLE 9.1

Effects of water activity in foods

Water activity limit	Effect	Examples of foods
0.95–0.91	*Pseudomonas, Salmonella, Bacillus, Vibrio parahaemolyticus, Clostridium perfringens, Cl. botulinum, Lactobacillus,* some molds, and some yeasts inhibited	Some cheeses, some fruit juice concentrates, cooked sausages and bread, cured meat
0.91–0.87	Most yeasts and micrococcus inhibited	Dry cheeses, margarine, fermented sausage, sponge cake
0.87–0.80	Most enzymes, *Sacharomyces* spp., molds (mycotoxigenic penicillia), and *Staphylococcus aureus* inhibited	Most fruit cakes, fruit juice concentrates, cooked rice, sweetened condensed milk, pulses
0.80–0.75	Most halophilic bacteria, aspergilli inhibited	Jam, marmalade, most marshmallows
0.75–0.65	Most Xerophilic fungi (e.g., *Aspergillus candidus*), *Sacharomyces bisporus* inhibited	Jelly, molasses, some dried fruits, nuts, raw cane sugar
0.65–0.60	Almost all osmophilic yeasts and a few molds (e.g., *Monascus bisporus*) inhibited	Dried fruits, some toffees, honey
0.55	Deoxyribonucleic acid (DNA) becomes disorderly	Noodles, spaghetti, dried spices
0.50	No microbial activity	Dried grains, dried spices, whole egg powder

For ideal solutions, $a_w = X_w$ (9.2)

where, X_w is the mole fraction of water in the solution.

For non-ideal solutions, the above equation is modified by including another factor known as the *activity coefficient* (γ) as

$$a_w = \gamma X_w \qquad (9.3)$$

The *Norrish equation* (for water activity for high moisture content foods) is given as

$$\log \frac{a_w}{X_w} = -k(1 - X_w)^2 \qquad (9.4)$$

where, X_w = mole fraction of water in the food/solution. k is a constant. The values of k for different solutes have been found. For example, $k_{sucrose} = 2.7$ and $k_{glucose/fructose/invertase} = 0.7$.

Example 9.1 What is the water activity of a 30% sucrose solution?

SOLUTION:

The molecular weight of sucrose ($C_{12}H_{22}O_{11}$) is 342.

Thus, the mole fraction of water in the 30% solution is:

$$X_w = \frac{(70/18)}{(70/18) + (30/342)} = 0.978$$

Substituting the values in Norrish's equation and taking $k_{sucrose} = 2.7$,

$$\log a_w = \log X_w - 2.7(1 - X_w)^2 = \log(0.978) - 2.7(0.978)^2$$

$$= -9.66 \times 10^{-3} - 1.3068 \times 10^{-3} = -0.01097$$

Hence, $a_w = 0.975$

A mixture involving several components will have its water activity the same as the product of the water activities of all components.

$$a_w = (a_{w1})(a_{w2})$$

Example 9.2 A liquid food constitutes 60% solids and remaining 40% water. The solids have been found to be 40% sucrose and 60% hexose. What will be the water activity of the food? Take the k value for hexose as 0.7 and that for sucrose as 2.7. The molecular weights of sucrose and hexose are 342 and 180, respectively.

SOLUTION:

If 100 g of the food is taken, the amount of sucrose and hexose will be 24 g and 36 g. The amount of water is 40 g.

If the sucrose dissolves completely in the whole water,

$$X_{sucrose} = (24/342)/(40/18 + 24/342) = 0.0306$$

$$X_w = 1 - 0.0306 = 0.9694$$

Using a k value for sucrose as 2.7, $\log(a_{w1}) = \log(0.9694) - 2.7(1 - 0.9694)^2$

$$or, (a_{w1}) = 0.9637$$

Considering the situation if only hexose is dissolved,

$$X_{hexose} = (36/180)/(40/18 + 36/180) = 0.08257$$

$$X_w = 1 - 0.08257 = 0.91743$$

Using 0.7 as the k value for hexose, $\log(a_{w2}) = \log(0.91743) - 0.7(1 - 0.91743)^2$

$$or, (a_{w2}) = 0.9074$$

Therefore, a_w of the mixture is $0.9637 \times 0.9074 = 0.8745$

The presence of undissolved solids does not affect the water activity. If the food contains some undissolved solids, they can be simply ignored in the calculations except in finding the fractions of different dissolved solids and water.

The *Brunauer-Emmett-Teller (BET) equation*, used for determining the water activity of low moisture content foods, is given as:

$$\frac{a_w}{M(1-a_w)} = \frac{1}{M_1 C} + \frac{C-1}{M_1 C} a_w \qquad (9.5)$$

where M is the moisture of the commodity (per cent dry basis), M_1 the moisture (percent dry basis) of a monomolecular layer (the surface of the solid particles is coated by water molecules in a mono-molecular layer), and C, a constant (related to heat of adsorption and dependent on temperature). The BET equation has been found particularly suitable for foods having a_w between 0 to 0.45.

BET plot is a plot of $a_w/M(1-a_w)$ against a_w. Since the plot is linear, the constant M_1 can be determined from the slope and intercept of the line.

There are also other equations, namely, the GAB (Guggenheim-Anderson-de Boer) equation, Caurie equation, Chen equation, Chung and Pfost equation, Day-Nelson equation, Halsey equation, Henderson equation, Iglesias-Chirife equation, Kuhn equation, Oswin equation, Smith equation, etc., which are used to represent water activity for different types of commodities at different ranges of moisture contents. These equations are mentioned in Toledo (2007).

As discussed above, water activity is also simply given as the ratio of the vapor pressure of water in a solution (p_s) to the vapor pressure of pure water (p_{ws}) at the same temperature. Thus, a_w is related to moisture content in a non-linear relationship known as a moisture sorption isotherm curve. These curves will be different for different commodities and for different temperatures.

9.1.3 Measurement of Water Activity

Water activity values can be measured by either a resistive electrolytic or a capacitance method or by measuring the dew point with a dew point hygrometer.

In the *resistive electrolytic hygrometers*, the sensing element is a liquid electrolyte, which is held in between two small glass rods by capillary force. The resistance of the electrolyte changes when it absorbs or loses moisture, and, thus, it varies directly with the relative air humidity. Once the vapor-liquid equilibrium is established, the resistance is directly proportional to the water activity of the sample. The volatiles in the surrounding environment may affect the performance of the electrolyte. Hence, chemical protection filters are used to absorb the volatile compounds and prevent their contact with the sensor.

In the *capacitance hygrometers*, there are two charged plates separated by a polymer membrane, which serves as the dielectric. The increase in moisture content of the membrane increases its capacity to hold a charge, and, thus, the measured capacitance is approximately proportional to the water activity. The observations are not affected by most volatile chemicals. However, they are affected by residual water on the polymer membrane. The instrument is less accurate than dew point hygrometers (+/- 0.015 a_w). It has to be calibrated regularly to get a better result.

The *dew point hygrometers* have a mirror over a closed sample chamber. The mirror is cooled during which an optical sensor is used to measure the dew point temperature. The relative humidity of the chamber is obtained from the measured dew point temperature by using a psychrometric chart. This method is accurate ($\pm$ 0.003 a_w) and often the fastest.

9.2 Classification of Microorganisms on the Basis of Temperature Range of Growth

Based on the temperature range of growth, the microorganisms can be categorized as psychrotrophic, psychrophilic, mesophilic, thermophilic, and thermoduric. Their characteristics are given in Table 9.2.

This classification is important to know the potential organisms present in a food processed and stored at different temperature conditions.

9.3 Growth of Bacteria and Generation Time

In autecological studies, the growth of bacteria (or other microorganisms) can be modeled in batch culture with four phases, i.e., lag phase, log or exponential phase, stationary

TABLE 9.2

Categorization of different microorganisms based on the temperature range of growth

Class	Optimum temperature for growth	Remarks
Psychrotrophic (cold tolerant)	12–18°C	Some microorganisms of this group can reproduce even at as low as 4°C. These are easily destroyed by heat.
Psychrophilic (cold loving)	20°C	These are of concern if the food is processed and stored in this temperature range.
Mesophilic (medium range)	20–44°C	The minimum temperature for growth is 5–10°C. Most of the packed foods are stored in this temperature range.
Thermophilic (heat loving)	45–65°C	The minimum temperature for growth varies between 30–40°C. Thus, these are of concern only if the food is prepared and kept in this temperature range.
Thermoduric (heat enduring)	Can survive above 70°C, but cannot reproduce at these temperatures	

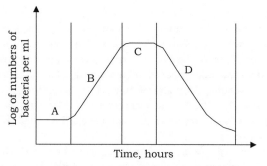

A: Lag phase;   B: Exponential phase;
C: Stationery phase; D: Death phase

FIGURE 9.1   A typical growth curve of bacteria in a media

phase, and death phase. Figure 9.1 shows typical growth curve for bacteria.

> *Autecology* or *population ecology* is a sub-field of ecology relating to the dynamics of species populations and how these populations interrelate with the environment.

In microbiology, inoculation is introducing microorganisms into a suitable culture in which they can grow and reproduce. Thus, when a bacterium is inoculated or newly introduced in a specific medium, they try to adapt themselves to the medium and there is limited cell division during the period of adaptation. This is known as the *lag phase*. During this phase, there may be increase in mass and volume of the bacterial cells and increased metabolic activity. There may also be synthesis of enzymes, proteins, and RNA. After the lag phase, the bacteria have adjusted themselves to the medium and start cell division, which is known as the *log phase*. Thereafter, a stage comes when the medium may not be able to supply the nutrients for the increased number of bacteria. Besides, the formation of organic acids and accumulation of wastes causes a break in the growth of bacteria. The growth and death rates become almost equal and, hence, the population of bacteria becomes somewhat constant. This phase is known as the *stationary phase*. In this stage, the *secondary metabolites*, such as antibiotics, are produced by the bacteria and the genes that are responsible for the sporulation process may start their activity.

Then the bacteria die in the *death phase* (decline phase). The number of viable cells decreases exponentially during this phase, which is essentially a reversal of the growth that occurred during the log phase. Microorganisms capable of forming resistant but dormant spores could remain alive during the death phase, even under certain unfavorable conditions.

Under favorable conditions, the bacterial population doubles at regular time intervals by binary fission. The increase in a bacterial population, in other words, is by geometric progression. If it is started with one cell, there are 2 cells after division in the first generation, increasing to 4, 8 and so on in the subsequent generations, i.e., the number will grow as 2^0, 2^1, 2^2, 2^3, ..., 2^n (where n = the number of generations of the cell population, doubling during each generation). Thus, in the log

phase, the growth is exponential in nature and a straight line is obtained when cell number against time data is plotted on a semi-log graph paper. The organisms' specific growth rate is represented by the slope of this line. The growth rate is influenced by the environmental conditions, among other factors.

The time interval required for the cells to divide is termed *generation time* (G) or, simply, it is the time per generation.

$$G = \text{total time for growth } (\theta)/n \qquad (9.6)$$

The generation time is also termed the *time for doubling* the bacterial population.

If B_i is the number of bacteria at the beginning and B_f at the end of the time duration, then

$$B_f = B_i \times 2^n \qquad (9.7)$$

(This equation expresses the growth by binary fission)

$$\text{Thus,} \quad \log(B_f) = \log(B_i) + n \log(2)$$

$$n = \frac{\log(B_f) - \log(B_i)}{\log(2)} = \frac{\log(B_f) - \log(B_i)}{0.301}$$

$$\text{i.e.} \quad n = 3.3 \log\left(\frac{B_f}{B_i}\right)$$

$$\text{As,} \quad G = \frac{\theta}{n}, G = \theta \bigg/ 3.3 \log\left(\frac{B_f}{B_i}\right) \qquad (9.8)$$

The generation time varies from as little as about 12 minutes to 24 hours or even more for bacteria, depending on its species. In laboratory studies, the generation time has been observed to be 15–20 minutes for *E. coli*. For most known culturable bacteria, generation times vary from about 15 minutes to 1 hour. The *Mycobacterium tuberculosis*, a common pathogenic bacterium, has a quite long generation time.

The main factors affecting the growth of microorganisms are the temperature, pH, water activity, as well as the type of substrate.

9.4 Thermobacteriology

9.4.1 Effect of Heat on Microorganisms– D, z, and F Values

Heat processing causes destruction of microorganisms. The rate of destruction of microorganisms during the thermal processing follows a first-order reaction (discussed in Chapter 4), i.e., if N is the number of microorganisms, then the change in number of microorganisms with time can be given as:

$$\frac{dN}{d\theta} \alpha - N \qquad (9.9)$$

In other words, the rate of change in the number of microorganisms at any time under a specific processing condition is proportional to the number of microorganisms present at that time. The negative sign indicates that the number will decrease with respect to time. The death of a microorganism is defined as when it has lost its ability to reproduce.

The above equation can be written as:

$$\frac{dN}{d\theta} = -kN \tag{9.10}$$

Here k is a constant, which can be called the *death rate constant*. Heat tolerance of each species is different and, hence, the above relationship will hold good only for the specific microorganism for the specified processing condition.

If we consider that the number of microorganisms is N_0 at time 0 and N at time θ, integrating within limits,

$$\int_{N_0}^{N} \frac{dN}{N} = -k \int_{0}^{\theta} d\theta \tag{9.11}$$

$$\text{or,} \quad \ln\left(\frac{N}{N_0}\right) = -k\theta \tag{9.12}$$

$$\text{or,} \quad \ln\left(\frac{N_0}{N}\right) = k\theta \tag{9.13}$$

$$\text{i.e., } \ln N = \ln N_0 - k\theta \tag{9.14}$$

The equation 9.13 can also be written as:

$$\log\left(\frac{N_0}{N}\right) = \frac{k\theta}{2.303} \tag{9.15}$$

$$\text{i.e. } \log N = \log N_0 - \frac{k\theta}{2.303} \tag{9.16}$$

The equations 9.13 and 9.15 indicate that microorganisms follow logarithmic order of death. It means, if we heat a food to a high enough temperature, the same percentage of microorganisms will be destroyed during any specific time interval regardless of their initial number.

The *death rate curve* of a particular microorganism at a specified temperature is shown in Figure 9.2(a). The *thermal death time* or *decimal reduction time* is defined as the time needed to destroy 90% of the microorganisms (i.e., to bring their number down to 10% of the number present initially). It is referred to *D value*, usually expressed in minutes, and it is different for different types of microorganisms. Higher D value indicates greater heat resistance.

The D values for most of the microorganisms (related to food spoilage) and other food ingredients have been found out at 121.1°C (250°F), which is known as D_{121} (or if expressed in °F as D_{250}). D_{121} values are used as a basis for comparison of heat resistance of microorganisms and food nutrients.

If we consider Figure 9.2(a), the slope of the death rate curve is -1/D.

Similarly, if we consider the equation 9.15, the slope of the curve is -k/2.303.

$$\text{Thus, D} = 2.303/k \tag{9.17}$$

$$\text{i.e., } \log N = \log N_0 - \frac{\theta}{D} \tag{9.18}$$

A 1D process reduces cell numbers by 90%, i.e., $\log(N_0/N) = 1$, i.e., if the initial number of microorganisms is 10, the final number after a 1D process will be 1. This is called *one log cycle reduction*. If the initial number is 10^7, then one log cycle reduction will bring the number to 10^6. This states that for reducing the microbial population from 10^7 to 10^6 the time taken is the same as that for reducing the number from 10 to 1.

It may also be looked at from another angle. One log cycle reduction is 90% reduction. Two log cycles reduction is 99% reduction. Similarly, 3 log cycles reduction will be 99.9% reduction and 5 log cycle reduction will be 99.999% reduction.

A 10D process causes 10 decimal reductions. For example, if the initial number of cells is 10^6, then a 5D process will result in cell numbers 10 and a 10D process would reduce cell numbers to 10^{-4} per container, i.e., one cell can be found in every ten thousand containers. The number of log cycle reduction, i.e., log (N_0/N_f) is also known as *spore log reduction* and is denoted as S or Y_N.

The equation 9.15 can also be written as:

$$\theta = D \log (N_0/N) \tag{9.19}$$

If the final microorganism population is given as N_f, the process time at any temperature t can be written as:

$$F_t = D_t \log (N_0/N_f) = D_t \times S \tag{9.20}$$

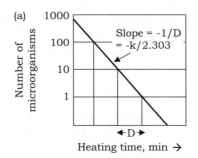

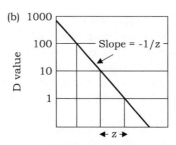

FIGURE 9.2 (a) Death rate curve, (b) Thermal death time curve

i.e., $S = \dfrac{F_t}{D_t}$ (9.21)

The time required to reduce the microbial load from an initial value to a specific desirable value at a specific temperature is defined as *F value*. The F value is the product of D value for the target microorganism at the specific temperature and the desired log cycle reduction.

If the number of residual microorganisms are observed as N_1 and N_2 after heating times θ_1 and θ_2, respectively, then the D value can be determined as follows.

$$D = \frac{\theta_1 - \theta_2}{\log(N_1) - \log(N_2)}$$ (9.22)

The above equation can be used to experimentally to determine the D value for a specific microorganism at a specific temperature by observing the time required for reducing the number of bacteria to two different levels at the same temperature.

As the microorganisms follow a first order reaction, the shape of the curve will be as shown in Figure 9.2(a). However, certain variations in the shape of the microbial destruction curves are also observed. The number of microorganisms may increase for a while before the first order reaction behavior is obtained. For mixed cultures, the curve may not be exactly of first order.

The destruction of microorganisms is also dependent on the temperature used for processing and the rate of destruction is more at higher temperatures. Figure 9.2(b) represents the D values plotted against the temperature, which is known as the *thermal death time* (TDT) curve. A parameter z is defined as the change in temperature required to bring about a tenfold change in decimal reduction time, i.e., the increase in temperature necessary to obtain the same lethal action or the same effect (i.e., same N_0/N) in one-tenth of the time. It represents the change in the death rate when the temperature changes. The slope is -1/z for the TDT curve.

The Figure 9.2(b) can be used to obtain the z-value by taking two D values at two different temperatures and calculating the slope of the line. The following formula can also be used.

$$z = \frac{t_2 - t_1}{\log(D_1) - \log(D_2)}$$ (9.23)

where both t_1 and t_2 should be taken in °C or K.

The D and z values indicate the heat resistance of microorganisms or enzymes, or chemical constituent of the food. As can be seen from the figure, if the z values are more, then the slope will be less and, to reduce the time of reaction, larger changes in temperature would be required. For a spore forming bacteria, the usual z value is 10°C.

As discussed above, the D and z-values indicate the resistance of microorganisms. Thus, both can be used to find an equivalent thermal process for a microorganism if a standard process is known.

From Figure 9.2(b),

$$\log D_t = -t/z + c$$ (9.24)

Where D_t is the D value at any specific temperature t°C. We can also write the above equation as:

$$\log D_{121} = -121/z + c$$ (9.25)

Hence, from equations 9.24 and 9.25,

$$\log (D_t/D_{121}) = (121-t)/z$$ (9.26)

or, $D_t = D_{121}\, 10^{(121-t)/z}$ (9.27)

As $F_t = D_t \times S$, the F value at any temperature F_t can also be written as:

$$F_t = F_{121}\, 10^{(121-t)/z}$$ (9.28)

It is mentioned earlier that the D_{121} (D value at 121°C) values have been found for most microorganisms. So from equation 9.27, D value at any temperature t can be found. Unless mentioned, z value is usually taken as 10°C. D_{121} at z =10°C is usually denoted as D_0 and, accordingly, F_{121} value at z =10°C is denoted as F_0.

The D and z values of some spoilage microorganisms and nutrients in different types of food have been given in Annexure XI. As can be seen, the D and z values of microorganisms are usually lower than the vitamins and enzymes. Thus, the processing conditions usually adopted for destroying microorganisms have little effect on the nutrients available in food.

For example, *Cl. botulinum* spores have a D value of 0.1 to 0.3 min, whereas the D values of thiamine and peroxidase are 158 min and 3 min, respectively. Thus, for a 10 log cycle reduction of *Cl. botulinum* spores, it will require a processing of three minutes. However, the processing of three minutes has a negligible effect on the thiamine and there will be only one log cycle reduction of peroxidase.

A close examination of the above table also indicates that the chemical changes have larger z values than that of the microorganisms. The color and flavor components also have larger z values. Thus, the vitamins, color, flavor, etc. will be relatively unaffected in a process intended for the destruction of microorganisms. This also explains why the high temperature short time (HTST) method of processing helps in better retention of vitamins and sensory quality as compared to the long hold type methods.

The type of food also affects the heat resistance of microorganisms. *Bacillus stearothermophilus* sourced from spinach has been found to have more D and z values than that sourced from green beans.

Example 9.3 If a can has a spore load of 100, calculate a target process time such that the final spore load will be 10^{-5} (1 in 100000). The D value of the specific spore at the specific temperature is 1.6 min.

SOLUTION:

$$S = \log \frac{N_0}{N} = 7$$

$$F = 7 \times 1.6 = 11.2 \text{ min}$$

Example 9.4 If under the same conditions, another spore has a D value of 0.2 min, what would be the process time for a 12D process? The initial spore load was found to be 10 per can.

SOLUTION:

In this case, S= 12

$$F = 12 \times 0.2 = 2.4 \text{ min.}$$

Example 9.5 What will be the process time at 121°C for 10 log cycle reduction of *Clostridium botulinum* if the D_{121} value of the microorganism is 0.216 min?

SOLUTION:

Given, $D_{121} = 0.216$ min

$$\theta = D_t \times \log(N_0/N) = D_t \times S,$$
where S is the number of log cycle reduction.

Thus, $\theta = 0.216 \times 10 = 2.16$ min

or, it can be simply written as:
$F_{121} = D_{121} \times 10 = 0.216 \times 10 = 2.16$ minutes.

Example 9.6 In the above example, find out the process time at 130°C when z= 10°C

SOLUTION:

D_{130} can be calculated using the equation

$$D_{130} = D_{121}10^{(121-130)/z}$$

or, $D_{130} = 0.216 \times 10^{(121-130)/10} = 0.0272$

$$F_{130} = S \times D_{130} = 10 \times 0.0272 = 0.272 \text{ min}$$

Otherwise also we can find F_{130} as:

$$F_{121} = 2.16 \text{ min}$$

$$F_{130} = F_{121} \times 10^{(121-130)/z} = 0.272 \text{ min}$$

Check that when the process temperature is increased by just 9°C, the process time for the same bacterial reduction is remarkably reduced.

Example 9.7 The initial bacterial load of a can is 10^4. Calculate the process time so that the final spore load will be one in a thousand cans. The D_0 for spores = 2 min. If under the same condition, the D_0 value is 0.2 min, what would be the process time?

SOLUTION:

Given, $D_{121} = 2$ min (Remember that D_0 means D_{121})

$$N_0 = 10^4$$

$N = 10^{-3}$ (1 in a thousand samples means the spore load is 10^{-3})

$$\theta = D_t \times \log(N_0/N) = 2 \times \log (10^4/10^{-3})$$
$$= 2 \times 7 = 14 \text{ min}$$

i.e., The process time will be 14 min.
In the second case if the D_0 value is 0.2 min,

$$\theta = D_t \times \log(N_0/N) = 0.2 \times \log (10^4/10^{-3})$$
$$= 0.2 \times 7 = 1.4 \text{ min}$$

i.e., The process time will be 1.4 min.
Check that if the D value is less for the same bacterial reduction, the process time is less.

Example 9.8 If the process time F_0 for 99.9999% reduction of a particular bacterial spore is 1.3 min, calculate the process time for 10 log cycle reduction.

SOLUTION:

Given, $F_0 = 1.3$ min.

99.9999 % reduction means 6 log cycle reduction (i.e. S = 6)

Thus, $1.3 = D \times 6$, i.e. $D = 1.3/6$

For 10 log cycle reduction (S = 10)

$$F = D \times S = (1.3/6) \times 10 = 2.16 \text{ min}$$

Remember that when we say F_0, it is the reference process time (F_{121}), i.e. the process operates at 121°C.

Example 9.9 During pasteurization of milk, the same reduction of the target microorganism was observed maintaining the milk at 63°C for 30 min or at 72°C for 15 sec. Determine the z value of the microorganism.

SOLUTION:

$$D_t = D_{121}10^{(121-t)/z}$$

For the first situation $D_{63} = D_{121}10^{(121-63)/z}$

i.e., $30 = D_{121}10^{(121-63)/z}$ (A)

For the second situation $D_{72} = D_{121}10^{(121-72)/z}$

i.e. $0.25 = D_{121}10^{(121-72)/z}$ (B)

Dividing equation (A) by (B), we get

$$120 = 10^{(9/z)}$$

$$\log(120) = 9/z$$

$$z = 4.33°C$$

Example 9.10 A spherical shaped food having 6 mm diameter is blanched for 4 min at 105°C. Calculate the percent reduction of a particular vitamin that has z = 18°C and D_{121} = 2.7 minutes.

SOLUTION:

$$D_t = D_{121}10^{(121-t)/z}$$

$$D_{105} = D_{121}10^{(121-105)/18} = 2.7 \times 7.74 = 20.905 \text{ min}$$

$$\log(N_o/N) = t/D_t$$

$$\log(N_o/N) = 4/20.905 = 0.19134$$

$$N_o/N = 1.554$$

Thus, the number of log cycle reduction is 1.554. But the question asks to find the percent reduction in the nutrient. For that, $(N_o-N)/N_o$ has to be calculated in percent.

$$(N_o-N)/N_o \times 100 = ((1.554N-N)/1.554N) \times 100 = 35.65 \%$$

Example 9.11 During an experiment to decide the retorting time, two samples were drawn after 10 and 20 minutes of heating at 121°C. If the numbers of residual microorganisms were 3600 and 24, then find out the D value? The lag time was earlier found to be only 0.6 min for heating the tubes to 121°C, and thus it can be ignored. If the number of microorganisms after 20 minutes is 18, what will be the D value?

SOLUTION:

As the lag time is ignored, the equation 9.22 can be used.

$$D = \frac{20-10}{\log(3600)-\log(24)} = \frac{10}{2.176} = 4.59 \text{ min}$$

For the second case

$$D = \frac{20-10}{\log(3600)-\log(18)} = \frac{10}{2.30} = 4.34 \text{ min}$$

Check that there is more reduction of microorganisms in the second case as the D value is lower.

As mentioned at a later stage, the type of substrate also affects the microbial resistance to heat.

9.4.2 Factors Affecting Microbial Resistance to Heat

The resistance of microorganisms (as well as of enzymes) to heat is affected by the type of microorganism (or enzyme), incubation conditions (age of culture, temperature, type of culture medium, etc.), and the condition of food during the heat treatment, among the many factors. The vegetative cells have lower heat resistance than spores, and, hence, foods with only vegetative spores need less severe processing conditions than those with spores. The characteristics of the food, such as the pH, composition, and water activity, are some important factors that affect the heat resistance of microorganisms.

The effects of temperature and pH on the heat resistance of microorganisms (indicative only) are shown in Figure 9.3.

As the temperature increases, the D value decreases and that is why the processing time is shorter at higher temperatures. When the pH is lower the D value will be smaller, i.e., the microorganisms are less heat resistant in acidic foods.

The microorganisms find it more comfortable to grow on low acid foods than on high acid foods, and, hence, the low

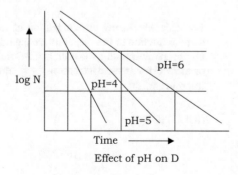

Effect of pH on D

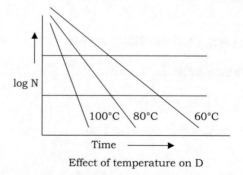

Effect of temperature on D

FIGURE 9.3 Effect of temperature and pH on the heat resistance of microorganisms

acid foods usually have a higher initial number of microorganisms than high acid foods. Therefore, low acids foods require more heat treatment than high acid foods. Accordingly, for consideration of heat processing requirements, the food products have been classified as high acid foods (pH < 4.5) and low acid foods (pH ≥ 4.5).

Cl. botulinum comfortably grows in low acid foods (for pH >4.5) stored in cans, i.e., under anaerobic conditions. Thus, for thermal processing of low acid foods, *Cl. botulinum* is the main target microorganism. However, the chance also exists of the presence of more heat-resistant spoilage bacteria. So the cans are given more intense processing than usually required for *Cl. botulinum*. For pH between 4.5 and 3.7, the target organisms are the yeasts and fungi or heat resistant enzymes. Very highly acidic foods (pH<3.7) do not normally support growth of microorganisms, and, hence, the thermal processing mostly aims at enzyme inactivation. Thus, the processing condition is less severe.

9.4.3 Relationship between Death Rate Constant and Activation Energy

For the design of thermal processes, it is necessary to know which changes take place in the food and at what rate the changes of individual components take place. The reaction rates depend on the components and concentration (C) of food and also on process parameters such as temperature and water activity, a_w.

$$r = \frac{dC}{d\theta} = -kC^n \tag{9.29}$$

C can be the concentration of a substance or microorganism, or some measure of a quantity of any parameter, such as color and taste.

The reaction rate, k, is not a constant and varies with temperature. As per the Arrhenius equation, the relationship between k and absolute temperature (T) is represented as follows.

$$k = Ae^{-\left(\frac{E_a}{RT}\right)},$$

where E_a is the activation energy. \tag{9.30}

If the reaction rates are taken at two temperatures, T_1 and T_2

$$k_1 = Ae^{-\left(\frac{E_a}{RT_1}\right)} \text{ and } k_2 = Ae^{-\left(\frac{E_a}{RT_2}\right)}$$

$$\ln k_2 - \ln k_1 = -\frac{E_a}{R}\left(\frac{1}{T_2} - \frac{1}{T_1}\right)$$

$$\ln\left(k_2/k_1\right) = -\frac{E_a}{R}\left(\frac{1}{T_2} - \frac{1}{T_1}\right) \tag{9.31}$$

$$\text{Or, } \log\left(k_2/k_1\right) = -\frac{E_a}{2.303 \times R}\left(\frac{1}{T_2} - \frac{1}{T_1}\right) \tag{9.31a}$$

Since k = 2.303/D, the equation 9.31 indicates that D value decreases in terms of activation energy. This does not agree with the Arrhenius equation. The numerical value of E_a is variable, but for both enzyme reactions and the growth of microorganisms, it generally lies between 3×10^7 and 12×10^7 kJ.mol^{-1}.

Similarly, for death rates at two temperatures,

$$D_{t1} = D_{121}10^{(121-t1)/z} = \frac{2.303}{k_1}$$

$$D_{t2} = D_{121}10^{(121-t2)/z} = \frac{2.303}{k_2}$$

$$\text{Thus, } \left(k_2/k_1\right) = 10^{-\frac{T_1-T_2}{z}}$$

$$\ln\left(k_2/k_1\right) = -\frac{T_1-T_2}{z}\ln(10) \tag{9.32}$$

Combining the two equations 9.31 and 9.32,

$$-\frac{E_a}{R}\left(\frac{1}{T_2} - \frac{1}{T_1}\right) = -\frac{T_1-T_2}{z}\ln(10)$$

$$\frac{E_a}{R}\left(\frac{T_1-T_2}{T_1T_2}\right) = \frac{T_1-T_2}{z}\ln(10)$$

$$\frac{E_a}{R} = \left(\frac{T_1T_2}{z}\right)\ln(10) \tag{9.33}$$

$$z = 2.303\frac{R}{E_a}T_1T_2 \tag{9.34}$$

If the Arrhenius equation is considered to be more accurate, it will mean that z is not constant but a function of temperature.

Example 9.12 Considering the example 9.9, find out the activation energy for the process.

SOLUTION:
We will use equation 9.31 to obtain the activation energy.

The temperatures are given as 72°C (=345.15 K) and 63°C (=336.15 K)

$$\log\frac{D_{72}}{D_{63}} = -\frac{E_a}{2.303 \times R}\left(\frac{1}{336.15} - \frac{1}{345.15}\right)$$

$$\ln\frac{0.25}{30} = -\frac{E_a}{2.303 \times 8.31434}\left(7.157\times10^{-5}\right) = -E_a \times 4.051\times10^{-6}$$

$$E_a = 513230.61 \text{ J.mol}^{-1}$$

Example 9.13 In a milk processing plant milk is pasteurized for an 8-log cycle reduction by

keeping it at 78°C for 15 s or at 88°C for 4 s. Find the decimal reduction times for these two methods. Calculate the z value of the target microorganism. Also calculate the activation energy required for the pasteurization process.

SOLUTION:

In this situation, $\log (N_0/N) = S = 8$
For the first case:

$$F_1 = D_1 \log_{10} (N_0/N_1)$$

$$\text{or, } 4 \text{ s} = D_1 \times (8)$$

$$\text{Thus, } D_1 = 0.5 \text{ s}$$

For the second case:

$$F_2 = D_2 \log_{10} (N_0/N_2) = D_2 \times 8$$

$$\text{or, } 15 \text{ s} = D_2 \times 8$$

$$\text{Thus, } D_2 = 1.875 \text{ s}$$

Considering that the z value is constant for both methods.

$$D_t = D_{121} 10^{(121-t)/z}$$

Thus, for the first method, $4 = D_{121} 10^{(121-88)/z}$ (A)

and, for the second method, $1.875 = D_{121} 10^{(121-78)/z}$ (B)

Dividing (A) by (B) $\dfrac{0.5}{1.875} = 10^{(-88+78)/z}$

$$-0.574 = -10 / z$$

$$\text{or, } z = 17.42°C$$

To find out the activation energy, the equation 9.33 is used.

$$\frac{E_a}{R} = \left(\frac{T_1 T_2}{z}\right) \ln(10)$$

Thus,

$$E_a = \frac{(351.15)(361.15)\ln(10)}{17.42} \times 8.31434$$

$$= 139372.02 \text{ kJ.}\left(\text{kg mole}\right)^{-1}$$

(Here we take R as 8.31434 J. (mol. K)$^{-1}$ and the temperatures in K. But remember that z is a temperature difference, and, hence, it is same for both Celsius and Kelvin. In the MKS system the value of R is 1.987 cal.mol^{-1}.K^{-1})

Example 9.14 The target microorganism in a thermal process has the D_0 and z values as 3 min and 10°C. What will be the process time at 140°C for 10D inactivation? The z value was determined by taking observations at two temperatures, namely, 110°C and 121°C.
Use both the thermal death time model and the Arrhenius equation to compare the results.

SOLUTION:

Using the thermal death time model:
  Here the D_0 value is 3 min and there will be 10D inactivation.

$$\text{So, } F_0 = 10 \times 3 = 30 \text{ min}$$

Now,
$$F_{140} = F_0 10^{(121-140)/z} = (30)10^{(121-140)/10} = 0.377 \text{ min}$$

Here the z value has been found by taking the two temperatures as 121°C (=394.15 K) and 110°C (=383.15 K)
  Hence, as per equation 9.33,

$$E_a = R\left(\frac{T_1 T_2}{z}\right)\ln(10) = 8.31434\left(\frac{394.15 \times 383.15}{10}\right)(2.303)$$

$$= 289169.23 \text{ J.(gmol)}^{-1}$$

As per equation 9.31, $\ln\left(\dfrac{k_2}{k_1}\right) = -\dfrac{E_a}{R}\left(\dfrac{1}{T_2} - \dfrac{1}{T_1}\right)$

i.e., $\left(\dfrac{k_2}{k_1}\right) = e^{-\frac{E_a}{R}\left(\frac{1}{T_2} - \frac{1}{T_1}\right)}$

Since k is inversely proportional to D or F

$$\left(\frac{F_1}{F_2}\right) = e^{-\frac{E_a}{R}\left(\frac{1}{T_2} - \frac{1}{T_1}\right)}$$

Or, $F_1 = F_2 e^{-\frac{E_a}{R}\left(\frac{1}{T_2} - \frac{1}{T_1}\right)}$

In this case $T_1 = 140 + 273.15 = 413.15$ K and $T_2 = 121 + 273.15 = 394.15$ K

$$F_{140} = F_{121} e^{-\frac{289169.23}{8.31434}\left(\frac{1}{394.15} - \frac{1}{413.15}\right)} = F_{121} \times e^{-4.057967}$$

$$= 30 \times 0.01728 = 0.5185 \text{ min}$$

The above example indicates that use of the Arrhenius equation gives a higher time of processing than the thermal death time method. The difference in results from both methods may be negligible at temperatures below 120–125°C. But in UHT processing, higher temperatures are used, and, hence, more accurate calculations as the Arrhenius model are preferred. Some other modern calculation techniques are also available for the purpose.

9.4.4 Reaction Quotient

The *reaction quotient* is an index to show how a temperature change of 10 K (or 10°C) changes the reaction rate. It is denoted as Q_{10}.

$$Q_{10} = \frac{\text{Reaction rate at } (t+10)^0 C}{\text{Reaction rate at } t^0 C} = \frac{k_{t+10}}{k_t} \quad (9.35)$$

Now, $\dfrac{k_t}{2.303} = \dfrac{1}{D_{121} 10^{\frac{121-t}{z}}}$

$$\frac{k_{t+10}}{2.303} = \frac{1}{D_{121} 10^{\frac{121-(t+10)}{z}}}$$

$$\frac{k_{t+10}}{k_t} = \frac{D_{121} 10^{\frac{121-(t)}{z}}}{D_{121} 10^{\frac{121-(t+10)}{z}}} = 10^{\frac{121-t}{z} - \frac{121-t-10}{z}} = 10^{10/z}$$

$$\log Q_{10} = \frac{10}{z} \quad (9.36)$$

Example 9.14 The D and z values for a particular microorganism are 3 min and 12°C. The microorganism can degrade a particular food component in 200 hours at 25°C. What will be the time of degradation of the same food component when stored at 35°C?

SOLUTION:

$$\log Q_{10} = \frac{10}{z}$$

Thus, $\log Q_{10} = 10/12$,

i.e., $Q_{10} = 6.813$

$$k_{t+10} = Q_{10} \times k_t = 6.813 \times k_t$$

Hence, the reaction rate will be faster by 6.813 times at 35°C that that at 25°C.
Thus, the shelf life of the food = 200/6.813 = 29.355 hours

In this chapter, we discussed the importance of water activity in food preparation and storage. We also distinguished between the moisture content and water activity and studied the different models to express water activity. The different types of microorganisms have different ranges of water activity for growth and reproduction. Similarly, the behavior of microorganisms during thermal processing was also discussed and the calculation of the process time for specific situations was analyzed, which form the basis of design of thermal processing parameters and equipment. The concepts discussed in this chapter will find applications in food processing operations involving moisture removal and heat processing.

Check Your Understanding

1. Milk stored for 18 hours at 20°C results in a 200 times increase in bacterial count. What will be the increase in bacterial count if the same milk is stored at the same temperature for 6 hours?

   Answer: 5.85 times

2. What is the doubling time (in minutes) of a bacterium with a specific growth rate of 2.3 h^{-1} in 500 ml of growth medium?

   Answer: 18.1 minutes

3. Milk pasteurization is carried out either at 85°C temperature for 4 s or at 71°C for 40 s. In both cases the sterilizing value is 8. What are the decimal reduction times for the two processes? Calculate the z values for reference temperatures of 71°C and 85°C. Also obtain the activation energy value for the pasteurization process.

   Answer: z = 14°C, E_a = 168405.74 kJ.(kg mole)$^{-1}$

4. The D_0 value and z value of a specific microorganism are 2.55 min and 10°C. What will be the process time for 8D inactivation at a temperature of 130°C? Use both the thermal death time model and the Arrhenius equation to compare the results. The z value was determined by taking observations at two temperatures, namely, 115°C and 121°C.

   Answer: 2.77 min

5. The shelf life of a food is 7 days when stored at 30°C. Assuming that Q_{10} value is 1.4 for deteriorative reactions that occur in the food, estimate the shelf life when the storage temperature is 10 °C.

   Answer: 13.72 days

Objective Questions

1. Define water activity and state its importance in food preservation. What are the factors that affect the water activity of a food?

2. Classify the microorganisms on the basis of their optimum growth temperatures.

3. Explain the growth phases of bacteria in a growing medium.

4. Derive a relationship between D value at any temperature and D value at 121°C.

5. Low acid foods require more severe thermal processing than high acid foods. Explain why?

6. Derive a relationship between the death rate constant and activation energy.

7. Define (a) generation time, (b) TDT curve.

8. Differentiate between

   a. D value and F value;

b.   D value and k value;

c.   Reaction quotient and z value.

BIBLIOGRAPHY

Adams, M.R. and Moss, M.O. 2008. *Food Microbiology*, 3rd ed. RSC Publishing, Cambridge, UK.

Cleland, A.C. and Robertson, G.L. 1986. Determination of thermal process to ensure commercial sterility of food in cans. In S. Thorn (ed.) *Development in Food Preservation*, Vol. 3. Elsevier Applied Science Publishers, London.

David, J. 1996. Principles of thermal processing and optimization. In J.R.D. David, R.H. Graves and V.R. Carlson (eds.) *Aseptic Processing and Packaging of Food*. CRC Press, Boca Raton, 3–20.

Fellows, P.J. 2000. *Food Processing Technology*. Woodhead Publishing, Cambridge, UK.

Hallstrom, B., Skjoldebrand, C., Tragardh, C. 1988. *Heat Transfer and Food Products*. Elsevier Applied Science, London.

Holdsworth, S.D. 1985. Optimization of thermal processing, a review, *J. Food Eng.* 4: 89.

Lund, D.B. 1975. Heal processing. In: M. Karel, O. Fennema and D. Lund (eds.) *Principles of Food Science. Vol. 2, Principles of Food Preservation*. Marcel Dekker, New York, 32–86.

Palaniapan, S. and Sizer, C.E. 1997. Aseptic process validated for foods containing particulates. *Food Technol.* 51(8): 60–68.

Pflug, I.J. 2010. *Microbiology and Engineering of Sterilization Processes*, 14th ed. Environmental Sterilization Laboratory, Otterbein, IN.

Rai, B. and Bhunia, A.K. 2007. *Fundamental Food Microbiology*. 5th ed. CRC Press, Boca Raton.

Ramesh, K.V. 2019. *Food Microbiology*. MJP Publishers, Chennai.

Ramesh, M.N. 1999. Food preservation by heat treatment. In: M.S. Rahman (ed.) *Handbook of Food Preservation*. Marcel Dekker, New York, 95–172.

Reed, J.M., Bohrer, C.W. and Cameron, E.J. 1951. Spore destruction rate studies on organisms of significance in the processing of canned foods. *Food Res* 16: 338–408.

Stumbo, C.R. 1973. *Thermobacteriology in Food Processing*, 2nd ed. Academic Press, New York.

Toledo, R.T. 2007. *Fundamentals of Food Process Engineering*, 3rd ed. Springer, New York.

Troller, J.A. and Christian, J.H.B. 1978. *Water Activity and Food*. Academic Press, London.

Van Den Berg, C. 1986. Water activity. In: D. Mac Carthyu (ed.) *Concentration and Drying of Foods*. Elsevier Applied Science, Barking, Essex, 11–36.

Section B

Unit Operations

B1

Primary Processing, Separation, Size Reduction and Mixing

10

Primary Processing

Some primary processing operations are carried out in the food production/processing areas before the food is actually processed. Unit operations include cleaning/washing, sorting, and grading. Peeling of fruits and vegetables is also a common unit operation in food processing plants. These unit operations help to improve the quality of the raw materials and to prepare the product for further processing. In this chapter, we will discuss the principles and different equipment used for these operations and their working principles.

10.1 Cleaning, Sorting, and Grading

10.1.1 Methods and Principles

Removal of undesirable materials including dirt, husk, straw, and other foreign matters from the raw material at the beginning of the processing operations is known as *cleaning*. The process helps to upgrade the quality of raw materials and subsequent finished products. In the case of fruits and vegetables, the surface of the food needs to be cleaned or removed before the subsequent operations. Peeling or descaling fish is, hence, considered as cleaning.

There may be different types of contaminants present in the raw materials received in a food processing plant, which may be categorized as metals, minerals, plant and animal parts, chemicals, and microbial cells and products. The microbial products can be in the form of discolored products, off-flavors, and toxins. The process of cleaning is based on the product and the type and nature of the admixtures/ dirt.

The cleaning methods can be broadly classified under two categories as:

- *Dry cleaning* (mechanical separation based on differences in size, shape, color, magnetic, frictional, and other such characteristics);
- *Wet cleaning* or washing (soaking, washing assisted with rotating drums, spraying and brushing of the commodity or shuffling of the wash water, flotation washing, ultrasonic cleaning, etc.).

As the raw material received in a food processing plant can have different types of admixtures, more than one cleaning procedure (or both dry and wet cleaning methods) may be followed.

Equipment, such as the screens, winnowers, destoners, color sorters, spiral separators, and magnetic separators, is used for separation of undesirable materials from the grains, dried peas, etc., that are characterized by their low moisture content,

small size, more mechanical strength, and comparatively freer flow than the fruits and vegetables.

As compared to wet cleaning devices (also known as washers), the cleaners involve smaller, cheaper equipment. The dry effluent can be discarded easily and cheaply, though there is capital expenditure for pollution control. Failure to remove dust from the work area also creates health and explosion hazards and may recontaminate the product.

Cleaning should be done as soon as the materials are received in the plant to reduce wastage of food, minimize pollution in the plant, and protect the machines for subsequent processing. It also protects the operators and helps to improve the economics of processing.

Scalping is a cleaning operation but refers to the initial removal of large particles prior to the actual cleaning operation. This is usually carried out before the raw materials enter the mill premises. It improves the capacity and efficiency of the cleaners installed in the mills and also reduces pollution in mill premises.

Classification or separation of products is based on various characteristics and is termed *grading*. *Sorting* refers to selection of a fraction of the product from the whole mass for specific use and/or according to a specific criterion. For example, the potato can be graded on the basis of size into many fractions. However, for preparing potato chips, we sort or pick the large size potatoes of the processable variety from the whole mass, i.e., sorting involves acceptance of a fraction as per the desired criterion/use and rejection of the remainder. The leftover material after sorting can be used otherwise. The most important initial sorting in a processing plant is for variety and maturity. Sorting and grading not only increase the market value, but also facilitate subsequent peeling operations, packaging, storage, and processing (heat transfer to the thermal center depends on size) operations.

Sorting and grading involve separation on the basis of different physical and other characteristics, such as size, shape, density, surface texture, color, total soluble solids, moisture content, freshness, ripeness, etc. Sorting is done by hand in small-scale operations. However, for commercial purposes, different types of sorters are available for different types of commodities and on the basis of separation. Sorting is also done on the basis of biochemical characteristics, such as TSS, acidity, moisture content, freshness, microbial load, etc.

Table 10.1 shows some examples of the properties used for cleaning and separation processes, including sorting and

TABLE 10.1

Common properties used for separation processes in food grains/horticultural products

Properties of input	Type of separator
Width and thickness	Sieves, sifters, thickness graders, inclined sifters, grading reels, etc.
Length	Indented type of disc type separator
Aerodynamic properties	Pneumatic separator, husk aspirator, cyclone separator
Form and state of surface	Spiral separator, belt type separator
Specific gravity and coefficient of friction	Separating tables, stone separator
Ferromagnetic properties	Magnetic and electromagnetic separators
Electrical properties	Electrostatic separators
Color	Electronic separators

grading in food grains/ horticultural products and the type of equipment used.

Effectiveness of separation

Let us consider a screen separator, which separates the feed on the basis of size. The undersize particles (particles having a size smaller than the opening of the screen) pass through the screen and the oversize particles (particles with a size larger than the screen openings) move on the screen surface and are discharged at the end of the screen. A 100% efficient operation would mean that there is no undersize material in the overflow and no oversize material in the underflow. However, this rarely happens as some small-size materials flow along with the large-size materials and are collected with the larger materials. Similarly, the collection below the screen may contain some oversize materials, which should have been ideally retained on the top of the screen. Thus, a term *separation effectiveness* is used to explain the ability of the separator to divide the feed into the two fractions. Separation effectiveness can also be termed *effectiveness of sorting* or *effectiveness of cleaning* on the basis of application.

Let us consider a screen separator, where F, O, and U are the mass flow rates of feed, overflow, and underflow, in $kg.h^{-1}$. The overflow can be either the desirable product or the reject (or byproduct). For example, if large-size stones are separated from rice on a screen, then the underflow is the desirable material and the overflow is the reject. Similarly, if broken rice is sorted from head rice on the screen, then the overflow is the main product.

Let X_f, X_o, and X_u be mass fractions of the desirable material in feed, overflow, and underflow, respectively.

$$\text{Then, } F = O + U$$

$$F.X_f = O.X_o + U.X_u$$

$$U = F - O, \text{ and } O = F - U$$

$$\text{Hence, } \frac{O}{F} = \frac{X_f - X_u}{X_o - X_u} \text{ and } \frac{U}{F} = \frac{X_o - X_f}{X_o - X_u} \quad (10.1)$$

The effectiveness of separation of the oversize materials in the overflow is

$$E_o = \frac{O.X_o}{F.X_f} \quad (10.2)$$

and the effectiveness of separation of the undersize materials in underflow is

$$E_u = \frac{U \times (1 - X_u)}{F \times (1 - X_f)} \quad (10.3)$$

The overall screen effectiveness is given as:

$$E = E_o \times E_u = \frac{X_o(1 - X_u)(X_o - X_f)(X_f - X_u)}{X_f(1 - X_f)(X_o - X_u)^2} \quad (10.4)$$

The above equation can also be used for determination of separation effectiveness in other types of solid separators (cleaners, sorters, graders, etc.), where the two fractions can be designated as desirable fractions and undesirable fractions. For convenience, the O and U can also be denoted as P and R to indicate the product (desirable fraction or main fraction) and reject, respectively.

Example 10.1 One single sieve was used to separate small stones from rice. After sieving 200 g of the grain, it was observed that 30 g of the sample passed through the sieve. On further analysis, it was observed that the materials retained on the screen had 10 g stones and the materials that passed through the sieve had 18 g rice grains with them. Find the effectiveness of the screen.

SOLUTION:

Here F = 200 g, O = 170 g, U = 30 g

$$X_o = 160/170 = 0.94$$

$$X_u = 18/30 = 0.60$$

$$X_f = (160+18)/200 = 0.89$$

$$E = \frac{X_o(1 - X_u)(X_o - X_f)(X_f - X_u)}{X_f(1 - X_f)(X_o - X_u)^2}$$

$$= \frac{(0.94)(0.4)(0.29)(0.05)}{(0.89)(0.11)(0.34)^2} = 0.4817$$

Example 10.2 An air screen grain cleaner having two screens was evaluated and the following observations were noted for a 200.0 g sample

Fractions	Clean grain outlet	Outflow of blower	Screen overflow	Screen underflow	Total
Clean grain	191.5 g	0.3 g	0.1 g	0.1 g	192.0 g
Impurities	0.4 g	5.0 g	1.5 g	1.1 g	8.0 g
Total	191.9 g	5.3 g	1.6 g	1.2 g	200.0 g

Compute the cleaning efficiency of the cleaner.

SOLUTION:

The different fractions are given in the problem; Thus, we get,

impurities present in feed: 4%
impurities present in clean grain: 0.21%
clean seed in outflow of blower 5.7%
clean seed in overflow of 1st screen 6.25%
clean seed in underflow 8.3%

Fraction of clean seed in feed 100−4 = 96% or 0.96

Clean seed fraction in clean grain outlet 100−0.21
= 99.8 %, or 0.998

Clean seed fraction in foreign matter outlet
= (5.7 + 6.25 + 8.3)/ 100 = 0.2025

$$\text{Cleaning effectiveness} = \frac{X_p(1-X_r)(X_p-X_f)(X_f-X_r)}{X_f(1-X_f)(X_p-X_r)^2}$$

$$= \frac{0.998(0.797)(0.038)(0.757)}{0.96(0.04)(0.7955)^2}$$

$$= 0.9416$$

The cleaning effectiveness is 94.16%.

10.1.2 Equipment

Screen Cleaners and Sorters

Construction and Operation

- Screening is the method of size-based separation of feed into different fractions. One screen can separate a mass into two fractions. A number of screens of different sized perforations can be used one above the other (as in a multi-deck flatbed screen) or in a series to obtain more than two fractions. The latter is commonly used for size grading. The schematic diagram of a two-screen grain cleaner is given in Figure 10.1.

- Perforated sheet metal or woven wire mesh are used to construct the screen cleaners. The openings are square or rectangular in the wire mesh. The openings may be triangular, oblong, or round in perforated metal sheets.

- The common arrangements for screening include shaking, gyrating, and vibrating the flat screens or rotating a cylindrical screen.

- A *trommel* or *revolving screen* rotates on its longitudinal axis. The cylinder axis is inclined downward.

- The material retained on the screen is known as the *oversize* or *plus* material and the material passing the screen is known as *undersize* or *minus* material. If two screens are used, then the material that passes through a particular screen and retained over the subsequent screen is the *intermediate* material.

- The openings on the screens are usually specified by "mesh," which is the number of openings per linear inch, counted from the center of any wire to a point exactly 1 inch away, or by the size of the clear opening between the wires. When the opening size is equal to or more than ½ inch, the screen is also called a *space cloth*. *Aperture* is the minimum clear space between the edges of the opening on the screening surface.

- Screens used for cleaning are often used with an air blast (*air screen cleaners*) for improving the effectiveness of separation. These are known as the air screen cleaners. Some accessories, such as screen brushes, oil cloth covers, screen knockers, and rubber balls, are provided with the screen cleaners to reduce the blockage of the pores and, thus, to improve the efficiency. Dash et al. (2012) have discussed in detail the types of screen cleaners used in grain processing industries.

- Large-size screens are also used for sorting and grading fruits and vegetables on the basis of size. If this is used for sorting on the basis of size, then the size of the openings of the screen becomes progressively larger from the feed end to the discharge end;

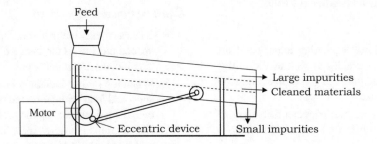

FIGURE 10.1 Schematic diagram of a two-screen grain cleaner

the smaller objects fall down near the feeding end of the screen and the larger ones move longer distances before they fall down. By having partitions at suitable locations, the material is graded on the basis of size.

- *Drum screens*, used for peas and beans, are either concentric (one drum inside another) or parallel (materials leave one drum to move to the other) or series types (a single drum constructed from sections with different size apertures).

- *Grizzly* offers a sloping grid made of metal bars. The bars may be of gradually reducing diameter, thus having a gradual increase in the gap between the bars, or of same diameter diverging type. The aperture can be either continuously diverging type or step wise increasing type. Cables or flatbed conveyor belts can also be used for continuous diverging systems in place of rollers. As the fruits flow parallel to the length of bars, smaller fruits fall down first and the larger ones move a longer distance and fall down as the gap between the bars increases more than the size of the fruits. The grizzly is often in the form of a short endless trough. In another type, the bars are fixed at right angle to the path of materials.

- A *rotary cylinder sizer* is a size-grading device for large fruits that is composed of three to five hollow cylinders, which are rotated by an electric motor. Each cylinder is perforated, and the size of the holes gradually increase from the feed end. Thus, fruits of a size smaller than the holes pass through and the large-size ones move to the next cylinder. Oversized fruits are collected as the overflow of the last cylinder.

Salient Features

- Different factors affecting the throughput of sieves are the physical nature and tension of the sieve material, the frequency and amplitude of shaking, the nature and shape of the particles, and the sticking of particles in the sieve holes (and the mechanisms employed to prevent this).

- For effective separation of grains, the rectangles and oblong openings in the wire mesh are oriented parallel to the direction of particle flow.

- In case of a grizzly or rotary cylinder sorter, used for sorting of large fruits, care should be taken to minimize the drop of the product from the screen to the collection chute to avoid bruising the fruits.

Applications

- The method is used for separating impurities from grains, fruits, and vegetables as well as for sorting and grading different types of agricultural commodities on the basis of size.

- Larger-size particles, i.e., those greater than approximately 50μm (0.05 mm), can be separated easily by sieves and screens.

- The equipment, like the grizzly and rotary cylinder sorter, work best with round commodities.

Disc Separator

Construction and Operation

- A *disk separator*, also known as an *indented disk separator*, separates different fractions of a mixture on the basis of differences in *length*.

- There are a number of disks inside a close housing, which are fitted in a series on a horizontal shaft. There are pockets or indentations on the disk surfaces, which are slightly undercut on each disk. As the disks are made to revolve through a mixture of grains or seeds, short grains or undersized particles are picked up by the disks. As the disk revolves, a condition will come when the pockets will face downward, the grains will then fall down and be carried into a trough located by the machine's side.

- Larger or whole grains are not lifted by the disks. These are moved by the disk spokes to the end of the machine and are discharged. The tasks of moving the grains through the machine, mixing them, and bringing them in contact with the pockets are performed by the vanes fitted with the spokes of the disks.

- There are three basic shapes of disk pockets of various sizes. The "R" pockets are used for separating broken pieces from rice; the label "R" refers to rice. The "V" pocket picks up and removes round-shaped grains; its label "V" comes from "Vetch."

- By using different combinations of the pockets of different shapes and sizes, a number of grades of the grains with different lengths can be obtained; in such situations the pocket size increases progressively from inlet to discharge.

Salient Features

- Density, moisture content, or seed coat texture do not affect the separation.

Applications

Separating small weed seeds from grains, size grading for precision planters, and removing brokens from whole grains can be conveniently carried out using a disk separator.

Indented Cylinder Separator

Construction and Operation

- It has a slightly inclined rotating cylinder with closely spaced indents on the inside of the cylinder surface. The indents are of hemispherical shape.

- When the grains, including brokens (or a mixture of short and long grains), are fed into the cylinder at one end, the indents pick up the brokens and short grains. These grains fall down from the inner surface of the cylinder as the indents face down during the rotation.

The magnitude of the centrifugal force determines the distance that grains travel before they fall.

- A longitudinal trough collects the short grains and brokens. The position of the trough can be adjusted to control the size of separation. There is a screw conveyor in the trough, arranged such that these materials are moved and collected at the upper end of the cylinder.
- The long grains are not picked by the indents and gradually move toward the other end of the cylinder due to the inclination.
- The cylinders with indents of different sizes are available; a particular cylinder has all indents of the same size.

Salient Features

- The separation is influenced by the shape and size of the indents, seed coat texture, seed density, and moisture content.
- The factors affecting separation include the speed of operation of the cylinder and the position of the adjustable trough. Too high speed of the cylinder will cause centrifuging and will not allow the grains to fall as desired. Too low speeds are also undesirable.

Applications

The cylinder separator has similar applications as the disk separator.

Gravity Separator/Destoner

Construction and Operation

- The separator permits the grains to flow down due to gravity over an inclined perforated surface and upward movement of air causes grain flotation.
- The separator has a slightly inclined deck with triangular-shaped perforations. Air is blown from the bottom through the perforations. There are baffles below the deck to have uniform distribution of air through it as the air is blown.
- There is a feed box at the upper end of the deck from which the grain is uniformly fed on to the deck. The air velocity should be sufficiently high so as to partially lift the granular material from the deck floor. It causes the materials to vertically stratify in layers as per density.
- The deck is also given an oscillating movement to force the heavy seeds in contact with the deck floor to move uphill while the air keeps the light seeds in floating condition to eventually move downhill.
- Ultimately, the light materials move to the lower end of the discharge edge and the heavy materials move to the upper end.
- A *destoner* is a special case of specific gravity separator to separate stones of the same size from seeds.

- The weight-based vertical stratification of the seed mixture combined with the separation of different strata by the oscillating deck leads to the successful operation.

Salient Features

- The pressure of the air rising through the deck can be controlled precisely, facilitating the handling of materials with different densities.
- The mixture can be graded into a number of fractions by putting cross slats at several locations on the discharge edge of the deck.
- The separation can be controlled by adjusting the feed rate, slope of the deck, deck vibration, and air flow rate.

Applications

- Gravity separators are very effective in removing badly damaged, diseased, deteriorated, and insect-infested seeds as well as weed seeds.
- This is used for removal of materials of the same size but different densities, such as stones of the same size from the grains.

Pneumatic Separator/ Aspirator

Construction and Operation

- In pneumatic separation system, air is blown to remove the lighter materials from the seeds, fruits, and vegetables.
- When the air stream crosses the mixture of seeds and lighter materials, the materials having terminal velocity lower than the air velocity (such as lighter seeds and impurities) are blown away.
- A typical equipment consists of a channel in which the mixture of materials is allowed to fall. Air is directed upward at a specified flow rate to enable the lighter materials to rise through the column and get collected at the top through a discharge port. The materials with higher terminal velocities are unaffected and fall down the channel and get collected at the bottom.
- In the case of aspirators, the fan is at the discharge end, which creates negative pressure (vacuum) to allow the atmospheric air to enter into the channel through a specified inlet close to the inlet of materials. The air velocity passing through the bed should be more than the terminal velocity of the lighter ones and less than the heavier ones.
- A *scalping aspirator* is a device in which rough separation of a feed mixture is obtained by dropping it into a rising air column.
- A *fractionating aspirator* is a device in which the air column expands gradually from the lower to upper end (Figure 10.2), and, thus, the air velocity gradually decreases from the lower section to the upper

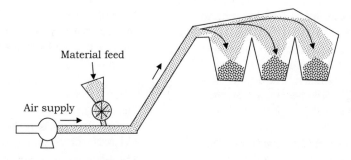

FIGURE 10.2 Fractionation of materials by using air stream

section. The seed mixture is introduced near the lower end. Heavier materials having higher terminal velocities than the air velocity at the inlet section fall down. The remaining materials are lifted by the air column and as the air velocity gradually reduces, these materials are collected in separate spouts as per their terminal velocities.

Applications

- Air separation is used for separation of lighter materials, chaffs, leaves, straw, seed coats, etc. from the seeds and grains. Air separation is often used in combination with screening to improve the effectiveness of separation.
- The scalping aspirator is used for initial cleaning and scalping operations in food processing plants to remove lighter chaff, straw, leaves, immature seeds, etc. from the raw materials.
- Fractionating aspirators can classify the materials on the basis of their terminal velocities, thus helping separation of materials on the basis of their moisture content.

Floatation Sorter

- Some products, such as peas, beans, etc., are sorted by flotation in brine. This causes the sound, mature, and denser seeds to sink and the immature ones to float.
- The method is also used to clean small seeds, such as mustard and pulses and vegetable seeds, from lighter impurities.

Roll Mill/ Velvet Roll Separator

Construction and Operation

- The velvet roll classifies seeds on the basis of their shape and surface texture.
- It consists of two parallel inclined rolls covered with velvet cloth and placed side by side close to each other. The rolls rotate in opposite directions.
- The seed mixture is fed at the upper end of the rollers.
- As the rollers rotate, the rough surfaced grains or the grains having sharp or broken edges are caught in the velvet and are thrown outside. The smooth grains are not held by the velvet and move toward the lower

end of the machine in the V-shaped channel formed by the rolls.
- There is an adjustable cover or shield above the rolls to guide the rough and flat seeds to be properly thrown over the sides.

Salient Features

- There are several categories in which the discharge from the side of the rolls is caught. The roughest seeds get ejected first.
- In a typical equipment, the feed rate, roll inclination, speed of rollers, and cylinder roughness can be adjusted to obtain the desired separation.
- A number of roll pairs can be used one above another to increase throughput capacity.

Applications

- This separator effectively separates the grain having a rough seed coat or sharp angles from smooth surface grains.

Magnetic Separator

Construction and Operation

- The magnetic separator can be used to separate metallic (magnetic) impurities from the grains or to separate different components of a grain mixture on the basis of stickiness properties and surface texture.
- The schematic diagrams of two types of magnetic separators are given in Figure 10.3. In cases where the separator is used to remove the metallic impurities, the magnet may be housed within a drum or below a flat belt, on which the particles are allowed to move continuously. The non-magnetic materials fall due to gravity at the end pulley of the belt or from the surface of the drum. The magnetic materials stick to the magnet and stay on the belt or move on the drum to a longer distance to the end of the magnetic field. Alternatively, a scrapper can be used to remove these materials from the drum/ belt.
- For separation of a seed mixture based on the stickiness properties and surface texture, a small amount of water and fine iron powder is mixed with the feed materials in a mixing device. The iron powder sticks to the rough, cracked, and sticky seeds. The seeds with a smooth surface are not affected.

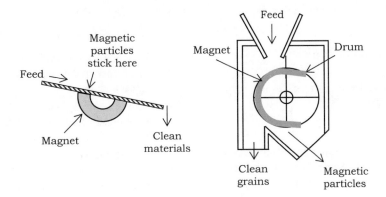

FIGURE 10.3 Schematic diagrams of two types of magnetic separators

This mixture is fed at the top of a horizontally revolving magnetic drum. Smooth seeds, relatively free of the iron powder, fall from the drum.

The seeds, on which some iron powder has stuck, move on the drum until the magnetic field exists and are collected separately. Otherwise, a brush can be employed to remove these seeds from the surface of the drum.

Salient Features

- The effectiveness of separation depends on the magnet's force of attraction. The degree of separation is also affected by such factors as difference in seed coats, amount of iron powder, amount of water mixed, and efficacy of powder-water mixing operation.
- As compared to ordinary permanent magnets, electromagnets are preferred where the magnetic field strength can be controlled by changing the strength of the electric current.

Applications

- Magnetic separators are used for removing magnetic impurities, such as pig iron, steel, nickel, and cobalt, to protect subsequent processing machineries and to improve food quality and safety.
- The separator can be used for separating weed seeds and seeds with rough surfaces from smooth surface seeds.

Inclined Belt Separator

Construction and Operation

- The differences in surface texture of the material and/or shape are the basis of separation.
- There is an inclined belt, which is moved slowly against the slope. A mixture of seeds is fed at the midpoint along the length of the belt. The round or smooth seeds roll / slide down the belt at a speed higher than the belt's speed in opposite direction, whereas the flat and rough seeds stick to the belt and move on the belt against the slope.

- Two hoppers, at the two ends of the belt, collect the separated materials.
- Belts of different degrees of roughness, made up of canvas or smooth plastic, are available for specific needs.
- Several belts simultaneously, one above the other, may be used in a single machine to have higher capacity.

Salient Features

- Separation is controlled through adjusting feed rate, angle of inclination, and the belt speed.
- High feed rates may not cause effective separation; very low feed rates cause reduced capacity.

Applications

- The separator is used to classify smooth or round seeds from rough and flat seeds.

Frictional Separator

Construction and Operation

- Angles of friction of different types of grains of even similar size differ; this forms the basis of the operation of a frictional separator.
- It consists of an inclined plane surface on which the grains are allowed to move.
- As the frictional forces on each constituent will be different, they will move at different velocities along the surface. Heavier particles will form the lower layer during the movement and move at a lower velocity; lighter particles will also move downward but at a higher velocity as the upper layer.
- The particle velocity at any point on the inclined plane (i.e., the particle moving in the lower layer) can be represented as follows.

$$v_1 = (2g.K_1. x_1)^{1/2},$$

where $K_1 = \sin \alpha - \mu_1 \cos \alpha$, $x_1 =$ distance travelled by the particle, which has initial velocity zero, $\alpha =$ angle of inclination, $\mu_1 =$ coefficient of friction

The particle velocity in the upper layer is given as follows:

$$v_u = v_l (1 + (K_u/K_l))$$

v_u = absolute velocity of particles in the upper layer ($v_u > v_l$ by the factor K_u/K_l)

Applications

- The equipment can be used for separating husked grains from the unhusked ones because the frictional angles for the two components differ considerably.
- The equipment is used for separation of similar size seeds as oats and hulled oats, millets, etc.

Fluidized Bed Cleaner/ Separator

Construction and Operation

- Differences in density and size of different types of grains in a mixture are exploited in a fluidized bed cleaner/ separator.
- It constitutes a perforated deck on which the granular materials are put in a thin layer. Air is passed vertically from below, at a controlled rate so as to create boiling or bubbling of the bed. This causes layering of the particles as per their densities and particle size.
- The mixture of grains, in a continuous system, falls from the hopper onto the deck with or without the help of a vibratory feeder.
- The deck inclination is 10–20° so that the fluidized grains move down the slope resulting in their classification.
- Toward the lower end of the deck, a clear separation of light materials (like fine chaff and dust etc.) at the top and the heavier materials (good seeds) in the lower layers is obtained.
- The channel is built in the shape of a converging taper with the bed becoming deeper at the discharge end and helping in better separation. A splitter assists in the collection of upper and lower layers in different outlets.

Salient Features

- The air flow rate has to be precisely controlled, which depends on the relative terminal velocities of the constituents to be separated. High air flow rates may not cause proper separation.
- Proper adjustment of the splitting device at the outlet is also important.

Applications

This equipment permits effective cleaning of lighter seeds, such as cabbage, carrot, onion, radish, lettuce, grass seeds, etc.

Vibratory Separator

Construction and Operation

- The vibratory separator makes separation of two constituents on the basis of differences in shape and surface texture.
- There is a rectangular deck, inclined both sideways and endwise. It is vibrated by an electromagnetic vibrator.
- When the mixture to be separated is fed near the center of the upper end of the deck the flat or rough seeds move to the higher side of the deck on the discharge edge. The round and smoother seeds tumble and slide to the lower side.
- Suitable dividers positioned along the discharge edge help to collect the different fractions.
- The position of dividers can be changed or additional number of dividers can be placed as per the requirement of separation.

Applications

This is used in rice mills for separation of paddy from brown rice.

Cyclone Separator

The cyclone separator is a device commonly used for separation of dust and light materials from air. This is a commonly used device with the spray dryers for collection of dried powders from drying air. Two common applications during grain processing are the collection of husk after husk aspiration in rice mills and the collection of light particles in an air screen cleaner. The cyclone separator is also used to separate solids from the discharge of a pneumatic conveyor. The cyclone separation principle is discussed in more details in Chapter 12.

Spiral Separator

Construction and Operation

- It separates the grains as per their shape (roundness) and ability to roll.
- There is a stationary, open screw conveyor standing vertically on one end, with inner and outer helices.
- The mixture is fed at the top of the unit. The round materials of the mixture quickly roll down the inclined surface and, as the velocity increases, the centrifugal force throws them into the outer helix. The non-round and flat materials move down slowly in the inner helix. The materials of the inner and outer helices are collected separately.

Salient Features

- The spiral separator has no moving part, and it does not require any external power supply, except for feeding the materials at the top of the unit.

- The feeding rate for effective separation needs to be controlled such that each grain/ particle rolls independently.

Applications

This separator is used for removing round seeds, such as mustard, rape, soybean, etc., from non-round seeds, such as wheat, flax, oats, etc.

Weight Sorter

Construction and Operation

- Large-size fruits and vegetables, eggs, etc. are sorted on the basis of weight.
- The equipment consists of a conveyor that carries the commodity above a series of counterbalanced arms.
- On the basis of weight, the arms release the heavy materials, which fall down and are collected. The light materials move on the conveyor to the next weigher.

  (One of the earliest devices developed to sort food items on a weight basis had a central rotating hub carrying a number of radially disposed cups balanced on single leaf spring arms. The weight of the food object would make the relevant cup to tip and discharge the object through a cam rail tipping arrangement. In some cases, spiral rollers consisting of aluminum pipes guided the input objects to the weight-sensing component where the lighter objects would continue to move forward while the heavier objects would drop down through a padded chute.)

Applications

- Such sorters have been developed to sort different types of fruits and vegetables.
- Weight sorting is more accurate than other methods and is, therefore, used for more valuable foods.

Automated Electronic Sorters

Some physical or biochemical parameters, such as color, shape, size, and structure, are used to differentiate between the objects in a mixture with the help of one or more digital and/or analog signal processors and I/O devices. Automatic digital sorters are non-destructive devices. These intelligent sorters improve the quality of the output, increase capacity, and improve the performance.

In optical sorters, reflected visible, hyperspectral, or x-ray radiation from the objects to be sorted is analyzed to separate them out. Incident radiation may be narrowed down to specific wavebands to enhance the separation efficiency. There are now specialized cameras or receivers available to receive the images or signals, process them rapidly, and control the sorters for best results. The sequence of actions is as flows: (1) entry of the objects to be sorted, (2) aligning of the material for processing by vibratory feeders, (3) movement and orientation of the material in a single line along the specified path, (4) analysis of the reflected light from the singulated input material for its rejection or acceptance, and (5) either rejection or acceptance of the material.

Color Sorter

Construction and Operation

- A color separator separates the raw food materials, such as peas or grains, on the basis of their differences in color or brightness.
- It consists of an optical chamber; there are two photocells fixed at a particular angle on one side and there is an electrically charged needle on the other side. On the lower side there are two electrodes that are maintained with a high potential difference between them; the gap between these electrodes forms the grain outlet.
- As the grain mixture is fed uniformly through the chamber along the specified path, the photocells direct their beams to one point along the path of the grains (Figure 10.4). When the beam encounters a dark object, it actuates the needle to impart a charge to the object. The particles pass the photo detector one at a time.
- The grains then move between the two electrodes, where the charged objects are segregated from those that are not charged.
- In another design, the machine picks up the seeds on a series of suction fingers and takes them past a phototube. They are ejected into separating containers, one at a time, on the basis of their color or brightness.
- In yet another design, the photodetectors measure the reflected color of each piece and compare it with preset standards. The defective pieces are separated by a short blast of compressed air.

Salient Features

- Small-particulate foods are automatically sorted at very high rates.
- The angle, shape, and lining material of the chute can be varied to control the velocity of pieces as they

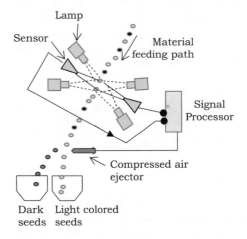

FIGURE 10.4 Schematic representation of a color sorting device

pass the photodetector. The color of the background, type, and intensity of light used for illuminating the food are also closely controlled.

Applications

The separator is used for large seeds and small freely moving vegetables for removal of off-colored ones. The color within the same crops and variety may be due to differences in maturity and disease.

Electrostatic Separator

Construction and Operation

- Electrostatic separators separate seeds on the basis of differences in their electrical properties.
- When different seeds are charged with static electricity by a specific source, the magnitude and nature of the charge imparted to the particles vary depending on the electrostatic characteristics of the seeds. Some seeds retain the charge for a longer period as compared to others. When these seeds are passed through another electric field then the action of the outer electric field produces some mechanical work on the charged particles, which is used for separation.
- The materials are moved on a belt and are charged. The seeds gradually lose their charge. At the end of the belt, the seeds, which have quickly lost their charge, fall in a normal manner from the belt. However, the poorly conducting seeds and the seeds that lose their charge slowly, adhere to the belt and move to a slightly longer distance along the pulley before falling down.
- Suitably placed chutes collect the different fractions of materials.

Salient Features

- The degree of separation depends on the relative ability of different fractions in the seed mixture to conduct electricity or to hold a surface charge.
- The magnitudes of the charge imparted to the seeds depend on their electrical conductivity and dielectric constant. The other properties as particle size and shape; specific gravity and design of the separators also affect the magnitude of the charge.

Applications

This separator can separate different seeds of the same size having different electrical properties and also same seeds having different moisture levels.

Picker Belt

Separations that cannot be made by machines may be made by hand, or hand picker belts. There is a moving belt on which the materials to be sorted move. The operators stand on both sides of the belt and manually sort and remove the undesirable materials from the belt. The device helps to reduce drudgery and improve capacity of manual sorting.

Various small-size tools are also available for manual sorting of the large size fruits. Besides, many different types of equipment are also available for sorting and grading specific commodities, such as an onion grader, coffee grader, etc.

10.1.3 Washing

The wet cleaner or washer is used for soft fruits and vegetables for removing soil and other sticking contaminants, pesticide residues, etc. Different types and combinations of detergents and sterilants can be used. This method is specifically used for root crops and fruits, such as oranges and mangoes, in which dirt often sticks to the surfaces. Washing is also often helpful in initial cooling of the materials as they are received in the processing plant. In addition, it reduces the microbial load and the residues of fungicides, insecticides, etc. from the surface of the fruits. Often the fruits are washed in a continuous tunnel while they are conveyed from the receiving yard to the initial processing section.

Washing with warm water may accelerate chemical and microbial spoilage; this is promoted if the materials are not immediately cooled after they are washed with warm water. There is also a large volume of effluent, which has to be properly disposed. Wash water after suitable filtration and chlorination is often recirculated.

Methods of Washing

Major modes of washing are discussed below.

  Soaking. It involves soaking the fruits, vegetables, and root crops in still or moving water. Often this is used as a preliminary step before other washing modes, such as spraying or shuffling, to loosen the mud on the surface of the commodity. Often two to three drums are used, and the materials move in a series from one drum to other for step-wise soaking and washing.
  The drums may be made of galvanized sheet metal or concrete fitted with drain holes. Steel drums, cut in half, can be used for the purpose. The metal edges should be suitably covered with split rubber or plastic hose. There is a provision for adding fresh water and removing the floating produce from the drum. If

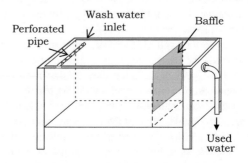

FIGURE 10.5 A simple soaking and washing tank

concrete tanks are used, they should be tiled. Figure 10.5 shows a simple soaking and washing tank made of sheet metal.

Floatation cleaning. The lighter impurities float in water and the heavier materials remain at the bottom. This is helpful for peas and some small vegetables.

Shuffle washing. In such devices, the wash water is given a vigorous reciprocating motion. This is used for difficult cleaning jobs; however, it should not be used if the product may become damaged by shuffling.

Spray washing. It consists of using sprays to wash away the dirt. The intensity (pressure of the spray) and the type of distribution of sprays are decided on the basis of the product and contaminant characteristics. Modern installations use a pressure of 15 atm or more. This method requires a large quantity of water.

Rubbing by brushes. It consists of rotating brushes to rub the surfaces of the commodities while the product moves on a specified path. The relative movement of the brushes moves the materials from one end to the other end. The brushes are made up of fiber, rubber, sponge, or other materials.

The washing time is controlled primarily by the relative motion of the brushes and also by the flow of wash water if the product is guided in a water bath. It may be combined with sprays or soaking. The brushes may have to be replaced periodically.

Ginger is often brushed without using water to remove the clods of soil.

Rotary drum washing. It consists of a perforated long drum with its axis slightly inclined to horizontal (Figure 10.6 and 10.7 b). The drum may be made of wire mesh or of a metal sheet with a large area of perorations fastened onto a suitable frame. The drum is rotated along the axis with a suitable drive mechanism. The drum is partially immersed in a tank of water. The materials are fed from one end of the drum and as the drum rotates slowly, the materials move slowly to the lower end being immersed in water. There may be screw flights inside the drum to guide the movement of the materials from one end to other. During this movement, the materials tumble, rub against the inner surface of the drum and also against each other so that the sticking dirt is loosened and removed.

There may be auxiliary spray nozzles near the axis of the drum for better washing effectiveness.

The time of washing can be controlled by the inclination of the drum and by the helical flights on the inner surface of the drum. The efficiency of cleaning depends on the machine parameters, such as roughness or amount of corrugations on the inside surface of the drum, the speed of rotation, and the duration that the materials stay in the drum. The other characteristics of this device include high capacity, simplicity in operation, and almost no damage to the product.

This washer is suitable for the materials those can easily roll down the drum, such as potatoes, large fruits, etc.

This type of washer is effective for removing sandy or loamy soil and for removing spray residues from fruits.

Ultrasonic washing. Ultrasonic cleaning is a process that uses ultrasound (usually between 20 and 40 kHz) to agitate a fluid during washing and is used for cleaning equipment parts, scientific samples, and process equipment as heat exchangers, pipes, etc. There is a chamber to hold the materials and an ultrasound generating transducer built into the chamber (or lowered into the fluid as and when required). Transducers are usually piezoelectric (e.g., made with lead zirconate titanate (PZT), barium titanate, etc.), but they are sometimes magnetostrictive. During the process, the ultrasonic waves produce cavitation in the fluid, i.e., many millions of partial vacuum bubbles are formed, which collapse with enormous energy. The agitation of the water causes high forces on contaminants adhering on the surfaces of the

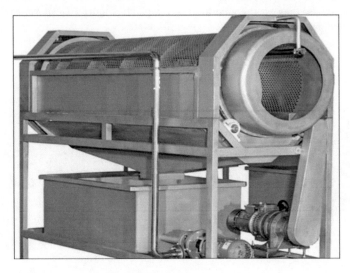

FIGURE 10.6 Rotary drum washer (Courtesy: M/s Shiva Engineers, Pune)

product, and can separate oil, rust, grease, bacteria, lime scale, etc. The materials to be cleaned should be non-absorbent (metals, plastics, etc.) and inert to the cleaning fluid. The higher the frequency, the smaller the nodes between the cavitation points, which helps better cleaning. It also helps to remove the dirt from cracks and blind holes. Three stage cascading tanks are also available, where the fluids can be reused.

The fluid is usually water, although solvents of appropriate types are also used depending on the dirt. Cleaning solutions also can contain detergents, wetting agents, etc. and the selection of the solution depends on the item cleaned. A good wetting agent may be added to reduce the surface tension to increase the cavitation levels. The solutions are usually used at a temperature of 50–65°C. The time of washing is usually 3–6 minutes, though more time can be expended, even up to 20 minutes for difficult cleaning.

Commercial Washers

In commercial washers usually more than one washing mode is involved. Spray nozzles in combination with other washing devices help to improve performance.

The commercial washers can also be classified as batch type or continuous types. In continuous types, a conveyor carries the commodity under sprays of water. To reduce water consumption, the water is recirculated through a filter. Washing is also sometimes integrated with the movement of raw commodities into the plant. Figure 10.7 shows two such commercial washers.

Some points, which should be kept in mind for washing fruits and vegetables, are as follows.

- The type of washer should not damage the fruits. The rotary washer should not be used for fragile vegetables.
- Washing should be done as soon as the commodity is received in the plant to remove the "field heat" and also to reduce the load and growth of microorganisms.
- Spoiled, skin-damaged, and rotten fruits should be removed before washing, or else they cause contamination of fruit and the equipment.

- Washing should be done before cutting and peeling to prevent loss of solids and nutrition.
- Pre-washing can be done by dipping in cold or warm water (up to 50°C) with or without detergents. Compressed air may be supplemented for bubbling of the water. If warm water is used for washing or pre-washing, the product needs to be immediately cooled to normal temperature by air currents, which also helps surface drying. For fruits and vegetables, rapid washing and a short re-drying help to better retain the quality.
- Often chlorinated water is used for washing to reduce microbial levels. Alternatively, permissible detergents or 1.5% HCl solution could be added to the washing water. The level of chlorine should be monitored to avoid excess chlorine smell in the washed product. Special disinfecting/cleaning agents are also available, which work better than the chlorinated water. These also incorporate wetting agents that give better washing.
- Water needs to be changed frequently if the soil and other contaminants are more. To reduce the amount of water consumption, it is advisable to recirculate water through a filter. As discussed earlier, a series of tanks may be used, the first to collect most of the dirt and the subsequent ones for step-wise final washing.
- Washing reduces the microbial load and better washing causes higher reduction of microorganisms from the surface of the fruits. Therefore, the ratio of the total number of microorganisms present on the fruit surface after and before washing subtracted from 1.0 will be a measure of the washing efficiency. Usually, a sixfold reduction indicates good washing. The final wash water should have a very low bacteria load and no mold and yeast.

In addition, commercial washers are also available for washing of crates, cans, bottles, and utensils used in food processing plants. Figure 10.8 shows one such washer and dryer used for crates.

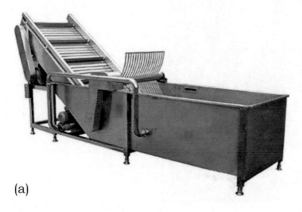

(a)   (b)

FIGURE 10.7   (a) A continuous immersion type vegetable washer (Courtesy: M/s Shiva Engineers, Pune), (b) A rotary drum type commercial vegetable washer (Courtesy: M/s Bajaj Processpack Limited, Noida)

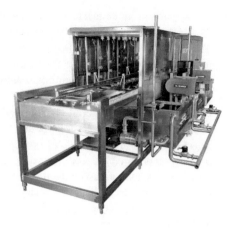 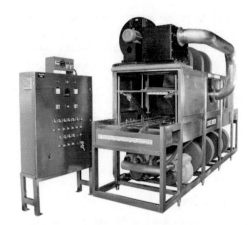

FIGURE 10.8   Crate washer and crate dryer (Courtesy: M/s Bajaj Processpack Limited)

10.2 Peeling

10.2.1 Methods and Principles

Peeling or removing the inedible peel or skins of fruits and vegetables is an important primary processing operation. The objective, ideally, is to remove the peel only and retain the edible portion intact. However, depending on the peeling method and the precision of peeling, there may be some damage to the edible portion. The effort is to minimize the damage.

Peeling is done primarily by manual, mechanical, chemical, and thermal processes. However, infrared peeling, ultrasonic peeling, ohmic peeling, and enzymatic peeling methods have also been developed more recently.

Manual peeling with sharp, stainless steel knives is used only when the other methods are either not advisable or sometimes as a complementary operation to the other three ways. Peelers for specific commodities are available.

10.2.2 Equipment

Mechanical Peelers

Construction and Operation

- In the case of mechanical peelers, the surface of the fruit is cut/rubbed for removing the skin. These can be grouped on the basis of their working principle as a knife peeler, abrasion peeler, rotating drum peeler, and brush peeler. This process may be either a batch or a continuous one.

- The knife peeler has knives for peeling commodities, such as apples, pears, potatoes, etc.

- The abrasion peeler has an abrasion device, such as carborundum (silicon carbide) coat or grinder, to grind off the peel of crops. Carborundum stones of varying degrees of roughness are also available.

- The rotating drum peeler has a rotating drum made of a sieve material to peel the root vegetables, turmeric, etc.

- The brush peeler features brushes for rubbing and removing the skin. Brushes having varying degrees of stiffness are available.

Salient Features

- The process should remove only the outer skin of the product, leaving no material on the surface for later removal by hand trimming. Peeling works better for uniform and round potatoes, for which preliminary sorting is done.

- In most cases the shapes of the products are uneven, and, hence, there is loss of edible portions during peeling, which can be expressed as peeling loss. The peeling loss is affected by the (1) commodity characteristics, namely, the size and shape of the commodity, and (2) equipment and operational characteristics as the depth of the cut by a knife or the roughness of an abrasive surface, load going to the peeler, time of peeling, and the rotation of the drum, etc.

- Usually, small-size fruits loose more edible portions as compared to large ones because of their comparatively higher surface area per unit weight.

- Proper loading is important. The peelers can perform well when operated at full capacity, which ensures that all surfaces of the commodity are equally exposed to the rasping action. A lower load will cause the same surface to be more exposed to the peeling surface and will cause greater loss of edible material.

- Keeping the commodity for more time in the peeler or using higher speeds of the drum or knives than recommended will cause higher loss.

- A small amount of water is added during the peeling for easy removal of the peel fragments and better polish.

- Washing of the raw material should precede peeling or else the peeled tubers may be contaminated. For some commodities, blanching precedes peeling.

Applications

- The knife peeler is used for apples, pears, potatoes, etc.

- The abrasion peeler is used for crops like potatoes, carrots, sweet potatoes, and other root vegetables.

For preparation of potato chips, abrasive peelers are preferred over chemical peelers. Abrasion peeling results in about 10% loss of the original tuber weight.

- The rotating drum peeler is commonly used to peel root vegetables, turmeric, etc.
- The brush peeler is used for products with soft skins, often also used as a secondary step after chemical peeling.

Chemical Peelers

Construction and Operation

- The fruits and vegetables are dipped in a caustic chemical solution (most commonly, sodium hydroxide (NaOH) or potassium hydroxide) in a range of 76–100°C. The solution is known as lye and, thus, chemical method is also known as *lye peeling*. The hot alkali solution dissolves the pectic and hemicellulosic materials in the cell walls and loosens/separates skin from the inner tissues.
- The usual concentration of lye is about 0.5–2% though it can go up to 15% and the time of dipping is about 0.5–5 minutes depending on the type of commodity.
- The chemical peeler thus consists of a tank to hold the lye solution and the product may be immersed in a batch wise or continuous manner. After treatment, the vegetables are tumbled in a wash to separate the loosened skin.
- For a continuous method, the speed of the conveyor is adjusted to maintain the length of time of the product in the tank. The vegetables with loosened skins are conveyed from this tank to another section, where water jets remove the skin and residual lye.
- In some cases, the lye solution is sprayed on the commodity. Wetting agents are also added for better contact of the solution and the commodity.

Salient Features

- The concentration of the lye must be properly maintained throughout. A significant amount of caustic is carried out by the product; thus, an adequate amount of makeup caustic has to be added to maintain the concentration of lye. Some water may also be added to maintain the level of water in the tank. The concentration can be measured by measurement of toroidal conductivity of the solution and using the calibrated curve. Routine titration can be used to crosscheck the information.
- Peeling time and peeling quality depends on the concentration of the alkaline solution.
- NaOH is preferred to KOH as KOH is generally expensive than NaOH.
- With a view to prevent enzymatic browning, chemically peeled material is often boiled in water for a short duration or immersed in diluted citric acid solutions.

- In comparison to the mechanical peeler, the chemical peelers involve less cost, less time, and less waste.
- The disadvantage is the treatment with the chemical and the possible residue after the treatment. From a food safety standpoint, the minimum recommended concentration of lye is advisable.
- Proper selection of materials for the peeler is also important as the chemicals are eroding.
- The timing and concentration of the lye have to be properly monitored as over-exposure affects the inner tissues and causes oxidation. The concentration of the lye has to be periodically checked and corrected as the lye concentration reduces during the process.
- There should be proper disposal of skins and chemicals.

Applications

Lye peeling is used with commodities like peaches, tomatoes, kiwifruit, and potatoes. Only the recommended lye concentration should be used. Recent trends are gradually aiming to reduce the use of chemicals.

Thermal Peelers

The thermal peelers may be hot air/hot water or steam peelers. These are also known as wet heat and dry heat peelers.

Dry heat peeler or flame peeler. Vegetables can be peeled by exposing them to hot gases or direct flame (about 1 minute at 1000°C). The most commonly used peelers are the rotary tube type. The intense heat from the flame or the hot gas causes steam to develop under the skins of vegetables such as potatoes, loosening them, and subsequently washing the skin off by a water jet.

Wet heat (steam) peeler. The vegetables are exposed to steam under pressure (about 10 atm) in a closed chamber. The skin and the underlying tissue of the vegetable are softened by the high-pressure steam. Subsequently, the steam under the skin inflates as the external pressure is suddenly released, allowing the skin to expand and get cracked. The cracked skin is then easily washed away with the help of a jet of water at high pressure (up to 12 atm). Thick-skinned vegetables, such as beets, carrots, potatoes, and sweet potatoes, can be peeled this way. Such high-pressure peelers are batch processes, and the resultant capacity is lower.

In either the chemical or the thermal peeling methods, which involve application of heat, the commodity must be cooled quickly or else the product quality is impaired.

Infrared Peeler

Lye and steam peeling methods require the use of chemicals or heat. Lye peeling effluent has high pH and poses disposal problems. Steam peeling requires intensive use of thermal energy. A peeling method requiring neither chemicals nor steam would be preferable.

Infrared thermal radiation is in the bandwidth of 760 nm to 1.0 mm. This radiation is easily absorbed by the organic compounds and moisture in the peel. Surface temperature of a 50 mm diameter tomato increases by 10°C when exposed to IR radiation for 60 s while the tomato flesh temperature remains close to the initial fruit temperature. IR heat causes the internal vapor pressure below the tomato skin to rise rapidly, causing the skin to rupture. IR peeling has been found to give better peeling performance as compared to the mechanical, chemical, or thermal methods. IR peeling could save about 25% energy as compared to lye and steam peeling. Tomatoes, pears, peaches, and apricots have been successfully peeled using IR peeling method. The uniqueness of IR peeling is the chemical and water less peeling resulting in better economy and environmental sustainability.

Ultrasonic Peeler

Ultrasonic peeling utilizes the low frequency ultrasound (20–100 kHz) to treat the fruits and vegetables submerged in hot water. The ultrasonic cavitation effect through compression and rarefaction of high frequency sound waves results in detachment of the skin from the flesh. It is the synergistic effect of the power ultrasound and hot water that causes the peeling. Ultrasonic peeling has been successfully tested with the tomato. Tests have revealed that the application of ultrasound for tomatoes increased the content of secondary metabolites in the fruit without affecting its quality.

Ohmic Peeler

Peeling is achieved by electro-heating the fruit surface by controlling the electrical conductivity of the peeling medium. It has been tested for tomatoes where the fruits are immersed in a salt or sodium hydroxide solution. During the peeling process, the fruit acts as an electrical resister where the skin is heated when a current is passed through it. Ohmic heating could also be used along with lye peeling to achieve better results.

Enzymatic Peeler

Enzymatic peeling uses the power of enzymes, such as pectinases, hemi-cellulases, and cellulases, to hydrolyze polysaccharides in the peel of the fruits and vegetables. These enzymes are allowed to infuse through the peel to loosen the attachment between the peel and the flesh. Two commonly used enzymes, poly-galacturonase (PG) and pectin methylesterase (PME), hydrolyze pectin in the plant cell wall and the middle lamella. For a specific fruit or vegetable, it is necessary to select the appropriate enzyme, its concentration, temperature, and pH to obtain successful peeling. Peeling of citrus fruits, grapefruits, stone fruit, and many other commodities has been successfully carried out using enzymes.

In addition to the above-mentioned innovative peeling methods, such methods as vacuum peeling, cryogenic peeling, peeling with calcium chloride, acid peeling, etc. have been studied. The commercialization of these methods will depend on scaling up of their capacities and lowering their costs.

10.3 Blanching

10.3.1 Methods and Principles

Blanching is the process of dipping vegetables and some fruits in hot water or exposing them to live steam for a defined time period prior to further processing to destroy the activity of enzymes present in these raw materials. As discussed in Chapter 9 under the section on thermo-bacteriology, the enzymes have specific D and z values and the rate of inactivation of enzymes is dependent on the treatment time and temperature. In addition, blanching leads to a reduction in the number of microbes on food surfaces. As a consequence of the blanching, the tissues of the vegetables and fruits undergo softening, which, in turn, facilitates the filling of blanched material into containers, removal of air from intercellular spaces, and formation of head space vacuum in the cans. The time and temperature combination chosen for blanching is such that it ensures adequate enzyme inactivation and prevents excessive softening of tissues while retaining food flavor. Blanching in certain cases helps in peeling commodities like beet root and tomatoes by loosening the skin.

Blanching is done before freezing and dehydration to prevent degradation of color. Also, blanching is done prior to canning to prevent any deterioration in the canning process before it is actually thermally processed. As the time and temperature of treatment are not enough for the desired bacterial reduction for safe storage and consumption, it is not an independent method of preservation.

For adequate inactivation of enzymes and protection of the texture and nutrients, the following sequence of operations is needed.

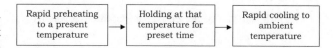

10.3.2 Equipment

Blanching can be done with steam, hot water, or a combination of both. The blanching can be done either in batch or in continuous manner. Accordingly, the blanchers can be primarily grouped as steam or water blanchers.

Continuous Steam Blancher

Construction and Operation

- It consists of a mesh conveyor on which the pieces of commodities are moved through a steam chamber.
- The steam chamber may be about 15 m long, 1-1.5 m wide, and up to 2 m high.
- A schematic diagram is shown in Figure 10.9(a). There may be another chamber for pre-conditioning before the steam chamber and one for cooling after the steam chamber.
- The cooling can be done by fog sprays, resulting in reduction in evaporative losses from the food as well as the amount of effluent.

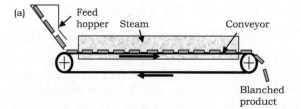

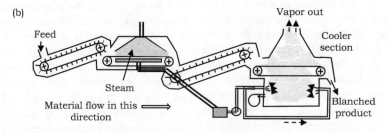

FIGURE 10.9 Schematic diagram of (a) a simple steam blancher, (b) a typical IQB continuous steam blancher and cooler (after Fellows, 2000, with permission)

- The residence time is decided by the conveyor speed.
- To prevent the steam from escaping out of the chamber, water sprays, rotary valves, or hydrostatic seals are used at the inlet/outlet.
- In case of individual quick blanching (IQB), the food, in a single layer, is heated to the desired temperature and held there for the desired time. Figure 10.9(b) is a schematic of a typical IQB continuous steam blancher and cooler.

Salient Features

- The steam can be reused by passing through venturi valves.
- High throughput capacities of up to 4000 to 5000 kg.h⁻¹ can be obtained.
- Conventional steam blanching usually gives a nutrient loss of about 5%. A short preconditioning of the commodities by warm air (65°C) reduces the loss of nutrients. However, the cost of preconditioning must be balanced against the cost of effluent treatment.
- IQB helps in reduced heating time and increased energy efficiency up to 85% and more. As an example, for 1 cm diced carrot, the heating and holding times are 25 s and 50 s, respectively. IQB also helps in increasing the amount of the product that can be blanched per kg of steam. The amount of small particulate foods per kg of steam varies, in general, from 0.5 kg to 6–7 kg.

Batch Type Fluidized Bed Blancher

Construction and Operation

- This consists of a chamber in which an air and steam mixture is used for blanching.
- The air velocity is approximately 4–5 m.s⁻¹ and the materials should be in a fluidizable form.

Salient Features

- The food needs to be circulated uniformly and continuously.
- It helps to produce faster and more uniform heating.
- The loss of soluble heat-sensitive components, such as vitamins, and the volume of the effluent are reduced.

Hot Water Blancher

- In the hot water blanchers, the water temperature is kept between 85° and 100°C. There is provision for heating the water in the tank and maintaining its temperature at the desired level.
- After the food is heated for the desired time, it is immediately taken out and cooled quickly. Cooling is usually done by water at room temperature.
- The *batch type hot water blancher* consists of a large tank in which there is provision for heating water and maintaining it at the desired constant temperature. This is used for small establishments. In some designs, there is a provision for cooling the produce immediately after blanching in a separate tank by immersing in cold water. Figure 10.10 a shows a batch type blancher-cooler.
- A *conveyor type blancher* (Figure 10.10 b) carries the materials by a conveyor through the hot water kept in a tank. The time of travel is adjusted with the time of blanching.
- A *reel blancher/ rotary blancher* has a slowly rotating and slightly inclined cylindrical drum wrapped with a mesh submerged partly in hot water. Internal flights of the drum guide the movement of materials within the drum. The commodities remain immersed in the hot water and get blanched as they move in the drum from the input end to the discharge end.
- A *pipe blancher* is essentially an insulated metal pipe having feeding and discharge ports. Hot water

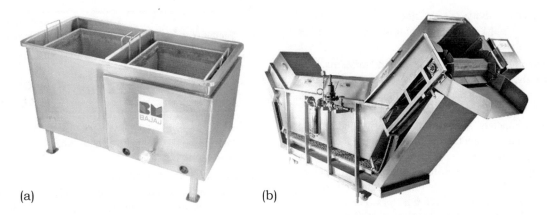

FIGURE 10.10 (a) Batch type blancher cooler (b) Conveyor type blancher (Courtesy: M/s Bajaj Processpack Limited, Noida)

is moved in the pipe, which also carries the commodities with it. The residence time for the material to be blanched is determined by the pipe length and the velocity of the water. Such blancher offers higher capacity in a smaller floor space. Another important aspect is that the conveying and blanching can be integrated here.

- The blancher cooler shown in Figure 10.11 is based on the individual quick blanching principle. A single conveyor carries the commodity through the different sections as pre-heating, blanching, and cooling. There is heat regeneration for preheating (and also the water which loses heat is used for precooling). It requires less water (approx. 1000 liter per 10 MT of product) and less energy (since up to 70% of heat can be recovered). More product can be blanched per kg of steam (up to 20 kg commodity in a blancher cooler as compared to only 0.25–0.5 kg in conventional hot water blanchers). It also reduces the amount of effluent.

Steam + Hot Water Blancher

This involves heating the commodity in steam for achieving a rapid temperature rise and then there is a hot water spray to maximize contact and, thus, the heat transfer. There is also immersion in water for direct heat transfer between hot water and individual product pieces.

Other methods, i.e., vacuum steam blanching, hot air blanching, microwave blanching as well as in-can blanching, are also available.

10.3.3 Considerations during Blanching

- The effectiveness of blanching depends on the method, treatment parameters, size of food pieces, and type of fruit or vegetable.

- The effectiveness of blanching or *degree of blanching* is determined by testing the residual catalase/peroxidase activity after blanching. Peroxidase is more heat resistant than catalase and the inactivation of peroxidase activity implies destruction of other less heat-resistant enzymes. The severity of blanching is also indicated by the loss of ascorbic acid.

- Under-blanching has been observed to cause more damage in comparison to no blanching. If the heat treatment disrupts the tissues only and does not inactivate enzymes, it will cause more degradation due to more exposure of the substrates to the enzymes. If only some enzymes are destroyed, it may cause

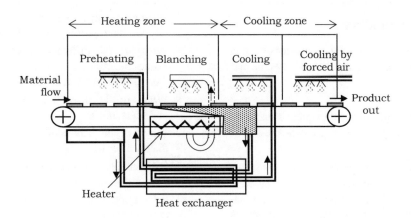

FIGURE 10.11 Schematic diagram of a continuous blancher cooler (after Fellows, 2000, with permission)

increased activity of the others. Under-blanching may lead to the development of off flavors in dried or frozen foods during storage.

- Blanching causes loss of vitamins and minerals, specifically, the water-soluble ones. The loss takes place due to leaching in blanch water, thermal destruction, and oxidation. The extent of losses depends on several parameters, including the maturity and variety of the food, the method of blanching, the method of food preparation (e.g., extent of slicing, cutting, or dicing), the surface area-to-volume ratio of the pieces of materials, the blanching time and temperature, the method of cooling, and other factors. There is lower loss in high temperature short time (HTST) methods. The commodity-water ratio in water blanching as well as cooling also affects the loss of water-soluble constituents.

- In general, steam blanching leads to smaller loss of water-soluble constituents, reduced wastes (effluents), reduced cost of waste disposal, and reduced capital as well as running cost. It is easy to keep the commodities and equipment clean and sterile. In comparison to water blanching, steam blanching is more uniform.

- During blanching, the color of some foods is brightened primarily due to the removal of hair and dust from the surface. There is destruction of some food pigments depending on their D value. The green color of the vegetables can be retained by adding calcium oxide or sodium carbonate (0.125% w/w) to the blanched water.

- Enzymatic browning can be prevented by immersion of cut potatoes and apples in dilute salt solution (NaCl- 2% w/w) prior to blanching.

- As discussed earlier, blanching softens the commodities and, therefore, helps filling the blanched product in cans. However, excess softening needs to be prevented for some foods. Insoluble calcium pectate complexes are formed by the addition of calcium chloride to blanched water. It helps in the maintenance of tissue firmness.

Check Your Understanding

1. If a square wire mesh has a wire diameter of 1 mm and the size of the opening is 2 mm, what is the percentage of open area?

    Answer: 44.44%.

2. What will be the open area in a square mesh wire cloth with a size of opening 2 mm, and the mesh is 8?

    Answer: 44.44%.

3. During evaluation of a grain cleaner with two screens, one above the other, the 200 g samples were collected each from different points and were analyzed for the desirable and undesirable fractions, as given in the following table.

Sample fraction	Desirable material, g	Undesirable material, g
Feed, g	188.0	12.0
Clean grain outlet, g	198.7	1.3
Oversize outlet, g	8.51	191.49
Undersize outlet, g	36.36	163.64

Determine the cleaner's effectiveness.
Answer: 0.88

4. In a screen separator, the mass flow rates of feed, overflow, and underflow are 150, 140, and 10 kg h^{-1}, respectively. Mass fraction of material in the feed and overflow are 0.9 and 0.96, respectively. Find the effectiveness of separation.

    Answer: 62.38%

Objective Questions

Cleaning, Sorting, and Grading

1. Differentiate between cleaning, sorting, and grading.
2. What are the advantages of cleaning fruits and vegetables? What are the common impurities in fruits and vegetables?
3. What are the relative advantages and disadvantages of wet and dry cleaning?
4. Why is sorting by size and shape important before further processing operations?
5. State the principles of separation in the following equipment with examples of fruits and vegetables, which are sorted/ graded with these.

    (i) inclined draper, (ii) spiral separator, (iii) disk separator, (iv) cyclone separator, (v) pneumatic separator, (vi) air screen cleaner, (vii) destoner, (viii) indented cylinder separator, (vi) roller separator, (vii) color sorter, (viii) electrostatic separator, (ix) image processing method, (x) aspiration sorting, (xi) floatation sorting, (xii) weight sorter.

6. Explain the sorting principles of tomatoes, apples, peas, and dehusked coconuts for different applications.
7. What are the different types of magnetic separators? Explain their working principles.
8. What are the different classifications of screens? Name the factors affecting screen effectiveness. Derive an expression of screen effectiveness.
9. What are the different methods for washing fruits and vegetables?
10. Why is drying fruits and vegetables necessary immediately after washing?
11. Differentiate between (a) spray washer and shuffle washer, (b) rotary drum washer and brush washer.
12. Why is detergents / hydrochloric acid solution added to the cleaning water? What are the concentrations of the chemical used?

13. Is it advisable to wash fruits and vegetables after peeling? Justify your answer.

14. Name the washing / cleaning devices most suitable for peas, potatoes, tomatoes, carrots, cauliflower, oranges, mangoes, pineapple.

15. What is ultrasonic cleaning? What are its application areas?

Peeling

16. Name the different methods of peeling.

17. What are the different types of equipment used for mechanical peeling? Name suitable peelers for commodities such as apples, carrots, potatoes, onions, and sweet potatoes.

18. What are the different factors affecting peeling loss?

19. State the relative advantages and disadvantages of chemical and mechanical methods of peeling.

20. Explain the working principle of chemical peeling.

21. Differentiate between wet heat peeling and dry heat peeling.

22. Why is lye peeling not commonly recommended for potatoes?

23. Explain how thermal energy is used for peeling fruits and vegetables.

24. What precautionary measures should be taken to reduce recontamination of fruits during the peeling operation?

Blanching

25. Explain the objectives of blanching. State the processing parameters for blanching vegetables.

26. Why is the blanching operation carried out for most vegetables even before other preservation methods, such as drying or freezing?

27. What are the factors affecting blanching operations?

28. Explain enzymatic browning and non-enzymatic browning.

29. Name some vegetables that are not blanched, and explain why.

30. Differentiate between the hot water blanching and steam blanching processes and explain their relative advantages and disadvantages.

31. Write notes on (i) continuous steam blancher (ii) pipe blancher (iii) blancher-cooler, (iv) fluidized bed blancher.

32. Explain the difference between a reel blancher and a pipe blancher.

33. What are the devices used to prevent loss of steam in a steam blancher?

34. Explain the IQB hot water blancher with figure. Explain the advantages of IQB over conventional steam blanching.

35. Does blanching cause some loss of vitamins? What are the factors affecting such losses during blanching? What method is adopted in an IQB steam blanching system to minimize losses?

BIBLIOGRAPHY

Bhatti, S. and Varma, U. 1995. *Fruit and Vegetable Processing: Organisations and Institutions.* CBS Publishers, New Delhi.

Brennan, J.G., Butters, J.R., Cowell, N.D. and Lilly, A.E.V. 1976. *Food Engineering Operations*, 2nd ed. Elsevier Applied Science, London, UK..

Chakravorty, A. 2019. *Post-harvest Technology of Cereals, Pulses and Oilseeds.* 3rd ed. Oxford and IBH Publishing Co., New Delhi.

Charm, S.E. 1978. *The Fundamentals of Food Engineering*, 3rd ed. AVI Publishing Co., Westport, Connecticut.

Cruess, W.V. 1938. *Commercial Fruit and Vegetable Products.* McGraw Hill, New York. .

Dash, S.K., Bebartta, J.P. and Kar, A. 2012. *Rice Processing and Allied Operations*, Kalyani Publishers, New Delhi.

Earle, R.L. and Earle, M.D. 2004. *Unit Operations in Food Processing*, Web Edn. The New Zealand Institute of Food Science & Technology (Inc.), Auckland https://www.nzifst .org.nz/resources/unitoperations/index.htm

Fellows, P.J. 2000. *Food Processing Technology*, 2nd ed. Woodhead Publishing, Cambridge, UK.

Heldman, D.R. and Singh, R.P. 1981. *Food Process Engineering.* AVI Publishing Co., Westport, Connecticut.

Henderson, S.M. and Perry, R.L. 1976. *Agricultural Process Engineering.* AVI Publishing Co., Westport, Connecticut.

https://www.colorsortergroup.com/what-is-color-sorter-machine .html

Nelson, P.E. and Tressler, D.K. 1980. *Fruit and Vegetable Juice Processing Technology*, 3rd ed. AVI Publishing Co., Westport, Connecticut. .

Pandey, P.H. 1994. *Principles of Agricultural Processing.* Kalyani Publishers, New Delhi.

Pillaiyar, P. 1988. *Rice: Post Production Manual.* Wiley Eastern Limited, New Delhi.

Sahay, K.M. and Singh, K.K. 1994. *Unit Operations in Agricultural Processing.* Vikas Publishing House, New Delhi.

Srivastava, R.P. and Kumar, S. 2008. *Fruit and Vegetable Preservation: Principles and Practices.* Kalyani Publishers, New Delhi.

Sudheer, K.P. and Indira, V. 2016. *Post-Harvest Technology of Horticultural Crops. Horticultural Science*, Vol VII, 1st ed. New India Publishing Agency, New Delhi. pp. 420.

Sudheer, K.P. and Indira, V. 2017. *Entrepreneurship Development in Food Processing.* 1st ed.. New India Publishing Agency, New Delhi. pp. 356.

Sudheer, K.P. and Indira, V. 2018. *Entrepreneurship and Skill Development in Horticultural Processing*, 1st ed. New India Publishing Agency, New Delhi. pp. 340.

Thompson, A.K. 2003. *Fruit and Vegetables harvesting, Handling and Storage.* . Blackwell Publishing, Oxford, UK.

Urschel, J.R. 1988. The science and art of cutting food products. In: A. Turner (ed.) *Food Technology International Europe.* Sterling Publications International, London, 87–91.

Watson, E.L. and Harper, J.C. 1989. *Elements of Food Engineering*, 2nd Edn. Van Nostrand Reinhold, New York.

11

Size Reduction

Size reduction, also known as *comminution*, is an essential unit operation in food processing. Many products, such as meal, powder, flour, and split, are obtained through size reduction. Fruits and vegetables are cut and sliced before processing or cooking. Polishing of rice and homogenization of milk also involve size reduction operations. In addition to adding variety to the products, reducing particle size also increases the surface area and reactivity of the solids. It reduces the bulk of fibrous material for easier handling, permits the separation of unwanted materials by mechanical methods, and also helps in waste disposal.

The purpose of size reduction and the requirement of end products may differ according to the situation. The raw material in most of the cases is irregular in size and shape and, hence, the size reduction operations and the equipment differ for different cases. There are two major categories of size reduction on the basis of the raw material being either solid or liquid. Size reduction of solids can be achieved through impact, compression, abrasion, and cutting. For liquids, size reduction is achieved by homogenization or atomization.

11.1 Size Reduction of Hard Solids

11.1.1 Methods and Principles

The size reduction of solid and hard agricultural and food products, such as grains, can be achieved by several means; the main principles are compression or crushing, impact, shearing/ attrition, and cutting. These methods can be used individually or in combination.

A material undergoes *crushing* and is ruptured when the magnitude of an external force applied on the material exceeds its strength. Sugarcane crushers, oil expellers, roller flour mills, etc. use compression for size reduction. *Impact* is subjecting a material to an abrupt force in excess of its strength, causing the material's failure. The hammer mill and leg pounder for rice use impact force for size reduction. *Shearing* is essentially a combination of cutting and crushing. The shearing units consist of a knife and a bar. If the edge of knife or shearing edge is very thin and sharp, the size reduction process can be approximated as cutting, whereas that with a thick and dull shearing edge can be approximated as crushing. *Cutting* occurs when a sharp and thin knife is forced through the material undergoing size reduction. It is typically used for soft solids. Cutting gives a definitive particle size and sometimes a definite shape, with few or no fines. Sometimes size reduction occurs from *attrition* by rubbing a particle against a rough surface and sometimes by rubbing with other particles as happens in a rice

polisher. Milk homogenization involves reducing the size of fat globules from intense shear in the supporting fluid.

Energy Requirement

Size reduction of solid particles takes place by mechanical action causing stress; fracturing of materials leads to the reduction in size. As discussed above, the force applied may be either compression, impact, cutting, or shear or a combination thereof. During the process, mechanical moving parts of a size reduction machine act on the material to create stresses. The stress is initially absorbed internally by the material as strain energy. When the local strain energy exceeds a critical level, fracture takes place along lines of weakness and the stored energy is released. The critical level of the strain energy at which the fracture occurs is a function of the material. The creation of new surfaces requires only a part of the energy applied to the material and is known as the *surface energy*. A larger part of the applied energy is, however, dissipated as heat, raising the temperature of the product and the equipment. Some energy may be lost due to vaporization of moisture from the material without any increase in temperature. The time duration for which the stress is applied is important since lower stress concentrations for longer periods could also cause the material to disintegrate. The extent of size reduction is, thus, affected by the magnitude of the force as well as the time duration of the force application.

The hardness of the material and its tendency to crack, its friability, affect the material's energy requirement for size reduction. In addition, the amount of power necessary for the size reduction depends on the feed rate, moisture content, product particle size requirement, and type and condition of equipment. As indicated above, energy in excess of that actually required for size reduction is converted into heat; this conversion should be kept as low as practicable. Hence, for an efficient size reduction process, the energy applied to the material should exceed the minimum energy needed to rupture the material by as small a margin as possible.

Size reduction is a very energy inefficient process, and, thus, it is important to use energy as efficiently as possible. Unfortunately, it is not easy to calculate the minimum energy required for a given size reduction process; some general laws have been proposed for estimating the energy requirement for size reduction of hard solids. These laws are based on the principle that when a particle is reduced to symmetrical small sized particles then the energy requirement is a function of the size of the original and the final products. As it is assumed that the feed and product are symmetrical in shape, a common dimension can be used to calculate the energy requirement.

Let us consider the particle of dimension L is reduced to a symmetrical product with dimension ΔL.

Relating the required energy to some function of the size of the initial and final particles, an equation can be written as:

$$\Delta E \alpha \frac{\Delta L^l}{L^m}$$

$$\text{or, } \Delta E = C \frac{\Delta L^l}{L^m}$$

A simplified form can be written as:

$$\Delta E = C \frac{\Delta L}{L^n} \tag{11.1}$$

In the above equation, ΔE is the energy requirement for size reduction, L and ΔL are the original and final sizes of the particles (both are symmetrical), and C, l, m, and n are constants. Since the particles are assumed symmetrical, a common dimension can be used for representation of energy.

Actually, the reduced particles usually resemble polyhedrons with nearly plane faces and sharp edges and corners. After smoothening by abrasion, they may take other shapes. For compact particles in which the length, breadth, and thickness are nearly equal, the largest diameter or apparent diameter is taken as the particle size. For plate-like or needle-like shapes, other dimensions are used to express the size.

The equation 11.1 states that the energy required to reduce the size of a material is proportional to a dimension of the reduced material relative to a similar dimension of the original material raised to a power n.

Thus, the energy necessary to reduce a specific mass of particles from one size to another is:

$$E = -C \int_1^2 \frac{dL}{L^n} \tag{11.2}$$

Based on this generalization, some laws have been developed for energy requirement for size reduction and these laws have been named after the corresponding scientists.

Kick's Law

This law assumes that the forces acting on the material cause plastic deformation in it and the resultant stresses are analyzed within the elastic limit of the deformation. It assumes that the energy requirement for size reduction is a function of a common linear dimension of the material; therefore, the value of "n" is 1. Thus, the energy requirement can be given as follows:

$$E = -C \int_1^2 \frac{dL}{L}$$

$$E = -C.\ln\left(\frac{L_2}{L_1}\right) = C.\ln\left(\frac{L_1}{L_2}\right)$$

Here $E = P/\dot{f}$ = (Power/feed rate) and has a unit of kW.h.ton^{-1}. L_1 and L_2 are the characteristic dimensions of the feed and product, respectively. Denoting the constant as C_K, the Kick's constant, the above equation can also be written as:

$$E = C_K.\ln\left(\frac{L_1}{L_2}\right) \tag{11.3}$$

Kick's law considers that the energy needed for size reduction of a specific quantity of material within its elastic limit is constant for the same reduction ratio, and it is independent of the original sizes. The law, thus, suggests that a material's crushing energy requirement would be the same whether the size reduction is from 12 cm down to 4 cm (threefold) or from 6 mm to 2 mm. The feed and discharge openings in a crusher often decide the maximum dimensions (diameters) of feed and product. Thus, the *reduction ratio* of a crusher is often expressed as the ratio of the feed opening to the discharge opening.

Rittinger's Law

This law assumes that size reduction is a shearing procedure, and, hence, the energy required in crushing is proportional to the change in surface area, i.e., the energy requirement is proportional to the square of the common linear dimension. The value of "n" in equation 11.2 becomes 2 in this case.

$$E = -C \int_1^2 \frac{dL}{L^2}$$

$$E = -C.\left[\frac{L^{-1}}{-1}\right]_1^2 = C\left(\frac{1}{L_2} - \frac{1}{L_1}\right)$$

Denoting the constant as C_R, the Rittinger's constant, the above equation can also be written as:

$$E = \frac{P}{f} = C_R\left(\frac{1}{L_2} - \frac{1}{L_1}\right) \tag{11.4}$$

Equation 11.4 represents Rittinger's law. The surface area per unit mass, known as the specific surface of a particle, is proportional to $1/L$, the energy required to reduce the size of particles from 10 cm to 5 cm would be the same as that required to reduce the size of particles from 5 mm down to 3.33 mm.

Both Kick's law and Rittinger's law can be used over limited ranges of particle sizes, provided C_K and C_R are obtained experimentally with the material being crushed and specific equipment. Kick's law has been found to be reasonably good for the grinding of coarse particles in which the increase in surface area per unit mass is relatively small. Rittinger's law fits the experimental data better for the creation of fine powders, large areas of new surfaces, through size reduction. These laws have limited utility and are infrequently applied today.

Bond's Law

It is a more realist law than Kick's law and Rittinger's law. This law states that the energy required to form particles of size "D_p" from very large feed is proportional to the square root of the surface-to-volume ratio of the product.

The sphericity (φ) of a non-spherical particle can be given as:

$$\varphi = \frac{6/D_p}{S_p/V_p} \tag{11.5}$$

where D_p is the equivalent diameter of the particle, S_p is the surface area of one particle, and V_p is the volume of one particle.

$$\text{Thus,} \quad \frac{S_p}{V_p} = \frac{6}{\varphi.D_p} \qquad (11.6)$$

$$\text{As per Bond's law,} \quad \frac{P}{f} = E = \frac{C_B}{\sqrt{D_p}} \qquad (11.7)$$

C_B is known as the *Bond constant*, which is also dependent on the type of machine and material being handled as in the cases of C_R and C_K.

For use of Bond's law, the *work index* (W_i) is defined as the gross energy requirement (kW.h.ton^{-1}) to reduce a very large feed to a size such that 80% of the product passes through a 100 μm screen. As per the above definition, W_i is specific for the material being reduced and can be found experimentally from laboratory crushing and grinding tests.

If, D_p = diameter of screen hole in which 80% of the product passes (mm) and D_f = diameter of screen hole in which 80% of the feed passes (mm), P is in kW, and f is feed in tons per hour, then

$$C_B = \sqrt{100 \times 10^{-3}} \times W_i = 0.3162 \times W_i$$

$$\frac{P}{f} = E = 0.3162 \times W_i \left(\frac{1}{\sqrt{D_p}} - \frac{1}{\sqrt{D_f}} \right) \qquad (11.8)$$

W_i includes the friction in the crusher and the power given by the above equation is gross power.

Integrating the equation (11.1) within limits gives a generalized relationship as follows:

$$E = \frac{C}{1-n} \left(L_2^{1-n} - L_1^{1-n} \right) \qquad (11.9)$$

This equation has been mostly used for grinding small grains. The exponent is slightly less than 2 for starchy grains and may be larger than 2 for fibrous materials.

Example 11.1 A 5 hp motor is used to reduce the size of a food particle from 8 mm to 0.008 mm. Assuming Rittinger's law, find the size of the motor to reduce the particle size to 0.0008 mm.

SOLUTION:

We will use the Rittinger's equation.

$$E = C_R \left(\frac{1}{L_2} - \frac{1}{L_1} \right)$$

To produce particles of 0.008 mm

$$E_1 = C_R \left(\frac{1}{0.008 \times 10^{-3}} - \frac{1}{8 \times 10^{-3}} \right)$$

$$= C_R \left(125000 - 125 \right) = 124875 \, C_R$$

Energy required to reduce particle size to 0.0008 mm

$$E_2 = C_R \left(\frac{1}{0.0008 \times 10^{-3}} - \frac{1}{8 \times 10^{-3}} \right)$$

$$= 1249875 \, C_R$$

$$\text{Now,} \quad \frac{E_2}{E_1} = \frac{1249875 C_R}{124875 C_R} = 10.01$$

$$\text{Thus,} \quad E_2 = 10.01 \, (5) = 50 \text{ hp}$$

Example 11.2 During a milling experiment, it was observed that to grind 3.25 mm sized grains to IS sieve 40 (0.42 mm opening) the power requirement was 7 kW. Calculate the power requirement for milling the same grain by the same mill to IS sieve 20 (0.211 mm opening) using (1) Kick's law, and (2) Rittinger's law. The feed rate of milling is 250 kg.h^{-1}.

SOLUTION:

(i) According to Kick's law,

$$\frac{P}{f} = C_K \ln \left(\frac{L_f}{L_p} \right)$$

$$\frac{7}{0.25} = C_K \ln \left(\frac{3.25}{0.42} \right),$$

$$\text{or, } 28 = C_K \times 2.046$$

$$C_K = 13.684$$

Substituting C_K for the second condition,

$$\frac{P}{0.25} = 13.684 \times \ln \left(\frac{3.25}{0.211} \right)$$

$$P = 0.25 \times 13.684 \times 2.734 = 9.355 \text{ kW.}$$

(ii) According the Rittinger's law,

$$\frac{P}{f} = C_R \left(\frac{1}{L_p} - \frac{1}{L_f} \right)$$

$$\text{or, } \frac{7}{0.25} = C_R \left(\frac{1}{0.42} - \frac{1}{3.25} \right)$$

$$\text{or, } 28 = C_R \, (2.073), \text{ or } C_R = 13.505$$

$$\text{Now,} \quad \frac{P}{0.25} = 13.505 \times \left(\frac{1}{0.211} - \frac{1}{3.25} \right)$$

$$\text{or, } P = 0.25 \times 13.505 \times 4.432 = 14.96 \text{ kW}$$

Example 11.3 How much power is required to crush 1.2 t of a material per hour if 80% of the

feed passes through IS sieve no. 340 (3.25 mm opening) and 80% of the product passes through IS sieve no. 40 (0.42 mm opening). Take the work index of the material as 8.40.

SOLUTION:

According to Bond's law,

$$\frac{P}{\dot{f}} = 0.3162 \times W_i \left(\frac{1}{\sqrt{D_p}} - \frac{1}{\sqrt{D_f}} \right)$$

Here, $W_i = 8.40$, $f = 1.2$ t.h^{-1}, $D_p = 0.42$ mm, $D_f = 3.25$ mm

$$\frac{P}{1.2} = 0.3162 \times 8.4 \times \left(\frac{1}{\sqrt{0.42}} - \frac{1}{\sqrt{3.25}} \right)$$

$$= (1.54 - 0.555) \times 0.3162 \times 8.4 = 2.6242$$

or, $P = 3.149$ kW.

Example 11.4 A hammer mill is used to grind one metric ton of granular material having 3 mm mean diameter so that 80% of the ground particles pass through a sieve of 260 μm opening. The median diameter of the ground particles has been found to be 180 μm. If the hammer mill has the Bond's energy constant as 0.06 kW.h.(mm)$^{1/2}$. kg^{-1}, what is the energy required in grinding?

SOLUTION:

Given, weight = 1000 kg

$$D_f = 3 \text{ mm}, D_p = 260 \text{ μm} = 0.26 \text{ mm}$$

$$C_B = 0.06 \text{ kW.h.(mm)}^{1/2}.\text{kg}^{-1}$$

$$\frac{P(kWh)}{f(kg)} = E = C_B \left(\frac{kWh.\sqrt{mm}}{kg} \right) \left(\frac{1}{\sqrt{D_p}} - \frac{1}{\sqrt{D_f}} \right) \left(\frac{1}{\sqrt{mm}} \right)$$

or,

$$P = 1000 \times 0.06 \times \left(\frac{1}{\sqrt{0.26}} - \frac{1}{\sqrt{3}} \right) = 1000 \times 0.06 \times 1.3838$$

$$= 83.028 \text{ kW.h.kg}^{-1}$$

Example 11.5 In a hammer mill, if the feed has the size of 20 mm and the product size is 1.2 mm (i.e., 80% of the product is less than 1.2 mm), the power required is 5.6 kW. How much power would be required for the same feed if 80% of the product is needed to be less than 0.8 mm.

SOLUTION:

For the 1st case:
Given, $D_f = 20$ mm, $D_p = 1.2$ mm,

Power = 56 kW

$$E = \frac{P}{\dot{f}} = C_B \left(\frac{1}{\sqrt{D_p}} - \frac{1}{\sqrt{D_f}} \right)$$

or, $$\frac{5.6}{\dot{f}} = C_B \left(\frac{1}{\sqrt{1.2}} - \frac{1}{\sqrt{20}} \right)$$

$$= C_B (0.9128 - 0.2236) = 0.6893 \, C_B$$

or, $C_B = 8.124 /f$
For the 2nd case:

$$P = \dot{f} . \frac{8.124}{\dot{f}} \left(\frac{1}{\sqrt{0.8}} - \frac{1}{\sqrt{20}} \right)$$

$$= 8.124 \times (1.118 - 0.2236)$$

$$= 8.124 \times (0.8944)$$

$$= 7.27 \text{ kW}$$

Example 11.6 During the size reduction process, a crystal of 0.5 mm size (80% of the feed passes through 0.5 mm sieve) is ground to fine powder so that 80% of the product passes through a 0.088 mm sieve (170 mesh Tyler sieve). The power requirement was found to be 2 hp. If the desired final product should be such that 80% would pass through a 250 mesh sieve (0.063 mm) and the throughput has to be increased by 20%, what will be the power requirement? Use Bond's equation.

SOLUTION:

For the first case, feed size = 0.5 mm, product size = 0.088 mm, feed rate = $\dot{f}$

Thus, $$\frac{2}{\dot{f}} = 0.3162 \times W_i \left(\frac{1}{\sqrt{0.088}} - \frac{1}{\sqrt{0.5}} \right)$$

i.e. $$\frac{2}{\dot{f}} = 0.3162 \times W_i (1.9568) \qquad (A)$$

For the second case, feed size = 0.5 mm, product size = 0.063 mm and feed rate = $1.2(\dot{f})$
If P is the power requirement,

$$\frac{P}{1.2 \times \dot{f}} = 0.3162 \times W_i \left(\frac{1}{\sqrt{0.063}} - \frac{1}{\sqrt{0.5}} \right)$$

i.e. $$\frac{P}{1.2 \times \dot{f}} = 0.3162 \times W_i (2.57)$$

i.e. $$\frac{P}{\dot{f}} = 1.2 \times 0.3162 \times W_i (2.57) \qquad (B)$$

Dividing equation B with equation A,

$$P/2 = 1.2 \times 2.57/ 1.9568 = 0.9494 \text{ hp},$$

or, $P = 3.152$ hp

Crushing Efficiency

The ratio of the surface energy created by crushing to the energy absorbed by the solid is referred to as *crushing efficiency* (η_c).

$$\eta_C = \frac{e(A_p - A_f)}{E_a} \qquad (11.10)$$

where e = surface energy per unit area, N.m.m^{-2}, A_p and A_f are areas per unit mass of product and feed, respectively, m² and E_a is the energy absorbed by unit mass of material, N.m. The surface energy created by fracture is small in comparison to the total mechanical energy stored in the material at the time of rupture and most of the latter is converted to heat. Therefore, the crushing efficiency is low.

Usually, during the size reduction process, some part of the input energy is lost due to friction, and, hence, the ratio of the energy absorbed (i.e., actually used for size reduction) to the applied energy is known as the *mechanical efficiency* 'η_m'.

$$\eta_m = E_a/E \qquad (11.11)$$

The power required by the machine can be calculated as:

$$P = E.\dot{f} = \dot{f}.e\frac{(A_p - A_f)}{\eta_a \eta_c} \qquad (11.12)$$

$$\text{or,} \quad P = \frac{6\dot{f}e}{\eta_m \eta_c \rho_p}\left(\frac{1}{\varphi_p D_p} - \frac{1}{\varphi_f D_f}\right) \qquad (11.13)$$

where D_f and D_p are the volume surface mean diameter of the feed and product, respectively, φ_f and φ_p are the sphericities of the feed and the product, ρ_p is the particle density, and $\dot{f}$ is the feed rate.

Effectiveness of Grinding

The effectiveness of grinding of solid materials can be expressed in terms of specific power consumption, degree of grinding, and the specific load that the initial product would exert on the working tool of the crusher/grinder.

The following equation represents the degree of grinding (I).

$$I = S_a/S_b \qquad (11.14)$$

Where the overall surface areas of the product are S_a and S_b after and before grinding.

The moisture content of the product, in addition to affecting the power consumption, also affects the efficiency of grinding.

Overall extraction (Ex) efficiency is given as:

$$Ex\ (\%) = U_f - U_i \qquad (11.15)$$

where the percentages of undersize particles are U_f and U_i for the final (ground) product and the initial product (feed), respectively.

During the process of crushing, the product from a size reduction machine consists of a mixture of coarse, fine, and even dust particles of various sizes; the ratio of diameters of the largest to the smallest particle could be of the order of 10^4. The size of the fine particles will not change much as the grinding continues; however, the coarser particles will be further reduced.

In real situations, the fines can be minimized in some size reduction processes but cannot be eliminated. In particular, grinders can control the largest particles sizes but not the fines.

During size reduction, the surface area of a particle increases, which also increases the reactivity of food during processing and storage. During mixing also, the surface area is an important parameter that affects the mixing index. Thus, the estimation of surface area is often required. The *specific surface* represents the surface area per unit mass. The specific surface can be estimated by knowing the shape factor and the size distribution of the particles.

As simple approximations, the characteristic particle dimension, D_p, can be used to determine the volume and surface area as follows:

$$V_p = \gamma D_p^3 \qquad (11.16)$$

$$\text{and} \quad S_p = 6\tau D_p^2 \qquad (11.17)$$

where V_p is the volume of the particle, S_p is the particle surface area, D_p is the typical dimension of the particle, and γ, τ are factors related to the particle geometries.

For example, for a cube, the volume is D_p^3 and the surface area is $6D_p^2$, i.e., γ, τ are 1 each. For a sphere, $\gamma = (\pi/6)$ and $\tau = (\pi/6)$. Thus, the ratio of surface area to volume is $6/D_p$ in each case.

A *shape factor* can be defined as $\tau/\gamma = \lambda$. For the cube and sphere that were discussed, $\lambda = 1$. It has been experimentally observed that for many grinding processes, this shape factor (surface area to volume ratio) of the resulting particles is approximately 1.75, indicating that the surface area is nearly twice that for a cube or a sphere.

The general expression for the ratio of surface area to volume is:

$$\frac{S_p}{V_p} = \frac{6\lambda}{D_p} \qquad (11.18)$$

Thus, if there is a mass m of particles of density ρ_p, the number of particles is $m/(\rho_p V_p)$ each of area S_p.

So total surface area, $S_t = \frac{m}{\rho_p V_p}\cdot\frac{6\lambda V_p}{D_p} = \frac{6\lambda m}{\rho_p D_p}$

$$(11.19)$$

where S_t represents the total surface area of all the particles contained in the mass of the material. Total surface area of a powder can be estimated by combining the results of the sieve analysis with equation 11.19.

Sieve Analysis

The particle size distribution of a granular material can be studied by a procedure known as *sieve analysis*. In food processing, specifically during the size reduction process, it is important to know the average size and the size distribution of powders and granular materials as the size also decides the surface area, and the surface area affects the rate of reaction on surfaces of the product. The size also affects performance of machines during heat transfer, cyclone separation, drying, diffusion, extrusion cooking, etc. Sieve analysis is often referred to as a *gradation test*.

Standard screens are used for measurement of size range of particles. Internationally, two scales commonly used are the US sieve series and the Tyler standard sieve series. The Bureau of Indian Standards has also standardized mesh sizes for screen analysis. The US-sieve mesh designation has been accepted by the International Standards Organization. The ASTM standard sieve numbers are almost similar as the mesh numbers of the Tyler sieves. Annexure XII displays a list of the most common sieve sizes.

As discussed in Chapter 10, the mesh number gives the number of openings per linear inch in a screen. In the standard sieves, the opening sizes vary by a factor of $\sqrt{2}$. In some cases, intermediate sieves are also available with the ratio $\sqrt[4]{2}:1$. Though the sieve sizes are expected to confirm to the sizes specified by the numbers assigned to them, minor variations from the specifications are also expected due to the wear and tear and deformation of the wires.

During sieve analysis, a set of sieves with specified openings are used. The top-most screen has the largest opening, and the opening size reduces successively for the lower screens. At the bottom there is a pan, which has no opening and is known as the receiver. There is a top cover. The complete set of sieves is fixed to a mechanical sieve shaker. As the material of specified quantity is put on the top screen, due to the shaking of the screens the material falls down and is collected on different sieves as per their sizes. Then the material retained on each sieve is weighed and the percentage retained on each sieve is calculated. These values are then put to a standard analysis procedure to obtain the average size of the particles. Often a semi-log plot (sieve size scale is logarithmic) is also used to represent the percent passing vs. opening size in screens.

The cumulative percent of materials retained up to a particular sieve is also used to determine the percentage of materials that can pass through the screen.

$$\% \text{ passing} = 100\% - \% \text{ cumulative retained} \quad (11.20)$$

With a view to achieving reproducible results in screen analysis, it is important to standardize the procedure. As an example, for analysis of food grains, the set of IS sieves recommended are nos. 100, 70, 50, 40, 30, 20, and 15 and pan. The sample is oven dried to constant weight and 250 g of it is put on the top-most sieve. Then the sieve set is shaken for 5 minutes and the material retained on each sieve is weighed for analysis as outlined earlier.

Some common terms associated with particle size distribution analysis are as follows.

Fineness modulus (FM). It represents the mean size of the particles of the aggregate. It is calculated by adding the product of percentage weight of material retained on each of the standard sieves with the number of sieves from bottom (pan = 0, bottom-most sieve = 1 and so on) and then dividing it by 100. The fineness modulus is an index to indicate the uniformity of grind in the resultant product.

The *average particle size* is obtained as:

$$D_{avg} = 0.135(1.366)^{FM} \quad (11.21)$$

Effective size. The effective size (in microns) is the maximum particle size of the smallest 10% of the materials. It also denotes the sieve opening through which only 10% of the total collection of materials can pass. It is designated by the symbol D_{10}.

Uniformity coefficient. It is the ratio of the maximum size of the smallest 60% particles (the opening allows 60% materials to pass through that) to the effective size.

$$\text{Uniformity coefficient} = D_{60}/D_{10} \quad (11.22)$$

Example 11.7 and 11.8 illustrate the screen analysis for determining the fineness modulus, average particle size, effective size, and uniformity coefficient of the particles.

Example 11.7 In a screen analysis with 250 g of a flour sample, the following observations were taken. Find the fineness modulus and average particle size.

IS sieve no.	100	70	50	40	30	20	15	Pan
Weight of material retained, g	0.0	1.2	14.3	30.5	85.0	94.8	18.0	6.2

SOLUTION:

For the above data, the following table is constructed.

IS sieve no.	Weight of material retained	Percent material retained	Multiply with	Obtained value
100	0.0	0	7	0
70	1.2	0.48	6	2.88
50	14.3	5.72	5	28.6
40	30.5	12.2	4	48.8
30	85	34.0	3	102
20	94.8	37.92	2	75.84
15	18.0	7.2	1	7.2
Pan	6.2	2.48	0	0
Total	*250*	*100*		*265.32*

Fineness modulus = 265.32/100 = 2.653
Average particle size = 0.135(1.366)$_{2.653}$= 0.309 mm

Example 11.8 In a sieve analysis, the following results were obtained.

Tyler sieve no.	3.5	10	20	30	40	60	80	Pan
Weight of material retained, %	5	7	40	16	10	15	5	2

Find out the effective size and uniformity coefficient of the particles.

SOLUTION:

From the above data and Annexure XII, the following table is obtained.

Tyler sieve number	Size of opening, mm	% passing
3.5	5.66	95
10	2	88
20	0.841	48
30	0.595	32
40	0.42	22
60	0.25	7
80	0.177	2

We plot the above data on a semi-log paper as shown in Figure 11.1. From the plot, the effective size of the particles is obtained as $D_{10} = 0.36$ mm and the uniformity coefficient is obtained as $D_{60}/D_{10} = 1.2/0.36 = 3.33$.

The weighted average will be taken to represent a mixture of different sized particles.

An important assumption in the above analysis is that all particles are spherical or nearly spherical, so that the particles, with diameters lower than the size of the screen, can pass through. Obviously, the analysis may not give reliable results if the particles are flat and elongated. Microscopic examination of the fractions in such situations can corroborate the result.

The above analysis gives accurate and consistent results for materials with size more than 100 mesh (150µm). For still finer materials, it may not be so accurate as the mechanical energy needed for the particles to pass down the opening, the surface attraction between the particles and screen and in between the particles increase with reduction in particle size.

Wet sieve analysis consists of suspending particles and allowing them to pass through the sieve; however, the liquid should not affect the particles. This helps for a more efficient movement of fine material through the sieve than by shaking the dry material.

Laser diffraction and image analysis methods are some advanced systems for analysis of particle size distribution. Digital image processing has now become the quickest and easiest method for determining the particle size distribution and is being extensively applied for analyzing rock masses. Its use in food powders is still limited. Centrifuge methods and sedimentation are also used.

11.1.2 Equipment

Classifications

The general categorization of the size reduction equipment (McCabe et al. 1993) is given in Table 11.1.

Compressive forces are used in crushers to break large pieces of solid materials into small lumps. The crushers can be primary crushers or secondary crushers depending on the particle sizes. In the case of grinders, impact and attrition are used with compression. The ultrafine grinders mostly work on attrition. Cutters are used for soft solids and give products of definite size and shape. The size reduction machines are also characterized on the basis of the final particle size (approx.) as primary crusher (150–250 mm), secondary crusher (6 mm), intermediate grinder (40 mesh), fine grinder (200 mesh), and ultrafine grinder (1–50 µm).

Grinding by standard equipment generates a significant amount of heat, which may cause the loss of flavor and taste of the foods. Specifically, spices, which are valued for their characteristic flavor and aroma, degrade in quality. The product loses volatile oil even up to 30–40% (in some studies it has

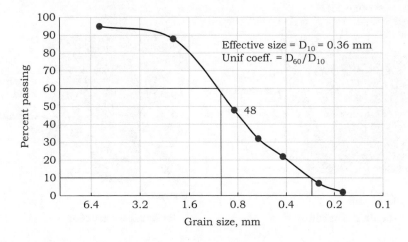

FIGURE 11.1 Determination of effective size and uniformity coefficient of particles

TABLE 11.1

Categories of size reduction equipment

Broad category	Typical Examples
Crushers	Jaw crusher, gyratory crusher, roll crushers (smooth roll type and toothed roll type)
Grinders	Hammer mill, attrition mill, bowl mill, roller mill, tumbling mill, compartment mill
Ultrafine grinders	Fluid energy mills, agitated mills, hammer mills with internal classification
Cutters, dicers, and slicers	Knife cutters, dicers, slicers

been reported to be as high as 60% also) and original color. Generation of high temperature during grinding also causes melting of fat, which causes the ground powder to stick on the grinder surface. Grinding of spices at cryogenic conditions can avoid such problems.

Jaw Crusher

Construction and Operation

- It consists of two jaws, made up of heavy materials, opening at the top like a "V" (Figure 11.2 a).
- The jaw faces are either flat or slightly bulged, sometimes with horizontal grooves.
- One of the jaws is nearly vertical and is fixed, and is called the anvil jaw, while the other one is made to reciprocate in a horizontal plane by an eccentric device and is called swinging jaw.
- The angle between the two jaws is kept at 20–30°. The number of strokes given to the swinging jaw by the eccentric unit ranges from 250 to 400 per minute.
- When the feed is dropped between two jaws, the large-sized materials are broken by compression (when the movable jaw moves toward the

stationary jaw) into smaller pieces. Subsequently, the reduced materials fall down into the narrower gap between the jaws at lower part of the machine and are further reduced by the closing and opening of the jaws into still smaller pieces. The final broken materials drop out at the discharge opening at the bottom.

- In another design, the pivot point is at the top of the movable jaw or above the top of the jaws on the center line of the jaw opening. Thus, the bottom of the movable jaw, which is close to the discharge end and is narrowest, has the maximum to and fro movement. Hence, there is almost no chance of choking of materials in this crusher.

Applications

These are used for coarse reduction of large amount of solids and for primary crushing of hard materials. It is not extensively used in the food industry.

Gyratory Crusher

Construction and Operation

- The gyratory crusher consists of an inverted cone crushing head carried on a heavy shaft pivoted at the top (Figure 11.2 b). It gyrates in an open-top funnel-shaped casing. An eccentric unit fitted at the bottom rotates the crushing head.
- During operation, the bottom of the crushing head moves toward and then away from the stationary wall. Thus, the material gets crushed when it is fed between the circular jaws, trapped between the outer fixed and the inner gyrating cones and forced into a gradually narrowing space. The crushing operation continues until the material is sufficiently small to pass out from the bottom.
- The crushing head is free to rotate on the shaft and turns slowly because of friction with the material

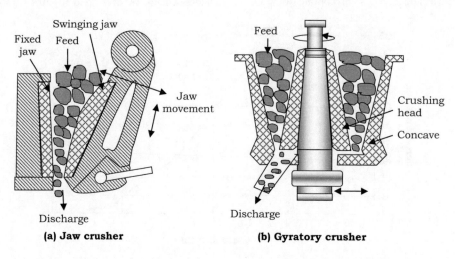

(a) Jaw crusher   **(b) Gyratory crusher**

FIGURE 11.2   Schematic representation of difference between the working elements of a jaw crusher and gyratory crusher

being crushed. The per minute crushing head speed is about 125 to 425 gyrations.

Salient Features

- Gyratory crushers, unlike the jaw crushers, have continuous discharge, maintaining uniform load on the driving motor.
- It needs less power than the jaw crusher and the maintenance requirement is lower.
- The capacity of the machine varies with the jaw setting, the strength of the material being crushed, and the speed of gyration of the machine.

Applications

This is applied for large hard-ore and mineral crushing applications and has found limited use in food applications.

Smooth-Roll Crusher

Construction and Operation

- It consists of two heavy smooth-faced metal cylinders or rolls kept close to each other on a parallel horizontal axis (Figure 11.3).
- The rollers used in wheat flour milling are usually 25 cm in diameter and either 80 cm, 100 cm, or 150 cm long.
- The rolls are made to rotate toward each other, usually at the same speed, which generally ranges from 50 to 300 rev.min^{-1}. The rolls may be driven either at differential speeds or one roll may be stationary.
- The particles to be broken fall between the two rolls, trapped and nipped between them where they break by compression.
- The rolls are frequently of large diameter with relatively narrow faces so that they can handle moderately large lumps.
- With a view to avoiding any damage to the roll surface, at least one of the rolls is spring loaded since it is possible that there is some unbreakable material (stones, nuts, bolts, etc.) in the feed.

Salient Features

- Spacing between the rolls controls the product size and capacity of the machine.
- The size of the largest particle that would be nipped between the rolls can be determined if the geometry of the particles and friction coefficient between the rolls and the feed material are known.
- Normally for crushers with very big rolls (as compared to the gap), the following equation can be used.

$$D_p = 0.04\ R + G/2 \qquad (11.23)$$

where R = roll radius, D_p = maximum size of particle, and G = gap between the rolls.

Applications

The smooth-roll crushers make meal or flakes from food grains or grits.

Serrated or Toothed-Roll Crusher

Construction and Operation

- As the name suggests, in such crushers the rolls are serrated as per need. The equipment may contain two rolls, through which the material would pass, or only one roll working against a stationary curved breaker plate (Figure 11.4).
- The two rolls may rotate at differential speeds, which help to tear the materials fed through them.
- Another type of equipment contains a small high-speed roll with transverse breaker bars on its face turning toward a large slow-speed smooth roll.

Salient Features

- These serrated-roll machines operate by compression, impact, and shear (the smooth roll crushers work by compression alone).
- Some crushing rolls contain heavy pyramidal teeth to take care of coarse feed.
- The main factors that affect the effectiveness of the roller mills (both smooth and toothed rolls) include the clearance between the rolls and the geometric

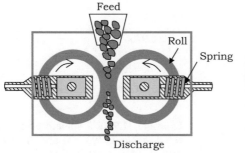

FIGURE 11.3 (a) Working of a smooth roll crusher, (b) a roller flour mill

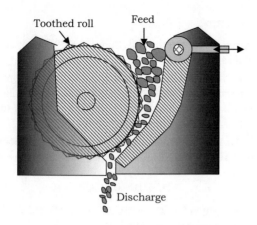

FIGURE 11.4   Working of a single serrated roll crusher

and kinematic parameters of the rolls. The geometric parameters include the roll diameter, clearance, and initial size of the particle. Shape, number, slope, mutual position, and shape of the cross section of corrugation of rolls significantly affect the quality and yield of the reduced material, the total output, and the specific power consumption of the roller mills.

- The efficiency of grinding depends on the ratio of speeds, v_f/v_s, which is an important kinematic parameter.

- A pair of rolls has the capacity as follows:

$$Q_t = l\, bv\rho\, \Psi \qquad (11.24)$$

where l and b are the length of the rolls and clearance between the rolls; v is the mean product velocity in the grinding zone; ρ is the bulk product density before grinding; Ψ is the filling coefficient in the grinding zone ($\Psi = Q_a/Q_t$, where Q_a and Q_t are the actual and theoretical capacities of the rolls).

The capacity also depends on the type and moisture content of the grain.

Applications

- Toothed-roll crushers are more versatile than smooth-roll crushers. These are used for flour, sugarcane, soybeans, etc.

- The serrated-roll crushers, unlike smooth-roll crushers, can accommodate larger particles. The sugarcane industry is an example where multi-stage rolls crush the cane.

- In a wheat mill, there are usually two types of rolls, namely the *break rolls* and the *reduction rolls*. The break roll system usually consists of about four pairs of corrugated steel rolls. In each pair, one roll revolves faster than the other, usually in a ratio of 2.5:1. The break rolls are used to produce middlings from wheat. After each reduction of endosperm (middlings), the flour is sifted away from the bigger size middlings and the remaining middlings are passed to the next

set of rolls. Subsequently, there is a series of pair of *reduction rolls,* which have smooth-faced rolls. These are used for further size reduction and for preparing the ground flour of desired size. The reduction rolls are also further divided into *coarse rolls* and *fine rolls* depending on the clearance between the rolls. Each set of rolls takes stock from the preceding one. As many as 12 to 14 reduction rolls are used in most flour mills, but all of these may not be used for all size reductions. In addition, a stand-by system known as a *scratch system* is also available in most of the flour mills, which is usually an extension of the break roll system. It is possible to grind flour into very fine particles by single grinding under high grinding pressure. But the practice of single grinding may rupture the starch and, hence, should be avoided.

Example 11.8 In a roll mill, soybean seeds of approx. 9 mm diameter are reduced to an equivalent diameter of 2.5 mm. The rolls rotate toward each other at equal speed. If the coefficient of friction between the soybean seeds and the roll material is 0.22, find the diameter of the rolls.

SOLUTION:

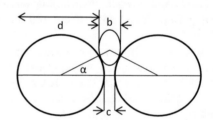

In the present situation,

$$b = \text{maximum diameter of seed} = 9 \text{ mm}$$

$$c = \text{diameter of product} = 2.5 \text{ mm}$$

$$\mu = 0.22 = \tan \alpha$$

$$\text{or, } \alpha = 12.41°$$

$$\text{Now, } \cos\alpha = \frac{r + c/2}{r + b/2} = \frac{d + c}{d + b}$$

$$\text{or, } \cos(12.41) = \frac{d + 2.5}{d + 9}$$

$$\text{or, } 0.9766 = \frac{d + 2.5}{d + 9}$$

i.e., $d = 268.8$ mm

The roll diameter will be 26.88 cm.

Example 11.9 In a smooth crushing roll, if the roll diameter is 300 mm, and the gap between the rolls is 10 mm, what is the largest particle size that can be crushed?

SOLUTION:

Using equation 11.23,

$$D_p = 0.04\, R + G / 2,$$

$$= 0.04 \times 150 + 5 = 11\ mm$$

Example 11.10 Wheat is being milled in a roller mill with 30 cm long rolls and 0.4 mm clearance between the rolls. The mean calculated velocity of the product is 3 m.s⁻¹ in the grinding zone, and the coefficient of filling of the grinding zone is 0.5. Find the milling capacity. The bulk density of wheat before grinding may be taken as 720 kg.m⁻³.

SOLUTION:

As per equation 11.24, $Q = l\, bv\rho\, \Psi$

Here, $\rho = 720\ kg.m^{-3}$, $l = 0.3\ m$, $v = 3\ m.s^{-1}$, $b = 0.4\ mm = 0.0004\ m$, and $\Psi = 0.5$

Thus, the capacity is obtained as:

$$Q = (720)(0.3)(3)(0.0004)(0.5) = 0.1296\ kg.s^{-1}$$

$$= (3.6)(0.1296) = 0.467\ t.h^{-1}$$

Example 11.11 Spherical grains of 9.5 mm diameter are crushed into 2 mm diameter particles using a 8 kW roller mill. The feed rate is 5 kg.s⁻¹.

(a) Find the diameter of the rolls if the coefficient of friction is 0.24.

(b) Find the power consumption if the required capacity is 8 kg.s⁻¹ and the final particle diameter is 3 mm.

SOLUTION:

(a) The particles are to be reduced from 9.5 mm diameter to 2 mm diameter.

As we have discussed in the previous example, b = 9.5 mm and c = 2 mm

$\mu = 0.24$, i.e., the angle of friction, $\alpha = \tan^{-1}(0.24) = 13.49°$

Now, $\cos\alpha = \dfrac{r + c/2}{r + b/2} = \dfrac{d + c}{d + b}$

or, $\cos(13.49) = \dfrac{d + 2}{d + 9.5}$

or, d = 262.24 mm

The roll diameter will be 26.22 cm

(b) As per Bond's law

$$\frac{8}{5} = 0.3162 \times W_i \left(\frac{1}{\sqrt{D_p}} - \frac{1}{\sqrt{D_f}} \right)$$

$$1.6 = 0.3162 \times W_i \left(\frac{1}{\sqrt{2}} - \frac{1}{\sqrt{9.5}} \right)$$

$$1.6 = 0.3162 \times W_i (0.38266)$$

or $W_i = 13.223\ kW.s.mm^{1/2}.kg^{-1}$

For the 2nd case,

$$\frac{P}{8} = 0.3162 \times 13.223 \times \left(\frac{1}{\sqrt{3}} - \frac{1}{\sqrt{9.5}} \right)$$

$$P = 8 \times 4.18 \times (0.2529)$$

$$= 8.459\ kW$$

Hammer Mill

Construction and Operation

- Hammer mills consist of a high-speed rotor, rotating at 1500 to 4000 rev.min⁻¹ in a cylindrical hard metal casing (Figure 11.5). Usually, the rotor shaft is horizontal.

- A screen with chosen perforation size makes up the lower portion of the casing whereas the top half is a solid one. The whole unit is kept in another casing to collect the ground material in a specified outlet and, thus, keeping the environment within the plant clean.

- The rotor has pinned to it a number of fixed or swinging hammers. The hammers are straight metal bars with plain, enlarged, or sharpened ends, which hit the materials fed to the chamber.

- In many cases, several rotor disks, mounted on the same shaft, carry four to eight swing hammers each.

- It is from the top of the casing that the material is fed into the mill. As the material falls into the chamber, it is ground by the hammers fixed on the rotor shaft until it is fine enough to pass through the screen at the bottom.

- Thus, the screen size controls the fineness of grinding, and the particle size of the final product can be changed just by changing the screen.

Salient Features

- Impact and some amount of shear between the feed and screen are responsible to a large extent for the size reduction.

- Hammer mills are versatile and rugged in design and operation, in addition to being simple, producing no damage to the machine while running at no load.

- There is little probability of damage of mill due to foreign objects. In particular, in the case of a swinging hammer mill, the hammers are not damaged even if some unbreakable solid material is fed into the milling chamber by chance.

- Hammers become worn during the course of operation; however, the worn hammers do not significantly reduce the efficiency of the mill.

(a)

(b)

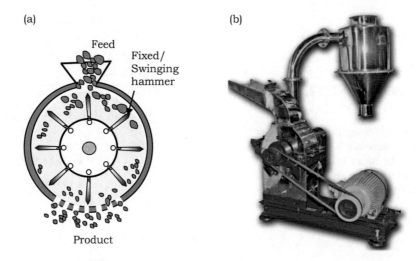

FIGURE 11.5 (a) Working of a hammer mill, (b) a hammer mill with cyclone separator for dust recovery (Courtesy: M/s DP Pulverizer Industries, Mumbai)

- Hammer mills can handle both brittle and fibrous materials. In the case of fibrous materials, a cutting action is given by the projecting sections.
- The high power requirement of a hammer mill is its main disadvantage. The type of feed is responsible for deciding the capacity and power requirement of a hammer mill. Energy consumption of commercial mills ranges can be 30–75 kJ per kg of biomass or even more. The specific energy consumption will depend on the reduction ratio.

Applications

Hammer mills are widely used for spice and poultry feed grinding.

Classifying Hammer Mill

A classifying hammer mill consists of a set of swing hammers between two rotor disks. In addition to the hammers, the rotor shaft carries two fans to draw air through the mill and discharge into ducts used for collection of the product. There are short radial vanes on the rotor disks for separating oversize particles from the acceptable size.

During operation, the particles are given a high rotational velocity in the grinding chamber, which causes the coarse particles to concentrate at the wall of the chamber due to the centrifugal force acting on them. The air stream carries finer particles inward from the grinding zone toward the shaft. The particles are thrown outward by the separator vanes. Thus, acceptable fine particles are carried through and the large size particles are thrown back for further reduction in the grinding chamber.

Rotor speed or the size and number of the separator vanes control the achievement of the maximum particle size.

High-speed hammer mills provided with internal and external classification are used for ultrafine grinding to particle sizes 1–20 µm. Fluid or jet mills are also used for ultrafine grinding. Agitated mills are used for ultrafine wet grinding.

Attrition Mill

Construction and Operation

- An attrition mill, also known as a *plate* or *burr mill,* consists of two rotating circular disks kept close to each other on the same horizontal or vertical axis. The surfaces of the disks are roughened or grooved.
- One disk is stationary while the other one is rotating in a single runner mill. Both disks are driven at high speeds but in opposite directions in case of a double runner mill.
- It is near the axis of rotation that the material to be milled is fed into the gap between the disks through the hub of one of the disks. The feed material moves in an outward direction, owing to the rotation of the disks where it is crushed and sheared as it makes its way to the edge of the discs. The milled material is discharged from the periphery into a stationary casing.
- The gap between the disks can be adjusted within a narrow range to control the product particle size and the rate of milling.
- At least one disk is spring loaded to avoid damage due to over loading or entry of any unbreakable foreign material along with the feed.
- The burr mill discs are usually 20 to 140 cm in diameter and operate at 350 to 700 rev.min^{-1} (Figure 11.6).

Salient Features

- As compared to other size reduction machines, a burr mill requires lower power and investment.

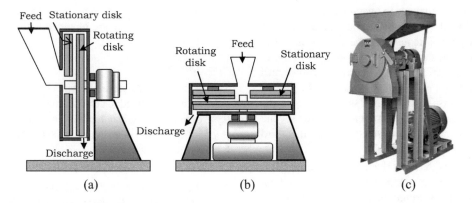

FIGURE 11.6 (a) and (b) Schematic diagrams of horizontal axis type and vertical axis type attrition mills, (c) a horizontal axis type attrition mill (Courtesy: M/s DP Pulverizer Industries, Mumbai)

- The material is reduced by crushing and shear.
- Type of disks and the gap between them control the fineness of grinding.
- There are different patterns of grooves and corrugations on the disc surfaces for performing a variety of operations, such as granulating, cracking, grinding, and shredding.
- Unbreakable foreign matter in the feed may result in damage/breakage to the plates.
- Materials are needed to be fed slowly in attrition mills; overfeeding leads to reduction in mill performance and undue heat generation.

Applications

- Burr mills have been used to produce whole grain and dehusked grain flour. Heat generation during grinding is a serious disqualification for their use in spice grinding.
- Grinding of soft materials is carried out using double-runner disks type attrition mills.
- Plate mills having very fine clearances and very high speeds have led to the development of *colloid mill*s to produce particles of colloidal dimensions.

Ball Mill

Construction and Operation

- A ball mill consists of a cylindrical or conical shell in which some solid grinding balls are kept. The shell slowly rotates about a horizontal axis.
- The shell is usually made of steel lined with high carbon steel plate, porcelain, or silica rock. The balls are made up of steel, 25–125 mm in diameter.
- With the rotation of the ball mill, the balls are carried up the side of the shell by centrifugal force almost near to the top. Then from those elevated positions the balls drop by gravity on the materials kept below. Thus, size reduction takes place by impact of the balls on the material. Grinding utilizes the energy imparted to the balls while raising them.

- A large ball mill may be up to 4.25 m in length and 3 m in diameter. Thus, the material to be ground is impacted by the balls from a height of more than 2.5 m.
- The balls usually occupy half the volume of the shell and the void fraction of the ball mass at rest is typically 0.4.
- A ball mill comes under a group of size reduction machines known as *tumbling mills*. *Rod mill* and *pebble mill* are the other types of tumbling mills. Metal rods are there in a rod mill, instead of balls, for grinding. Pebble mills have flint pebbles or porcelain or zircon spheres. Usually, the pebbles are 50–175 mm in size. The tumbling mills can be batch or continuous type.
- A *conical ball mill* is an improved type of mill in which there is segregation of the grinding zones. It usually consists of a 60° cone at the feed entering section and a 30° cone at the exit section (Figure 11.7). It also contains balls of different sizes. As the balls go on wearing and become smaller as the mill is operated, new large balls are added periodically. As the shell rotates, the large balls move toward the point of maximum diameter, and the smaller balls move toward discharge. Thus, the largest balls dropping from the maximum height carry out the initial size reduction; smaller balls falling from a smaller height grind the smaller particles.

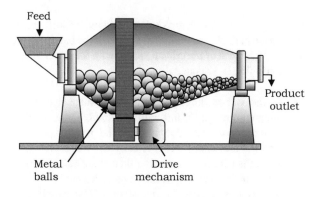

FIGURE 11.7 Schematic diagram of a conical ball mill

- A *tube mill* is continuous in operation and has a long cylindrical shell where the material is ground for 2–5 times as long as it remains in the shorter ball mill. It produces very fine powders in a single pass when the power availability is not a constraint. A tube mill can be converted into a *compartment mill* by placing slotted transverse partitions in it such that large balls are contained in one compartment while the second compartment would have smaller balls, the third compartment may have pebbles.

- In ball mills or tube mills the material is often lifted by scoops and dropped into a cone that directs it out through a hollow trunnion, known as a *trunnion discharge*.

- In rod mills much of the grinding is by rolling compression and attrition. Rods are usually 25–125 mm diameter. The materials escape through the openings in the chamber wall at one end of the cylinder, which is known as *peripheral discharge*.

Salient Features

- The grinding may be accomplished with dry solids; however, the feed is usually a suspension of particles in water. The mill efficiency and capacity are increased in this way.

- In case of wet grinding, the discharge openings maintain the water level such that the suspension just fills the void space in the mass of balls.

- While the balls are in contact with the wall, there is some grinding by slipping and rolling over each other, but most of the grinding is by impact.

- Higher rotational speed of the mill leads to higher height of balls inside the mill and higher power consumption. Falling from more height also increases impact force.

- However, *centrifuging*, i.e., balls sticking to the mill wall during rotation and not falling from the top, will cause little or no grinding. The speed of rotation should, thus, be controlled to evade centrifuging. The critical speed, i.e., the rotational speed at which centrifuging occurs, is determined as follows.

$$\eta_c = \frac{1}{2\pi}\sqrt{g\big/(R-r)} \qquad (11.25)$$

where, η_c = critical speed, rev.s^{-1}; R = radius of the mill, m; r = radius of the ball, m. The rotational speeds of the ball mills are kept at 65% to 80% of the critical speed, with the lower values for wet grinding in viscous suspension.

- In the wet milling type ball mill, water may be pumped through the mill with the solid. Screens are used to separate the product from the oversize particles, but they cannot economically make separations when the particles are smaller than about 150 to 200 mesh. In that case the reduced particles are carried by water to a centrifugal classifier. The classifier separates the undersize and oversize particles; the oversize particles are repulped with more water and returned to the mill.

Applications

Size reduction of abrasive materials into medium and fine particles is recommended to be done in ball mills. The food industry uses ball mills only to a limited extent; one such application is grinding of food coloring materials. The rod mills are used for intermediate grinding.

Example 11.12 Find the speed of rotation of a ball mill having a diameter of 1 m and charged with 40 mm balls. The ball mill is grinding solid matter. Also find the critical speed and speed of rotation for wet grinding in viscous suspension.

SOLUTION:

The critical speed, $n_c = \dfrac{1}{2\pi}\sqrt{\dfrac{g}{R-r}}$

or, $n_c = \dfrac{1}{2\pi}\sqrt{\dfrac{9.81}{(1.0-0.04)/2}} = 0.72$ revolutions per second

$= 43.17$ rev. min^{-1}

The operating speed for solid matter grinding = $n_c \times 0.8 = 34.54$ rev. min^{-1}

For the second case, the critical speed remains same.

Operating speed for grinding in suspension = $43.17 \times 0.65 = 28.06$ rev. min^{-1}

Agitated Mill

Construction and Operation

- The mill has a stationary vertical vessel of 4-1200 *l* capacity filled with liquid and the grinding medium suspended in it. Hard solid elements like sand grains, balls, or pellets are examples of grinding medium.

- The bed of grinding media is set into motion by an agitator. A multi-arm impeller is employed in some designs to agitate the charge. In some designs, there is a reciprocating central column for grinding hard materials, vibrating the vessel contents at about 20 Hz.

- The slurry is fed at the top of the mill while the product is collected at the bottom through a screen.

- Different mill designs make both dry and wet grinding possible. As discussed in the case of wet grinding, the powder is mixed beforehand in a liquid and then fed to the mill as a concentrated suspension or slurry.

- Due to the relative movement between the grinding media and the feed material, the material is comminuted as the result of impact and shear forces.

Applications

Particles of 1 μm size or finer are produced by agitated mills, which find their use in ore industries.

Fluid Energy Mill

Construction and Operation

- Inter-particle attrition is the mechanism of size reduction for particles suspended in high velocity streams of either compressed air or superheated steam.

- A typical chamber, as shown in Figure 11.8, is an oval loop of pipe that is 25–200 mm in diameter and 1.2–2.4 m in height. The air is admitted into the chamber at a high pressure (7 atm) through nozzles, which then move at a high speed in a circular or elliptical path.

- Feed enters the chamber at the bottom. As it is moved by the compressed air, the inter-particle attrition and rubbing of the particles on the chamber wall cause the size reduction.

- As these particles move, because of spinning along the upper bend the particles are separated as per size and the larger particles are thrown toward the upper wall, whereas the smaller (reduced) particles remain close to the inner wall of the loop.

- There is an opening along the inner wall of the loop to collect the reduced particles and send them subsequently to a cyclone separator and bag filter for collection. The larger particles continue to move in the loop for further size reduction.

- In some designs, the gas jets carrying the particles move in opposite directions to cause the particles to strike each other and in others the high velocity air vigorously agitates a fluidized bed.

Salient Features

- As there is no moving part in the mill, the cost of maintenance is low and also due to absence of abrasion of the mill, there is no contamination of the feed. It is also very easy to clean the mill.

- The classification is supported by the complex pattern of the swirl generated in the gas stream at the bend in the loop of the pipe.

- The air requirements per kg of product is about 6–9 kg. Inert gas can also be used for air-sensitive feed.

- The process does not generate heat and, hence, it is used for milling of heat sensitive particles.

- The surfaces of the ground particles are smoother and less angular in shape; repeated reduction of feed produces finer particles of spherical shape.

- Such mills work best for feed with particle size less than 100 mesh though they can be used for particles as large as 12 mm. Non-sticky particles can be reduced to ½ to 10 μm in diameter.

- The loop mills can have a capacity as high as 6000 kg.h^{-1}.

- The device requires high pressure and energy consumption is high.

Applications

- The fluid energy mill is used for size reduction of non-sticky solids. It is not suitable for soft and fibrous materials as well as materials of larger sizes. It can be used for heat sensitive solids.

- The use of inert gas also permits use of materials that are more reactive to oxygen.

- The device can also be used for drying of moist feed.

Fixed Head mill

The process of shearing a material between a fixed casing and a rotating head, with very narrow gap between them, is accomplished using various forms of mills. One such type is a pin mill in which pins are attached on the surfaces of both the static and the moving plates and the material is sheared between the pins.

Edge Runner Mill

Construction and Operation

- The *edge runner mill*, also known as a *roller stone mill*, crushes the materials into fine powders by rotating stones or grinding wheels. Usually, there are two large stones that turn slowly in a large bowl.

- The mill rotates on a central shaft by an electric motor provided at the basement and rotates opposite to the direction of rolling of wheels.

- The material to be ground is put on the bed and it is kept in the path of the stone wheels with the help of a scrapper.

- The size reduction is done by crushing due to the heavy weight of the stones and the shearing force which is involved during the movement of these stones.

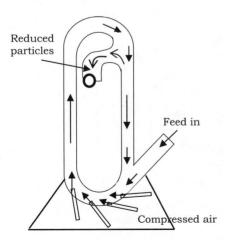

FIGURE 11.8 Working of a fluid energy mill

- The material is ground for a pre-set duration before it is passed through the sieves to get the required size powder.

Salient Features

- This machine operates without the danger of clogging or jamming by accumulation of the material and offers greater efficiency and high comparative quantities.
- Very fine particle sizes of the feed materials can be obtained.

Applications

The edge runner mill is commonly used for grinding most of the drugs into fine powder. It is also used for grinding chocolate and confectionery flour.

Rietz Mill or Disintegrator

Construction and Operation

- It consists of a rotor inside a circular screen enclosure. The rotating shaft is usually vertical.
- The rotor includes a number of hammers running at a fairly close clearance. The hammers, in general, are fixed to the shaft rigidly; however, swing hammers are also used in some cases.
- Feed enters the milling chamber parallel to the axis as in disk type attrition mill.
- The product is discharged radially out through a perforated sizing screen, which surrounds the rotor.
- Rietz machines are normally supplied in rotor diameters from 10 to 60 cm. The hammer tip speeds are in the range of 5.2 to 11.1 m.s^{-1} and power ranges from 0.3 to 150 kW.

Salient Features

- The Rietz mill is mostly used on wet materials. The advantage of this mill is found in those cases where solid content is in the range of 40% to 80%.

- It keeps running because the close hammer clearance keeps the sizing screen open. Therefore, more fine and uniform grinding is possible.
- It is able to grind materials to below 15 µm size.
- The disintegration-resistant and tramp materials can also be separated by providing a differential discharge system for the final product.

Applications

The Rietz disintegrator is used to pulverize and shred difficult to grind, dry materials as well as materials high in moisture and/or oil to coarse or fine particle sizes.

Concentric Cylinder Abrasive Mill

Such mills are mostly used for scouring of husk or seed coverings of pulses and cereals. These mills work on the principle of friction. Inside a larger drum (preferably made up of perforated metal sheet), an abrasive roller rotates and the frictional forces cause the scouring/separation. The outer metal cylinder may also be fabricated as the bottom half is perforated, whereas the upper half portion is made of plain mild steel sheet.

Cryogenic Grinder

Construction and Operation

- The two basic components of a cryogenic grinding system are the precooling and grinding units (Figure 11.9). Figure 11.10 shows a commercially available small capacity low temperature pulverizer.
- Heat content of the material is considerably reduced by the precooler before it is fed to the grinder. The components of the cryogen grinding unit include a properly insulated barrel enclosing a screw conveyor and with a provision to introduce liquid nitrogen into the barrel. The other components are a liquid nitrogen dewar, an air compressor, an arrangement for power transmission, and a control panel.
- In cryogenic grinding, usually liquid nitrogen is used. The other cryogens include hydrogen, methane, helium, krypton, neon, argon, and liquefied natural

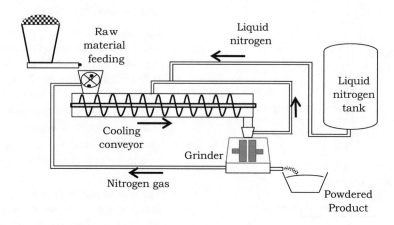

FIGURE 11.9 Schematic diagram of a cryogenic grinding system (after https://www.aiche.org, with permission)

FIGURE 11.10 A low temperature pulverizer (Courtesy: M/s Bajaj Processpack Limited, Noida)

gas. Nitrogen has an added advantage of creating an inert atmosphere.

- The loading into a hopper is done through a vacuum conveyor, after which the material to fed in the cryogenic screw conveyor. After the material is cooled, it is fed to the grinder.
- The grinding operation is usually by impact or attrition. A hammer mill, ball mill, or pin mill is commonly used.
- The ground powder and the nitrogen vapor are collected from the grinding unit. The nitrogen vapor is recycled.

Salient Features

- The objective of precooling the material, before it enters the grinder, is to make it brittle by lowering its temperature below its brittle point as well as the freezing point of oil in the material. Specifically, the moisture and the oil in the spices are changed to crystalline form such that clogging during the process of grinding is effectively avoided. As a result, the energy consumption for size reduction is reduced considerably. It also gives a higher production rate in the size reduction machine.
- Another key advantage of cryogenic grinding is the higher retention of etheric oils.
- The liquid nitrogen quickly evaporates at ambient temperature. Therefore, it is important that the liquid nitrogen consumption is economized for acceptable operating cost. The loss of liquid nitrogen is minimized through insulation and proper design of the precooler.

- The efficiency of cryogen utilization is improved by proper maintenance of retention time of the material in the gaseous zone and liquid nitrogen.
- Use of liquid nitrogen during grinding eliminates rancidity, by obviating the chances of oxidation, and by creating an inert atmosphere.
- Lower cost, finer particle size, and more uniform particle distribution are other attributes of cryogenic grinding,
- However, special care should be taken to prevent exposure of cryogens to skin and eyes.

Applications

Cryogenic machining can be ideally utilized for those materials that are elastic in nature, possess low combustion temperatures, have low melting points, and are sensitive to oxygen. Besides, as it reduces loss of flavor and taste, it is very widely accepted for use in the spices industry.

11.1.3 Accessories in Size Reduction Units

- The size reduction machine performs well only if the feed is of suitable size, enters at a uniform rate, and the product is removed as soon as possible after the particles have attained the desired size. The unbreakable materials should be separated from the feed by suitable separators.
- It is also important that the heat generated during milling is removed, in particular, for the heat sensitive products.
- Thus, the important accessories in a size reduction unit are the metal separators, constant rate feeders, heaters, coolers, pumps, and blowers.
- Cooling water or refrigerated brine may be circulated through coils or jackets in the mill to keep the temperature low during grinding. Sometimes chilled air is blown through the mill. Dry ice may also be admitted with the feed. Cryogenic grinding involves using liquid nitrogen to have a grinding temperature below -75°C. Such a low temperature changes the breaking characteristics of the solid, usually making it more friable.

11.1.4 Important Considerations during Size Reduction

The following are some important points to know while using the size reduction machines.

- Size reduction requires a huge amount of energy and is perhaps the most inefficient of all unit operations as less than 1% of the energy supplied is used for creating new surfaces and over 99% of the energy is utilized in operating the equipment, producing undesirable heat and noise.
- In many mills the feed is reduced by passing it once through the mill. When the oversize particles are not returned to the mill then it is said to be an *open circuit*

operation. When the oversize particles are returned to the mill after separation from the desired size particles, it is said to be a *closed-circuit operation.*

- Over-grinding of the fine particles consumes excess energy and closed-circuit operation is preferred for reduction to fine and ultrafine sizes. However, there is a necessity of wet classifiers or air separators to separate the recommended size particles from the large-size particles and conveying system to take the large-size materials to the top of the machine, which require energy. Energy must also be supplied for the conveyors and separators in a closed-circuit system. The energy requirement in a closed-circuit system is usually lower by 25% than the open circuit system.

- In some situations, the discharge of the reduced particles is blocked deliberately to crush the particles many times before they are discharged. This is known as *choke crushing.* When there is no choke crushing, i.e., in normal situations, the crushed particles are allowed to freely go out of the size reduction machine, this is known as *free-discharge crushing* or *free crushing.*

- Determination of power requirement for a particular grinding job is difficult. The exact amount of power requirement depends on type of material, moisture content of feed, material feed rate, type and condition of mill, product particle size, specification (number) of abrasive surface, etc.

- Depending on the intended use, the powders of ground spices have particle sizes, generally, in the range of 50 to 850 μm. Accordingly, a hammer mill, an attrition mill, or a pin mill is used for spice grinding.

11.2 Size Reduction of Soft Solids

11.2.1 Methods and Principles

Soft solids, such as fruits and vegetables, are cut into pieces, varying from very small to large sizes. Such operations can be termed as *disintegration* or breakup of fruits and vegetables. These breakup operations can be carried out either with minimal changes in form or with considerable changes. The general size reduction laws such as those of Bond, Kick, or Rittinger cannot be applied to these size reduction operations.

Some examples of size reduction or disintegration operations with little change in form include pitting, coring, destemming, husking, shelling, etc. The main objective of these operations is to prepare the commodity for processing by separating the undesirable components without damaging the main edible component. The process must be consistent and effective in producing uniform units of the given commodity and, of course, should not impair the final product quality.

Breakup operations with considerable change in form include operations such as cutting, shredding, crushing and/or comminuting, homogenizing, sheeting, extracting, and/or juicing. In these operations the form of the commodity is changed, and varieties of products of different sizes and shapes become available.

11.2.2 Equipment

Cutting Machines

Construction and Operation

- The cutting machines may be of different configurations to produce slices, dices, or cubes of different sizes. The slices may be flat, wavy, or wrinkle types.

- A rotary knife cutter is a common cutting machine with a horizontal rotor having several knives fixed to it and rotating at 200 to 900 rev.min^{-1} within a cylindrical chamber. Figure 11.11 (a) and (b) show the working of a rotary knife cutter and a commercial slicer working on this mechanism.

- The knives are made up of tempered steel or stellite edges and they pass with close clearance over approximately a similar number of stationary bed knives.

- Flying knives are sometimes parallel to the bed knives and cut at an angle depending on the properties of the feed.

- Feed enters the chamber from the top and are cut by the rotating knives and then discharged from the bottom of the equipment.

- Devices having knives attached on a horizontal disc are also available.

- Design of rotary cutters and granulators is similar. While a rotary cutter gives products of regular shapes as cubes, thin squares, etc., the granulator yields materials with irregular shapes. Other devices, such as a *bowl chopper* with vertical rotating cutting knife, are available.

Salient Features

- Different types of knife cutters have been developed for cutting slices/chips of various commodities. The selection of any particular equipment is mostly on the basis of the type of product.

- There are some products that become very sticky during cutting because of the presence of sugar. There are also fibrous products that cause accumulation of fiber in the cutter.

- Some products must be moisturized prior to cutting to make them wet and prevent overheating.

- The food-cutting machines are available in different capacities, the capacities can be very high, such as 3500 kg potatoes per hour or more.

- Sharpness of the knives needs to be maintained so that they cut instead of tearing.

Applications

- Knife cutters are used to make thin slices of fruits and vegetables.

- Special knife cutters have been developed for cutting slices/chips of potatoes, cassava, bananas, etc.

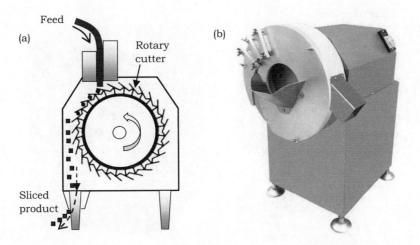

FIGURE 11.11 (a) Working of a slicer (b) a commercial ginger slicer (Courtesy: M/s Shiva Engineers, Pune)

Shredding, Crushing, Chopping, and Juice Extraction Machines

Shredding can be done on similar machines as used for cutting but here the product is also torn apart while it is being cut.

Crushing and chopping are the operations carried out on fully ripened and too soft fruits, which cannot be sliced or diced. These types of products are used for preparation of products such as soups, pies, side dishes, and puddings.

Crushing of products by chopping them into small pieces (0.025–0.075") prior to heating speeds up enzyme activity, which is often prevented by introduction of superheated steam into the device.

At the time of crushing and immediately after crushing, the air must be removed to prevent oxidation and other chemical reactions in tissues. The use of a vacuum helps to eliminate the air and to create a better consistency in the final product.

Different types of pulpers and juice extractors are available, which primarily work on the principle of compression.

11.3 Size Reduction of Liquid Foods

11.3.1 Methods and Principles

Reduction in the solid or liquid particle size in a dispersed phase is achieved through the application of intense shearing forces, known as *homogenization*. For example, milk is an emulsion of fat in water and the size of fat globules is reduced to less than about 2 μm during homogenization. As the fat globules (or other solids/liquids in other materials) are broken, it also increases the number of particles in the liquid. Homogenization involves breaking of solids/liquids at a high shearing force and is, thus, a more severe operation than emulsification.

11.3.2 Equipment

There are different types of homogenizers, i.e., high-speed mixers, pressure homogenizers, colloid mills, ultrasonic homogenizers, hydro-shear homogenizers, and microfluidizers.

Pressure Homogenizer

Construction and Operation

- The pressure homogenizers involve forcing the liquid through a very small gap (orifice) by a high-pressure pump.

- The main parts of a homogenizer are the pump, valve, valve seat, breaker ring, and the tension spring.

- The pump usually operates at 10,000–70,000 kPa. The small adjustable gap (up to 300 μm) is created by a homogenizing valve resting on the valve sheet on the discharge side. The tension spring is used to hold the valve in position.

- As the liquid is forced to move through the narrow gap, it attains very high velocity (80–150 m.s⁻¹) and hits the breaker ring, which is a hard surface in the path of the liquid. After that the liquid emerges from the valve and the velocity is reduced almost instantaneously.

- The size reduction or breaking of the particles dispersed in the continuous phase takes place by two means, namely, by powerful shearing action when the liquid passes through the narrow orifice at a high velocity and then by disruption due to impact at the breaker ring. There is further breakage of particles due to a sudden drop in pressure or explosive effect as the fluid leaves the valve.

- In foods like milk and milk products, the distribution of the emulsifying agent may not be proper over the newly formed surfaces causing fat globules to clump together. Therefore, two-stage homogenization is used for breaking up the clusters of globules (Figure 11.12 a). Using a two-stage homogenizer also helps to maintain the effectiveness of homogenization even if there is slight wear and tear in the first valve.

Salient Features

- The shearing effect of the valve changes with the velocity of the fluid and for a certain valve, shearing effect can be acceptable only at a given velocity.

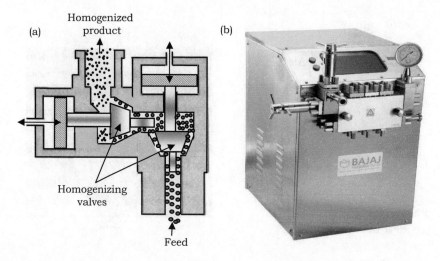

(a) Homogenized product

(b)

Homogenizing valves

Feed

FIGURE 11.12 (a) Schematic representation of size reduction in a two-stage homogenizer (after Fellows, 2000 with permission), (b) a homogenizer (Courtesy: M/s Bajaj Processpack Limited, Noida)

- With fluctuating pressure, the velocity will fluctuate, thus causing non-uniformity in the size reduction and decreasing efficiency. Therefore, positive displacement pumps, preferably three-plunger positive displacement pumps with overlapping strokes, are used for the purpose.

Applications

Homogenization is a common unit operation in milk processing; the average size of fat globules is reduced to less than 2 μm from the initial size of 4–10 μm and helps the dispersion of fat globules uniformly in milk. The milk viscosity increases as a result of the formation of higher number of fat globules and adsorption of casein onto the globule surfaces. The color of some foods, like milk, is affected by homogenization because the larger number of globules causes greater reflectance and scattering of light. A commercial homogenizer is shown in Figure 11.12 (b).

Colloid Mill

Construction and Operation

- A colloid mill is a device that produces hydraulic shear to reduce either the particle size of a solid in suspension in a liquid or the droplet size of a liquid suspended in another liquid.
- This is essentially a disc mill with a disc rotating at 2000–18000 rev.min⁻¹ having a small clearance (0.05–1.3 mm) between a stationary disc and the rotating disc (Figure 11.13). Discs are available in flat, conical, and corrugated shapes for different applications.
- The material, after being fed through the hopper, is passed through the narrow gap between the rotor and stator and, thus, reduced to fine particle size due to intense hydraulic shear. Higher shear rates lead to smaller droplets, down to approximately 1 μm, which are more resistant to emulsion separation.
- Design variants include using either two counter-rotating discs or intermeshing pins on the disc

surfaces to increase the shearing action. Like in a *paste mill*, the discs may be mounted horizontally for highly viscous foods, such as peanut butter or fish pastes.

- The high levels of applied shear to the process liquid in a colloid mill help to form stable suspension or emulsion by dispersing the particles or liquid droplets in addition to disrupting structures in the fluid.
- The rotor and stator surfaces may be smooth or rough. Rough-surfaced mills produce the shearing action by creating intense eddy current, turbulence, and the impact on the particles.
- The capacity of the machine can be 2–3 l.min⁻¹ to 400 l.min⁻¹.

Salient Features

- To prevent damage to a colloidal mill, the material to be ground should be pre-milled as finely as possible.
- Higher shear rates lead to smaller droplets, the particle size can be controlled by modifying the gap between the rotor and the stator.
- Often there is little size reduction in the mill, and the principal action is the disruption of the lightly bonded clusters or agglomerates.
- Particles size of as small as 3 microns can be obtained by the colloidal mill.
- Colloidal mills are more effective for high-viscosity fluids as compared to pressure homogenizers because colloid mills produce high shearing forces. Colloid mills with intermediate-viscosity liquids tend to produce larger droplet sizes as compared to those from pressure homogenizers
- The greater friction created in viscous foods may require these mills to be cooled by recirculating water.
- Fibrous materials can be milled using rough surfaced rotor and stator.

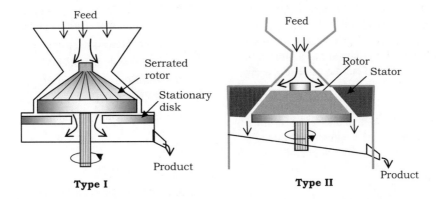

Type I   **Type II**

FIGURE 11.13   Working of colloid mill

Applications

- A colloidal mill reduces the size of particles of solid materials present in suspensions or emulsions. It also increases the stability of emulsions and suspensions. The most efficient system for dispersion and homogenization of raw materials is a colloid mill.

- Colloid mills are used in food, pharmaceutical, paint, and many other industries. Soy milk, peanut butter, fruit jam, cream, and fish liver oil are some of the products manufactured in colloidal mills. Colloid mills are also used for syrups, milk, and purees.

Ultrasonic Homogenizer

Construction and Operation

- Ultrasonic homogenizer uses high-frequency sound waves (18–30 kHz) to cause alternate cycles of compression and tension in low-viscosity liquids and cavitation of air bubbles; all these contribute to form emulsions with droplet sizes of 1–2 μm.

- A vibrating metal blade is used to produce the ultrasonic energy and the blade's clamping position is adjusted to control blade's vibration frequency.

Applications

Ultrasonic homogenizers produce baby foods, cream for salads, synthetic creams, ice creams, and essential oil emulsions. Dispersal of powders in liquids is also achieved through this equipment.

Hydro-Shear Homogenizer

A hydro-shear homogenizer consists of a double-cone shaped chamber with a tangential feed pipe at the center and outlet pipes at the end of each cone.

The liquid feed enters into the chamber at high tangential velocity and swirls in, like a cyclone chamber, until it reaches the center and is discharged. During the swirling, the differences in velocity between adjacent layers of liquid cause high shearing forces, which together with cavitation and ultra-high

frequency vibration, break droplets in the dispersed phase to within a range of 2–8 μm.

A *microfluidizer* is another similar type of equipment where a fluid mixture is pumped into a chamber, creating shear and turbulence when the mixture components intermingle. The dispersed phase is reduced to droplets of less than 1 μm size within a narrow size range.

Check Your Understanding

1. A hammer mill for reducing the size of Bengal gram from 6.25 mm to 0.75 mm requires 3.3 kW-h energy per ton of material. Considering that Bond's law represents the size reduction, what would be the amount of energy required for reducing the size to 0.1 mm?

   Answer: 12.077 kWh.ton^{-1}

2. The work index of a material is 6.25. If 80% of the feed and 80% of the product pass through IS Sieve No. 340 (3.25 mm opening) and IS Sieve No. 40 (0.42 mm opening), respectively, what will be the power consumed to crush 5000 kg.h^{-1} of the material?

   Answer: 9.77 kW

3. A fine powder of sugar is prepared from the sugar crystals originally of 500 μm size (80% of the feed pass through this size sieve) to a size such that 80% of the product pass through US standard sieve number 170 (88 μm size). It was observed that a 3 hp motor was adequate for the required throughput. If the final product has to be such that 80% would pass through a No. 120 sieve (125 μm size) and the throughput will be increased by 30%, what will be the power requirement? Use Bond's equation.

   Answer: 2.8183 hp

4. Corn, having an equivalent diameter of 8 mm, is crushed using a pair of rolls to prepare grits (equivalent diameter of 3 mm), raw material for making corn flakes. Both the rolls rotate convergent at equal speed. Calculate the diameter of each roll if the coefficient of friction is 0.25 between the corn and the material of the roll surface.

   Answer: 15.8 cm

5. Energy required to grind a given mass of particles from a mean diameter of 12 mm to 4 mm is 12 kJ. kg^{-1}. If energy consumed to grind the same mass of particles of 2 mm mean diameter to x mm mean diameter is 252 kJ.kg^{-1}, find out the value of x using Rittinger's law.

   Answer: 0.25 mm

6. Spherical grains of 25 mm diameter are crushed using a 10 kW roller crusher having a feed rate of 5 kg. s^{-1} into 5 mm diameter particles. Find the diameter of the rolls if the coefficient of friction is 0.268. Find the power consumption if the required capacity is 2.5 kg. s^{-1} and the final particle diameter is 2.5 mm.

   Answer: 1125 mm; 8.746 kW

7. Sorghum is ground from the average particle size of 4 mm to 0.4 mm. Feed rate of the mill is 100 kg.h^{-1} and power requirement is 5 kW. What is the Rittinger's constant for the size reduction process?

   Answer: 22.22

8. Find the operating speed of rotation of a ball mill having a diameter of 1600 mm and charged with 60 mm balls. The ball mill is grinding solid matter. What would be the operating speed if the equipment with the same specifications is used for wet grinding in viscous suspension?

   Answer: 27.36 rev. min^{-1}, 22.23 rev. min^{-1}

Objective Questions

1. State the different laws of size reduction related to solid grinding with their mathematical expressions. Explain the different situations under which these laws are used for determining the energy required for size reduction.

2. Define (a) work index, (b) crushing efficiency, (c) effectiveness of grinding, (d) fineness modulus, (e) effective size in size reduction, and (f) uniformity coefficient

3. Differentiate between the working principles of
   a. Crushers and grinders;
   b. Jaw crusher and gyratory crusher;
   c. Smooth roll crusher and toothed roll crusher;
   d. Hammer mill and attrition mill;
   e. Roller mill and rod mill;
   f. Agitated mill and fluid energy mill.

4. Write a short note on cryogenic grinding. State the basic components of a cryogenic grinding system.

5. What are the different size reduction processes of fruits and vegetables (a) which do not change the form of the product appreciably, and (b) which change the form of the product appreciably?

6. Why is the crushing of fruits recommended with super-heated steam or in a vacuum chamber?

7. Differentiate between pitting and coring, snipping and destemming, crushing and shredding.

8. Describe the working principles of
   a. Pressure homogenizer;
   b. Ultrasonic homogenizer;
   c. Colloid mill;
   d. Hydroshear homogenizer.

BIBLIOGRAPHY

Ahmad, T. 1999. *Dairy Plant Engineering and Management*. 4th ed. Kitab Mahal, Delhi.

Brennan, J.G. (Ed.). 2006. *Food Processing Handbook*, 3rd ed. WILEY-VCH Verlag GmbH & Co. KGaA, Weinheim.

Brennan, J.G., Butters, J.R., Cowell, N.D. and Lilly, A.E.V. 1976. *Food Engineering Operations*, 2nd ed. Elsevier Applied Science, London.

Chakravorty, A. 2019. *Post-harvest Technology of Cereals, Pulses and Oilseeds, 3rd* Edn.. Oxford and IBH Publishing Co, New Delhi.

Charm, S.E. 1978. *Fundamentals of Food Engineering*, 3rd ed. AVI, Westport, Connecticut.

Dash, S.K., Bebartta, J.P. and Kar, A. 2012. *Rice Processing and Allied Operations*. Kalyani Publishers, New Delhi.

Earle, R.L. and Earle, M.D. 2004. *Unit Operations in Food Processing*, Web Edn. The New Zealand Institute of Food Science & Technology, Inc., Auckland, https://www.nzifst .org.nz/ resources/unitoperations/index.htm.

Fellows, P.J. 2000. *Food Processing Technology*. Woodhead Publishing, Cambridge, UK.

Heldman, D.R. and Singh, R.P. 1981. *Food Process Engineering*. AVI Publishing Co., Westport, Connecticut.

Henderson, S.M. and Perry, R.L. 1976. *Agricultural Process Engineering*. The AVI Pub. Co., Inc., Westport, Connecticut.

https://www.aiche.org/resources/publications/cep/2015/september/cool-down-liquid-nitrogen.

Junghare, H.K., Hamjade, M., Patil, C.K., Girase, S.B. and Lele, M.M. 2017. A Review on Cryogenic Grinding. *International Journal of Current Engineering and Technology*. Spl Issue 7: 420–423

Leniger, H.A. and Beverloo, W.A. 1975. *Food Process Engineering*. D. Reidel, Dordrecht, 169–188.

Lewis, M.J. 1990. *Physical Properties of Foods and Food Processing Systems*. Woodhead Publishing, Cambridge, 184–195.

Loncin, M. and Merson, R.L. 1979. *Food Engineering*. Academic Press, Cambridge, MA, pp.246–264.

McCabe, W.L., Smith, J.C. and Harriot, P. 1993. *Unit Operations of Chemical Engineering*, 5th ed. McGraw-Hill, Inc., New York.

Nelson, P.E. and Tressler, D.K. 1980. *Fruit and Vegetable Juice Processing Technology*, 3rd ed.. AVI Publishing Co., Westport, Connecticut268–309.

Pandey PH. 1994. *Principles of Agricultural Processing*. Kalyani Publishers, New Delhi.

Pillaiyar P. 1988. *Rice: Post Production Manual*. Wiley Eastern Ltd., New Delhi.

Rao, D.G. 2009. *Fundamentals of Food Engineering.* PHI Learning Pvt. Ltd. , New Delhi.

Sahay, K.M. and Singh, K.K. 1994. *Unit Operations in Agricultural Processing.* Vikas Publishing House, New Delhi.

Singh, R.P. and Heldman, D.R. 2014. *Introduction to Food Engineering.* Academic Press, San Diego, CA.

Urschel, J.R. 1988. The science and art of cutting food products. In: A Turner (ed.) *Food Technology International Europe.* Sterling Publications International, London, 87–91.

Watson, E.L. and Harper, J.C. 1989. *Elements of Food Engineering*, 2nd ed.. Van Nostrand Reinhold, New York.

12

Mechanical Separation

Separation of components is often needed in food industries, for sorting, cleaning, classification, etc. The separation of impurities from water, fat from milk, pulp from juices, oil from oil-bearing materials, and powder from air/gas coming out of pneumatic conveyors/spray dryers are all mechanical separation operations. The main processes used for the separation of different components of food are screening, filtration, settling and sedimentation, centrifugation and expression. Although the processes of evaporation and drying also separate water from food, the objective is to preserve the resultant product. In contrast, the mechanical separation processes are just used to separate the different fractions and are not essentially food preservation operations.

12.1 Methods of Mechanical Separation

On the basis of materials separated, different mechanical separators can be classified as given in Table 12.1.

The methods of separation of solids from solids have been discussed in Chapter 10 and we will not elaborate further in this chapter. The main principles of separations involving separation of solids from liquids or gases, liquid from liquid or liquid from solid, etc. can be classified under different categories as filtration, settling and sedimentation, centrifugation, pressing, and other combination methods.

Filtration is the separation of insoluble solids from a fluid by guiding the mixture to pass through a bed of porous material. In *settling* and *sedimentation*, the particles are separated from the fluid due to the difference in their densities. The particles of different densities settle at different rates under the force of gravity. The application of centrifugal force to separate solids from liquids or immiscible liquids is known as *centrifugation*. Cyclone separators also employ centrifugal force to separate solids from fluids. *Expression* or pressing separates liquids from solids by applying pressure on the solids as is done in oil expellers or in juice presses. Innovative separation techniques, such as membrane separation and supercritical fluid extraction, are also discussed in this chapter.

12.2 Filtration

12.2.1 Methods and Principles

During filtration, the fluid (usually called the feed slurry) is passed through a bed of porous material (called the *filter medium* or *septum*) to separate the insoluble solids from it. The fluid may be a liquid or a gas. The pressure difference causes

the fluid to pass through the pores whereas the solid particles are held onto the pores. Thereafter, as the filtration continues and more solids are held on the surface of the septum as a porous cake (called the *filter cake*) that acts as the medium for continued filtration.

Filtration is usually practiced when the fluid has a relatively small amount of solid content in it. For large amount of suspended solids, initial settling or sedimentation followed by filtration is helpful. Either the filtrate or the cake may be important to us, or, in some cases, both are important as in filtration of fruit juices or in the filtration of miscella obtained from oil mills. Sometimes, neither the filtrate nor the cake is important, but separation is required, such as in the partition of waste solids from the waste liquid prior to disposal.

Rate of Filtration and Resistances to Flow

As mentioned above, during the process of filtration, the solid particles are initially trapped in the septum and then a cake of solids is formed on the septum as the filtration continues.

The *rate of filtration* through a filter medium is given as:

$$\text{Rate of filtration} = \frac{\text{driving force(pressure difference across filter)}}{\text{resistance to flow}} \quad (12.1)$$

As the cake thickens during filtration, the area of fluid flow reduces. It, in turn, increases the total resistance and decreases the filtrate flow. The pressure required to maintain the same flow rate of the filtrate increases.

There are two types of resistances, namely, the *medium resistance* due to the filter medium and the pre-coated layer of filtrate cake, and the *specific cake resistance*, which is the resistance to the filtrate flow across the cake. The total pressure drop through the filter is given as:

$$\Delta p = \Delta p_m + \Delta p_c \quad (12.2)$$

where, Δp_m is the pressure drop across the filter medium, and Δp_c is the pressure drop across the cake.

Let us represent the flow process during filtration as shown in Figure 12.1. In this figure, t_c (m) is the thickness of the cake that is formed after time t (s). The filter cross-sectional area is A (m²) and v (m.s⁻¹) is the linear velocity of the filtrate perpendicular to the cake. Let m (kg) be the mass of the filter cake and μ (N.s.m⁻²) the viscosity of the filtrate.

TABLE 12.1

Different types of mechanical separators

Materials to be separated	Type of separator
Solids from solids	Based on differences in size: screens
	Based on differences in other characteristics: air classifiers, centrifugal classifiers, specific gravity separation tables, spiral separators, magnetic separators, electrostatic separators, electronic color separators, floatation separators, etc.
Solid from liquid	Filters, centrifugal filters, clarifiers, thickeners, sedimentation tanks, sedimentation centrifuges, hydroclones (liquid cyclones), membrane separation
Liquid from solid	Presses, centrifugal extractors
Liquid from liquid	Settling tanks, liquid cyclones, centrifugal decanters, coalescers
Solid from gas	Settling chambers, cyclones, air filters, bag filters, impingement separators, electrostatic and high-tension precipitators
Liquid from gas	Settling chambers, cyclones, impingement separators, electrostatic precipitators
Gas from liquid	Still tanks, deaerators, foam breakers

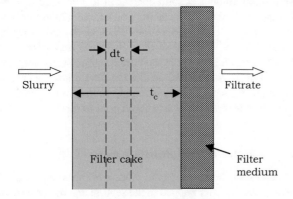

FIGURE 12.1 Schematic diagram of flow across a filtering medium

Now the medium resistance, $R_m (m^{-1})$ and the specific cake resistance, α $(m.kg^{-1})$ can be written as:

$$R_m = \frac{\Delta p_m}{\mu v} \quad (12.3)$$

$$\text{and} \quad \alpha = \frac{\Delta p_c}{\mu v (m/A)} \quad (12.4)$$

Thus, substituting the values in equation 12.2,

$$\Delta p = \alpha \mu v \frac{m}{A} + \mu v R_m \quad (12.5)$$

The mass of the filter cake could be determined if the volume of the filtrate along with the concentration of cake solids is known. Thus, if V (m^3) is the volume of filtrate and c_s $(kg.m^{-3})$ is the concentration of cake solids in the suspension, then

$$m = V. c_s \quad (12.6)$$

V can also be related to m, as follows:

$$m = V.c_s = \frac{\rho c_x}{1 - Mc_x} V \quad (12.7)$$

where M is the mass ratio of wet cake to dry cake, c_x $(kg.m^{-3})$ is the mass fraction of solids in the slurry and ρ $(kg.m^{-3})$ is the density of filtrate.

The filtrate velocity, v, can be given as $\dfrac{(dV/d\theta)}{A}$

$$(12.8)$$

Substituting in equation 12.5,

$$\Delta p = \frac{\mu(dV/d\theta)}{A}\left(\frac{\alpha.V.c_s}{A} + R_m\right) \quad (12.9)$$

$$\text{Or,} \quad \frac{d\theta}{dV} = \frac{\mu}{A.\Delta p}\left(\frac{\alpha V.c_s}{A} + R_m\right) \quad (12.10)$$

The above equation represents a straight line of the form, y = mx + C,

$$\text{where,} \quad y = dq/dV, m = \frac{\mu \alpha c_s}{A^2 \Delta p} \quad \text{and} \quad c = \frac{\mu.R_m}{A.\Delta p}$$

Thus, it can be used to obtain the specific cake resistance and the medium resistance from filtration data. The steps are as follows:

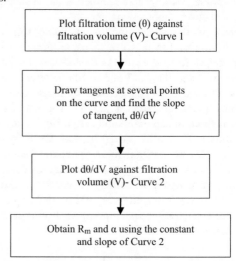

Steps to obtain specific cake resistance and the medium resistance

The equation 12.10 is known as the *Sperry equation*.

The volume of filtrate (V) and the cake thickness (t_c) can be related by a material balance as follows:

$$t_c.A.(1-\varepsilon).\rho_p = c_s (V+\varepsilon.t_c A) \quad (12.11)$$

In this equation, ρ_p (kg.m^{-3} of solid) is the density of solid particles in the cake and ε is the void fraction or the porosity of the cake. The volume of filtrate held in the cake is the final term in the above equation. This is usually small and is neglected.

The specific cake resistance, α can also be defined as:

$$\alpha = \frac{G(1-\varepsilon)S_p^2}{\rho_p \varepsilon^3} \tag{12.12}$$

where, S_p (m^2 of particle area per m^3 volume of solid particle) is the specific surface of particles, and G is a constant. The value of G is 4.17 for particles of definite shape and size.

The above equation indicates that void fraction ε and S_p affect the specific cake resistance. As pressure can affect ε, specific cake resistance is also a function of pressure.

The following empirical equation is used for relating α with the pressure

$$\alpha = \alpha_0 (-\Delta p)^\sigma \tag{12.13}$$

Here α_0 and σ are empirical constants. For incompressible cakes, $\sigma = 0$. Usually, the value of σ is between 0.1 and 0.8.

Sometimes, the following equation is also used.

$$\alpha = \alpha_0' \left[1 + \beta(-\Delta p)^\sigma \right] \tag{12.14}$$

Here, α_0', β and σ are empirical constants.

Constant Pressure Filtration

Constant pressure filtration occurs when the pressure difference across the filter remains almost constant. This situation can be observed when a centrifugal pump is used as the feed pump.

In this situation, the equation 12.10 can be integrated (with Δp as a constant value) as:

$$d\theta = dV\left(\frac{\mu \alpha c_s V}{A^2 \Delta p} + \frac{\mu R_m}{A.\Delta p}\right) \tag{12.15}$$

$$\text{or, } \int_0^\theta d\theta = \left(\frac{\mu \alpha c_s}{\Delta p A^2}\right)\int_0^V V.dV + \frac{\mu R_m}{A.\Delta p}\int_0^V dV \tag{12.16}$$

$$\text{Or, } \theta = \left(\frac{\mu \alpha c_s}{\Delta p}\frac{V^2}{2A^2} + \frac{\mu R_m}{\Delta p}\frac{V}{A}\right) \tag{12.17}$$

$$\frac{\theta}{V} = \left(\frac{\mu \alpha c_s}{2\Delta p A^2}V + \frac{\mu R_m}{A.\Delta p}\right) \tag{12.18}$$

If we denote, $\zeta_p = \dfrac{\mu \alpha c_s}{\Delta p.A^2}$ and $\chi = \dfrac{\mu R_m}{A.\Delta p}$, the above equations can be written in a simplified form as:

$$\theta = \zeta_p \frac{V^2}{2} + \chi.V \tag{12.19}$$

$$\frac{\theta}{V} = \zeta_p \frac{V}{2} + \chi \tag{12.20}$$

$$\text{and } \frac{d\theta}{dV} = \zeta_p V + \chi \tag{12.21}$$

Here the units of ζ_p is s.m^{-6} and that of χ is s.m^{-3}.

Equation 12.18 (or 12.20) suggests that, if we plot θ/V against V, we get a linear plot with the intercept χ and slope $\zeta_p/2$. Then from ζ_p and χ values, the values of α and R_m can be found.

While solving the equations 12.10 and 12.18, the values of R_m may be negative. It will happen specifically when the R_m is considerably smaller than α; this situation occurs when finely suspended materials present in the fluid quickly reduce the porosity of the medium, though the deposit as cake is very less.

Here the use of equation 12.17 can avoid negative R_m values. If the amount of suspended solids is zero, i.e., $c_s = 0$, then equation 12.17 becomes

$$\theta = \frac{\mu R_m V}{\Delta p.A} \tag{12.22}$$

As $R_m = \dfrac{\Delta p_m}{\mu v}$, we can write the equation 12.17 as:

$$\theta = \frac{\mu \alpha c_s}{2\Delta p A^2}V^2 + \frac{\mu \Delta p_m}{\Delta p.\mu.v}\frac{V}{A} \tag{12.23}$$

Substituting q for A.v (q = volumetric flow rate), the above equation can be modified as:

$$\theta = \frac{\mu \alpha c_s}{2\Delta p A^2}V^2 + \frac{V}{q} \tag{12.24}$$

$$\text{or, } \theta = \zeta_p \frac{V^2}{2} + \frac{V}{q} \tag{12.25}$$

The pressure difference (Δp) across the pre-coated filter and q will depend on the type of filter aid and applied pressure as well as the thickness of the cake. If the filter and filtrate remain the same, the volumetric flow rate will be proportional to the thickness and applied pressure. Thus, knowing q, α can be found from the slope of a regression equation for ($\theta - V/q$)-vs-V^2. A log-log plot of ($\theta - V/q$)-vs-V can also be used for the purpose.

Washing of Filter Cakes and Cycle Time

The *total cycle time* in a filter includes the actual filtration as well as the time for washing of the cake. The cake is washed by diffusion and displacing of the filtrate. In addition, additional time is required for removing the cake, cleaning the filter, and reassembling the filter. The sum of the times needed for cleaning, filtration, and washing account for the total cycle time for the filter.

To calculate the washing rates, it is assumed that the conditions during washing and those existing at the end of the filtration are the same, and that the cake structure remains unaltered when the slurry liquid is replaced by the wash liquid.

The final rate of filtration for constant pressure filtration using the same pressure in washing and filtering (which normally occurs in leaf filters) is the reciprocal of equation 12.21.

$$\left(\frac{dV}{d\theta}\right)_f = \frac{1}{\zeta_p V_f + \chi} \qquad (12.26)$$

where, $\left(\dfrac{dV}{d\theta}\right)_f$ is the rate of washing in m³.s⁻¹ and V_f is the volume of filtrate until the end of filtration, m³.

The wash liquid has to travel through an area only half of that during filtration and a cake twice as thick in case of plate and frame filter press. Hence, ¼ of the final filtration rate is the predicted washing rate.

$$\left(\frac{dV}{d\theta}\right)_f = \frac{1}{4}\left(\frac{1}{\zeta_p V_f + \chi}\right) \qquad (12.27)$$

Actually, it is because of cake consolidation, formation of cracks, and channelling that the washing rate may be lower than the predicted value. Studies have indicated that the washing rates were 70% to 92% of the predicted values in a small plate and frame filter press.

Rate of Filtration in Continuous Filters

An example of continuous filtration is a rotary drum filter. Here there is continuous movement of the feed and filtrate; the cake is also moved at a steady, continuous rate. The pressure drop is maintained at a constant value. Resistance of the filter medium in continuous filtration is generally ignored because it is very small in comparison to the cake resistance.

Therefore, in the equation 12.21, $\chi = 0$

$$\int_0^\theta d\theta = \zeta_p \int_0^V V.dV \qquad (12.28)$$

$$\theta = \zeta_p \frac{V^2}{2} \qquad (12.29)$$

In a continuous filter, if the total cycle time is given as θ_c and θ is the time when actual cake formation takes place, we may correlate them with a factor f (fraction of cycle used for cake formation) as:

$$\theta = f.\theta_c \qquad (12.30)$$

In a rotary drum filter, f is the fraction of total time when a point on the surface of the drum is submerged in the slurry. In the equation 12.30, θ is the time during which the cake is formed.

Now substituting the value of ζ_p in equation 12.29, we get

$$\theta = \frac{\mu\alpha c_s}{\Delta p A^2}\frac{V^2}{2} \qquad (12.31)$$

$$\frac{V^2}{A^2\theta} = \frac{2\Delta p}{\mu\alpha c_s} \qquad (12.32)$$

$$\frac{V^2}{A^2\theta_c} = \frac{2f\Delta p}{\mu\alpha c_s} \qquad (12.33)$$

$$\frac{V^2}{A^2\theta_c^2} = \frac{2f\Delta p}{\theta_c\mu\alpha c_s} \qquad (12.34)$$

Thus, the flow rate is given as:

$$\frac{V}{A\theta_c} = \left[\frac{2f\Delta p}{\theta_c\mu\alpha c_s}\right]^{1/2} \qquad (12.35)$$

For conditions when the specific cake resistance varies with pressure, we can use equation 12.13 to get the value of α to be used in equation 12.35. Thus, we find that the square root of viscosity and cycle time have inverse proportionality with the flow rate.

In a continuous filtration system, the filter resistance term B has to be included in the calculations if the filter medium resistance is relatively large. Also, B should not be ignored if short cycle times are used.

In that case

$$\theta = f.\theta_c = \zeta_p\frac{V^2}{2} + \chi.V \qquad (12.36)$$

Flow rate can be given as:

$$\frac{V}{A\theta_c} = \frac{-R_m/\theta_c + \left[\dfrac{R_m^2}{\theta_c^2} + \dfrac{2c_s\alpha(\Delta p)f}{\mu\theta_c}\right]^{1/2}}{\alpha c_s} \qquad (12.37)$$

Constant Rate Filtration

If we use a positive displacement pump, then the filtration occurs at a constant rate. In that situation, the equation 12.10 can be rearranged for the constant rate filtration.

$$\frac{d\theta}{dV} = \frac{\mu}{A.\Delta p}\left(\frac{\alpha V.c_s}{A} + R_m\right) \qquad (12.38)$$

$$\text{or, } \Delta p = \left(\frac{\mu\alpha c_s}{A^2}.\frac{dV}{d\theta}\right)V + \frac{\mu R_m}{A}.\frac{dV}{d\theta} \qquad (12.39)$$

$$\text{Substituting, } \zeta_v = \left(\frac{\mu\alpha c_s}{A^2}.\frac{dV}{d\theta}\right), \text{ and } \chi' = \frac{\mu R_m}{A}.\frac{dV}{d\theta} \qquad (12.40)$$

The above equation can be written as:

$$\Delta p = \zeta_v V + \chi' \qquad (12.41)$$

Here the ζ_v has a unit of N.m⁻⁵, and χ' has a unit of N.m⁻².

We assume that ζ_v and χ' are constant characteristics of the cake and rate of filtrate flow while the cake is incompressible. We get a straight line for a constant rate $dV/d\theta$ if Δp is plotted against V (the total volume of collected filtrate). The slope of the line is ζ_v and intercept χ'. The pressure increases and the volume of collected filtrate decreases with the increase in cake thickness.

Substituting $V = \theta(dV/d\theta)$, the equation can also be written as:

$$\Delta p = \left(\frac{\mu\alpha c_s}{A^2}\cdot\left(\frac{dV}{d\theta}\right)^2\right)\theta + \frac{\mu R_m}{A}\cdot\frac{dV}{d\theta} \qquad (12.42)$$

This equation is also used when the specific cake resistance varies.

Centrifugal Filtration

In the case of centrifugal filtration, the specific cake resistance may change markedly while the area for flow and driving force increase from the axis toward the periphery. For analysis of centrifugal filtration, we will start from the equation of constant pressure filtration with an assumption that the cake is already deposited.

As shown in Figure 12.2, let r_C be the inner radius of the basket and r_I the radius up to the face of the cake. Let r_L be the inner radius of the liquid surface.

The following assumptions are made for the analysis.

The flow is considered to be laminar. An average value of specific cake resistance (α) can be used by assuming the cake to be incompressible. It is also assumed that in a large diameter centrifuge there is a thin cake. Thus, the flow area (A) is nearly constant. The liquid velocity is as follows:

$$v = \frac{q}{A} = \frac{dV}{Ad\theta} \qquad (12.43)$$

Here, the filtrate flow is q (m³.s⁻¹) and v (m.s⁻¹) is the velocity.

Now, we reproduce the equation 12.10 below.

$$\frac{d\theta}{dV} = \frac{\mu}{A.\Delta p}\left(\frac{\alpha V.c_s}{A} + R_m\right) \qquad (12.44)$$

This can in another way be written as follows:

$$\frac{dV}{Ad\theta} = \frac{\Delta p}{\mu\left(\dfrac{\alpha V.c_s}{A} + R_m\right)} \qquad (12.45)$$

Substituting equation 12.43 in equation 12.45 and substituting $m_c = V.c_s$, we get

$$\Delta p = \mu q\left(\frac{\alpha m_c}{A^2} + \frac{R_m}{A}\right) \qquad (12.46)$$

In this equation m_c (kg) is the cake mass deposited on the filter.

For a hydraulic head of dz (m), the pressure drop is given as, $dp = \rho.g.dz$.

Acceleration due to gravity, g, is replaced by $\omega^2 r$ in a centrifugal field, and dz by dr so as to get

$$dp = \rho\,\omega^2 rdr \qquad (12.47)$$

Integrating between r_L and r_C,

$$\Delta p = \frac{\rho\omega^2\left(r_C^2 - r_L^2\right)}{2} \qquad (12.48)$$

Thus, $q = \dfrac{\rho\omega^2\left(r_C^2 - r_L^2\right)}{2\mu\left(\dfrac{m_c\alpha}{A^2} + \dfrac{R_m}{A}\right)}$ $\qquad (12.49)$

The following equation has been developed for a special case where a change in radius changes the flow rate considerably.

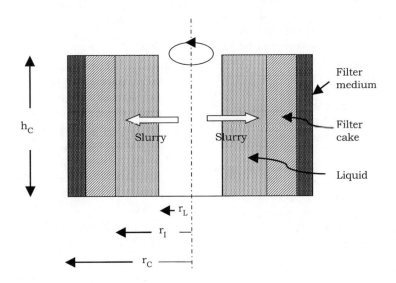

FIGURE 12.2   A representation of centrifugal filtration

$$q = \frac{\rho\omega^2\left(r_2^2 - r_1^2\right)}{2\mu\left(\dfrac{m_c\alpha}{\overline{A}_L\overline{A}_a} + \dfrac{R_m}{A_2}\right)} \qquad (12.50)$$

where, area of filter medium, $A_2 = 2\pi r_c h_c$, $\overline{A}_L = \dfrac{2\pi h_C\left(r_C - r_L\right)}{\ln\left(r_C/r_L\right)}$,

the logarithmic mean cake area, $\overline{A}_a = \left(r_C + r_L\right)\pi h_C$ is the arithmetic mean cake area. The equation is valid at a given time for a given cake mass. It does not cover the whole filtration cycle.

12.2.2 Equipment

Many types of filters are used in the food industry as well as other process industries to cater to the needs of the whole array of materials and diverse process conditions.

Classifications of Filters

The equipment used for commercial filtration is classified in a number of ways. The common type of classification can be as batch type or continuous type systems. In batch type filters, the cake is removed periodically from the equipment by stopping the filtration for a while. In continuous type filters, the cake is continuously removed while the filtration is continuing. Depending on whether the clarified filtrate or the filter cake or the outlet liquid is the desired product, the filters may be classified accordingly. They can also be classified as gravity, pressure, or vacuum type depending on the type of force employed to have the flow through the septum.

The arrangement of the filter media is another important criterion of classification. The filter cloth can be arranged as individual leaves dipped in a slurry or in series as flat plates in an enclosure or on rotating mesh drums in the slurry.

The classification of filters can broadly be as clarifying filters, cake filters, and cross flow filters. *Cake filters* have a thinner filter medium than the clarifying filter. Solid particles enter the pores of the filter medium and fill them during the initial period of filtration. The extra solids added subsequently during the filtration process are retained on the septum surface in the form of a cake. Thus, after the initial operation, the cake performs the job of actual filtration. However, the filtration rate reduces as the depth of cake continues to increase; thus, the cake has to be removed from time to time. The cake filters are used when there are large amounts of solids in the liquid. Filter press used in the oil mills or the filter candle used in household filters are cake filters. The surface of the candle of household filter is often cleaned when the rate of filtration reduces.

The *clarifying filters* produce clear liquids, such as beverages or clean gas, by removing small amounts of solids. The filter medium is thick and, thus, the small amount of solids present in the fluid is deposited within the pores and is not visible as a cake on the septum surface. This is also known as a *deep bed filter*. Examples are Cartridge filter and Ultra filter.

The feed suspension in *cross flow filters* flows across the filter medium at a very high velocity under pressure. Thus, the flows of the fluid and the filtrate are in a cross-wise manner. There may be a thin deposit of the cake on the surface of the medium, but it does not grow to appreciable thickness because of the high velocity fluid flow.

The cake filters can be categorized as continuous or discontinuous types on the basis of whether the discharge of the cake (filtered solids) is continuous or intermittent. In continuous filters, the discharge of liquid and cleaning of the septum surface (i.e., removal of the cake) are continuous and there is no need to stop the filtration to discharge the cake. But in discontinuous filters, the flow of fluids is interrupted in between to remove the cake. Pressure filters are usually discontinuous as pressure has to be applied on the upstream side where the cake is deposited. Since a vacuum has to be maintained on the downstream side, the vacuum filters are usually continuous.

The pressure filters, depending on the type of force applied, can be either gravity filters or centrifugal filters. The filter press, shell and leaf filter, and automatic belt filter come under the category of discontinuous type pressure filters using gravity force. The suspended batch centrifuge and automatic batch centrifuge are centrifugal filters. The continuous conveyor type centrifuge is an example of a continuous pressure filter. The rotary drum filter, horizontal belt filter, and rotary vacuum disc filter are continuous type vacuum filters. The vacuum nutsch is a discontinuous type vacuum filter.

Feed can be modified by such means as addition of filter aids, heating, or recrystallization to improve the filtration rate. Filter aids are explained at the end of this section.

Gravity/Pressure Filters

Under this group of equipment, we will be discussing about the bed filter, plate and frame filter press, shell and leaf filter and the belt filter.

Bed Filter

Construction and Operation

- This is the simplest type of filter, which consists of an open tank with layers of sand and gravel as the filter medium (Figure 12.3).
- The bottom layer is often gravel, there are aggregates above the gravel, and there is fine sand layer at the top. In water filtration applications, sometimes there is a layer of anthracite coal above sand and gravel layers. The depth of filter bed can be as high as 3 m.
- The water/slurry slowly filters through the layers and the filtrate goes out at a specified outlet at the bottom.
- There is provision for back washing to clear the precipitates of filtered articles clogged in the sand.

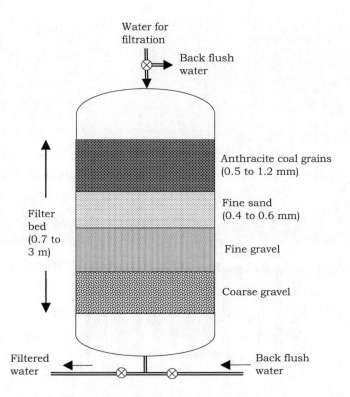

FIGURE 12.3 Schematic diagram of a gravity filter with backwash system

Salient Features

- The flow rate drops when the sand is clogged. Back washing by forcing water in the opposite direction can improve the flow rate.
- It can be used only when the precipitates can be easily removed by back washing in situations when the precipitates do not adhere strongly to the sand.

Applications

The bed filter is commonly used for clarification of potable water, for filtering of rainwater, or for treatment of waste water. This works well for relatively small amounts of solids.

Plate and Frame Filter Press

Construction and Operation

- The plate and frame filter press comprises a number of plates having corrugated surfaces or channels cut into them and there are filter cloths over each side of the plates. The plates are held by frames such that together they make a series of chambers. The frames are supported on a pair of rails (Figure 12.4).
- The frames can be brought close to each other to make the chambers or can be separated apart with the help of a hand screw or a hydraulic press by moving the frame on the rail.
- The plates may be 6 to 50 mm thick placed vertically or horizontally having square or rectangular shape. Usually, square plates are used and the size may be 0.15 m ×0.15 m to 2 m × 2 m.

- As the frames are closed and slurry is fed to the chambers from the top, the filter cloth does not allow the solids to pass through and the filtrate passes through the cloth to the plate side which then moves down along the corrugations (or channels) on the plate. Thus, the frame side of the cloth is where the cake forms.
- There are openings in each plate and frame and these openings, registering together, form the feed channel. Slurry enters the chamber through these openings and after filtration the filtrate passes down the channels on plates to the discharge outlet at the bottom. All discharge outlets connect to a common header.
- Solids form the cake on the cloth covered faces of the plates and the pressure maintained is usually low (1.5-3.2 kg.cm^{-2} gauge pressure).
- As layer of cake builds up, the flow rate is decreased. The filtration is continued until the frames are completely full; at this stage usually the flow of filtration is almost stopped. This is when the press is said to be *jammed* or known as *solid filling*.
- Usually, the cake is washed before removing from the frame. However, there may be a blowing of air before wash is applied to recover some solution. As the chamber is completely filled with solids, no displacement of cake occurs at that moment. Then behind the filter cloth of each alternate plate, wash solution or water is applied under pressure. To remove the wash and to obtain a more dry cake an air blow may be given.

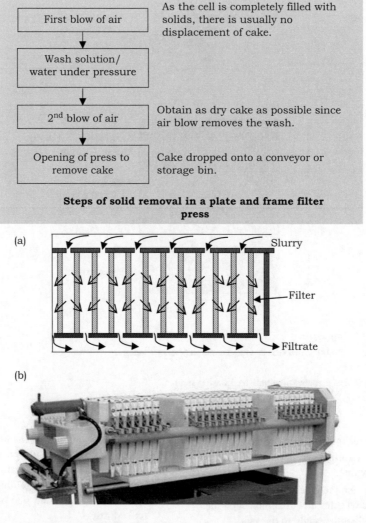

FIGURE 12.4 (a) Flow of materials in a plate and frame filter press, (b) outer view of a plate and frame filter press

- There may be separate channels for the wash water. Thus, the plates can be of two types, namely, plates having no such ducts or plates provisioned with ducts to allow wash water behind the filter cloth.
- The press is opened, the cake removed, and it is deposited onto a conveyor or in a storage bin.
- The cycle of operation is continued after reassembling the system.

Salient Features

- A plate and frame filter press has high flexibility for different types of products, involves low capital cost, and can be easily maintained.
- It operates in a batch-wise manner and requires dismantling the press and cleaning at the end of each cycle manually. Thus, of the total operating cost, the cost of labor and the downtime cost together may become a major component. Some designs of the filter press have spare sets of frames, which are fitted on a shaft to reduce the down time.
- The major advantage is the dryness of the cake obtained from the plate and frame press.

- It can be used under moderately high pressure when necessary, for example, for filtering viscous products.
- In many cases there are separate discharges for each frame, so that if the filtrate from a specific outlet is observed to be cloudy, it indicates that there may be damage in the filter cloth; it helps for better maintenance.

Applications

This is widely used for oil filtration and for squeezing of juices from apples and other commodities.

Shell and Leaf Filter

Construction and Operation

- The filter has a set of vertical leaves made up of a hollow wire framework covered by the filter cloth. A retractable tract holds the leaves.
- A hollow frame connects the leaves, and it also acts as the outlet channel for the filtrate.
- The above arrangement is stacked within a pressure vessel. There is an inlet pipe to allow the slurry into the pressure vessel and there is also a connection from the outlet channel of the leaves to the outside of the pressure vessel (Figure 12.5).

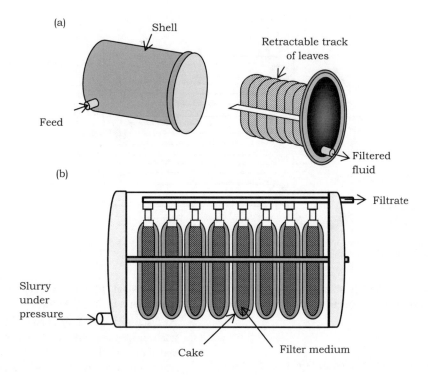

FIGURE 12.5 Schematic of (a) shell and leaves assembled on a retractable tract (b) flow streams in the filter

- During operation, at a pressure of approx. 400×10^3 Pa, the feed slurry is pumped into the shell. Thus, through the filter medium, the filtrate passes into the discharge manifold.

- After the filtration operation or after the cake has built up to such an extent that the rate of filtration has slowed down, the filtration is stopped, the pressure is released, and the shell is opened for blowing or washing away the cake.

- Sometimes, to dislodge the cakes, air is blown into the leaves in the opposite direction. To wash away the cakes without opening the shell, water jets, in place of air, are also employed.

- In some designs, the uniformity of the cake build-up is improved by rotating the set of leaves at about 1-2 rev.min^{-1}.

Salient Features

- It operates under pressure higher than the filter press and involves a cost higher than the plate filters.

- It is not economical to handle a large quantity of sludge.

Applications

For filtration applications involving the filter aids and for filtration of small amounts of suspended solids from liquids, shell and leaf filters are widely used.

Belt Filter

Construction and Operation

- An automated belt filter consists of a movable belt made of filter cloth that moves through a horizontal chamber or a number of horizontal chambers arranged one above the other. There may be up to 20 chambers in a system.

- During the filtration, the belt remains stationary, and the top surface of the belt collects the solids.

- After the chamber is filled with solids, high pressure water is pumped from the top through a diaphragm to press the cake and express some of the liquid from it. Thereafter, wash water may be passed through the cake and the cake is recompressed if desired. Another blow of air through the cake removes more liquid.

- The chamber is then opened hydraulically, and the belt is moved a distance, which is little more than the length of the chamber while the filter cake is scrapped off. At the same time a portion of belt passes between spray nozzles for washing.

- The belt is stopped after all the cake has been discharged. Thereafter, the chamber is again closed, and the filtration cycle is repeated.

- In a multi-chamber system, as shown in Figure 12.6, the liquid slurry is fed to the chambers through individual inlets. The continuous belt passes sequentially from one chamber to the other. Each chamber has separate scrapper for collection of the cake and provision for directing the cake to a common outlet. After passing through all the chambers, the belt passes through the washing zone before re-entering the top chamber.

- The size of the filter can be upto 30-40 m^2.

Salient Features

- It is a discontinuous type of filter; however, the capacity can be high as a number of chambers can be used simultaneously for filtration.

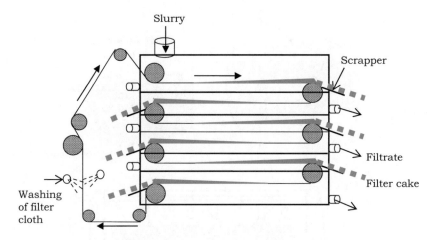

FIGURE 12.6   Schematic diagram of an automatic belt filter

- The overall cycle is relatively short, typically 10 to 30 minutes. All the steps as mentioned above can be computer controlled.

Applications

The belt filter can be used for filtration of liquid having high solid contents.

Three-Belt Gravity-Pressure Type Belt Filter

Construction and Operation

- A three-belt type filtration device involves filtration in two stages, namely, the gravity drainage stage and the compression stage. During compression, the slurry first goes through a low-pressure compression followed by high-pressure compression. There are different belts or the filter cloth arrangements for the gravity drainage stage and compression stage.
- Before the filtration process, the feed slurry is preconditioned; it is coagulated or flocculated depending on the process as well as the feed characteristics. A flocculator fitted with an agitator adds a polymer (usually a high molecular weight polymer) to the slurry.
- A distributor spreads the conditioned slurry as an even mat on the belt of gravity drainage zone.
- The gravity zone comprises a single flat or inclined belt that is used for drainage of water from the slurry by gravity. Most of the free water is drained during this stage.
- In the pressure zone, the feed is compressed in between two belts (filter cloths) moving close to each other. The belt is supported by a series of perforated rollers; these rollers increase the pressure on the slurry, improve the drainage by producing discontinuity and breaking the water surface tension.
- The pressure zone, as mentioned, can be further differentiated as low-pressure and high-pressure zones. In the low-pressure zone, also termed *wedge zone*, upper and lower belts gradually and slightly increase specific pressure applied to the slurry. The high-pressure zone

is also called "S" zone as the two belts form "S" configuration rounding the rollers.

- As discussed, there is one belt in the gravity discharge zone and separate two belts in the pressure zones. The speed of the belts can be different in different zones. Thus, this device is a three-belt system.
- In the pressure zone, the feed is compressed in between two belts (filter cloths) moving close to each other. The belt is supported by a series of perforated rollers; these rollers increase the pressure on the slurry, improve the drainage by producing discontinuity and breaking the water surface tension.
- Two doctor blades discharge the dewatered cake.
- The belts are continuously cleaned under the washing manifold which has spraying nozzles.
- Additional devices for washing of the cake and the belt and drying of the cake with infrared/hot gas/microwaves can be attached depending on the required process parameters.

Salient Features

- The belt types for the gravity and pressure zones can be different for different applications. Double weave woven wire belts with different yarn types are commonly used. These also give a better durability as compared to conventional wire belts.
- The mass flow rate (kg.s⁻¹) on a belt filter press is obtained by multiplying the mass loading (kg.m⁻²) with the belt speed (m.s⁻¹) and the initial width of slurry across the belt (m). The input to a belt press filter is often expressed as the mass of dry solids per unit time per unit belt width. The feed solids concentration is usually in 1–10% range. After dewatering, the cake may contain up to 12–50% solids.
- The formation of a thick cake in gravity drainage zone may affect the rate of filtration. The amount of water drained in the gravity zone depends on the conditioning of the slurry/sludge in addition to the type of solids and the filter media. Too much polymer

and more mixing are undesirable. Heating the slurry reduces its viscosity and improves filtration rate. Low pH also decreases flocculation, and thus, the conditioning parameters should be properly maintained. Addition of a surfactant also improves dewatering.

- In the pressure zone, the optimum number of rollers need to be used. Further increase in the number of rollers does not help in auxiliary dewatering. Reducing the belt speed is more advantageous than increasing the time of pressing for the water removal. However, lowering the belt speed reduces the throughput capacity.

- Generally, the minimum design discharge cake thickness is 3–5 mm to ensure proper discharging of the cake.

- The filtrate obtained by this press is usually not fully clear; this necessitates for its further treatment before it is reused or discharged as a waste. Further clarification or chemical treatment is often practised.

- The fraction of dry solids in the cake is used as a measure of efficiency of a belt filter press. The other efficiency parameters are solids recovery and lateral migration of slurry on the belt. Solids recovery is the percentage of dry solids recovered out of the solids originally present in the feed slurry.

- Belt filters use relatively lower pressures as compared to other compression filters as filter press. Thus, they also have low initial costs and low energy requirement. They offer a long life and easy maintenance. The adjustment and monitoring of the operations are also easy. The other advantages associated with the belt filter press include high throughput capacity and ability to use higher hydraulic loadings with dilute feed slurry (feed solid concentration below 1.5%).

- The cost of flocculant is often a key operating cost during dewatering. The belt press filters in general have lower flocculant consumption than other types of thickeners, though it is higher than the membrane filter presses and centrifuges. The overall cycle is relatively short, typically 10 to 30 minutes. All the steps as mentioned above can be computer controlled.

Applications

- The presses are designed for dewatering different kinds of slurry/sludge (sludge is usually more than 50% solids) by means of progressive compression between two permeable belts. This is primarily used for urban sewage and waste water treatment, dyeing, tanneries, and breweries, as well as chemical and paper industries.

- In food industry, this is used in apple juice, cider, and wine making.

- The V-fold belt can be used for small-scale applications. As the maximum belt size is 0.75 m, it can treat up to approximately 3000 liter.h^{-1} of slurry.

Centrifugal Filters

In the case of centrifugal filters, the slurry is kept under centrifugal force in a drum, the surface of which is made up of wire mesh or a slotted sheet covered with the septum. As the drum rotates at a high speed, the filtrate tends to move away through the septum and the solids remain behind.

The arrangement gives a much drier cake than that possible with the plate and frame press or the belt filter. Rotating the drum for a small time after the feeding has been stopped can remove more water from the cake. Thus, the energy required in subsequent drying of the cake can be reduced (in case the objective is to obtain a cake as dry as possible).

The main types of centrifugal filters are suspended batch centrifuge, automatic short-cycle batch centrifuge, and continuous conveyor centrifuge.

Suspended Batch Centrifuge

Construction and Operation

- A common type of batch centrifuge, called the *top suspended centrifuge*, consists of a perforated basket 75–120 cm in diameter and 45–75 cm in depth that rotates at 600–1800 rev.min^{-1}. The filter medium is generally made of canvas or other fabric, or there is a wire cloth lining on the perforated wall of the basket. The rotating basket is held inside a solid cylindrical casing (Figure 12.7).

- A free-swinging vertical shaft driven from the top holds the basket from bottom.

- There is an inlet pipe into the basket for entry of feed slurry. As the basket rotates at a high speed, the filtrate passes through the filter medium to the casing and is discharged from the casing with the help of a discharge pipe.

- During filtration, wash liquid may be spurted through the cake for removal of the entrained soluble material.

- Usually a maximum cake thickness of 5–15 cm is allowed inside the basket. After filtration, the cake is further rotated at a high speed to make it as dry as possible, sometimes this speed is more than that used in the loading and washing stages.

- Subsequently, the speed of the basket is reduced to about 30–50 rev.min^{-1}, after which the cake is scrapped off with a knife. The cakes in the form of flakes are guided to move into a basket through a specific outlet.

- The cycle is repeated after the filter medium is rinsed clean.

- Automatic controls are often provided for some or all of the steps in the cycle.

Salient Features

- In the case of sugar refining, the suspended batch centrifuges have a cycle of about 2–3 minutes per load and the capacity can be up to 5 tph of crystals. In other cases, suspended centrifuges usually operate at cycles of 10 to 30 minutes per load and the capacity can be 800–1800 kg.h^{-1}.

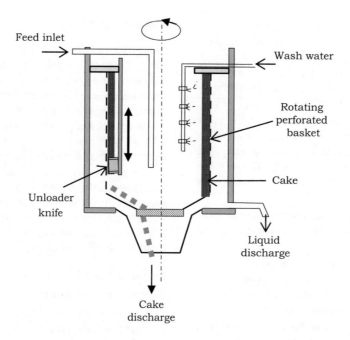

FIGURE 12.7   Schematic diagram of a suspended batch centrifuge

- Another type of batch centrifuge is driven from the bottom. The solids are unloaded by hand through the top of the casing or ploughed out through the openings in the floor of the basket, such as in top suspended machines.

Applications

As mentioned above, top suspended centrifuges have been used extensively in sugar refining. For still higher capacities, other automatic centrifuges or continuous conveyor centrifuges are used.

Automatic Batch Centrifuge

Construction and Operation

- The automatic batch centrifuge consists of a basket with 50–110 cm diameter that rotates about the horizontal axis. Unlike the suspended batch centrifuges, fine metal screens are used as filter medium in these machines (Figure 12.8).
- Cycle timers and solenoid-operated valves control the different operations as feeding, washing, spinning, rinsing, and unloading. As practiced for other centrifuges, the feed slurry, wash liquid, and screen rinse are sprayed into the basket at specified time intervals. Any part of the cycle may be shortened or lengthened as desired.
- With coarse crystals the total operating cycle ranges from 35 to 90 seconds, so that the hourly throughput is large.
- A heavy knife scraps the cake while the basket runs at full speed.

Salient Features

- Because of the short cycle and small amount of holdup required for feed slurry, filtrate and

discharged solids, the automatic centrifuges are easily integrated into continuous processes.

- Automatic centrifuges have high productive capacity with free draining crystals. They are not used for solids, which do not move freely through the chute, or for solids draining slowly as it would increase the cycle time and would be uneconomical.
- The small batches of solids can be effectively washed with small amounts of wash liquid, and, as in any batch machine, the amount of washing can be temporarily increased to clean up off-quality material, if required.
- Crystals experience considerable degradation or breakage by the unloading knife.

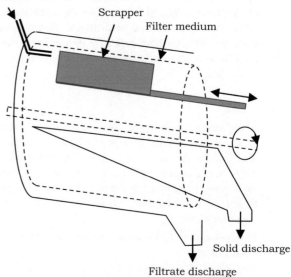

FIGURE 12.8   Schematic diagram of an automatic batch centrifuge

Applications

These are used for high-capacity applications; however, these are typically not used when most of the particles in the feed are finer than 150 mesh.

Continuous Filtering centrifuge

Construction and Operation

- A continuous centrifugal separator for coarse crystals is the *reciprocating-conveyor centrifuge* shown in Figure 12.9.
- It comprises of a rotating basket of 30 to 120 cm diameter. The basket has a slotted wall covered with a woven metal cloth.
- There is a revolving funnel for feeding the liquid; the funnel is revolved to agitate the feed slurry gently. There is a stationary pipe at the small end of the funnel to allow the feed to go inside.
- As the feed moves toward the large end of the funnel due to the rotation of the funnel, it gains speed and is thrown onto the wall of the basket and on which it rapidly moves along with the wall. Due to the centrifugal force, the filtrate goes out through the metal cloth and the crystals are collected on the upstream (inner) side.
- There is a reciprocating pusher to move the solid layer deposited over the filtering surface. The crystals are moved a small distance with each stroke of the pusher toward the edge of the basket, after which they are discharged into a large casing and then to a collector chute. The maximum thickness of the crystal built up is usually 25–75 mm.
- Filtrate and any wash liquid sprayed on the crystals during their travel leave the casing through separate outlets.

Salient Features

- The moderate acceleration of the feed slurry and gentle removal of the discharged solids minimize breakage of the crystals.
- Multi-stage units that minimize the distance of travel of the crystals in each stage are used for solid cakes that do not convey properly in a single-stage machine.

Applications

Reciprocating centrifuges can be effectively used for removal of solids in which materials finer than 100 mesh is less than 10%.

Discontinuous Vacuum Filters

As discussed, the vacuum filters are usually discontinuous. Vacuum nutsch is one such piece of equipment.

Vacuum Nutsch

Construction and Operation

- A vacuum nutsch employs a vacuum on the downstream side and has a size little more than a large Buchner funnel, 1–3 m.
- The thickness of the layer of solids can be about 10–30 cm.
- It is operated in a batch-wise manner.

Salient Features

It can be used for corrosive materials provided the nutsch is made up of a corrosion resistant material.

Applications

Nutsches are not recommended for large-scale production operations because of the high labor cost involved in digging out the solid cake.

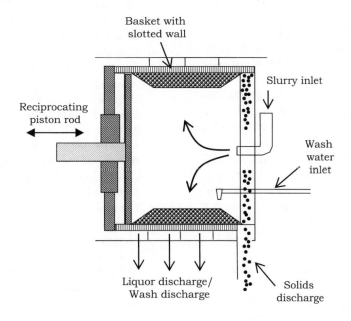

FIGURE 12.9   Schematic diagram of a reciprocating conveyor centrifuge

Continuous Vacuum Filters

In continuous vacuum filters, there is negative pressure on the downstream side of a moving septum, so the liquor is sucked through the septum and the cake is deposited on the upstream side. The filter medium continuously moves and the removal of the cake from the filtration zone, washing, drying, and its removal take place continuously. The filter medium then re-enters the slurry and another layer of cake is deposited on it. In a continuous vacuum, the pressure differential is ordinarily between 25 and 500 mm Hg across the septum.

Rotary Drum Filter

Construction and Operation

- The rotary drum filter, which continuously filters, washes, and removes the cake, consists of a horizontal drum having a slotted wall on which there is a lining of the filter medium (Figure 12.10). The drum is partially submerged in the liquid (to be filtered) kept in a tank and is rotated at 0.1-2 rev.min^{-1}. Standard drum sizes range from 0.3 m diameter 0.3 m face to 3 m diameter with a face of 4.3 m.

- There is another smaller drum with a solid surface under the main drum and some radial partitions between these two drums divide the annular space into separate compartments or channels.

- Each compartment is connected by an internal conduit to a hole in the rotating plate of the rotary valve (or valve mechanism), which creates either suction or positive air pressure in the compartment during the operation.

- During the operation, at any time a part of the drum is dipped in the slurry tank and vacuum is applied to the compartments with faces in the slurry. Thus, filtrate passes through the septum and then through the compartments. The solid cake is deposited on the surface of the septum on the upstream side.

- There is an agitator provided at the lower section of the slurry tank to keep the solids in suspension.

- As the compartment moves to a position such that it is no longer dipped in the slurry, washing the cakes starts, followed by drying. The wash liquid is usually sprayed directly onto the surface of the cake.

- A vacuum is applied to the panel from a separate system, which sucks the liquid and air through the cake layer. There is a separate tank to collect the wash water.

- After the cake of solids on the face of the compartment has become as dry as possible, the vacuum is cut off and the cake is loosened by blowing compressed air under the cake. Subsequently, there is a horizontal knife, known as a *doctor's blade*, at the end of the cycle to scrap off the cake. The cake is collected in a trough.

- After the cake has been removed, the compartment reenters the slurry and a new cycle starts.

- As discussed above, some part of the septum is in the filtering zone at all times, a part is in the washing zone, and a part is used for drying and discharging the cake. Thus, the discharge of both solids and liquids from the filter is uninterrupted. Automatic controls are used to sequentially open and close the vacuum and the auxiliary systems for filtration, drying, washing, and discharge of the cake.

Different variations of the rotary drum filter are commercially available. In some designs the drum is not divided into compartments and vacuum is applied to the entire inner surface of the filter medium. The filtrate and wash liquid are removed together, and the solids are loosened before discharge by blowing air through the cloth from a stationary shoe inside the drum. In another method, the removal of the cake is done by separating the filter cloth from the drum surface and passing it around a small diameter roller. The cloth may be washed as it returns from the roller to the underside of the drum.

A *precoat filter* is a modified form of rotary drum filter, in which a layer of porous filter aid, such as diatomaceous earth, is coated on the filter medium. This is specifically advantageous when small amounts of fine or gelatinous solids are to be removed. With an ordinary filtration process, these solids would choke the filter cloth. Both the cake and the precoated materials are scrapped together after filtration and, hence, the cake is not pure. Usually, this cake has to be discarded; thus, if the cake is a valuable product, such filters should be avoided.

Salient Features

- The amount of submergence of the drum is variable. Most bottom feed filters operate with about 30% of the filter area submerged in the slurry. A high submergence filter with 60%–70% of its filter area submerged may be used when no washing and a high filtration capacity are desired.

- The cake formed on industrial rotary vacuum filters is 3–40 mm thick.

- The maximum pressure differential is about 1 atm for vacuum filter. Hence, for viscous liquids, this filter is not suitable.

- They are sometimes adopted under positive pressures up to about 15 atm for situations in which vacuum filtration is not feasible or economical. This situation may arise when the solids are very fine and filter very slowly or when the liquid has high viscosity, more than 1 poise, or is a saturated solution that will crystallize on cooling. In that case, the drum is enclosed in another shell to maintain the pressure. Pressure type filters cost about two times as much as vacuum filters.

- The system is compact and can give high throughput capacity. The capacity of any rotary filter depends strongly on the characteristics of the feed slurry and particularly on the thickness of the cake, which may be deposited in practical operation.

- If the solid particles in the slurry are coarse enough so that they settle quickly, then they may not come in contact with the filter cloth and the rotary drum

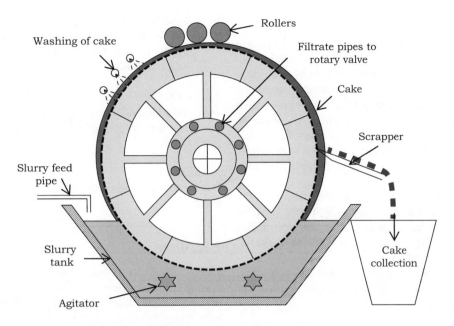

FIGURE 12.10   Schematic diagram of a rotary drum filter (After Fellows, 2000 with permission)

filter will not work satisfactorily. Therefore, it is very important to timely or continuously agitate the slurry.

Applications

It is the most common type of continuous vacuum filter. However, its high cost and complexity limits its application only for high capacity and special problems.

Rotary Disk Filter

Construction and Operation

- It consists of a horizontal hollow rotating shaft supporting concentric vertical disks.
- The disks are also hollow and are covered with the filter cloth.
- Each disk is divided into partitions, which have individual outlets to the central shaft.
- As the rotary drum filter, the complete system is partly submerged in the slurry.
- After the filtrate has been sucked and the cake has been deposited, the cake undergoes washing, drying, and scrapping off at the upper portion of the assembly.

Salient Features

- As compared to the rotary drum filter, washing in this filter is less efficient.
- When the rotating annular filtering surface of a vacuum filter is divided into sectors, it is termed a *continuous rotary horizontal filter*. As it rotates, the processes of filtration, washing of the cake, drying, and removal of the cake take place in a cyclic manner. The washing efficiency of this filter is superior to that of a rotary drum filter.

Horizontal Vacuum Belt Filter

Construction and Operation

- As the name indicates, there is a horizontal belt, which is made up of the filter medium. In a continuous system, known as the *moving belt filter*, the belt conveyor has a perforated belt or there are transversely ridged supports on which the filter cloth rests.
- There is a longitudinal vacuum box connected to the downstream side of the filter cloth. The liquid fed above the belt is sucked through the filter cloth and the belt due to the vacuum created by the box.
- Feed slurry from a distributor at one end of the unit flows onto the belt and is filtered while the other end discharges the washed cake.
- In some models known as *indexing belt filters*, the vacuum is intermittently cut off and reapplied; the belt is moved onward half a meter or so, when the vacuum is off and is held stationary while the vacuum is applied. This helps in maintaining a good vacuum seal between the vacuum box and the moving belt.
- Such belt filters can have a very large surface area up to 130 m^2 filtration area with 4.8 m to 38 m length and 0.6 m to 3.6 m width.

Applications

- This filter is used for feed containing fast settling solid and coarse particles. As mentioned previously for the fast settling feeds the rotary drum filter may not be that useful.
- Belt filters are especially useful in waste treatment; since the waste frequently contains a very wide range of particle sizes.

Clarifying Filters

The basic difference in between the clarifying filter and cake filter is that the pores of the clarifying filter have a much larger diameter than the particles that are intended to be removed. The retention of particles is initially by electrostatic attraction or by surface forces so the particles retained in the pores could even be smaller than the pore size. The particles trapped in the pores reduce the effective diameter of the channels, but normally they do not block them completely. There is no formation of cake on the septum surface. Thus, clarifying filtration is also known as *depth filtration*.

Ultimately, a situation is attained when the solids block the pores entirely, which causes the filtrate flow to stop. The filter medium has to be replaced at that time. Thus, this type of filter is not recommended when the solids concentration in the liquid to be filtered is very high.

The gravity-bed filters for water treatment and very small cartridge and edge filters of various designs are examples of clarifying filters for liquids.

Cartridge Filter

Construction and Operation

- This consists of a series of thin metal discs 7.5 to 25 cm in diameter very closely arranged in a vertical stack on a vertical hollow shaft. The complete set is fitted in a closed cylindrical chamber.
- When the liquid is fed to the chamber under pressure, it tries to move to the hollow shaft through the narrow spaces between the disks. During this, most of the particles, which are smaller than the diameter of disks, are trapped on the disk surfaces. The hollow discs guide the filtrate to the exit at the top of the chamber.
- This type of filter is also known as an *edge filter* as most particles are separated at the edges of the disks.
- The solids accumulated on the disc surfaces are discharged from the cartridge from time to time. In the type of cartridge filter shown in Figure 12.11, the cartridge is given a half-turn and some stationary

teeth of a comb cleaner pass between the disks wiping the solids. The solids drop to the bottom of the casing, from where they may be removed at fairly long intervals.

Salient Features

Some cartridge filters contain tubular elements of porous metal or ceramic material that trap the particles inside the pores.

Applications

It is used for removal of small amounts of solids from process fluids.

Gas Filters

Filters are used for separating dust from air (gas) in food processing industries. Some such applications are separation of dust or powders escaping with air from the cyclone separator after spray drying, cleaning of process air, etc. The common filters used for gas filtration include pad filters, granular beds, and bag filters. In all of these filters most of the separation is by impingement.

Pad Filter

- It consists of a frame holding the pads of cellulose pulp, cotton, felt, glass fiber, or metal screen and the air is passed through the pads. The cleaned air is collected on the other side of the pads.
- The pad material may be dry or coated with a viscous oil to improve its dust-holding capacity.
- Disposable pads are used for light duty applications, but the pads are frequently cleaned, rinsed, and recoated with oil in large-scale gas cleaning.

Granular Bed Filter

The granular bed filter contains stationary or moving beds of granules of varied sizes, which may be as small as 30 mesh to as high as 40 mm. The collected dust can be removed by back washing.

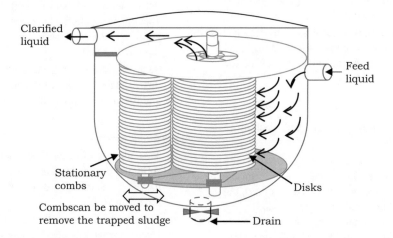

FIGURE 12.11 Working of a cartridge filter

Bag Filter

- A bag filter contains one or more large bags of felt or thin woven fabric, mounted inside a metal housing.

- The air to be cleaned is usually introduced at the bottom of the bag and allowed to pass outward, while the bag retains the dust.

- In some systems, the flow is cut off periodically and clean air is blown back to remove the dust. In some other systems the bag is mechanically shaken for the purpose.

- In most cases, bag filters serve as clarifiers with particles trapped within the fabric of the bag, but with a greater amount of dust, a thin cake of dust may be allowed to build up before it is discharged.

12.2.3 Filter Media

Duck weave or twill is a heavy cloth that is commonly used filter medium; other examples are felted pads of cellulose, woolen, metal, nylon, or other synthetic cloth. In the case of corrosive liquids, materials such as woolen cloth, metal cloth (usually made up of stainless steel), glass cloth, or paper are used as the filter media. Synthetic fabrics like polypropylene, various polyesters, and nylon can also be used where resistance to chemicals is required.

The filter media should have the porosity to remove solids to give a clear filtrate, and, at the same time, the pores should not be plugged. The media should have sufficient tear strength and must possess chemical resistance to the solutions to reduce the degradation rate. The filter cloth should also allow for easy and clean removal of the cake.

12.2.4 Filter Aids

Common filter aids are inert porous solids, perlite, wood cellulose, ground chalk, and diatomaceous earth. These aids are used as a precoat on the filter media before filtration; these are capable of preventing the plunge of gelatinous type solids in the filter medium. However, it is more commonly added to slurry to increase the porosity of the cake. The filter aid may be in a dry form or in the form of slurry, which is continuously metered into the feed slurry. In continuous vacuum filters the precoating is done once at the start of the filtration process; the filter aid can be added to the pan directly.

The concentration of the filter aid should be controlled such that the deposited cake would have satisfactory porosity for the filtration to continue. The usual concentration of the filter aid is one to two times that of suspended particles.

Filtration with a filter aid is usually employed when the filtrate is the valuable component. If the particles are to be recovered in pure form, then a filter aid should not be used. Filter aids are also not used when the filter aid and the precipitate cannot be chemically separated.

12.3 Settling and Sedimentation

12.3.1 Methods and Principles

Gravitational force drives settling of particles in a fluid for separating them. The particles can be solid particles (as solids suspended in water settle at the base of a bottle after a long time) or liquid drops (as fat globules or cream float in milk). The fluid can be a liquid or a gas.

When there is only a small amount of solid particles in the fluid, then each particle can be considered to be at an adequate distance from other particles and the vessel walls, so that the settling of the solid particle is not affected either by other particles or by the vessel walls. This process is known as *free settling*. However, when the concentration of particles is higher, the settling is retarded and is known as *hindered settling*.

Sedimentation is the process of separation of dilute slurry or suspension into a clear liquid and slurry of higher solid content by using gravity force.

Forces Acting on a Particle and Settling Velocity

When a particle of higher density than that of the surrounding fluid moves in the fluid, it is under the influence of gravity force, buoyancy force, and resistance or drag force acting in direction opposite to the motion of the particle.

Assume that a particle of mass "m" (kg) is falling at a velocity "v" (m.s^{-1}) relative to the fluid. Assume further that densities of the liquid and the particle are ρ and ρ_p (kg.m^{-3}). V_p (m^3) is the volume of particle and g (m.s^{-2}) is the acceleration due to gravity.

Then the magnitudes of the forces are as follows:

$$\text{Gravitational force, } F_g = mg \quad (12.51)$$

$$\text{Buoyancy force } F_b = \frac{m\rho g}{\rho_p} = V_p \rho g \quad (12.52)$$

$$\text{Drag force } F_d = C_d \frac{\rho v^2}{2} A \quad (12.53)$$

The frictional resistance or the drag force, F_d is proportional to $v^2/2$.

The drag coefficient C_d is dimensionless and given as

$$C_d = \frac{F_d/A}{\rho v^2/2} \quad (12.54)$$

Thus, the force due to the acceleration is the resultant force on the particle.

$$\text{Hence, } m\frac{dv}{d\theta} = F_g - F_b - F_d = mg - \frac{m\rho g}{\rho_p} - C_d \frac{\rho v^2}{2} A \quad (12.55)$$

When the particle begins moving from its rest position, the fall consists of a very short initial acceleration period (say 1/10 seconds) followed by a period of constant velocity. The constant velocity (zero acceleration) is denoted as the free settling velocity, v_t or the *terminal velocity* and the period of fall under terminal velocity is important.

At terminal velocity,

Gravity force - Buoyancy force – Drag force = 0

$$\text{i.e. } g - \frac{\rho g}{\rho_p} - C_d \frac{\rho v_t^2}{2m} A = 0 \tag{12.56}$$

$$\text{or, } g\left(\frac{\rho_p - \rho}{\rho_p}\right) = C_d \frac{\rho v_t^2}{2m} A \tag{12.57}$$

If the diameter of the particle is D_p(m),

$$\text{then, } A = \left(\frac{\pi}{4}\right)D_p^2 \text{ ,and mass m} = \left(\frac{\pi}{6}\right)D_p^3 \rho_p$$

$$\text{therefore, } g\left(\frac{\rho_p - \rho}{\rho_p}\right) = C_d \frac{\rho v_t^2}{2\frac{\pi}{6}D_p^3 \rho_p} \cdot \frac{\pi}{4}D_p^2$$

$$\text{or } g\left(\frac{\rho_p - \rho}{\rho_p}\right) = C_d \frac{3}{4}\frac{\rho v_t^2}{D_p \rho_p}$$

$$\text{or, } v_t = \sqrt{\frac{4}{3}\frac{(\rho_p - \rho)gD_p}{\rho C_d}} \tag{12.58}$$

The C_d used in the above derivation is appropriate only when v_t is small and the flow condition is laminar at the fluid-particle interface. The drag coefficient for rigid spheres is dependent on the Reynolds number as shown in Figure 12.12.

The drag coefficient of rigid spheres in the laminar flow region for Re <1.9, *Stokes' law* region, is given as:

$$C_d = \frac{24}{Re} = \frac{24\mu}{\rho v D_p} \tag{12.59}$$

$$\text{Thus, } v_t = \frac{gD_p^2(\rho_p - \rho)}{18\mu} \tag{12.60}$$

Because v_t also affects C_d, the terminal velocity for a particular particle using the equation 12.58 or 12.60 is obtained by trial and error.

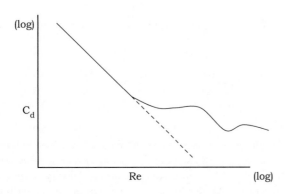

FIGURE 12.12 Relationship between drag coefficient and Reynolds number for a rigid sphere

The terminal velocity for other Re values can be found as follows:

For Re values between 1.9 and 500

$$v_t = \frac{0.153a^{0.171}D_p^{1.14}(\rho_p - \rho)^{0.71}}{\rho^{0.29}\mu^{0.43}} \tag{12.61}$$

For Re between 500 and 2×10^5

$$v_t = 1.75\sqrt{\frac{aD_p(\rho_p - \rho)}{\rho}} \tag{12.62}$$

In general, an index K can be used to find the value of C_d, where

$$K = d\left[\frac{\rho g(\rho - \rho_p)}{\mu^2}\right]^{0.33} \tag{12.63}$$

Where, d represents the projected particle diameter.

A general form of equation can be written for the relationship between C_d and Re as:

$$C_d = \frac{b}{Re^n} \tag{12.64}$$

The values of the constants b and n are given in Table 12.2.

The value of C_d for other particle shapes will be different from those shown in Figure 12.12.

If Re at fluid particle interface exceeds 1.9,

$$mg\left[1 - \frac{\rho}{\rho_p}\right] = C_d A_p \rho \frac{v_t^2}{2} \tag{12.65}$$

Substituting the values of A_p and m,

$$v_t = 1.155\left[\frac{gD_p(\rho_p - \rho)}{\rho.C_d}\right]^{0.5} \tag{12.66}$$

Example 12.1 It is required to separate green peas from lighter foreign matter like stalks and leaves by blowing air. The projected diameter and sphericity of green peas are 7.2 mm and 0.7. The bulk density and true density are 570 kg.m^{-3} and 980 kg.m^{-3}. The density and viscosity of air may be taken as 1.165 kg.m^{-3} and 0.0185 cP at this temperature.

SOLUTION:

Given density of air = 1.165 kg.m^{-3}

Acceleration due to gravity = g = 9.81 m. s^{-2}

TABLE 12.2

Values of constants b and n to be used in equation 12.64

		K	Re	b	n
Stokes' law region	less than 0.33	less than 1.9	24	1
Intermediate flow	0.33< K <44	1.9–500	18.5	0.6
Newton's law	K > 44	500–2×10^5	0.44	0

For the peas, $K = 0.0072 \left[\dfrac{(1.165)(9.81)(980-1.165)}{(0.000185)^2} \right]^{0.33}$

$$= 0.0072 \times (3.26 \times 10^{11})^{0.33}$$

$$= 0.0072 \times 6305.78 = 45.40$$

Since $K > 44$, $C_d = 0.44$

Thus, $v_r = 1.155 \left[\dfrac{(9.81)(0.0072)(980-1.165)}{(1.165)(0.44)} \right]^{0.5}$

$$= 1.155 \times (134.875)^{0.5}$$

$$= 13.41 \text{ m. s}^{-1}$$

The air velocity above 13.41 m.s^{-1} will blow away the peas with the lighter materials. Thus, the air velocity should be less than 13.41 m.s^{-1}.

In this case it is assumed that the lighter materials have lower terminal velocities than the peas. In the case where it is required to separate two individual seeds by air classification, then the terminal velocities of the two seeds have to be found and the air velocity should be kept in between these two terminal velocities.

Air classification is commonly used in husk separators in rice mills and in different types of grain cleaners. It is also used for fractionation of particles of different sizes.

These equations also hold good for sedimentation, floatation, centrifugation, and fluidization applications.

In the case of very small particles, *Brownian motion* affects the settling. Brownian motion is the random movement of particles by molecular collision between the particles and fluid molecules. As the movement of the particles is random, the effect of gravity is alleviated and the settling rate is reduced. The particles may not even settle at all. When the sizes of particles are a few μm, Brownian motion becomes important and affects the settling process. To reduce this effect during separation, centrifugal force is applied.

Drag Coefficient for a Non-Rigid Sphere

When particles are not rigid as in the case of settling of cream in milk, there may be deformation of the particles and there may also be some internal circulation within the particle. Both these phenomena affect the drag coefficient and terminal velocity.

- For Re less than 50, the drag coefficients for air bubbles rising in water are same as those for rigid spheres in water.
- For Re up to about 100, the drag relationship for liquid drops in gases is the same as that for solid particles. Large drops deform under increased drag.
- For Re up to 10, the drag coefficient curve for small liquid drops in immiscible liquids follows that for rigid spheres. For Re between 10 and 500,

the terminal velocity for liquid drops is higher than that for solids, which is primarily due to the internal circulation in the drop.

Hindered Settling

As we discussed earlier, when the number of particles in a fluid is large so that the surrounding particles influence the settling of the particular particle then it is called hindered settling. As the flow is obstructed, the equation for free settling cannot be applied. The drag force will be more in the suspension as other particles also interfere. Thus, the effective viscosity of the mixture will be more, which can be written as:

$$\mu_m = \frac{\mu}{\psi} \tag{12.67}$$

where, μ_m = effective viscosity of the slurry, μ = actual viscosity of the fluid, and Ψ is an empirical correction factor depending upon the concentration.

The factor Ψ depends on the fraction of volume the liquid occupies in the slurry (ε).

$$\psi = \frac{1}{10^{1.82(1-\varepsilon)}} \tag{12.68}$$

The bulk density of the liquid slurry,

$$\rho_m = \varepsilon \rho + (1-\varepsilon)\rho_p \tag{12.69}$$

where, ρ_m is density of slurry in kg.m^{-3}.

Thus, the difference in density between slurry and particle is

$$\rho_p - \rho_m = \rho_p - [\varepsilon \rho + (1-\varepsilon)\rho_p] \tag{12.70}$$

$$= \varepsilon(\rho_p - \rho) \tag{12.71}$$

The value of v_t in the container will be ε times the value obtained from Stokes' law.

Thus, starting with equation 12.60,

$$v_t = \frac{gD_p^2(\rho_p - \rho)}{18\mu_m} \tag{12.72}$$

By modifying μ and $(\rho_p - \rho)$ for this condition, we rewrite the equation as:

$$v_t = \frac{gD_p^2(\rho_p - \rho)\varepsilon}{18\mu / \psi} \tag{12.73}$$

$$\text{or, } v_t = \frac{gD_p^2(\rho_p - \rho)\varepsilon\psi}{18\mu} \tag{12.74}$$

If we multiply ε again to take care of the relative velocity effect, the above equation becomes

$$v_t = \frac{gD_p^2(\rho_p - \rho)}{18\mu}\varepsilon^2\psi \qquad (12.75)$$

The Reynolds's number, in this case, can be found as:

$$Re = \frac{\rho_m D_p v_t}{\mu_m \varepsilon} \qquad (12.76)$$

$$= \frac{\rho_m g D_p^3(\rho_p - \rho)\rho_m}{18\mu^2}\varepsilon\psi^2 \qquad (12.77)$$

When the Re is less than 1, then the settling is in Stokes' law region. The effect of concentration will be more for non-spherical particles and angular particles.

Effect of Wall During Settling

When the particle diameter D_p becomes significant as compared to the diameter of the settling tank, then the wall may cause retardation for settling. The terminal velocity is reduced. A correction factor, ζ_w is multiplied with the terminal velocity to take care of the wall effect.

$$\zeta_w = \frac{1}{1 + 2.1(D_p / D_w)}, \text{ for } D_p / D_w > 0.05 \quad (12.78)$$

In case of a completely turbulent regime, the correction factor is

$$\zeta_w = \frac{1 - (D_p / D_w)^2}{[1 + (D_p / D_w)^4]^{1/2}} \qquad (12.79)$$

Classification of Solids through Settling

A *classifier* is a piece of equipment in which the solid particles can be separated into different fractions depending upon their settling velocities in a fluid. Usually two methods, namely, the *differential settling method* and the *sink-and-float method* are in vogue. In the first case, the liquid used for settling has a density in between the densities of the particles to be separated. For example, if the fraction of particles that are to be separated have densities of 1030 kg.m^{-3} and 1200 kg.m^{-3}, respectively, then the density of liquid can be, say, 1100 kg.m^{-3}. Thus, the particles of higher density (1200 kg.m^{-3}) will settle down and those of 1030 kg.m^{-3} will float.

The differential settling can also separate solid particles into several size fractions on the basis of their settling velocities. Here, densities of the two substances to be separated are higher than that of the fluid medium. All the materials settle here, whereas in the previous case, one component settles and the other floats.

The sink and float method is independent of particle size; only densities are important here. Since most solids have high densities, the liquids must have densities more than water. Not many natural liquids have this property and, thus, pseudo-liquids (a suspension of very fine materials in water) are used. Settling velocity of fine solid materials in the medium is negligible because their diameters are very small. The settling is mostly hindered type.

Suppose there are two materials A and B, with densities ρ_A and ρ_B, and $\rho_A > \rho_B$.

The terminal velocities of A and B can be written as:

$$v_{tA} = \sqrt{\frac{4}{3}\frac{(\rho_A - \rho)gD_A}{C_{dA}\rho}} \qquad (12.80)$$

$$v_{tB} = \sqrt{\frac{4}{3}\frac{(\rho_B - \rho)gD_B}{C_{dB}\rho}} \qquad (12.81)$$

If the settling velocities for both particles are the same, then $v_{tA} = v_{tB}$

$$\text{or, } \frac{4}{3}\frac{(\rho_A - \rho)gD_A}{C_{dA}.\rho} = \frac{4}{3}\frac{(\rho_A - \rho)gD_B}{C_{dB}.\rho}$$

$$\text{or, } \frac{D_A}{D_B} = \left(\frac{\rho_B - \rho}{\rho_A - \rho}\right)\left(\frac{C_{dA}}{C_{dB}}\right) \qquad (12.82)$$

Case 1: Settling of spherical particles in Newton's law turbulent region

The drag coefficient for essentially spherical particles at very high Re remains almost constant in the turbulent Newton's law region.

So, taking $C_{dA} = C_{dB}$,

$$\frac{D_A}{D_B} = \frac{\rho_B - \rho}{\rho_A - \rho} = \left(\frac{\rho_B - \rho}{\rho_A - \rho}\right)^{1.0} \qquad (12.83)$$

For particles where $D_A = D_B$, combining equations 12.80 and 12.81 gives

$$\frac{v_{tA}}{v_{tB}} = \left(\frac{\rho_A - \rho}{\rho_B - \rho}\right)^{0.5} \qquad (12.84)$$

If both particles are allowed to settle in the same medium, as per the equations 12.83 and 12.84, as the diameter of the particle increases, the terminal velocity increases at a different rate for two particles (which is dependent on the density of particle). Also, if the density of the medium is increased, then the gap between D_A and D_B increases because the numerator reduces more as compared to the denominator.

Case 2: For laminar Stokes' law settling,

$$C_{dA} = \frac{24\mu}{D_A.v_{tA}.\rho}, \quad C_{dB} = \frac{24\mu}{D_A.v_{tB}.\rho}$$

$$\text{If } v_{tA} = v_{tB}, \text{ then } \frac{C_{dA}}{C_{dB}} = \frac{D_B}{D_A} \qquad (12.85)$$

Thus, $\dfrac{D_A}{D_B} = \dfrac{\rho_B - \rho}{\rho_A - \rho}\left(\dfrac{D_B}{D_A}\right)$

$$\left(\dfrac{D_A}{D_B}\right)^2 = \dfrac{\rho_B - \rho}{\rho_A - \rho}$$

$$\left(\dfrac{D_A}{D_B}\right) = \left(\dfrac{\rho_B - \rho}{\rho_A - \rho}\right)^{0.5} \tag{12.86}$$

Case 3: In the transition flow when it is neither laminar nor turbulent,

$$\left(\dfrac{D_A}{D_B}\right) = \left(\dfrac{\rho_B - \rho}{\rho_A - \rho}\right)^{n} \quad \text{where } 0.5 < n < 1 \tag{12.87}$$

One disadvantage associated with this differential settling method is that the terminal velocity of smaller heavy particles and larger light particles may be the same and they may settle at the same time.

Sedimentation

Sedimentation, also known as *thickening*, is the process in which a dilute slurry undergoes the process of settling under gravity, resulting in a slurry with higher concentration and a clear fluid. Accordingly, the equipment is called a settling tank or thickener. At the start of the process, all particles settle by free settling. With time, as the materials settle, the liquid column is divided into different zones, as shown in Figure 12.13 (a). Initially after some time, the slurry is classified into two zones, the zone of clear liquid and the zone of slurry or suspension. As time proceeds it is further classified as clear liquid (1), suspension (2), dense slurry (3) and compacted solids (4). Actually, the zone 3 is a transition phase in between the suspension and settled solids. During the process, the height of the interface layer between the clear zone and suspension (z) reduces at a constant rate. The settled particles also gradually become more compact by removal of liquids; thus, its thickness also decreases. At the end, the transition and the suspension zones disappear, and we get the clear liquid at the top and settled solids at the bottom. The moment at which the compression starts is called the *critical point*.

Usually, a graduated cylinder is used to conduct a batch settling test. If we plot the interface height, z-vs-time, we get a curve, as shown in Figure 12.13(b). The velocity of settling is constant up to the critical point, and then reduces asymptotically. The settling velocity is, hence, found by drawing a tangent to the curve at any time because $v = dz/d\theta$.

For such a process, $c_\theta z_\theta = c_o z_0$ (12.88)

where, c_θ and z_θ are the average concentration of the suspension and the height of slurry, respectively, at any time θ, and the corresponding values at time 0 are c_0 and z_0.

$$\text{Thus,} \quad c_\theta = \left(\dfrac{z_0}{z_\theta}\right)c_0 \tag{12.89}$$

12.3.2 Equipment

Gravity Settling Tank

Construction and Operation

- A simple gravity settler is essentially a horizontal tank or box that separates the suspended solid materials from a fluid by settling.
- The fluid (liquid or gas) with dispersed solids is made to enter from one end of the tank at the top and is moved at a very slow velocity toward the other end so as to allow adequate time for settling of the dispersed solids (Figures 12.14 (a) and (b)).
- Specific designs are used for settling of liquid-liquid dispersion, i.e., to separate liquids based on their densities. In this case, the liquid velocity should be such that the smallest droplets should be able to move downward from the top or to move upward from the bottom to the interface.

Salient Features

- Particles attain their terminal settling velocities and settle at the floor.
- The settling time obtained as a result of dividing the height of the settling chamber by the terminal velocity should be smaller than that of air retention in the chamber; it puts a restriction on the height of the chamber.

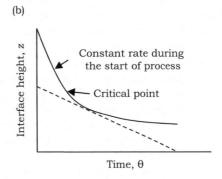

FIGURE 12.13 (a) Schematic of a batch sedimentation process, (b) change in interface height during settling

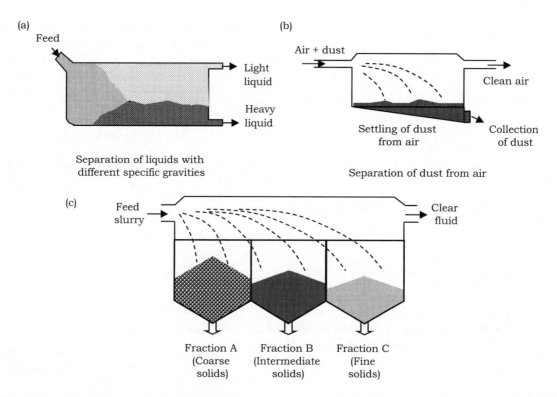

FIGURE 12.14   (a) and (b) Schematic diagram of gravity settling tanks for separation of two liquids with different specific gravities and for separation of dust from air, (c) a gravity settling classifier

• The residence time of the air in the settler can be calculated by dividing volumetric air flow through the chamber by the chamber size.

Applications

This is mostly used for settling of dust from process air or exhaust from an industry and for separating the liquid based on densities.

Gravity Settling Classifier

Construction and Operation

• The classifier works on the principle that when a liquid containing suspended solids of different sizes is moved slowly in a horizontal direction, the larger particles settle fast and the smaller particles take more time to settle.

• The equipment consists of a large tank subdivided horizontally into several sections, each section having a separate outlet at the bottom. The separate sections can be made simply by placing three baffles along the tank, as shown in Figure 12.14(c).

• As the slurry is fed to the tank at one end and it moves to the other end at a very slow velocity, the larger particles, due to their higher terminal velocities, settle to the bottom near the entrance and the smaller particles, with lower terminal velocities, move along the tank and settle to the bottom near the exit.

• The *Spitzkasten classifier* is a gravity-settling classifier in which there are a number of conical containers

with a successively increasing diameter in the direction of flow. The slurry is moved sequentially from one container to the other. In each container, there is a water inlet at the bottom that compels water to move in an upward direction. The desired size range for each container is achieved by controlling the velocity of the water moving upward. The largest and fastest-settling particles are collected in the first vessel and the slowest settling particles settle in the last.

Applications

This is used for settling of suspended solids from slurry and classifying them as per their particle sizes.

Sedimentation Thickener

Construction and Operation

• The sedimentation thickener consists of a large tank with slowly revolving rake.

• The slurry is fed at the center of the tank at a sufficient height from the surface of the liquid. The movement of the rake spreads the slurry radially and, due to the density difference, the liquid moves upward and the solids tend to settle.

• The gentle stirring helps to remove water from the sludge and at the same time the rake guides the sludge toward the center of the bottom, where it is removed from the tank.

• The settling of the solids can be divided into three zones: the upper free settling zone, the middle transition zone, and the lower compression zone. The settling of the solids in the upper zone is by free settling. In the transition zone, the concentration of solids is more and there is hindered settling. Thereafter, the lower compression zone is where the sludge settles and that is an extreme case of hindered settling.

• If the upward velocity of the fluid at the upper portion of the tank is lower than the minimum terminal settling velocity of the solids, then it is possible to expect the clear liquid to exit at the top edge of the tank.

Applications

This is used for obtaining a clear fluid and a slurry of higher solids concentration through fractionation of a dilute slurry by gravity settling.

12.4 Centrifugal Separation

12.4.1 Methods and Principles

In the case of settling or sedimentation the gravitational force acting on the particles causes the separation. The process is very slow for very small particles. Hence, centrifugal force can be employed to increase the rate of separation. The equipment used for the purpose is known as a centrifuge.

Principle of Separation

The magnitude of centrifugal force, in a separation process, applied on the particles is usually many times that of gravity. Hence, the particles that take a very long time to settle under gravity or do not settle at all could be separated very easily. With particles of different sizes and densities, relative settling velocities are not changed, but the main benefit of centrifugal force is to overcome the disturbing effects of free convection currents and Brownian motion. The schematics of the centrifugal separation processes for separation of solids from a liquid, and for the separation of liquids on the basis of difference in specific gravities are shown in Figure 12.15.

Suppose a cylindrical bowl rotating along its vertical axis receives slurry (with solid particles) at its center. If the bowl rotates at a reasonable speed, the slurry will be thrown toward the periphery by centrifugal force. In consequence, both the liquid particles and the suspended solid particles are influenced by the horizontal centrifugal force as well as the downward gravitational force due to their mass. However, the gravitational force is very small as compared to the centrifugal force, and, hence, it can be considered negligible. The particles will experience the centrifugal force, but, as the liquid particles have lower densities (mass), they will be subjected to lower centrifugal force. The solid particles with higher densities will experience higher centrifugal force and will be thrown toward the wall of the bowl.

The acceleration due to centrifugal force, a (m.s^{-2}), is given as follows:

$$a = \omega^2 r \qquad (12.90)$$

where r (m) is the radius of rotation and ω (rad.s^{-1}) is the angular velocity.

The centrifugal force, F$_c$ (N) acting on a particle is given as:

$$F_c = m.a = m.\,\omega^2 r \qquad (12.91)$$

$$\text{Or, } F_c = m.\,(v/r)^2 r = m.v^2/r \qquad (12.92)$$

The gravitational force is F$_g$ =m.g; it means that the centrifugal force is $\omega^2 r/g$ or v^2/rg times more than the gravitational force.

$$\omega = \frac{2\pi N}{60}, \text{ where N is the rev.min}^{-1} \qquad (12.93)$$

Thus, we can represent the ratio between the centrifugal force and gravitational force as:

$$\frac{F_c}{F_g} = \frac{\omega^2 r}{g} = \frac{r}{g}\left(\frac{2\pi N}{60}\right)^2 = 0.001118 \times rN^2 \qquad (12.94)$$

Settling in a Centrifuge

Let us take the case where feed enters at the bottom of a tubular bowl centrifuge and exits at the top. The centrifuge

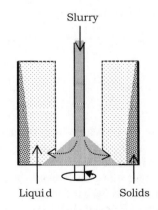

Slurry

Liquid   Solids

(a)Separation of solids from liquid

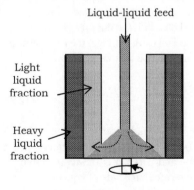

Liquid-liquid feed

Light liquid fraction

Heavy liquid fraction

(b) Separation of liquid-liquid fractions

FIGURE 12.15 Schematic of the centrifugal separation process (a) separation of solids from a liquid, (b) separation of liquids on the basis of difference in specific gravities

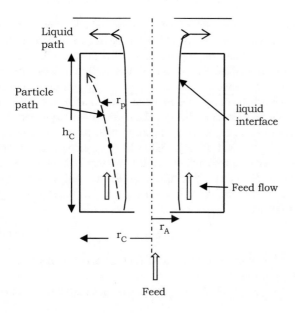

Liquid path

Particle path

h_C

r_p

liquid interface

Feed flow

r_C

r_A

Feed

FIGURE 12.16   Fluid flow and settling of particles in a centrifuge

rotates along its own axis. The slurry is subjected to centrifugal force while it moves from the feed end to the discharge end. Heavier particles tend to move towards the periphery of the centrifuge. If the residence time is enough such that the particles strike the wall of the centrifuge before escaping with the liquid from the chamber, they will be retained and collected along the wall.

Therefore, the diameter of the smallest particle that can be removed in a centrifuge can be calculated as the particle moves radially at its terminal settling velocity.

Let the radius and height of the centrifuge be denoted as r_C and h_C. It is assumed that all solid particles are carried by the liquid moving upward at a uniform velocity. Let v_t be the terminal settling velocity of the solid particles. The path of the particles can be shown as in Figure 12.16.

Suppose the particle at the end of the residence time is at distance r_p from the vertical axis. If more residence time is allowed, i.e., if the r_p is allowed to be same as r_C, then the particle will hit the wall of the bowl and be deposited on the wall, and, thus, it will be separated from the liquid. However, the particle would leave the bowl with the fluid if $r_p < r_C$.

If the settling is in the Stokes' law range, then the terminal settling velocity at radius r can be written as follows:

$$v_t = \frac{a D_p^2 (\rho_p - \rho)}{18\mu} = \frac{\omega^2 r D_p^2 (\rho_p - \rho)}{18\mu} \qquad (12.95)$$

In this equation v_t (m.s^{-1}) is the radial settling velocity, D_p (m) particle diameter, ρ_p (kg.m^{-3}) is the particle density, ρ (kg.m^{-3}) is the fluid density, and μ (Pa.s) is the liquid viscosity. For hindered settling, the right hand side of equation 12.95 is multiplied by a factor $\varepsilon^2 \psi_p$. Here, ε is the fractional volume that the liquid occupies in the slurry mixture, as discussed in equation 12.75 and ψ_p is a correction factor.

Since $v_t = \dfrac{dr}{d\theta}$,

$$d\theta = \frac{18\mu}{\omega^2 D_p^2 (\rho_p - \rho)} \cdot \frac{dr}{r} \qquad (12.96)$$

Integrating within limits, at time $\theta = 0$, the particle is at r_A, and at time $\theta = \theta_r$, the particle is at r_C.

$$\text{Or,} \quad \int_0^{\theta_r} d\theta = \frac{18\mu}{\omega^2 D_p^2 (\rho_p - \rho)} \int_{r_A}^{r_C} \frac{dr}{r} \qquad (12.97)$$

$$\theta_r = \frac{18\mu}{\omega^2 D_p^2 (\rho_p - \rho)} \ln\left(\frac{r_C}{r_A}\right) \qquad (12.98)$$

The residence time θ_r can be written as V/q, where V(m^3) is the volume of liquid in the bowl and q (m^3.s^{-1}) is the volumetric feed flow rate.

$$V = \pi h_C \left(r_C^2 - r_A^2\right) \qquad (12.99)$$

$$\text{Thus,} \quad \frac{V}{q} = \frac{18\mu}{\omega^2 D_p^2 (\rho_p - \rho)} \ln\left(\frac{r_C}{r_A}\right) \qquad (12.100)$$

$$q = \frac{\omega^2 D_p^2 (\rho_p - \rho)}{18\mu \ln\left(\dfrac{r_C}{r_A}\right)} \pi h_C \left(r_C^2 - r_A^2\right) \qquad (12.101)$$

From this equation, D_p can be calculated. The particles with diameter D_p or more will touch the wall of the centrifuge and will be removed from the feed. Particles having a diameter smaller than that calculated in the above equation will not settle and will move out.

If D_{pc} is defined as a critical diameter, i.e., the diameter of a particle that reaches half the distance between r_A and r_C, then we can find D_{pc} by integrating the equation within limits as follows:

$$\text{at } \theta = 0, \; r = \frac{(r_A + r_C)}{2}, \text{ and at } \theta = \theta_r, \; r = r_C$$

$$q_c = \frac{\omega^2 (\rho_p - \rho) D_{pc}^2}{18\mu \ln\left[2r_C / (r_A + r_C)\right]} (V)$$

$$= \frac{\omega^2 (\rho_p - \rho) D_{pc}^2}{18\mu \ln\left[2r_C / (r_A + r_C)\right]} \left[\pi h_C \left(r_C^2 - r_A^2\right)\right] \qquad (12.102)$$

Thus, the particles settling at the wall will have a diameter more than D_{pc} at the flow rate of q_c. The smaller particles will remain in the liquid and escape along with it.

In a case with a small thickness of liquid layer as compared to the radius of the bowl, modifying the equation 12.95 by substituting r_C for r and D_{pc} for D_p, we get

$$v_t = \frac{\omega^2 r_C D_{pc}^2 (\rho_p - \rho)}{18\mu} \qquad (12.103)$$

The time of settling θ_r for the critical D_{pc} is

$$\theta_r = \frac{V}{q_c} = \frac{(r_C - r_A)/2}{v_t} \qquad (12.104)$$

Substituting equation 12.103 into 12.104 and arranging,

$$q_C = \frac{\omega^2 r_C (\rho_p - \rho) D_{pc}^2}{18\mu[(r_C - r_A)/2]}(V) \qquad (12.105)$$

Expressing the volume V as:

$$V = 2\pi r_C (r_C - r_A) h_C \qquad (12.106)$$

$$q_c = \frac{2\pi h_C \omega^2 r_C^2 (\rho_p - \rho) D_{pc}^2}{9\mu} \qquad (12.107)$$

The above equations are applicable for liquid-liquid systems where liquid droplets migrate in the continuous phase and join together.

Example 12.2 A centrifuge is used to separate suspended solids (with particle density 1800 kg.m⁻³) from water. The centrifuge has a bowl with a working diameter of 5.6 cm and height of 20 cm. The diameter of the liquid exit path at the top of the bowl is 1.6 cm. If the flow rate is 0.004 m³.h⁻¹ and N = 18000 rev.min⁻¹, find the particles' critical diameter to enable them to move with the exit stream. The density and viscosity of water may be taken as 1000 kg.m⁻³ and 1 centipoise.

SOLUTION:

The working diameter is 5.6 cm, thus $r_c = 2.8$ cm, $h_C = 20$ cm, $r_A = 0.8$ cm

The equation 12.102 will be used here.

$$q_c = \frac{\omega^2 (\rho_p - \rho) D_{pc}^2}{18\mu \ln[2 r_C / (r_A + r_C)]} [\pi h_C (r_C^2 - r_A^2)]$$

Here,

$$\omega = \frac{2\pi N}{60} = \frac{2\pi(18000)}{60} = 1884.96 \text{ rad.s}^{-1}$$

The bowl volume is

$$V = \pi h_C (r_C^2 - r_A^2) = \pi(0.20)\left[0.028^2 - 0.008^2\right]$$

$$= 4.52 \times 10^{-4} \text{m}^3$$

Viscosity, $\mu = 1$ cP $= 0.001$ kg.m⁻¹.s⁻¹ $= 0.001$ Pa.s
The flow rate q_c is

$$q_c = \frac{0.004}{3600} = 1.11 \times 10^{-6} \text{ m}^3.\text{s}^{-1}$$

Substituting into the above equation

$$1.11 \times 10^{-6} = \frac{(1884.96)^2 (1800 - 1000) D_{pc}^2}{18(0.001) \ln[2 \times 0.028 / (0.008 + 0.028)]}(4.52 \times 10^{-4})$$

or $\quad 1.11 \times 10^{-6} = \dfrac{1284791.631 \times D_{pc}^2}{7.953 \times 10^{-3}}$

$$D_{pc} = 8.289 \times 10^{-8} \text{ m or } 0.0829 \text{ μm}$$

Example 12.3 A centrifuge rotating at 1440 rev .min⁻¹ separates oil from a dispersion. The oil is present in the form of spherical globules of 45 μm diameter. The density of oil is 900 kg.m⁻³. Separation occurs at an effective radius of 4.2 cm. What will be the velocity of oil through water? Take the viscosity of water as 0.79 cP and the density of water as 1000 kg m⁻³.

SOLUTION:

We use equation 12.103.

$$v = \frac{\omega^2 r_C D_{pc}^2 (\rho_p - \rho)}{18\mu} = \frac{(2\pi N)^2 r d^2 (\rho_p - \rho)}{18\mu}$$

or, $\quad v = \dfrac{\left(\dfrac{2\pi \times 1440}{60}\right)^2 \times 4.2 \times 10^{-2} \times \left(45 \times 10^{-6}\right)^2 (1000 - 900)}{18 \times 0.79 \times 10^{-3}}$

$$= 0.0138 \text{ m.s}^{-1} = 13.8 \text{ mm.s}^{-1}$$

Scaling up of Centrifuges

If we multiply and divide equation 12.102 by 2g and then substituting equation 12.60 with D_{pc} into equation 12.102, we obtain

$$q_c = \frac{2(\rho_p - \rho) g D_{pc}^2}{18\mu} \frac{\omega^2 V}{2g \ln[2 r_C / (r_A + r_C)]} \qquad (12.108)$$

Where the particle's terminal settling velocity is v_t.
Considering Σ as a physical characteristic of the centrifuge as:

$$\Sigma = \frac{\omega^2 V}{2g \ln\left[2 r_C / (r_A + r_C)\right]} = \frac{\omega^2 \left[\pi h_C (r_C^2 - r_A^2)\right]}{2g \ln\left[2 r_C / (r_A + r_C)\right]} \qquad (12.109)$$

Then, the above equation 12.108 can be written as:

$$q_c = 2 v_t . \Sigma \qquad (12.110)$$

Using the equation 12.110 in equation 12.107 for settling in thin layer as a special case

$$\Sigma = \frac{\omega^2 \pi h_C 2 r_C^2}{g} \qquad (12.111)$$

In fact, Σ is the area of a gravitational settler that, for the same feed rate, has the same sedimentation characteristics as those of the centrifuge. Scaling up from a laboratory test of q_1 and Σ_1 to q_2 (for $v_{t1} = v_{t2}$) requires

$$\frac{q_1}{\Sigma_1} = \frac{q_2}{\Sigma_2} \qquad (12.112)$$

Such a type of scaling up is possible when the centrifuges are of a similar type of geometry and the centrifugal forces are within a factor of 2 from each other.

A factor E may be incorporated, for different configurations, in the above equation

$$\frac{q_1}{\Sigma_1} = \frac{q_2}{E.\Sigma_2} \qquad (12.113)$$

Efficiency values for different types of centrifuges, determined experimentally, are available in the literature.

Separation of Liquids in a Tubular Bowl Centrifuge

The situation occurs when the immiscible liquids are finely dispersed as an emulsion. The separation occurs due to the difference in specific gravities of the liquids. A common example of separation of liquids in a centrifuge is the separation of cream and skim milk from an emulsion of milk. This is also used for concentration of fruit juices.

Let us consider the case of a tubular bowl centrifuge (Figure 12.17), which is used to separate two liquids with different densities. Let the density of the lighter liquid be ρ_L and that of the heavier liquid be ρ_H.

When the liquid is fed into the centrifuge, the heavier liquid will move toward the wall because of the centrifugal force and the lighter liquid will move along the axis of rotation. Thus, a vertical weir is put somewhat at an intermediate position so that the heavier and lighter liquids exit through different outlets. It is important to know the location of the interface of heavy-light liquids to decide on the position of the vertical weir. Ultimately, it is the position of the weir that controls

the volumetric hold up V in the centrifuge and determines the effectiveness of separation.

Let r_L = radius up to surface of light liquid layer(m)

r_I = radius up to liquid-liquid interface (m)

r_H = radius up to surface of heavy liquid downstream (m)

h_C = height of the bowl (m)

Pressure exerted by the light liquid and heavy liquid counterbalance each other at the interface as shown in Figure 12.18. We will find the pressures for both liquids and equate them.

First considering the light liquid, the force on a liquid particle at distance r is given as:

$$F_c = m\omega^2 r \qquad (12.114)$$

where, the mass of the liquid particle is m.
The differential force across a thickness dr is

$$dF_c = dm\, \omega^2 r$$

Substituting, $dm = (2\pi rh_C).dr.\rho$
we get $dF_c = (2\pi rh_C).dr.\rho_L\omega^2 r$

$$\text{or,} \quad \frac{dF_c}{A} = dp = \frac{(2\pi rh_C)dr.\rho_L.\omega^2 r}{2\pi rh_C}, \text{ as } A = (2\pi rh_C)$$

$$(12.115)$$

$$\text{Hence, } p_2 - p_1 = \frac{\rho_L\omega^2}{2}\left(r_I^2 - r_L^2\right) \qquad (12.116)$$

Similarly, the heavy phase of thickness r_I-r_H exerts the pressure at the interface r_I as follows.

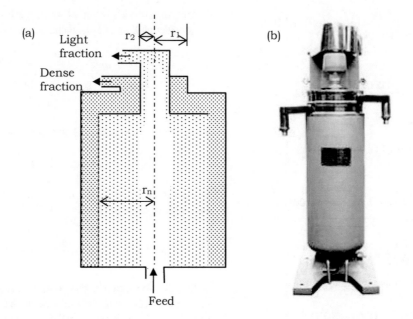

FIGURE 12.17   (a) Separation of light and dense fractions of a liquid in a tubular bowl centrifuge, (b) outer view of the tubular bowl centrifuge

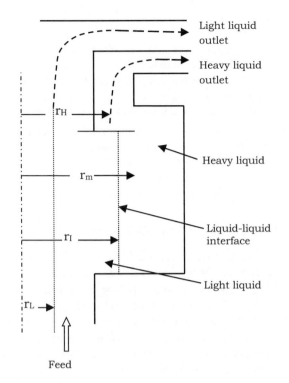

FIGURE 12.18 Calculation of interface position for a tubular bowl centrifuge

$$p_2 - p_1 = \frac{\rho_H \omega^2}{2}\left(r_I^2 - r_H^2\right) \quad (12.117)$$

At the interface layer

$$\frac{\rho_H \omega^2}{2}\left(r_I^2 - r_H^2\right) = \frac{\rho_L \omega^2}{2}\left(r_I^2 - r_L^2\right) \quad (12.118)$$

$$\text{or,} \quad r_I^2 = \frac{\rho_H r_H^2 - \rho_L r_L^2}{\rho_H - \rho_L} \quad (12.119)$$

Thus, the interface position can be found as:

$$r_I = \sqrt{\frac{\rho_H r_H^2 - \rho_L r_L^2}{\rho_H - \rho_L}} \quad (12.120)$$

The interface should be at a position smaller than the distance r_m so that the lighter liquid does not go out with the heavier liquid.

Centrifuges can develop forces of 13000 times gravity (13000×g). Super centrifuges have narrow diameter 100–150 mm and they are tall. Some narrow centrifuges, known as *ultra-centrifuges,* have only about 75 mm diameter and rotational speeds of up to 60000 rev.min^{-1}.

> **Example 12.4 A bowl centrifuge having an internal diameter of 230 mm is used to separate water from an oil. Let the density of the oil be 900 kg.m^{-3} and that of water can be taken as 1000 kg.m^{-3}. The water (i.e., the heavy liquid) port is at 5 cm from the axis of rotation and the oil port is at 3 cm. Find the position of the interface, i.e., the radius of the neutral zone.**

SOLUTION:

In a centrifuge the radius of the neutral zone can be given as:

$$r_I = \sqrt{\frac{\rho_H r_H^2 - \rho_L r_L^2}{\rho_H - \rho_L}}$$

$$\text{Thus,} \quad r_I = \sqrt{\frac{1000(0.05)^2 - 900(0.03)^2}{1000 - 900}}$$

$$= \sqrt{\frac{2.5 - 0.81}{100}} = 0.13 \text{ m} = 13 \text{ cm}$$

12.4.2 Equipment

Forces on particles in centrifugal separators are multiple and, thus, the particles that do not settle easily in gravity settlers and can often be separated from fluids by centrifugal force. Besides, it is possible by the centrifugal force to separate much smaller particles than is possible in gravity settlers. The high centrifugal force assists in overcoming the disturbing effects of Brownian motion and free convection currents, although it does not change the relative settling velocities of small particles. Centrifugal separation is also much faster than the gravitational separation when the two fractions to be separated do not differ much in their densities.

Classifications

The centrifugal separators, also known as *centrifuges,* are commonly classified under the category of separators for immiscible solids (e.g., tubular bowl centrifuge and disc bowl centrifuge), clarifiers (for removing small amounts of solids, e.g., solid bowl clarifier, nozzle centrifuge, and valve discharge centrifuge) and desludging or dewatering centrifuges for removing large amounts of solids from liquid (e.g., basket centrifuge, screen conveyor centrifuge, conveyor bowl centrifuge, etc.).

Tubular Bowl Centrifuge

- It consists of a vertical cylinder (or bowl), typically 10–15 cm in diameter and 75 cm long, which rotates inside a stationary casing at 15000 to 50000 rev.min^{-1} depending on the diameter. The smaller the diameter, the higher the revolutions per unit time.

- It is at the base of the bowl that feed liquor is introduced (Figure 12.17). Due to the centrifugal force, the liquid particles with larger densities move toward the periphery of the bowl. whereas those with smaller densities remain close to the axis. A vertical weir is fixed at an intermediate position to direct the denser and lighter liquids through two separate outlets.

- Centrifugal forces developed in super centrifuges are about 13,000 times the force of gravity.
- There are *ultra-centrifuges* having a small diameter of about 7.5 cm and very high speeds of 60000 rev.min^{-1}.

Disc Bowl Centrifuge

- A stack of inverted metal cones is contained in a cylindrical bowl, 0.2–1.2 m in diameter (Figure 12.19).
- The gap between the cones is about 0.5–1.3 mm and there are matching holes on the cones forming flow channels for liquid movement.
- The stack of cones rotates at 2000–7000 rev.min^{-1}.
- It is at the base of the disc stack that the feed is introduced. The denser fraction moves toward the wall of the bowl due to the centrifugal force on the liquid along the underside of the discs. The lighter fraction, on the other hand, moves toward the center along the upper surfaces.
- A weir system at the top removes both liquid streams continuously in a similar way as the tubular bowl system.
- Capacities can be up to 15,0000 l.h^{-1} (for both disc and tubular bowl types).
- Better separation is obtained by the use of disc bowl centrifuge due to the thinner layers of liquid formed in between the cones.
- Periodic cleaning of the cone assembly is required to remove solids deposited on the cone surfaces.

Applications

Disc bowl centrifuges are used to separate cream from milk and to clarify oils, coffee extracts and juices and for starch-gluten separation.

Solid Bowl Clarifier

- A solid bowl clarifier consists of a rotating cylindrical bowl, 0.6–1.0 m in diameter.
- The liquor with the suspended solids is fed into the bowl. Due to rotation of the bowl, the solids are thrown outward and form a cake on the bowl wall.
- The bowl is drained when the cake has reached a predetermined thickness and the cake is automatically removed through an opening in the base.
- This type of clarifier is normally used with a maximum of 3% w/w solids in the liquor.

Nozzle Centrifuge/ Valve Discharge Centrifuge

- These centrifuges are similar to disc bowl types except that the bowls have a bi-conical shape.
- Solids are continuously discharged through small holes at the periphery of the bowl in a nozzle centrifuge and are collected in a container.
- Holes in a valve discharge centrifuge are fitted with valves that open periodically for a fraction of a second to discharge the accumulated solids.
- These centrifuges are used for feeds with higher solids. Feed liquor is separated into three streams: a light phase, a dense phase, and solids. The valve type centrifuge gives lower liquor wastage and produces drier solids.
- Capacities up to 300 000 l.h^{-1} are achievable.

Applications

Centrifugal clarifiers are used to treat oils, juices, beer, and starches and to recover yeast cells.

Conveyor Bowl Centrifuge

- A conveyor bowl centrifuge consists of a rotating solid bowl in which the sludge is separated after

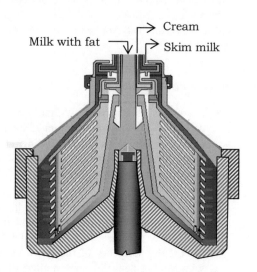

FIGURE 12.19 (a) Schematic diagram of flow occurring in a disc bowl centrifuge (b) a disc bowl type cream separator

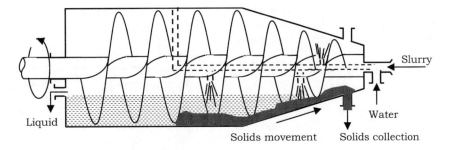

FIGURE 12.20 Schematic diagram of a conveyor bowl centrifuge (after Leniger and Beverloo, 1975)

being deposited on the bowl wall (Figure 12.20). The screen conveyor centrifuge has a similar design, but the bowl is perforated to remove the liquid fraction.

- Solids are deposited on the wall and conveyed to one end of the centrifuge in the solid bowl centrifuge, and the liquid fraction moves to the other, larger-diameter, end.

- As compared to other types of equipment, the solids to be discharged are relatively dry.

- In comparison to the screw conveyor centrifuge, the solid bowl rotates up to 25 rev.min^{-1} faster.

Reciprocating Conveyor Centrifuge

- It consists of a rotating basket, 0.3–1.2 m in diameter. The feed enters into the basket through a funnel, which also rotates at the same speed.

- A layer of cake is formed on the bowl as the liquid passes through perforations in the bowl wall.

- A reciprocating arm pushes the cake a little forward when the layer of cake has become 5–7.5 cm thick. The area, thus, is made ready for receiving more feed.

- Fragile solids, such as crystals from liquor, can be separated by reciprocating conveyor centrifuges.

Basket Centrifuge

- It has a perforated metal basket lined with a filtering medium.

- It rotates at up to 2000 rev.min^{-1} and the rotation is controlled by automatic devices.

- Usually, depending on the feed material, the cycles are of 5–30 minutes duration.

- Initially the bowl is rotated slowly, when the feed liquor is introduced.

- The speed is then increased to separate solids.

- Finally, the bowl is slowed, and the cake is discharged through the base by a blade.

- Capacities of these dewatering centrifuges are up to 90,000 l.h^{-1}.

Applications

Feeds with high solid contents are separated using de-sludging centrifuges. They are used to separate coffee, cocoa, and tea slurries.

12.5 Separation of Fine Solids from Gas

12.5.1 Principle of Separation

A cyclone separator is a device for collection of dust from gas (air) by the application of centrifugal and gravity force. The equipment consists of a vertical cylinder with a conical bottom. During the operation, the air carrying the fine solid particles at a very high velocity enters the cyclone separator tangentially at the top. As the velocity is very high, the air forms a vortex around the center of the chamber. The centrifugal force, thus generated, enables the particles to move radially and hit the wall of the chamber. After hitting the chamber wall, the particles move down along the wall and are collected at the bottom. The air, without the suspended particles, exits the cyclone from the top after it reaches the bottom of the cone and then moves upward in a smaller spiral in the center of the cone and cylinder. Thus, there is a double vortex.

In a cyclone separator, a particle is acted upon by two forces, the centrifugal force and the weight of the particle. The particle moves toward the cylinder wall in a helix along the resultant of the centrifugal force and particle weight. The centrifugal force can be described as:

$$C_f = Wv^2/gR \qquad (12.121)$$

where, C_f (kg) = centrifugal force, W (kg) = weight of particle, R (m) = radius of rotation, and v (m.s^{-1}) linear or tangential velocity.

The separating force can be given as:

$$F = W\sqrt{\frac{v^2}{g^2R^2}+1} \qquad (12.122)$$

The centrifugal force is many times the force of gravity. The performance factor or separation factor of cyclone is given by

$$S = C_f/W = v^2/gR \qquad (12.123)$$

With the increase in S, the separation becomes more effective. Cyclone separator separates particle having size 5–200 μm. For particles over 200 μm, gravity settling chambers are used.

The pressure drop through a cyclone can be given by the following equation.

$$dp = \frac{12Wh}{KE^2\left(\dfrac{L}{d}\right)^{1/3}\left(\dfrac{H}{d}\right)^{1/3}} \qquad (12.124)$$

where, dp= pressure drop, number of inlet velocity heads, d = diameter of cylinder, m; L=height of cylinder, m; H=cone height, W=entry width, h=entry height, E=exit duct diameter, and K=vane constant, 0.5–0.7.

The smallest particle size that can be removed by a cyclone can be estimated as:

$$D_p = \sqrt{\frac{9\mu E}{2\pi Nv(\rho_p - \rho_a)\left(\dfrac{4R}{E}\right)^K}}$$ (12.125)

where, D_p(m) = smallest particle diameter, μ (kg.m^{-1}s^{-1}) = viscosity of air, E (m) =exit duct diameter, N=number of effective turns of air stream in inner spiral, v (m.s^{-1}) = entrance velocity, ρ_p (kg.m^{-3}) and ρ_a(kg.m^{-3}) are densities of particle and air, R (m) = radius at which the spiral velocity is equal to v, and K= a constant, 0.5-0.7.

The terminal settling velocities to cause effective separation can be found out as follows.

Let us assume that the particle sizes are small enough for the Stokes' law to be applicable.

If the particles entering the cyclone reach the terminal settling velocities very quickly, the terminal radial velocity, v_{tr} can be given as:

$$v_{tr} = \frac{\omega^2 r D_p^2 (\rho_p - \rho)}{18\mu}$$ (12.126)

If, v_{tan} = particle's tangential velocity at radius r, then $\omega = v_{tan}/r$

$$\text{Hence, } v_{tr} = \frac{D_p^2 g (\rho_p - \rho)}{18\mu} \cdot \frac{v_{tan}^2}{gr} = v_t \frac{v_{tan}^2}{gr}$$ (12.127)

where, v_t = gravitational terminal settling velocity

With the increase in the terminal velocity v_t, the radial velocity v_{tr} increases and the particles settle more easily along the walls. Since it is difficult to find out the radial velocity, the following empirical equation is often used.

$$v_{tr} = \frac{K D_p^2 (\rho_p - \rho)}{18\mu r^M} (\text{K and M are constants})$$ (12.128)

The *efficiency of cyclone separation* is represented as the mass fraction of particles of a given size that are collected relative to the particles of the same size present. As the particle size increases, the cyclone efficiency rises rapidly. The diameter, D_{pc}, for which 50% of the mass of entering particles is retained is known as the *cut diameter*.

12.5.2 Cyclone Separator

Construction and Operation

- It consists of a conical bottomed vertical cylinder as shown in Figure 12.21.
- There is a tangential inlet at the top through which the gas-solid particle mixture enters at a very high velocity. Thus, the mixture has a rotating motion and

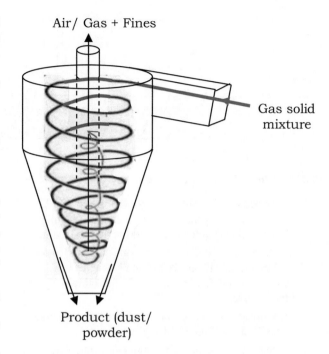

Air / Gas + Fines

Gas solid mixture

Product (dust/ powder)

FIGURE 12.21   Working of a cyclone separator

the air flows spirally in a vortex downward adjacent to the wall.

- The centrifugal force throws the particles radially toward the wall. The particles after hitting the wall of the cyclone give away the kinetic energy and fall downward due to gravity along the wall. These are collected at the bottom of the cone.
- The air, free from suspended solids, exits at the center of the cyclone top but not before reaching the bottom of the cone and then moving upward in a smaller spiral in the center of the cyclone.

Salient Features

- Cyclone separator provides more effective separation in comparison to gravity settling chambers because centrifugal force in cyclone separator is 5 to 2500 times more than the force of gravity and, therefore, the resultant outward force on the particles creates high tangential velocities. The centrifugal force in a cyclone ranges from about 5 times gravity in large, low velocity units to 2500 times gravity in small, high-resistance units.
- As mentioned above, the equipment generates one downward and another upward vortex both in the same direction.
- Cyclones where water is sprayed inside are known as *wet-scrubber cyclones* for making the removal of the solids easier.
- The settling rate of the particles depends upon their size and weight. Lighter particles take longer time to settle.
- The number of effective turns inside the cylinder is also important but, in practice, it is usually very

difficult to estimate. For a common cyclone, the number of effective turns is taken to be two.

- The cyclone separator has no moving part and practically no maintenance is required. Cyclone separators are effective for gas-particle separation with the least cost.

Applications

- A cyclone separator is used in the collection of dried particles from air. Some common applications are the separation of milk solids from air from the exit of a spray dryer or separation of rice bran from air coming out of a polisher. It is also commonly used for collection of dust and wastes during processing of grains and to separate out airborne material from the discharge of a pneumatic conveyor.

- As discussed earlier in this section, particles over 5 μm in diameter are generally removed by cyclones. Gravity settling chambers are adequate for particles over 200 μm in size.

Example 12.5 Particles of an average diameter of 50 μm and having a density of 1100 kg.m⁻³, suspended in air, are needed to be separated in a cyclone having 750 mm diameter. The entry velocity is kept at 14 m.s⁻¹. What will be the centrifugal force acting radially in the cyclone and the separation factor of the cyclone?

SOLUTION:

The centrifugal force, $C_f = mv^2/r$ or Wv^2/gr

The weight of the particle can be given as:

$$W = \frac{4}{3}\pi r^3 \rho = (4/3)(3.1416)(25 \times 10^{-6})^3 (1100)$$

$$= 7.199 \times 10^{-11}$$

Hence, $C_f = \dfrac{7.199 \times 10^{-11} \times 14 \times 14}{9.81 \times 0.375} = 3.836 \times 10^{-9}$ kgf

The separation factor, $S = \dfrac{C_f}{W} = \dfrac{v^2}{gr} = \dfrac{14 \times 14}{9.81 \times 0.375} = 53.28$

Example 12.6 The dimensions of a cyclone separator are as follows: diameter- 175 mm, air inlet diameter- 25 cm, separating height- 2.5× diameter of inlet, helix pitch 15°, inlet width- 10 cm. If it is used for separate particles of specific gravity 1.28 and the entry velocity is maintained at 14 m.s⁻¹, estimate the pressure drop through the unit. Also find the smallest particle which can be collected.

SOLUTION:

The pressure drop through the cyclone can be given by

$$dp = \frac{12Wh}{KE^2 \left(\dfrac{L}{d}\right)^{1/3}\left(\dfrac{H}{d}\right)^{1/3}}$$

Here, dp= pressure drop, number of inlet velocity heads
d = cyclone diameter = 0.175 m
L = cyclone height = 0.75 m
H=cone height = 1.8 m
W=entry width = 0.1m
E=exit duct diameter = 0.3m
K=vane constant, 0.5-0.7
h = entry height = 0.3 +0.9×tan 15° = 0.54

Substituting K=0.5,

$$dp = \frac{12 \times 0.1 \times 0.54}{0.5 \times 0.3^2 \left(\dfrac{0.75}{1.8}\right)^{1/3}\left(\dfrac{1.8}{1.8}\right)^{1/3}} = \frac{0.648}{0.0336} = 19.27$$

(The entry height was calculated at the center line of cyclone considering 15° helix pitch.)

The smallest particle size removed by the unit can be estimated by the equation 12.125.

Taking the viscosity of air as 5×10^{-5} kg.m⁻¹.s⁻¹, and the density of air as 1.293 kg.m⁻³,

$$D_p = \sqrt{\frac{9\mu E}{2\pi Nv(\rho_p - \rho_a)\left(\dfrac{4R}{E}\right)^K}}$$

$$= \sqrt{\frac{9 \times 5 \times 10^{-5} \times 0.3}{2 \times \pi \times 2 \times 15(1200 - 1.293)\left(\dfrac{4 \times 0.9}{0.3}\right)^{0.5}}}$$

So $D_p = \sqrt{\dfrac{13.5 \times 10^{-5}}{7.817 \times 10^5}} = 1.314 \times 10^{-5}$m or 13 μm.

Example 12.7 The solid particles suspended in air have an average diameter of 18 μm and density of 760 kg.m⁻³. The air is entered into a cyclone of 0.72 m diameter at a tangential velocity of 28 m.s⁻¹ at 0.36 m. The density and viscosity of air are 1.16 kg.m⁻³ and 1.85×10⁻⁵ Pa.s, respectively. What is the terminal radial velocity of the particle?

SOLUTION:

The terminal velocity is found as:

$$v_t = \frac{gd^2(\rho_p - \rho_a)}{18\mu} = \frac{9.81 \times (18 \times 10^{-6})^2 \times (760 - 1.16)}{18 \times (1.85 \times 10^{-5})}$$

$$= 0.00724 \text{ m.s}^{-1}$$

The terminal radial velocity of the particle is given by

$$v_r = \frac{v_t}{g} \times \frac{v_{tan^2}}{r} = \frac{0.00724}{9.81} \times \frac{28^2}{0.36}$$

$$= 1.607 \text{ m. s}^{-1}$$

12.6 Membrane Separation

12.6.1 General Principles

Principle of Separation

Membrane separation is a process of utilizing semi-permeable membranes along with the differences in molecular weight and size of the components for separating components of a liquid food. The feed suspension is moved under pressure at a reasonably high velocity across it (Figure 12.22). Thus, it is a system of cross flow filtration and the separation takes place due either to the pressure difference or to the concentration difference between the two sides of the membrane material.

Membranes are available in different pore sizes. Depending on the pore size, a specific membrane material restricts the movement of materials and molecules having a size larger than that of the pores and the smaller size molecules and solvent move through to the downstream. Thus, the feed is separated into two streams, namely, the *retentate* or the high molecular weight fraction retained on the upstream side, and the *permeate* or low molecular weight fractions that move across the membrane. Usually there is no buildup of the layer of solid on the upstream side due to the high velocity of the flow.

On the pore sizes and pressures used for separation, membrane filtration systems have been classified as reverse osmosis (RO), nanofiltration (NF), ultrafiltration (UF), and microfiltration (MF).

Membrane filtration/separation has a wide range of applications in industry to concentrate food solutions. As foods are complex mixtures of various organic and inorganic compounds having varying molecular weights, membrane separation is often used to separate the food components for new food preparations or to obtain high value food ingredients. They can be applied to fractionate solutions of macromolecules, selectively removing water and some solutes from a solution.

Movement of Molecules Through a Membrane

The properties of solute and suspended solids influence the membrane separation process. The membrane attributes affecting the transmembrane flux and the system performance are thickness, mean pore size, pore size distribution, configuration of pores, membrane *tortuosity* (or the twisted path across the membrane thickness for fluid or particle flow) and the solute rejection properties.

The motion of molecules through the barrier takes place mainly due to permeation, Knudsen diffusion and/or convection. Permeation involves the diffusion of molecules through the membrane; the subsequent desorption of the diffused molecules in the downstream also affects the permeation process. *Knudsen diffusion* or *Knudsen flux* occurs when the mean pore diameter of the porous medium is smaller than the mean free path of the particles (i.e., $d/\lambda < 0.2$). In this mechanism, the molecules bounce back and forth between the walls of the porous medium without colliding with each other. The partial pressure gradient is the main driving force for the flow. Convection takes place during microfiltration and ultrafiltration involving viscous flow ($d/\lambda > 20$).

Osmosis essentially means craving for water. In the case of osmosis, there is a semi-permeable barrier separating a solution having the solute on one side and the pure solvent on the other side. The concentration difference causes the pure solvent to pass across the membrane. Even though the chemical potential across the membrane is the same at equilibrium condition, there is a difference in hydrostatic pressure. In this case the hydrostatic head is the osmotic pressure.

Reverse osmosis is opposite of osmosis; the solvent in a solution is forced through a semi-permeable membrane toward the pure solvent side by application of pressure. In this process the applied pressure should be more than the osmotic pressure. The molecules dissolve at the surface of the membrane on the upstream side and are diffused through the membrane, which is removed at the other surface on the downstream side. Thus, diffusion drives molecules through the membrane in reverse osmosis. Osmotic energy is required to overcome the diffusion resistance.

Osmotic pressure is directly proportional to concentration, hence, a colligative property.

$$\Pi \propto \frac{1}{M_w} \text{ and } \Pi \propto c$$

$$\text{and, it is obtained as, } \Pi = \frac{RTc}{M_w} \qquad (12.129)$$

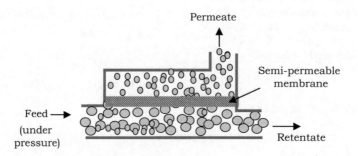

FIGURE 12.22   Separation by membrane processing

where, Π (kPa) = osmotic pressure, R (kPa.m^{-3}.mol^{-1}.K^{-1}) = gas constant, T is the absolute temperature and c is the concentration.

The above equation, known as *Vant hoff's equation*, is applicable for dilute solutions.

During the separation, the osmotic pressure reduces the pressure differential across the membrane. For forcing the solvent across the membrane, the pressure on the upstream side must be higher than the osmotic pressure. The osmotic pressure can also be expressed as:

$$\Pi = \frac{RT}{V}\ln(x.\gamma) \tag{12.130}$$

where, γ = activity coefficient, V = molar volume, and x is the mole fraction of water.

Since the product x.γ is the water activity, the above equation can be written as:

$$\Pi = \frac{RT}{V}\ln(a_w) \tag{12.131}$$

Example 12.6 Sucrose solution is being concentrated to 45% at 30°C by reverse osmosis. What will be the minimum pressure required across the membrane?

The formula that will be used is

$$\Pi = \frac{RT}{V}\ln(a_w),$$

thus, the values of V and a_w need to be found for this condition.

The water activity of a 45% sucrose solution can be found as follows.

The mole fraction of solvent,

$$X_w = \frac{55/18}{55/18 + 45/342} = \frac{3.0555}{3.0555 + 0.1315} = 0.9587$$

$\log(a_w) = \log(X_w) - k(1-X_w)^2$ (as equation 9.4)
For sucrose, K = 2.7

So, $\log(a_w) = \log(0.9587) - 2.7(1-0.9587)^2$

$$= -0.01832 - 4.60 \times 10^{-3} = -0.0229$$

Thus, $a_w = 0.9486$

Taking R = 8315 N.m.(kg mole)$^{-1}$K^{-1} and T = 303.15 K and the density of water = 1000 kg.m^{-3},

$$V = (18 \text{ kg. (kg-mole)}^{-1}) \times (1/1000 \text{ m}^3.\text{kg}^{-1}) = 0.018 \text{ m}^3. \text{(kg-mole)}^{-1}$$

$$\Pi = \frac{(8315)(303.15)}{0.018}\ln(0.9486) = 140038458.3 \times \ln(0.9486)$$

$$= -7389558.547 \text{ Pa} = 7389.56 \text{ kPa}$$

Thus, the osmotic pressure is 7389.56 kPa. Thus, the minimum pressure required is 7389.56 kPa.

As discussed above, the pressure difference across the membrane is the driving force for the flow across the membrane. The total pressure difference can be written as:

$$\Delta p = \Delta p_m + \Delta p_b \tag{12.132}$$

where, Δp_m is the pressure drop across the medium and Δp_b is the pressure drop due to the boundary layer of the retentate suspension flowing across the membrane surface and due to fouling (solid deposits) on the membrane surface.

The Sperry equation can be modified for the flux across the membrane with a compressible deposit on the surface of the membrane by using $\alpha = \alpha_0.\Delta p^n$

$$\frac{dV}{d\theta} = \frac{Aq\Delta p}{\mu\alpha_0\Delta p^n c.q(V/A) + A.\Delta p} \tag{12.133}$$

where, n = compressibility factor, q = permeability for pure water corresponding to the applied pressure, Δp, and c = concentration of solids, n is taken as 0 for a completely incompressible cake and as 1 for a completely compressible cake.

The second term in the denominator of the above equation can be ignored for situations when the solid deposits and the fluid boundary layer at the membrane surface mostly control the flux. Thus, the above equation can be modified as:

$$\text{i.e. } \frac{dV}{d\theta} = \frac{A^2\Delta p^{1-n}}{\mu\alpha_0 c.V} \tag{12.134}$$

For a completely compressible solid deposit, if the value of n is taken as 1 in the above equation, the equation reduces to

$$\frac{dV}{d\theta} = \frac{A^2}{\mu\alpha_0 c.V} \tag{12.135}$$

Equation 12.135 states that for this situation, the transmembrane pressure does not affect the flux. UF and RO are examples in which such conditions prevail.

Even in liquids without any suspended solid, the flux across the membrane for solutions is lower than that for pure water. This phenomenon is known as *polarization concentration*. The reduction in flux is affected by the operating conditions, the solid concentration in feed, and the flow rate of the solution. Ideally, the flux remains constant if transmembrane pressure is maintained constant. Fouling on the membrane surface may cause reduction of flux under constant operating conditions.

Considering that the difference between the applied pressure and the osmotic pressure is the driving force for the flow across the membrane, a simple equation can be written for the differential pressure as:

$$p = \frac{p_f + p_r}{2} - p_p \tag{12.136}$$

where, p (kPa) is the transmembrane pressure, p_f, p_r, and p_p are the pressures of feed, retentate, and the permeate (all in kPa).

In addition to the driving force, the other parameters affecting the flow include the concentration of solute within the membrane and the motion of the components across the membrane, etc. The mobility, in turn, depends on the membrane structure and solute size. The transmembrane flux will increase with increase in applied pressure. The flux will also increase with increased membrane permeance and decreased solute concentration in feed. The flow velocity, viscosity, temperature of the feed liquid and presence of other components of higher and lower molecular weights also affect the flux.

The flux (J) through the membrane can also be given as:

$$J = k.A.(p - \Pi) \tag{12.137}$$

where, p (kPa) is the applied pressure, Π (kPa) is the osmotic pressure, k (kg.m^{-2}.h^{-1}.kPa^{-1}) is the mass transfer coefficient and A (m^2) is the membrane area.

$$\text{For dilute solutions, } \Pi = MRT \tag{12.138}$$

where, T is the absolute temperature, M (mol.m^{-3}) is the molar concentration, and R (kPa.m^{-3}.mol^{-1}.K^{-1}) is universal gas constant.

Many foods have high osmotic pressure (e.g., fresh fruit juices can have osmotic pressures in the range of 6–10×10^5 Pa) and, hence, a high applied pressure is required to maintain the flow.

The separation is also affected by the fluid's boundary layer at the surface of the membrane. If the thickness of this laminar boundary layer is δ, solid concentrations at the boundary layer and in the liquid bulk are c and c_l, then the following equation can be used to relate the mass of solids reaching the boundary layer along with the solvent and those leaving the boundary layer by diffusion.

$$\left(\frac{\partial V}{\partial \theta}\right)c = -D\left(\frac{\partial c}{\partial x}\right) \tag{12.139}$$

In this equation, D is the mass diffusivity of solids. Denoting $(\partial V/\partial \theta) = J$, the flux across the membrane and integrating with respect to x with the boundary conditions, $c = c_l$ at $x = \delta$ and $c = c_s$ at $x = 0$ (at the surface)

$$\ln\left(\frac{c}{c_l}\right) = \frac{J}{D}(\delta - x) \tag{12.140}$$

$$\ln\left(\frac{c_s}{c_l}\right) = \frac{J}{D}(\delta) \tag{12.141}$$

The equation 12.141 states that at lower δ there would be reduced polarization concentration even at maximum flux. The thickness of the boundary layer (δ) can be reduced by maintaining high fluid velocities.

From equation 12.141,

$$J = \frac{D}{\delta}\ln\left(\frac{C_s}{C_l}\right) \tag{12.142}$$

If D/δ is substituted by k_s, a mass transfer coefficient for the solids, equation 12.142 can be modified as:

$$J = k_s(\ln c_s - \ln c_l) \tag{12.143}$$

$$\text{or, } \ln c_l = -J/k_s + \ln c_s \tag{12.144}$$

In the above equation, if the c_s is constant, the equation is of the form, y = mx + C, i.e., a plot of bulk concentration c_l against the flux (J) across the membrane on a semi log paper will be a straight line with a slope of $-(1/k_s)$; in other words, the rate of reduction of flux (J) varies inversely with the k_s. Thus, the general equations for evaluating mass transfer coefficients are useful in explaining the effects of fluid velocity on the flux across the membrane, when the flux is controlled only by polarization concentration.

Mass transfer in turbulent flow can be estimated by the *Dittus-Boetler equation*.

$$Sh = 0.023 \, Re^{0.8}Sc^{0.33} \tag{12.145}$$

Where, Re is Reynolds number, Sh is the Sherwood number, and Sc is Schimdt number.

Considering that

$$Sh = \frac{k_s d}{D}, Sc = \frac{\mu}{\rho D}, \text{ and } Re = \frac{\rho v d}{\mu} \quad (D = \text{mass diffusivity,}$$

m^2.s^{-1} and d is the diameter of the pipe), the above equation can be modified as:

$$k_s = 0.023 \frac{D}{d} \cdot \frac{d^{0.8}v^{0.8}\rho^{0.8}}{\mu^{0.8}} \cdot \frac{\mu^{0.33}}{D^{0.33}\rho^{0.33}}$$

$$\text{or, } k_s = 0.023 \cdot \frac{D^{0.67}v^{0.8}\rho^{0.47}}{d^{0.2}\mu^{0.47}} \tag{12.146}$$

The Dittus-Boetler equation applies for tube flow. For non-circular cross sections, hydraulic radius should be used in place of the d.

In the case of laminar flow, an expression for the k_s can be developed starting from the *Sieder-Tate* equation for heat transfer.

$$Sh = 1.86\left[Re.Pr.\frac{d}{L}\right]^{0.33} \tag{12.147}$$

Here, L is the length of the channel.

Substituting the values of these dimensionless numbers,

$$k_s = 1.86 \cdot \frac{D^{0.67}v^{0.33}}{d^{0.33}L^{0.33}} \tag{12.148}$$

If the diffusivity D of the molecular species and surface concentrations in the separation process are known, then the equations 12.146 and 12.148 can be used to determine the transmembrane flux. However, the concentration at the membrane surface (c_s) cannot be predicted and, hence, these are not adequate for determining the actual transmembrane flux.

The *Einstein-Stokes equation* can also be used to calculate D as follows.

$$D = \frac{1.38 \times 10^{-23}\,T}{6\pi\mu r} \qquad (12.149)$$

Where T is in K, r is the molecular radius in m, μ in Pa.s, and D is in $m^2.s^{-1}$

Selectivity of a Membrane

The selectivity of a membrane is an expression for the extent of separation of the solute from the feed. Thus, if c_f and c_p are the concentrations of solute in feed and permeate, respectively, then the extent of retention of solutes can be given as:

$$R = \frac{c_f - c_p}{c_f} = 1 - \frac{c_p}{c_f} \qquad (12.150)$$

Another term *real retention* is defined as:

$$R_r = \frac{c_m - c_p}{c_m} = 1 - \frac{c_p}{c_m} \qquad (12.151)$$

where, c_m is the concentration of solute at the membrane solution interface. This corresponds to the differences in solute concentration at the membrane-solution interface and that in the downstream permeate.

The solute concentration at the membrane surface at the upstream side will always be higher than that of the bulk fluid, and thus the real retention is always more than the observed retention. Please note that the c_p values of zero in the above equations will yield the R and R_c values as 1. The R and R_r values are almost the same in the absence of a polarized layer. These values can be obtained by experimental measurements. For obtaining R_r, batch experiments are conducted with high stirring speed and low operating pressure. The feed concentration is also kept low.

The type of membrane material and operating conditions affect the selectivity of the membrane and solute retention. The solutes with low molecular weight would pass more easily than the high molecular weight molecules. Plotting the retention factor against the molecular weight of the solute for a particular membrane gives a curve of sigmoidal shape.

Some solutes may be repelled by specific membranes, e.g., the cellulose acetate membranes may repel the mineral salts. The rejection is also affected by interactions between solutes.

Usually, the rejection properties of membranes are defined in terms of *molecular weight cutoff* (MWCO). For determining the MWCO, experiments are conducted with neutral solutes of various molecular weights at low feed concentration, low transmembrane pressure drop, and high turbulence. Retention values at steady state are measured when experiments, using each of these solutes, are conducted. A plot of the measured retention values vs. molecular weight of solutes on a semi-log paper will be as shown in Figure 12.23 (two curves for two situations are shown). The molecular weight cut off of the membrane is the molecular weight of the solute at 90% retention.

The ability of the membrane to remove water while retaining the solutes increases with an increase in applied pressures at lower pressures and decreases with an increase in applied pressure at higher pressures.

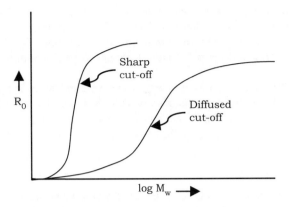

FIGURE 12.23 Sharp and a diffused molecular cut-off curves for a membrane

The retentate is often recycled to obtain both the desired final solid concentration of the product and high cross membrane fluid velocities.

A molecular weight cut-off (MWCO) curve may be either a diffused cut off or a sharp cut off. If the R value increases quickly to 90% level without much increase in the molecular weight values, it is called a sharp cut-off curve or else it is a diffused type cut-off curve. There should be very accurate control over the operating conditions for obtaining sharp cut-off characteristics. Commercial membranes are generally diffused type.

Membrane Permeability

The membrane permeability (ψ) can be represented as:

$$\psi = \frac{J^0}{\Delta p} \qquad (12.152)$$

Where, J^0 is the pure water flux and Δp is pressure drop across the membrane. ψ has the unit of $m^3.m^{-2}.(Pa\text{-}s)^{-1}$ or $m.(Pa\text{-}s)^{-1}$.

The membrane permeability does not depend on either the cross-flow velocity or the stirring speed. The membrane permeability can be obtained by conducting experiments using distilled water at different pressures. The water flux is measured at various pressure drops. A straight line passing through the origin is obtained when permeate flux is plotted against the operating pressure. The slope of this curve is the membrane permeability.

Sterilizing Filtration

Sterilizing filtration employs a micro-filtration process with pore size of less than 0.2 μm and aims to reduce the microbial load. The relationship for the rate of micro-filtration process can be given as follows.

$$\frac{dV}{d\theta} = k\frac{V^{-n}}{S} \qquad (12.153)$$

where, S = load of suspended material and microorganisms removed by the filter. The characteristics of fluid, filter medium, suspended solids, and the fluid velocity across the membrane are given by the two constants, k and n. The constant k also includes the effect of pressure drop across the membrane. The value of n is more than 1, thus, the rate of filtration decreases rapidly with an increase in infiltrate volume.

The micro-porous membranes are preferred to cartridge filters for sterilizing filtration as there is a need for a much smaller pressure drop and more efficient removal of microorganisms as compared to the latter.

Cycle Time

In the case of membrane filtration, the fluid moves at a very high velocity, thus causing reduced cake formation. However, backwashing or flushing is still required to remove the cake. The membranes that are fragile to resist back-washing pressure use flushing for cake removal. During the flushing process, the filtration is periodically stopped by reducing the transmembrane pressure, but the velocity of fluid flow is not reduced. The flushing fluid may be discarded, but it is usually filtered by a coarser filter, or else it would mix with the feed and reduce the filtration rate.

The filter assemblies are often pre-sterilized with high-pressure steam or chemical sterilants. The treatment solution must be compatible with the filter material. A test known as a *bubble point test* is often used to determine the condition of the filter after pre-sterilization. The test involves noting the pressure required to dislodge the liquid from the membrane after wetting the membrane and introducing sterile air. Hydrogen peroxide, chlorine solutions, and iodophores are the common chemical sterilants used for the purpose. The time needed for pre-sterilization, backwashing, or flushing, etc. should be considered for estimating the total cycle time.

The membrane separation system can be operated either batch wise or in a continuous manner. In a batch system, the liquid is recirculated until the desired concentration is achieved.

In the case of continuous operation, after the steady state is achieved, the feed rate will be the same as the sum of the permeate and concentrate flow rates. The degree of concentration is calculated with the help of the concentrate flow rate and feed rate.

12.6.2 Methods and Equipment

Methods

As discussed earlier, movement of various chemical species is restricted through a membrane. Different membrane processing methods are categorized as follows depending on the pore size and the pressure used (Figure 12.24).

* Reverse osmosis;
* Nano filtration;
* Ultra Filtration;
* Microfiltration.

There are different applications for different permeabilities. *Reverse osmosis* (RO) can separate solutes with low molecular weights, such as salts and monosaccharides, having high osmotic pressure from the solvent; hence, the method is so named. The *nanofiltration* (NF) can remove monovalent ions like sodium and chlorine and can be used for partial demineralization. The *ultra-filtration* (UF) can retain the large and macromolecules as colloids and proteins. The *microfiltration* (MF) can remove the bacteria and macromolecules.

These semipermeable membranes are arranged in different ways in an assembly, which, together with the inlet and outlet pipes, pressurizing device, and other accessories, make up the equipment. We will first discuss the features and applications of the different types of membranes and then the arrangements of the membranes in the assembly.

Reverse Osmosis

Characteristics

* The basic nomenclature comes from the fact that the solvent flows in the direction opposite to that of normal osmotic flow. The separation is caused by a solution-diffusion mechanism in the polymer and not by screening at the membrane surface.
* The pores in the membrane are in the range of 2–10 A° and, thus, very low molecular weight solutes in only very small amounts can pass through them. The

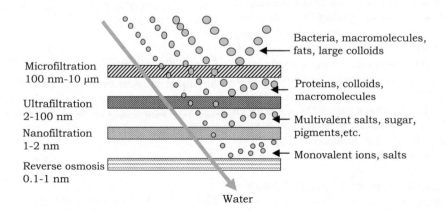

FIGURE 12.24 Pore sizes of different types of membranes and their separation capabilities (after Yang et al. 2019 with permission)

RO is often known as *hyper-filtration* when it is used to separate small molecules and ions.

- Polymers used as RO materials have a high rejection for soluble salts while they have high water permeability.
- The operating pressure is usually more than 2500 kPa.
- Chemical composition, molecular structure, and thickness are the important factors that are responsible for the performance of RO membranes.
- RO has low operating cost and there is no need for heat treatment during or before RO.

Applications

- RO is used to separate low molecular weight solutes (such as aroma compounds, salts, monosaccharides, etc.), which exert high osmotic pressure, from water.
- The common applications include purification of water and desalinization of sea water. It is also used for purification of wastewater.
- RO membranes are commercially used for concentration and purification of fruit juices, dairy products, vegetable oils, enzymes, and fermentation liquors. It is also widely applied for concentration of whey. It is also used in industry for preconcentrating juices before evaporation, which helps in reducing heat treatment.
- It is used to concentrate wheat starch, natural extracts and flavors, citric acid, egg white, coffee, milk, and syrups.
- RO system clarifies wine and beer. It can also de-alcoholize the low alcohol wines and beers.
- RO is the most economical when treating dilute solutions.

Nano-Filtration

Characteristics

- It stands in between UF and RO with a pore size of 5–20 A° and pressure range of 1500–2500 kPa. As it operates at a lower pressure than reverse osmosis, it is also known as *loose reverse osmosis*.

Applications

- Materials with molecular weights ranging from 200 to 1000 Da can be removed. Thus, it is used for partial demineralization of solutions; it also helps in removal of inorganic salts, e.g., urea, lactic acid, Na, Cl, K. It can remove the salts and thus it is used for reducing salt in cheese making.
- It is used for acid removal.
- It can be used for filtration of dyes and low molecular weight organics.
- Nano filtration is used for pre-concentration of materials to be treated with RO and as pre-treatment for electro dialysis and ion exchange.

Ultra-Filtration

Characteristics

- Unlike reverse osmosis, ultra-filtration operates at low pressure (600–800 kPa) and the pore size range is from 20 to 1000 A°.
- UF membranes are thicker than RO membranes (usually 0.1–0.5 μm thick).
- The transport of solutes and materials is mainly due to convection and diffusion.
- A specialized ultra-filtration process involving the dilution of the retentate with water for repeat filtration through the ultra-filtration membrane is termed as *Dia-filtration*. As seen in Figure 12.25, the permeate would be the low molecular weight compounds as salts and sugar. The retentate, which is mostly fats and proteins, is recycled, diluted if needed, and fed into the system along with the feed. It helps in further increasing the concentration of retained components and reducing the concentrations of soluble permeates.

Applications

- It allows larger molecules like lactose and minerals to pass through and retains molecules like proteins and colloids, which have a low osmotic pressure. Thus, UF is commonly used for pre-concentrating milk for cheese making and for making other milk products.

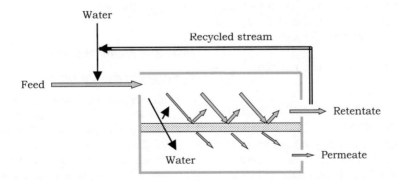

FIGURE 12.25 Schematic of diafiltration process

- It is also used for selective removal of salts and lactose from whey, concentration of enzymes, pectin, and recovery of proteins from cheese whey.
- UF is widely employed for the separation of enzymes from the fermentation process. UF is also useful as biochemical reactors where the reaction products may reduce the reaction rate such as microbial or enzyme conversion processes.
- It is used for concentration of sucrose and tomato paste.
- UF is used for clarification of pectinase enzyme-treated fruit mash.
- UF is used as a pre-treatment before RO so that organic and colloidal materials do not cause fouling.
- It is used for water treatment for the removal of contaminants and bacteria of sizes smaller than 1μm down to molecules of about 1nm in size (M≅300).

Microfiltration

Characteristics

- Microfiltration membranes have a pore size of more than 1000 A°.
- Materials with molecular weights more than 1×10^5 are separated with microfiltration. In fact, it lies between UF and conventional filtration systems.
- The operating pressure is usually 200–400 kPa.

Applications

- MF is used for producing low-heat sterile milk in the dairy industry.
- MF can separate the fat globules and colloids, which are dispersed in milk.
- For the size of particles removed by MF, cake filtrations could be used; however, the cake of 1 μm particles would offer a high resistance to flow.

Membrane Materials

Cellulose acetate (CA) is the most commonly used polymer as it has the desirable characteristics and at the same time it is inexpensive. The modified natural cellulose polymers, synthetic polymers, inorganic ceramic materials as well as metals are also used for preparation of membranes. Polyacrylonitrile, polyamides, and polyurethanes are also used for low pH materials and low temperature resistance applications. Polysulphones and ceramic materials are better suited for high temperatures and wider pH range.

Desirable characteristics of membrane materials can be listed as follows.

- It should have high porosity and there should not be much variation in the pore-sizes.
- The molecular structure of the membrane controls the diffusion of solutes and, hence, the membrane material should have the desirable permeability and solute rejection properties.

- The membrane should allow high permeate flows without being distorted. They should have high mechanical strength so as not to lose their characteristic porous structure during the operation.
- They should be resistant to heat, abrasion, chemicals, bacteria, and decomposition by water.
- The material should be able to be cleaned and sanitized easily.
- They should be durable and inexpensive.

Depending on the process, the membranes can be formed and classified as *symmetric* or *asymmetric*, depending on whether the pores are uniform or not across the thickness. They are also known as *isotropic* and *anisotropic* membranes. Usually, we obtain an asymmetric membrane when a polymer film is casted on another polymeric support. The anisotropic membranes provide extra strength to the membrane. The membranes are also classified as *homogeneous* or *heterogeneous*. The former are isotropic membranes having a pore size of 2–10 A° and the main mechanism for separation is diffusion. The membranes used for RO come under this group. But in case of microfiltration, the pore sizes can range between 10 and 1000 A°; thus, this is an example of heterogeneous membrane and the separation takes place both by diffusion and by convection. A heterogeneous membrane can be used as a support for the homogeneous membrane for some specific applications. Membranes having a continuous gradation in pore size are also available.

Membrane Modules

Membrane module is the arrangement of membranes in the equipment.

Usually, the membranes are formed by mixing a polymer solution in a volatile solvent such as acetone, and then subsequently evaporating the solvent, which gives a porous sheet. Membranes with larger pores are prepared by exposing the membranes to high energy radiation. When the radiation enters the polymer, it dissolves material it contacts and when this part is removed, the pores are created.

There are four types of membrane modules:

- Plate and frame;
- Tubular;
- Hollow-fiber;
- Spiral wound.

Plate and Frame Module

Construction and Operation

- Like a plate and frame filter press, membranes are stacked together with intermediate spacers in the plate and frame design. The membranes in forms of flat sheets are sandwiched between the membrane support plates. Commercial plate-frame units have membrane plates mounted vertically (Figure 12.26a).

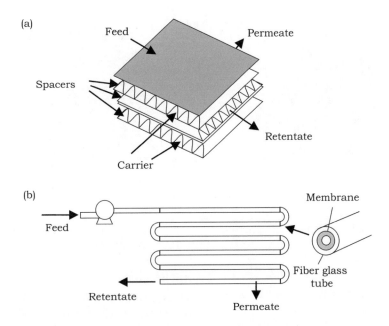

FIGURE 12.26 (a) Plate and frame membrane system, and (b) Tubular membrane system

- Gaskets with locking devices and glue are used to seal the membranes to the plates. Membranes could also be directly bonded to the plates.
- The permeate is collected through tubes at the bottom of the plates and comes from the porous plates that have internal flow channels. The ribs or grooves on plates form the side flow channels.
- The flow can be either in parallel or in series.

Salient Features

- The arrangement uses a compact space to pack a large membrane surface area.
- The replacement of individual plates is easy and, hence, it is less expensive.

Applications

The system can be used for both UF and MF.

Tubular Module

Construction and Operation

- The membrane is formed on the inner surface of a porous tube and thus the tube acts as the support for the membrane layer (Figure 12.26b).
- The feed enters the tube at high velocities (as high as 6 m.s^{-1}).
- The permeate moves through the membrane and comes out radially from the membrane system. It may be recycled back depending on the application.
- Some tubular designs are mechanically cleaned with sponge balls. Sponge balls are also used to enhance cleaning by chemicals by reducing time and quantity of chemicals.

Salient Features

- Tubular MF, UF, and NF systems do not require significant pre-filtration. The high cross-flow velocities minimize the formation of a cake on the surface of the membrane and, thus, high flux rates are achieved. The cleaning of the membrane surface is easy even for applications with high suspended solids.
- For juice clarification applications, it is required to maintain high yield rate as well as high concentration levels of suspended solids in the final product. Tubular membranes are better suited for such applications.

Applications

Such modules can be used advantageously for feed with a high amount of suspended solids. Juice clarification and treatment of paper industry waste are successful examples of tubular membranes.

Hollow Fiber Module

Construction and Operation

- It consists of a number of hollow fibers (in which each fiber is a tubular module) and kept in a large pipe like a shell and tube heat exchanger. Thus, a compact module with a large membrane surface area is created, which gives a high capacity with minimum space (Figure 12.27).
- Depending on the application, up to 20 tubes can be connected using special end caps in series or in parallel.
- The membranes could be arranged such that the module could be operated as a single pass system or the feed could be circulated within.

(a)

(b)

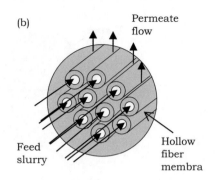

FIGURE 12.27   (a) Hollow fiber modules, (b) flow streams in hollow fiber module

- The system is constructed to withstand the permeate back pressure.

Salient Features

- The advantages of such modules include high throughput in less space.
- Back-flushing periodically can remove solids from the membrane inside the surface, thus reducing the need for frequent chemical cleaning. Thus, it also helps in lowering labor and chemical costs.
- Membrane fouling is reduced due to the tangential flow along the membrane surface as in the tubular module.
- Future extension of the plant is easy due to the modular construction of the system.

Applications

Applications of hollow fiber membranes include dairy processing, juice clarification, potable water treatment, wine filtration, etc.

Spiral Wound Module

Construction and Operation

- The membrane is formed as a film on a flat sheet and then several membranes are sandwiched together with permeate carrier and feed spacers in between. The spacers are usually of 0.07 to 0.25 cm (Figure 12.28).
- The membrane assembly is looped around a perforated tube and sealed at each the edges. The overall diameter is about 6 cm to 45 cm and length 75 cm to 150 cm.
- The feed liquid flows tangentially to the membranes after entering the cartridge. The permeate flows into the central tube and the concentrate exits the module at the other end.
- Separator screens are provided to create turbulence, which eventually helps in increasing the flux.

Salient Features

- The turbulent flow and the low volume of liquid with a large membrane area reduce the requirement for large pumps.

- The spiral wound system involves lower cost and is, thus, popular.

Applications

This type of module is used for dairy processing, potable water treatment, protein separation, whey protein concentration, brackish water treatment, seawater desalination, etc.

12.6.3 Membrane Concentration vs. Concentration by Evaporation

The general advantages and limitations of the membrane separation process over the conventional evaporation process are as follows.

Advantages	Limitations
• Membrane concentration is carried out at ambient temperature, thus eliminating the energy requirement for heating. The additional investments required for generating heat/steam for the evaporation process is not required. • The sensory and nutritional qualities of foods are better retained as the food is not heated. • As there is no heat requirement during the separation process and the maintenance requirement is negligible, it involves low operational cost and less labor.	• It involves higher capital investment than the conventional evaporation system. • Along with the applied pressure, the membrane pore size and the change in feed concentration affect the separation and the flux. • The maximum concentration that can be achieved by membranes is about 30% total solids, whereas the evaporators have no such limitation. • There may be cake formation on the membranes, which affect the capacity and efficiency.

12.6.4 Pervaporation

Pervaporation is a method where the separation of a liquid feed mixture is done by partial vaporization through a selectively permeable membrane. The partial vaporization is achieved by reducing the pressure on the permeate side. The vapor passes through the membrane as the permeate and the liquid remains

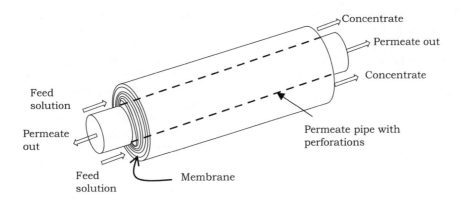

FIGURE 12.28 Flow pattern in a spiral wound membrane system

as the retentate. Polyvinyl alcohol and cellulose acetate are the hydrophilic polymers that permit water permeation. These can be used for reducing the alcoholic content of wines and beers. Polydimethyl siloxane or polytrimethyl silylpropyne are hydrophobic polymers, which are also used in pervaporation to concentrate aroma compounds.

Check Your Understanding

1. A centrifuge is used to separate solids from water having a density of 2050 kg.m^{-3}. The density and viscosity of water may be taken as 1000 kg.m^{-3} and 1 centipoise. The centrifuge has a bowl with r_C = 2.5 cm and r_A = 0.76 cm and height h_C = 20 cm. If the flow rate is maintained at 0.0035 m³.h^{-1} and N=24000 rev.min^{-1}, Determine the particles' critical diameter characterizing the exit stream.

   Answer: 0.0563 µm

2. What will be the minimum pressure required across the membrane during reverse osmosis to concentrate sucrose solution to 40% at 30°C?

   Answer: 5859.7 kPa.

3. What will be the osmotic pressure of salt solution made of one liter water and 10 g sodium chloride at 30 °C?

   Answer: 4.13 bar

4. What will be the centrifugal constant if the basket centrifuge of 0.5 m diameter is rotating with 2000 rev.min^{-1}?

   Answer: 1117.86

   (Hint: Centrifugal constant = v^2/gr)

5. Pineapple fibers are settling in its juice due to gravity. The fiber particles are of 100 µm diameter having a mass density of 1065 kg.m^{-3}. The density and viscosity of juice are 1020 kg.m^{-3} and 0.5 cP, respectively. Calculate the terminal velocity of the fiber.

   Answer: 4.91 ×10^{-1} mm.s^{-1}.

6. A high-speed tubular ultracentrifuge with a bowl radius of 100 mm and a height 500 mm rotates at 20,000 rpm and settles starch particles (average diameter of 20 µm) on the wall. What will be the ratio of centrifugal force to the gravitational force acting on the particle?

   Answer: 44760:1

7. How many folds would the g-number of a centrifuge increase by doubling both the spinning speed and bowl diameter?

   Answer: 8 folds.

Objective Questions

1. Explain the principles of mechanical separation for different types of fractions or impurities.

2. What are the factors those affect the rate of filtration?

3. Enumerate the forces acting on a particle during settling and determine the terminal settling velocity.

4. How does the filter help in improving the rate of filtration?

5. Explain the differences in the principles of separation in tubular bowl centrifuge and disk bowl centrifuge.

6. Explain the process of separation in a cyclone mentioning the forces acting on a particle.

7. Write notes on
   a. Centrifugal filtration;
   b. Filter press;
   c. Shell and leaf filter;
   d. Automatic belt filter;
   e. Rotary drum filter;
   f. Vacuum nutsch;
   g. Edge filter;
   h. Precoat filter;
   i. Rotary disc filter;
   j. Solid bowl clarifier;
   k. Conveyor bowl centrifuge;
   l. Basket centrifuge;
   m. Nozzle discharge centrifuge;
   n. Cyclone separator;
   o. Sterilizing filtration;
   p. Membrane materials.

8. Define (a) Sperry equation, (b) cross flow filter, (c) cycle time during filtration, (d) Stokes' law, (e) Drag coefficient, (f) Ultra centrifuge, (g) Dittus-Boetler equation, (h) selectivity of a membrane, (i) membrane permeability, (j) molecular weight cutoff, (k) pervaporation.

9. Differentiate between

   a. Medium resistance and specific cake resistance during filtration;

   b. Constant rate filtration and constant pressure filtration;

   c. Continuous filter and discontinuous filter;

   d. Cake filter and clarifying filter;

   e. Pressure filter and vacuum filter;

   f. Suspended batch centrifuge and automatic batch centrifuge;

   g. Rotary drum filter and cartridge filter;

   h. Pad filter and bag filter;

   i. Settling and sedimentation;

   j. Free settling and hindered settling;

   k. Settling tank and settling classifier;

   l. Reverse osmosis, nano-filtration, ultrafiltration and microfiltration;

   m. Tubular module, hollow fiber module and spiral wound module;

   n. Membrane concentration and concentration by evaporation.

BIBLIOGRAPHY

Ahmad, T. 1999. *Dairy Plant Engineering and Management* 4th ed. Kitab Mahal, New Delhi.

Brennan, J.G. (ed.) 2006. *Food Processing Handbook* 3rd ed. WILEY-VCH Verlag GmbH & Co. KGaA, Weinheim.

Brennan, J.G., Butters, J.R., Cowell, N.D. and Lilly, A.E.V. 1976. *Food Engineering Operations* 2nd ed. Elsevier Applied Science, London.

Castel, C. and Favre, E. 2018. Membrane separations and energy efficiency. *Journal of Membrane Science* 548: 345–357.

Charm, S.E. 1978. *Fundamentals of Food Engineering* 3rd ed. AVI Publishing Co., Westport, Connecticut.

Couto, C.F., Lange, L.C. and Amaral, M.C.S. 2018. A critical review on membrane separation processes applied to remove pharmaceutically active compounds from water and wastewater. *Journal of Water Process Engineering* 26: 156–175.

Dash, S.K., Bebartta, J.P. and Kar, A. 2012. *Rice Processing and Allied Operations*. Kalyani Publishers, New Delhi.

Earle, R.L. 1983. *Unit Operations in Food Processing* 2nd ed. Pergamon Press, Oxford.

Earle, R.L. and Earle, M.D. 2004. *Unit Operations in Food Processing*, Web Edn. The New Zealand Institute of Food Science & Technology, Inc., Auckland. https://www.nzifst.org.nz/resources/unitoperations/index.htm

Fellows, P.J. 2000. *Food Processing Technology*. Woodhead Publishing, Cambridge, UK.

Geankoplis, C.J.C. 2004. *Transport Processes and Separation Process Principles* 4th ed. Prentice-Hall of India, New Delhi.

Glover, F.A. 1971. Concentration of milk by ultrafiltration and reverse osmosis. *Journal of Dairy Research* 38: 373–379.

Heldman, D.R. and Hartel, R.W. 1997. *Principles of Food Processing*. Aspen Publishers, Inc., Gaithersburg.

Heldman, D.R. and Singh, R.P. 1981. *Food Process Engineering*. AVI Publishing Co., Westport, CT.

Hemfort, H. 1984. *Centrifugal Clarifiers and Decanters for Biotechnology*. Westfalia Separator AG, 4740 Oelde 1, Germany.

Kessler, H.G. 2002. *Food and Bio Process Engineering* 5th ed. Verlag A Kessler Publishing House, Munchen, Germany.

Leniger, H.A. and Beverloo, W.A. 1975. *Food Process Engineering*. D. Reidel, Dordrecht, 498–531.

Lewis, M.J. 1996. Ultrafiltration. In: A.S. Grandison and M.J. Lewis (eds) *Separation Processes in the Food and Biotechnology Industries*. Woodhead Publishing, Cambridge, 97–140.

McCabe, W.L., Smith, J.C. and Harriot, P. 1993. *Unit Operations of Chemical Engineering* 5th ed. McGraw-Hill, Inc., New York.

Porter, M.C. 1979. Membrane filtration. In: P.A. Schweitzer (ed.) *Handbook of Separation Techniques for Chemical Engineers*. McGraw Hill Book Co., New York.

Prabhudesai, R.K. 1979. Leaching. In: P.A. Schweitzer (ed.) *Handbook of Separation Techniques for Chemical Engineers*. McGraw-Hill Book Co., New York.

Saldaña, M.D.A., Gamarra, F.M.C. and Siloto, R.M.P. 2010. Emerging technologies used for the extraction of phytochemicals from fruits, vegetables, and other natural sources; In L.A. de la Rosa, E. Alvarez Parrilla and G.A. Gonzalez-Aguilar (eds) *Fruit and Vegetable Phytochemicals: Chemistry, Nutritional Value, and Stability*. Blackwell Publishing, Iowa, 235–270.

Singh, R.P. and Heldman, D.R. 2014. *Introduction to Food Engineering*. Academic Press, San Diego, CA.

Sourirajan, S. 1970. *Reverse Osmosis*. Logos Press, New York.

Strathmann, H. 1981. Membrane separation processes. *Journal of Membrane Science* 9(1–2): 121–189.

Toledo, R.T. 2007. *Fundamentals of Food Process Engineering* 3rd ed. Springer, New York.

Watson, E.L. and Harper, J.C. 1989. *Elements of Food Engineering* 2nd ed. Van Nostrand Reinhold, New York.

Yang, Z., Zhou, Y., Feng, Z., Rui, X., Zhang, T. and Zhang, Z. 2019. A review on reverse osmosis and nanofiltration membranes for water purification. *Polymers* 11: 1252. doi:10.3390/polym11081252

13

Extraction

The dictionary meaning of extraction is to remove some material of interest from a solid or liquid matrix. The material of interest may itself have economic value or its extraction increases the economic value of the remaining feed material or both the extract and the remaining feed have increased economic value. For example, lycopene extracted from the tomato peel is a valuable biomolecule for pharmaceuticals. However, when toxins are extracted out of soybeans, the remaining material after extraction has enhanced food value. When caffeine is extracted from coffee beans, not only the caffeine but also the decaffeinated coffee beans have enhanced economic values. Extraction, as a separation process, has found applications in such diverse fields as mining, metallurgy, chemical engineering, medicine, food, and agriculture.

13.1 Methods and Principles

13.1.1 Methods of Extraction

In food processing, extraction is applied on the raw materials and/or intermediate products to improve the quality of the finished products. It is defined as the process where one or more components from a feed material are removed with the use of a fluid phase followed by the recovery of the separated components from the fluid phase. Obviously, a fluid is used as the agent to extract the targeted components. The feed material could be a solid or a liquid matrix. Extraction is a preferred process in many food operations because of its simplicity and low energy expenditure. Thermally sensitive components can be conveniently extracted in comparison to some other separation techniques.

The edible oil industry has been a major user of extraction technology to extract oil from oilseeds and other feeds. Soybean, corn, and rice bran contain less than 20% oil and mechanical expression does not yield adequate oil. Therefore, these oilseeds are suitably prepared and subjected to solvent extraction for satisfactory oil recovery. Groundnuts have about 45% oil and a major fraction of this oil is obtained through mechanical expression leaving about 6–7% oil in the cake. This cake can now be subjected to solvent extraction to recover almost 90% of the oil in the cake. Extraction is the main process to obtain essential oils and flavors from spices and herbs, pigments from horticultural products, caffeine from tea and coffee, and many other similar applications. Extraction finds applications in several domains, such as fermentation, pharmaceuticals, fragrances, food products, and agriculture chemicals. A few examples of industrially practiced extraction processes are given in Table 13.1.

The feed may be a solid, a semi-solid, or a liquid. The solute in the context of the extraction could either be a solid or a liquid. If feed is a solid, the extraction is termed as *solid-liquid extraction* or *leaching*. If the feed is a liquid, the extraction is called *liquid-liquid extraction* or *solvent extraction* or *partitioning*. Liquid-liquid extraction is based on the solute's solubility being different in the two liquids, which themselves are immiscible.

Extraction can be either a batch or a continuous process. The continuous extraction processes could use either co-current or counter-current flow of the materials.

Since extraction in general is carried out at low temperatures, it is used for extracting heat-sensitive or non-volatile components.

13.1.2 Principles of Different Extraction Methods

Extraction is a composite process consisting of mixing, diffusion, phase equilibrium, and solubility.

Molecular diffusion coefficients, solubility, and mixing define the rate at which the process occurs, and the equilibrium is reached. Solute molecules move in a process of diffusion through a continuum in one phase as explained by Fick's first law of diffusion. In addition, the molecular movement could also be through the interface between the two phases or into the solid. For the solute to dissolve in the solvent, the solute must diffuse out of the solid into the solvent phase. The time period needed to achieve equilibrium between the phases is determined by the rate of diffusion. The theory of mass transfer using Fick's principles can very well be used to analyze the solute transfer. One result of the analysis clearly indicates that time for solute transfer can be reduced by increasing the surface area of the feed. However, the size should not be too small to create the problem of separation.

The final extract should have the highest solute concentration approaching the saturation concentration. Therefore, the solvent to feed ratio is maintained such that the equilibrium reached is lower than the saturation concentration.

Even in supercritical fluid extraction, the supercritical solvent is recycled for maximizing the extraction. However, the desired degree of solute recovery is achieved in fewer cycles because of high solute solubility of the supercritical solvent.

Besides the selection of the right solvent, there are several considerations to successfully and efficiently carry out the extraction. The objective is to extract as much solute from the feed as possible with minimum time, amount of solvent, and operating expenses. This requires that the operating conditions should be tailored to achieve the objective while maintaining

TABLE 13.1

Examples of extraction processes in food industries

Feed material	Solvent	Product	Co-products
Crushed sugarcane	Water	Sugar	Bagasse, molasses
Roasted and crushed coffee beans	Hot water	Decaffeinated beans	Caffeine
Flaked soybean	Hexane	Soybean oil	De-oiled soymeal
Green coffee beans	Supercritical CO_2	Decaffeinated coffee	Caffeine
Ginger rhizomes	Supercritical CO_2	Ginger extract	Gingerol
Egg yoke	Supercritical CO_2	De-cholesterolized Egg yoke	Cholesterol
Hops	Supercritical CO_2	Hops extract	Hops essential oils

the economics and sustainability. The time duration for the solute transport depends on the solute and solvent properties, feed characteristics, agitation, and temperature. Obviously, increasing the surface area and temperature would accelerate the extraction process.

When the quantity of solvent is adequate for the solid to satisfy the solute solubility requirement, solute concentrations in both the solid and the solvent phases are equal. In other words, the solution adhering to the solids and the liquid or solvent phase will have the same concentration. However, when the amount of solvent is not sufficient to dissolve all the solute present in the feed, equilibrium is considered to have been achieved when no further changes in solute concentrations in either phase occur even with prolonged contact duration. Sufficient contact time between the two phases needs to be provided for the equilibrium to occur. Estimation of the number of extraction stages requires the determination of the equilibrium relationships between the feed and the solvent for any set of flow conditions and the degree of separation.

It is based on laboratory data that the equilibrium conditions are usually determined because measurements are more accurate than the predictive equations-based values. There are analytical methods also to determine the equilibrium conditions. The correlation of equilibrium data in terms of *activity coefficients* can be obtained from laboratory measurements using semi-empirical equations available for this purpose. In the absence of experimental data, however, one must turn to purely predictive models.

It is useful to define certain terms frequently mentioned in relation to extraction. The raw material from which one or more components need to be extracted is called *feed*. The material that needs to be extracted is termed *solute* and the material that is added to the feed for the extraction is called the *solvent*. The solvent forms a phase different from that containing the solute. The two phases for extraction could be either solid – liquid or two immiscible liquids. The solvent together with the solute, when the solute from the feed moves to the solvent, is called the *extract*. The feed bereft of the solute is now called the *raffinate*. As such, the concept of extraction is quite simple; bring the feed and the solvent in contact, allow time for the extraction to occur, separate the extract from the raffinate, and separate the solute from the solvent, if necessary.

Conservation laws of mass and energy need to be observed for each of the extraction stages. In the case of steady-state processes there is no accumulation.

13.1.3 Determination of Number of Stages Required for a Process

Energy and Material Balance Method

The schematic representation of a typical leaching process in a single stage is shown in Figure 13.1.

Assume that a leaching process consists of the extractable solute (F_e), the liquid solvent (S), and the insoluble matrix (F_{in}). Thus, if feed and extract are denoted as F and E,

$$F = F_e + F_{in} \tag{13.1}$$

$$E = F_e + S \tag{13.2}$$

Let Y be the fraction of the extractable solute left in the solid matrix (slurries) (the same as X, the fraction of the extractable solute in the extract), and N is defined as the insoluble solid component as a fraction of the extract, then

$$Y = F_e/E \text{ and } N = F_{in}/E \tag{13.3}$$

In a situation where the quantity of insoluble solids in the feed is zero, N will be zero. If the amount of solvent is zero, then the value of N is equal to B/C and Y = 1.0.

As indicated earlier, it is not possible to achieve almost complete extraction in a single stage, in a finite time with finite solvent. Therefore, the process is carried out in multiple stages. The multistage extraction processes could be either continuous or semi-continuous. The slurry and the liquid streams could

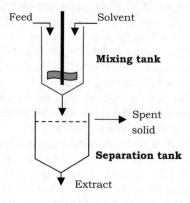

FIGURE 13.1 Schematic representation of a typical single-stage leaching process

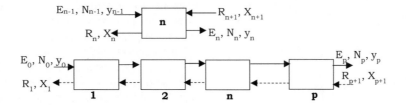

FIGURE 13.2   Material flows in a multistage leaching process

be either counter-current or co-current. The most effective flow arrangement has been found to be counter-current for the extraction process.

For a counter-current leaching operation with a number of stages, the different components and their fractions will be as shown in Figure 13.2.

In this situation,

$$E_{n-1} + R_{n+1} = E_n + R_n \qquad (13.4)$$

$$\text{or } E_{n-1} - R_n = E_n - R_{n+1} \qquad (13.5)$$

If there are p stages in a leaching process, the simultaneous algebraic equations can be expressed as follows.

$$E_0 - R_1 = E_1 - R_2 = E_2 - R_3 = \ldots = E_n - R_{n+1} = \ldots$$
$$= E_p - R_{p+1} = \text{Constant} = \Delta \qquad (13.6)$$

The inherent assumption in writing the above equations is that the flows into and out of a stage are in equilibrium, i.e., for any given stage, the extract and the solution in the slurry leaving the stage are in equilibrium or $Y^* = X$ (equilibrium concentration fraction is indicated by Y^*).

Stage-by-stage material balance for a multistage extraction can be carried out to determine the number of stages. Because the conditions of extraction are known only at the entrance and the outlet, the stage-by-stage material balance will give rise to a set of simultaneous material balance equations. The set of equations is then solved to determine the solute concentrations in the underflow and the overflow leaving each stage. Each stage is assumed to be ideal to solve the equations, comparing the calculated solute concentrations with the specified values. Alternatively, a graphical method can be employed to determine the unknown quantities.

Graphical Method

Consider a counter-current extraction process requiring p stages. The function $N = f(y)$ is assumed to be known. The equilibrium function $Y^* = f(X)$ is assumed to be known. The extracts do not contain any solid particles. The solid and the solvent compositions at the beginning of the process are specified and the expected solute recovery is also given. Evaluate p, i.e., the number of extraction stages, for the extraction process.

The steps involved in a graphical method, known as the *Ponchon–Savarit diagram,* used for finding the solution are as follows.

a)   Draw a slurry line, i.e., a line representing $N = f(y)$. It is also termed the *overflow line.*

b)   Locate the four points on this line for the two entering and two exiting streams, i.e., E_0, E_p, R_1, and R_{p+1}.

   (The point representing $E_0 - R_1 = \Delta$ lies on the line connecting E_0 and R_1. Similarly, Δ lies on the line connecting E_p and R_{p+1}. The point Δ then lies at the point where these two lines intersect.)

c)   The point E_1 is found on the slurry line from the point R_1 because of the equilibrium condition. R_2 is located on the x, y line found by connecting the E_1 with Δ. For clear extracts, $N = 0$.

d)   The two-step construction is repeated until R_{p+1}; the number of such two-step constructions obtained will be the number of stages required for the leaching.

The above graphical method has been illustrated using the following example.

Example 13.1 A variety of microbe produces 0.15 g of pigment from 1.0 g of dry microbial mass. A solvent has been used to extract the pigment from the microbial biomass. It is expected that 90% of the pigment could be recovered using a counter-current multistage leaching process. The recommended ratio of solvent to feed is 1.0. It has been determined that, after draining, 0.6 g of liquid is retained per gram of lycopene-free fungus tissue. The extracts do not contain any traces of the biomass. How many stages are required?

SOLUTION:

Using the above-mentioned information, we will construct the Ponchon-Savarit diagram (Figure 13.3). Since for all x and y, $N = 1.0 / 0.6$, the slurry line is $N = 1.667$. The extract line is $N = 0$ since the extracts do not have any biomass traces.

The feed has no solvent, therefore, $y = 1.0$ and N_0 is 6.7 (1.0 / 0.15), that represents E_0.

At stage 1, clear solvent is fed, indicating $N_p = 0$ and $x = 0$, establishing point R_{p+1}.

One kg of feed contains 0.15 kg pigment and 90% of the pigment is expected to be extracted through leaching. Thus, 0.135 kg pigment will be extracted and 0.015 kg will remain with the spent feed. The solvent used is 1.0 kg.

Therefore, the amount of raffinate is equal to 0.85 + 0.85 × 0.6 = 1.36 kg and the clear extract is 2.0 − 1.36 = 0.64 kg.

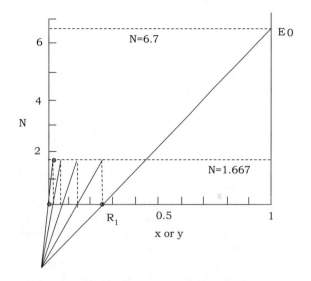

FIGURE 13.3 Determination of number of extraction stages through graphical method

$$Y_p = (0.15 \times 0.10) / (0.85 \times 0.6) = 0.0294$$

$$X_1 = 0.135 / 0.64 = 0.211$$

The four points E_o, E_p, R_1, and R_{p+1}, plotted on the graph, represent the streams entering and exiting a stage. The difference point Δ has been marked where the straight lines, joining points E_o with R_1 and R_{p+1} with E_p, intersect. Use $Y^* = X$ to mark the equilibrium stages. The results indicate that 4 to 5 stages would be required to achieve the desired level of leaching.

13.1.4 Stage Efficiency

Stage efficiency reached in an extraction stage is defined as the ratio of concentration of solute in the solvent actually achieved in a stage and the equilibrium concentration of solute in the solvent phase. The efficiency, also called *Murphy efficiency*, for the extraction process is represented as follows:

$$\eta_M = \frac{X_0 - X}{X_0 - X^*} \tag{13.7}$$

where, X is the solute concentration in the extract leaving the stage, X_0 is the solute concentration in the extract entering the stage, and X^* is the equilibrium concentration of the solute.

An extraction stage is designated as an "ideal stage" if equilibrium is reached in the stage, i.e., it is 100% efficient.

The overall efficiency, achieved in a multistage extraction process, is also the Murphy efficiency and can be estimated as follows:

$$\eta_M \approx \eta_{overall} = \frac{N_{theoretical}}{N_{real}} \tag{13.8}$$

13.1.5 Unsteady Mass Transfer in a Stage

Each single stage in the overall leaching process can be analyzed as a transient phenomenon and the time for the mixture

to approach equilibrium could be determined. It is also equivalent to a batch process where an agitated mixing vessel is used to mix solids with the solvent, followed by separation of solids from the liquid.

If M and C are the mass of the solute transferred in the solution and the solute concentration in the bulk liquid in time θ and V is the volume of the mixture, then

$$M = CV$$

$$dM = V \, dC$$

$$\frac{dM}{d\theta} = V \frac{dC}{d\theta} = K_L A (C_S - C) \tag{13.9}$$

where, K_L is the mass transfer coefficient in the liquid phase, A is the surface area of the solids participating in the mass transfer and C_S is the solute concentration at the solid surface.

$$\text{Thus, } \frac{dC}{d\theta} = K_L \frac{A}{V} (C_S - C) \tag{13.10}$$

Integration of the differential equation yields the following equation for the solute concentration at any time θ.

$$\int \frac{dC}{C_S - C} = \frac{K_L A}{V} d\theta$$

$$\ln \frac{C_S - C}{C_S} = \frac{K_L A}{V} \theta$$

$$\frac{C_S - C}{C_S} = e^{-\frac{K_L A}{V} \theta}$$

$$C = C_S \left(1 - e^{-\frac{K_L A}{V} \theta} \right) \tag{13.11}$$

Example 13.2 A solid matrix weighing 500 kg contains 22.5% of the water-soluble component by mass. It is agitated with 100,000 liters of water for 500 s. The raffinate contains 25% of the solution after each decanting. The saturation concentration of water with the solute is 2.3 kg.m^{-3}. Find the concentration of the solute in the solution after the leaching. In a preliminary trial using a vessel of 1000 liter volume, a solute from an inert solid leached for 10 s and the resulting saturation of water was 70%. It is assumed that the conditions are equivalent to those in the trial.

SOLUTION:

From the data for the trial, the value of the mass transfer coefficient, $K_L.A$, is evaluated. This value of $K_L.A$ is then utilized for evaluating the unknown concentration at the end of the full-scale leaching operation.

For the trial, V = 1000 liters = 1.0 m^3, θ = 10 s, C$_s$ = 2.3 kg.m^{-3} and C = 0.7 C$_s$

$$C = C_S\left(1 - e^{-\frac{K_LA}{V}\theta}\right), \text{or } 0.7 \times C_S = C_S\left(1 - e^{-\frac{K_LA}{1}10}\right)$$

or, K$_L$.A = 0.12

For the full-scale leaching operation, V = 100,000 liter = 100 m^3, θ = 500 s

$$\text{Thus, } C = (2.5)\left(1 - e^{-\frac{0.12}{100} \times 500}\right) = 1.12 \text{ kg.m}^{-3}$$

The initial amount of the solute in the solid = (500) (0.225) =112.5 kg

The maximum concentration of the solute that can be removed = 112.5 / 100 = 1.125 kg.m^{-3}.

It indicates that the extraction process is highly efficient.

13.2 Leaching (Solid-Liquid Extraction)

Leaching is defined as a mass transfer process when a liquid solvent is used to recover a desired component from a solid matrix. The process of leaching happens when a tea bag or coffee beans are put in a cup of hot water or when crushed sugarcane is added with hot water. In these cases, the material of interest is embedded in a solid. To take this material of interest out of the solid, a liquid is mixed with the solid mass and a certain amount of time is allowed for the material of interest to leave the solid mass and become a component of the liquid. If required, the liquid solution could be further processed to obtain pure material of interest such as sugar in the case of sugarcane. Else, the liquid solution itself, tea and coffee, can be the final product.

Clearly, the solvent must have a high solubility for the solute. Besides, the solid needs to have a large surface area for the mass transfer to occur faster and there should be sufficient time for the process to conclude. Leaching is sometimes also referred as *percolation*. *Lixiviation* is the term used when alkali is leached out of wood ashes. *Elutriation* is specifically used when the solute is present on the surface of an otherwise insoluble solid and the solvent just washes it off the solid surface. *Decoction* is the process of leaching when the solvent is used at its boiling temperature. When leaching is carried out to remove an undesirable material from the solid, it is better known as *washing*. An example is to wash away the lactose and other solutes from cheese curd.

13.2.1 Principle

As discussed, due to several limitations in respect of perfect mixing of the feed and the solvent, diffusion coefficients, and solubility constraints, leaching may not be complete in a single stage. Therefore, a multistage process is required to maximize the solute recovery.

The following affect the effectiveness of the process.

1. Solid structure;
2. Solid particle size and its distribution;
3. Solute solubility of the solvent and its dependence on temperature;
4. Diffusion behavior of the solute from the solid to the solvent.

Like conduction and convection heat transfer, the diffusion rate of the solute from within the solid to its surface may be the slower process as compared to the mass transfer from the solid surface to the solvent.

The solute, during leaching, moves from the solid matrix to the solvent through (1) molecular transport within the solid, and (2) convective transport from the solid surface to the bulk solvent. Each leaching stage consists of bringing the two phases in contact for a specific time duration to facilitate the solute transfer and then the two phases are separated.

The net molecular transport in the process ceases to exist once the equilibrium is reached between the solid and the liquid phases, i.e., the chemical potential is the same everywhere. It would take a very long time to reach true equilibrium. However, for practical purposes, equilibrium is assumed to exist after a finite duration in the process. The real process is assumed to be a number of equilibrium stages and represents the deviations between the theoretical and actual processes through efficiency factors obtained experimentally.

13.2.2 Equipment

Leaching could be an *unsteady* or a *steady-state* process. Unsteady state leaching processes are those where the state of equilibrium is not achieved. Such situations normally occur in mining, metallurgical, and many chemical processes. In the food processing domain, it is essentially a steady state continuous leaching process. In batch leaching processes, carried out at small scales, transient conditions are assumed and the time durations required for effective leaching are determined. Continuous leaching processes are further classified as co-current and counter current processes. Counter current leaching processes are sub-divided as variable underflow and constant underflow processes.

Different multistage leaching methods differ in the way the solids and the solvents flow through the stages. The main multistage mechanisms are as follows.

Fixed Bed Extractor

As the name indicates, the solid beds are fixed. Consider a battery of six columns containing the solid matrix, also called percolators. Consider the solid matrix to be roasted and ground coffee beans.

In the beginning of the leaching process, though all the percolators are filled with the solid matrix, only five percolators are used and the sixth one is kept as a stand-by bed. Solvent, i.e., hot water, is added to bed no. 1 and moves down in the coffee bed, extracting the coffee. The system is operated at high pressure in accordance with the high extraction temperature. The extract from the bed is pumped to the next bed for further enrichment with the solute, and the process goes on in the sequence 1-2-3-4-5 through the beds. Clearly, the extract received from the fifth bed is the most concentrated with the coffee.

When the contents of the bed no. 1 are thoroughly exhausted, it is disconnected and the bed no. 6 at the end of the battery, containing fresh feed, is connected. The fresh solvent is added to the bed no. 2 and the flow sequence of the extract becomes 2-3-4-5-6. The bed no. 1 is emptied, cleaned, and refilled with fresh material. In the next stage, the bed no. 2 is disconnected and the liquid flow sequence becomes 3-4-5-6-1, and so on. This mechanism of sequentially shifting the extract flow through the battery of beds effectively simulates the counter-current effect without moving the solids from one bed to another. Thus, it is often categorized as counter-current flow extractors.

Belt Extractor

The solid matrix with appropriate size reduction is continually fed through a hopper on a perforated slowly moving belt. Liquid is sprayed on the belt in such a way that the fresh solvent is applied on the last section of the belt before the spent solids are discharged. The extract from this section is sprayed on the belt section just before the last section. In other words, the extract sprayed at the belt section containing fresh solid matrix is received from the next belt section. The most concentrated extract is thus received by the fresh solid matrix and is called full miscella. The belt extractors are high-capacity extractors. These could be two stage extractors where two belts exist in a series.

Carrousel Extractor

The main component of the carrousel extractor is a vertical cylindrical chamber with a perforated floor, in which there is a slowly revolving rotor. The rotor has segments created by radial partition walls and containing the solid material from which the solute needs to be extracted. The solvent flows from the top to the bottom through the solid matrix and is collected at the bottom for recirculation through the solid bed at the next section.

Auger Extractor

A screw conveyor carries the solid matrix up in a cylindrical enclosure and the solvent flows down the cylinder. The cylinder and the screw could also be in an inclined position.

Bucket Extractor

The material containing the solute is filled in the buckets that have perforated bottoms. These buckets move through the solvent stream either horizontally or vertically. In the case of a bucket elevator, there is vertical movement of buckets and the solvent flows down through the buckets. The collection of the extract is at the bottom.

13.3 Liquid-Liquid Extraction

13.3.1 Principle

The proposition of pulling a component from the feed solvent and adding it to the extracting solvent is termed *liquid-liquid extraction*. It is assumed that the two solvents are not miscible. There are compounds that are very poorly miscible in organic solvents. But water is a good solvent for such compounds. Such compounds could be extracted into organic compounds by repetitive extraction using a liquid-liquid extractor.

Extraction of a solute from a two component solution is accomplished by mixing the solution with a solvent that is immiscible in the feed liquid mixture but the solute is more soluble in the solvent. This is *single-contact batch extraction*.

On an industrial scale, the common practice is to use more than one stage for the extraction and is normally carried out on a continuous basis.

The equipment consists of discrete mixers and settlers. Alternatively, some forms of column contactors, where the feed and solvent phases flow counter-currently by virtue of the density difference between the phases, may be employed. An adequate settling volume needs to be provided for complete phase disengagement when it is time for final settling or phase separation under gravity.

As discussed, the result of an extraction operation consists of spent feed slurry, more usually termed the raffinate, and the solvent containing extracted solute, termed the extract.

There are four separate requirements for carrying out an efficient extraction:

a) Dispersion of solvent into the feed;
b) Creation of large interfacial area for diffusion from one phase to the other;
c) Allowing an acceptable level of solute diffusion by providing adequate holding or retention time for the diffusion to take place;
d) Separation of the raffinate and the extract.

Thus, steps involved in a liquid-liquid extraction include (1) mixing (contacting) of the feed and extracting liquids and (2) phase separation. The selection of solvent and the operating conditions need to be considered in both the steps indicated above. A case in point is that of vigorous mixing that is carried out to make the extraction faster. However, the formation of emulsion impairs the phase separation.

Molecular and/or eddy diffusional mechanisms govern the complex process of diffusion of a solute from one liquid phase to the other. Mass transfer flux is proportional to the driving force, instantaneous concentration in this case. The ratio of mass transfer flux and the driving force is termed the mass transfer coefficient, in terms of either the continuous or the dispersed phase driving force.

The extraction for neutral (i.e., neither acidic nor basic) organic compound requires simply carrying out the extraction repeatedly using an organic solvent. In the case of acidic compounds, an aqueous base is used to selectively deprotonate the compound for its release into the aqueous phase. There are only a few organic compounds that are soluble in water, the remaining organic phase containing any by-products and/or unreacted starting materials could be discarded. The desired product can be precipitated by acidification of the aqueous phase. If the product is basic, a sequence very similar to the acidic compound can be performed. The basic compound can be protonated by using an aqueous acid, releasing the protonated compound into the aqueous phase, and discarding the organic phase. The basic compound will be deprotonated through neutralizing the acidic phase leading to precipitation. In some cases, second extraction may be required using an organic solvent.

As discussed, liquid-liquid extraction is typically carried out using an aqueous phase (either pure water or aqueous solution) and an organic phase. The desired compound (an organic molecule) can theoretically be in either phase depending on the nature of the compound and the nature of the aqueous phase.

Choice of Solvents

The solvent should be such that it is selective for the solute being extracted. Besides, it should have good recoverability in addition to being of low cost, noncorrosive, non-inflammable, and with low solubility in the feed phase. The ease of attracting the solute from the feed stream to the solvent reflects the selectivity of the solvent. To avoid subsequent phase separation, the interfacial tension between the two phases should be sufficiently low. Density difference between the two phases should be sufficiently large such that counter-current flow of the phases could be maintained under gravity. Diethyl ether is the most commonly used solvent in liquid-liquid extraction. Ethanol and acetone mix with water and, therefore, are not recommended as solvents in liquid-liquid extractions. Some other recommended solvents are hexane, diethyl ether, dichloromethane, and toluene.

Equilibrium is reached after mixing and it is indicated by equal chemical potentials of the extractable solute in the two liquids. If C_1 and C_2 are the equilibrium concentrations of the solute in the two liquids, a distribution factor K is defined as follows.

$$K = (C_1/C_2) \tag{13.12}$$

Solutes relative preference is represented by the distribution coefficient for the two solvents. In the context of the extraction process, the chemical potential and concentration for an ideal solution are proportional. Under this assumption, the distribution coefficient at a given temperature is independent of the concentration and, thus, practically constant.

13.3.2 Equipment

Contactors bring the feed and solvent phases in intimate contact to let the transfer of the solute from one phase to the other occur at a fast pace. A contactor may just be a stirred tank connected to a settling chamber, termed a *mixer-settler*. A number of mixer-settler units may be assembled to form a multistage arrangement to obtain the required degree of extraction.

13.4 Supercritical Fluid Extraction

Supercritical fluid extraction (SFE) uses supercritical fluids as the extracting solvent for separating the extractant from the feed matrix. The extraction temperature and pressure, having a direct bearing on fluid density, transport properties, extraction yield, and extraction time, can be controlled.

Supercritical fluid extraction may be carried out in high-pressure equipment in a batch or continuous manner on both solid and liquid feeds with or without recycling of the solvent. The supercritical fluid as the solvent is brought in contact with the feed from which a desirable product is required to be extracted. The supercritical fluid saturated with the extracted product is subsequently expanded to atmospheric conditions and the solubilized product is recovered in the separation vessel. The supercritical fluid could be recycled for further use. Another method to separate the solute from the SCF is to reduce the temperature. The SCF could be reheated for recycling without the need for recompression.

13.4.1 Principle

Supercritical Fluid

The formation of a supercritical fluid is the result of a dynamic equilibrium. A single-component fluid becomes supercritical when its temperature and pressure exceed its critical temperature and pressure, respectively. The phase diagram for a pure material (Figure 13.4) shows how the material exists in the form of solid, liquid, and gas at different temperature and pressure values. The solid line delineating the solid-gas regions on the left indicates sublimation of the solid into gaseous form. There exists a point where all three phases exist in equilibrium and this unique point is called the triple point.

There exists a situation called *critical point*, beyond which the fluid cannot change from a gas to a liquid or from a liquid to a gas phase. This condition is called the supercritical condition for the fluid. Increasing temperature cannot result in turning to gas, and increasing pressure cannot result in turning to liquid at this point.

The critical point is also defined by a critical temperature (CT) and critical pressure (CP). Any further increase in pressure or temperature, beyond the critical point, does not cause any change in the phase of the fluid. In other words, in the phase diagram, the region above the CT and CP values is defined as the *supercritical region*.

Properties of Supercritical Fluids

The characteristic properties of a supercritical fluid (SCF) are density, diffusivity, and viscosity. Supercritical values for these properties lie between liquids and gases. For example, carbon dioxide becomes a supercritical fluid above its critical temperature (31.1°C) and pressure (73 bar) with the density closer to that of liquid. It exhibits properties that lie between liquid and gaseous phases of CO_2. The density of a supercritical fluid is almost that of a liquid, but it is not a liquid. In addition, the solubility of solutes in a supercritical fluid approaches the solubility in a liquid. Thus, the principle of solute extraction from solids using a supercritical fluid is very similar to that

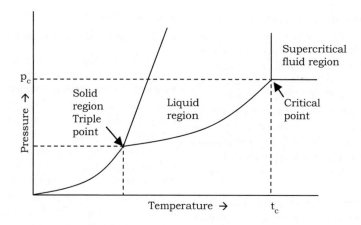

FIGURE 13.4   An idealized phase diagram

for solid-liquid extractions. Supercritical fluids have negligible surface tension and viscosity, exhibit gas-like viscosities and diffusion coefficients, and can easily penetrate into the solid matrix. Table 13.2 gives the critical properties of some solvents used in supercritical fluid extraction (Ruhan et al. 2009).

The most commonly used SCF is carbon dioxide because it is a readily available, inexpensive, non-flammable, and non-toxic material. In some cases, the addition of ethanol in small amounts changes the polarity of supercritical carbon dioxide to improve extractions. One of the most important advantages of using SCFs as a solvent in extraction is the possibility of selective extraction of compounds.

Supercritical solubility of a substance is given as the mole fraction or the concentration (w/w) of the substance in the supercritical phase at equilibrium with the pure fluid. Solute solubility in SCF can be modified by the use of entrainers. Lecithin solubility, for example, is enhanced when ethyl alcohol is used as an entrainer with supercritical carbon dioxide. There are various co-solvents and modifiers, such as isopropyl alcohol and ethanol, n-hexane, heptane, pentane, toluene, methanol, acetone, and formic acid with ammonium formate in methanol. A supercritical carbon dioxide-ethanol mixture has shown higher solubilities for limonoids from citrus seeds and phosphatidyl choline from dried egg yolk.

Pressure and temperature generally influence the solubility of solutes in the supercritical solvent. Therefore, the same super-critical fluid at a different temperature-pressure combination can extract a different compound. A supercritical fluid extraction (SCFE) process may seek to separate multiple

solutes from a solid matrix. A two-stage process may be used to extract more than one group of compounds present in a solute. The SCF may easily dissolve one group of compounds under a given set of conditions, while a group of the less soluble compounds is left back in the solid matrix. The SCF under a different set of conditions, favorable for the dissolution of the remaining compounds in the solid, may now be used for their extraction.

Critical Points of Mixtures

Critical temperature and pressure of a mixture of n compounds may be estimated as weighted averages of mole fractions of the n compounds present in the mixture.

$$T_c = (x_1 T_{c1} + x_2 T_{c2} \ldots + x_n T_{cn})/n \qquad (13.13)$$

$$P_c = (x_1 P_{c1} + x_2 P_{c2} \ldots + x_n P_{cn})/n \qquad (13.14)$$

13.4.2 Equipment

Construction and Operation

- A SCFE system is depicted in Figure 13.5. It basically consists of a pump to force the super critical fluid through the sample, a chamber for holding the sample, and a collecting vessel. The sequence of operations are as follows:

TABLE 13.2

Critical properties of some solvents

Solvent	Molecular weight (g.mol^{-1})	Temperature (K)	Pressure (MPa)	Density (g.cm^{-3})
CO$_2$	44.01	304.1	7.38	0.469
Water	18.02	647.3	22.12	0.348
Methane	16.04	190.4	4.60	0.162
Ethane	30.07	305.3	4.87	0.203
Ethanol	46.07	516.2	6.38	0.276
Propane	44.09	369.8	4.25	0.217
Hexane	86.18	507.4	3.03	0.233
Butane	58.12	425.2	3.80	0.225

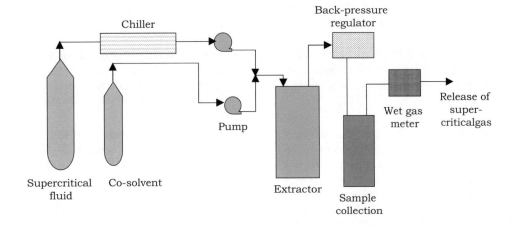

FIGURE 13.5   Schematic diagram of a supercritical fluid extraction system

- The liquid CO_2 is metered by flow meter and then pumped.
- After pumping the liquid is heated (by an in-line heat exchanger) to attain supercritical conditions. Co-solvents may also be treated in the similar manner.
- The supercritical fluid then goes into the extraction vessel and dissolves the targeted material within the solid matrix.
- The SCF then passes into a separator at lower pressure, where it turns into a gas. The reduction in the SCF pressure is achieved by throttling the flow through an orifice or a needle valve before the separator tank. The solute separates from the gas and settles out since the solute has much lower solubility in the gas as compared to the liquid SCF. Temperatures in the process are maintained and excessive cooling at the throttling valve, due to the Joule-Kelvin effect, is prevented with the help of heat exchangers.
- A series of vessels at reducing pressures can be used to further fractionate the collection.
- The supercritical fluid can be recycled after cooling and recompressing or can be discharged into the atmosphere.
- Charging and emptying the extractor is a batch operation.
- The carbon dioxide is maintained at less than 5°C in liquid form and the pressure is about 50 bars for pumping. At laboratory-scale, reciprocating or syringe pumps are used. Diaphragm pumps are common for large scale extractions.
- Another pump is often provisioned to deliver co-solvents or entrainer, such as food grade ethanol, for faster extraction and automated cleaning of system fluidics.
- The sample holding chamber, which is also a pressure vessel, can be a simple tubing or large vessel equipped with suitable fittings. The vessel equipped

with a heating device is usually supposed to be under 74 bars to 800 bars.
- The vessel incorporates a heating device as there is significant cooling due to the adiabatic expansion of the CO_2.
- A capillary tube, cut to length, or a needle valve is used in small systems to accurately maintain the pressure at different flow rates in the vessel. A back-pressure regulator is provided in large systems.

Salient Features

- The major SCF extraction process parameters are SCF flow rate, pressure, and temperature in the extraction vessel.
- Temperature levels in both the tanks are controlled.
- The fluid must be at low temperature before pumping to remain as liquid. Then it is pressurized to achieve supercritical conditions. Further, as mentioned above, the expansion of fluid in the separator causes adiabatic cooling and, thus, heating is required to prevent excessive cooling. In view of this, there are cooling and heating requirements at different sections, for which proper devices are integrated.
- Extraction efficiency of a SCFE process could be optimized by selecting an appropriate temperature and pressure combination.
- The following are the major parameters in deciding the efficiency of the process.
  - Extraction temperature;
  - *Miscibility pressure*, also known as *threshold pressure*, i.e., the pressure at which the solute begins to dissolve in the SCF;
  - Pressure at which the solute solubility in the SCF is the maximum;
  - Pressure range in which the solute solubility in the SCF would vary between zero and its maximum value.

- Each supercritical fluid has its own limitations for applications due to the variable properties and other restrictions.
- Channelling of SCF flow through the bed of solids and entrainment of the non-extractable component in the SCF are the two major problems in supercritical fluid extraction.
- There is a need to allow adequate contact time for the solvent to a) let it penetrate solid particles, and b) permit diffusion of solute from the solid interior into the solvent. SCF flow rate in the extraction process not only controls the contact time but also makes enough solvent available for the concentration of dissolved solutes in the SCF to remain below the solute solubility. It follows that a larger rate of SCF flow will be required if a large quantity of solute is to be extracted within a reasonable length of time.
- Prolonged contact time is needed for the feed and the SCF in the extraction chamber because solute diffusion from the solid into the supercritical fluid may be slow although the SCF penetrates into the interior of a solid rapidly due to the high SCF diffusivity.

The time of solid-solvent contact is the quotient of extraction vessel volume divided by the solvent volumetric flow rate. The volume is calculated at the temperature and pressure inside the extraction vessel. Normally, volume of the solvent is measured at atmospheric pressure after the gas exits the expansion tank. From this measured volume, the number of moles of gas is calculated and the volume of the supercritical fluid in the extraction vessel is calculated using the equations of state for gases.

SCF vs. Traditional Extraction Processes

- In addition to the advantages of lower consumption of solvents and greater selectivity, SCF results into faster extractions and higher yields in comparison to hydro-distillation for essential oil extraction.
- The low viscosity and high diffusivity of the super critical fluid accelerates the extraction process. Thus, SCF extraction process takes just 10 to 60 minutes, while the traditional extraction process continues for several hours.
- The oils extracted using traditional processes are generally refined and some valuable compounds are often lost during the process. However, SCFE process can avoid the refining process and can also be used to extract only the specific desirable compounds.
- There is no degradation of temperature-sensitive materials as there is no heating to remove the solute from solution.
- Supercritical fluids are chemically inert, non-polluting, non-flammable, harmless, and inexpensive, and they possess greater environmental sustainability. SC-CO$_2$ is preferred over CFC compounds due

to environmental considerations. There is also the promise of recovery of useful compounds from processing by-products.

- SCFE produces higher phenolic concentrations from grape pomace as compared to the traditional extraction techniques as normal solvent extraction, solid-liquid extraction, soxhlet extraction, and hydro-distillation.
- Unlike traditional methods, SCFE can be used to remove undesirable compounds from foods. For example, supercritical CO$_2$ with water as a co-solvent applied to green tea can remove caffeine, but not the antioxidants.

Applications

- SCFE is used for
  - high value processes as extraction of oils from hop and decaffeination of coffee and tea (supercritical carbon dioxide in coffee industry has replaced other solvents such as trichloroethylene and ethyl acetate);
  - Preparation of high-quality cocoa butter from cocoa beans (using carbon dioxide as the fluid);
  - extraction of oils from nuts, seeds or grains, and fragrant roots for food and pharmaceutical applications (in addition to allowing more efficient extraction, process also yields oils with bioactive compounds such as carotenoids, plant sterols, vitamin E, etc.);
  - extraction of essential oils and its derivatives, tocopherols and phytosterols, etc.
- Supercritical fluid extraction has been reported to lower the cholesterol content of food products of animal origin, such as dry egg yolk, dried chicken, beef patties, milk, etc. by 65% to 98%. Phospholipids from triglycerides are separated in case of soybean using SCF to minimize degumming or ginger flavor is separated from dried ginger.
- Supercritical CO$_2$ could be used for effective microbial inactivation for sterilization of foods that are sensitive to temperature and pressure variations.

The supercritical fluid extraction can be used as a technique for new food preparations. However, the creation of supercritical fluids by increasing pressure and/or temperature is an expensive affair. In such cases as water, there are corrosion problems. However, these constraints have now been overcome and supercritical fluids are being used for extraction applications, on an industrial scale, for a wide range of extractants as well as food matrices.

13.5 Advanced Extraction Technologies

There have been a number of advancements in the domain of food science and engineering that have impacted the extraction technology as well. These new advancements have permitted better contact between the feed and solvent, enhanced solubility, increased diffusion, efficient resource utilization, replacement of hazardous chemicals with eco-friendly ones for environmental sustainability, etc. Some of the recent advancements in the context of improving the extraction processes are briefly explained below.

13.5.1 Pulsed Electric Field-Assisted Extraction

Effectiveness of solid-liquid extraction in the food industry is considerably enhanced by subjecting the feed to pulsed electric field (PEF). Short impulses (microseconds to milliseconds) of high voltage amplitude electric waves directed to the product enhance the release of intra-cellular compounds without any significant temperature rise. Therefore, PEF has been observed to significantly increase the extraction of anthocyanins and phenolics in red wine must; betanine from beetroot and carrot juice, and many other similar products.

13.5.2 Microwave-Assisted Extraction

Microwave-assisted extraction technology is gaining increased acceptability on account of savings in energy, water, and solvents. Microwaves target the energy in the biological feed at the moist cells, increasing the pressure on the cell walls, rupturing them, and releasing the phytoconstituents. Microwave-assisted hydro-distillation (MAHD) has been observed to improve the quality of the extract, enhance the yield of the targeted compounds, and reduce the process time. It is truly a green and effective alternative to conventional leaching techniques. Further improvement in microwave-assisted extraction is possible by operating it under a vacuum. Vacuum microwave-assisted extraction (VMAE) operates at low oxygen sub-pressure, improves extraction efficiency, and the extraction time is reduced.

13.5.3 Ultrasound-Assisted Extraction

Highly energetic ultrasound waves travel through the feed-liquid mass generating alternating high and low pressure cycles and resulting into acoustic cavitation. Acoustic or ultrasonic cavitation, in turn, raises at the local level, extreme temperatures, pressures, and high shear forces in the extraction mass. Implosion of cavitation bubbles in the solvent creates macro-turbulences and micro-mixing. As a result, the extraction performance is greatly enhanced. A few examples are extraction of vanillin, polyphenols, herbal extracts, and caffeine from green tea. It has been observed in recent studies that the solvent circulation while subjecting the feed-solvent mass to ultrasound, termed *ultrasound-assisted dynamic extraction*, reduced the solvent consumption and extraction time.

13.5.4 Subcritical Water Extraction

Pressurized (1000–6000 kPa) low-polarity water is termed *subcritical water*. This water at 100–374°C is used as the solvent with the resultant higher extraction yields as compared to conventional solvents. Examples are the extraction of carotenoids from microalgae and rice bran oil through simultaneous lipase inactivation. Advantages of this intervention include a less costly extracting solvent, shorter extraction times, higher-quality extracts, and an eco-friendly process.

13.5.5 High Pressure Assisted Extraction

Hydrostatic pressures in the range of 100 to 800 MPa, even up to 1000 MPa, have been advantageously utilized in the extraction of ginsenosides from ginseng, polyphenols from green tea leaves, anthocyanins from grape skin pomace, and grape waste and flavonoids from propolis at room temperature. High hydrostatic pressure essentially destroys the cell membranes in the feed and, thus, the bioactive compounds are extracted easily. In the process, degrading enzymes may also be inactivated.

13.5.6 Aqueous Two-Phase Extraction

The two aqueous phases could be two water-soluble polymers or a polymer and a salt. It is a liquid-liquid extraction method. Consider the simplest one-step batch extraction. The two-phase system is prepared, and the feed is added to this two-phase system. After mixing, the phase separation is carried out by either gravitational settling or centrifugation. Extraction of betalains (a natural colorant from beetroot) is an example of this eco-friendly process.

13.5.7 Enzyme-Assisted Aqueous Extraction

Enzyme treatment along with an aqueous extraction process have proved to increase the oil yield from oilseeds to about 90%. Removal of lipids from oilseeds is facilitated by decreased surface activity of oleoresins, which, in turn, is the result of the enzymatic activity to break down the walls of cotyledon cells and lipid bodies. Enzyme-assisted aqueous extraction has been demonstrated to extract oils from oilseeds and some fruits. The extraction process is carried out at relatively low temperatures as compared to other methods and the solvent used is water.

Check Your Understanding

1. What are the common solvents used for the leaching process?
2. What are the standard operating parameters in a supercritical fluid extraction process?
3. What is stage efficiency in an extraction process?
4. Write notes on
    a. Super critical fluids;
    b. SCFE equipment;
    c. Different types of extractors for leaching;
    d. Fixed bed extractor;
    e. Bucket extractor;
    f. Auger extractor;
    g. Carrousel extractor;

    h. Contractor;

    i. PEF extraction;

    j. Microwave assisted extraction;

    k. Enzyme assisted extraction;

    l. Ultra sound assisted extraction.

5. Differentiate between

    a. Leaching and solvent extraction;

    b. Leaching and washing;

    c. SCF extraction and traditional extraction.

BIBLIOGRAPHY

Barba, F.J., Zhu, Z., Koubaa, M., Sant'Ana, AS. and Orlen, V. 2016. Green alternative methods for the extraction of antioxidant bioactive compounds from winery wastes and by-products: A review. *Trends Food Sci Technol.* 49: 96–109.

Berk, Z. 2018. *Food Process Engineering and Technology, 3rd* ed. Elsevier Applied Science, London

Brennan, J.G. (Ed.) 2006. *Food Processing Handbook*, 3rd ed. WILEY-VCH Verlag GmbH & Co. KGaA, Weinheim.

Brennan, J.G., Butters, J.R., Cowell, N.D. and Lilly, A.E.V. 1976. *Food Engineering Operations*, 2nd ed. Elsevier Applied Science, London.

Brunner, G. 2003. Supercritical fluid extraction of ethanol from aqueous solutions. *J. Supercrit Fluids*, 25(1): 45–55.

Charm, S.E. 1978. *Fundamentals of Food Engineering*, 3rd Edn. AVI Publishing Co., Westport, Connecticut.

Daintree, L., Kordikowski, A. and York, P. 2008. Separation processes for organic molecules using SCF Technologies. *Adv Drug Del Rev.* 60(3): 351–72.

Earle, R.L. and Earle, M.D. 2004. *Unit Operations in Food Processing*, Web Edn. The New Zealand Institute of Food Science & Technology, Inc., Auckland, https://www.nzifst.org.nz/resources/unitoperations/index.htm

Fellows, P.J. 2000. *Food Processing Technology*. Woodhead Publishing, Cambridge, UK.

Geankoplis, C.J.. 2004. *Transport Processes and Separation Process Principles*, 4th ed. Prentice-Hall of India, New Delhi.

Hamm, W. 1992. Liquid-liquid extraction in the Food Processing. In: J. D. Thorton (ed.) *Science and Practice of Liquid-liquid Extraction, Vol. 2*, Clarendon Press. Oxford.

Heldman, D.R. and Hartel, R.W. 1997. *Principles of Food Processing.* Aspen Publishers, Inc. Maryland.

Heldman, D.R. and Singh, R.P. 1981. *Food Process Engineering*, AVI Publishing Co., Westport, Connecticut.

Herrero, M., Mendiola, J., Cifuentes, A. and Ib´a˜nez, E. 2010. Supercritical fluid extraction: Recent advances and applications. *J Chromatogr.* 16: 2495–511.

Johnston, K.P. and Penninger, J.M.L. (Eds.) 1989. *Supercritical Fluid Science and Technology. ACS Symposium Series 406.* American Chemical Society, Washington, DC.

Jumaah, F., Sandahl, M. and Turner, C. 2015. Supercritical fluid extraction and chromatography of lipids in bilberry. *J Am Oil Chem Soc.* 92: 1103–11.

Kessler, H.G. 2002. *Food and Bio Process Engineering*, 5th ed. Verlag A Kessler Publishing House, Munchen, Germany.

Leniger, H.A. and Beverloo, W.A. 1975. *Food Process Engineering.* D. Reidel, Dordrecht. .

McCabe, W.L., Smith, J.C. and Harriot, P. 1993. *Unit Operations of Chemical Engineering*, 5th ed. McGraw-Hill, Inc., New York.

McHugh, M. and Krukonis, V. 1994. *Supercritical Fluid Extraction.* 2nd ed. Butterworth Heinemann, Boston.

Prabhudesai, R.K. 1979. Leaching. In: P.A. Schweitzer (Ed.) *Handbook of Separation Techniques for Chemical Engineers*, McGraw-Hill Book Co., New York.

Rizvi, S.S.H., Daniels, J.A., Benado, A.L. and Zollweg, J.A. 1986. Supercritical fluid extraction: Operating principles and food applications. *Food Technol.* 40 (7): 56–64.

Rizvi, S.S.H., Mulvaney, S.J. and Sokhey, A.S. 1995. The combined application of supercritical fluid and extrusion technology. *Trends Food Sci Technol.* 6: 232–240.

Ruhan, A., Motonobu, G. and Mitsuru, S. 2009. Supercritical fluid extraction in food analysis. In: S. Otles (ed.) *Handbook of Food Analysis Instruments.* CRC Press, Boca Raton, 25–54.

Safapuri, T.A., Massodi, F.A., Rather S.A., Wani, S.M., Gull, A. 2019. Supercritical fluid extraction: A review. *J. Biol. Chem. Chron.* 5(1): 114–122.

Saldaña, M.D.A., Gamarra, F.M.C. and Siloto, R.M.P. 2010. Emerging technologies used for the extraction of phytochemicals from fruits, vegetables, and other natural sources. In: L.A. de la Rosa, E. AlvarezParrilla, G.A. Gonzalez-Aguilar (eds) *Fruit and Vegetable Phytochemicals: Chemistry, Nutritional Value, and Stability.* Blackwell Publishing: Iowa, 235–270.

Schwarztberg, H.G. 1987. Leaching. Organic materials. In: R.W. Rousseau (ed.) Handbook of Separation Process Technology, Wiley, New York, 540–577

Sihvonen M., Jarvenpaa, E., Hietaniemi V. and Huopalahti R. 1999. Advances in supercritical carbon dioxide technologies. *Trends Food Sci Technol.* 10(6–7): 217–22.

Singh, R.P. and Heldman, D.R. 2014. *Introduction to Food Engineering.* Academic Press, San Diego, CA.

Temelli, F. 2009. Perspectives on supercritical fluid processing of fats and oils. *J. Supercrit. Fluids* 47(3): 583–590.

Thornton, J.D. (Ed.).1992. *Science and Practice of Liquid-liquid Extraction).* Clarendon Press, Oxford.

Thornton, J.D. 2006. *Extraction, Liquid-Liquid.* https://doi.org/10.1615/A to Z. extraction_liquid-liquid (Accessed 3rd July, 2022)

Toledo, R.T. 2007. *Fundamentals of Food Process Engineering*, 3rd ed. Springer, New York.

Treybal, R.E. 1963. *Liquid Extraction.* McGraw-Hill, New York.

Wang, L. and Weller, C. 2006. Recent advances in extraction of nutraceuticals from plants, *Trends Food Sci Technol.* 17(6): 300–12.

Wang, L., Weller, C., Schlegel, V., Carr, T. and Cuppett, S. 2008. Supercritical CO_2 extraction of lipids from grain sorghum dried distillers grains with solubles. *Bioresour Technol.* 99(5): 1373–82.

Watson, E.L. and Harper, J.C. 1989. *Elements of Food Engineering*, 2nd ed. Van Nostrand Reinhold, New York.

14

Mixing and Emulsification

14.1 Mixing

Mixing is the random distribution or dispersion of one or more components in another (or others). The phases that are initially separate cannot be differentiated after proper mixing. In the food industry, mixing is used for altering the quality of a product and/or to develop new products. The larger component in mixing is said to be the *continuous phase* and the smaller component, which has almost lost its identity in the mixture, is said to be the *dispersed phase*.

The mixing operation should not affect the nutritive value or the shelf life of the components used in mixing or the final product. However, the components can have their own effects, e.g., addition of salt or sugar can reduce the water activity of the mixture. During mixing the interaction between the components may cause some chemical changes in the food. Mixing operation may also cause generation of heat, which may accelerate the rate of chemical reactions.

14.1.1 Methods and Principles

The mixing operations used in the food industry can be broadly discussed under three categories as follows.

- mixing of dry materials as powders, granular solids,
- mixing of liquids as concentrated juice and sugar, water and milk powder, etc.
- mixing of pastes, e.g. salt in ginger paste, salt/ sugar with dough, etc.

Mixers are usually specifically designed for the type of products, e.g., powders, pastes, or liquids, etc. as per the requirement of power and degree of uniformity required. Often the mixers commonly used for pastes can be used for mixing dry powders also.

The performance of a mixer is characterized by the properties of the product, the time required for mixing, and the power requirement. Sometimes, we require a very high degree of uniformity, sometimes we desire quick mixing, and sometimes the most important concern is the power requirement.

A mixing process basically starts by placing the two separate materials (to be mixed) in the mixing vessel. If small samples are taken, just before the mixing operation, from different locations in the vessel, nearly all samples will consist of either one or the other of the pure components. As mixing proceeds the variations in the concentrations of any sample within the whole mass will tend to diminish and will approach the overall fraction of the components in the vessel. Mixing is said to be complete if one material is uniformly distributed through the other, i.e., a state in which samples taken from any location will have the components in the same proportion as in the complete mixture.

Thus, the mixing effectiveness can be determined by taking some random samples after or during mixing and finding out how the fraction of a component varies in these samples. However, the result of analysis of mixing effectiveness depends, among other factors, on sample size also. If the sample is very small, it may represent only one or the other pure component. A very large sample size may not effectively represent the composition. Thus, the proper selection of the sample size is important. For example, for preparation of salty dried ginger grits, the salt content should be 1% in dried ginger, then 10 kg salt has to be added to 990 kg dried ginger grits. But if the product is needed to be sold in 20 g sachets, each packet should have 0.2 g salt. Thus, in this case the sample size can be taken as 20 g.

Mixing of Powders and Granular Materials

Principle

Some typical examples are the mixing of sugar with wheat flour, oil with pulses or talc powder with rice (as done during glazing operations). The mixing of dry materials is usually done by slow speed agitation by impeller or blades, by tumbling, or less frequently by centrifugal smearing and impact.

The degree of mixing in the case of powders and granular materials depends primarily on the following factors.

- Characteristics of materials being mixed (particle size, shape and density of each component, surface characteristics and tendency of the materials to aggregate); mixing is better if the components are smaller in size and are uniform in shape.
- Operational conditions of individual materials during mixing, such as moisture content and flow characteristics
- Type of mixer

The solids do not aggregate easily and cannot be mixed as uniformly as can be done with liquids.

Mixing Effectiveness

As discussed above, the mixing effectiveness can be determined by taking some random samples after or during mixing and finding out the variations in concentrations of a particular sample. In case of granular solids, the standard deviation (σ_m) of the fractions of any component from the mean value (that would be observed in a sample from a completely random,

DOI: 10.1201/9781003285076-17

blended mixture) is taken. In case of pastes, this σ_m value is zero; however, with granular solids it is not zero.

Let the number of samples be n and the average value of the measured concentrations of a component in the samples be $\bar{x}$. If n is very large, $\bar{x}$ will be equal to the overall average fraction of the component in the mix.

If the solids were perfectly mixed, then every measured value of x_i would be equal to $\bar{x}$. If the mixing is not complete, x_i will differ from $\bar{x}$ and their standard deviation about $\bar{x}$ can be used as an expression for the effectiveness of mixing.

The standard deviation at any time during mixing can be given as:

$$\sigma_m = \sqrt{\frac{\sum_{i=1}^{n}\left(x_i - \bar{x}\right)^2}{n-1}} \quad (14.1)$$

In this equation, σ_m is the standard deviation, x_i is the concentration of the component in i^{th} sample, n is the number of samples and $\bar{x}$ is the mean concentration of the samples.

The standard deviation can also be written as:

$$\sigma_m = \sqrt{\frac{\sum x_i^2 - \bar{x}\sum x_i}{n-1}} \quad (14.2)$$

Initially the compositions of different fractions in samples drawn from different locations of the mixture are very different from each other. Thus, the standard deviation will be more. As mixing proceeds, the uniformity of the mixture increases and standard deviation will reduce and tend toward zero and, hence, low standard deviation signifies good mixing.

> In the above discussions and also in practical applications, the standard deviation is a relative term and its significance varies with the fractions of different materials in the mix. A standard deviation of 0.001 would be more significant if the overall average fraction of the component in the mix is 0.01 rather than when it is 0.1.

If two components are being mixed, before mixing has begun, the components in the mixer remain separately (or as separate layers); thus, the samples in the first layer will have $x_i = 0$; in the other layer it will be $x_i = 1$.

Suppose the two constituents of the mixture are A and B. The fraction of A in the mixture is assumed to be "a" and that of B as "b". At the start of mixing, a sample drawn from the bulk would be either A or B. For this two-component mixture, a + b = 1

$$\sigma_0^2 = \frac{1}{n}\left[a.n.(1-a)^2 + (1-a)n(0-a)^2\right] \quad \left(\text{for n samples}\right)$$

$$\sigma_0^2 = a.(1-a) \quad (14.3)$$

i.e., under these conditions, the standard deviation is given by

$$\sigma_0 = \sqrt{a(1-a)} \quad (14.4)$$

where a is the overall fraction of a component in the two component mix.

Let us consider a completely blended mixture of sugar and flour and from that we take "n" samples, each containing "N" particles. Let the overall fraction (by number of particles) of sugar in the total mix be a_p. If N is small (say about 100), then the fraction of sugar in each sample will vary and it will not be same even if we consider it as 100% mixing.

For a completely random mixture, there is a theoretical standard deviation (σ_e), which can be given by

$$\sigma_e = \sqrt{\frac{a_p\left(1-a_p\right)}{N}} \quad (14.5)$$

Thus, a *mixing index* (MI), the ratio of standard deviation of a random mixture at zero mixing to the standard deviation at any time, σ_m, can be given as:

$$MI = \frac{\sigma_e}{\sigma_m} = \sqrt{\frac{a_p\left(1-a_p\right)\left(n-1\right)}{N\sum_{i=1}^{n}\left(x_i - \bar{x}\right)^2}} \quad (14.6)$$

The mixing index before the start of mixing can be given as:

$$MI_0 = \frac{\sigma_e}{\sigma_0} = \frac{1}{\sqrt{N}} \quad (14.7)$$

The standard deviation is a relative measure of mixing and is applicable for tests of a specific material in a specific mixer. With the help of the measured standard deviation values at different times, some other mixing indices have been proposed to explain and monitor the mixing operations and the performance of the mixers.

For mixing of nearly equal masses of the constituents and/or at relatively low mixing rates, the following expression for the mixing index (MI) can be used.

$$MI_1 = \frac{\sigma_m - \sigma_\infty}{\sigma_0 - \sigma_\infty} \quad (14.8)$$

When one of the components is much smaller than the other, as in the mixing of a small quantity of salt in large bulk of material and/or at higher mixing rates, the following expression is used.

$$MI_2 = \frac{\log\sigma_m - \log\sigma_\infty}{\log\sigma_0 - \log\sigma_\infty} \quad (14.9)$$

For mixing of solids or liquids in a similar way as the first one,

$$MI_3 = \frac{\sigma_m^2 - \sigma_\infty^2}{\sigma_0^2 - \sigma_\infty^2} \quad (14.10)$$

In the above equations, σ_0 the standard deviation of a sample, just at the start of mixing, σ_∞ is the standard deviation of a "perfectly mixed" sample, and σ_m the standard deviation of a sample taken at any time during mixing. σ^2 denotes the variance. It may be noted that the above equations are of the similar form as the moisture ratio or the temperature ratio during a drying or heating/cooling process.

Even though each mixing index has been proposed for separate applications, usually all three are examined and the equation that most fits the mixer type and the particular situation is selected.

> If we take a completely uniform mixture of granular material and try to further mix in an equipment, it may cause some separation of the mixed parts, which we can call *un-mixing*. Thus, during any mixing operation, there is also certain un-mixing and the homogeneity of the final product is decided by the balance between the mixing and un-mixing. The above discussion also implies that proper time is important for uniform mixing. The time required to reach this equilibrium stage is determined by the type of ingredients, type of equipment and operating conditions. Further during mixing of dry solids, un-blending forces, usually electrostatic in nature, are present, which prevent complete blending and, hence, long mixing times may rather cause un-mixing and segregation.

As discussed, the perfect mixing (where, $\sigma_\infty = 0$) can never be achieved for powders and particulate solids, but in good mixers the value of the standard deviation can be very low. Further, even from a well-mixed solid sample, small random samples will differ distinctly in composition, and, hence, a large number of samples is to be taken for analysis.

The mixing index, MI, indicates how far the mixing has continued toward equilibrium (the mixing rate is proportional to driving force). The equilibrium value of MI is 1. Thus, in analogy with heat or moisture transfer, for small time of mixing, the rate of change in MI is directly proportional to 1-MI.

$$\frac{dMI}{d\theta} = k(1 - MI) \quad (14.11)$$

where k is a constant. The rearrangement and integration between limits yields

$$\int_0^\theta d\theta = \frac{1}{k} \int_{MI_0}^{MI} \frac{dMI}{1 - MI} \quad (14.12)$$

$$\theta = \frac{1}{k} \ln \frac{1 - MI_0}{1 - MI} \quad (14.13)$$

$$\theta = \frac{1}{k} \ln \frac{1 - \frac{1}{\sqrt{n}}}{1 - MI} \quad (14.14)$$

The equation can calculate the time required for any desired degree of mixing as long as un-blending forces can be ignored.

In a simplified form, the time of mixing (θ_m), expressed in seconds, has been proposed to be related to mixing index as:

$$\ln (MI) = -k.\theta_m \quad (14.15)$$

where k is the *mixing rate constant*. It varies with the type of components and the mixer.

Mixing of Pastes

Principle

For the pastes or highly viscous materials, the mixing involves kneading, folding, and shearing. Kneading involves pressing and squeezing the material (such as dough) against the vessel wall. The other two mechanisms are folding unmixed food into the mixed part, and applying shear to stretch the material. The majority of kneading machines split the mass apart and shear it between a moving blade and a stationary surface. A good mixing operation requires creating and recombining more and more new surfaces in the food. The pastes do not flow easily; thus, the mixer blades must move throughout the vessel, or the food should be directed to the mixer blades. A substantial amount of energy is required even with fairly thin materials and as the same mass becomes stiff and rubbery, power requirements increase.

Usually in mixing of solids and pastes, a pure homogenous mixture is not achieved (in liquids it is possible). The mixing of pastes and solids also requires more power than the mixing of liquid(s); often the power needed to drive a mixer for pastes and deformable solids is many times higher than that needed by a mixer for liquids. Some equipment normally used for blending pastes can also be used for solid mixing.

The properties of the materials may vary widely, which affect the mixing process. Even the properties of a single material may change during the course of mixing. Some materials, which are initially in the form of free and dry material, may become pasty during the process (say by addition of water), and they may change to free and granular again. Thus, the different properties as viscosity, density, stiffness, tackiness, and wettability can affect the mixing process.

During mixing, only a part of the energy supplied to the mixer is used for mixing and the remaining is dissipated as heat. In most cases the actual percentage of utilization of energy for mixing is very small. That is why the product after mixing has a higher temperature than the raw materials. The heat generated during mixing is often desired to be removed to protect the machine and the quality of the food material. Other things being equal, if the mixer takes less time to bring the material to the desired degree of uniformity, the useful fraction of energy for mixing will be more.

Mixing Effectiveness

Let us consider a pasty solid mass to which some kind of tracer material has been added. To find the mixing effectiveness, after the tracer (discontinuous phase) material is added, a number of samples are taken from various locations in the mixer and the fraction of tracer x_i is determined as we have done for analysis of powder/particulate materials mixing. In this case also the standard deviations are observed and the mixing index is calculated as in case of powders and particulate materials. Initially the standard deviation will be higher and as mixing proceeds, the uniformity of the mixture would increase and the standard deviation will diminish to zero.

As we discussed before, with particulate solids the equilibrium standard deviation for complete mixing σ_e is used as a reference. With pastes, we take the reference at zero mixing, i.e. σ_0. These two are closely related (equations 14.4 and 14.5). In fact, if the value of N is set equal to 1 in equation 14.5, the equations become identical. In other words, if the sample contains one particle then the analysis will indicate $x_i = 1$ or 0 and nothing in between.

$$MI_p = \frac{\sigma_0}{\sigma_m} \qquad (14.16)$$

$$\text{Thus,} \quad MI_p = \frac{\sigma_0}{\sigma_m} = \sqrt{\frac{(n-1)a(1-a)}{\sum x_i^2 - \bar{x}\sum x_i}} \qquad (14.17)$$

In any batch process, the value of MI_p is initially 1, and it increases as mixing improves. Theoretically, MI_p should be ∞ after a long time. But unlike liquids, solids or pastes do not mix very well and the measured values of x_i will never be same as $\bar{x}$. So MI_p is finite even with perfectly mixed material.

Mixing of Liquids

Principle

For mixing of liquids of low or medium viscosities (which can easily flow) or for dispersing small amount of solids in a liquid, the liquid is agitated usually by mechanical agitators. The agitation in a vessel causes entrainment of slow-moving parts within the faster moving parts. In a vessel using mechanical agitator(s), the type of flow is affected by the impeller type, fluid characteristic, size of the tank, and the arrangement of baffles and agitator.

The velocity components in the fluid during mixing by an agitator can have different components, as shown in Figure 14.1. In simplified form, the velocity of the fluid at any point in the tank has three components, namely, the longitudinal velocity parallel to the impeller shaft, the radial velocity perpendicular to the impeller shaft, and the rotational velocity, which is tangential to the impeller shaft in a circular manner. These three velocity components govern the spatial variations in the overall flow patterns in the vessel.

In case of a vertical shaft, the radial and longitudinal components of the flow cause the mixing. For a vertical and centrally located shaft, the tangential component of the flow causes the liquid to move in a circular path around the shaft. The adjoining layers of circulating liquid move at the same speed, which causes the formation of a vortex in the liquid and there is no mixing. It also throws the solids or denser materials toward the wall of the vessel due to centrifugal force, from where they fall along the wall and are collected at the bottom. In other words, it helps concentration of the product rather than mixing. If the impeller speeds are too high, the vortex may be so profound that it tends to reach the impeller, and this causes the air from above the liquid to move down into the liquid. It is generally undesirable as incorporation of air into the product causes ancillary problems, such as oxidation, and may cause discoloration. Hence, the speed of the agitator should not be so high as to create a vortex. Use of off-center or angled mixer shafts or provision of baffles or angled blades prevents the above problems (Figure 14.2).

The above problems can also be avoided, specifically in small vessels, by keeping the shaft away from the centerline of the vessel, then tilting toward the wall. In the case of larger vessels, mounting the agitator on the side of the tank, with the shaft in a horizontal plane, but at an angle with the radius, can elude the vortex formation.

In the case of large tanks with vertical agitators, baffles are provided to avoid swirling. In un-baffled vessels, the mixing is not perfect (Figure 14.3). The baffles obstruct rotational flow, but the radial or longitudinal flow components are not affected. Vertical strips fixed perpendicular to the wall of the mixing vessel serve the best. For large tanks 3–4 baffles are provided. The width of baffles is kept less than one-twelfth and one-eighteenth of the vessel diameter for turbines and propellers, respectively. The baffles can be avoided for side-entering, inclined, or off-center propellers. These will of course add to the complexity of flow.

In the case of propeller agitators, the liquid moves down from the suction of the propeller, then spreads radially and is subsequently lifted along the wall due to the action of propellers. Thereafter the liquid moves to the suction of propeller for the next cycle of movement. These are used when heavy solid particles are to be kept in suspension and for liquid viscosities less than 50 poise.

There is good radial flow in the plane of the impeller for paddle agitators and flat-blade turbines. The flow divides at

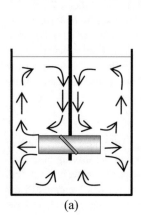 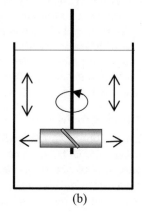

(a)	(b)

FIGURE 14.1 (a) Velocity components during liquid mixing process with an agitator (b) the longitudinal, radial, and tangential velocity components during agitation

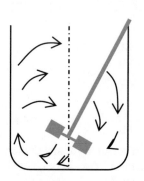 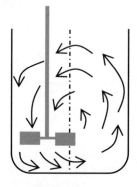

FIGURE 14.2 Flow patterns observed with off-center propellers

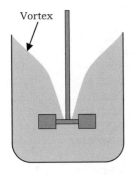
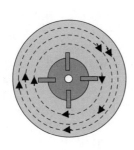

Vortex

FIGURE 14.3 Swirling flow pattern in an un-baffled vessel (after Oldshue, 1969 with permission)

the wall and forms two detached motion patterns. One moves down alongside the wall and back to the center of the impeller from below; the other rises to the surface and moves back to the impeller from above. For a vessel without baffles, there are vortex formations at moderate stirrer speeds; however, for vessels with baffles, there is more rapid mixing of the liquid.

In tall tanks, often more than one impellers are fixed on the same shaft and, in that case, each impeller acts as a separate mixer.

Most liquid foods are non-Newtonian, coming under the categories of pseudoplastic foods (e.g., sauces), dilatants foods (e.g., corn flour and chocolates), or visco-elastic foods (e.g., bread dough). In case of pseudoplastic foods, there is formation of a zone around a small agitator where a partial material discontinuity develops as mixing proceeds and bulk of the food does not move. Such a zone becomes clearer at higher speeds of the agitator. Dilatant foods show a shear thickening behavior, and the agitator may stop functioning if adequate power is not available. Visco-elastic foods behave like pastes and require folding and stretching action.

Mixing Effectiveness

The mixing effectiveness can be found out by taking standard deviations of a particular fraction and using a suitable mixing index as discussed earlier. The mixing rate constant, k, as mentioned above is governed by the characteristics of the mixer and the constituent liquids.

The relationship between k and the mixer characteristics is given by

$$k \alpha \frac{D^3 N}{D_v^2 h} \quad (14.18)$$

where D (m) is the agitator diameter, D_v (m) the vessel diameter, N (rev. s^{-1}) the agitator speed and h (m) the height of liquid in the vessel.

The following relationship of some dimensionless numbers can be used to express the power requirement during mixing operations.

$$Po = K(Re)^p(Fr)^q \quad (14.19)$$

Here Po is the Power number, Re is the Reynolds number, Fr is the Froude number and K, p and q are factors linked to the geometry of the agitator. These factors are usually determined by conducting experiments.

$$Re = \frac{\rho_m D^2 N}{\mu_m} \quad (14.20)$$

$$Fr = \frac{DN^2}{g} \quad (14.21)$$

$$Po = \frac{P}{\rho_m N^3 D^5} \quad (14.22)$$

P (kW) is the power transmitted via the agitator, μ_m (N.s.m^{-2}) the viscosity of the mixture, and ρ_m (kg.m^{-3}) the density of the mixture.

The Froude number is important only when a vortex is formed in an un-baffled vessel. In other cases, this term is insignificant and, hence, ignored.

The density of a mixture for use in the above equations can be obtained as follows:

$$\rho_m V_m = \rho_1 V_1 + \rho_2 V_2 \quad (14.23)$$

where the subscripts 1 and 2 refer to the two components of the mixture, and subscript m is for the mixture. ρ and V represent the density and volume.

The viscosity of a mixture is found using the following equations.

$$\mu_m (\text{unbaffled mixer}) = \mu_1^{V_1} \mu_2^{V_2} \quad (14.24)$$

$$\mu_m (\text{baffled mixer}) = \frac{\mu_1}{V_1} \left(\frac{1 + 1.5\mu_2 V_2}{\mu_1 + \mu_2} \right) \quad (14.25)$$

Here 1 and 2 represent the two fractions.

The power requirement of a mixer varies with the amount, nature, and uniformity of foods in the mixer and the type, size, speed as well as the position of the agitator. Fellows (2000) describes the details of the characteristic changes in power consumption Po of propellers at different Re.

Example 14.1 During a food preparation, 100 kg flour was mixed with 1.2 kg of salt. After mixing for one minute, 10 g samples from different locations were analyzed and the weights of salt in samples were observed to be 0.14, 0.098, 5.8, 0.08, 0.004, 3.6, 0.09, 6.4, 0.008, and 0.98 g. Find the standard deviation of the sample compositions from the mean composition.

After 2 minutes, again 10 samples were taken and the weight of salt in these samples were found as 0.05, 0.12, 0.17, 0.046, 0.08, 0.0078, 0.096, 0.092, 0.09, and 0.014 g. Check whether the subsequent mixing for one minute achieved any improvement in mixing or not.

SOLUTION:

The overall mean composition of the salt in flour is 1.2/101.2 = 0.011858.

First we need to find out the sum of $\left(x_i - \overline{x}\right)^2$ to obtain the standard deviation. For the first case, the table will be as follows.

Sample no.	Weight in g per 10 g	Value of x_i	$\bar{x}$	$x_i - \bar{x}$	$\left(x_i - \bar{x}\right)^2$
1	0.14	0.014	0.011858	0.002142	4.5894E-06
2	0.098	0.0098	0.011858	-0.00206	4.2342E-06
3	5.8	0.58	0.011858	0.568142	0.32278566
4	0.08	0.008	0.011858	-0.00386	1.4882E-05
5	0.004	0.0004	0.011858	-0.01146	0.00013128
6	3.6	0.36	0.011858	0.348142	0.12120306
7	0.09	0.009	0.011858	-0.00286	8.1665E-06
8	6.4	0.64	0.011858	0.628142	0.39456274
9	0.008	0.0008	0.011858	-0.01106	0.00012227
10	0.98	0.098	0.011858	0.086142	0.00742049
Total					0.84625738

Thus, $\sigma_m = \sqrt{\dfrac{\sum\limits_{i=1}^{n}\left(x_i - \bar{x}\right)^2}{n}} = \sqrt{\dfrac{0.84625738}{10}} = 0.29$

For the second case, the table will be as follows.

Sample no.	Weight in g per 10 g	Value of x_i	$\bar{x}$	$x_i - \bar{x}$	$\left(x_i - \bar{x}\right)^2$
1	0.05	0.005	0.011858	-0.00686	4.7028E-05
2	0.12	0.012	0.011858	0.000142	2.0247E-08
3	0.17	0.017	0.011858	0.005142	2.6443E-05
4	0.046	0.0046	0.011858	-0.00726	5.2674E-05
5	0.08	0.008	0.011858	-0.00386	1.4882E-05
6	0.0078	0.00078	0.011858	-0.01108	0.00012272
7	0.096	0.0096	0.011858	-0.00226	5.0972E-06
8	0.092	0.0092	0.011858	-0.00266	7.0634E-06
9	0.09	0.009	0.011858	-0.00286	8.1665E-06
10	0.014	0.0014	0.011858	-0.01046	0.00010936
Total					0.0003945

$\sigma_m = \sqrt{\dfrac{\sum\limits_{i=1}^{n}\left(x_i - \bar{x}\right)^2}{n}} = \sqrt{\dfrac{0.0003945}{10}} = 0.00628$

The standard deviation from the mean has reduced, which indicates better mixing as compared to the first observation. Thus, the subsequent mixing for one minute was desirable. It can be further improved also.

Example 14.2 Four kg powdered sugar is to be mixed with 50 kg flour. During the mixing process ten gram samples were taken from 10 locations after 2 and 4 minutes. The percentages of sugar in the samples are as below.

After 2 minutes
0.4, 13.2, 3.2, 12.4, 10.4, 2.0, 12.2, 0.9, 2.2, 0.2
After 4 minutes
3.2, 6.2, 7.4, 9.2, 9.8, 10.4, 11.4, 6.8, 7.8, 9.6
Find the mixing index after 2 minutes and 4 minutes.

SOLUTION:

The following table is prepared from the observations.

| Sample no. | $\bar{x}$ (= 4/54) | After 2 min | | | After 4 min | | |
		Percentage of sugar	Value of x_i	$\left(x_i - \bar{x}\right)^2$	Percentage of sugar	Value of x_i	$\left(x_i - \bar{x}\right)^2$
1	0.074074	0.4	0.004	0.00491038	3.2	0.032	0.00177023
2	0.074074	13.2	0.132	0.00335541	6.2	0.062	0.00014578
3	0.074074	3.2	0.032	0.00177023	7.4	0.074	5.487E-09
4	0.074074	12.4	0.124	0.0024926	9.2	0.092	0.00032134
5	0.074074	10.4	0.104	0.00089556	9.8	0.098	0.00057245
6	0.074074	2	0.02	0.00292401	10.4	0.104	0.00089556
7	0.074074	12.2	0.122	0.00229689	11.4	0.114	0.00159408
8	0.074074	0.9	0.009	0.00423464	6.8	0.068	3.6894E-05
9	0.074074	2.2	0.022	0.00271171	7.8	0.078	1.5413E-05
10	0.074074	0.2	0.002	0.00519467	9.6	0.096	0.00048075
Total				0.03078609			0.0058325

Thus, the variance after 2 min,

$$\sigma_m^2 = \frac{0.03078609}{10} = 0.00308$$

and, the variance after 4 min,

$$\sigma_m^2 = \frac{0.0058325}{10} = 0.00058$$

The value of $\sigma_o^2 = (4/54)(50/54) = 0.0686$

As the number of particles in the samples is very large, σ_r^2 can be taken as 0. Thus, the mixing indices are obtained as:

$$\text{After 2 min}, (MI)_{2\,\text{min}} = \frac{(0.0686 - 0.00308)}{(0.0686)} = 0.955$$

$$\text{After 4 min}, (MI)_{4\,\text{min}} = \frac{(0.0686 - 0.00058)}{(0.0686)} = 0.9915$$

Example 14.3 In a food processing plant, finger millet flour is mixed with corn flour in a ratio of 30:70. Both powders are of the same size. After 2 minutes of mixing, the variance of the sample compositions of corn flour was found to be 0.08. Each sample contains 60 particles. What will be the time required for thorough mixing, if the variance of the sample composition is to be 0.01 after thorough mixing?

SOLUTION:

Taking a = 0.30 as the fractional content of finger millet flour,

$$(1 - a) = (1 - 0.30) = 0.70$$

$$\sigma_o^2 = 0.3 \times 0.7 = 0.21$$

$$\sigma_\infty^2 = \sigma_o^2 / N = 0.21/60 = 0.0035$$

In this case, the mixing index is obtained as:

$$MI = \frac{\sigma_m^2 - \sigma_\infty^2}{\sigma_0^2 - \sigma_\infty^2}$$

After 2 minutes, $MI = (0.08 - 0.0035)/(0.21 - 0.0035)$

$$= 0.0765 / 0.2065 = 0.3704$$

Using equation 14.15 to find out the mixing rate constant,

$$\ln(MI) = -k.\theta_m$$

Here, the time is 2 min, thus, $\theta_m = 120$ s

$$\text{or, } \ln(0.3704) = -k(120)$$

$$\text{or, } k = 0.0083$$

For the 2nd case, i.e. for $\sigma_m^2 = 0.01$,

$$MI = (0.01 - 0.0035)/(0.21 - 0.0035)$$

$$= 0.0065 / 0.2065 = 0.0315$$

Substituting the value of k, the time can be found out as

$$\ln(0.0315) = -0.0083 (\theta_m)$$

$$\text{or, } \theta_m = 416.68 \text{ s} = 6.94 \text{ min}$$

Example 14.4 Water is mixed with a concentrated syrup (60°Brix) in a ratio of 4:1 in a cylindrical tank 1.5 m in diameter. The temperature is 30°C. A propeller agitator of 18 cm diameter is used for mixing, which operates at 860 rev.min⁻¹. What will be the effective Reynolds number for the mixing operation?

At 30°C, the viscosities of syrup and water are 0.036 N.s.m⁻² and 0.001 N.s.m⁻². The densities at that temperature are 1060 kg.m⁻³ for the syrup and 1000 kg.m⁻³ for water.

SOLUTION:

The density of the mixed liquid can be obtained as:

$$\rho_m = \rho_1 V_1 + \rho_2 V_2$$

where, ρ_m is the density of the mixture, ρ_1 and ρ_2 are the densities of the materials and V_1 and V_2 are the fractional volumes.

In this case, ρ_1 and ρ_2 are 1000 and 1060 kg.m⁻³.

$$V_1 = 4/5 = 0.8 \text{ and } V_2 = 1 - 0.8 = 0.2$$

$$\rho_m = (1000)(0.8) + (1060)(0.2) = 1012 \text{ kg.m}^{-3}$$

As per equation 14.24, for unbaffled agitation

$$\mu_m(\text{unbaffled}) = \mu_1^{V_1} \mu_2^{V_2}$$

Thus $\mu_m = (0.001)^{0.8}(0.036)^{0.2} = 2.048 \times 10^{-3}$ N.s m⁻²

Here D = 0.18 m and N = 860 rev.min⁻¹ = 14.333 rev. s⁻¹

Thus, the Reynolds number can be given as:

$$Re = \frac{\rho_m D^2 N}{\mu_m} = \frac{(1012)(0.18)^2(14.333)}{2.048 \times 10^{-3}} = 229473.6$$

Example 14.5 In the above problem, if Power number corresponding to the Re is given as 1.4, what will be the size of the motor required?

SOLUTION:

$$Po = \frac{P}{\rho_m N^3 D^5}$$

Thus

$$P = (Po)\rho_m N^3 D^5 = (1.4)(1012)(14.333)^3(0.18)^5 = 788.28 \text{ J.s}^{-1}$$

i.e., P = 0.788 kW = 1.05 hp

14.1.2 Equipment

Mixers for Powders and Granular Materials

Different types of mixing equipment, such as a tumbling mixer, ribbon mixer, and vertical screw mixer, are used for mixing powders and granular/ particulate materials. Extruders and some size reduction equipment also serve as mixers in some instances.

Tumbling Mixer

Construction and Operation

- It basically consists of a drum, inside which the components to be mixed are kept. Mixing takes place with the rotation of the drum along its own axis.
- The drum may be of different configurations as cylindrical, double cone, Y-cone, and V-cone types. The twin shell blender shown at (b) is made from two cylinders joined to form a V and rotated about a horizontal axis.
- It is filled to approximately 50–60% of its volume and is rotated at 20–100 rev.min^{-1}, for say 5–20 minutes.
- The drum may have provision of some devices for spraying small amount of liquids into the mix.
- Baffles or mechanically driven counter-rotating arms are provided in the drum for breaking up lumps of solids and to improve the mixing efficiency.
- The drum also has an arrangement to tilt it for easy discharge of the materials after mixing. A conveyor may be fitted to convey the mixed materials to the next section.

Salient Features

- Optimum mixing for specific components depends on the speed of rotation and shape of the vessel.
- The speed of rotation should be lower that the critical speed (it is the speed at which centrifugal force exceeds gravity force), or else the material will not fall from the periphery of the drum and there will be no mixing.
- The shells are more effective than double cone for some blending operations.
- The tumbling mixer normally requires a little less power than a ribbon blender for the same volume of operation (Figure 14.4).

Applications

These mixers are also used for mixing of granular materials with salt, sugar, etc. They are also used for coating applications. The ball mills without balls can also be used as tumbling mixers, which can effectively mix suspensions of dense solids in liquids and heavy dry powders.

Ribbon Mixer

Construction and Operation

- It consists of a closed hemi-spherical horizontal trough containing a shaft and a helical ribbon agitator (two or more narrow thin metal blades or ribbons, which are in the form of helices). Usually the counter rotating ribbons are mounted on the same shaft (Figure 14.5).
- The pitches of the ribbons are made differently so that when one ribbon moves the material rapidly forwards through the trough, the second one moves the material slowly backwards. This creates relative motion within the feed and mixing takes place as the material is conveyed from the feed end to the discharge end, i.e., there is a continuous mixing and conveying action.
- The counteracting agitators also cause turbulence in the food that helps better mixing.
- Ribbons may be continuous or interrupted.

Salient Features

- The mixing trough is kept covered to avoid contamination of the product and to prevent the operational area from becoming dusty. However, during inspection, the top can be opened.
- The mixer can be designed either for continuous or for batch-wise operation.
- It requires moderate power.

Applications

The mixer can be used for pastes and for powders that do not flow readily, although it is commonly used for dry ingredients and small-particulate foods.

Vertical Screw Mixer

Construction and Operation

- It is constructed of a conical tank, which has a rotating vertical screw within it. The screw orbits around the central axis (Figure 14.6), which helps to mix the

(a)

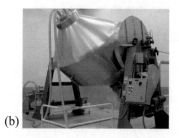

(b)

FIGURE 14.4   (a) Simple tumbling mixer (b) a tumbling mixer connected to vacuum (Courtesy: M/s Shiva Engineers, Pune)

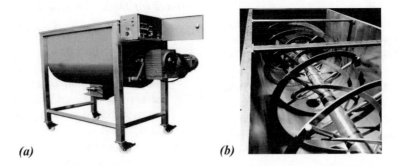

FIGURE 14.5 (a) Ribbon mixer (Courtesy: M/s Shiva Engineers, Pune) (b) Inner view showing the ribbon (Courtesy: M/s Arvinda Blenders, Ahmedabad)

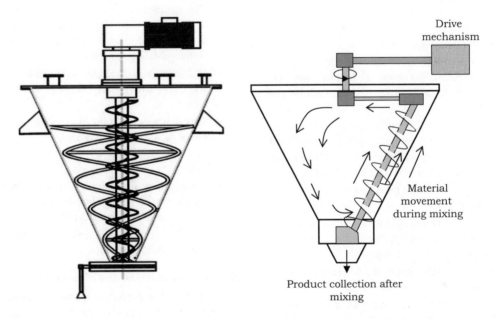

FIGURE 14.6 Schematic diagram of a vertical screw mixer (a) vertical fixed screw (courtesy: Arvinda Blenders, Ahmedabad, India), and (b) an orbiting screw

contents thoroughly. There may be two screws opposite to each other also.

- The screw or the helical conveyor elevates and circulates the material.
- Many different designs are available. In the design shown in Figure 14.6(b), the double motion helix orbits about the central axis of a conical vessel, touching all parts of the mix.

Salient Features

- Mixing is slower than in a ribbon blender, but the power requirement is also less.
- In some designs, the central screw is fixed, which is less effective but involves less capital cost.

Applications

Applications for such an equipment include assimilation of small amounts of a material into a bulk of other material, such as grains, light solids, etc.

Impact Wheel

Construction and Operation

- It consists of a high-speed spinning disc 25–70 cm in diameter in a stationary casing. The disk may be either vertical or horizontal.
- Premix of dry ingredients is fed continuously near the center of the disc, which is thrown outward in the casing as a thin layer.
- Mixing takes place with the help of centrifugal action. The high shearing forces acting on the materials during their movement over the disk surface also helps thorough blending of the materials.
- Some types of attrition mills can also be designed only for mixing, and not for size reduction. In that case the materials are dropped onto a horizontal double rotor. The rotor has short vertical pins near its periphery, which help to increase the mixing effectiveness.

Salient Features

- The premix fed to an impact wheel must be fairly uniform as the product has no hold up in the mixer and there is no chance of recombining material that has passed through.

- Sometimes several passes through the same machine or through a series of machines are necessary.

- The revolution of the disc is determined depending on the type of material. The revolutions per minute will be less for easy mixing materials.

- For light free-flowing powders, the capacity of impact wheels can be up to 25 tonnes.h^{-1}.

Applications

These are used for fine, light powders.

Muller Mixer

Construction and Operation

- Mulling is a smearing or rubbing action similar to that in a mortar and pestle. Thus, there are wide and heavy wheels in the mixer to cause such type of action (Figure 14.7).

- The pan may be stationary or rotating. In a particular design with stationary pan, the central vertical shaft is driven, causing the muller wheels to roll in a circular path over a layer of solids on the pan floor.

- The rubbing action takes place due to the slip of the wheels on the solids.

- Some plows (or blades) guide the solids under the muller wheels or to an opening in the pan for discharge of product after mixing.

- In another design the axis of the wheels is held stationary, and the pan is rotated.

- In some cases, the wheels are placed in an offset manner, and both the pan and the wheels are driven.

- If the muller wheels are replaced with mixing plows, it is called a *pan mixer*.

- Continuous muller mixers with two mixing pans connected in series are also available.

Applications

The mixer is used for small batches of heavy solids and pastes; they are especially effective in uniformly coating the particles of granular solid with a small amount of liquid.

FIGURE 14.7   Working elements of a muller mixer

Mixers for High Viscosity Liquids and Pastes

Slow-Speed Vertical Shaft Impeller

Construction and Operation

- The most common examples are the multiple-paddle (gate) agitator, counter-rotating agitator, and anchor and gate agitator, which agitate and mix the high viscosity material in a tank with suitable arrangements.

- The anchor agitator is a type of paddle agitator, commonly used in heated mixing vessels (Figure 14.8). In this case, there is a very small clearance between the blades and the inner wall of the vessel, and, hence, the blades continuously scrap the heat transfer surface and prevent fouling on the hot surface.

- The agitator usually operates in combination with a paddle or other agitator, usually rotating in the opposite direction at higher speed, which are called counter-rotating agitators.

- In some mixers, the arms on the gate intermesh with stationary arms on the anchor to surge the shearing action.

- Radial movement in the food mix is promoted by the use of inclined vertical blades.

Applications

- These mixers develop high shearing forces and are used for more viscous liquids. However, the mixing is not as good as other type of agitators.

- The anchor and gate agitator is often used with heated mixing vessels.

Kneaders, Dispersers, and Masticators

The kneaders, dispersers, and masticators basically consist of two heavy blades rotating on parallel horizontal shafts, which are fixed in a short trough. The bottom of the trough is saddle

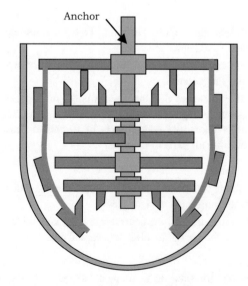

FIGURE 14.8   Schematic diagram of an anchor and gate agitator (after McDonagh, 1987)

shaped. The blades turn toward each other at the top, which forces the dough to move downward over the point of the saddle, then the dough is sheared between the blades and the wall of the trough.

The blades usually turn at different speeds, usually with a speed ratio of about 1.5:1. In some machines the blades overlap and turn at the same speed or with a speed ratio of 2:1 (i.e., one rotation of one blade corresponds to two rotations of the other blade in the same time interval).

Different designs of blades are available. The common *sigma blade* is used for general-purpose kneading. The *double-naben* or *fishtail blade* is used for heavy plastic materials. The *dispersion blade* is used for situations where high shear forces are required to disperse powders or liquids into plastic or rubbery masses. The *disperser* is thus heavier in construction than a kneader and wants more power.

A *masticator* is still heavier with heavier blades and requires even more power. Both shafts are independently driven. These are often called *intensive mixers* and are used in the plastic industry. In many kneading machines the trough is open, but, in some designs, known as *internal mixers*, the mixing chamber is closed during the operation.

Sometimes a vessel, in which the blades can completely sweep the inner surface of the pan, can be used for kneading.

The sigma blade mixer is common in the food industry, and we will discuss it below.

Sigma Blade Mixer

Construction and Operation

- This is also known as a *Z-blade mixer* (as the blades have a Z-shape) or a *twin shaft horizontal blade mixer*. It consists of a metal trough in which two heavy-duty blades are fixed on two horizontal axes (Figure 14.9).
- The Z-shaped blades intermesh as they rotate toward each other at the same or different speeds. It produces shearing forces between the two blades and between the blades and the specially designed base of the trough. The blades usually rotate at 14–60 rev. min^{-1}.

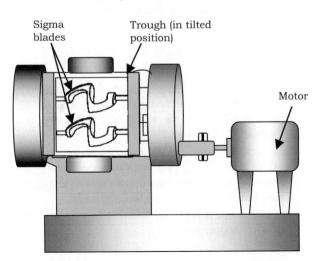

FIGURE 14.9 Schematic diagram of a two-arm kneader

- The time required may vary from 5 to 20 min or longer.
- The trough normally has a provision of tilting it for unloading. There may be a screw conveyor below the trough to carry the mixed material to the next processing section.

Salient Features

- A substantial amount of power is used by these mixers and it is dissipated as heat in the product. Thus, the mix may require to be cooled, and, hence, the wall of the trough may be jacketed for the flow of cooling medium.
- As the amount of power required is high, the mixing time should be less, which is possible only if high mixing efficiency is attained.
- Sometimes the mixture constituents may be heated while in the machine.
- The blade edges may be serrated to give a shredding action.

Applications

The mixer is used for mixing of dough and high viscous pastes.

Change Can Mixer

Construction and Operation

- The equipment is so named because the can and agitator assembly can be separated easily. It basically consists of a removable can of 20–400 l size that holds the material to be mixed. The complete agitator assembly is mounted concentrically with respect to the axis of can (Figure 14.10).
- During operation, after keeping the ingredients in the can, the agitator assembly is lowered into the can. After mixing, the assembly is raised and the can is emptied. Thus, such an arrangement helps in easily putting the load and discharging as well as cleaning the different parts of the assembly.
- In a specific design known as a *pony mixer*, the agitator consists of several vertical blades or fingers held on a rotating blade and positioned near the wall of the can.
- The blades are slightly twisted.
- The can may rest on a turntable driven in a direction opposite to that of the agitator. It helps in guiding all the mass to the blades for mixing.
- In the *beater mixer*, the can or vessel is kept stationery, and the agitator has a planetary motion, which helps it to reach all parts of the vessel. The shape of the beaters is designed so that there is less clearance between the beater and wall as well as the bottom of the mixing vessel.

Applications

These are used for small amounts of viscous pastes.

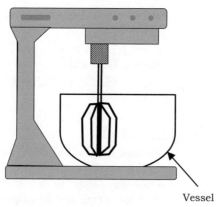
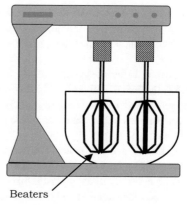

Vessel   Beaters

(a) Beater mixer with planetary
gear and one beater

(b) With two beaters

FIGURE 14.10   Schematic diagram of a beater mixer

Pug Mill

Construction and Operation

- The mixing chamber consists of a horizontal open trough or closed cylinder, inside which blades or knives set in a helical pattern rotate on a horizontal shaft.
- The solids continuously enter one end of the mixing chamber, and they are mixed as they move forward to the discharge end.
- The mill can be a single shaft or a double shaft type. The double shaft type is used when more rapid or more thorough mixing is required.
- The mixing chamber is usually cylindrical, but in some specific designs it may be polygonal in cross section to prevent sticky solids from being carried around with the shaft.
- Sometimes the pug mill is operated under a vacuum to remove air from clay or other materials. They may have jacketed walls for heating or cooling.

Applications

These are used to mix liquids with solids to form thick, heavy slurries.

Continuous Kneader

Construction and Operation

- In a continuous kneader, there is usually a single horizontal shaft having rows of teeth arranged in a spiral pattern. As it rotates slowly in the mixing vessel, the material is mixed and moved from one end to the other in the chamber.
- The shaft turns and also reciprocates in the axial direction.
- There are stationary teeth fitted in the wall of the casing and, when in operation, the teeth on the rotor pass with close clearance between stationary teeth set.
- For light solids the chamber is usually open and for plastic masses it is closed.

Applications

The machine is used for heavy, stiff, and gummy materials and when the capacity required is very high, i.e., several tons per hour.

If the discharge opening of a continuous kneader is restricted with the help of an extrusion die and the pressure and other parameters are controlled, then it works as an extruder.

Screw Conveyor Mixer

Construction and Operation

- The screw conveyor mixer is a continuous mixer, in which there may be single or twin screws fixed closely in a stationary casing (or barrel). The casing may be slotted in some applications.
- The clearance between the screw and the barrel wall is kept very small to enable the desired shearing and kneading action.
- The screws move the materials through the barrel and, at the same time, cause the mixing action.
- To increase the shearing action, the screw may be interposed with pins.

Applications

This kind of equipment is useful for mixing viscous and pasty materials, and it is also used for extrusion and butter margarine manufacture.

Planetary Mixers

Construction and Operation

- In this machine the rotating blades follow a planetary path inside the vessel, and, hence, the mixing takes place in all parts of the vessel. The blades usually rotate at 40–370 rev.min^{-1}.
- In another type of design, there are fixed rotating blades kept in an offset manner from the center of a revolving vessel. The vessel may revolve either counter-currently or co-currently with the blades.
- The clearance between the blades and the vessel wall is less in both types.

Applications

These are useful for blending ingredients and mixing pastes.

Static and Motionless Mixer
Construction and Operation

- These are mixing devices installed in the processing line and consist of a series of static mixing elements within a housing. Capital costs and maintenance requirements are considerably reduced since there is no requirement for agitators, tanks, and moving parts.
- The mixing is in the form of radial mixing, division of flow, and transient mixing.
- In radial mixing, the mixing elements deflect the fluid through a series of 180° rotations. The fluid is forced from the center to the wall of the housing and then again forced back to the center.
- In flow division, the material is continuously split into an increasing number of components (e.g., first split into two components by the first mixing element, then split into four components by the second element, and so on). In between each splitting, the material is rotated through 180°, which gives thorough mixing.
- The space between the elements allows relaxation of viscous materials after radial mixings.

Applications
Such mixers are used in chocolate manufacture.

In addition, equipment like butter churns, bowl choppers, and rollers also do the mixing operation while carrying out simultaneous homogenization or size reduction. Mixing of high viscosity materials is carried out in roller and colloid mills along with size reduction.

Mixers for Low Viscosity Liquids
For mixing of low viscosity liquids usually a vertical axis cylindrical tank is used in which an impeller is fixed onto a shaft (Figure 14.11). The impeller is driven by a motor. The top of the tank may be open or closed. The tank bottom is rounded as the fluid currents would not penetrate into sharp corners or regions. The round bottom also helps subsequent cleaning operations. The liquid is fed into the tank from the top with a suitable arrangement. There is also a pipe at the bottom for discharge of the mixed product. There may be provisions for heating and cooling of the product mix during mixing.

In a vertical cylindrical tank, the depth of the liquid should be almost the same or a little higher than the tank diameter. If the depth of the liquid is more, then two or more impellers are fixed on the same shaft.

Impeller agitators can be classified as either axial-flow or radial-flow impellers. In the axial- flow ones, the flow currents are parallel with the axis of the impeller shaft. In the other type, the flow currents are in a tangential or radial direction. The three main types of impellers are paddles, propellers, and turbines. The paddle agitators consist of flat blades. A propeller is a device with a rotating hub; it works similar to an Archimedes' screw and, when rotated, it transforms rotational

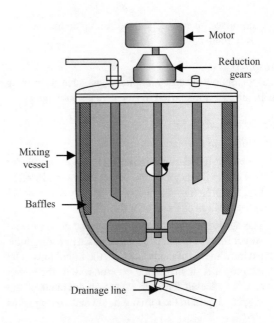

FIGURE 14.11 Schematic diagram of a vessel with agitator

power into a linear thrust by acting upon a working fluid such as water. Some common types of propellers are three-blade marine impellers, open straight blade turbines, vaned disk turbine, and pitched blade turbines.

Paddle Agitator
Construction and Operation

- The paddle agitator consists of wide flat blades fixed on a vertical shaft. The assembly is fitted inside a big tank.
- The paddle agitators usually have two or four bladed paddles. The blades are usually vertical; however, they may be pitched to improve the mixing.
- Usually the blade diameters are 50%–80% of the vessel diameter. They rotate at a slow to moderate speed (20–150 rev.min^{-1}). The width of the blade is typically one-sixth to one-tenth of its length.
- In the case of deep tanks, proper mixing is achieved by mounting several paddles one above the other on the same shaft.
- The movement of the blades creates turbulence and shear in the liquid and thus mixing takes place.
- Baffles are provided in the tank to improve the relative motion within the liquid.
- The tanks, which do not have baffles, have pitched blades to maintain longitudinal flow.

Salient Features

- The liquid is moved radially and tangentially due to the push of the paddles. In the case of pitched blades, there is a vertical component of flow also.
- In an un-baffled vessel, the slow speed of the paddle gives mild agitation; but higher speeds can cause swirling and, hence, baffles become necessary.

- The mixer gives excellent performance for low viscosity fluids.

Applications

It is used for low viscosity applications, such as mixing of sugar in juice. It is also used in water treatment plants.

Propeller Agitator

Construction and Operation

- *Propeller agitators* are impellers having short blades (1/4 of the diameter of the vessel or lower). In practice, the propeller diameters are usually less than 45 cm regardless of the size of the tank.
- The agitator is located in a position so as to prevent vortex formation and to promote radial and longitudinal movements of the liquids. As the liquid moves by the propeller, it strikes the floor or wall of the vessel and is deflected there. It causes entrainment of the stagnant liquid in the moving liquid and mixing takes place.
- Vessel wall is often fitted with baffles to break off rotational flow and to improve mixing.
- Propellers are operated at 400–1500 rev.min^{-1}, the higher speeds corresponding to smaller diameter ones.
- In the case of deep tanks, two or more propellers are fixed on the same shaft. Usually in such cases the propellers move the liquid in the same direction, but some may be fixed in a manner to work in the opposite direction to improve the turbulence.
- Toothed propellers are used for special purposes.

Applications

- Propeller agitators are used for liquids of low viscosity, specifically for preparing syrups or brines, diluting concentrated solutions, blending miscible liquids, and dissolving ingredients.
- These are effective in very large vessels.

Turbine Agitator

Construction and Operation

- The turbine agitator is similar to a multi-bladed paddle agitator, but it has shorter blades. The blades are attached to a rotating central shaft in the mixing tank.
- The diameter of the impeller ranges from 30% to 50% of the diameter of the vessel. They usually rotate at 30–500 rev.min^{-1}.
- Usually there are more than four blades. The blades may be straight or curved; they may also be pitched or vertical.
- A special type turbine agitator, namely the *vaned disc impeller*, has the blades mounted on a flat disc. The disk is fitted on a vertical shaft. The tank has baffles for aiding the mixing operation.

Special Features

- In low viscosity liquids, turbines generate strong currents for thorough entrainment in the liquid.

There is a zone of high turbulence and shear near the impeller.

- The flow has mainly radial and tangential components. As discussed earlier, vortexing and swirling may be induced by the tangential components of the flow, which should be avoided by the baffles or by using a diffuser ring.

Applications

Turbine agitators can be used for mixing liquids over a wide range of viscosities. These can be used for pre-mixing emulsions due to the high shearing forces generated at the edges of the impeller blades.

Powder-Liquid Mixers

Construction and Operation

- A uniform stream of powder can be mixed with liquid sprays in a chamber.
- Blades or rotors may carry out the subsequent mixing.
- Pipes having some stationary mixing blades fitted inside can also be used for mixing powders with liquids. The ingredients are pumped through the pipe for mixing to take place (Figure 14.12).
- As the liquid is moved in a pipe, these types of mixers involve a very short residence time.

Applications

As mentioned above, these are used for mixing powders with liquids.

Draft Tubes

The return flow to an impeller of any type approaches the impeller from all directions. In most cases, this is not considered a problem. However, when the direction and velocity of flow to the suction of the impeller must be controlled, for instance, when the solid particles that have a tendency to float on the liquid surface are to be dispersed in the liquid, draft tubes are used (Figure 14.13). They maintain high shear at the impeller itself. For propellers, the draft tubes are mounted around the impeller, and, for turbines, they are fixed immediately above the impeller. The draft tubes also help in the manufacture of certain emulsions.

The draft tubes increase the fluid friction and reduce the rate of flow for a given power input. So these are not used unless they are highly essential.

14.1.3 Factors Affecting the Selection of Mixing Equipment

The following factors are considered for the selection of mixing equipment.

Commodity Factors

- Type of food(s) and ingredients to be mixed;
- Amount of food constituents that need to be mixed;
- Desired degree of mixing.

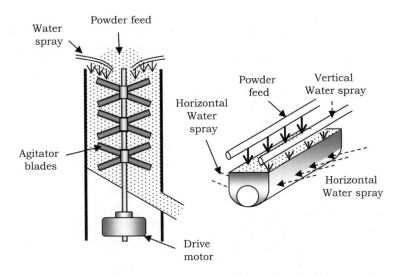

FIGURE 14.12 Schematic of different types of powder-liquid mixing methods

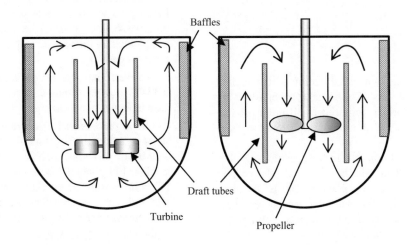

FIGURE 14.13 Flow pattern in draft tubes in baffled tanks

Machine Factors

- Energy consumption;
- Speed;
- Any special requirement during mixing for better mixing of the constituents.

14.2 Emulsification

14.2.1 Principle

Achieving a stable emulsion of two or more immiscible liquids by thorough mixing is the objective of emulsification. A smaller fraction, known as the dispersed phase, is distributed as minute droplets within the larger or the continuous phase. Butter (an emulsion of water in fat), low-fat spreads (an emulsion of water in fat), milk (an emulsion of fat in water), margarine, salad cream, and mayonnaise are a few examples of emulsions. Cakes, ice cream, and sausage meat are examples of more complex emulsions.

The formation and maintenance of a stable emulsion is difficult due to higher interfacial tension between the continuous and dispersed phases. Use of emulsifying agents helps the formation of emulsions by reducing the energy required to form an emulsion. The type and extent of the emulsifying agent affect the stability of emulsions. Stokes' law explains the different physical factors influencing the stability of an emulsion. As discussed in Chapter 12,

$$v = \frac{gD_p^2\left(\rho_p - \rho_s\right)}{18\mu} \qquad (14.26)$$

In this equation, v (m.s^{-1}) is the velocity of separation of the phases, D_p (m) is the diameter of droplets being dispersed, ρ_p (kg.m^{-3}) and ρ_s(kg.m^{-3}) are the densities of dispersed phase and continuous phase, and μ (N.s.m^{-2}) is the viscosity of continuous phase.

The emulsion will be more stable when the velocity of separation is small, i.e., when the droplet sizes are small, the densities of the two phases are very different and the continuous phase is of high viscosity. The droplet sizes normally range between 1μm and 10 μm. The stability of emulsions is greatly influenced by interfacial forces acting at the globule surfaces.

Naturally occurring phospholipids and proteins work as emulsifying agents. However, many synthetic agents, such as esters of glycerol and esters of fatty acids, are normally used in food processing, which have proved to be more effective. Synthetic emulsifying agents are further classified as polar (which bind to water and produce o/w emulsions) and non-polar types (which adhere to oils producing w/o emulsions).

> The hydrophile-lipophile balance (HLB) value is used to characterize the emulsifying agents. The lipophilic emulsifying agents have low HLB values (below 9) and are used for water in oil emulsions. High HLB values (11 to 20) correspond to hydrophilic characteristics and such agents are used for oil-in-water emulsions, solubilizers, and detergents. The intermediate group emulsifying agents (HLB values between 8 and 11) are used as wetting agents.

The ionic and non-ionic type classification is also applied to polar emulsifying agents. The ionic type emulsifying agents have different surface activities at different pH values; however, the activity of non-ionic emulsifiers is independent of pH.

Phospholipids (e.g., lecithin), protein (e.g., gelatin, egg albumin), potassium, or sodium salts of oleic acid, and sodium stearyl-2-lactylates are ionic emulsifiers having HLB values between 18–20. Examples for non-ionic emulsifiers having 2–5 HLB values are glycerol monostearate, polyglycerol esters, and sorbitol esters of fatty acids.

Hydrocolloids, such as alginates, carboxymethyl cellulose (CMC), guar, gum arabic, and locust bean come under the group of non-ionic polar agents. Pectin and methyl cellulose are also commonly used non-ionic emulsifying agents. The purpose of stabilizers, when dissolved in water is to form viscous solutions. The stabilizers, which are polysaccharide hydrocolloids, increase the viscosity when used in oil-in-water emulsions and prevent coalescing of droplets by forming a 3-D network in the emulsion. Microcrystalline cellulose and related cellulose powders are examples of stabilizers for water-in-oil emulsions.

The selection of a proper emulsifying agent and stabilizer during food preparation is as important as the control over the homogenization conditions, as these affect the texture and mouthfeel of the food.

14.2.2 Equipment

Emulsions of low viscosity liquids can be prepared by using high-speed mixers, such as turbines and propeller agitators. The mixers create shearing action for emulsification. Homogenizers permit size reduction of droplets in the dispersed phase and those of emulsifying agents. Homogenization also prevents coalescing of the fine droplets by reducing the interfacial tension among the phases, imparting greater stability to the resultant emulsion.

Check Your Understanding

1. Flour (99.2 kg) was mixed with 0.80 kg of salt. After mixing the bulk for a while, samples, each weighing 10 g, were taken for analysis. The weights of salt in samples were 0.12, 0.092, 5.6, 0.074, 0.003, 3.8, 0.11, 6.2, 0.004, and 0.92 g. From the mean composition, compute the standard deviation of the sample compositions.

   After some time again 10 samples were taken and the weights per 10 g sample were found as 0.07, 0.09, 0.12, 0.006, 0.05, 0.0078, 0.088, 0.09, 0.094, and 0.01 g. See whether the mixed condition is better than the previous case or not.

   Answer: The standard deviations are 0.2872 and 0.00432. Lower standard deviation in the second case indicates good mixing.

2. In a food processing unit operation, 5 kg powdered sugar is mixed to 45 kg flour and it is desired to have a uniform mixture. To observe the uniformity of mixing, 100 g samples were taken from 10 locations (10 samples were taken) after 5 minutes and 10 minutes and were analyzed. The percentages of sugar in the samples were as below.

   After 5 min: 0.2, 12.5, 2.4, 14.8, 11.2, 1.8, 13.4, 0.9, 6.2, 2.6

   After 10 min: 4.8, 5.6, 6.8, 8.9, 10.1, 10.6, 11.2, 11.8, 12.1, 12.6

   Find the mixing index after 5 minutes and 10 minutes.

   Answer: 0.963 and 0.991

3. In a food processing operation, dried mushroom in powder form is mixed with starch in a ratio of 35:65. After three minutes of mixing, the variance of the sample compositions of starch was found to be 0.078. If the sample contains 30 particles and the dried mushroom and starch are almost the same size, find the time required for mixing with a value of 0.01 for the maximum sample composition.

   Answer: 11.88 min

4. During preparation of a food material, 800 kg flour is mixed with 80 kg of sugar. Ten 100 g samples are taken for analysis of the sugar percentage after 1, 5, and 10 minutes. The results are as follows:

	Percentage of sugar				
After 1 min	0.3	3	4.2	7.4	9.4
	13.2	16.4	7.2	2.6	0.5
After 5 min	3.2	5.4	8.5	9.4	13.2
	10.8	12.2	8.6	7.6	6.2
After 10 min	8.8	9.5	9.8	10	10.2
	11.4	10.8	9.4	9.1	7.2

Assuming that $\sigma_\infty = 0.01$ can express the perfect mixing condition, calculate the mixing index for each mixing time and the time required for thorough mixing.

Answer: Mixing indices are 0.501, 0.336, and 0.042 after 1, 5 and 10 minutes. The total time required for thorough mixing will be 14.48 minutes.

5. Concentrated apple juice (60°Brix) is needed to be diluted by mixing water in a ratio of 1 to 4 in a cylindrical tank 1.2 m in diameter using a propeller agitator of 15 cm diameter. The agitator operates at 900 rev.min^{-1}. At 30°C operating temperature, the viscosities of apple juice and water are 0.03 N.s.m^{-2} and 0.001 N.s.m^{-2}. The densities at that temperature are 1044 kg.m^{-3} for apple juice and 1000 kg.m^{-3} for water. What will be the effective Reynolds number for the mixing operation?

Answer: 145817

6. In a mixing tank operating at very high Reynolds number ($>10^4$), if the diameter of the impeller is doubled (other conditions remaining constant), how many times will the power required increase?

Answer: 32 times

(Hint: Power number, $P_0 = \dfrac{P}{\rho N^3 D^5}$)

Objective Questions

1. What are the factors those that affect the degree of mixing for powders and granules? What will happen if a perfectly mixed product is further stirred?

2. Explain the mixing indices for mixing of solids and powders. State the situations in which these mixing indices are used.

3. What are the factors that affect the rate of mixing in the case of solids?

4. State the relationship between the time of mixing and mixing effectiveness.

5. Explain the different velocity components in liquid mixing.

6. What are the modes of mixing high viscosity materials?

7. How are the mixing effectiveness of pastes and liquids expressed? What are the factors that affect the mixing effectiveness during mixing liquids and pastes?

8. Explain the dimensionless numbers that can be used to express the power requirement during mixing liquids.

9. Differentiate between the working principles of
   a. mixers for liquids and mixers for pastes;
   b. ribbon mixer and screw conveyor mixer;
   c. sigma blade mixer and change can mixer;
   d. paddle agitator, propeller agitator and turbine agitators.

10. Write short notes on the following
   a. Emulsification;
   b. tumbling mixer;
   c. factors affecting selection of mixing equipment;
   d. planetary mixer;
   e. slow speed liquid mixers.

BIBLIOGRAPHY

Brennan, J.G. (Ed.) 2006. *Food Processing Handbook*, 3rd ed. WILEY-VCH Verlag GmbH & Co. KGaA, Weinheim.

Brennan, J.G., Butters, J.R., Cowell, N.D. and Lilly, A.E.V. 1976. *Food Engineering Operations*, 2nd ed. Elsevier Applied Science, London.

Cullen, P.J. (Ed.) 2009. *Food Mixing: Principles and Applications.* Wiley-Blackwell, UK..

Earle, R.L. and Earle, M.D. 2004. *Unit Operations in Food Processing.* Web Edn. The New Zealand Institute of Food Science & Technology Inc., Auckland. https://www.nzifst.org.nz/ resources/unitoperations/index.htm

Fellows, P.J. 2000. *Food Processing Technology.* Woodhead Publishing, Cambridge, UK.

https://arvindablenders.com/vertical-cone-screw-blender-dryers.htm

Kessler, H.G. 2002. *Food and Bio Process Engineering*, 5th ed. Verlag A Kessler Publishing House, Munchen, Germany.

Leniger, H.A. and Beverloo, W.A. 1975. *Food Process Engineering.* D. Reidel, Dordrecht. .

Lewis, M.J. 1990. *Physical Properties of Foods and Food Processing Systems.* Woodhead Publishing, Cambridge.

Lindley, J.A. 1991a. Mixing processes for agricultural and food materials. 1: Fundamentals of mixing. *Journal of Agricultural and Engineering Research* 48: 153–170.

Lindley, J.A. 1991b. Mixing processes for agricultural and food materials. 2: Highly viscous liquids and cohesive materials. *Journal of Agricultural and Engineering Research* 48: 229–247.

Lindley, J.A. 1991c. Mixing processes for agricultural and food materials. 3: Powders and particulates. *Journal of Agricultural and Engineering Research* 48: 1–19.

Matz, S.A. 1972. *Bakery Technology and Engineering.* AVI Publishing Co., Westport, Connecticut.

McCabe, W.L., Smith, J.C. and Harriot, P. 1993. *Unit Operations of Chemical Engineering*, 5th ed. McGraw-Hill, Inc., New York.

Mcdonagh, M. 1987. Mixers for powder/ liquid dispersions. *The Chem. Engr.* March 28–32.

Oldshue, J.Y. 1969. Suspending solids and dispersing gases in mixing vessel. *Industrial and Engineering Chemistry* 61(9): 79–89.

Rielly, C.D. Mixing in food processing. In: Fryer P.J., Pyle D.L., Rielly C.D. (eds) *Chemical Engineering for the Food Industry. Food Engineering Series.* Springer, Boston, MA. https://doi.org/10.1007/978-1-4615-3864-6_10

Watson, E.L., and Harper, J.C. 1989. *Elements of Food Engineering*, 2nd ed. Van Nostrand Reinhold, New York.

B2

Processing Involving Application of Heat

15

Heat Exchange

Heat exchange is a common unit operation in food processing industries. The liquid foods are pasteurized and sterilized by heating them in different types of heat exchangers. Hot water blanching involves dipping food in hot water and the hot water is produced by heat exchangers. The heat exchangers are also integral parts of evaporation systems.

A heat exchanger is needed when thermal energy is required to be transferred between (a) two or more fluids, (b) between a fluid and a solid surface, or (c) between a fluid and solid particulate material. A wide variety of heat exchangers are employed in the food and allied industries. They are available in many shapes and forms; the selection depends mainly on the characteristics of the heat exchanging fluids and the desirable characteristics of the finished product. Examples of heat exchangers are air heaters, feed heaters, evaporators, boilers, condensers, and cooling towers. Usually, a heat exchanger does not have any moving part in it. There are, however, exceptions where some mechanical rotation or motion is necessary, such as a scraped-surface heat exchanger or a rotary regenerative exchanger.

15.1 Methods and Principles

The heat exchangers can be classified in many ways depending on the type of heat transfer process, such as direct type or indirect type, or based on construction or fluid flow patterns.

15.1.1 Indirect Heat Exchange

In indirect type heat exchangers (for example, the plate heat exchanger or the tubular heat exchanger), heat transfer occurs through the metal wall/plate/tube that separates the heating/cooling medium from the fluid to be heated or cooled. Temperatures continuously change at the exchanger surface and in the two fluids.

Rate of Heat Transfer

In these cases, the heat transfer is by conduction and convection in series. In Chapter 5, we discussed the heat transfer between two fluids through a flat surface/wall and between two fluids through a metallic pipe/cylinder and discussed the governing equations as follows:

$$\frac{q}{A} = \frac{\Delta t}{\dfrac{1}{h_i} + \dfrac{x_1}{k_1} + \dfrac{1}{h_o}} \tag{15.1}$$

$$q = \frac{\Delta t}{\dfrac{1}{h_i A_i} + \dfrac{\ln\left(\frac{r_2}{r_1}\right)}{2\pi Lk} + \dfrac{1}{h_o A_o}} \tag{15.2}$$

The following example describes the use of these equations for calculating heat transfer in heat exchangers.

Example 15.1 In a tubular heat exchanger, steam flowing on the outside of the exchanger tubes is heating the milk flowing through the tubes. Initial experiments show that, based on the outside area, the overall coefficient of thermal energy exchange is 1365 W.m^{-2}.K^{-1}. Fouling deposit on the inside surface of the tubes, after the milk is processed for some time, causes the overall coefficient to drop to 255 W.m^{-2}.K^{-1}. Determine the value of the coefficient of thermal energy exchange for the fouling. The outside and inside diameters of the tubes are 33 mm and 26 mm, respectively.

SOLUTION:

Let U_1 and U_2 be the resistances without fouling and with fouling. Resistance to heat transfer is taken as negligible for the outside surface of the tube since steam is condensing. Therefore, the heat transfer resistance in the beginning is written as follows:

$$\frac{1}{U_1 A_o} = \frac{\ln\left(r_o/r_i\right)}{2\pi Lk} + \frac{1}{A_i h_i}$$

i.e. $\dfrac{1}{U_1} = \dfrac{A_o \ln\left(r_o/r_i\right)}{2\pi Lk} + \dfrac{A_o}{A_i h_i} = \dfrac{r_o \ln\left(r_o/r_i\right)}{k} + \dfrac{r_o}{r_i h_i} = \dfrac{1}{1365}$

Now, when fouling occurs, the resistance is

$$\frac{1}{U_2 A_i} = \frac{1}{A_i h_f} + \frac{\ln\left(r_o/r_i\right)}{2\pi kL} + \frac{1}{A_i h_i}$$

i.e. $\dfrac{1}{U_2} = \dfrac{A_i}{A_i h_f} + \dfrac{A_i \ln\left(r_o/r_i\right)}{2\pi kL} + \dfrac{A_i}{A_i h_i}$

i.e. $\dfrac{1}{U_2} = \dfrac{1}{h_f} + \dfrac{2\pi r_i L \ln\left(r_o/r_i\right)}{2\pi kL} + \dfrac{1}{h_i} = \dfrac{1}{h_f} + \dfrac{r_i \ln\left(r_o/r_i\right)}{k} + \dfrac{1}{h_i}$

It can also be written as:

$$\frac{1}{U_2} = \frac{1}{h_f} + \left(\frac{r_i}{r_o}\right)\left(\frac{r_o \ln\left(r_o/r_i\right)}{k} + \frac{r_o}{r_i h_i}\right)$$

DOI: 10.1201/9781003285076-19

Thus, $\dfrac{1}{255} = \dfrac{1}{h_f} + \dfrac{1}{1365} \times \dfrac{r_i}{r_o}$

or $\dfrac{1}{h_f} = \dfrac{1}{255} - \dfrac{1}{1365} \times \dfrac{13 \times 10^{-3}}{16.5 \times 10^{-3}}$

or $h_f = 299.4 \; W.m^{-2}.K^{-1}$

Mean Temperature Difference

As mentioned earlier, the basic equation for heat transfer by convection is as follows:

$$q = U.A.\Delta t,$$

Here, temperature difference Δt refers to two fluids or a fluid and a surface.

It is important that the mean temperature difference, Δt_m, is correctly determined for use in the thermal exchange equations.

The *log mean temperature difference*, Δt_{lm} (also referred as LMTD in this text), for heat exchangers can be given as follows:

$$\Delta t_{lm} = \dfrac{\Delta t_1 - \Delta t_2}{\ln\left(\dfrac{\Delta t_1}{\Delta t_2}\right)} \qquad (15.3)$$

where, Δt_1 is temperature differences between the hot and cold fluids at section 1 and Δt_2 is the temperature differences at section 2. The entry point of either hot or cold fluid can be taken as section 1 and the other as section 2.

Figure 15.1(a and b) shows the temperature profiles for concurrent and counter-current heat exchangers. t_{hi} and t_{ho} denote the temperatures of the hot fluid at inlet and exit, and t_{ci} and t_{co} denote the temperatures of the cold fluid at inlet and exit, respectively.

In the 1st case (concurrent flow),

$$\Delta t_1 = t_{hi} - t_{ci} \text{ and } \Delta t_2 = t_{ho} - t_{co}$$

Thus, $\Delta t_{lm} = \dfrac{(t_{hi} - t_{ci}) - (t_{ho} - t_{co})}{\ln\left(\dfrac{t_{hi} - t_{ci}}{t_{ho} - t_{co}}\right)} \qquad (15.4)$

In the 2nd case (counter current flow),

$$\Delta t_1 = t_{hi} - t_{co} \text{ and } \Delta t_2 = t_{ho} - t_{ci}$$

Thus, $\Delta t_{lm} = \dfrac{(t_{hi} - t_{co}) - (t_{ho} - t_{ci})}{\ln\left(\dfrac{t_{hi} - t_{co}}{t_{ho} - t_{ci}}\right)} \qquad (15.5)$

Figure 15.1(c) shows a situation where the temperature of the heating/cooling medium practically remains constant. For example, in evaporators, the heating is usually done by the condensation of steam, and practically there is no change in the temperature of the heating medium. The mean temperature difference in this case can be obtained as equation 15.4.

Here, $t_{hi} = t_{ho}$ (we can denote it as t_h)

Thus, $\Delta t_{lm} = \dfrac{(t_h - t_{ci}) - (t_h - t_{co})}{\ln\left(\dfrac{t_h - t_{ci}}{t_h - t_{co}}\right)} \qquad (15.6)$

Example 15.2 Milk temperature in a counter-current heat exchanger rises from 4°C to 60°C and the temperature of the heating medium comes down from 80°C to 70°C. Assume the coefficient of overall heat transfer (U) and the heat exchanger area to be 800 W.m^{-2}.K^{-1} and 60 m^2, respectively. Find the heat supplied to the milk.

SOLUTION:

$$\Delta t_1 = 70 - 4 = 66°C$$

$$\Delta t_2 = 80 - 60 = 20°C$$

$$\Delta t_{lm} = (66 - 20) / \ln(66 / 20) = 46 / \ln(66 / 20)$$

$$= 38.53°C$$

Now, $q = U.A.\Delta t$

$$= 800 \left(W.m^{-2}.K^{-1}\right) '' 60 \left(m^2\right) \times 38.53 (K)$$

$$= 1849440 \; W = 1849.44 \; kW$$

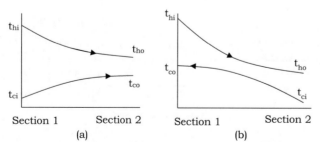

FIGURE 15.1 Temperature profiles of heat exchangers: (a) parallel flow type, (b) counter flow type, and (c) temperature of heating medium remains constant

Example 15.3 For the previous problem, find the milk flow rate into the system. Make any other assumption if necessary.

SOLUTION:

We know that heat is supplied to raise the milk temperature to the desired extent.

$$\text{Thus, } q = m\, c_p\, \Delta t$$

Specific heat of milk is taken as 4 kJ.kg^{-1}. K^{-1}.

$$1849440 \text{ J. s}^{-1} = m \text{ (kg. s}^{-1}) \times 4000 \text{ (J.kg}^{-1}.\text{K}^{-1}) \times (60\text{-}4) \text{ (K)}$$

$$\text{Therefore, } m = 8.256 \text{ kg. s}^{-1}$$

Example 15.4 The temperature of milk increases from 4°C to 64°C in a continuous heating system where hot water enters concurrently into the heat exchanger at 80°C. Find the temperature change for the hot water assuming that the hot water flow rate is 50 kg. s^{-1} and the milk flow rate is 6 kg.s^{-1}. The coefficient of overall heat transfer is 860 W.m^{-2}.K^{-1}. Take the specific heats of milk and water as 4 and 4.2 kJ.kg^{-1}.K^{-1}, respectively. Also determine the exchanger's heat transfer area.

SOLUTION:

For the milk, $q_m = m \times c_p \times \Delta t$

$$= 6\left(\text{kg.s}^{-1}\right) \times 4\left(\text{kJ.kg}^{-1}.\text{K}^{-1}\right) \times 60\left(\text{K}\right)$$

$$= 1440 \text{ kW}$$

For the water, $q_w = m \times c_p \times \Delta t$

or, $50 \text{ (kg.s}^{-1}) \times 4.2 \text{ (kJ.kg}^{-1}.\text{K}^{-1}) \times \Delta t \text{ (K)} = 1440 \text{ kW}$

$$\text{or, } \Delta t = 6.86 \text{ K (or °C)}$$

Outlet temperature of heating water is

$$80 - 6.86 = 73.14°\text{C}$$

The rate of heat transfer is calculated as follows:

$$\text{U.A. } \Delta t_{lm} = 1440 \text{ kW}$$

Inlet temperature difference, $\Delta t_1 = 80 - 4 = 76°\text{C}$;

Outlet temperature difference, $\Delta t_2 = 73.14 - 64 = 9.14°\text{C}$

Thus, LMTD, $\Delta t_{lm} = \left(\Delta t_1 - \Delta t_2\right) / \ln\left(\Delta t_1 / \Delta t_2\right)$

$$= \left(76 - 9.14\right) / \ln\left(76 / 9.14\right) = 31.57°\text{C}$$

Heat transfer

area, $A = Q / \left(U.\Delta t_{lm}\right)$

$$= 1440 \times 10^3 \left(\text{W}\right) / \left(\left(860 \text{ W.m}^{-2}.\text{K}^{-1}\right) \times \left(31.57 \text{ K}\right)\right)$$

$$= 53.04 \text{ m}^2.$$

Example 15.5 For the previous problem, if steam at atmospheric pressure is utilized for heating instead of hot water, find the amount of steam required. Assume that only the steam's latent heat is used to heat the milk.

SOLUTION:

Latent heat of steam at atmospheric pressure is 2257 kJ.kg^{-1}

$q = m \times \lambda$ (m = mass of steam condensed for heating)

$$\text{or, } 1440 \text{ (kJ. s}^{-1}) = m \text{ (kg.s}^{-1}) \times (2257 \text{ kJ.kg}^{-1})$$

$$\text{Thus, } m = 1440/2257 = 0.638 \text{ kg. s}^{-1}$$

Check that if the heat of condensation is used for heating, then the amount of water/steam required is considerably reduced.

Example 15.6 Milk is being processed at the rate of 2000 kg.h^{-1} and its post-processing temperature is 73°C. This milk requires cooling to 5°C for which refrigerant water is available at -5°C. Employing a counter-current heat exchanger, the water leaves the heat exchanger at 10°C. A 3 mm stainless steel pipe with internal diameter of 2.5 cm is used for the thermal exchange. The milk side and the water side convective heat exchange coefficients are given as 1260 and 3050 W.m^{-2}.K^{-1}, respectively. Thermal conductivity of the steel used for the pipe is 21 W.m^{-1}K^{-1}. Determine the length of the heat exchanger pipe.

SOLUTION:

Thermal resistance for a heat exchanger surface is found as follows:

$$\frac{1}{U} = \frac{1}{h_a} + \frac{x}{k} + \frac{1}{h_b}$$

$$\frac{1}{U} = \frac{1}{1260} + \frac{3 \times 10^{-3}}{21} + \frac{1}{3050} = 1.26 \times 10^{-3} \text{ m}^2.\text{K.W}^{-1}$$

Therefore, U = 790.90 W.m^{-2}.K^{-1}
    Now, q = U.A.Δt_m

Here $\Delta t_{lm} = \dfrac{\left(73-10\right) - \left(5 - (-5)\right)}{\ln\left[\left(73-10\right) / \left(5 - (-5)\right)\right]} = 28.796°\text{C}$

Thus, q = (790.90) × (π.0.025.L)
    Again, heat removed from the milk q = m.c_p.($t_a - t_b$)

$$\text{or } q = (2000/3600) \times (3.9 \times 10^3) \times (73\text{-}5)$$
$$= 147333.3 \text{ J.s}^{-1}$$

Hence, A = 147333.3 / (790.90 ×28.796)
= 6.47 m²

As, A = π D L,

Hence, the length of the pipe, L = 6.47 /
(π ×0.025) = 82.38 m.

Example 15.7 Determine the initial steam requirement to heat 50 kg of cream at 25°C in a jacketed pan having a heating surface area of 0.8 m². Steam with latent heat content of 2197 kJ.kg⁻¹ is available at saturation temperature of 121.8°C and 2.1 bar absolute pressure. The overall heat transfer coefficient for the pan surface is 400 W.m⁻²K⁻¹.

SOLUTION:

Initial temperature of cream = t_c = 25°C
   Therefore, initial heat transfer rate, q = 400 × 0.8× (121.8-25) = 30976 W
And the amount of steam = q / λ = 30976/ (2197×10³) = 0.0141 kg.s⁻¹

Example 15.8 Find the amount of steam required in a heat exchanger of shell and tube type where milk at 10°C is pumped into the tube side to heat the milk to 70°C. Assume that only latent heat of steam, supplied in the shell at 130°C, is used for the heating.

In this system, if the tube side is maintained such that the evaporation of milk takes place at 70°C, then what will be the steam required per kg of evaporation per hour. The heat of vaporization at 70°C may be taken as 2332 kJ.kg⁻¹. Milk has 12% total solids.

SOLUTION:

For the 1st situation:
   We take 1 kg.s⁻¹ milk as the basis for calculation.
   The steam at 130°C has latent heat of condensation as 2173.7 kJ.kg⁻¹ (from steam table).

By heat balance, $m \times c_p \times \Delta t = W_s \times \lambda$

or, 1 (kg. s⁻¹) × 4000 (J.kg⁻¹.K⁻¹) × (70 - 10)(K)
= W_s (kg.s⁻¹) ×2173.7×10³ (J.kg⁻¹)

or, W_s = 0.11 kg. s⁻¹

Thus, we require 0.11 kg.s⁻¹ of steam per 1 kg.s⁻¹ of milk flow.
   For the 2nd situation:
   The heat balance is $m \times c_p \times \Delta t + V \times h_{fg} = W_s \times \lambda$

Where, V is the amount of vapor removed and h_{fg} is the latent heat of vaporization.
   From 1 kg of milk heated, the potential evaporation will be 0.88 kg. Thus,

$$1\left(kg.s^{-1}\right) \times 4000\left(J.kg^{-1}.K^{-1}\right) \times (70-10)(K)$$

$$+ 0.88 \times 2332 \times 10^3 \left(J.kg^{-1}\right)$$

$$= W_s\left(kg.s^{-1}\right) \times 2173.7 \times 10^3 \left(J.kg^{-1}\right)$$

or, W_s = 1.05 kg.s⁻¹
The amount of steam is 1.05 kg.s⁻¹ for evaporation of 0.88 kg water from milk. So for 1 kg.h⁻¹ evaporation, the steam required will be 1.05/0.88 = 1.19 kg.s⁻¹

Example 15.9 A stainless steel tube (k_a = 19 W.m⁻¹.K⁻¹) of 4 cm outer diameter and 2 cm inner diameter is covered with 3 cm thick insulation (k_b = 0.2 W.m⁻¹.K⁻¹). The inner wall of the pipe is at a temperature of 120°C. The temperature is 40°C outside the insulated pipe. At what rate is the heat lost per meter length of the tube?

SOLUTION:

Given:

Internal diameter of tube = 20 mm,
i.e. r_1 = 0.01m
Outer diameter of tube = 40 mm, i.e. r_2 = 0.02 m
Insulation thickness = 30 mm, i.e. r_3 = 0.05 m,
Inside temperature, t_1 = 120°C
Outside temperature, t_2 = 40°C
The tube length is 1 m.

The heat transfer rate through a composite cylinder is given as follows:

$$q = \frac{\Delta t}{\Sigma R}$$

Here, $\Sigma R = \dfrac{r_2 - r_1}{k_1 A_{lm_1}} + \dfrac{r_3 - r_2}{k_2 A_{lm_2}}$

Hence, the log mean area for each case has to be found.
   Pipes inside surface area = A_1 = $2\pi r_1 L$ = 0.0628 m²
   Pipes outside surface area = A_2 = $2\pi r_2 L$ = 0.1257 m²
   Insulated pipes outside surface area = A_3 = $2\pi r_3 L$= 0.314 m²

$$A_{lm_1} = \frac{A_2 - A_1}{\ln\left(A_2 / A_1\right)} = \frac{0.1257 - 0.0628}{\ln\left(0.1257 / 0.0628\right)} = 0.0906 \ m^2$$

Similarly,
$$A_{lm_2} = \frac{A_3 - A_2}{\ln\left(A_3/A_2\right)} = \frac{0.314 - 0.1257}{\ln\left(0.314/0.1257\right)}$$
$$= 0.2057\ m^2$$

The resultant heat transfer rate is calculated as follows:

$$q = \frac{120 - 40}{\dfrac{0.02 - 0.01}{(19)(0.0906)} + \dfrac{0.05 - 0.02}{(0.2)(0.2057)}} = 108.84\ W$$

Example 15.10 Milk stored at 4°C is pasteurized at 78°C in a plate type heat exchanger and is then cooled to 4°C. There are three heat exchangers: one for regenerative heating, one for heating, and the other for cooling. The milk flow rate is 4000 l.h⁻¹. In the heating section, the milk is heated by hot water, which enters the heat exchanger at 90°C. The flow rate of hot water is 7200 l.h⁻¹. In the cooling section, the chilled water enters at 1°C and, after receiving heat from milk, exits at 6°C. 70% of the heat exchange takes place in the regeneration section. The overall heat transfer coefficients for the heating, cooling, and regeneration sections are calculated as 2700 W.m⁻².K⁻¹, 2600 W.m⁻².K⁻¹ and 2500 W.m⁻².K⁻¹. If each plate of the heat exchanger has the heat transfer area of 0.6 m², calculate the number of plates required in each section.

Assume that the density of milk is 1030 kg.m⁻³, the density of water is 960 kg.m⁻³ at 90°C and 1000 kg.m⁻³ at 1°C/6°C. The specific heat of water and milk may be taken as constant at 4.2 kJ.kg⁻¹.K⁻¹ and 3.9 kJ.kg⁻¹.K⁻¹. Take any other assumption if necessary.

SOLUTION:

With the given values in the problem, the flow chart is drawn as below.

There is 70% heating in regenerator, i.e., temperature change of milk = 70% of (78 - 4) = 51.8°C

Cold milk leaves the regeneration section at 4 + 51.8 = 55.8°C

Milk is cooled in the regeneration section to 78 – 51.8 = 26.2°C

Specific heat of milk = 3.9 kJ.kg⁻¹.K⁻¹

Volumetric flow rate of milk =

$$4000\ l.h^{-1} = \frac{4000}{3600} = 1.11\ l.s^{-1} = 1.11 \times 1030/1000\ kg.s^{-1}$$
$$= 1.1433\ kg.s^{-1}$$

Volumetric flow rate of water =

$$7200\ l.h^{-1} = \frac{7200 \times 10^{-3}}{3600} = 2.0 \times 10^{-3}\ m^3.s^{-1}$$
$$= 2.0 \times 10^{-3} \times 960 = 1.92\ kg.s^{-1}$$

Amount of heat required to heat milk from 4°C to 78°C
$$= m\ c_P \Delta T$$
$$= 1.1433 \times 3.9 \times 1000 \times (78 - 4)$$
$$= 329956.38\ W$$

The heat balance and calculations for different sections are as below:

For the regeneration stage

Heat supplied = 70% of 329956.38 W = 230969.466 W

Temperature difference across the heat exchanger plates = 78 - 55.8 = 22.2°C

As pasteurized milk is used for heating raw milk, the temperature difference between two streams will remain the same throughout.

$$q = UA\ \Delta T$$

Thus, in this section,
$$A = \frac{q}{U \times \Delta T} = \frac{230969.466}{2500 \times 22.2} = 4.1616\ m^2$$

Each plate has an area of 0.6 m²

$$So, no\ of\ plates = \frac{4.1616}{0.6} = 6.93 \approx 7$$

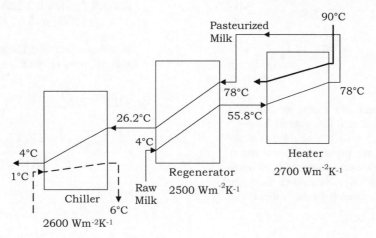

Heating Section

q = 30% of total heat supplied = $0.3 \times 329956.38 = 98986.914$

For hot water, $q = m\,c_p\,\Delta T$

$$\Delta T = \frac{q}{m.c_p} = \frac{98986.914}{1.92 \times 4200} = 12.275^0C$$

Temperature of hot water leaving the heating section is $90 - 12.275 = 77.72°C$
Thus, in the heater, $\Delta T_i = 90 - 78 = 12°C$ and ΔT_o $= 77.72 - 55.8°C = 21.92°C$

$$\Delta T_{LMTD} = \frac{21.92 - 12}{\ln\frac{21.92}{12}} = 16.46^0C$$

$$q = UA\,\Delta T_{LMTD}$$

i.e. $A = \frac{q}{U.\Delta T_{LMTD}} = \frac{98986.914}{2700 \times 16.46} = 2.23\ m^2$

Thus, no of plates $= \frac{2.23}{0.6} = 3.71 \approx 4$

Cooling Stage (For milk)

$$q = m\,c_p\Delta T$$
$$= 1.1433 \times 3900 \times (26.2 - 4)$$
$$= 98986.914\ W$$

In this case, $\Delta T_i = 26.2 - 6^0C = 20.2^0C$

$$\Delta T_o = 4 - 1 = 3^0C$$

$$\Delta T_{LMTD} = \frac{20.2 - 3}{\ln\left(\frac{20.2}{3}\right)} = 9.07^0C$$

As done above, $A = \frac{q}{U\Delta T_{LMTD}} = \frac{98986.914}{2600 \times 9.07} = 4.197\ m^2$

No of plates $= \frac{4.197}{0.6} = 6.96 \approx 7$

Example 15.11 Saturated steam at 205 kPa (absolute), 121°C and at a flow rate of 2 kg. s⁻¹ is carried through a tube having 50 mm outer diameter and 30 mm inner diameter. The steam tube has a 60 mm insulation layer of silica foam with k= 0.055 W.m⁻¹.K⁻¹. How much heat per meter length of the insulated tube is lost if the

ambient temperature is 30°C? Neglect the tube's thermal resistance.

SOLUTION:
Given:

$$OD = 50\ mm,\ ID = 30\ mm,$$
$$m = 2kg.s^{-1},\ k_2 = 0.055\ W.m^{-1}.K^{-1}$$
$$r_1 = 0.015\ m,\ t_3 = 30°C,\ r_2 = 0.025\ m,\ r_3 = 0.025 + 0.06 = 0.085\ m$$
$$t_1 = 121°C$$

Heat transfer is taking place in a cylinder.

$$\text{Thus, } q = \frac{2\pi L(t_1 - t_3)}{\frac{1}{k}\ln\left(\frac{r_2}{r_1}\right) + \frac{1}{k_2}\ln\left(\frac{r_3}{r_2}\right)}$$

The first term of the denominator is 0 since the heat resistance of tube wall can be neglected

or, $\frac{q}{L} = \frac{2\pi(121 - 30)}{\frac{1}{0.055}\ln\left(\frac{0.085}{0.025}\right)} = 25.41\ W.m^{-1}$

Thus, the rate of heat loss per m length of the insulated tube is 25.41 W.

Example 15.12 In case of a sterilization system involving indirect heating, saturated steam at 150°C is utilized for heating milk from 85°C to 140°C. Heating occurs in such a way that only latent heat of condensation is supplied to milk. If the milk flow rate is 2.6 m³.s⁻¹, find the amount of steam required for heating. Assume there is no contact between the heating medium and milk. Assume that the steam at 150°C gives away 2112 kJ.kg⁻¹ for condensation and the bulk density of milk is 1.03 kg.m⁻³.

SOLUTION:

The milk flow rate is 2.6 m³.s⁻¹ = 2.6 (m³.s⁻¹) × 1.03(kg.m⁻³) = 2.678 kg.s⁻¹

If m is the mass of milk to be heated, the quantity of steam, W_s, to be supplied is calculated as follows:

$$W_s = \frac{(m \times c_p \times \Delta t)}{\lambda}$$
$$= \frac{(2.678(kg.s^{-1}) \times 4000(J.kg^{-1}.°C^{-1}) \times (140 - 85)(°C))}{(2112 \times 10^3)(J.kg^{-1})}$$
$$= 0.279\ kg.s^{-1}$$

Heat Flow through a Tube with Constant Heating/Cooling Medium Temperature

The temperature of the cooling or heating fluid in many heat exchangers remains unchanged even though heat is exchanged with the other stream. For example, if steam is used for heating by using only the latent heat and the condensate temperature remains the same as that of the steam (as happens usually during evaporation of food), the heat transfer analysis is carried out as follows.

Let, as shown in Figure 15.2, the fluid flowing over the outside of the heat exchanger tube heats the liquid flowing inside it.

The temperatures of the liquid and the outside temperature are t_0 and t', respectively, at the initial stage.

Suppose L is the required length of the heat exchanger tube to raise the temperature of the liquid from t_0 to t. Considering a small element of liquid dx at a distance x from the liquid inlet, the liquid temperatures at x and x+dx are assumed to be t and t+dt, respectively.

If $\dot{m}$ = mass flow rate, heat gained by the liquid is $\dot{m}c_p\left\{(t+dt)-t\right\}$.

Heat transferred from outside through dx = $U.A.(t'-t)$

$$= U.(\pi D.dx)(t'-t)$$

Thus, the energy balance under steady state is presented as follows:

$$\dot{m}c_p\left\{(t+dt)-t\right\} = -U.(\pi D.dx)(t'-t)$$

$$\text{or, } \frac{dt}{t'-t} = \frac{-U.\pi D.dx}{\dot{m}.c_p}$$

$$\text{or, } \int_{t_0}^{t}\frac{dt}{t'-t} = \frac{-U.\pi D}{\dot{m}.c_p}\int_{0}^{L}dx$$

$$\text{or, } \log_e\frac{t'-t}{t'-t_0} = -\frac{U.\pi D}{\dot{m}.c_p}L$$

$$\text{or, } \frac{t'-t}{t'-t_0} = e^{-\frac{U.\pi D}{\dot{m}.c_p}L} \tag{15.7}$$

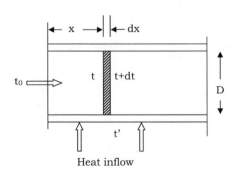

FIGURE 15.2 Analysis of heat flow through a tube with constant heating/cooling medium temperature

where, t is the fluid temperature at distance L from the inlet.

The above equation can be satisfactorily used for tubular heat exchangers and is valid for steady state conditions. Remember that $\pi DL = A$ is the total area available for thermal exchange in the heat exchanger.

In such cases, either the heating source or the heated fluid may flow within the tube.

> **Example 15.13** Milk is cooled from 60°C initial temperature using direct-expansion type surface cooler having 1.8 m² heat transfer area. The refrigerant is supplied at -4°C. Determine the final milk temperature when the overall thermal conductance is 800 W.m⁻².K⁻¹ and the milk flow rate is 36 kg.min⁻¹. Take any other assumptions necessary.

SOLUTION:

We assume the milk's specific heat as 4 kJ.kg⁻¹.°C⁻¹

The following steady state equation is used.

$$\frac{t-t'}{t_0-t'} = e^{\left(\frac{-UA}{c_p\dot{m}}\right)}$$

In the above equation we have multiplied both numerator and denominator on LHS of the equation 15.7 by -1. For use in this equation, the values are, A = 1.8 m², t_0 = 60°C, t' = -4 °C, U = 800 W.m⁻².K⁻¹, and the mass flow rate of milk = 36 kg.min⁻¹ = 0.6 kg.s⁻¹.

Substituting the values:

$$\frac{t-(-4)}{60-(-4)} = \exp\left(\frac{-(800)(1.8)}{(4000)(0.6)}\right)$$

$$t+4 = (64)e^{-0.6}$$

$$t+4 = 64\times0.5488 = 35.124°C$$

$$t = 31.124 °C$$

The final temperature of milk will be thus 31.124°C.

Heat Exchanger Effectiveness

Analysis of the heat exchanger is made convenient by the LMTD approach provided the inlet and outlet temperatures are known or can be determined. The analysis is easier with the effectiveness concept, E, when the heat exchanger temperatures at inlet and outlet need to be determined.

Heat exchanger effectiveness is defined as follows:

$$E_o = \frac{\text{Actual heat transfer}}{\text{Maximum possible heat transfer}}$$

$$\text{or, } E_o = \frac{t_{hi}-t_{ho}}{t_{hi}-t_c} \tag{15.8}$$

It may be mentioned that $t_{hi} - t_{ho} = (t_{hi} - t_c) - (t_{ho} - t_c)$

Substituting the values

$$E_o = 1 - e^{-\frac{U.A}{w_h.c_{ph}}} \qquad (15.9)$$

Heat Exchanger Efficiency

The efficiency is the ratio of heat transfer that actually occurs in the heat exchanger to the ratio of the maximum possible heat transfer that could occur. In a heat exchanger under steady state, heat absorbed by the cold fluid is equal to the heat given off by the hot fluid. If the cold and the hot fluids are indicated by subscripts c and h, respectively, then

$$q = m_c c_{pc}(t_{co} - t_{ci}) = m_h c_{ph}(t_{hi} - t_{ho}) \qquad (15.10)$$

Considering a counter flow heat exchanger (Figure 15.1(b)), let us denote $m_c.c_{pc}$ as C_c and $m_h.c_{ph}$ as C_h (the heat capacities). If the change of temperature for cold fluid is more than that in case of hot fluid, then $C_h > C_c$. Thus, let us denote C_h as C_{max} and C_c as C_{min}. If the heat exchanger had an infinite area, the temperature of hot fluid would coincide with that of the cold fluid at some point, whereby t_{co} would be equal to t_{hi}.

The exchanger efficiency is written as follows:

$$\varepsilon = \frac{\text{actual heat transfer}}{\text{maximum possible heat transfer}} = \frac{C_h(t_{hi} - t_{ho})}{C_c(t_{hi} - t_{ci})}$$

$$= \frac{C_{max}(t_{hi} - t_{ho})}{C_{min}(t_{hi} - t_{ci})}$$

$$(15.11)$$

Similarly, for situations where $C_h < C_c$, the heat exchanger efficiency would be

$$\varepsilon = \frac{C_c(t_{co} - t_{ci})}{C_h(t_{hi} - t_{ci})} = \frac{C_{max}(t_{co} - t_{ci})}{C_{min}(t_{hi} - t_{ci})} \qquad (15.12)$$

From equations 15.10-12,

$$q = \varepsilon.C_{min}(t_{hi} - t_{ci})$$

Equations 15.11 and 15.12 can be combined and written as:

$$\varepsilon = \frac{C_h(t_{hi} - t_{ho})}{C_{min}(t_{hi} - t_{ci})} = \frac{C_c(t_{co} - t_{ci})}{C_{min}(t_{hi} - t_{ci})} \qquad (15.13)$$

The rate of heat transfer in terms of LMTD and coefficient of overall heat transfer (U) is given as follows.

$$q = C_c(t_{co} - t_{ci}) = UA\frac{(t_{ho} - t_{ci}) - (t_{hi} - t_{co})}{\ln\frac{(t_{ho} - t_{ci})}{(t_{hi} - t_{co})}}$$

At any instant of time, temperatures t_h and t_c correspond to hot and cold fluids, respectively. Heat transfer change (dq) taking place across a heat exchanger's differential area (dA) results in a small temperature change in hot (dt_h) and cold (dt_c) fluids.

$$dq = C_c dt_c = C_h dt_h = U.dA.(t_h - t_c) \qquad (15.14)$$

The enthalpy balance at any instant of time is written as follows:

$$C_h(t_{hi} - t_h) = C_c(t_{co} - t_c) \qquad (15.15)$$

The equation 15.15 can be rearranged to be written as:

$$t_h - t_c = t_{hi} - \frac{C_c}{C_h}t_{co} + t_c\left(\frac{C_c}{C_h} - 1\right) \qquad (15.16)$$

Combining equations 15.14 and 15.16

$$\frac{U.dA}{C_c} = \frac{dt_c}{t_{hi} - \frac{C_c}{C_h}t_{co} + t_c\left(\frac{C_c}{C_h} - 1\right)} \qquad (15.17)$$

Equating the equation 15.17 with boundary conditions at A = 0, $t_c = t_{ci}$ and at A=A, $t_c = t_{co}$,

$$\frac{U}{C_c}\int_0^A dA = \int_{t_{ci}}^{t_{co}} \frac{dt_c}{t_{hi} - \frac{C_c}{C_h}t_{co} + t_c\left(\frac{C_c}{C_h} - 1\right)}$$

Solving the above equation, we get

$$-UA\frac{C_c - C_h}{C_c C_h} = \ln\left[\frac{(t_{hi} - t_{ci}) - \frac{C_c}{C_h}(t_{co} - t_{ci})}{(t_{hi} - t_{ci}) - (t_{co} - t_{ci})}\right] \qquad (15.18)$$

From equation 15.13, $\quad \varepsilon\frac{C_{min}}{C_c} = \frac{(t_{co} - t_{ci})}{(t_{hi} - t_{ci})} \quad (15.19)$

From equations 15.18 and 15.19,

$$\varepsilon = \frac{1 - \exp\left[-UA\left(\frac{C_c - C_h}{C_c C_h}\right)\right]}{\frac{C_{min}}{C_h} - \frac{C_{min}}{C_c}\exp\left[-UA\left(\frac{C_c - C_h}{C_c C_h}\right)\right]} \qquad (15.20)$$

Now, considering the two specific cases that were discussed earlier,

Case 1: $C_h < C_c$, i.e. $C_h = C_{min}$ and $C_c = C_{max}$, the above equation 15.20 will be

$$\varepsilon = \frac{1 - \exp\left[-\frac{UA}{C_{min}}\left(1 - \frac{C_{min}}{C_{max}}\right)\right]}{1 - \frac{C_{min}}{C_{max}}\exp\left[-\frac{UA}{C_{min}}\left(1 - \frac{C_{min}}{C_{max}}\right)\right]} \qquad (15.21)$$

Case 2: $C_h > C_c$, i.e., $C_c = C_{min}$ and $C_h = C_{max}$. In this case also we get the equation 15.21.

Thus, equation 15.21 is used to obtain the counter current heat exchanger efficiency.

Heat exchanger efficiency under parallel flow condition can also be obtained as follows.

$$\varepsilon = \frac{1 - \exp\left[-\dfrac{UA}{C_{min}}\left(1 + \dfrac{C_{min}}{C_{max}}\right)\right]}{1 + \dfrac{C_{min}}{C_{max}}} \quad (15.22)$$

For a given process, the number of required heat transfer units (NTU) are obtained as follows:

$$NTU = \frac{UA}{C_{min}} \quad (15.23)$$

Substituting the above and by using $\beta = C_{min}/C_{max}$,
For counter flow condition

$$\varepsilon = \frac{1 - \exp\left[-NTU(1-\beta)\right]}{1 - \beta\exp\left[-NTU(1-\beta)\right]} \quad (15.24)$$

For parallel flow condition,

$$\varepsilon = \frac{1 - \exp\left[-NTU(1+\beta)\right]}{1 + \beta} \quad (15.25)$$

For a given NTU, if the efficiency of a heat exchanger is determined under parallel flow condition, it will be lower than that for a similar heat exchanger with counter flow.

Example 15.14 Fruit juice having a specific heat of 3.51 kJ.kg⁻¹.K⁻¹ is flowing in a double tube concentric counter current heat exchanger at the rate of 2000 kg.h⁻¹. The juice is to be cooled from 90°C to 60°C using cold water entering at 30°C with a flow rate of 3500 kg.h⁻¹. The specific heat of water is 4.2 kJ.kg⁻¹.K⁻¹ and there is 10% heat loss from fruit juice to the surroundings. The overall heat transfer coefficient is 558 W.m⁻².K⁻¹ and the area of heat exchanger is 2.48 m². What will be the outlet temperature of cooling water? Also what is the NTU of the heat exchanger?

SOLUTION:

Only 90% of the heat supplied by the milk is taken by the cold water as 10% is lost to the surroundings. Making the heat balance,

$$3500 \times 4.2 \times (t - 30) = 0.9 \times 2000 \times 3.5 \times (90 - 60)$$

Thus, $t = 42.86\ °C$

$$C_{min} = \frac{(3.5 \times 1000 \times 2000)}{3600} = 1.94 \times 10^3\,W.K^{-1}$$

$$NTU = \frac{UA}{C_{min}} = \frac{(558 \times 2.48)}{1.94 \times 10^3} = 0.71$$

Thickness of Insulation

In different heating and cooling applications insulation materials are provided to reduce heat flow, i.e., to prevent heat loss (or gain). The wall of cold stores and the tubes of heat exchangers exposed to the outside are some common examples. In general, the insulating materials have a network of pores (or air spaces). The more pores, the more efficient is the insulator. Common insulating materials include silica, glass wool, glass, foamed polystyrene, cellulose, etc.

The surface area for heat transfer increases as the insulation thickness on a cylindrical object increases. On the other hand, the amount of outward conductive heat flow is restricted by the insulating material due to its low thermal conductivity. Thus, there will be a certain minimum insulation thickness up to which the heat loss will continue to increase. This minimum thickness of insulation is known as the *critical thickness*.

Assume that r_i and r_o are inner and outer radii of the insulated pipe. Let the temperatures of medium (say gas or liquid) inside the pipe be t_i and that flowing outside pipe be t_a. Here, r_o is required to be determined. Assume that the insulation has thickness t and thermal conductivity of insulator k_i.

R, the resistance to thermal energy flow, is represented as follows:

$$R = R_{conduction} + R_{convection}$$

$$R = \frac{\ln\left(\dfrac{r_o}{r_i}\right)}{2\pi k_i} + \frac{1}{2\pi h r_o} \quad (15.26)$$

The driving force for heat flow $= \Delta t = t_i - t_o$
The heat transfer rate out of the pipe $= q = \Delta t/R$

$$\text{i.e.,}\quad q = 2\pi(t_i - t_a) \bigg/ \left(\frac{\ln\left(\dfrac{r_o}{r_i}\right)}{k_i} + \frac{1}{h r_o}\right) \quad (15.27)$$

The minimum or critical radius is obtained by equating dq/dr_o to zero to obtain the condition where the heat transfer rate is the maximum. The final expression will be

$$r_{critical} = \frac{k_i}{h} \quad (15.28)$$

Adding insulation to the pipe would be counter-productive as long as $r_{critical}$ remains larger than the insulated pipe's outer radius.

15.1.2 Direct Heat Exchange

In the case of direct type heat exchangers, the two fluids intended for exchange of heat are mixed directly. For example, for heating milk, superheated steam is directly mixed with the milk in a chamber. The product temperature is increased as the steam supplies the heat. During the process, the steam condenses and the condensate mixes with the product, i.e., some extra water is added to the product. This extra water added during the process is removed in an expansion chamber. Mass and heat balances in direct heating systems are discussed below. Later in this chapter, we will discuss the details of the construction and operation of direct heating systems.

Example 15.15 In a direct steam injection type heat exchange system, steam at 150°C is injected in milk to heat it from 80°C to 145°C. Direct steam injection also results in the addition of some extra water into the milk. Determine the additional water quantity added to the milk in the present situation. The steam at 150°C has 2112.2 kJ.kg⁻¹ latent heat.

SOLUTION:

In this case, the steam is directly mixed with the milk to increase its temperature. Condensation of steam provides the necessary heat to increase the milk's temperature. The temperature of the condensate drops down to the final milk temperature. The above situation can now be translated into the following equation.

Heat gained by milk = Heat lost by steam

i.e.$(m)(c_p)(t_2-t_1) = (W_s)(\lambda) + W_s.c_p.(t_s-t_2)$

Assume 1 kg.s⁻¹ is the milk flowing in the system. Let specific heats of milk and steam (condensate) be 4000 and 4200 J.kg⁻¹.°C⁻¹, respectively.

$$1\left(kg.s^{-1}\right) \times 4000\left(J.kg^{-1}.°C^{-1}\right) \times \left(145-80\right)\left(°C\right)$$

$$= W_s\left(kg.s^{-1}\right) \times 2112.2 \times 10^3\left(J.kg^{-1}\right)$$

$$+ W_s\left(kg.s^{-1}\right) \times 4.2 \times 10^3\left(J.kg^{-1}.°C^{-1}\right) \times \left(150-145\right)\left(°C\right)$$

i.e., 260000 J.s⁻¹ $= W_s\,(2112.2 + 21) \times 10^3$ J.s⁻¹
$= W_s \times (2133.2) \times 10^3$ J.s⁻¹

or, $W_s = 0.1218$ kg.s⁻¹

Thus, 0.1218 kg extra water is added per kg of milk.
The total heat supplied by the steam includes both latent and sensible components with the latent heat component being much higher than the sensible heat component. Thus, the sensible heat component of steam may be ignored in cases where very precise calculations are not required.

Example 15.16 Steam is utilized to heat 100 kg of food material by direct injection from 4°C to 80°C. The food has a specific heat of 3.5 kJ.kg⁻¹. K⁻¹. Steam is available at 120°C with an enthalpy

of 2700 kJ.kg⁻¹. Enthalpy of the steam condensate is 344 kJ.kg⁻¹. Find the steam quantity required for the heating application. Heat loss from the system to the atmosphere may be neglected.

SOLUTION:

Heat balance for direct steam injection system of heat exchange can be written as follows.

Heat going in with steam and milk = Heat going out with condensate and heated milk

If W_s (kg) is the amount of steam required to heat 100 kg of food material, then

heat going into the system with steam and milk

$= m\,c_p t_1 + W_s h_s$

$= \left(100\right)\left(3.5\right)\left(4\right) + \left(W_s\right)\left(2700\right)$ kJ

Heat out from the system with milk and condensate

$= m\,c_p t_2 + W_s h_{sc}$

$= \left(100\right)\left(3.5\right)\left(80\right) + \left(W_s\right)\left(344\right)$

Solving the above,

$(100)(3.5)(4) + (W_s)(2700) = (100)(3.5)(80) + (W_s)(344)$

or, $W_s = (100)(3.5)\,(80-4)/(2700-344) = 11.29$ kg

Thus, the enthalpy balance in direct heat transfer can also be simply written as in the above problem:

$(\text{Heat utilized})_{\text{to increase the temperature of milk}}$
$= (\text{Heat supplied})_{\text{by steam's condensation}}$

(Note: If the enthalpy of condensate is not given, then the product of $c_p.t_2$ (= 4.187×80) can be used as the heat of condensate at 80°C. In this example, the enthalpy of the condensate at 80°C is given; thus, it has considered the sensible heat loss by the condensate.

15.2 Heat Exchange Equipment

15.2.1 Classifications

The classification of heat exchange equipment (heat exchangers) is based on such factors as number of fluids, mechanism of heat transfer, heat transfer process, flow arrangements, and construction features.

On the basis of heat transfer processes, heat exchangers are either *indirect contact* or *direct contact* types. As the name indicates, the product and the heat transfer medium in an *indirect contact heat exchanger* are kept physically separated by employing a thin wall of a highly conducting material. Also called *surface heat exchangers*, indirect contact heat exchangers are further classified as *storage*, *direct transfer*, and *fluidized bed* types. A few examples of indirect transfer type heat exchangers, where the fluids are separated by a heat transfer surface, are plate-type, tubular, and extended surface exchangers.

There is direct physical contact between the product and the heating medium in *direct contact heat exchangers*. A common application is evaporative cooling, which involves both heat and mass transfer. These types of heat exchangers offer much higher heat transfer rates than the indirect types. There is no fouling problem as there is no heat transfer surface. Compared to the indirect contact types, the construction of direct physical contact types is relatively inexpensive.

The direct contact heat exchangers may be further classified as immiscible fluid exchangers, gas-liquid exchangers (spray dryers, spray tower, etc.), and liquid-vapor exchangers (steam injection and steam infusion heat exchangers). In particular, heat transfer rate is very high due to phase change when steam is the heating medium. Such heat exchangers are also known as *direct steam heaters*.

Classification on the basis of construction features leads to the heat exchangers being plate-type, tubular, extended surface, and regenerative. There are some other distinct heat exchangers also, such as a scraped surface exchanger. Though the scraped surface heat exchanger can be categorized as a tubular type, there are modifications over the conventional tubular exchangers to make them distinct from others in the group.

Heat exchanger classification based on the number of fluids has two, three, or more fluid types. There may be several flow arrangements depending on the fluids and the operational requirements.

Classification is based on the mechanism of heat transfer (such as single- or two-phase convection on both sides or one side) as well as according to the functions as heaters, coolers, chillers, condensers, etc. It is a sensible heat exchanger if no phase change occurs in any of the fluids during the heat exchange.

Some heat exchangers, such as fired heaters and boilers, have internal thermal energy sources as electric heaters. The fluidized bed exchangers may have combustion and chemical reaction within them.

15.2.2 Indirect Type Heat Exchangers

Plate Heat Exchanger

Construction and Operation

- There are a number of parallel stainless-steel plates in a plate heat exchanger (PHE) and the plates are fitted on a frame such that they can move toward or apart from each other. The usual plate thickness is 2.5–3 mm and spacing is 1.25–7.75 mm. The plates are usually arranged in a vertical rather than in a horizontal position to save space. Figure 15.3 (a) and (b) present the schematic diagram of plates used in the heat exchanger, and the arrangement of plates in a plate type heat exchanger. Figure 15.4 shows an industrial plate heat exchanger used for pasteurization of milk.

- A jack or screw device is used to tighten the plates on the frame. When tightened, the plates form a number of narrow chambers separated by the metallic plates.

- There are four corner ports on each plate; the corner ports forming the distribution channels for the two fluids, when the plates are assembled.

- The frame has two horizontal bars, one at the top and the other at the bottom to hold the plates, to ensure proper alignment and to allow the plates to move closer and apart from each other. The top and bottom edges at the center of each plate are notched. When the movable end cover is removed, the carrying bars are of such appropriate length that the plates may be slid away from each other for inspection and cleaning.

- The frame has a fixed end cover (which is known as *head piece*) and a movable end cover (called *tail piece or* follower). As the plates are tightened by applying pressure on the latter, this is also known as the *pressure plate*. Connected to external piping, the end covers have nozzles for the incoming and the outgoing fluids, which line up with the distribution headers (ports).

- Each plate is stamped or embossed a corrugated (or wavy) surface pattern on it. These corrugations form the spacing between two plates and also help in providing mechanical support to the plate stack. Further, when the plates are kept close to each other,

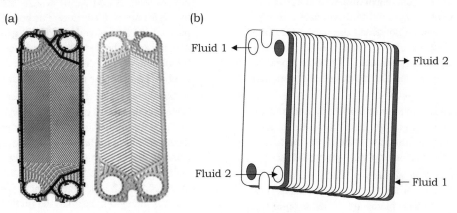

(a)   (b)

Fluid 1   Fluid 2

Fluid 2   Fluid 1

FIGURE 15.3 (a) Plates used in the heat exchanger, (b) schematic diagram of arrangement of plates in a plate type heat exchanger

FIGURE 15.4   A plate heat exchanger used for pasteurization of milk

it reduces the equivalent diameter to have a high level of turbulence and high heat transfer rate.

- The shapes of the channels depend on the kind and shape of corrugations. There are more than 60 patterns of corrugations worldwide. Plates, depending on whether the generated turbulence is of a high or low intensity, are designated as hard or soft.

- As the fluid enters the system at one end in series or in parallel manner, the fluid passes through alternate chambers (channels). Product flow through alternate channels in the pack of plates is facilitated by the varying patterns of open and blind holes.

- There is no mixing of the heat exchanging fluids as they move in different chambers. The metallic plates on both sides of each liquid allow only the exchange of heat.

- The most conventional flow arrangement is 1pass–1pass counter flow. The 1-pass-1-pass U configuration design has all fluid passages on the fixed end cover that permits convenient disassembly for cleaning/repair without any need to disconnect any piping. The flow can be prevented through some ports using suitable gasketing to permit more than one pass for either one or both fluids. The flow pipes are connected to both fixed and movable end covers in a multi-pass heat exchanger.

- Depending on the manner of fabrication and desired leak tightness, PHEs can be classified as *welded*, *gasketed*, or *brazed*. Spiral plate, lamella, and plate coil exchangers are some other plate-type exchangers.

- *Gasketed plate heat exchanger.* The gasketed PHE is common in food processing, in which, using rubber gaskets, the metal plates are sealed around the edges. In fact, this is the most common type of PHE

and, unless specifically mentioned, the PHE means gasketed plate heat exchanger.

- Elastomeric molded gaskets are employed for sealing purposes and the gasket is generally 5 mm thick. The gaskets are compressed by about 25% of thickness while tightening them in peripheral grooves. Double seals around the port sections act as an additional safety measure to prevent intermixing of fluids in the unlikely event of gasket failure.

- The two fluids that would exchange heat are directed to flow through alternate chambers by the holes made in the plates as well as gaskets in the corner of the plates.

- Typical gasket materials for high temperature applications as in sterilization and pasteurization are butyl and nitrile rubber.

- *Welded plate heat exchanger.* Use of corrosive fluids is not permitted in the gasketed PHE. Thus, to alleviate this limitation as well as to the work under higher temperatures and pressure, welded plate heat exchangers are used. A number of designs are available, and the plates are welded on one or both fluid sides. The plate size in case of welded PHE is generally larger in comparison to the plate size in gasketed PHE to reduce the overall welding cost. However, it cannot be disassembled easily on the welded fluid sides.

- *Extended surface heat exchangers* use extended surface elements (or commonly referred to as fins) on one or both fluid sides. The fins permit expansion of the surface area for heat transfer and, thus, result in a heat exchanger that is more compact. Fins are capable of increasing the heat transfer surface area by as much as 5 to 12 times the primary surface area. An effort is made to maximize the fin density expressed in terms of fins per unit length (fin frequency).

Salient Features

- Plate-type exchangers are simple and economical in operation and occupy less floor space. They have good heat exchange efficiency and can be used for varying tasks, such as heating, cooling, regeneration, and holding.

- The heat exchanger can be easily disassembled for cleaning, inspection, and maintenance.

- In comparison to other non-contact type heat exchangers, it requires lower capital investment to heat or cool a product to a temperature very close to the heating/cooling medium temperature (even up to 1°C difference or so is achievable). Thermal effectiveness of such a heat exchanger is very high (up to about 93%), resulting in economical low-grade heat recovery.

- Flexibility in the selection of plate size, corrugation patterns, and flow arrangements helps to easily change the heat transfer surface area or to rearrange them for different tasks or capacities. Also, the heat exchanger capacity can be varied conveniently by addition or removal of plates. Product flow rates in industrial plate heat exchangers are high, i.e., 5000–20,000 kg.h^{-1}.

- Small hydraulic diameter flow passages, swirl or vortex flow generation, and breakup and reattachment of boundary layers make it possible to attain very high heat transfer coefficients. A multi-pass flow arrangement is also preferred for obtaining a higher coefficient of heat transfer.

- High coefficients of heat transfer in addition to the possibility of utilizing counter flow of the fluids result in a considerable reduction in the surface area requirement for a heat duty PHE application (one-half to two-third) as compared to a shell-and-tube exchanger. Thus, the overall volume, space requirement, and cost for the exchanger are reduced.

- There is less fouling in PHEs, which also helps in maintaining high heat transfer coefficient throughout the operation.

- Usually there is no leakage from one fluid to the other through the plates. In case there is any leakage through the gasket, the flow of fluid will be from inside of the heat exchanger to outside as the fluid streams are under pressure. This prevents any sort of contamination of the fluids.

- The limiting pressure of operation in a PHE is decided by the gasket materials. It is usually operated below 1.0 MPa, although it is capable of handling pressures up to about 3 MPa. Similarly, the long gasket periphery of the gasketed PHEs does not allow them to be used in high-vacuum applications.

- The gasket materials can allow up to 260°C operating temperatures. However, operating temperatures are usually below 150°C with a view to enhance the life of the gasket materials and also to use comparatively inexpensive gasket materials.

- Pressure drop in a PHE, as compared to a shell-and-tube exchanger, is very high for equivalent flow velocities. The pressure drop, however, does not create a problem because of short plates and not so high flow velocities.

- The pressure within the heat exchanger has to be kept constant. A change in pressure changes the handling time and, thus, the heating temperature. At low pressures and low fluid velocity (1.5-2 m.s^{-1}), there may be uneven heating and *burn on* (fouling or burning of solids on heat exchanger surfaces). It causes retarded heat transfer or increased pressure drop over a period of time. Under such conditions, more frequent cleaning is required by stopping the process and cleaning the plates.

A common example of fouling is that when we boil milk at home, a layer of solid is deposited on the container surface. In industrial heat exchangers, the formation of such solid deposits causes reduced heat transfer rate, solids' loss, and the loss of overall economy.

- The overall heat transfer coefficient ranges from 8000–20000 kJ.m^{-2}.h^{-1}.°C^{-1} in these heat exchangers.

- A single frame can support more than one exchanger. For example, in the case of milk pasteurization, three or more heat exchangers or sections for regeneration, heating, and cooling can be arranged on a single frame.

Fluid velocity as well as the pressure drop should be high for increasing the handling capacity (amount of fluid flow in the heat exchanger).

Alternatively, the design is made such that the fluid moves through the heat exchanger in a number of parallel passages. For example, the product can move in two parallel flows, which change direction four times in the heat exchanger. The heating fluid changes direction twice when it moves in four parallel flows. The flow is designated as 4×2/ 2×4; the first number in the numerator indicates the number of parallel flows, the second indicating the number of passes for the hot fluid. In the denominator, the first number is the number of parallel flows and the second number is the number of passes for the other medium. Figure 15.5 shows the different flow patterns in a plate type heat exchanger.

Applications

- Plate heat exchangers can successfully process low-viscosity (less than 5 Pa.s) liquid foods.

- These can be operated under fully sanitized conditions, and, hence, they are suitable for food industries. These are very common in the dairy industry for milk pasteurization. It is also commonly used in juice, beverage, and alcoholic drink processing industries.

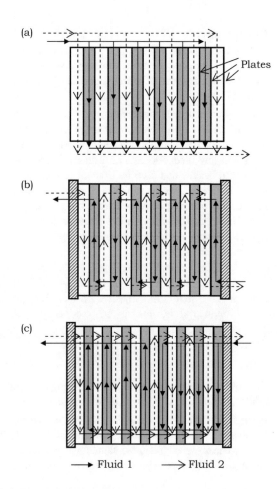

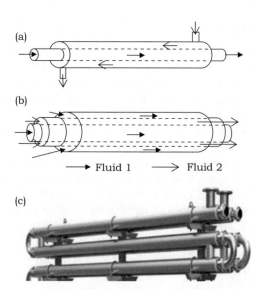

FIGURE 15.6 (a) and (b) Schematic of a double and a triple tube tubular heat exchanger, (c) outer view of a tubular heat exchanger (Courtesy: M/s Bajaj Processpack Limited)

FIGURE 15.5 Flow patterns in a plate type heat exchanger (a) parallel flow (b) counter flow (c) a 4×2/ 2×4 plate heat exchanger

- The PHEs can handle suspensions provided the largest suspended particle size is smaller than one-third of the average channel gap to avoid clogging. Similarly, extremely viscous fluids may cause flow problems, especially in cooling.

Concentric Tube or Tubular Heat Exchanger

Construction and Operation

- Tubular heat exchangers, as discussed earlier, may be classified as concentric tube, spiral tube, and shell and tube exchangers. These are one of the simplest heat exchangers, though the designs may vary widely.

- The concentric tube heat exchangers used in food industries are commonly either *double tube* or *triple tube* types (Figure 15.6). The tubes are made of stainless steel.

- A pipe is positioned concentrically inside another pipe in a double tube heat exchanger. The flows of the two streams in the annular space and in the inner pipe are either in a concurrent or in a counter-current manner. The inner tube may contain obstructions or fins for increasing heat transfer by creating

turbulence. The fins may be placed either radially or longitudinally.

- It is in the inner annular space that the product in a triple-tube heat exchanger flows, and the heating/cooling fluid flows in the inner tube and outer annular space. Clearly, the rate of heat transfer is almost doubled. The innermost tube may contain fins.

- If a large heat transfer surface is required, banks of sections are used because the length of each section is usually limited to standard pipe lengths.

Salient Features

- There is no problem for the flow distribution. As the tubular heat exchanger has a few seals as compared to the PHE, the cleaning and maintenance of aseptic conditions are easier; cleaning in place (CIP) is also possible. But it is difficult to inspect the heat transfer surfaces for food deposits.

- These require higher pressures and occupy more space than plate heat exchangers of same capacity. As the pressure requirement is high and the flow velocity has to be maintained within the specified range, very large diameter tubes cannot be used.

- The high flow rates cause turbulence at the tube walls, which helps in uniform heat transfer and less product deposition.

- The production capacity cannot be changed. More equipment is required for increasing the production capacity.

- In the straight tube systems, turbulence is created by counter-current flow and helical corrugations, resulting in enhancement of heat transfer rate. Turbulence increase can also be achieved by altering the relative positions of the tubes.

- Tubular exchangers can be customized for practically any capacity as well as operating conditions. These exchangers have a wide range of operating conditions, limited only by construction materials, i.e., from high vacuum to over 100 MPa pressure. The equipment can be operated at pressures up to 2 MPa even for viscous fluids. It can be used for any temperature difference, i.e., from cryogenics to 1100°C, between the fluids and higher liquid flow rates (6m.s⁻¹) are also possible.

Applications

- This is used for low and medium viscosity products, such as orange juice, cottage cheese, wash water, ice cream mix, etc. Heat sensitive products and products that may cause fouling on the heat exchanger surfaces can also be handled by this exchanger. They are also especially suitable at high operating temperatures and pressures.

- All types of tubular exchangers find their use primarily for heat transfer applications involving liquid-to-liquid as well as liquid-to-phase change, e.g., condensation or evaporation processes.

- Triple tube heat exchangers find their use in those situations where a need exists for increased heat transfer area. Specifically, during cooling, the rate of cooling is limited by high product viscosity; hence, the surface area of the heat transfer should be more.

Shell and Tube Heat Exchanger

Construction and Operation

- This consists of a cylindrical casing (shell), with a number of stainless steel tubes inside in the form of a bundle. The axis of the tube bundle is parallel to that of the shell.

- Circular pipes of diameter less than 0.6 m and rolled and welded metal plate for higher diameters are used for the shell.

- Usually there are 5 to 7 tubes within the shell and the tube diameter is 10–15 mm. Generally circular elliptical, rectangular, round/ flat, or twisted tubes are used.

- Many different types of internal constructions are available. The shape of the tube bundles can be straight or in the form of a U. Some other shapes are helical, bayonet, serpentine, sine-wave bend, L-shape, or hockey sticks, etc. There are manifolds at each end of the shell where the tubes get connected. Figure 15.7 presents some arrangements of tubes in the heat exchangers of shell and tube type.

- The other major components are the front- and rear-end heads, tube sheets, and baffles. Front- and rear-end heads act as entrance and exit for the tube fluid. The rear-end heads are often provisioned to take care of thermal expansion of the tube. Figure 15.8 shows a commercial small capacity shell and tube heat exchanger.

- Through the shell, while one fluid flows over the tubes, the other flows in the tubes. To avoid high

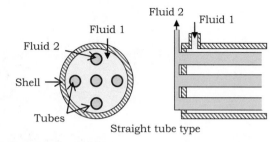
Straight tube type

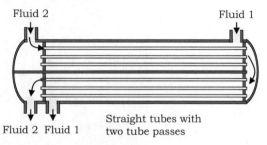
Straight tubes with two tube passes

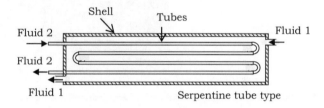
Serpentine tube type

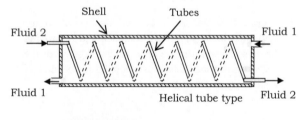
Helical tube type

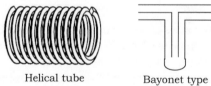
Helical tube   Bayonet type

FIGURE 15.7 Schematic diagram of arrangement of tubes in the heat exchangers of shell and tube type (after Ahmad, 1999 with permission)

pressure shells, it is preferable to pass high pressure fluids through tubes.

- For increasing the heat transfer, the fluid is made to pass more than once along the shell length and, hence, in more than one tubes. In other words, straight tubes are connected at the ends with bends and thus a "tube assembly" or "tube bundle" is formed. It is also possible to fold the stainless steel tubes into hair pin or "trombone" forms. Spiraling of tubes also helps to increase the travel time of the fluid in tube in the shell. However, the number of passes on the shell side is only one.

- For a two-tube-pass exchanger, the mean temperature difference and thus the heat transfer rate can be increased by a counter-flow arrangement.

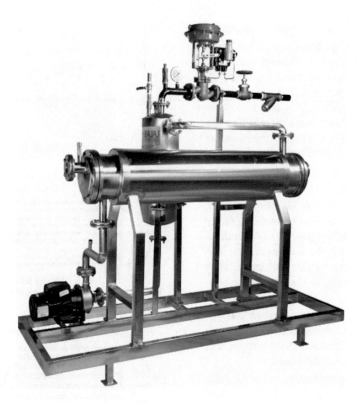

FIGURE 15.8   A shell and tube heat exchanger (Courtesy: M/s Bajaj Processpack Limited)

- A cross-flow pattern is created by baffles in the shell side, increasing the turbulence and improving the heat transfer. Baffles also prevent sagging of the tubes. Baffles are usually not needed in the shell if condensing steam is used.

Salient Features

- Higher rates of heat transfer can be achieved by maintaining the flow of fluids over the tubes instead of flow parallel to the tubes.
- Selection of the type of shell and tube exchanger is dependent on such factors as cost, cleanability, temperature, corrosion, operating pressure, pressure drop, and hazards. Selection of the heat exchanger is also dependent on the nature of the shell side fluid.
- It is difficult to clean the shell side without removing the tube bundles; thus, the fluid to be placed in the shell should be clean and non-corrosive or less corrosive. Also, by the same reasoning, to minimize corrosion, corrosive fluid should not be passed through the tubes. If it is necessary to use corrosive fluids, suitable alloys should be used for the tubes.

Applications

Heat exchangers of shell and tube type are employed where the requirement is of large surface area for heat transfer. The commonly used evaporation systems or boilers mostly have heat exchangers of the shell and tube type.

Spiral Tube Heat Exchanger

Construction and Operation

- Contained in a shell are spirally wound coils numbering one or more. Hence, the heat exchanger of the spiral tube type is also classified under the group of heat exchangers of shell and tube type.
- As the heat exchange surface area in a given space can be greatly increased by spiraling, it reduces the space requirement and cost.
- The coil helps to amplify turbulence at relatively low flow rates. The heat transfer rate for a spiral tube is almost two to four times of that for heat exchangers with straight tube-in-tube or shell-and-tube arrangements. It also reduces fouling.
- As the tube has the shape of a continuous helix or coil, the thermal expansion of the tube does not pose any problem. However, the coil is almost impossible to clean.

Applications

A uniform distribution of particles is achievable with the coil type heat exchanger. Therefore, it is suitable for high viscosity liquids, such as fruit purees, salad dressings, cheese sauce, foods containing a range of particle sizes, and fruit sauces/bases for yoghurts and pies.

Scraped-Surface Heat Exchanger

There is often solids deposition (burn on) on the walls of a tubular heat exchanger because of high amounts of dissolved solids present in the products. This deposition serves as an insulating layer and reduces the heat transfer. In addition, there is also hydraulic drag, which necessitates frequent cleaning of the heat exchanger surfaces. This problem is avoided by using a scraped surface heat exchanger in which the wall is continuously scrapped to reduce the thermal resistance.

Construction and Operation

- There are two tubes as in a double tube type heat exchanger with some mechanical means, such as rotating blades, to continuously scrap the inside surface of the inner tube. The range of rotor speeds is 150–500 RPM.
- While the food is allowed to pass through the inner tube, the heating/cooling fluid flows over the outer jacket.
- The blades scrap the deposits from the wall and mix them with the product and provide turbulence (Figure 15.9).
- The food contact parts are made up of corrosion-resistant material, such as hard chromium-plated nickel, high-grade stainless steel (SS 316), or pure nickel. The cylinder material may be chromium-plated nickel, stainless steel, or any other suitable alloy. The compatibility of the material with the product to be heated should be ensured. In addition, the materials used for the scraper blade and cylinder should also be compatible with each other, e. g., use of stainless steel blades with a stainless steel cylinder is not recommended as the blade may cause wear of the cylinder.
- The rotor blades are also often covered with plastic laminate or molded plastic.

Salient Features

- Heating fluid is usually hot water or steam. Brine or a refrigerant such as Freon can be used as the cooling medium.
- Although higher rotational speed would allow better heat transfer, there is a possibility of product damage due to maceration by too many blades and too high speeds. Therefore, it is necessary for any specific product to make a careful selection of rotor speed and the gap between the rotor and the cylinder.
- The heat transfer rate depends on rotor speed, number of blades fixed on the rotor, viscosity of product, temperature difference, and surface area of cylinder. The shaft speed is maintained such that the surface is properly wiped throughout to give the desired heat transfer coefficient.
- Since the product is constantly blended during the heat transfer process, the result is a uniform product.
- Such exchangers are suitable for applications where a high temperature difference may exist between the two fluid streams. Heat transfer coefficients are of the order of 8000–12000 $kJ.m^{-2}.h^{-1}.K^{-1}$ and vary with rotor speed.

Applications

Heating, sterilizing, pasteurizing, whipping, emulsifying, gelling, plasticizing, and crystallizing can be carried out by scraped-surface heat exchangers. However, a scraped-surface system involves higher cost of equipment and operation, and, hence, it is advantageous when the high viscosity of product makes other heat transfer methods impracticable.

A variety of pumpable fluids, such as soups, fruit juices, tomato paste, etc., can be processed using the heat exchanger.

Steam Jacketed Kettle

Construction and Operation

- This is a batch type system that comprises a vessel having a jacket. The jacket usually covers the lower portion of the vessel and extends up to half of the sides.
- The liquid food is kept in the vessel. The heating medium (hot water or steam) passes through the jacket. Steam condenses in the jacket after giving away heat. Thus, there is a pipe for collection of condensate and a drain for products at the lower side of the vessel. Near the top of the jacket there are two outlets, one for the steam and the other for non-condensed gases.
- The top of the vessel may be kept open or covered with a lid.

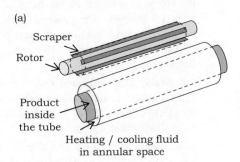

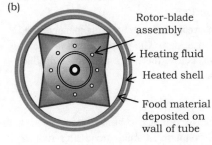

FIGURE 15.9 Schematic of two designs of scraped-surface heat exchanger

- There is provision for tilting the kettle for easy discharge of the product.

- Agitators in some designs are provided to ensure thorough mixing of the product during heating. Some coils or steam pipes may also extend through the vessel to provide additional heating surface.

- The other important accessories are safety valve, pressure gauge, steam inlet valve, steam trap, condensate drain valve, etc.

Salient Features

- It is desirable to have uniform temperature at all parts of the jacket, and, hence, the steam is moved rapidly over the heating surface, which also helps in attaining higher heat transfer coefficient (1000–8000 kJ.m^{-2}.h^{-1}.°C^{-1}).

- For efficient operation, a minimum amount of air needs to be allowed with steam in the jacket and superheating of steam is avoided. The steam trap helps in removing air and condensate adequately.

Applications

This is mostly used for heating milk, fruits and vegetables, juices, etc. As it is a batch type system, it can handle different types of foods, different ranges of viscosities, and different capacities. The holding times can also be varied as per requirements. Due to these flexibilities, the jacketed kettle is commonly seen in small/medium food processing enterprises.

Surface Heat Exchanger

Construction and Operation

- This is made up of a series of tubes or plates stacked vertically (Figure 15.10).
- The heating or cooling medium flows inside the plates or tubes; the product falls as a thin film outside.
- There is a trough at the top to distribute the product. The product is collected at the bottom.

Salient Features

- The flow rate of product should be maintained constant to maintain a uniform thickness of the tubes. Too slow movement causes improper coverage, too fast movement can move the product away from the heat transfer surface.

- This type of heat exchanger is inexpensive and very flexible in design. The product moves outside the tubes, and, hence, there is a chance of product contamination.

Applications

As there is chance of product contamination, such heat exchangers are not common in food processing industries. They may be used for heat recovery from process/waste materials.

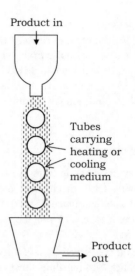

FIGURE 15.10 Schematic of a surface type heat exchanger

15.2.3 Direct Type Heat Exchangers

The steam infusion and steam injection heat exchangers are known as direct steam heaters, i.e., the heat transfer occurs by direct contact of steam and the product. Therefore, the heat transfer is instantaneous. The heating efficiency is almost 100%. The steam, after giving away the heat, condenses and is mixed with the product. Thus, there is an addition of moisture to the product. The system is known as *steam injection* or *steam-into-product* when the steam is injected through a nozzle into the product. In another method, a chamber is kept filled with pressurized steam at the heating temperature. As the product enters the chamber and falls through the steam, there is heat transfer. Such a system is called *steam infusion type* or *product-into-steam type*. The heating process in infusers is gentler than injectors.

Steam Injection Heat Exchanger

Construction and Operation

- This involves instantaneous heating of the product to the desired temperature by directly injecting high pressure and a high temperature steam.

- Usually, the product is preheated before sending into the direct type heat exchanger, into which steam is injected at a pressure of 965×10^3 Pa. For quick dispersion and mixing with the product, the steam is injected in the form of a thin sheet or small bubbles. The product is supplied to the heating chamber by a high-pressure pump. Boiling of the product is prevented by appropriately maintaining the pressure.

- The product is heated to the desired temperature (say 150°C) almost instantaneously due to the release of the steam's latent heat of vaporization. The condensate mixes with the product.

- The holding time in the chamber is very less, i.e., about 2 to 3 s after which the product and condensate mix is taken for immediate cooling. There may be some restrictors at the holding tube's outlet to control the fluid flow.

- An expansion vessel (a large chamber used to reduce the pressure of the condensate and product mix) is used to remove the extra water that is added to the product during the heat exchange.

    If the temperature of the product fed to the chamber is 80°C (the product is usually preheated before feeding into the heat exchanger), and it has to be heated to 140°C by steam condensing at 150°C, then the thermal energy required by per kg of the product to heat to the desired temperature, for instance, is about 60 × 4 kJ = 240 kJ. The energy released by the condensation is about 2.1 MJ per kg of steam. Therefore, about 0.11 kg per kg of product or 11% of the product will be the amount of steam needed to be condensed. However, this added volume is evaporated in the expansion cooling chamber.

- In some cases, the added water is acceptable and is allowed to remain as such in the food.

- There are different designs of steam injectors.

    - The product tube is in the form of a venturi. Four tubes with orifices surrounding the product tube supply the steam. With a view to encourage swirl and turbulences in the product, orifices are drilled either at an angle or radially.

    - The steam, in another design, is injected across the flow of the product at a sharp angle in the form of a thin cone directed inward. It helps in condensation and quick mixing.

- In yet another design the product is made to pass through a venturi, and it is in the expansion section of the venturi where the steam is injected, leading to increased product pressure. The steam is injected as a thin annulus around the product. Further increase and decrease of pressure within the injector is caused

by a second venturi section encouraging condensation and diffusion.

Salient Features

- Application of very high steam pressure adds to the cost of equipment. Hence, the lowest feasible pressure differential between product and steam is used.

- During the process of heating, a condensate film is formed first around the product, which subsequently mixes with the product due to turbulence. However, the film of condensate creates a barrier between the product and the injected steam, effectively preventing the product from overheating.

- The steam used for direct heat exchange should be "culinary" (food) steam, i.e., it should be free from damaging compounds. If the steam contains any additional compounds, they will be directly added to the product. The expansion vessel can remove only the extra water added to the product; however, these additional compounds may not be removed.

Applications

This is often used in the UHT sterilization of milk and other products.

Steam-Infusion Heat Exchanger

Construction and Operation

- This consists of a conical bottom steam pressure vessel. Here the steam pressure is about 450×10^3 Pa, which is much lower as compared to the injection system. A comparison of the steam injection and infusion type heat exchangers is shown in Figure 15.11.

- A series of parallel and horizontal distribution tubes are laid in the vessel for feeding the product. The tubes having thin slits along their bottom guide the product to fall as free-falling thin films in the laminar

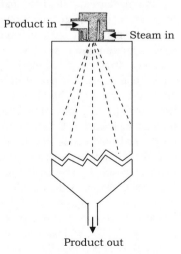

Steam injection type
heat exchanger

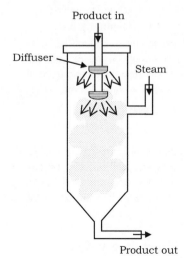

Steam infusion type
heat exchanger

FIGURE 15.11 Comparison of steam injection and infusion type heat exchangers

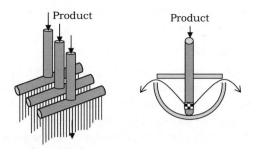

FIGURE 15.12 Product flow in steam infusion systems (after Ahmad, 1999 with permission)

range. The liquid viscosity determines the size of the spreaders. Figure 15.12 presents a schematic of the product flow in steam infusion systems.

- In another design, there is at the top a loose circular disc. A thin gap between the bowl and the loose disc allows the product to emerge in the form of a thin umbrella, permitting, thereby, heating of both sides of the umbrella sheet by steam.

- For special applications there may be more than one distributor in one chamber.

- Due to steam condensation, the product temperature rises very rapidly. The product, after heating, together with the condensed steam is collected and released at the conical bottom of the chamber.

- A strategy to achieve desired cooking is to retain a specific amount of liquid at the bottom of the chamber.

Salient Features

- The heat transfer rate is very high as the steam comes in contact with minute droplets of the food.

- Infusion heat exchangers may have fouling at the base and on the wall of the vessel but there is no cleaning problem and there is little effect on plant operation.

- Like in the steam injection system, the condensed steam dilutes the product, and, hence, the extra water added is removed by a vacuum cooling system.

- The steam infusers must be able to endure the operating pressure and must always be longer than the corresponding injectors. Because the temperature of steam may be higher, some products become overheated by the injectors.

Applications

This is used for sterilizing and/or cooking many types of products, e.g., milk, ice cream mixes, soups, chocolates, processed cheese, etc. Specially designed spreaders are used to handle the products that contain particulates such as diced vegetables and meat chunks.

15.2.4 Comparison of Direct Type and Indirect Type Heat Exchangers

- Heating and cooling occur at high rates when there is direct contact between the food and the heating medium (steam) in direct type heat exchangers.

Direct contact systems are appropriate for HTST and UHT applications because the heating and holding times are usually of a few seconds.

- In the indirect system, usually a single product pump and a homogenizer together are required. A direct heating system, however, needs (1) product pumps of different types, (2) a homogenizer, and (3) a pump to remove non-condensable gases from a vapor condenser. Thus, for same throughput capacity, the total operating cost for a direct type heat exchanger is almost double of that of an indirect one. The initial investment for a direct plant is also almost two times that for a PHE or tubular heat exchanger.

- There is no heat transfer surface and, thus, no fouling problem on heat exchanger surfaces in direct type heat exchangers, and, hence, the cleaning requirement is greatly reduced. Consequently, as compared to a PHE, a direct system can be run for almost twice the time duration without cleaning. The characteristics of a tubular heat exchanger are intermediate.

- Running cost in a direct plant is high because there is less regenerative heating; usually less than 60% in comparison to even 90% in indirect plants. The requirement for water (for condenser) is more, i.e., about 1.5 l of water per liter of product or more.

- The direct types can be used for handling more viscous products as compared to indirect types.

15.2.5 Other Special Types of Heat Exchangers

There are some other equipment as bulk milk coolers, spray coolers, cascade coolers, etc. where different types of heat exchange methods are used. In a bulk milk cooler, the liquid milk is stored in the tank and chilled water or refrigerant is circulated in coils fixed around the tank. An agitator may be provided for improving the heat transfer. The selection of the heat exchanger would broadly depend on the type of heating/cooling medium and the type and desirable qualities of the final product among other considerations.

In another system, such as a rock bed, which is a porous solid aggregate, the fluid flows in a zigzag fashion through the pores. Structure of the surface participating in heat transfer is cellular. Heat is stored in the solid matrix when the elevated temperature fluid is made to flow through the rock bed. Subsequently, when fluid at a lower temperature passes through the heated rock bed, it retrieves the stored heat. Thus, the thermal energy transfer is intermittent or cyclic. This is an example of a storage heat exchanger, also known as a regenerator or a regenerative heat exchanger.

A heat exchanger with a fluidized bed involves the heating of granular or finely divided materials while they are in fluidized state (Figure 15.13). Air at a slightly higher than the particulate bed's terminal velocity is allowed to flow upward from below. This brings the solid particles to a fluidized state and the behavior of the particulate material in the bed is that of a liquid. The particles are thoroughly mixed and there is a uniform temperature for the total bed (gas and particles). The fluidized bed heat exchangers are commonly used in drying,

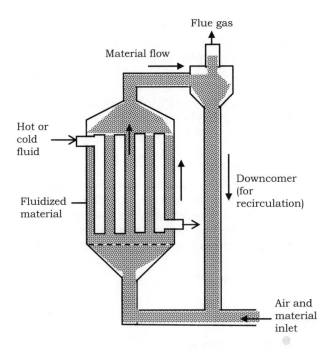

Flue gas

Material flow

Hot or
cold
fluid

Fluidized
material

Downcomer
(for
recirculation)

Air and
material
inlet

FIGURE 15.13 Schematic of a fluidized bed heat exchanger

mixing, adsorption, and waste heat recovery. The hot flue gases, exiting from the heat exchanger, may be sent to another heat exchanger for recovery of heat.

15.2.6 Heat Exchanger Considerations

- Heat exchanger capacity is decided by such parameters as the fluid velocity, area, viscosity, and the difference in temperatures of the two streams.

- Product quality may be impaired to an unacceptable level if difference between the temperatures of the heating medium and product is very high and the heating is being carried out at a very high temperature. A better strategy to increase the capacity of the system is to increase the heat transfer surface area instead of increasing the temperature of the heating medium.

- The mean temperature difference in parallel flow systems is lower than the counter flow systems. Thus, to have the same amount of heat transfer, a larger heat transfer is required. This adds to the initial cost. Hence the counter flow systems are preferred to concurrent flow systems.

- In a parallel flow system, the fluid to be heated or cooled cannot be brought close to the entering temperature of heating or cooling medium. However, in a counter flow system with adequate heat exchange surface area, the final temperature of the desired stream can approach quite close to the other stream's inlet temperature.

- In heat exchangers the film coefficients of the same order of magnitude are observed on both sides and the value remains close to the smaller of the two

film coefficients. Increasing one film coefficient by increasing the velocity of fluid without increasing the other does not result in an overall heat transfer coefficient increase. Therefore, to achieve higher overall coefficient of heat transfer, velocities of both the fluids involved in the heat exchange need to be increased.

Check Your Understanding

1. The temperature of milk at 4 kg.s^{-1} increases from 4°C to 60°C in a continuous pasteurization system where hot water enters the heat exchanger at 80°C. Find the temperature change for the hot water assuming that the hot water flow rate is 50 kg.s^{-1}. The coefficient of the overall heat transfer is 800 W.m^{-2}.K^{-1} in the concurrent heat exchanger and the specific heats of milk and water are 4 and 4.2 kJ.kg^{-1}.K^{-1}, respectively. Determine the exchanger's heat transfer area. If steam at atmospheric pressure is utilized for heating instead of hot water, find the amount of steam required. Assume that only the steam's latent heat is used to heat the milk.

   Answer: 4.267 K, 29.27 m², 0.396 kg.s^{-1}

2. Determine the cooler surface area needed in a counter flow heat exchanger to cool 250 kg.h^{-1} of milk from 26°C to 4°C. The surface cooler uses 2°C cold water at the rate four times that of milk. Assume 2400 W.m^{-2}. K^{-1} as overall thermal conductance and 4 kJ.kg^{-1}.K^{-1} as the milk's specific heat.

   Answer: 0.34 m².

3. In case of a sterilization system, saturated steam at 150°C is utilized for heating milk from 90°C to 140°C. Heating occurs in such a way that only latent heat of condensation is supplied to milk. If the milk flow rate is 3 m³.s^{-1}, find the amount of steam required for heating. Assume there is no contact between the heating medium and milk. Steam at 150°C possesses 2112.2 kJ.kg^{-1} latent heat of condensation. There is no direct contact of the heating medium and the milk.

   Answer: 0.2926 kg.s^{-1}

4. Milk is cooled from 40°C initial temperature using a direct-expansion surface cooler having 1.5 m² heat transfer area and -4°C refrigerant temperature. Determine the final milk temperature when the overall thermal conductance is 600 W.m^{-2}.K^{-1} and the milk flow rate is 40 kg.min^{-1}.

   Answer: 27.4°C.

5. A liquid food flows through the inside pipe of a double pipe heat exchanger in a turbulent regime. Steam is allowed to condense at the outside of the inner pipe and thus its wall temperature is maintained constant. At a certain flow rate of the liquid, the heat transfer coefficient at the inside of the inner pipe has been found to be 1200 W.m^{-2}.K^{-1}. Estimate the heat

transfer coefficient when the liquid flow rate through the pipe doubles.

Answer: 2089.32 W.m⁻².K⁻¹

6. Calculate the amount of steam at 120°C (enthalpy, h=2700 kJ.kg⁻¹) that must be added to 100 kg of food material with a specific heat of 3.5 kJ.kg⁻¹.K⁻¹ to heat the product from 40°C to 80°C by direct steam injection. Enthalpy of condensate (water) is 344 kJ.kg⁻¹. Heat loss to the steam may be neglected.

Answer: 11.3 kg

7. Consider a parallel flow heat exchanger with area A_p and a counter flow heat exchanger with area A_c. In both the heat exchangers the hot stream flowing at 1 kg.s⁻¹ cools from 80°C to 50°C. For the cold stream in both the heat exchangers the flow rate and the inlet temperature are 2 kg.s⁻¹ and 10°C, respectively. The hot and cold streams in both the heat exchangers are of the same fluid. Also both the heat exchangers have the same overall heat transfer coefficient. What is the ratio A_c/A_p?

Answer: 0.9278

8. It is envisaged to heat fruit juice to 75°C in a double pipe counter-current heat exchanger. Fruit juice is available at the rate of 68 kg.min⁻¹and the initial temperature of 35°C. The heat transfer fluid is an oil having a specific heat of 1.9 kJ.kg⁻¹.K⁻¹. Calculate the heat exchanger area when the oil temperature is 110°C at the entrance of the exchanger and 75°C at the exit. Fruit juice has a specific heat of 4.18 kJ.kg⁻¹.K⁻¹ and 320 W.m⁻².K⁻¹ is the coefficient of overall heat transfer for the heat exchanger surface.

Answer: 15.81 m²

9. It is proposed to use a double pipe type counter-current heat exchanger for preheating fruit juice from 5°C to 45°C. Hot water is the heating agent that enters the heat exchanger at 75°C and leaves at 65°C. The fruit juice is available at the flow rate of 1.5 kg.s⁻¹ and has 3.85 kJ.kg⁻¹.°C⁻¹ specific heat. Calculate the coefficient of overall heat transfer assuming the heat exchanger area to be 10 m².

Answer: 533.73 W.m⁻².K⁻¹

10. Calculate the amount of cooling water and heat transfer area required for a counter-current heat exchanger where milk at the rate of 50 kg.h⁻¹ and 73°C temperature after pasteurization needs to be cooled to 15°C. Cooling water has temperatures of 10°C at the entrance and 17°C at the exit. Coefficient of overall heat transfer is 600 W.m⁻².K⁻¹ on the inside surface of the heat exchanger pipe. Specific heat values in kJ.kg⁻¹.K⁻¹ are 3.9 for milk and 4.2 for water.

Answer: 0.248 m²

11. In a parallel flow heat exchanger, the heat capacity rates (product of specific heat at constant pressure and mass flow rate) of the hot and cold fluid are equal. The hot fluid flowing at 1 kg.s⁻¹ with $c_p = 4$ kJ.kg⁻¹.K⁻¹ enters the heat exchanger at 102°C while the cold fluid has an inlet temperature at 102°C. The overall heat transfer coefficient for the heat exchanger is estimated to be 1 kW.m⁻².K⁻¹ and the corresponding heat transfer surface area is 5 m². Neglect heat transfer between heat exchanger and the ambient. The heat exchanger is characterized by the relation, $2\varepsilon = -\exp(-2\ NTU)$. What is the exit temperature for the cold fluid?

Answer: 55°C

12. A fluid is flowing inside the inner tube of a double pipe heat exchanger with diameter "d." For a fixed mass flow rate under turbulent flow conditions, what will be the relationship between the tube side heat transfer coefficient and the diameter of the tube?

Answer: hα d⁻¹·⁸

Objective Questions

1. Differentiate between
   a. Direct type and indirect type heat exchanger;
   b. Heat exchanger effectiveness and heat exchanger efficiency;
   c. Direct transfer type heat exchanger and storage type heat exchanger;
   d. Tubular heat exchanger and shell and tube type heat exchanger;
   e. Steam injection type heat exchanger and steam infusion type heat exchanger.

2. Write short notes on (a) log mean temperature difference, (b) scraped-surface heat exchanger, (c) direct steam heaters, (d) spiral tube heat exchanger, (e) surface heat exchanger, (f) fluidized bed heat exchanger.

3. Which one out of the concurrent type or counter-current type flow gives a better heat exchange efficiency? Why?

BIBLIOGRAPHY

Ahmad, T. 1999. *Dairy Plant Engineering and Management*, 4th ed. Kitab Mahal, New Delhi.

Albright, L.D. 1990. *Environment Control for Animals and Plants*. ASAE, St. Joseph, MI.

Bennet, C.O. and Myers, J.E. 1962. *Momentum, Heat, and Mass Transport*. McGraw-Hill, Inc., New York.

Brennan, J.G., Butters, J.R., Cowell, N.D. and Lilly, A.E.V. 1976. *Food Engineering Operations*, 2nd ed. Elsevier Applied Science, London.

Bromley, L.A. 1950. Heat transfer in stable film boiling. *Chemical Engineering Progress* 46: 221-227.

Charm, S.E. 1978. *Fundamentals of Food Engineering*, 3rd ed. AVI Publishing Co., Westport, Connecticut.

Datta, A.K. 2001. *Transport Phenomena in Food Process Engineering*. Himalaya Publishing House, Mumbai.

Dickerson, R.W. 1968. Thermal properties of foods. In D.K. Tressler, W.B. Van Arsdel, and M.R. Copley (eds.).*The Freezing Preservation of Food*. AVI Publishing Co., Westport, Connecticut.

Earle, R.L. and Earle, M D. 2004. *Unit Operations in Food Processing* (Web Edn.) The New Zealand Institute of Food Science & Technology, Inc., Auckland. https://www.nzifst .org.nz/resources/unitoperations/index.htm

Fellows, P.J. 2000. *Food Processing Technology*. Woodhead Publishing, Cambridge, UK.

Geankoplis, C.J.C. 2004. *Transport Processes and Separation Process Principles*. 4th ed. Prentice-Hall of India, New Delhi.

Gupta, C.P. and Prakash, R. 1994. *Engineering Heat Transfer*. Nem Chand and Bros., Roorkee.

Heldman, D.R. and Singh, R.P.1981. *Food Process Engineering*.2nd ed. AVI Publishing Co., Westport, Connecticut.

Henderson, S.M. and Perry, R.L. 1976. *Agricultural Process Engineering*. The AVI Pub. Co., Westport, Connecticut.

Holman, J.P. 1997. *Heat Transfer*, 8th ed. McGraw Hill Book Co., New York.

Kumar, D.S. 2008. *Engineering Thermodynamics*. S.K. Kataria & Sons, New Delhi.

Rao, D.G. 2009. *Fundamentals of Food Engineering*. PHI Learning Pvt. Ltd., New Delhi.

Sahay, K.M. and Singh, K.K. 1994. *Unit Operations in Agricultural Processing*. Vikas Publishing House, New Delhi.

Singh, R.P. and Heldman, D.R. 2014. *Introduction to Food Engineering*. Academic Press, San Diego, CA.

Toledo, R.T. 2007. *Fundamentals of Food Process Engineering*, 3rd ed. Springer, New York.

Watson, E.L., and Harper, J.C. 1989. *Elements of Food Engineering*, 2nd ed. Van Nostrand Reinhold, New York.

16

Thermal Processing

In the food processing industry, heat processing or thermal processing methods, such as pasteurization, sterilization, and blanching, are very common. Heat processing offers the following advantages.

- It imparts preservative effect on foods by inactivation of enzymes and killing of microorganisms and other extraneous organisms. The destruction of microorganisms is also due to the inactivation of enzymes (due to denaturation of proteins) and enzyme-controlled metabolism in microorganisms.
- It helps in changing the texture, flavor, and eating quality of food.
- It facilitates better availability of some nutrients present in the food by improving their digestibility. For example, heat processing improves digestibility of proteins.
- It helps in the destruction of anti-nutritional constituents of foods.

The control of processing conditions in heat processing is very simple, and, hence, this is the most preferred method for killing microorganisms in food processing.

16.1 Methods and Principles

16.1.1 Heat Processing Methods

Blanching is a short time exposure of food to a relatively lower temperature to inactivate the enzymes and to prevent browning during storage. *Pasteurization* is the process of exposing the food to a specified temperature for a specific time period for killing selected pathogenic microorganisms. The specific time and temperature combination depends on the target microorganism that has to be destroyed. For example, the target microorganisms during milk pasteurization are the *Mycobacterium tuberculosis* and *Coxiella burnetti*. However, the pasteurization process cannot destroy the *thermoduric* or *thermophiles* groups of microorganisms. This classification is discussed in Chapter 9.

Sterilization is the process of exposing the food product to a high temperature for a long time period to destroy almost all microbial and enzyme activities. Even though, in some cases, sterilization is defined as the destruction of all microorganisms, we will learn at a later stage that, in reality, complete destruction is not achieved. As sterilization eliminates almost all microorganisms and inhibits enzyme activity, it is possible to store a properly packaged product for a long period at normal temperature, even up to six months or more. But pasteurization does not kill all microorganisms, and, hence, these residual microorganisms can grow under favorable conditions and can spoil the food. Therefore, the pasteurized food has to be stored in proper packages under refrigerated conditions. The shelf life of pasteurized food is limited to 7 to 15 days even under refrigerated conditions. The advantage of pasteurization is that, unlike sterilization, it does not notably change the organoleptic and eating qualities of food. As the process of sterilization was initially developed for treating products within cans, sterilization is also otherwise known as *canning*.

16.1.2 Processing Time

In Chapter 9, under the section on thermo-microbiology, it was noted that the process time would depend on the initial and final (desired) number of microorganisms and the temperature to which the microorganisms are exposed (which affects the D value of the microorganism) in addition to the thermal resistance of the target microorganism. Thus, the processing time required for sterilization will be much lower than sterilization. For example, if the decimal reduction time for the target microorganism at a specific temperature is 0.3 minute and the pasteurization and sterilization requirements are 6 and 12 D, respectively, then the processing times for pasteurization and sterilization would be 1.8 minutes and 3.6 minutes. Sterilization usually employs a much higher temperature than pasteurization. Similarly, the processing parameters for blanching are decided based on the target enzyme.

Also, we have discussed in Chapter 9 that the *F value* is the time required to reduce the number of microorganisms from an initial value to a desired final value at a specific temperature.

The F value is usually expressed with suffixes indicating the z value of the target microorganism and the processing temperature. For example, F_{121}^{10} represents a process that operates at 121°C and where z value of 10°C characterizes a microorganism. This value is taken as the reference F value or F_0. If z is not 10°C, then it is noted as F_0^z. Similarly, the D value at 121°C is known as D_0. A can of food is actually processed for more time than the F value to include the time required for the heat to penetrate to its thermal center.

Some examples related to the selection of process time have been discussed in Chapter 9. A few more examples are given below.

> **Example 16.1 If a can has a spore load of 100, calculate the minimum process time so that the final spore load will be 1 in a lakh cans. The D_0 for spores =1.6 min.**

SOLUTION:

Given, $D_0 = 1.6$ min (D_0 means D_{121})

$N_0 = 100$

$N = 10^{-5}$ (1 in 100,000 samples means the spore load is 10^{-5})

$\theta = D_t \times \log(N_0/N) = 1.6 \times \log(10^2/10^{-5}) = 1.6 \times 7 = 11.2$ min

i.e. The process time will be minimum 11.2 min.

Example 16.2 If the process time F_0 for 6 log cycle reduction of a particular bacterial spore is 3 min, calculate the process time for 10 log cycle reduction.

SOLUTION:

Given, $F_0 = 3$ min.

$S = 6$ (6 log cycle reduction means 99.9999 % reduction)

Thus, $3 = D \times 6$

$D = 3/6$

For 10 log cycle reduction ($S = 10$)

$F = D \times S = (3/6) \times 10 = 5$ min

Example 16.3 The F_0 (sterilizing time) for a process has been found to be 2.90 minutes. Determine the probability of spoilage by an organism if each can contains 12 spores of this organism ($D_0 = 1.6$ min). Assume the z-value remains unchanged for the organism.

SOLUTION:

$F = D \times S$

Thus $S = F / D$

i.e., the reduction in microorganisms at an F_0 of 2.90 min can be given as:

$$S = \log \frac{N_0}{N} = \frac{F}{D} = \frac{2.90}{1.6}$$

$$\log \frac{N_0}{N} = 1.8125$$

$$N_0/N = 64.94$$

$$N = 12/64.94 = 0.185$$

Or, the probability of spoilage will be 18 in 100 cans.

Example 16.4 A food sample has a bioburden of 640 cfu.ml^{-1}. If the D value of the microorganism is 2.5 minutes, what will be the total process time for the final microbial load to be 10^{-6} cfu.ml^{-1}?

SOLUTION:

To reduce the bioburden from 640 to 1, the log cycle reduction is log (640/1) = 2.806
Thus, total log cycle reduction = 2.806 + 6 = 8.806

Total process time = $F = D \times S = 2.5 \times 8.806 = 22.015$ min

F_t^z is also defined as the *accumulated lethality* to reflect the total lethal effect of heat applied. The concept of F was first introduced by Ball (1923). The F value is used for comparing the effectiveness of different heat sterilization methods. Mention is also made of the F_t^z as *F-biological*.

> Like the F values for sterilization (for heat resistant spores) are expressed at a reference temperature of 121.1°C (250°F), the F value for the pasteurization process is expressed at the reference temperature of 82.2°C (180°F); this is the temperature at which most heat-resistant organisms and vegetative cells are inactivated.

The F values are extensively used by food canning industry in comparing processes and in developing new schedules. When two processes are equally effective in destroying a given microorganism, they are considered to be equivalent.

16.1.3 Lethality Factor

The *lethal rate or lethality factor* is an index of expressing the degree of lethality at the retort temperature in relation to the lethality at the reference temperature. It is denoted as L. The subscript r refers to retort temperature.

$$L = \frac{F_0}{F_r} \tag{16.1}$$

Using equation 9.28, the above equation can be written as:

$$L = \frac{F_0}{F_r} = 10^{(t-121)/z} \tag{16.2}$$

If a food is processed at 110°C and the z value of the most heat resistant micro-organism is 10°C, then

$$\text{Lethal rate} = 10^{(110-121)/10} = 0.08 \tag{16.3}$$

If the temperature is 115°C, then lethal rate = 0.25.

The equation 16.2 states that the lethal rate is equal to D_{121}/D_t. It also says that as the temperature of processing increases the lethal rate increases. The F value of a thermal process can be obtained by plotting the lethal rates against the process time. The same z value (for the specific microorganism or food

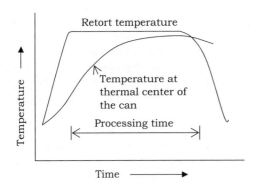

FIGURE 16.1 Change in temperature at the thermal center of a can during retort processing (after Fellows, 2000 with permission)

constituent) should be taken for calculating the F value, which was originally taken for calculating the D_t.

As the food is heated or cooled in a can, the temperatures within the can go on changing. In this case an integrated lethality is calculated. The temperature profile in a conduction heated food is shown in Figure 16.1. Initially, after the steam is introduced into the retort, it takes some time to reach the temperature needed for sterilization. But the temperature of the product (taken at the thermal center of the cans) lags behind and reaches the retort temperature after some time. In a batch retort, the time taken from the introduction of steam to the time when the processing temperature t_r has been attained is called the retort *come-up time*. Destruction of microorganisms begins when the food temperature goes above (say) 70°C. The processing time considers the lethal effect during the come-up time as well as the actual time the product is kept at the retort temperature. Thereafter, cooling is started by introducing cold water into the system. In a conventional canning process the temperature of the can during cooling also lags behind the retort temperature.

The lethal rate in equation 16.2 is calculated by taking small time increments ($\Delta \theta$) in the process and then using the average temperature at each time increment. Lethality is given as $\sum L_t.\Delta.\theta$. The z value is taken as 10°C in equation 9.28 (as also 16.2) to calculate L. Considering t = f(θ),

$$F = \int_0^\theta 10^{\frac{f(\theta)-121}{z}} d\theta \qquad (16.4)$$

The appropriate z value of a particular biological entity should be considered if the F_0 value is subsequently used to define the lethality of that entity.

The lethal rate depends on temperature and z; tabulated lethal rates are available. Tables for F_0 values are also available. As an example, the F_0 values remain within 3–6 minutes for vegetables in brine and within 12–15 minutes for meat in gravy.

16.1.4 Commercial Sterility

From the death rate curve of microorganisms, it is observed that to destroy all microorganisms, theoretically, the time required is infinite. Similarly, for obtaining a very low level of residual microorganisms (say 10^{-15}), the processing conditions

will be very severe and this would, in addition to affecting the sensory and nutritional quality, also add to the cost. Hence, the processing conditions are kept such that it kills the microorganisms to a considerably safe level and at the same time causes minimum damage to other attributes, i.e., we target at a specified number of microorganisms in the food, which can be considered safe for consumption without much loss of the essential components of food. This is known as the level of *commercial sterility*. It means that the majority of containers are sterile but there is a chance that non-pathogenic cells survive in a predetermined number of containers. The objective is to substantially inactivate all microorganisms and spores in the food, which would have otherwise grown during subsequent storage conditions. As was discussed earlier, the death of a microorganism means it has lost its ability to reproduce. A commercially sterile product may be defined as the condition in which the microorganisms present in a container are unable to reproduce under normal storage conditions.

As per the regulations, the thermal process for low acid canned foods should target at 12 D reduction of Cl. botulinum. If it is assumed that there are 10 surviving spores in one can, after 12 D process the number of microorganisms will be 10^{-11}. However, the calculation techniques as 12 log cycle reduction in number of spores neither takes into account the initial number of microorganisms nor does it define the final spore count in the product. Thus, the interpretation has now changed to the final microbial population (probability of survival) of 10^{-12} (i.e. one microorganism in 10^{12} cans). The final number of microorganisms for public health systems is kept at 10^{-9}, that for mesophilic spoilage is kept at 10^{-6} and that for thermophilic spoilage is kept as 10^{-2}.

In microbiology, *sterility assurance level* (SAL) defines the probability that a device is not sterile. The SAL of 10^{-6} (SAL6) is normally required for medical devices. If the initial microbial load is 10^6, to achieve SAL6, a total 12 log reduction is required.

The time-temperature combinations used in canning affects the quality of many foods. Thus, combination of other nonthermal processing methods with the thermal processing can be used, and canning parameters (time and temperature) need not have to be so severe.

Another approach (Pflug, 2010) considers the absolute spore concentration in the heated product. The residual microorganisms are denoted by a term PNSU- *Probability of a nonsterile unit* (Hallstrom et al. 1988). Pflug (2010) suggested that the PNSU for Cl. Botulinum should be 10^{-9}. Non-pathogenic mesophilic spore forming organisms should have a PNSU of 10^{-6} and thermophilic spore forming microorganisms should have PNSU 10^{-3}.

16.1.5 Considerations during Heat Processing

- As a food may contain many different types of microorganisms (or enzymes), the process conditions are decided on the basis of the most heat resistant microorganism (or enzyme) likely to be present in the food. The process time is kept more than the time calculated from the above equations. The initial heating period does not significantly contribute to total lethality, most of the lethal effect is achieved during the last few minutes of heating.

- The processing time and the spore log reduction depend on the initial load of microorganisms. The canning process normally requires heating of the food material to 121°C for complete commercial sterility of low-acid foods by eliminating all mesophilic microbes and spores of *Clostridium botulinum*. For achieving this target, the heating treatment is given for the duration equal to 12 D (termed "12 D processing") of *Clostridium botulinum* spores. The D value for *Clostridium botulinum* spores is 0.21 minutes. Therefore, 12 D is equal to 2.52 minutes and at the end of 2.52 minutes of treatment the microbial population would get reduced by a factor of 10^{12}. It has been found to be adequate for stability of low-acid foods at room temperature.

- But the severe heat treatment of the 12 D process may affect the sensory qualities and nutritional value extensively. Thus, a 5D or 8D process is used as a compromise between food safety and quality. In addition, in commercial processing, usually a specific time and temperature combination is used. Therefore, care is required to have a satisfactory and uniform microbiological quality of the raw products before processing. Reducing the initial microbial load by washing the commodity and maintaining proper hygiene in the food collection area as well as blanching are some steps in this regard. If the initial microbial load is kept at a low level, then the difference in effects of 5D or 8D process will be more negligible than the 12D process.

- Most of the enzymes are inactivated during normal processing. However, some enzymes, specifically important in acidic foods, are very heat resistant. Relatively short periods and lower temperatures of heat treatments may completely denature the enzymes.

- During in-container processing, the type of food, its physical state, pH, etc., and the rate of heat penetration in food, affect the time of processing. The heat penetration in different stages of the sterilization process (heating, holding, and cooling) is to be considered for the estimation of microbial destruction and nutritional losses.

- Heating of products within the can takes place by either conduction or convection or both. As liquids permit a better heat transfer than the solids due to the convection currents, normally foods are suspended in syrup or brine inside the can during the process of sterilization.

- During convection heating, in addition to the product viscosity and the presence of solid particles, the dimension of the container also affects the heat transfer. As per the general principles, the tall containers support better convection currents in liquid foods or food suspensions. The containers are also agitated to help in increasing heat transfer.

- Inside the retorts, the food materials are kept in cans, glass bottles, or other types of packages. The heat conduction characteristics of the package material have to be considered when calculating the heat transfer during the process. In flexible packages, the geometry of the product decides the geometry of packaging.

- The heat penetration and temperature of processing are considered at the thermal center of the cans. Thus, it is often required to monitor the temperature at the thermal center of the cans, which can be done with the help of a thermocouple. For conductive heating, the geometric center of the container is the thermal center. But usually the cans in retorts contain liquid foods or foods in suspension and there is convective heating. In such cases, the thermal center is usually at approximately one-third up from the container's base. However, the exact position of the thermal center varies in convective heating and should be found experimentally.

- If the spoilage microorganism is more heat resistant than *Clostridium botulinum*, then the spoilage of the food cannot be restricted even if the food is passed through a 12 D process based on *Cl. botulinum*. Thus, for food safety, the food product is recommended to be inoculated with an organism of known heat resistance to compare the actual spoilage with the theoretical spoilage, which is used in further designing the thermal process.

- Under-processed and leaking cans both pose health threats. If *Clostridium botulinum* spores are noticed, the health hazard is real and the material may need to be discarded. Intact cans containing only mesophilic, gram positive, spore-forming rods should be considered under-processed and may be further processed to impart safety.

- The routine quality assurance in a canning industry includes observing swollen and bloated cans in accelerated storage trials. If using a flat sour organism, samples for safety assessment can be taken after opening the can. It is important to maintain the level of spoilage and level of inoculation to easily evaluate the spoiled cans.

Example 16.5 A process is used for 6 log cycle reduction for an organism having 2.55 minutes as the D_0 value. An inoculated pack is to be made to verify the effectiveness of the process and tested with 100 cans. If the inoculated organism has D_0 as 2.2 min and the final population of the organism should be 2 in 100, what should be the level of inoculation?

SOLUTION:

Here the process time is 6×2.55 = 15.3 minutes

For the inoculums, the D value is 2.2 minutes, thus the number of cycles for the microorganism will be 15.3/2.2 = 6.95

Thus, $\log \dfrac{N_0}{0.02} = 6.95$

$$\dfrac{N_0}{0.02} = 10^{6.95} = 9006280$$

i.e., $N_0 = 180125$

Thus, the level of inoculation should be more than 180125.

16.1.6 Calculation of Process Time

The thermal process calculations and knowledge of their effectiveness are required for assessing the severity of process time and temperature or for determining the time or temperature needed to achieve the desired lethality.

As discussed earlier, lethality may be a single point value at thermal center of the food or as the integrated value. The temperature at the thermal center of food and that in other parts of the can are different, and, hence, it becomes difficult to calculate the processing times. The actual estimate of processing time should consider this spatial variation in temperatures in a can.

For microbial inactivation, it is adequate to consider the single point lethality. However, integrated lethality is essential for evaluating quality factor degradation.

There are two methods for determining the processing time.

1. *General (graphical) method*, which involves the graphical interpretation with lethality-time curve;
2. *Formula method*, in which an equation is employed to find out the process lethality or time using previously calculated tabulated values of parameters.

General (Graphical) Method

We have discussed the lethal rate or the lethal factor, which is given as:

$$L = 10^{(t-121)/z} \qquad \text{(as in 16.2)}$$

We have also discussed how the lethal rate increases with an increase in the processing temperature. The graphical method is based on the understanding that different time and temperature combinations can have similar lethality. The process lethality for a process is determined by graphical integration of the lethality value using the time-temperature data (equation 16.2).

$$F_0^z = \int_0^\theta L_t d\theta \qquad (16.5)$$

Process lethality can also be determined using the equation $S = F_t/D_t$. However, the sterilizing value is calculated from the following integral since D is not constant due to varying temperature

$$S = \int_0^\theta \dfrac{d\theta}{D_t} \qquad (16.6)$$

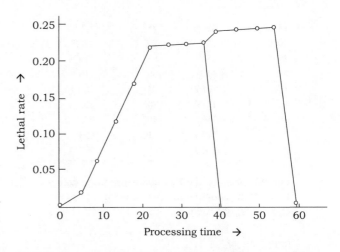

FIGURE 16.2 Lethal rate curve (after Fellows 2000 with permission)

Thus, the process involves the following steps.

1. First, the heating and cooling curves are drawn from the time temperature data (as shown in Figure 16.2).
2. The curve is then graphically integrated, i.e., the area is calculated to obtain either F or S.
3. For the process to be acceptable, the process lethality must be equal to the specified value for F or S. If it is more, then the processing is more severe than that desired and, if less, it does not give the desired inactivation of microorganisms. Thus, in both these cases, the time has to be readjusted for the specific temperature condition.
4. If the process lethality is more, then, the heating time is scaled back and the end point of the heating curve is obtained. The cooling curve is drawn for the new heating curve and the total area is recalculated. The process is repeated until we obtain the desired F and S.

The Simpson's rule can be used for integration selecting a time increment, $\delta\theta$, such that $\theta/\delta\theta$ will have an even number at the end of process time θ. If i is the number of increments,

$$A = \left(\dfrac{\delta\theta}{3}\right)\left[L_0 + 4L_1 + 2L_2 + 4L_3 + 2L_4 + \dots 2L_{i-2} + 4L_{i-1} + L_i\right]$$

$$(16.7)$$

The above process can be understood from the following example.

Example 16.6 During a thermal process having the retort at 121.1°C, the time temperature data at the thermal center were obtained as follows.

Time, min	Temp, °C	Time, min	Temp, °C
0	60	60	115.8
5	60	65	116.9
10	62	70	118.0

15	68	75	119.1
20	73.2	80	120.2
25	85.5	85	119.5
30	94.3	90	118.0
35	101.2	95	106.1
40	107.0	100	79.1
45	110.5	105	67.3
50	113.4	110	62.0
55	114.7		

For this process, the F_0 value needs to be calculated. Determine the process time for a lethality equal to F_0 of 11min.

SOLUTION:

We have to take an even number of increments. It is observed that the cooling starts from 80 minutes. From $\theta = 0$ to $\theta = 80$ min, $\delta\theta = 5$ min can give 16 increments.

$$t_0 = 121.1°C$$

Step I: From the equation, $L = 10^{(t-121.1)/10}$, the lethality values are found as follows:

$$L_0 = 10^{(60-121.1)/10} = 7.76 \times 10^{-7} \quad L_9 = 10^{(110.5-121.1)/10} = 0.0871$$

$$L_1 = 10^{(60-121.1)/10} = 7.76 \times 10^{-7} \quad L_{10} = 10^{(113.4-121.1)/10} = 0.1698$$

$$L_2 = 10^{(62-121.1)/10} = 1.23 \times 10^{-6}$$

$$L_3 = 10^{(68-121.1)/10} = 4.89 \times 10^{-6} \quad L_{11} = 10^{(114.7-121.1)/10} = 0.229$$

$$L_4 = 10^{(73.2-121.1)/10} = 1.622 \times 10^{-5} \quad L_{12} = 10^{(115.8-121.1)/10} = 0.295$$

$$L_5 = 10^{(85.5-121.1)/10} = 2.75 \times 10^{-4} \quad L_{13} = 10^{(116.9-121.1)/10} = 0.380$$

$$L_6 = 10^{(94.3-121.1)/10} = 2.089 \times 10^{-3} \quad L_{14} = 10^{(118-121.1)/10} = 0.489$$

$$L_7 = 10^{(101.2-121.1)/10} = 0.0102 \quad L_{15} = 10^{(119.1-121.1)/10} = 0.631$$

$$L_8 = 10^{(107-121.1)/10} = 0.0389 \quad L_{16} = 10^{(120.2-121.1)/10} = 0.813$$

Step II: The area under the heating curve is found as follows:

$$A = \left(\frac{\delta\theta}{3}\right)\left[L_0 + 4L_1 + 2L_2 + 4L_3 + 2L_4 + ...2L_{14} + 4L_{15} + L_{16}\right]$$

$$= \left(\frac{5}{3}\right)\left[8.1557\right] = 13.593$$

Step III: The area under the cooling curve is obtained as follows:

$$L_0 = 10^{(120.2-121.1)/10} = 0.813$$

$$L_1 = 10^{(119.5-121.1)/10} = 0.692$$

$$L_2 = 10^{(118-121.1)/10} = 0.490$$

$$L_3 = 10^{(106.1-121.1)/10} = 0.316$$

$$L_4 = 10^{(79.1-121.1)/10} = 6.309 \times 10^{-5}$$

$$L_5 = 10^{(67.3-121.1)/10} = 4.17 \times 10^{-6}$$

$$L_6 = 10^{(62-121.1)/10} = 1.23 \times 10^{-6}$$

$$A = \left(\frac{\delta\theta}{3}\right)\left[L_0 + 4L_1 + 2L_2 + 4L_3 + 2L_4 + 4L_5 + L_6\right]$$

$$= \left(\frac{5}{3}\right)\left[4.686\right] = 7.81$$

Step IV: Total area = 13.593 + 7.81 = 21.403

It may be observed that the cooling curve also contributes significantly to the total lethality. In this case, the calculated total lethality is more than the specified F value of 11 min. So the heating time needs to be reduced. This will also reduce the can temperature before cooling.

In this circumstance, suppose we reduce the heating time to 60 minutes. There will be 12 increments and the can temperature after 60 minutes will be 115.8°C. The cooling curve for the reset process will be parallel to that of the original process. The L values will be recalculated again, and the total area of the curve will be determined. If the calculated lethality is equal to the specified F, then the process is acceptable, or else, depending on whether the calculated lethality is more or less than the specified lethality, the heating time would be changed and the calculations would be done again.

In conductive heating foods, the thermal lag may cause an increase in the temperature at the thermal center of food even after the process of cooling is started. This necessitates conducting more trials and observing the temperatures at specified time intervals to determine the lethality.

Formula Method

The formula method is based on the tabulated values of parameter f_h/U expressed as lethality. For various conditions of heating and cooling, these values have previously been calculated by other researchers. The process is as follows.

If we plot the heating curve by taking the temperature at the thermal center of the food with time, the plot will be as shown in Figure 16.3.

A dimensionless temperature ratio, TR, can be expressed with the help of the following equation.

$$\frac{t_s - t}{t_s - t_0} = TR = erf\left[\frac{x}{(4\alpha t)^{0.5}}\right] \quad (16.8)$$

where, t is the temperature at a distance x from the surface of the can and erf is the error function. The error function values are given in standard mathematical tables.

The same equation can be used for thermal processing if t_s is replaced by t_r, the retort temperature. Here t_0 and t are the initial temperature and the temperature at any time at the thermal center of the can.

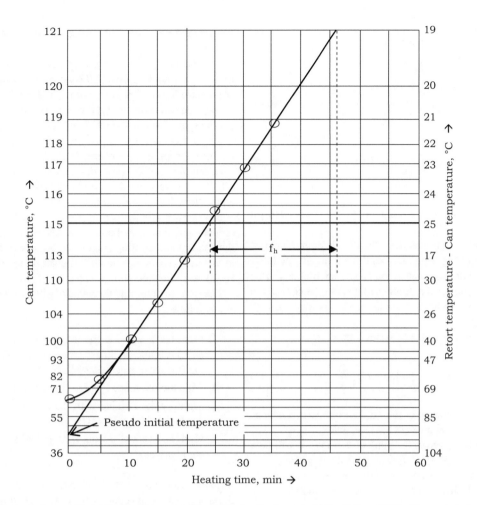

FIGURE 16.3   Change in temperature at the thermal center of the can by heating in a retort (after Fellows 2000 with permission)

Considering the Figure 16.3, if t_{ih} is the initial temperature at the thermal center of the can, and t_{pih} is the *pseudo-initial temperature* at the thermal center (the t_{pih} can be obtained by extending the curve to the y axis), then j_h, the *thermal lag factor*, can be defined as:

$$j_h = \frac{t_r - t_{pih}}{t_r - t_{ih}} \qquad (16.9)$$

The subscript h stands for heating.

j_h is also known as *intercept index* for the heating curve. Another intercept index j_c can, similarly, be defined for cooling. We will define some more terms.

I_h is the difference between the initial product temperature and the retort temperature.

$$I_h = t_r - t_{ih} \qquad (16.10)$$

$$\text{Thus, } j_h I_h = t_r - t_{pih} \qquad (16.11)$$

'f_h' is the time (minutes) for the heat penetration curve to cover one logarithmic cycle. It is also known as the *heating rate constant* or the *slope index* of the heating curve. If the heating curve consists of n line segments, then each line is represented by a slope index. As for f_h, the time required for

the thermal penetration curve to cover one logarithmic cycle of cooling, in minutes, is known as f_c.

In any heating process, even after the specified heating time, there is a gap between the retort temperature and the final product temperature (at the thermal center), this is denoted as "g." The value of g is affected by the D and z values of the microorganisms as well as the shape of the heating curve. The temperature of cooling water also influences "g." In conductive heating foods, the heat transfer is slow. Hence, if the proper rate of cooling is not maintained, the product may remain hot for more time even after the actual sterilization has been achieved.

In the case of unsteady state heat transfer, as happens in a can during the retorting process, the temperature ratio is given as:

$$\ln\left(\frac{t_r - t}{t_r - t_0}\right) = -\frac{hA}{mc_p}\theta \qquad (16.12)$$

From this equation, the slope,

$$f_h = \frac{m.c_p}{hA}$$

Considering $t_0 = t_{pih}$, from Figure 16.3,

$$\log\left(\frac{t_r - t}{j_h I_h}\right) = -\frac{\theta}{f_h} \qquad (16.13)$$

$$\log\left(t_r - t\right) = \log(j_h I_h) - \frac{\theta}{f_h} \qquad (16.14)$$

$$\text{or } \left(\frac{t_r - t}{j_h I_h}\right) = 10^{-\frac{\theta}{f_h}} \qquad (16.15)$$

$$\text{Thus, } t = t_r - j_h I_h \left(10\right)^{-\theta/f_h} \qquad (16.16)$$

The above equation indicates that when t_r-t is plotted against the time on a semi-log paper, the resulting straight line has a slope of -1/f_h and an intercept of log (jI). Also, higher j value indicates more time for the product to respond to changes in the temperature of the heating medium. This is why, j is also called *lag factor*.

The cooling curve can also be written as

$$\log\left(t - t_c\right) = \log(j_c I_c) - \frac{\theta_c}{f_c} \qquad (16.17)$$

where, t_g is the temperature of the product (at thermal center) at the end of the heating process, $t_g = t_r - g$, and $I_c = t_g - t_c$.

If we plot log $(t - t_c)$ versus θ_c (time of cooling), it will be a straight line with slope -1/f_c. $\theta_c = 0$ at the start of cooling.

The above equation can be modified to explain the temperature at any time during the cooling process, as follows:

$$t = t_c + j_c I_c \left(10\right)^{-\theta_c/f_c} \qquad (16.18)$$

Total lethality occurs during the initial part of the cooling curve just after the cooling water is introduced. However, the above equation represents only a portion of the cooling curve; it does not represent the complete initial segment that has the major impact for total lethality. Thus, the curved portion of the curve at the initiation of cooling has to be expressed mathematically to properly predict the total process lethality. The initial curved segment has been presented through hyperbolic, circular, and trigonometric functions by different researchers. The point where the curved and linear segments meet is also important.

From equation 16.17, the time of heating (θ) can be written as:

$$\theta = f_h \log\left(\frac{j_h I_h}{g}\right) \qquad (16.19)$$

All values excepting g can be obtained from the above heating curve.

We have discussed that in the formula method, lethality value expressed as the parameter f_h/U is tabulated. There are two common methods for calculation of g, namely the Hayakawa's (1970) and the Stumbo's (1973) methods. The Stumbo's method is more common; however, one can use

any method depending on the conditions existing during the process.

Stumbo (1973) has defined tables in which the g values have been found against f_h/U values for different j_c values (Annexure XIII). If the value of j_c is not available, the j values for heating can be used. But the j values are usually lower than j_c values, and, hence, the result will give a longer process time (a longer process time means a safer process since the destruction of microorganisms is greater).

In Stumbo's (1973) procedure a major assumption is that $f_h = f_c$. Thus, in the case the f_h and f_c values cannot be considered equal, then the general method is preferred.

Hayakawa's (1970) method can be used for different rates of heating and cooling because in this method, the lethality of the heating and cooling stages has been presented separately in different tables. Toledo (2007) gives more details about the Hayakawa's method and the tables.

The steps in obtaining the value of g by the Stumbo's method are as follows:

1. First, the F value at the retort temperature (denoted as F_1) is found. Standard tables are available to obtain the F value at selected z values (Annexure XIII);
2. Take the reference F value (10 min at 121°C);
3. Multiply the F value with the F_1 value and take the product as U, i.e. U = F.F_1;
4. Find the value of g from f_h/U and g tables for different j_c values.

As shown in Figure 16.3, the data for temperatures at different times are plotted directly onto a semi-log graph paper. The numbers are marked as t_r-t on the log scale and on the opposite side of the graph paper is the can temperature. It is possible to draw a straight line connecting most of the observation points and extend it to the axis ($\theta = 0$).

During processing in a batch retort, usually, the first 60% of the retort come-up time does not contribute to heating (and thus thermal processing). Hence, the actual process time is calculated by deducting 60% of the come-up time from the total calculated process time.

$$\text{Actual process time} = \theta - 0.6(\theta_{\text{come up}}) \qquad (16.20)$$

Another term, *pseudo-initial time* (θ_{pi}), may be defined as the time when the heating of the contents is assumed to start, i.e., after $0.6 \times \theta_{\text{come-up}}$ from the start of heating. If the curve is extended to the line $\theta = \theta_{pi}$, we get the pseudo-initial temperature, t_{pi}.

$$(t_r - t_{pi}) = jI \qquad (16.21)$$

As we have discussed earlier, the equations 16.9 and 16.10 can be used to determine j (intercept index) and I. On the semi-log plot, the time taken by the heating curve to cover one log cycle is known as the *slope index* f_h.

In a similar manner, the cooling curve can be plotted and f_c and j_c can be found. In this curve at time 0 ($\theta_c = 0$), the supply of steam is stopped and cooling water is introduced. The retort is assumed to immediately reach the cooling water temperature. It is at $\theta_c = 0$ that the intercept of the cooling curve is evaluated.

If the unaccomplished temperature differences are $t_1\text{-}t_c$ and $t_2\text{-}t_c$ at times θ_{c1} and θ_{c2}, respectively, then

$$f_c = \frac{\theta_{c1} - \theta_{c2}}{\log(t_1 - t_c) - \log(t_2 - t_c)} \quad (16.22)$$

In cases where a broken heating curve is displayed by the product, the calculations become more complicated and more complex formulae are used for determining the process times.

Example 16.7 A food is heated using a process based on $F_0 = 6$ min at 118°C temperature. The heat penetration data (curve) give t_{ih}=74°C, f_h= 20 min, j_c = 1.8, f_c=24 min, t_{pih}=40°C. The retort reached the process temperature in 8 min. Calculate the processing time.

SOLUTION:

As per the given data,

$$j_h = \frac{t_r - t_{pih}}{t_r - t_{ih}} = \frac{118 - 40}{118 - 74} = \frac{78}{44} = 1.77$$

$$I_h = 118 - 74 = 44°C$$

We know that F_0 is the same as F_{121}^{10}

From Annexure XIII for $121\text{-}t_r = 8°C$ and $z = 10°C$, $F_1 = 6.35$ (by interpolation)

Thus, $U = F.F_1 = (5.8)(6.35) = 36.83$

$$f_h/U = 20/36.83 = 0.543$$

Again, from $f_h/U = 0.543$ and $j_c = 1.6$, g is obtained as 0.094°C (by interpolation)

i.e., the thermal center reaches $113 - 0.094 = 112.91°C$.

$$B = f_h \log \frac{j_h I_h}{g} = 20 \log \frac{(1.83)(41)}{0.094} = 58.04 \, \text{min}$$

Process time = $58.04 - (0.6)(8) = 53.24$ minutes

16.1.7 Spoilage Probability Evaluation for a Process

When for a given process the probability of spoilage is determined, it is useful in deciding if the process needs any adjustments. The process time and the initial can temperature are known for a constant temperature process. We calculate g value, and U (at the specified z value) is determined from standard tables to calculate F_0^z. Then by substituting F_0^z for θ, the probability of spoilage is calculated. The general method can be used for determining process lethality. As the process temperature changes with time, to determine the temperature at the critical point, the finite difference methods can be used.

In general, a higher probability of spoilage is obtained by Hayakawa's method as compared to Stumbo's method. The process time obtained from Stumbo's method is higher than that obtained from Hayakawa's method.

16.2 Unit Operations in a Canning Process

The heat processing, or thermal processing or canning, involves exposing the food cans or packages to a high enough temperature for a specified duration of time. Besides metal cans, glass jars/bottles, flexible pouches, and rigid trays are also used for thermal processing. Sterilized or canned products can be stored for long periods at ambient temperatures. The normal flow of the food materials (fruits or vegetables) for canning is given in Figure 16.4.

Not all fruit varieties are suitable for canning. Thus, use of suitable varieties and adhering to the established commercial size and quality grades are important in the commercial canning process. The fruit must be harvested at the proper stage of maturiy, i.e., it should not be so ripe or so soft that it could be eaten fresh, yet it should have the flavor characteristics of the ripe fruit. Overripe fruits are infected with microorganisms and underripe fruits shrivel and toughen on canning.

The raw materials should arrive at the canning plant as early as possible, preferably within one hour of harvesting. After proper sorting of the materials, they are washed, peeled,

Receiving
↓
Sorting and grading
↓
Washing/ cleaning
↓
Peeling/coring/ pitting
↓
Curing/ Blanching
↓
Can filling
↓
Syruping or brining
↓
Lidding or clinching
↓
Exhausting
↓
Sealing
↓
Can washing
↓
Heat processing
↓
Cooling
↓
Drying
↓
Labelling
↓
Storage/ Marketing

FIGURE 16.4 Unit operations in a typical canning process

cored, and reduced in size as per requirements. The vegetables may be soaked in a potassium permangate solution to disinfect them. The vegetables may then be blanched. Use of hard water in blanching should be avoided as it toughens the tissues and destroys natural texture. The material is then filled in the cans and either syrup or brine is used to fill the can keeping some air space (called *head space*) at the top. Usually the syrup is added to fruits and the brine to vegetables. The temperature of the syrup/brine is kept at 79°–82°C and the head space is usually 0.3–0.4 cm.

The next step is the removal of air from the head space as well as the contents of the can. The process is known as *exhausting*. After exhausting, the cans are immediately sealed. The sealed cans are then washed with compressed air and/or water.

Then the actual heat treatment takes place for sterilization of the product. The heat processing can be done by saturated steam, by hot water, or by flames. After sterilization, the products are immedietely cooled to arrest further cooking of the product. After cooling, the cans are dried to remove moisture adhering on the outer surfaces, labelled, and stored/ sent to the market.

16.3 Exhausting

Exhausting is the process of removing air from the containers before sealing. It helps in the following ways.

- As the air (and oxygen) from the can is removed, it reduces the chances of corrosion and pin-holing of the inner surfaces of the cans.
- It helps to prevent the oxidative changes in food, such as discoloration, loss of vitamins, etc. It also reduces other types of chemical reactions.
- It helps to prevent strain on the can during sterilization. If the food product contains entrapped air, the air will expand during the heat processing and will cause bulging and strain on the container.
- The vacuum inside the cans keeps the ends of the can in concave shape. If the air is not removed and if there is formation of gas (due to decomposition of the can contents), the ends become convex. Thus, the concave shape of the can ends is used as a major indication of the quality and safety of food.
- Some fruits and vegetables expand or shrivel during heating. Hence, exhausting prevents overfilling of the cans. Simultaneously, it also allows a greater fill of soft fruits such as berries.

16.3.1 Methods

The exhausting process can be done by any one of the methods mentioned below or a combination thereof.

- filling the food into the container when it is hot and then sealing the container;
- filling the food at normal temperature and then heating it to 80°–95°C with the lid partially in place and then fully sealing the container when the food is still hot;
- using a vacuum pump;
- closing the can in an atmosphere of steam (a blast of steam is forced onto the head space of the can to replace the air, immediately after which the lid is sealed. When the can is cooled, the steam condenses and a partial vacuum is created in the head space. The steam is usually forced at 34–41.5 kPa).

Hot filling is the most common method as it also pre-heats the food and the actual processing time is reduced. The steam flow closing is applied to liquid foods (food kept in brine or syrup) because the liquid has less entrapped air and the liquid surface is flat so that the steam flow is uniform over the surface.

16.3.2 Equipment

Exhaust Box

As mentioned above, in this method the filled cans are exposed to a temperature of 80°–95°C. A long duration exhaust at a moderate temperature is more effective than a shorter duration exhaust at high temperature. Different types of exhaust boxes are used.

- *Disk exhaust box.* It consists of a rectangular metal box, in which there are three to four rows of large metal discs of 38–45 cm diameter fitted with cog gears that mesh together. There are curved iron rods above the discs, which guide the cans through the exhaust box. The cans travel down one row of discs and back the next. The number of rows and the length of the exhaust box vary depending on the desired capacity and the nature of the product.
- *Cable exhaust box.* It consists of a narrow, shallow, and rectangular metal chamber through which a steel cable or a chain conveyor passes (Figure 16.5). The cans are placed on the cable or conveyor to move through the chamber, during which they are heated or exposed to steam.
- *Rotary exhaust box.* The cans are carried on a reel through a steam-filled cylinder. The cans follow a circular or spiral path in the box.

Nowadays in many canneries, the exhaust boxes have been replaced by a steam-flow closure system or by a vacuum closing machine.

Vapor Sealing Equipment

Jars using the white vapo-seal caps (and certain other types of caps) are sealed in a current of steam. As the jars move in a metal box or chamber, a jet of steam is applied on the top layer of food to flush out most of the air in the head space. A lid falls in place, and it is forced onto the jar mechanically. After the steam in the head space of the jar condenses, a good vacuum is created.

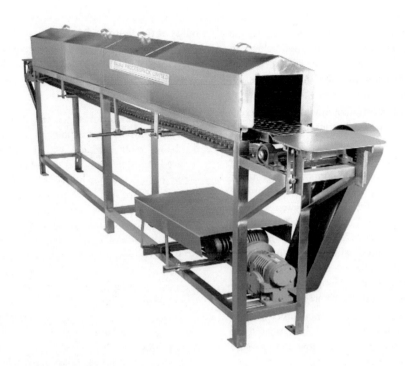

FIGURE 16.5 Exhaust box (Courtesy: M/s Bajaj Processpack Limited)

The "steam-flow closure" or "steam vac closure" in the double-seaming method of cans also employs a similar method. The air must be swept effectively from the head space by the steam flow. An atmosphere of steam is maintained around the can and cover assembly until the can is sealed to prevent the reentry of air into the can.

In this method, a good vacuum is obtained only if the head space is large. When the head space is small, almost no vacuum is obtained after sealing and cooling. Besides, the preparation of the food should be such that the food does not contain an excessive amount of air or gas, otherwise it could later fill the head space and reduce the vacuum.

Steam is in contact with the food for little time and there is neither any blanching effect nor sterilization. If blanching is necessary, then it must be done before canning.

Mechanical Vacuum Pump System

The glass jars are exhausted under a mechanically produced vacuum instead of employing heat.

Glass containers, sealed hot and allowed to cool, develop vacuum equal to approximately the theoretical value because the walls do not contract or expand as they do in the case of tin containers.

16.4 Processing of Cans

16.4.1 Methods

The heat processing is done usually by saturated steam, by hot water, or by flames. The first two are more common. The type of apparatus used for thermal processing can be broadly grouped depending on whether product is flowing (liquid and less viscous materials) or non-flowing type (solids or thick

pastes). The solids can also be processed either in packaged or in unpackaged form.

Heating by Saturated Steam

The containers are kept in saturated steam (under pressure) inside the retort (or a pressure vessel). The vessel is supplied with the steam with a pressure corresponding to the required treatment temperature. The cans on their outside surfaces experience steam condensation and the latent heat is transferred to the cans.

Some considerations are as follows.

- All parts of the retort should have even distribution of steam. In large-size retorts, steam should be applied through a large diameter pipe, running internally the whole length of the sterilizer if it is horizontal, with holes at frequent intervals for steam release.
- If the heating steam contains air or the retort has previously entrapped air, the air cannot condense and, therefore, it forms a layer between the steam and the can/bottle surfaces. This will lead to a lower temperature in the cans and under-processing of food. Hence, all air present in the retort should be removed by the incoming steam. The process is known as *venting*. Air vents of adequate size are an integral part of all batch sterilizers, which are fitted at points remote from the steam inlet. The air vents may be of thermostatic type working on the balanced pressure system, which automatically close when the temperature of the air-steam mixture passing through rises to about 99°C.
- Usually, the retorts have some system for agitating the cans to allow mixing of the product within the

can and thus uniform distribution of heat. In the absence of an agitation system, the foods closer to the inner surface of the cans may be over-processed and those at the center may be under-processed, which is more the case in in viscous foods.

- Since air is present in different amounts in different parts of a sterilizer, different bottles in the same batch will receive heat treatments of different intensities.

- A sterilizer should always be operated at a specific temperature (or pressure). A temperature recorder with suitable temperature range is fitted to record the variation of temperature with time inside the sterilizer. Pressure of the vessel is known by the pressure gauge fitted to it. An increase in the retort temperature can reduce the processing time, but it needs higher pressure and both the retort and the cans should be able to resist the higher pressure. All these also imply a higher cost of the retort and containers.

- After the desired degree of thermal processing is achieved, the cans (and contents therein) need to be cooled. If the cooling process is slow, the pressure in the containers remains high for a longer time. Therefore, the cooling air is also provided under pressure so that the strain on the container seams is minimized. It is called pressure cooling. As the temperature of food cools below 100°C, the pressure in the retort is reduced to normal and cooling is continued until the temperature drops down to 40°C. The moisture on the container surface dries at this temperature, and the setting of label adhesives is more rapid.

- Glass is a poor conductor of heat but has a high coefficient of thermal expansion. This will lead to mechanical stresses within the glass. Glass has a poor tensile strength and may break under the situation. This is called breakage by thermal shock. Thermal shock limits the rates of both heating and cooling. A good bottle will withstand a temperature difference of about 50°C if the outside is hotter than the inside, but at about 30°C only if the outside is cooler. Sudden temperature changes of this order should not occur at any time in the sterilizing process, or it will result in broken bottles. Thus, the cooling should also be controlled and controlled water cooling within the sterilizer is preferred to air cooling.

Heating by Hot Water

The glass containers or flexible pouches can be sterilized with hot water with high pressure air. Longer processing times would be required for glass containers than that for cans of the same size and shape due to the lower thermal conductivity of glass than metals.

The processing of retortable pouches is similar to that of glass in a hot water system. Heating of the flexible pouches is more rapid. A uniform thickness of food across the pouch is important for uniform heating; it is assured by keeping the liquid or semi-liquid foods horizontally. Vertical packs may bulge at the bottom. Often special frames are used to secure the products packed in flexible containers to prevent bulging or any other type of changes in package dimensions. It also prevents stresses at the seams, which may be caused by product expansion during heating.

Three important methods of heating retorts, namely, steam, water + air, and steam + air are in use. The first is used for metal cans and the other two for glass containers because the jar lids will be forced off at processing temperature and extra pressure is required to hold the lids in place on glass containers.

The process requirements for the come-up period, holding at the specified temperature, and cool-down period should be properly monitored. In the pure steam method, it is essential to vent the retort during the coming-up period so that all entrapped air is removed. Otherwise, heating will not be uniform. After the retort temperature is attained, the thermometer and pressure gauge should agree. If the pressure gauge shows more readings then there is entrapped air in the retort, it may lead to uneven heating. After the retort temperature is attained, the vents may be closed and only the small petcocks (bleeders) are left open to allow as much steam to escape as is necessary to remove non-condensable gases entering with the steam and to promote circulation.

Direct Flame Heating

Sterilization can also be done using direct flame heating. A conveyor moves the cans on an open flame. The flame temperature is usually at more than 1700°C and the air in contact with the cans is at 1200°–1400°C. The conveyor is arranged such that the cans also rotate as they move along the flame.

The product in the can is held in water, brine, or low sugar syrups so that there is rapid heat exchange between the contents and the can walls. There should be the maximum achievable vacuum in the cans without panelling at the time of filling. Otherwise, high sterilization temperature maintained in the process will cause excessive internal can pressure from expanding air and steam. The method is not suitable for small cans as the internal pressure may rise to about 275 kPa (at 130°C).

As the method uses a short processing time, it results in a food of high quality with less energy input (The energy saving may be about 20% when compared with conventional canning). An infrared controller is used to scan each can after processing. The sensor senses the amount of heat emitted by the can, which is an indication of the heat received. If the emitted heat is insufficient, the can is rejected.

16.4.2 Equipment

In general, a *cooker* is a device in which the temperature is maintained at 100°C or a little less for processing of fruits and acidic vegetables. A *retort* signifies an autoclave or closed processor in which the cans are heated in steam under pressure. In a retort, a temperature of 125°C or higher is easily attainable. Both manual and automatic control devices are available for maintaining the steam pressure and the process time.

Different types of heat exchangers, ohmic heating devices, etc. are also used for heat processing.

The type of equipment and the time and temperature of processing are decided depending upon the nature of fruits and vegetables. All fruits (except olives) are processed at 100°C. The acidity of fruits and acid vegetables (e.g., tomatoes) makes it possible to process safely at 100°C. But other vegetables, because of their higher microbial load due to low acidity, proximity to soil, hard texture, etc. need severe processing conditions. Temperature that would spoil the color, flavor, texture, and appearance of fruits in many cases improves the flavor and texture of vegetables.

Cookers

The cookers can be batch or continuous type.

Batch Type Open Cooker

- This consists of open insulated tanks filled with boiling water into which the cans are lowered in crates by means of a crane.
- The word open implies that it operates at atmospheric pressure. Open steam jets are used to maintain the temperature of water at the boiling point.
- The process is used only in small canneries.

Continuous Non-Agitating Open Cooker

- It has a long-insulated tank containing boiling water through which the cans move on metal baskets by an overhead conveyor.
- This cooker occupies a large floor space and a considerable amount of heat escapes from it through the evaporation of water, thus lowering the efficiency.
- It is also difficult to regulate and adjust the time of sterilization.

Continuous Agitating Cooker

- It consists of a long metal tank of heavy boiler plate inside which a spiral extends throughout the length.
- The cans enter a small porthole at one end and travel along the spiral with the help of a cylindrical reel that revolves inside the spiral.
- During the movement inside the cooker, the cans are rolled on the reel, thus thoroughly agitating the contents. This also improves the heat transfer into the cans.
- A number of outlet doors may be placed near the top of the processor so that the process time may be varied at will.
- The cross section of the cooker is conventionally cylindrical. However, the newer systems have the lower half of the shell as half circle and the upper half as a rectangle. This design permits lifting the flat lid easily during operation. It also helps to easily

remove the cans that have become crushed between the reel and the spiral and thus have caused the operation of the conveyor to be interrupted.

- Water is filled up to ½ – ¾ of its volume if the cans are to be processed in water, or else the cans can also be processed by live steam only.
- The use of water permits operation at any desired temperature. In most cases the sterilization temperature is kept at 1°–2°C below the boiling point of water, as it helps in avoiding heavy loss of heat. There is also no leakage of steam into the cook room and thus it helps in maintaining better working conditions. Operations slightly below the boiling point also permit use of a temperature controller that will close or throttle down the steam line during the frequent intervals when cans are not entering the processor. If the operation is at boiling point, then use of an automatic temperature regulator is not possible and manual operation of the steam valve may be necessary.
- The capacity of the processor (cooker) in cans/min varies with its length and diameter and with the length of the process. The output in cans/min is found by dividing the number of cans held by the processor with the reel completely filled by the length of the process in minutes.

Cooker-Cooler

- There is a tunnel that has three sections for preheating, sterilizing, and cooling, which are maintained at different temperature and pressures.
- The cans are carried on a conveyor through these three sections.

Retorts

Heat processing in a retort is often known as *retorting*. The retorts using steam can be classified as follows.

1. Batch retorts with or without agitation;
2. Rotating retorts without agitation;
3. Continuous process retorts;
4. Hydrostatic retorts;
5. Hydrolock retorts.

The first three categories are more common in the food industry.

Batch Retort

Construction and Operation

- It usually consists of a heavy steel pressure vessel of circular or rectangular shape.
- It may be horizontal or vertical. The vertical type sterilizers (Figure 16.6) are top loading types and require less floor space than horizontal (side loading types) retorts; however, the latter are more convenient to fill and empty and can have facilities for agitation.

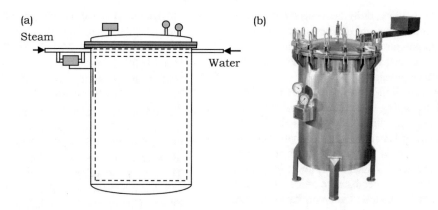

FIGURE 16.6   (a) Schematic diagram of a batch retort, (b) a batch retort (Courtesy: M/s Bajaj Processpack Limited)

FIGURE 16.7   Horizontal retort with trolleys to move the cans into the retort (Courtesy: M/s Bajaj Processpack Limited)

The horizontal types need more floor space; the typical length is 1.5 m–7 m.

- Usually, the bottles are loaded onto crates or small steel cars, which are then placed on a steel track in the retort (Figure 16.7). It helps in quickly and easily placing the cans inside the retort (the chamber is generally very hot during the operation). Perforated metal partitions are used to separate the layer of cans.

- Crates in horizontal retorts have a rectangular profile and the dimensions are such that they can also fit into cylindrical retorts.

- The vertical retorts can be placed in the form of a battery above which is a travelling crane. Cans are filled into circular crates that are lowered into the retort. Small vertical/horizontal sterilizers may be loaded manually.

- The vessel is fitted with a lid or door that is able to withstand the operating steam pressure. In special cases, both ends of the horizontal retort are fitted with heavy swinging doors so that after sterilization the crates of sterilized cans can be removed from one end of the retort while the crates of unsterilized cans enter at the opposite end.

- After the cans are put in the retort and the door is closed and sealed, the retort may be vented for some time and then steam is maintained in the vessel at

a pressure corresponding to the required treatment temperature. There may be provision to make the crates revolve during sterilization so as to agitate the can contents.

- After the desired processing time, the steam is vented into the atmosphere. Continuous venting of a small amount of steam is carried out by a steam bleeder, which also promotes steam flow within the retort.

- Compressed air is blown into the retort to prevent development of undue high pressure in the cans, which would otherwise deform the cans.

- The crates (bottles/cans) after sterilization are cooled immediately to prevent further cooking of the product. The cooling may be carried out within the retort itself. The cooling is achieved through water sprays. The retort should be opened only after the temperature has been reduced to less than 100°C. The crates of bottles are then taken out for either natural or forced air cooling.

- A system of hydraulically operated locks and pumps is used to open and close the retort as well as to supply steam and water.

- The final cooling of the cans may also be done in canals with the help of circulating chlorinated water. The cans are moved through them by overhead conveyors.

- Records are required to be maintained of the time and retort temperature for each batch. Thus, the two essential components of a retort include a recording device and an accurate temperature controller. The automatic control of operating steam pressure and the process time are more reliable than manual controls.

- During sterilization of such products as milk, a ring of white solids is formed at the edge of the milk surface on the bottle wall, which is caused by excessive foam formation during the bottle-filling operation. Hence, agitation by rotation or violent shaking of the bottles is one primary requirement for processing and cooling of milk in these retorts.

- It is also relatively easy to maintain them at a constant temperature. However, the batch retorts are labor intensive.

- The heat transfer is relatively slow in the absence of agitation, and, hence, these may not be suitable for viscous products.

- Batch sterilization systems are categorized based on the size of the pressure vessel, amount of thermal insulation, operating temperature, process time, and amount of steam used for venting. The consumption of steam varies accordingly.

- Usual process steam consumption is 0.2 and 0.6 kg of steam per liter of milk.

Applications

Batch sterilizers can be used for different package sizes. Usually, they are not used for viscous products unless there is agitation of cans in the retort.

Some special batch type retorts are discussed below.

Rotary Batch Sterilizer

The product is agitated by rotating the load of bottles/cans about a horizontal axis during heating.

Crate-Less Retort

In this case, the cans are dropped directly into the retort filled in water. As the name indicates, no crate is used. After processing, the initial cooling is done by introducing cold water into the retort. The final cooling is done outside by dropping the cans in a tank of cold chlorinated water. The hot water from one batch can be stored and reused in another batch. There is no need for venting of steam as the steam displaces water at an initial phase of the process.

Orbitort

It is a batch type retort, which is slowly rotated. It has two concentric cages. The loading of the cans is carried out in the annular space between the cages horizontally to the guide rails. With the retort rotation, the cans move along the guide rails, and, at the same time, the product is stirred.

Cascading Water Retort

Construction and Operation

- It consists of a huge chamber in which the cans are arranged in a stack and hot water is sprayed from the top. Steam spreaders heat the water internally. The source of overpressure is air (Figure 16.8).

- During processing, it is not the water but sprays or cascades (showers) that completely cover the containers.

- The spent water is collected at the bottom and then recycled into the system after passing it through another heat exchanger. There may also be another heat exchanger for cooling the used water so that it can be used subsequently for cooling of the cans.

Salient Features

- It uses a low volume of water;

- The energy requirement is also considerably lower than batch type systems.

Continuous Retorts

As in any other device, the continuous retorts have higher capacity and allow better monitoring of the process conditions. The final product is also more uniform. Different types

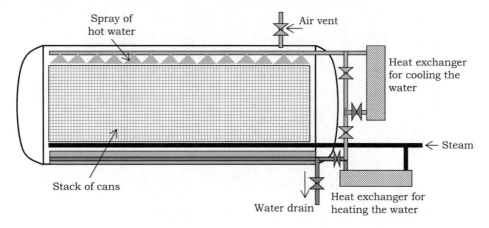

FIGURE 16.8 Schematic of a cascading water retort

of continuous retorts are available. The hydrostatic retort, hydrolock retort, and rotary sterilizer come under this category.

Hydrostatic Retort

Construction and Operation

- The hydrostatic retort is a large continuous sterilizer in which there is provision of steam injection in the sterilization chamber. Two water columns (barometric lock) are connected to the chamber, which eventually act as the feed and discharge gates.
- A conveyor carries the cans/packages through the steam chamber. As can be seen from Figure 16.9, the cans move through one water column into the steam chamber and remain there for a specified time for sterilization. Thereafter the cans move out of the chamber by passing through another water column.
- The temperatures of water columns are higher than the room temperature (due to condensation of steam) and, hence, the inlet column serves for the preheating of the cans and the outlet column does the initial cooling after sterilization. As the cans proceed for further cooling, they are exposed to sprays of cold water. Finally, the cans move through a tank of cold water for the final cooling process. The cans are then unloaded from the conveyor to make space for a new load of cans.
- A constant temperature is maintained in the processing chamber, known as a steam dome, by steam injection.
- The water column seals the steam in the chamber. The weight of water in the water columns counterbalances and maintains the pressure within the chamber. Change in the height of water column changes the steam pressure inside the chamber. For example, absolute pressure of steam to obtain a temperature of

121°C is 2.057 bar. The water column is required to be at least 10.7 m high. Pressure in the chamber must increase, necessitating an increase in the height of the water column, to increase the temperature.

- Hydrostatic cookers are accordingly large requiring tall structures, consequently, they are installed in the open.
- The temperature of water columns is also maintained by the condensation of steam in the steam dome.
- By varying the conveyor speed, the sterilization time can be varied. To increase the holding time in the steam dome, two, four, six, or eight passes are used.

Salient Features

- These retorts can handle cans of different sizes.
- The production rate can be very high, even up to 1,000 cans per minute.
- There is minimum container agitation during processing. However, the major limitation is the huge size of equipment and the high capital cost.

Hydrolock Retort

Construction and Operation

- These are continuous retorts with a rotating pressure lock, partly submerged in water, called *hydrolock* (Figure 16.10).
- First, the cans move through the hydrolock into a preheating chamber at the lower part of the retort. This section is filled with hot water. Then the cans move to the upper part of the retort containing steam where the actual sterilization takes place.
- A number of passes are possible to increase the holding time of cans during sterilization (Figure 16.10).

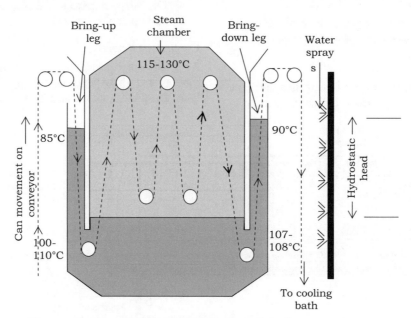

(Temperatures mentioned are indicative to explain the process.)

FIGURE 16.9 Schematic of a 6-pass hydrostatic retort (*After* https://www.barnardhealth.us; https://www.retorts.com)

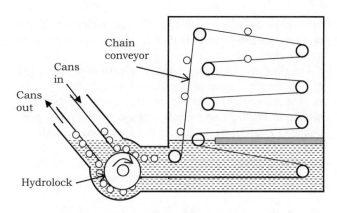

FIGURE 16.10 Schematic diagram of a hydrolock retort (after Hallstrom et al., 1988 with permission)

- After sterilization, the product passes through a cooling section.

Salient Features

- Hydrolock retorts can have different capacities and can accommodate varied can sizes.
- They are usually of large sizes and are costly.
- A common problem associated with this retort is leakage in the rotating valve.

Rotary Sterilizer

Construction and Operation

- It consists of a pressure vessel inside which there is a slowly rotating drum. Cans are moved along the periphery of the drum on can-shaped pockets guided by a helical track on the inner wall of the vessel (Figure 16.11).
- It may also consist of a series of cylindrical vessels called shells in which the cans move from one

shell to another. The shells are usually of 1.5 m diameter.

- There is a rotary valve at the top that allows continuous entry of the cans into the retort, but that prevents any loss of steam through it. In systems with more than one shell, there is a transfer valve that transfers cans from one retort to the other. There are revolving can-shaped pockets in some systems to carry cans into and out of the retort without loss of steam through self-sealing inlet and discharge valves.
- Cans move on the reels and roll along the cylindrical wall alternately when the reel rotates; the cans also rotate on their own axes. There is intermittent product agitation (during the rolling of the cans), which helps to mix the contents inside the can and improve heat transfer within the containers.
- The inside of the retort may be divided into different pressure sections, which are separated by pressure locks. In multi-shell systems, processing and cooling are done in separate shells.
- The cans attain the processing temperature immediately after entering a continuous retort. Thus, the whole time that the cans remain in the retort is the processing time.
- If N_t is the total number of cans in the retort when it is completely full, N_p is the number of pockets round the periphery of the reel, and Ω is the rotational speed of the reel, then the processing time is given as

$$t = \frac{N_t}{N_p \Omega} \qquad (16.23)$$

- Heat penetration parameters are determined by a simulator, called a *steritort*, at different reel rotational speeds.

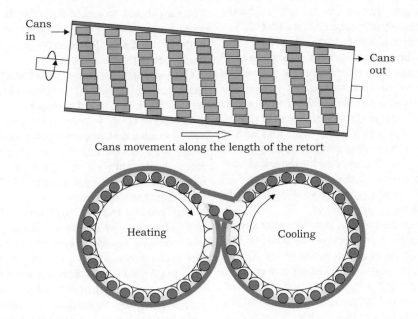

Cans movement along the length of the retort

FIGURE 16.11 Schematic of movement of cans in a continuous retort

Salient Features

- The floor space needed for continuous retort is only half that needed for a vertical retort installation of equal output. However, high initial investment is required and there are additional critical factors to monitor and control.

- The speed of rotation of the reel decides the residence time and the rate of heating. The speed of rotation should not exceed the critical limits beyond which soft products may become crushed, resulting in a homogenous mass.

- Agitation of the product increases convection inside the cans and thus the rate of heat penetration. This also allows the use of higher temperature (up to 138°C) and a shorter time (HTST) process. It increases the throughput capacity of the retort and helps in saving labor and steam. It also helps to improve product quality and uniformity.

- A continuous system can save as much as 60%–75% of the steam as compared with still retorts. It also reduces the water requirement for cooling of cans. However, the agitation is less for semi-solid products.

- There are restrictions imposed by reel steps, spacing of the spiral, and other features, limiting the range of both the length and the diameter of cylindrical container sizes that the retorts can accommodate.

Applications

As mentioned above, only specified sizes of cans can be used in the rotary sterilizer.

Important Accessories in Retort

The important accessories are temperature recorder/thermometer, air vent, and a pressure gauge in a retort. Steam vents are used during the coming-up period, which must be large enough to "purge" air from the retort quickly. The vertical retort should have a pressure gauge and mercury thermometer. In horizontal retorts, there should be two mercury thermometers one at each end.

The steam supply line that delivers steam to several retorts is called a header. The header must be large enough (about 2" diameter) to meet the peak demand for steam. There should be proper insulation on it to reduce heat loss by convection and radiation. Air is admitted so as to mix the water and maintain the additional pressure required to hold the lid of the jars in place.

In some systems, there may be a bypass line with a manual operated valve to allow entry of extra steam than that provided by the controller steam line. It is used during the initial heating period to reduce the come-up time. It must be turned off after the retort has attained the desired temperature, or else the extra steam could blow up (burst) the retort.

To uniformly distribute the steam in large-size retorts, some steam lines with perforations are connected to the main line. This is known as *steam spread*. The total cross section area of the holes should be 1.5 to 2 times the cross-section area of the steam inlet.

The water lines are attached along the inside top of the retort. On the outside, these are insulated to minimize the heat losses.

To protect the retort from very high pressure by expansion of the water and condensation of steam, there is an overflow valve or other protection device. For the vertical retort, the overflow line should be 2" in diameter and it is equipped with a controller.

For glass and retortable pouches, a mixture of steam and air is preferred to water or saturated steam as it eliminates the need of exhausting. In addition, sudden changes in pressure on heating or cooling cause glass breakage. To minimize rumbling in the retort, steam and air should be mixed outside the retort. A retort must have a blower system if it is designed for steam-air heating to allow proper movement inside the retort, which is required to maintain a uniform temperature.

16.4.3 Processing of Glass Containers

Usually, a stationary vertical retort is used for processing glass containers. It is very important to handle the glass containers with care because if a container breaks inside the retort, it will pollute the inside of the retort and the whole process has to be interrupted. As the glass cannot withstand sudden temperature changes, the heating and cooling should be gradual. Thus, the come-up time is lengthier. The retort is first filled with water, the containers are placed inside, and then steam is injected. Initially hot water can be filled into the system to save time and energy. Dry heat is not used for in-bottle sterilization to avoid thermal shock; cold water is introduced slowly for the cooling.

The water + air method gives more uniform heating of containers at different levels in the retort during the coming-up period than the steam alone or with steam + air methods. The water level should be well above the top-most containers and the top of the retort should be well vented.

If a horizontal retort is used for glass containers, it is important to have recirculation of water within the retort. An auxiliary small pump is used for the purpose. A petcock helps to remove any entrapped air and thus avoids an air lock in the line.

16.4.4 Cooling after Retorting

The cans should be cooled immediately after sterilization to avoid further cooking of the food. It may result in darkening of the food and may also cause flat sours in vegetables through the growth of thermophilic bacteria.

Cooling should not be carried out to a very low temperature as the cans will become wet and rusty; a final content temperature of 45°C is considered acceptable.

Cooling is done by sprays of cold water or by passing the cans through a tank of running cold water. Cooling by sprays offers more efficient use of water. The cooling water is often recycled after chlorination. The recommended level of free chlorine in the cooling water is about 1–2 ppm. In exceptional cases it may be 3 ppm.

The glass containers may break if they are directly exposed to the cooling water. Similarly, when the pressure of the retort is suddenly reduced, the lids may be forced off due to the pressure difference between the inside and the outside of the cans. Hence, the ingoing water is initially heated by live steam to a safe temperature. The temperature is then gradually reduced. At the same time air is blown into the chamber to maintain sufficient pressure. The air rising through the water mixes it and prevents blanketing.

Admitting cooling water at the top of the retort beneath the level of the water is also recommended.

For cooling of cans there is no need to temper the water; the cold water may be admitted directly. However, the air pressure should be sufficient to prevent buckling of large and medium-size cans.

16.5 Processing of Liquid Foods in Heat Exchangers

A temperature of 105°–120°C and a time of 10–30 minutes are needed for in-container sterilization of food. The processing should be such that the thermal center of the can receives this treatment. Often the nutritional and sensory characteristics of food degrade near the wall of the container. While reduction in processing times and protection of nutritional and sensory qualities accompany an increase in retort temperature, it requires using higher pressures. Ultimately there is a need for stronger containers and processing equipment. Thus, as an alternative, ultra-high temperature (UHT) sterilization involves keeping and packing sterilized food (processed under UHT condition for a shorter time) in a sterile atmosphere in pre-sterilized containers. The UHT processing is also termed *aseptic processing*.

16.5.1 Methods

Usually for UHT processing, the temperature is in the range of 135°–150°C and the holding time is just a few seconds. The processing time for *Cl. Botulinum* (F_0 = 3) at 141°C is only 1.8 s. From the basic understanding of the D_t and z values, we know that it is possible to achieve similar bacteriological destruction and much lower chemical change by this method as compared to the conventional sterilization method.

As the process time is of the order of a few seconds, it requires the use of a continuous flow heat exchanger. Time durations will be longer at temperatures below 135°C to be conveniently obtained in holding sections. Similarly, the time durations are too short at temperatures above 150°C, causing a problem in designing suitable heat exchangers. Hence, in UHT systems, the processing temperatures are selected within these limits. As a heat exchanger is used, the UHT sterilization system is suitable only for fluids and small particle suspensions.

It is important that the food should not boil in the heat exchanger even at the high temperatures maintained in the UHT system. Boiling will form vapor bubbles in the chamber, which will reduce the flow time since the bubbles displace the fluid. Specifically, in products like milk, a thick layer of precipitated solids is formed, restricting the product flow and reducing heat transfer. Thus, there is a need to prevent boiling, and, hence, the back pressure should be more than the vapor pressure of the product at the maximum temperature. As the food product in the UHT system has high water content, the pressure requirement equals the saturation water vapor pressure at the sterilization temperature.

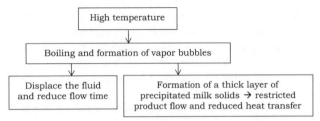

To avoid vaporization, maintain high back pressure in the chamber

For example, for the processing temperatures of 135°C and 150°C, the requirement of back pressures would be about 2.15 bar and 3.75 bar, respectively. However, a higher pressure of about 4 bar is recommended to prevent the formation of bubbles by the dissolved air and the separation of air at high temperature. Further, there will be a hydrodynamic pressure drop as the product is pumped through the heat exchanger, which may lead to higher pressure in other parts. The internal pressure may reach as high as 6–8 bar. Thus, a back-pressure valve or positive displacement timing pump is placed after the cooler to maintain the pressure within the system at a level at which there will be no product boiling in the chamber.

As discussed earlier, most of the lethality effect occurs toward the end of the heating stage and the beginning of the cooling stage. However, in UHT, the come-up time is negligible, and the cooling is also fast. Hence, the sterilization effect that occurs during heating to the necessary processing temperature and during the cooling process are ignored; in most cases, this is treated as a safety factor. The holding tube is the portion of the tube, after heating and before cooling sections, in which the product remains at the constant sterilization

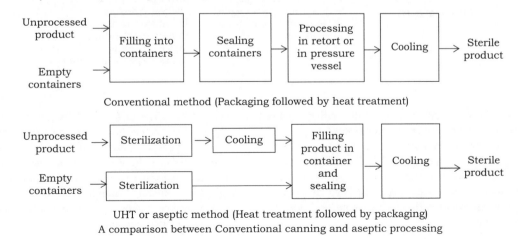

Conventional method (Packaging followed by heat treatment)

UHT or aseptic method (Heat treatment followed by packaging)
A comparison between Conventional canning and aseptic processing

temperature. In this section no further heat is supplied to the product, nor is any heat loss allowed by maintaining proper insulation. The residence time through the holding tube is the sterilization time.

The time of residence is decided by the fluid velocity in the heat exchanger. Usually, the velocity of fluid at the geometric center of the tube (i.e., for the fluid moving at maximum velocity) is considered for the holding time. A suitable pump is used to maintain a constant flow rate and thus the holding time. The velocity is very high and in the turbulent range. There should be a proper device/method to keep the heat transfer surfaces clean to reduce burn-on and to improve heat transfer.

UHT processing is being used successfully for liquid foods, such as milk, cream, yoghurt, fruit juices and concentrates, wine, etc. and also for foods with suspended solids, such as fruits and vegetables soups, tomato products, etc. As packaging is done after sterilization, very large bulk containers can be used for packaging the products. Cheaper packaging materials can also be used.

However, the plant is very costly and operation is much more complicated than the conventional retorts. The maintenance of complete aseptic conditions with sterile air and surfaces of filling machines, pre-sterilization of the packaging materials are the basic requirement of this system, which also adds to the cost.

16.5.2 Equipment

Classifications

The different systems used for UHT processing in food industry can be broadly classified as direct systems (using direct type heat exchangers, e.g., steam injection or infusion types) and indirect systems (using indirect type heat exchangers, e.g. plate or tubular types, etc.). In addition, UHT processing can also be done with other methods as microwave heating, dielectric heating, etc., though not common.

Steam Injection UHT Processing Equipment

Construction and Operation

- A steam injection type heat exchanger is used. The details of construction and operation have been discussed in Chapter 14. Steam at a high temperature is directly injected into the product for heating.

- First the liquid food is preheated to a temperature of 80°–85°C by indirect heating. This is usually achieved by regenerative heating utilized from sterilized product after the expansion cooling stage. For this purpose, the vapor released in the expansion vessel can also be used. The product temperature at the end of indirect heating must be closely controlled as it will affect the final temperature after mixing with steam. The holding time in the chamber is less, i.e., about 2–3 s after which the product and steam mix is taken for immediate cooling.

- The back pressure is maintained to prevent boiling in the heat exchanger. Usually, about 1 bar higher pressure than the pressure corresponding to sterilization temperature is maintained.

- After the liquid is sterilized, it is fed to a vacuum chamber, which is kept at 70°C for cooling. An expansion cooling vessel can also be used for faster cooling, to which the product is fed through a restrictor. The pressure in the cooling vessel is lower than atmospheric. It is also essential that the exact amount of water that was added due to condensation of steam is removed from the food during this stage. Hence, the vacuum is maintained corresponding to a boiling temperature of just 1°–2°C higher than that of the product temperature at the outlet of the final preheater.

- A temperature controller controls the steam supply to the injector by sensing the product temperature in the holding tube, and thus, controls the sterilization

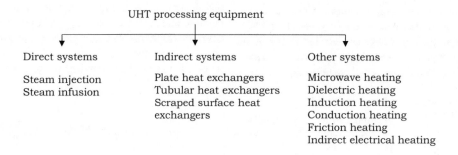

A comparison between the direct type and indirect type UHT sterilization systems is shown in Figure 16.12.

The temperature profile in a direct steam injection system is shown in Figure 16.13. Although the UHT sterilization systems in general use steam in different types of heat exchangers as the heating medium, systems have also been developed with hot water and electricity as the heating media for different types of commodities.

temperature. The difference between the temperature of the product before mixing with steam and that in expansion cooling is usually 1°–3°C. That is used to balance the water addition into the product as steam and water removal in the expansion chamber. The vacuum is maintained at a constant level in the expansion chamber by a suitable valve; it helps in maintaining constant temperature.

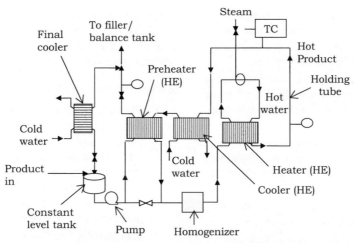

(a) Schematic of an indirect heating type UHT sterilization system

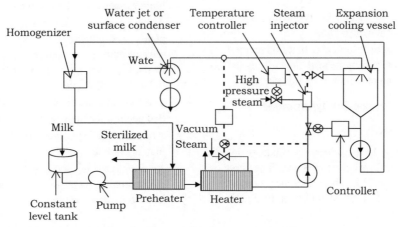

(b) Schematic of a direct heating type UHT sterilization system

FIGURE 16.12 Comparison between the indirect and direct heating type UHT sterilization systems

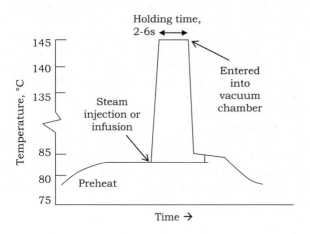

FIGURE 16.13 Temperature rise of a liquid food in a direct steam injection system (after Ahmad, 1999 with permission)

Suppose the feed temperature before entering the heat exchanger is 85°C and the sterilization temperature is 140°C. Then the heat required by the feed to reach 140°C temperature is approximately 55×4 = 220 kJ per kg. Similarly, 1 kg of steam by condensation at 150°C can give approximately 2115 kJ. Thus, the amount of condensed steam is about 0.10–0.11 kg per kg product.

This added water needs to be removed in the expansion vessel. Removal of the exact amount of water is important as this affects the product quality and the economics of the system.

• The cooled product from the lower part of the expansion chamber is pumped through the regenerator to the aseptic packaging section. In the case of milk, a homogenizer may be installed after the expansion chamber and before the regenerator.

Salient Features

• The added moisture due to steam condensation is removed from the finished product in the expansion vessel, and, hence, there is no change in the moisture content between the unprocessed and processed product.

• The sensory as well as nutritional properties are better retained because of instantaneous heating.

- The regenerative heating increases the system's thermal efficiency. However, the regeneration of energy in direct injection systems is less than 50% as compared to more than 90% in indirect systems.
- The uncondensed steam bubbles should not move into the holding tube as it would affect the flow by displacing the liquid and reduce the effective holding time.
- The processing conditions cannot be controlled as accurately as that with indirect systems. In addition, this method cannot be easily changed for different types of products.
- The method requires more expensive potable steam as compared to the normal processing steam.
- The production rates can be as high as 9000 kg.h^{-1}.

Applications

- As the heating and cooling are very rapid, the method can be used for heat sensitive foods.
- It is only for low-viscosity products that this method is suitable. It is particularly helpful for milk as the volatile materials can be removed by this method.

Steam Infusion UHT Processing Equipment

Construction and Operation

- Steam infusion type heat exchangers (Chapter 14) can be used for sterilization of products. The food is heated to 142°–146°C in about 2–3 s by contact with potable steam. The food is initially preheated by regenerative heating.
- The food and condensate mix is rapidly cooled to 65°–70°C in a vacuum chamber.

Salient Features

- The processing conditions are better controlled as compared to the steam injection types.
- In the steam injection system, the temperature of the injected steam is very high, which also increases the temperature of the surfaces of the chamber. Thus, there is a chance of contact of the food material with the hot surfaces and there may be burn-on and localized overheating of the product. However, in steam infusion types such problems are not observed.
- In this type, there may be blockage of the spray nozzles. For some foods, there may be separation of components.
- The production rates can be as high as 9000 kg.h^{-1}.

Applications

The method can handle higher viscosity foods as compared to steam injection systems.

Plate Heat Exchanger UHT Processing Equipment

The details of the plate type heat exchangers (PHEs) were discussed in Chapter 14. However, unlike the HTST process,

where the maximum product temperature is 72°–85°C, in the UHT, the product temperature is 135°–150°C, and thus, the heat exchangers must withstand higher temperature as well as internal pressures.

Salient Features of PHE for UHT Systems

- The seals should be perfect to avoid contamination of the food from the heating or cooling agents. Higher pressures may be observed in other parts of the system because of the pressure drop, which is due to the pumping of the product through the heat exchanger. As discussed earlier, the internal pressure of 6–8 bar may be observed. The gasketing materials used with plates must be able to withstand the severe conditions of temperature and pressure. High-quality rubbers are used for the purpose; nitrile rubber is used for temperatures up to about 138°C and resin-cured butyl rubber is used for temperatures up to 160°C.
- At high pressure and temperature, the stainless steel plates must withstand the flexion and distortion. To prevent damage to plates or seals, the mass of metal in the plate stack has to be carefully sterilized initially for uniform expansion.
- The food material is usually preheated by regenerative heating for improved energy efficiency.
- After the UHT treatment, the food has to be immediately cooled.
- As mentioned earlier, a sufficient back pressure is maintained to avoid formation of vapors within the heat exchanger chamber.

As seen from Figure 16.12, in the UHT indirect system for sterilization of milk, the feed is first pumped from the level-controlled balance tank to the heat exchanger for regenerative heating. Here, the milk temperature is increased to about 65°–85°C. Thereafter it is homogenized before sterilization. As the homogenization is done before sterilization, a non-aseptic homogenizer can be used. To bring the food temperature to 138°C, pressurized hot water is used in the heat exchanger using counter flow. The milk is held for 4 seconds in the holder and then cooled partly in a water-cooling section. Final cooling to about 20°C is done by regeneration. Milk then passes through a restrictor and desirable back pressure is applied on the milk after the heat exchanger. As the whole process takes place in a closed system, the holding tube dimensions, holding time, and time of flow in the other sections are controlled by the flow rate, which is ultimately controlled by the pumping rate of the homogenizer.

Tubular Heat Exchanger UHT Processing Equipment

The construction and operation of tubular heat exchangers have been discussed in Chapter 14. In the UHT systems, regenerative heating can be used in the heat exchangers to increase energy efficiency. The final heating is done by steam. The cooling is also done in two stages, the initial cooling by the raw product (in the regeneration section) and the final cooling by the cold water. The coil type tubular heat exchanger can be used for viscos foods and liquid foods containing particles of less than 1 cm size.

16.5.3 Reducing Fouling in Indirect Heat Exchangers in UHT Systems

The following points need consideration to reduce fouling in the indirect type of heat exchangers used for UHT processing.

- Large temperature differences in between the heating medium and the product (having tendency to foul) cause fouling. Maintenance of small temperature differentials help reduce fouling.
- The heat exchangers must have a high surface finish as rough surfaces provide more area for the materials to stick.
- Preheating of the feed reduces fouling. A recommended way for milk is to hold it for 10 minutes or more after heating it to 75°C or above. It is due to insolubilization of the milk salts and denaturation of proteins that the deposits are reduced.
- The use of pressurized hot water as the heating medium tends to reduce fouling as compared to steam heating.
- Higher flow velocities of both the feed and the heating medium reduce the amount of fouling. However, the velocity of the product is decided depending upon its holding time and pressure drop.
- Prevention of boiling within the heat exchanger and separation of dissolved gases in addition to the reduction in fouling is achievable by maintaining a back pressure of at least 1 bar more than the maximum product temperature employed in the heat exchanger. This method is commonly practiced in UHT heat exchangers to reduce fouling.

16.5.4 Packaging of UHT Processed Liquid Foods

The basic requirements of UHT processing include pre-sterilization of packaging materials and the surfaces of equipment. The different agents used for sterilization are moist heat, dry heat, high-intensity UV light, hydrogen peroxide, and ionizing radiation from either high-intensity electron beams or gamma rays. High-intensity and ionizing radiation generated from either high-intensity electron beams or gamma rays and UV light are not used in commercial food packaging applications.

Dry heat is less effective for the sterilization purpose than moist heat. Superheated steam or hot air could provide dry heat.

Hydrogen peroxide (H_2O_2) of 35% (w/w) concentration can be applied to the surface of packaging materials by either dipping, atomizing, or spraying. The materials are then passed through a heated chamber for vaporizing the hydrogen peroxide. The maximum residual hydrogen peroxide permitted is 0.1 ppm in the package as per US federal laws.

Laminated cartons are the preferred packaging materials for UHT processed foods due to lower cost of package, storage,

and transportation. The filling machines are kept in a sterile condition by UV light and filtered air. The inner side of the filling machine is kept at a positive air pressure to prevent any contaminant from entering inside.

16.6 Pasteurization of Foods

Pasteurization is a process named after Louis Pasteur, the French microbiologist who developed it in the 1860s. It is based on heating a product to a predetermined temperature and maintaining it for sufficient time so that nearly all objectionable microorganisms present in the product, are killed. It makes the food safe for human consumption, helps to retain the good flavor for a longer period, and improves the preservation quality. This process is commonly used for milk to increase its shelf-life and make it safer for human consumption. Other products that are also commonly pasteurized are eggs, juices, syrups, wine, beer, canned food, vinegar, and different dairy products.

16.6.1 Methods

Table 16.1 states the standardized time-temperature combinations used for pasteurization of milk and, accordingly, the methods and equipment have been named.

During initial stages of development of pasteurization process for milk, it was demonstrated that the process caused complete thermal death of *Mycobacterium tuberculosis* (TB germ) and the *Coxiella burnetii* (which then was considered to be responsible for the Q fever). At that time there was no facility to enumerate the actual reduction of bacteria. However, it was considered that the processing conditions, listed above, destroyed almost all common spoilage bacteria, including yeasts and molds and caused adequate destruction of common heat-resistant pathogenic organisms. However, now the Codex Alimentarius code states that milk pasteurization must achieve at least 5 log cycle reduction of *Coxiella burnetii*. The process kills the harmful bacteria *Escherichia coli O167:H7*, *Salmonella*, *Yersinia*, *Listeria*, *Campylobacter*, and *Staphylococcus aureus*.

TABLE 16.1

Different milk pasteurization methods

Method	Treatment	
	Temperature	**Duration**
Vat / long hold batch type pasteurization	63°C	30 min
High temperature short time (HTST) pasteurization	72°C	15 s
Higher heat short time (HHST) pasteurization	88°C	1 s
	90°C	0.5 s
	94°C	0.1 s
	96°C	0.05 s
	100°C	0.01 s
Ultra-pasteurization	138°C	2 s

The time and temperature combination to be maintained in a dairy plant depends on the initial microbial load and other operational considerations. Therefore, the combination for a dairy plant may vary slightly from the above tabulated values.

The fruit juice and beer, which come under the category of acidic foods (pH less than 4.6), are pasteurized for enzyme inactivation (e.g., polygalacturonase and pectin methylesterase in fruit juices) and for killing the yeasts and lactobacillus. Low acid foods such as liquid eggs and milk are pasteurized to destroy such spoilage organisms and pathogens as the yeast and molds.

Considering that pasteurization does not destroy all spoilage organisms, the product must be stored under refrigerated conditions. Many thermoduric bacteria such as *Enterococcus*, *Streptococcus thermophilus*, *Micrococcus*, *Lactobacillus*, and *Leuconostoc* species can survive pasteurization temperature.

16.6.2 Equipment

The equipment for pasteurization can be either a batch type using a vat or open jacketed kettle or a continuous type using a heat exchanger. Low-viscosity products, such as milk and juices, can be effectively pasteurized using plate heat exchangers. Scraped surface heat exchangers are useful for highly viscous material. Non-Newtonian fluids, e.g., dairy products, baby foods, and tomato ketchup, are pasteurized using shell or tube heat exchangers.

Long Hold or Vat Pasteurizer

Construction and Operation

- The product in vat or long hold pasteurization process, a batch process, is held at 63°C for 30 min.
- The operation is generally carried out in a jacketed kettle type heat exchanger. The vat pasteurizers can be either spray, flooded, or high-velocity flooded types, based on how the heating medium is applied in the external jacket.
- There are provisions for agitating the milk to reduce the temperature variations within the bulk.
- Controls for maintaining the temperature and holding time are two important accessories.
- There should be auxiliary heaters on the top of the vat for heating the air space above the liquid level (known as air space heaters).

Salient Features

- The basic requirement for effective pasteurization without deterioration of other characteristics are rapid product heating (achieved by circulating the heat transfer medium soon after the vat has begun

to be filled), maintenance of the pasteurization temperature for the desired time, and immediate cooling. Cold water is circulated in the jacket in some designs soon after the completion of the holding period. The cooling should be as fast as possible to prevent heating beyond the recommended time.

- The type of agitator will depend on the viscosity of the fluid and the tendency of the product to foam, as well as the effect on the cream line. Agitators required for highly viscous materials operate at slow speed and have large surface blades.
- Prevention of fouling on heating surfaces requires that the heat transfer medium should be just a few degrees warmer than milk. This condition also minimizes injury to cream line and flavor.
- Its installation cost is low and it has simple controls.
- The limitations include batch type and slow process. The running costs are also high as both heating and cooling are expensive.

Applications

- Vat pasteurizers suit the needs of small plants, i.e., processing of low volume products and different varieties of the product.
- It is also suitable for products such as bottle milk, sour cream, etc.
- Vat pasteurization for egg nog and frozen dessert mixes is done at 68°C for 30 minutes.

HTST Pasteurizer

Construction and Operation

- The high temperature short time (HTST) pasteurizers have a holding time of only 15 seconds and, thus, are usually constructed of continuous heat exchangers. Commonly used are the heat exchangers of the plate type.
- Figure 16.14 shows the basic HTST pasteurization process and its components. The process is summarized as follows.

  - Raw milk is transferred using a booster pump from a constant level tank to a regenerative heat exchanger where it is pre-heated with the help of the pasteurized milk to about 60°C.
  - The pre-heated milk enters the heater where its temperature is raised to the actual pasteurization temperature.
  - The milk then allowed to flow through the holder where the pasteurization temperature is maintained for the specified time.

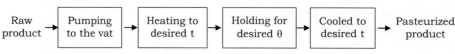

Basic operations in a vat pasteurizer

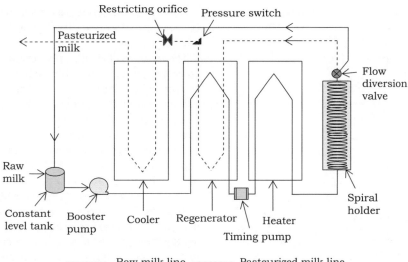

Raw milk line ——— Pasteurized milk line ·······

FIGURE 16.14 Process of HTST pasteurization

- The pasteurization is complete when the milk reaches the end of the holder.
- The pasteurized milk flows to the regenerator to give away some of its heat to the raw milk and simultaneously cool down to reduce its refrigeration requirement.
- The pasteurized milk from the regenerator flows to a chiller where the milk is cooled down to about 4°–5°C.

Salient Features

- The real heat requirement for pasteurization is considerably reduced by the use of the regenerator for enhancing the system's overall cost effectiveness.
- The major controls required for HTST pasteurization are the flow rate, temperature, and pressure. The pump must be able to maintain the desired flow rate in such a way that the resultant holding time meets the minimum legal requirement.
- There is a sensor actuated device at the end of the holder to ensure that the flow is diverted back to the system if the milk temperature is below the minimum legal value.
- It is essential that the pressure is well controlled in the regenerator, flow diversion valve, and the diverted milk lines (for homogenization, etc.). To ensure that the pressure on the pasteurized product side is always higher than that on the raw product side in the regenerator, the pump should have a pressure switch and a restrictor. This arrangement takes care of the leakage in the regenerator plates and the flow will be from pasteurized side to raw side.
- Pressure management is also required on the diverted milk line to take care of the holding time during diversion. In case the time is shorter during diverted flow than that in forward flow, a restricting orifice needs to be put in the diversion line.

- The HTST pasteurizer gives uniform treatment to the whole mass of product and the temperature is closely regulated. The regenerative heating system makes it more economical than the batch systems. However, the system is complicated and is designed for a specific product for a specific flow capacity and, thus, does not give enough flexibility in handling different types of products.

Higher Heat Short Time Pasteurizer

Construction and Operation

- The higher heat short time (HHST) pasteurization method involves heating milk at 88°C for 1 s or at 90°C at 0.5 s, 94°C 0.1 sec and 96°C for 0.05 sec or 100°C for 0.01 sec. The product after heating is immediately cooled and filled into hermetically sealed packages under aseptic conditions.
- The equipment and controls are similar to those of the HTST pasteurizers except that the UHT pasteurizers operate at higher temperatures. The holder required with UHT is of smaller size.

Salient Features

- Short holding time results in better texture of milk.
- It is possible to achieve greater bacterial destruction.
- The milk is shelf stable for at least six months and no refrigeration is required until the package is opened.
- The UHT treatment following regular pasteurization is capable of ensuring greater bacterial destruction together with beneficial effects on the body and texture of ice cream.

For the preparation of pasteurized milk, some other auxiliary steps are also carried out. The different steps in the process of preparation of pasteurized milk are shown in Figure 16.15.

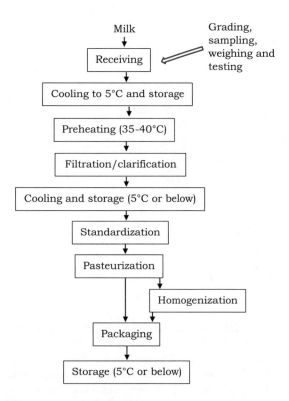

Milk → Receiving ← Grading, sampling, weighing and testing

Receiving → Cooling to 5°C and storage → Preheating (35-40°C) → Filtration/clarification → Cooling and storage (5°C or below) → Standardization → Pasteurization → Homogenization → Packaging → Storage (5°C or below)

FIGURE 16.15 Unit operations in preparation of pasteurized milk

Integration of Homogenizer with Pasteurization System

Since nearly all fluid milk and ice cream mixes, etc. are homogenized, homogenizers are integrated to the continuous pasteurization process. As the homogenization temperature must be at least 60°C, the homogenizer must be located either between the regenerator and heater or after the heater. The equipment, when installed, should not reduce the holding time below the legal minimum, either when it is operating or when it is at rest.

The capacity of the homogenizer can seldom be synchronized exactly with the timing pump unless a vented cover or other relief valve is employed, and then the pump operates at slightly greater capacity than the homogenizer. The usual practice is to use a homogenizer having 3%–8% greater capacity than the maximum flow rate of the system. It is equipped with a recirculation, by-pass loop from the discharge line to the suction feed line.

Ultra-Pasteurization Equipment

In *ultra-pasteurization*, milk is heated using high pressure steam at 180°–197°C. The milk temperature is raised to 138°C and held for at least 2 s before initiating rapid cooling. Although heated in commercially sterile equipment, the milk is not considered sterile because it is not hermetically sealed. The ultra-pasteurized milk needs to be refrigerated to give a shelf life of 30–90 days since it is not hermetically sealed.

Double pasteurization is a method in which there is secondary heating for increasing the degree of destruction of spoilage microorganisms.

The presence of alkaline phosphatase, denatured during the process of pasteurization, is used to determine the efficacy of pasteurization. The evidence of alkaline phosphatase destruction is proof of the destruction of common milk pathogens. Residual activity of α-amylase is the measure of the heat treatment effectiveness for liquid eggs.

Novel non-thermal methods such as high-pressure homogenization, pulsed electric field (PEF), pulsed high-intensity light, high voltage arc discharge, high-intensity laser, ionizing radiation, and microwave heating are also used for food pasteurization. A new method known as a *low temperature short time* (LTST) method has been patented, which involves exposing the food droplets to a high temperature, below the usual pasteurization temperatures, in a chamber. It is also termed *millisecond technology* (MST) because the liquid food is subjected to only a few thousandths of a second. Combined with HTST, it gives a significant increase in shelf life.

Some recently developed thermal processing equipment envisage separate treatments of the liquid and particulate components of a food. In addition, nowadays microwave heating and ohmic heating are also applied for heat processing. The microwave heating and ohmic heating have been discussed in detail in subsequent chapters.

Check Your Understanding

1. Eight log cycle reduction of *Clostridium botulinum* having z value of 9°C needs a process time of 1.5 minutes at 121°C temperature. What will be the process time for same degree of reduction at 130°C?

   Answer: 9 s

2. Decimal reduction times for *Bacillus subtilis* are 37 s and 12 s at temperatures of 120°C and 125°C, respectively. What is the temperature rise required to reduce the decimal reduction time at 120°C by a factor of 10?

   Answer: 10.22°C

3. A suspension contains 3.6×10^5 spores of *C. botulinum* having a D-value of 1.5 min at 121.1°C and 8.5×10^6 spores of *B. subtilis* having a D-value of 0.9 min at the same temperature. The suspension is heated at a constant temperature of 121.1°C. Find the heating time required for the suspension to obtain a survival probability of 10^{-3} for the most heat resistant organism.

   Answer: 12.83 min.

4. Pathogenic organism *Listeria monocytogen* in milk has a z value of 0.1°C at 71°C. Universal gas constant is 8.314 kJ.(kg mol)$^{-1}$.K^{-1} and decimal reduction time of *Listeria monocytogen* at 71°C is 5 s. Calculate the decimal reduction time of the same organism at 62°C.

   Answer: 149.4 s

   (Also solve the problems given in Chapter 15)

Objective Questions

1. Explain the differences between blanching, pasteurization, and sterilization.

2. State the relationship between the D values at any two temperatures t_1 and t_2.

3. Explain the basis of selection of temperature and time for pasteurization and sterilization of any food.

4. Why is the *Cl. botulinum* important from the canning point of view?

5. Is there any difference in consideration for thermal processing of low acid foods and high acid foods?

6. Will there be any advantage in thermal processing if the product is properly washed before canning?

7. Why are the filled cans exhausted before sealing during a canning process?

8. Explain the determination of thermal process time by the graphical method.

9. Draw the flow chart of HTST pasteurization system and explain the unit operations.

10. How are the temperature, flow rate, and pressure controlled in HTST pasteurizers?

11. What type of heat exchangers are used in UHT pasteurization systems? Explain briefly the working principle of these heat exchangers.

12. Differentiate between

    a. D value and z value;

    b. Log cycle reduction and spore log reduction;

    c. Cooker and retort;

    d. In-bottle sterilization and aseptic sterilization;

    e. Batch system and continuous system of pasteurization.

13. Explain

    a. Commercial sterility;

    b. Lethal rate;

    c. Come-up time during thermal processing;

    d. Pseudo-initial temperature;

    e. Unit operations in a canning process;

    f. Different components of an HTST pasteurizer.

14. Explain the working principles of

    a. Exhaust box;

    b. Direct flame heating method of sterilization;

    c. Batch type retort;

    d. Cascading water retort;

    e. Hydrostatic retort;

    f. Rotary sterilizer;

    g. UHT pasteurization;

    h. UHT sterilization system.

BIBLIOGRAPHY

Ahmad, T. 1999. *Dairy Plant Engineering and Management.* 4th ed. Kitab Mahal, New Delhi.

Ball, C.O. 1923. *Thermal Processing Time for Canned Foods.* Bull. 7–1 (37). National Research Council, Washington, DC.

Ball, C.O. and Olson, F.C.W. 1957. *Sterilization in Food Technology.* McGraw Hill Book Co., New York.

Brennan, J.G. (ed.) 2006. *Food Processing Handbook* 3rd ed. WILEY-VCH Verlag GmbH & Co. KGaA, Weinheim.

Brennan, J.G., Butters, J.R., Cowell, N.D. and Lilly, A.E.V. 1976. *Food Engineering Operations* 2nd ed. Elsevier Applied Science, London.

Burton, H. 1988. *UHT Processing of Milk and Milk Products.* Elsevier Applied Science, London.

Burton, H. and Perkins, A.G. 1970. Comparison of milks processed by the direct and indirect methods of UHT sterilization, Part 1. *Journal of Dairy Research* 37: 209–218.

Carlson, B. 1996. Food processing equipment: Historical and modern designs. In J.R.D. David, R.H. Graves, and V. R. Carlson (eds.) *Aseptic Processing and Packaging of Food.* CRC Press, Boca Raton, 51–94.

Carson, V.R. 1969. *Aseptic Processing* 3rd ed. Technical Digest CB 201, Cherry Burrel Corp., Chicago.

Charm, S.E. 1978. *Fundamentals of Food Engineering*, 3rd ed. AVI Publishing Co., Westport, Connecticut.

Cleland, A.C. and Robertson, G.L. 1986. Determination of thermal process to ensure commercial sterility of food in cans. In S. Thorn (ed.) *Development in Food Preservation* Vol.3, Elsevier Applied Science, New York.

Dash, S.K. and Sahoo, N.R. 2012. *Concepts of Food Process Engineering.* Kalyani Publishers, New Delhi.

David, J. 1996. Principles of thermal processing and optimization. In J.R.D. David, R.H. Graves, and V.R. Carlson, (eds.) *Aseptic Processing and Packaging of Food.* CRC Press, Boca Raton, 3–20.

Earle, R.L. and Earle, M.D. 2004. *Unit Operations in Food Processing.* Web Edn. The New Zealand Institute of Food Science & Technology, Inc., Auckland. https://www.nzifst.org.nz/resources/unitoperations/index.htm

Fellows, P.J. 2000. *Food Processing Technology.* Woodhead Publishing, Cambridge, UK.

Ford, J.E., Porter, J.W.G., Thompson, S.Y., Toothill, J. and Edwards-Webb, J. 1969. Effects of UHT processing and of subsequent storage on the vitamin content of milk. *Journal of Dairy Research* 36: 447–454.

Hallstrom, B., Skjoldebrand, C., Tragardh, C. 1988. *Heat Transfer and Food Products.* Elsevier Applied Science, London.

Hayakawa, K. 1970. Experimental formulas for accurate estimation of transient temperature of food and their application to thermal process evaluation. *Food Technology* 24(12): 1407–1417.

Hayakawa, K. 1977. Mathematical methods for estimating proper thermal processes and their computer implementation. *Advances in Food Research* 23: 76–141.

Heldman, D.R. and Hartel, R.W. 1997. *Principles of Food Processing.* Aspen Publishers, Inc., Gaithersburg.

Heldman, D.R. and Singh, R.P. 1981. *Food Process Engineering* 2nd ed.. AVI Publishing Co., Westport, Connecticut.

Holdsworth, S.D. 1985. Optimization of thermal processing, a review. *Journal of Food Engineering* 4:89.

Holdsworth, D. and Simpson, R. 2007. *Thermal Processing of Packaged Foods* 2nd ed. Blackie Academic and Professional, London.

https://www.barnardhealth.us/food-processing/info-vbz.html

https://www.retorts.com/white-papers/batch-versus-continuous-and-aseptic-sterilization/

https://fruitprocessingmachine.com/portfolio-items/retort-sterilizer/

https://www.hisaka.co.jp/english/food/product/

https://www.neelkanthretorts.com/hot-water-spray-retort.php

https://www.jbtc.com/foodtech/wp-content/uploads/sites/2/2021/09/JBT-White-Paper-ABRS.pdf

https://www.fsis.usda.gov/sites/default/files/media_file/2022-02/12-Hydrostatic-Retorts. pdf

Kessler, H.G. 2002. *Food and Bio Process Engineering*, 5th ed. Verlag A Kessler Publishing House, Munchen, Germany.

Leniger, H.A. and Beverloo, W.A. 1975. *Food Process Engineering*. D. Reidel Publishing Co., Boston, MA.

Lewis, M.J. 1993. UHT processing: Safety and quality aspects. In: A.Turner (ed.) *Food Technology International Europe*. Sterling Publications International, London, 47–51.

Lewis, M.J. and Heppell, N.J. 2000. *Continuous Thermal Processing of Foods: Pasteurization and UHT Sterilization*. Aspen Publishers, Inc., Gaithersburg, UK.

Lund, D.B. 1975. Heal Processing. In: M. Karel, O Fennema, and D. Lund (eds.) *Principles of Food Science. Vol.2, Principles of Food Preservation*. Marcel Dekker, New York, 32–86.

Lund, D.B. 1977. Maximizing nutrient retention. *Food Technology* 31(2): 71.

Ohlsson, T. 1992. R & D in aseptic particulate processing technology. In: A. Turner (ed.) *Food Technology International Europe*. Sterling Publications International, London, 49–53.

Palaniapan, S. and Sizer, C.E. 1997. Aseptic process validated for foods containing particulates. *Food Technology* 51(8): 60–68.

Pal, U.S., Das, M., Nayak, R. N., Sahoo, N. R., Panda, M. K. and Dash, S. K. 2019. Development and evaluation of retort pouch processed chhenapoda (cheese based baked sweet). *Journal of Food Science and Technology* 56: 302–309.

Pflug, Irving J. 2010. *Microbiology and Engineering of Sterilization Processes*. 14th ed. Environmental Sterilization Laboratory, Otterbein.

Ramesh, M.N. 1999. Food preservation by heat treatment. In: M.S. Rahman (ed.) *Handbook of Food Preservation*. Marcel Dekker, New York, 95–172.

Rao, D.G. 2009. *Fundamentals of Food Engineering*. PHI Learning Pvt. Ltd., New Delhi.

Reed, J.M. Bohrer, C.W. and Cameron, E.J. 1951. Spore destruction rate studies on organisms of significance in the processing of canned foods. *Food Research* 16: 338–408.

Sahay K.M. and Singh K.K. 1994. *Unit Operation of Agricultural Processing*. Vikas Publ. House, New Delhi.

Singh, R.P. and Heldman, D.R. 2013. Introduction to Food Engineering. 5th ed Academic Press, San Diego, CA.

Stumbo, C.R. 1973 *Thermobacteriology in Food Processing* 2nd ed. Academic Press, New York.

Toledo, R.T. 2007. *Fundamentals of Food Process Engineering* 3rd ed. Springer, USA.

Watson, E.L. and Harper, J.C. 1989. *Elements of Food Engineering* 2nd ed. Van Nostrand Reinhold, New York.

17

Evaporation

Evaporation is the process for concentration of solutions (often the solution of solids in water) by partial vaporization of the solvent. The difference between drying and evaporation is that the end product in drying is a solid whereas in evaporation the end product is a liquid of higher consistency. The difference between evaporation and distillation is that, in evaporation, the vapor is not separated further into fractions. Drying and distillation are discussed in subsequent chapters.

Some common evaporated products are condensed milk, fruit juice concentrates, jam, jelly, preserve, jaggery, etc. Evaporation of food helps in achieving the following.

- It enhances the shelf life of food by increasing the concentration of solids and reducing the water activity.
- The weight and volume of the liquid food are reduced, which helps in reducing cost involvement in storage, transportation, handling, and packaging. Some foods like milk, fruit juice, etc. are pre-concentrated before subsequent processing operations, such as drying, freezing, etc.
- Water from a food product may need to be partially removed for extending the shelf life and also to save energy in subsequent processing operations.
- It helps to develop new varieties of products for consumption, such as concentrated fruit juices, food drinks, tomato puree, etc., to add to taste and convenience in consumption.
- Some foods such as concentrated fruit juice are more convenient to use than the raw materials in further preparation processes. It is particularly helpful when the location of the processing unit is far away from the production area.
- Evaporation is also used in specialized cases for changing the flavor and/or color of foods (e.g., caramelized syrups)

The evaporator should give the concentrated product such that the original product will be obtained after dilution. In the process of evaporation, no pasteurization or sterilization can be assured as the holding time is not maintained.

17.1 Methods and Principles

17.1.1 Evaporation Process

The basic evaporation process consists of the heat transfer to the feed followed by vaporization of water from it. Thus, a heat exchange system is an essential part of the evaporator. Steam is the common heating medium in commercial evaporators. In the heat exchanger, first the feed temperature is raised to the level of its boiling point corresponding to the pressure within the evaporator chamber (sensible heating) and then the actual vaporization takes place (latent heat). Ideally, the feed is preheated close to the boiling point before feeding it into the evaporator. Thus, whatever heat is supplied in the evaporator, it acts as the latent heat of vaporization of water. Similarly, the steam also gives away only latent heat and the temperature of steam is almost same as that of the condensate. Any small reduction in the sensible heat of the heating medium (steam) is considered as negligible.

The evaporation rate is dependent on the rate of heat transfer to the food and the mass transfer from food as vapor. A higher temperature difference between the steam and the feed will increase the rate of evaporation. Therefore, the temperature differential is increased by increasing the steam temperature (by superheating of steam) and/or by reducing the boiling temperature of the feed in the evaporator. The latter is achieved decreasing the pressure. Evaporation at a low temperature is recommended as it helps to retain the sensory and nutritional quality of the food. Therefore, a *vacuum pump* needs to be connected to the feed chamber if the boiling temperature of the liquid has to be lowered. It can be either a mechanical vacuum pump or a steam ejector vacuum pump.

At high velocity, the fluid in the tube remains liquid practically until the end of the heat exchanger tube and flashes into a mixture of liquid and vapor as it exits at the end of the tubes. The place where the vapor is separated from the concentrated food is known as the vapor chamber. A huge quantity of vapor is produced, which is separated from the concentrated food. The *vapor separator* is a large chamber, whose diameter is more than that of a heat exchanger. Thus, the velocity of vapor greatly reduces. In fact, it is the most visible part of the evaporator. It also acts as a reservoir for the product.

During the process of separation, if the vapor comes out of the concentrate at a high rate and pressure, it tends to carry some concentrate droplets with it. This has to be prevented because it is loss of food material. Hence, in the vapor chamber, there is a device to arrest the concentrate droplets from the vapor before it leaves the separator. The losses of this type are known as *entrainment losses* and the device used is known as an *entrainment separator*. The recovered concentrate is returned back and the vapor leaving the evaporator is condensed in a condenser. The condensed vapor is often reused.

17.1.2 Components of an Evaporation System

As discussed above, an evaporator consists of a heat exchanger that transfers heat from steam to the feed; a chamber or other device for separating the vapor from the feed; and a vacuum pump. Some auxiliary parts are a boiler to produce steam; a condenser to condense the vapor evolved from the system; a device to remove non-condensable gases from the system, such as a steam jet ejector; pumps for feed; steam and condensate make up water and a preheater. The condensers and preheaters are also basically different types of heat exchangers. In some installations, the vapor is compressed with mechanical compressor or thermo-compression system and is mixed with the incoming steam to reduce the steam consumption. Figure 17.1 shows the basic parts of an evaporation system.

17.1.3 Performance of an Evaporator

The performance of an evaporator can be expressed in terms of its capacity and economy. The capacity is given in terms of kg of water vaporized per unit time. The *economy* (also known as *steam economy*) is given as the amount of vapor produced by the system per unit amount of steam utilized for the evaporation.

$$\text{Steam economy} = \frac{\text{amount of vapour produced}}{\text{amount of steam supplied}} \quad (17.1)$$

The steam economy is an index of the economy of evaporation. The lower the amount of steam consumed for each unit of vapor produced, the greater the above ratio and the greater steam economy.

The rated capacity of evaporator is often expressed as the amount of vapor produced per unit time and thus,

$$\begin{aligned}\text{Evaporator capacity} = \text{Steam consumption} \\ \times \text{steam economy} \quad (17.2)\end{aligned}$$

Example 17.1 The evaporation chamber is maintained so that the liquid food boils at 65°C in it. Find out the steam economy, if steam is supplied at 100°C.

SOLUTION:

The solution is obtained by assuming that the feed is preheated and admitted to the evaporator at the boiling point prevailing in the evaporation chamber (i.e., the heat required for evaporation is the latent heat at that temperature) and that only the steam's latent heat of condensation is used for heating.

From the steam table, latent heat values at 65°C and at 100°C are 2346 kJ.kg⁻¹, and 2257 kJ.kg⁻¹, respectively.

While 2346 kJ heat energy is required for evaporation of 1 kg water from food, the steam at 100°C supplies only 2257 kJ.kg⁻¹.

Thus, the amount of evaporation per kg of steam = 2257/2346 = 0.856 kg of water/ kg steam supplied. In other words, the steam economy = 0.856

Selection of Boiling Temperature of Food

To improve the rate of evaporation, the heat transfer rate should be high, which requires the use of a higher temperature difference between the heating and heated media. Thus, either the heating temperature (steam temperature) is increased or the boiling temperature of product is decreased or both are undertaken. Generally, it is preferred that the evaporators operate under vacuum because it is more economical to feed steam at modest pressure. In addition, use of high-pressure steam requires costlier construction. A vacuum in the evaporator reduces the boiling point of the liquid. Again, at higher pressure, the concentration of solutes is more, which, in turn, also increases the boiling point.

The maximum permissible boiling temperature and concentration of any specific liquid within the evaporator are also decided on the basis of the *temperature sensitivity of material*

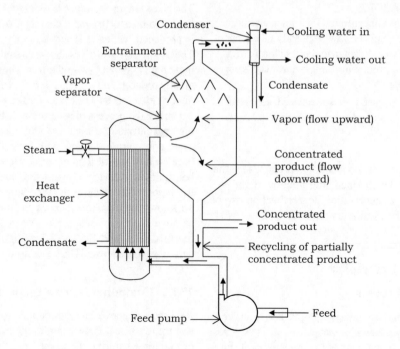

FIGURE 17.1 Basic parts of an evaporation system

and its solubility limit. As most of the food constituents are heat sensitive, to prevent loss of sensory and biochemical characteristics of the food, the boiling temperature of the food in the evaporator is kept low by maintaining a vacuum. The chemical composition of food influences the rate of fouling and possible changes in solute composition. Some products have the tendency to crystallize when taken out of the evaporator and cooled.

In the large evaporators and batch type evaporators, there may be variations in the temperature of liquid at different sections, the liquid near the heat exchanger surface will be at higher temperature than that at the thermal center. The product viscosity, velocity, and liquid side heat transfer coefficient are some other factors that affect the heat distribution in the food mass and, thus, heat damage at high temperatures.

In industrial evaporators the boiling point is not kept below 40°C as it would require a high cost of vacuum generation and the processing equipment will also require extra strength. In large evaporators, the boiling point of liquid at the base may be slightly higher because of increased pressure due to the weight of liquid above it.

Improving the Economy of Evaporation

The economics of evaporation can be improved by minimizing the loss of concentrates / solids, which tend to escape along with the vapor as well as by reducing the amount of energy consumed per unit evaporation.

Reducing the Loss of Product

The product losses are caused by foaming, frothing, and entrainment.

Foaming occurs due to proteins, carbohydrates, fatty acid solution, etc. present in the food. In addition to reducing the rate of heat transfer, foaming also causes inefficient separation of vapor and concentrate.

The loss of concentrate along with the vapor is known as entrainment losses. An entrainment separator or a disengagement space is used for minimizing this loss.

Improving Heat Exchange Efficiency

The extent of energy consumption is reduced by proper design and operation to minimize the various resistances to heat transfer. The heat exchange efficiency can be primarily improved by the following.

Reducing fouling and corrosion of heat transfer surfaces. If there is a wide temperature difference between the food and the heated surface, deposition of polysaccharides and denaturation of proteins cause fouling on the heat transfer surfaces. It acts as an insulating layer for heat transfer. If the scale is hard, the cleaning is difficult and expensive.

There is a chance of metal corrosion on the steam side also, which has to be prevented by using anti-corrosion chemicals, treatments, or surfaces. There may also be some ingredients in the food that are corrosive to heat exchanger surfaces.

During evaporation the fluid viscosity increases, which causes reduced heat transfer. Further, foods with high viscosity may cause more fouling and heat damage if not removed from the hot surfaces. Viscosity affects the rate of heat transfer as well as the handling of fluid.

Reducing the thickness of boundary films. Inducing turbulence in the fluid, agitating the fluid, and scraping the inner walls can reduce the thickness of boundary film and improve heat transfer.

Multiple Effect Evaporation

Evaporators may be single effect or multiple effect systems depending on the number of stages employed for evaporation. In a single effect evaporator, evaporation is achieved in a single stage with only one heat exchanger. As the vapor is removed from the food in the vapor separator, it is condensed into hot water. However, in multiple effect evaporation system, a number of evaporators are connected in series and the vapor removed from one evaporator is used as the heating medium in the next evaporator.

In a two-stage or two-effect evaporator the vapor produced in the first evaporator (first stage or first effect) goes to the steam chest of the second evaporator (second stage or second effect) and becomes the heating medium in the second evaporator. The feed is partially concentrated in the first effect, which then enters into the evaporation chamber of the second effect. The vapor produced in the second stage is condensed. Figure 17.2 compares the flow of feed and steam/vapor between a single effect and double effect evaporator.

The temperature and pressure of the vapor removed from the first stage (or effect) and the feed going to the second effect are the same, there is no driving force for vaporization in the second effect. Hence, either the boiling temperature of the liquid in the second effect is reduced or the temperature of the heating medium is increased so that the heat transfer can take place. The temperature of the vapor produced from the first effect can be improved either by mechanical or thermal compression before it enters the heating section of the second effect. The boiling temperature of the liquid in the second effect can be reduced by reducing the operating pressure in the second stage. The second option is more common.

Similarly, in a three-effect evaporator, the vapor produced in the second effect is used as the heating medium in the third effect. The third evaporator has the lowest pressure and the first one the highest. The vapor produced in the last effect is condensed and may be reused for some other applications.

Thus, in a multiple effect system, more than one evaporator is used. But as these evaporators are used in series, the throughput capacity (amount of feed handled) is not increased.

The heat transfer is given by:

$$q = U.A.\Delta t$$

where, Δt is the difference between the temperatures of steam and the feed in a single effect evaporator. For a multiple effect system, the difference between the steam temperature in the first effect and the boiling temperature of the feed in the last effect is the net temperature difference. Thus, for the same net temperature difference, the single effect and the multi-effect systems have the same heat removal/vapor removal capacity. Hence, the multi-effect system helps increase the steam economy (as the amount of steam required for the same amount of vaporization is reduced) at the expense of reduced throughput capacity.

(a)

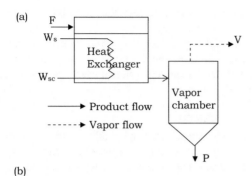

(b)

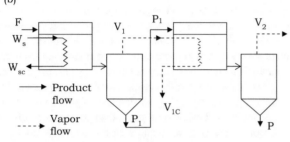

F: Feed; P: Final concentrated product; W_s: Steam into heat exchanger; W_{sc}: Condensate coming out of heat exchanger; V_1 and V_2 are the vapor removed from different effects, V_{1C} is the vapor condensate removed from first effect.

FIGURE 17.2   Flow of feed and steam/vapor in (a) single effect evaporator, and (b) double effect evaporator

Example 17.2 Find the steam economy in a triple effect evaporator if the steam entering into the first effect is at 100°C and the vacuum maintained in the evaporators are such that the boiling temperatures inside the first, second, and third evaporators are $t_1 = 85° \text{C}$, $t_2 = 70° \text{C}$, and $t_3 = 55° \text{C}$. The latent heat of vaporization corresponding to 100°C, 85°C, 70°C, and 55°C are 2257, 2296, 2333, and 2370 kJ.kg^{-1} (as obtained from a steam table).

SOLUTION:

The schematic diagram is shown with the corresponding temperatures.

Temperature of steam (W_s), $t_s = 100°C$,

In the first effect, condensate coming out (W_{sc}) will be at temperature t_s, because only latent heat is released, no sensible heat is given up.

Water evaporated in the first effect $= \dfrac{2257}{2296} = 0.983$ kg/ kg of steam supplied

Water evaporated in the second effect $= \dfrac{2296}{2333} \times 0.983 = 0.967$ kg/ kg of steam supplied to the 1st effect.

(The amount of vapor fed into the steam chest of 2nd effect is 0.983 kg for 1 kg of steam supplied to the first effect.)

Water evaporated in the third effect $= \dfrac{2333}{2370} \times 0.967$

$= 0.952$ kg/ kg of steam supplied to the 1st effect

(0.967 kg of vapor is fed into the steam chest of the third effect for each kg of steam supplied to the first effect.)

Thus, total water evaporated is 2.902 kg per kg of steam supplied to the system.

Steam economy = 2.902 kg.kg^{-1}

Example 17.3 In the above problem instead of a triple effect evaporator, if we take a single effect with boiling point of the food as 55°C, what will be the steam economy?

SOLUTION:

The amount of evaporation per kg of supplied steam will be

$$= \frac{2257}{2370} = 0.952 \text{ kg}$$

Thus, 0.952 kg/kg steam is the steam economy for the single effect system.

Clearly, the steam economies for the single effect and multiple effect systems having the same total temperature difference are 0.952 and 2.902 kg.kg^{-1}, respectively.

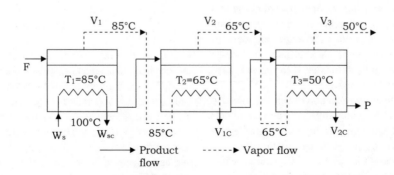

Steam (Vapor) and Product Flow in a Multiple Effect Evaporator

The condensate from the first effect is used as boiler feed. Condensates from subsequent effects are not used for the same because they may contain suspended solids.

Usually, the maximum boiling temperature in the first effect is kept as 70°C, and the minimum boiling point is in the last effect, which is about 40°C. Thus, a net temperature difference of 30°C is available. With three stages, the temperature difference in each effect will be approximately 10°C and with more effects, the temperature difference would decrease. As q = U. A. Δt, and Δt is reduced by increasing the number of effects, the heat transfer area, A, should be increased to maintain the same level of heat transfer. It requires more cost. Besides, the time taken by the product to pass the heat transfer area becomes longer, i.e., the holding time is increased. Therefore, in food industry, three to seven stages of evaporation are commonly used.

The multiple effect systems can be reverse feed, forward feed, parallel feed, or mixed feed types. In the reverse feed system, vapor passes in the direction opposite to that of the product and fresh steam is fed into the last effect. Since viscosity is reduced at high temperature, it is used when highly viscous products need to be concentrated. But forward feed is better if the feed is more sensitive to temperature. In another configuration, the fresh feed flow is parallel to all effects, but the vapor from each effect is used as the heating medium in the subsequent effect (Figure 17.3a). Figure 17.3b shows another possible arrangement in which the fresh steam is entered in the last effect and the vapor from the last effect is fed as the heating medium in the first effect.

Relationship between Overall Heat Transfer Coefficients and Temperatures in a Multiple Effect System

The heat transfer area of evaporators in all the stages in a multiple effect system remains the same as the heat exchangers have standard designs. In a steady operation, practically all of the heat expended in producing vapor in the first effect must be given up in the second effect as this vapor condenses. Thus, the rates of heat transfer in all the effects are nearly equal. Thus, the overall heat transfer coefficients in individual effects change depending upon the temperature differential. In other words, in a multiple effect evaporator the temperature drop is almost inversely proportional to the heat transfer coefficient.

Considering the triple effect evaporator as explained above

$$q_1 = U_1.A_1. \Delta t_1, \Delta t_1 = (t_s - t_1)$$

$$q_2 = U_2. A_2. \Delta t_2, \Delta t_2 = (t_1 - t_2)$$

$$q_3 = U_3.A_3. \Delta t_3, \Delta t_3 = (t_2 - t_3)$$

If the amount of steam condensing is denoted as W_s and the heat of condensation is λ, then $W_s. \lambda = q_1 = q_2 = q_3$ (assuming no boiling point rise)

Considering that the heat transfer areas are equal in the effects

$$\Delta t_i \alpha \frac{1}{U_i} \qquad (17.3)$$

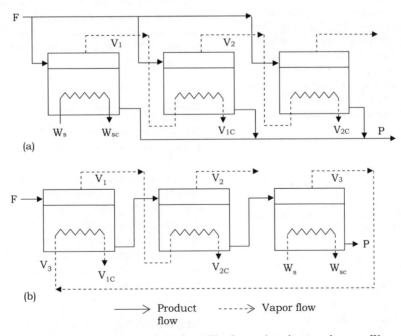

F: Feed; P: Concentrated product; W_s: Steam into heat exchanger; W_{sc}: Condensate coming out of heat exchanger; V_1, V_2 and V_3 are the vapor removed from different effects, V_{1C}, V_{2C} and V_{3C} are the vapor condensate removed from different effects.

FIGURE 17.3 Two possible flow arrangements in a multiple effect evaporation system

Total heat transfer, $q = q_1 + q_2 + q_3$

$$= U_1.A_1.\Delta t_1 + U_2.A_2.\Delta t_2 + U_3.A_3.\Delta t_3,$$

$$= U_{av}.A.\left(\Delta t_1 + \Delta t_2 + \Delta t_3\right)$$

$$= U_{av}.A.\Sigma\Delta t$$

$$(17.4)$$

As above, $\quad \sum \Delta t \alpha \dfrac{1}{U_{av}} \qquad (17.5)$

As the heat transfer resistances are in series,

$$\frac{1}{U_{av}} = \frac{1}{U_1} + \frac{1}{U_2} + \frac{1}{U_3}$$

By proportionality,

$$\frac{\Delta t_i}{\sum \Delta t} = \frac{1/U_i}{1/U_1 + 1/U_2 + 1/U_3} \qquad (17.6)$$

Calculations for multiple effect evaporators are made by trial and error. First, the temperature difference (Δt) is assumed for each effect and applying heat and material balance for each effect, the rate of heat transfer and the heat required to achieve the desired evaporation rate are compared for each effect. In case of differences, the values of assumed temperature differences are modified until the heat supply equals the heat requirement for each effect. Use of computers help to achieve the results of such iterative procedures in less time.

Example 17.4 Heat transfer coefficients for a triple effect evaporator in $W.m^{-2}.K^{-1}$ are 3000, 2400, and 1600. Steam temperature is 110°C for heating in the first effect. The boiling point of liquid in the last effect has been found as 55°C. What will be the liquid boiling points in the first and second effects? Heating surface areas in all the effects may be assumed to be equal.

SOLUTION:

The total temperature drop, $\Delta t = 110-55 = 55°C$

But total temperature drop is also written as the sum of individual differentials, i.e.,

$$\Delta t = \Delta t_1 + \Delta t_2 + \Delta t_3$$

Again, $q_1 = U_1 A_1 \Delta t_1 = U_2 A_2 \Delta t_2 = U_3 A_3 \Delta t_3$

$$U_1 \Delta t_1 = U_2 \Delta t_2 = U_3 \Delta t_3 = q/A$$

$$55 = \Delta t_1 + \Delta t_2 + \Delta t_3$$

$$= \Delta t_1 + (U_1 / U_2)\,\Delta t_1 + (U_1 / U_3)\,\Delta t_1$$

$$= \Delta t_1 (1 + (3000/2400) + (3000/1600)) = \Delta t_1 (4.125)$$

This gives, $\Delta t_1 = 13.33°C$

Now, $\Delta t_2 = (3000/2400)13.33 = 16.67°C$

$$\Delta t_3 = (3000/1600)13.33 = 25°C$$

Thus, the boiling point will be 110-13.33 = 96.67°C in the first effect.
For the second effect, the boiling point is 96.67-16.67 = 80°C
In the third effect, the boiling point is confirmed as 80-25=55°C

> In a multiple effect evaporation system, the net temperature difference can be controlled by controlling only two parameters, namely, the temperature of vapor in the last effect and temperature of steam in the first effect. Once the pressure in one effect is fixed, the temperature of steam and heat transfer coefficients establish the pressures in the remaining effects. For example, if the pressure p_3 becomes higher, it would raise the boiling point in the third effect. It decreases the temperature difference between vapor in heating coil and boiling point. Thus, the rate of vapor condensing in the coil is reduced. It causes more pressure build up in the second effect. In this way the pressures in all the effects are dependent on each other.

Vapor Recompression

The pressure and, consequently, the temperature of vapor are increased by using a mechanical compressor or by a venturi type steam jet (thermo-compression). The compressed vapor is mixed with the steam supplied as the heating medium to the same evaporator.

Thermal Vapor Recompression

In thermal vapor recompression (TVR), the vapor produced from an evaporator is passed through a nozzle, where it is compressed by high pressure steam and fed back into the system (Figure 17.4). The resulting steam mixture, as compared to the vapor produced, has a higher energy content. As it is mixed with the incoming fresh steam, the quantity of fresh steam needed is reduced.

Thermal recompression can be used with a multiple effect system at intermediate locations. Steam economies of up to 10 can be obtained in a combined thermal recompression multiple effect system as compared to about 6.0 to 6.5 for the multi-effect alone.

Mechanical Vapor Recompression

In mechanical vapor recompression (MVR), a mechanical compressor is used to compress the vapor produced from an evaporator, and the compressed vapor is then mixed with the fresh steam for heating. Actually, the evaporation is initiated with fresh steam and then the compressed vapor is added to

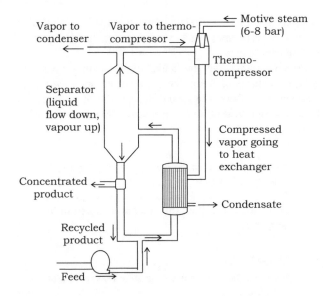

FIGURE 17.4 Schematic of thermal vapor recompression (After Singh and Heldman, 2014 with permission)

make up most of the steam addition. To make up the energy losses, just a small amount of fresh steam is required. The mechanical recompression systems can give steam economies up to 35 to 40. However, the extra cost of operation of the compressor needs to be considered to calculate the benefit cost ratio of the system.

In multiple effect system with MVR, vapor produced in the same effect and compressed to a higher temperature serves as the heating medium in the first effect.

Preheating of Feed

If the feed is introduced into the evaporator close to its boiling temperature (corresponding to the pressure in the evaporation chamber), then the complete heat (supplied by the heating medium) is used for producing vapor. But if the feed is at a lower temperature then some amount of heat is spent for sensible heating, which reduces the amount of vapor produced. Thus, often the vapor produced by the system is fed back to heat the incoming feed or the water used in the boiler to produce steam. It reduces the net energy used for evaporation, and improves steam economy. Preheating can also help in reducing the size of evaporator heating surface area.

17.1.4 Mass and Energy Balance in Evaporator

Mass Balance

In Chapter 3 on mass and energy balance, the mass balance equations for an evaporator have been discussed.

If F, V, and P are the amounts of feed entering the evaporator, vapor and concentrated products going out of the evaporator, X_f, X_p, and X_v are the fractions of solids in the feed, product and vapor, W_s and W_{sc} are the quantities of steam going into and condensate coming out of the evaporator, then the total mass balance and component (solid) balances are as follows:

$$F = V + P \tag{17.7}$$

$$F.X_f = P.X_p + V.X_v \tag{17.8}$$

If the fraction of solids in the vapor, X_v (i.e., the entrainment losses) is less and can be considered as negligible, the equation (17.8) reduces to

$$X_f . F = X_p . P \tag{17.9}$$

$$P = F\left(\frac{X_f}{X_p}\right) \tag{17.10}$$

$$V = F-P = F-F\left(\frac{X_f}{X_p}\right) = F\left(1-\frac{X_f}{X_p}\right) \tag{17.11}$$

The quantities F, V, and P may have the units of kg (total quantity in a batch system) or kg per unit time (the rate of flow for a continuous system). As the total amount of steam fed into the system goes out as condensate (without mixing with the solution), the terms are ignored while writing the equations 17.7 and 17.8.

The feed rate is generally constant for a particular evaporator (which is given as design feed rate).

Enthalpy Balance

As there are five streams contributing to the overall heat balance (Figure 17. 5), the equation can be written as:

$$F.h_f + W_s.h_s = V.h_v + P.h_p + W_s.h_{sc} \tag{17.12}$$

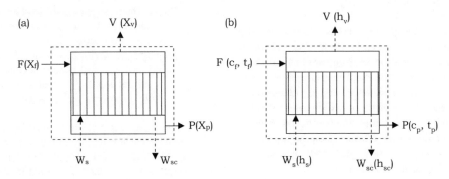

FIGURE 17.5 Parameters for (a) mass balance, and (b) enthalpy balance in an evaporation system

Here, h_f, h_p, and h_v are the enthalpies of feed, product, and vapor, respectively. h_s and h_{sc} are the enthalpies of steam and condensate, respectively.

So, h_s-h_{sc} = latent heat of evaporation, h_{fg}, and

$W_s = W_{sc}$ (as there is no loss of mass of steam)

If c and t are the specific heats and temperatures of different streams, then

$$F. c_f. t_f + W_s.h_{fg} = V.h_v + P.c_p.t_p$$

In the above equation t_p and t_v are equal and h_v is the saturation enthalpy after the boiling point rise (the boiling point rise is discussed after this section).

$$\text{Thus, } W_s.h_{fg} = V.h_v + P.c_p.t_p - F.c_f.t_f \qquad (17.13)$$

Replacing P and V in terms of F we get the final form of equation as:

$$W_s.h_{fg} = F.\left(\frac{X_f}{X_p}\right).c_p.t_p + F.\left(1 - \frac{X_f}{X_p}\right)h_v - F.c_f.t_f \quad (17.14)$$

$$\text{Also, } W_s. h_{fg} = q = U.A. \Delta t \qquad (17.15)$$

where, U is the overall heat transfer coefficient (OHTC), Δt is the net temperature difference in the evaporator, which is usually the log mean temperature difference in the heat exchanger.

The specific heat depends on the type of feed. It also changes during the process as its concentration increases. The specific heat of a liquid food with fraction of moisture m_w can be approximately taken as:

$$c_p \text{ (kJ.kg}^{-1}. \text{ K}^{-1}) = 1.25 \times (1 - m_w) + 4.18 \times (m_w) \quad (17.16)$$

Here, 1.25 kJ.kg^{-1}. K^{-1} and 4.18 kJ.kg^{-1}.K^{-1} are the specific heat values of dry matter and water, respectively.

The overall resistance to heat transfer between steam and boiling liquid in the evaporator is the sum of resistances offered by the steam film, scales on two sides, tube wall, and the resistance from the boiling liquid.

$$U = \frac{1}{\dfrac{1}{h_1} + \dfrac{1}{h_i} + \dfrac{1}{h_o} + \dfrac{x}{k} + \dfrac{1}{h_2}} \qquad (17.17)$$

Here, h_1 is the film coefficient for steam, which is usually 20000–40000 kJ.m^{-2}.h^{-1}.K^{-1}, h_i and h_o are fouling factors, h_2 is the film coefficient for the product. In most practical situations, the fouling factors and tube wall resistances are much smaller as compared to other resistances and are neglected in calculations. The steam film coefficient is high even when the condensation is film-wise. The liquid side coefficient depends on the viscosity of liquid. As the resistance on the liquid side controls the overall rate of heat transfer, forced circulation is preferred to reduce the thermal resistance.

The film thickness and viscosity increase with an increase in concentration during the process and cause a reduced heat transfer. Because it is difficult to measure the individual coefficients, experimental results are always expressed in terms of overall coefficients.

Promoters are sometimes added to the steam to give drop-wise condensation. Since the presence of non-condensable gases greatly affects the steam film coefficient, there should be a provision to vent non-condensable gases from the steam chest.

Example 17.5 A sugar solution with 12% solids concentration is to be evaporated to bring it to 60% solids concentration. Determine, per 100 kg of sugar solution, the amount of water to be removed.

SOLUTION:

We write the mass balance equation for evaporation as follows.

$$F = V + P$$

$$F \times X_f = P \times X_p$$

$$\text{Thus, } 100 \times 0.12 = P \times 0.6$$

$$\text{or, } P = 20 \text{ kg}$$

$$V = F - P = 80 \text{ kg}$$

Thus, 80 kg of water is to be removed per 100 kg of sugar solution.

Example 17.6 After the evaporation process a small amount of fresh juice, containing 10% solids, is mixed with the concentrated juice in order to improve the flavor of concentrated orange juice. If the fresh juice is concentrated to 45% solids in the evaporator and the final (desired) concentration is 40%, determine the weight fraction of the fresh juice required to be concentrated to final juice.

SOLUTION:

There are two unit operations as follows:

Step I: The fresh juice (with solid fraction 0.1) is evaporated to make concentrated juice (with solid fraction. 0.45).

Step II: The fresh juice (with solid fraction 0.1) is mixed with concentrated juice (with solid fraction 0.45) to make the final solid fraction (concentration) as 0.4.

Suppose, we have M kg of original juice (solid fraction 0.1), which is divided into two fractions M_1 and M_2. The first fraction M_1 (solid fraction 0.1) is concentrated to M_3, which has a solid fraction of 0.45. Then M_3 is mixed to M_2 to get the final solid fraction of 0.40.

We have to find out the ratio of M_1/M

Conducting the mass balance for the mixing operation:

$$M_2 \times (0.1) + M_3 \times (0.45) = (M_2+M_3) \times (0.4)$$

$$M_2 \times (0.4-0.1) = M_3 \times (0.45-0.4)$$

$$M_3 = 6\ M_2$$

Conducting the mass balance for the evaporation:

$$M_1 \times (0.1) = M_3 \times (0.45)$$

$$\text{or, } M_1 = 4.5\ M_3$$

$$\text{or, } M_1 = 27\ M_2$$

$$\text{As } M = M_1 + M_2,$$

$$\text{thus, } M_1/M = 27/28 = 0.9643 = 96.43\%$$

i.e., 96.43% of the fresh juice is concentrated. Please cross check the answer by starting your calculations from the fractionation operation to see that the final concentration is 40% solids.

Example 17.7 In an evaporator, the product boils at 90°C and steam used for heating is supplied at 120°C in the evaporator. Find the heat content of steam going into the system, condensate coming out of system, and the vapor going out of system. (Hint: Use steam table).

SOLUTION:

In this situation, the temperature of steam is 120°C.
The heat going into the system is the total heat of steam at 120°C.
Thus, from steam table, h_s =2706.3 kJ.kg^{-1}
The heat content of the condensate is h_{sc} = 504.7 kJ.kg^{-1}
(Check that h_s - h_{sc} is the latent heat that is used for the evaporation process.)
The evaporation takes place at 90°C, i.e., the vapor is at a temperature of 90°C.
So, the heat of vaporization at 90°C is h_v = 2661.3 kJ.kg^{-1}

Example 17.8 A fruit juice with 10% total soluble solids (TSS) is concentrated to 50% TSS in a single effect evaporator at atmospheric pressure (boiling point of juice can be taken as 100°C). Saturated steam at 2 bar is used for heating. If the feed rate is 10 kg.s^{-1}, what will be the amount of vapor and concentrated product? What will be the steam consumption if only the latent heat of condensation of steam is used for heating? Take the enthalpies for the saturated steam, condensate and vapor as 2706, 505, and 2676 kJ
.kg^{-1}, respectively. The specific heat for the juice (fresh and concentrated) can be taken as 4.18 kJ .kg^{-1}. K^{-1}.

SOLUTION:

It is assumed that the feed is heated using only the latent heat of condensation, and the feed enters at its boiling point at 100°C into the evaporator.
The latent heat of evaporation at 205 kPa pressure is 2706-505 = 2201 kJ.kg^{-1}.
Latent heat of vaporization at 100°C (which is also at pressure of 101.3 kPa or 1.01325 bar) is 2676-(100×4.18) = 2258 kJ.kg^{-1}.
First, carrying out the mass balance:

$$F \times X_f = P \times X_p$$

$$\text{Thus, } 10\times(0.1) = P \times(0.5)$$

$$\text{or, } P = 2 \text{ kg. s}^{-1}$$

$$V = 10 - 2 = 8 \text{ kg. s}^{-1}$$

The heat required for evaporation is
V×2258 kJ= 8 × 2258 kJ
(The feed enters at the boiling point and thus only latent heat is used for vaporization)
The latent heat of condensation of steam is 2201 kJ.kg^{-1}
Thus, steam required for evaporation = (8 × 2258)/2201 = 8.207 kg. s^{-1}

Example 17.9 Milk at 30°C and 10% soluble solids is fed to a single-effect evaporator for concentration, in which its vaporization temperature is maintained at 90°C. Temperature of steam utilized for heating is 120°C. The final concentration of milk is 45% (w/w). For the feed rate of 100 kg.h^{-1}, what will be the total amount of vapor produced and steam required for evaporation? Assume the specific heat of milk to be 4 kJ .kg^{-1}.K^{-1} during the process without any change. Assume any other value if needed.

SOLUTION:

Given t_f = 30°C, t_p = 90°C

$$F =100 \text{ kg.h}^{-1}$$

By mass balance, $F \times (0.1) = P \times(0.45)$

$$P = 100\times (0.1/0.45) = 22.22 \text{ kg.h}^{-1}$$

$$V = 100-22.22 = 77.78 \text{ kg.h}^{-1}$$

The heat balance in the evaporator can be written as:

$$W_s.h_{fg} = P.c_p. t_p + V.h_v - F.c_f.t_f$$

To use this equation, we need to know the h_{fg} and h_v, which are 2201.6 kJ.kg^{-1} at 120°C and 2661 kJ. kg^{-1} at 90°C (from steam table). We have assumed that the specific heat remains constant during the process at 4 kJ.kg^{-1}. °C^{-1}.

$$W_s \times (2201.6) = 22.22 \times 4 \times 90 + 77.78 \times 2661 - 100 \times 4 \times 30$$

$$= 202971.78 \text{ kJ.h}^{-1}$$

$$\text{Thus, } W_s = 92.19 \text{ kg.h}^{-1}$$

Example 17.10 Apple juice is concentrated from 10% solids to 40% solids in a continuous evaporator. The vacuum in the evaporator is maintained such that the juice boils at 60°C in the evaporator. The juice temperature at the inlet of the evaporator is 58°C. Determine the amount of juice that can be concentrated per hour. Take the enthalpy of water vapor as 2610 kJ.kg^{-1} at 60°C. Also take the specific heats of the fresh and concentrated juices as 3.89 and 2.85 kJ.kg^{-1}.K^{-1}, respectively. The heat transfer rate of the evaporator is 30 kW.

SOLUTION:

Given:

$$X_f = 10\%, \text{ i.e., } 0.1, X_p = 40\%, \text{ i.e., } 0.40$$

Taking the basis as 1 kg. s^{-1} of feed and if P and V are the amount of product and vapor in kg.s^{-1}, then,

$$(1)(0.1) = (P)(0.4)$$

$$\text{or } P = 0.25 \text{ F}$$

$$V = F - 0.25 \text{ F} = 0.75 \text{ F}$$

Heat balance around the evaporator gives,

$$W_s.h_{fg} + F. c_f. t_f = P. c_p. t_p + V.h_v$$

$$30 + (F)(3.89)(58-0) = (0.25)(F)(2.85)(60) + (0.75)(F)(2610)$$

$$30 = F(-194.5+42.75+1957.5) = F(1805.75)$$

$$\text{or } F = 0.0166 \text{ kg. s}^{-1}, \text{ i.e. } 59.8 \text{ kg.h}^{-1}$$

Example 17.11 Tomato juice is to be concentrated in an evaporator from 12% TSS to 30% TSS. The temperature of the feed at the inlet of evaporator is 40°C. The pressure in the evaporator is maintained such that the feed boils at 60°C. Steam used for heating is available at 1.7 atm absolute pressure. Determine the feed rate of tomato juice when U= 1240 W.m^{-2}.K^{-1} and the heat exchanger area (A) = 4.5 m^2. Assume the specific heat of the soluble solids as 0.3 kcal.kg^{-1}.°C^{-1}.

SOLUTION:

By mass balance

$$F \times (0.12) = P \times (0.3)$$

$$\text{Thus, } P = 0.4 \text{ F, and V} = 0.6 \text{ F}$$

The specific heat of soluble solids is 0.3 kcal. kg^{-1}.°C^{-1}.

Thus, the specific heat of feed,

$$c_f = 0.3 \times 0.12 + 1.0 \times (1-0.12) = 0.916 \text{ kcal.kg}^{-1}.K^{-1}$$

and the product's specific heat,

$$c_p = 0.3 \times 0.3 + 1.0 \times (1-0.3) = 0.79 \text{ kcal.kg}^{-1}.K^{-1}$$

The enthalpy balance equation is given as:

$$q = W_s.h_{fg} = P. c_p. t_p + V.h_v - F. c_f. t_f$$

Also, q = U.A. Δt
First, we have to find out the log mean temperature difference between the two streams.

The temperature of steam at 1.7 atm. absolute is 115°C. If we take that only the heat of condensation is used for heating and there is no temperature drop for steam,

$$\Delta t_1 = 75°C, \Delta t_2 = 55°C$$

$$\text{LMTD} = (75-55)/ \ln (75/55) = 64.48°C$$

Then, we have to know the heat of vapor. From steam table, h_v at 60°C is 2609 kJ.kg^{-1}

Substituting the values,

$$1240\left(W.m^{-2}.K^{-1}\right) \times 4.5\left(m^2\right) \times 64.48(K)$$

$$= 0.4F \times 0.79 \times 4.2 \times 1000 \times 60 + 0.6F \times 2609 \times 1000 -$$

$$F \times 0.916 \times 4.2 \times 1000 \times 40$$

Here, 1000 is multiplied with the terms in RHS to convert the specific heats to J.kg^{-1}.K^{-1}.

Now, 359798.4 (kg. s^{-1}) = 1798920 F
Thus, F = 0.204 kg. s^{-1}

Example 17.12 A sugar solution at 30°C needs to be concentrated to a level of 50% solids from initial concentration of 5%. Determine the amount of water required to be evaporated per kg of the feed. Also, calculate the specific energy requirement if 85% is the evaporator efficiency. Assume specific heat and boiling point of the solution as 3.95 kJ.kg^{-1}.°C^{-1} and 105°C, respectively. Latent heat of vaporization can be taken as 2502.3 kJ .kg^{-1}.

SOLUTION:

By mass balance

$$1 \times (0.05) = P \times (0.5)$$

Thus, P = 0.1 kg and V= 0.9 kg

The temperature of sugar solution has to be increased from 30°C to 105°C and then the vapor is removed.

Thus, sensible heat required =

$$m.c_p.\Delta t = 1\left(kg\right) \times 3.95\left(kJ.kg^{-1}.°C^{-1}\right)$$

$$\times \left(105 - 30\right)\left(°C\right) = 296.25 \text{ kJ}$$

Latent heat required for evaporation $\left(\text{of } 0.9 \text{ kg vapor}\right)$

$$= 0.9\left(kg\right) \times 2502.3\left(kJ.kg^{-1}\right)$$

$$= 2252.07 \text{ kJ}$$

Total heat = 296.25 + 2252.07 = 2548.32 kJ

Efficiency of evaporator = 85 %

Thus, actual heat requirement = 2548.32/0.85 = 2998.02 kJ

This is the heat requirement for evaporation of 0.9 kg vapor.

Thus, specific energy required = 2998.02/0.9 = 3331.13 kJ per kg of vapor

Example 17.13 A continuous triple effect evaporator is used to concentrate milk (1000 kg.h⁻¹) from 15% TSS to 50% TSS. Out of the total vapor produced, 50% is produced in the first effect and the other two contribute to 25% each. The boiling temperatures of milk in the three effects are observed to be 70°C, 60°C, and 50°C, respectively. The latent heats of vaporization at these temperatures are 2334 kJ. kg⁻¹, 2358.6 kJ.kg⁻¹, and 2383 kJ.kg⁻¹, respectively (as observed from steam table). Find out the total energy use in the evaporation process.

SOLUTION:

Milk feed rate = 1000 kg.h⁻¹

As per mass balance, 1000×(0.15) = P ×(0.5)

P = 300 kg.h⁻¹, V = 1000-300 = 700 kg.h⁻¹

Thus, evaporation in 1st effect = 350 kg.h⁻¹ and in 2nd and 3rd effects 175 kg.h⁻¹

Thus, total energy used for

evaporation $= 350 \times \left(2334\right) + 175 \times \left(2358.6\right) + 175 \times \left(2383\right)$

$$= 816900 + 412755 + 417025$$

$$= 1646680 \text{ kJ}$$

Example 17.14 Milk is being concentrated in an evaporator at 2500 kg.h⁻¹ feed rate from 12 to 50% solids. The heating steam is supplied at 2 bar (condensing temperature of steam is 120°C). The evaporation chamber is kept at 0.2 bar (boiling temperature in the evaporator may be assumed to be 60°C). The feed enters the evaporator at 21°C. The solution may be assumed to have negligible boiling point elevation. Take the specific heats of feed and the product as 4.0 and 3.9 kJ.kg⁻¹.°C⁻¹, respectively. If the overall heat transfer coefficient is 3000 W.m⁻².K⁻¹, what is the heating surface area required? Also obtain the steam economy for this process.

SOLUTION:

Given, concentration of feed and product are 12% and 50%

Thus for 2500 kg.h⁻¹ of feed,
the product = 2500 ×(0.12/.5) = 600 kg.h⁻¹
The water evaporated, V = 2500-600 = 1900 kg.h⁻¹
From steam table, total heat of vapor at 60 °C,
h_v = 2610 kJ.kg⁻¹
and, latent heat of steam at 120 °C, h_{fg} = 2201 kJ.kg⁻¹
Now the heat balance equation is
F. $c_f. t_f$+ $W_s.h_{fg}$ = P. $c_p. t_p$ + V.h_v
For the first situation, when the feed enters at 21°C (with all units in kJ.h⁻¹),

$$2500 \,(4.00)\,(21) + W_s\,(2201) = (600)\,(3.9)\,(60)$$
$$+ (1900)\,(2610)$$

$$210000 + W_s\,(2201) = 140400 + 4959000$$

$$W_s = 2221.44 \text{ kg.h}^{-1} = 0.617 \text{ kg. s}^{-1}$$

$$\text{Economy} = \frac{\text{vapour produced}}{\text{steam supplied}} = \frac{1900}{2221.44} = 0.855$$

The heating surface area, A can be found out from the equation

$$W_s.h_{fg} = q = U.A. \, \Delta t$$

Here the temperature is increased from 21°C to 60°C in the evaporator

$$\Delta t_1 = 120\text{-}21 = 99°C, \, \Delta t_2 = 120\text{-}60 = 60°C$$

$$\text{LMTD} = (99\text{-}60)/ \ln (99/60) = 77.88°C$$

Thus, (0.617) (kg.s⁻¹)×2201×10³ (J.kg⁻¹)
= 3000(W.m⁻².K⁻¹)×A (m²)×(77.88) (K)
Thus, A = 5.812 m²
The heat transfer area is 5.812 m².

Example 17.15 In a single effect falling film evaporator, the evaporation takes place at 60°C and

the temperature difference between the heating medium and evaporating liquid is 10°C. The length and diameter of the tubes in the evaporator are 6 m and 0.032 m, respectively and U = 2200 W.m⁻².K⁻¹. The vapor viscosity at 60°C is 11.15×10⁻⁶ Pa-s. Determine vapor velocity, Reynolds number, and pressure drop due to vapor flow. Take the latent heat of vaporization of water at 60°C as 2358 kJ.kg⁻¹ and specific volume of vapor at 60°C as 7.65 m³.kg⁻¹.

SOLUTION:

If m (kg. s⁻¹) is the vapor removed in the evaporator and h_{fg} is the latent heat of vaporization (J.kg⁻¹),

then the amount of vapor condensing in the evaporator is

$$m = U.A.\Delta t / h_{fg}$$

But, $A = \pi dL$, where L and d are the length and diameter of tube.

Similarly, the volumetric flow rate can be written as $Q = m.\nu$,

where ν (m³.kg⁻¹) is the specific volume of vapor at boiling temperature.

Thus, the velocity of vapor, v (m/s) in the evaporator tubes can be given by

$$v = Q / A_f = (m \nu) / A_f,$$

Here, A_f is the area of flow (which is same as the cross-sectional area of evaporator = $(\pi/4) d^2$)

Thus, velocity of flow, $v = \left(\dfrac{U(\pi dL)\Delta t}{h_{fg}} \right) \left(\dfrac{\nu}{(\pi/4)d^2} \right)$

Taking the values of ν at 60°C as 7.65 m³.kg⁻¹ and h_{fg} as 2358 kJ.kg⁻¹,

$$v = \left(\frac{2200(\pi)(0.032)(6)(10)}{(2358)10^3} \right) \left(\frac{7.65}{(\pi/4)(0.032)^2} \right) = 53.53 \text{ m.s}^{-1}$$

If the specific volume ν is 7.65 m³.kg⁻¹, the density is $1/\nu = 0.131$ kg.m⁻³

The Reynolds number is given by

$$Re = \left(\rho D v / \mu \right) = \frac{(0.131)(0.032)(53.53)}{11.15 \times 10^{-6}} = 20125.35$$

The pressure drop in the evaporator tube can be estimated by using equations applicable for friction losses in turbulent flow.

$$f = 0.079 \, (Re)^{-0.25} = 0.079 \, (20125.35)^{-0.25} = 6.633 \times 10^{-3}$$

The pressure drop is

$$F_f = \frac{\Delta p}{\rho} = \frac{4fLv^2}{2D} = \frac{(4)(0.006633)(6)(53.53)^2}{(2)(0.032)} = 7172.47$$

or $\Delta p = 7172.47 \times 0.131 = 933.699$ Pa

17.1.5 Boiling Point Rise

During the process of evaporation, the concentration of solids in the solution being evaporated increases, which causes a rise in the boiling point. This reduces the temperature gradient and, thus, the rate of heat transfer. Hence, for mass and energy balance, it is important to know the boiling point rise of the solution in the evaporator. Some methods for determination of boiling point rise are given below.

1. The boiling point of a solution at different pressures and concentrations of solid (better applicable for strong solutions) can be found out by the empirically derived Duhring's rule. As per this rule, the boiling point of a solution is stated to be a linear function of the boiling point of pure water at the same pressure. A plot of the above is known as a *Duhring's plot*. Standard plots are available for some common solutions as salts and sucrose. The trend of the lines is shown in Figure 17.6. Different lines are obtained for different concentrations (Earle, 2003).

2. Boiling point rise can also be found as (Heldman, 1997):

$$\ln(X_w) = -\frac{\lambda'}{R}\left(\frac{1}{T} - \frac{1}{T_b} \right) \tag{17.18}$$

where, X_w is the mole fraction of solvent, λ' is molal latent heat, R is Universal gas constant = 8.31434 J.mole⁻¹.K⁻¹, T and T_b are boiling points of pure water and solution under consideration (in K), respectively.

$\lambda = 2257.2$ kJ.kg⁻¹, 1 kg mole of water = 18 kg,

Hence, $\lambda' = 2257.2 \times 18$ kJ. (kg mole)⁻¹

X_w is found out by knowing the molecular weights of the solvent and solute.

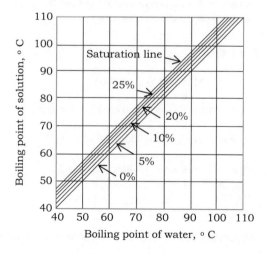

FIGURE 17.6 Example of a Duhring plot for a solution with different solute concentrations for a specific solute (The lines will be different for different solutes as the composition and molecular weights differ.)

Example 17.16 Find the boiling point for a 30% sucrose solution at normal atmospheric pressure?

SOLUTION:

For a 30% sucrose ($C_{12}H_{22}O_{11}$) solution,

molecular weight = 12×12 + 1×22 + 16×11 = 342

Thus, $(X_w)_{\text{30% sugar solution}} = \dfrac{0.70/18}{(0.70/18)+(0.30/342)} = 0.98$

Substituting the values in equation 17.18

$$\ln(0.98) = -\frac{2257.2 \times 18}{8.31434}\left(\frac{1}{373.15} - \frac{1}{T_b}\right)$$

$$4.13423 = \left(\frac{1}{373.15} - \frac{1}{T_b}\right)$$

$$\frac{1}{T_b} = \left(\frac{1}{373.15} - 4.13423\right) = 2.6757 \times 10^{-3}$$

$$T_b = 373.73 \text{ K} = 100.57 \text{ °C}$$

3. Soluble solids in most food products are mainly organic compounds, thus, the following equation expresses the boiling point rise in degrees Celsius.

$$\Delta t_b = K_b\, m \qquad (16.19)$$

Here, the boiling point of solution with molality "m" is Δt_b more than that of pure water at the given absolute pressure. K_b is a constant, often referred as the molal boiling point elevation constant.

The value of K_b for water is 0.512 °C.(mole)$^{-1}$. For acetic acid (normal boiling point 118.1°C) the value of K_b is 3.07 °C. (mole)$^{-1}$ and for chloroform (normal boiling point 61.3°C) the value of K_b is 3.63 °C.(mole)$^{-1}$.

It must be remembered from the basic knowledge of chemistry (also discussed in section 3.3) that

molality = moles of solute per kg of solvent

molarity = moles of solute per litre of solution

Example 17.19 Find the boiling point rise for a solution with 30% soluble solids (molecular weight= 342) at atmospheric pressure using the molality of the solution.

SOLUTION:

First the molality is found as follows:

300 g of solids with molecular weight of 342 means, the number of moles = 300/342, i.e., there are (300/342) moles in 700 g solvent.

Number of moles in 1000 g solvent

$$= m = \frac{(300/342)}{700} \times 1000 = 1.25$$

Thus, Δt_b = 0.51 x 1.25 = 0.64°C

The boiling point elevation, as obtained above, has to be added to the boiling point of water in the evaporator at the operating pressure (can be obtained from the steam table) to get the boiling point for the solution. For instance, in example 17.9, the t_p will be taken as 90°C+boiling point rise.

As the boiling point rises with the increase in concentration of solutes in a solution, the freezing point also depresses. The depression in a freezing point is given as 1.86×m, where m is the molality of the solution. 1.86 is the *Cryoscopic constant*.

Thus, the ratio of freezing point depression to boiling point elevation for any solution =1.86/0.51 = 3.64

As discussed earlier, in addition to the pressure or vacuum in evaporating space and concentration of solution, hydrostatic pressure of column of liquid also affects its boiling point. If the liquid column is high, then the boiling point will vary across the column and, hence, the boiling point is determined at half the depth of liquid in the evaporator.

17.2 Equipment

17.2.1 Classifications

In the food industry, evaporators can be broadly classified as *natural circulation evaporators* and *forced circulation evaporators*. In forced circulation evaporators, the liquid moves in the system by a pump. The high heat transfer rates achievable in forced circulation evaporators help to reduce the residence times and yield better product quality. The throughput capacity of the equipment is also increased.

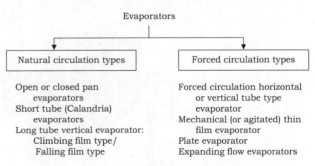

Evaporators are also classified as *once-through* or *circulation type,* depending on whether or not the feed is recirculated within the evaporation chamber. In circulation evaporators, a part of the concentrated liquor is mixed with fresh liquor and is fed back to the heat exchanger tubes. Since the liquid entering the tubes has a high proportion of the thick liquor, the heat transfer coefficient is low and the viscosity is high. Hence, for heat sensitive liquids, these are not appropriate. The forced circulation evaporators as well as climbing film evaporators are usually circulation types. The agitated film evaporators are always once through, which are suitable for heat sensitive materials.

Another classification can also be made depending on the number of effects used in the system as *single effect* or *multiple effect systems*. The multi-effect systems are once through evaporators. They can also be classified as *atmospheric type* or *vacuum type* depending on whether a vacuum pump is used to reduce the boiling point of the product or not. They can also be either *batch* or *continuous* type.

17.2.2 Open or Closed Pan Evaporator

Construction and Operation

- It is basically a pan kept on a fire/heat source. Common examples are the boiling pans used in jam/ jelly preparation. In such systems, the food has to be properly stirred/ mixed during the evaporation process. Another example is the open pan solar evaporator for salt water.

- However, in the food industry an open pan evaporator essentially means a jacketed kettle type heat exchanger (Figure 17.7 a), which consists of a hemispherical pan with an external jacket. The exterior jacket covers the curved bottom and about half way up to the cylindrical wall.

- It is a closed pan if it is fitted with a lid for making it airtight (Figure 17.7 b). The closed pan can be connected to a vacuum pump for vacuum operation.

- Steam is allowed to pass from the boiler into the external jacket for heating the liquid food material in the pan. After giving away the heat energy the condensed steam is collected at a specified outlet. To improve the rate of heat transfer, there may

be a stirrer or paddle in the pan. The agitation also reduces the chances of food burning on the heat exchange surface. It may also have internal tubes or coils through which the heating medium passes. These spiral coils are usually made up of heavy copper or stainless steel tubing (3") and are suspended in a horizontal plane at a few centimeters from the bottom of the kettle.

- In some models, the coil revolves around a horizontal shaft. It results in turbulence that improves heat transfer and decreases fouling.

Salient Features

- Characteristics of these evaporators include low capital cost, easy construction and maintenance, low energy efficiencies, and low heat transfer rates.

- If the food is not properly stirred and the heat transfer is not properly controlled, then there are chances of burning on the heating surfaces.

- If these evaporators are used for viscous products, the deposits on the heat transfer surfaces must be frequently cleaned.

Applications

- These evaporators are used for low or variable production rates or when frequent changes of product are required.

- These are used widely in small and medium scale industries for varieties of processed foods, such as sauces, gravies, jam, and other preserves.

- The jacketed open kettle type evaporators with a suitable agitating device and internal tubes/coils are used in the tomato industry for the manufacture of products such as tomato puree, tomato sauce, ketchup, etc. However, for very viscous products such as tomato paste or for heat sensitive pulps and juices, these evaporators are not recommended.

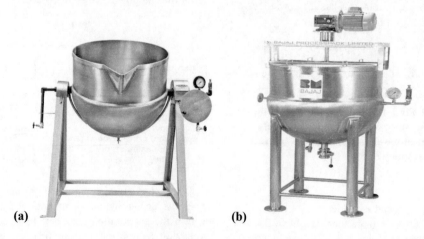

(a)   **(b)**

FIGURE 17.7   Jacketed kettle evaporator (a) open pan type, (b) closed pan type connected to vacuum (Courtesy: M/s Bajaj Processpack Limited, Noida)

17.2.3 Vacuum Pan Evaporator

The interior pressure of the evaporation chamber is reduced below atmospheric pressure to reduce the boiling point of the liquid to be evaporated (Figure 17.8). This also consequently reduces the need for heat in both boiling and condensation processes. The chamber is connected to a vacuum pump and the vapor removed is taken out and released into atmosphere. Evaporation at low temperatures maintains the quality of the food product in a better way.

17.2.4 Short-Tube Evaporator

Construction and Operation

- This comprises a shell and tube heat exchanger in which there is a vessel (or shell) normally having a vertical bundle of tubes (known as tube bundles). The tubes are usually of 50–100 mm diameter and 1–2 m length. There is a large diameter tube at the center of the shell (Figure 17.9 a). The shell and tube heat exchanger is also known as a *calandria* in the food industry.

- As the feed liquor flows inside the tube, it is heated by steam condensing on the outside of the tubes. The product boils upon heating. Heating of the fluid induces a density gradient. As a result, the warmer liquid rises through the small diameter tubes, and it comes down (recirculated) through the central large diameter tube (in which the temperature of the fluid is usually lower due to the more heat transfer path). The movement of the liquor occurs only due to natural convection.

- The product is recirculated repeatedly until the desired concentration is achieved and, after concentration, it is collected at the bottom. The vapor is released at the top.

- The fluid flow can also be enhanced by providing agitation.

- In some cases, the steam flows inside the tubes, and the product flows around the tubes.

- The horizontal tube bundle types are also available, but they are less common (Figure 17.9 b). In this case the tubes are at the lower part of the chamber. The product boils outside the tubes and steam flows through the tubes from the steam chest.

- In many designs, the tubes are fitted as a basket, which can be easily removed for cleaning. Hence, these are also known as *basket evaporators*.

- If there is an external pipe for recirculation of the feed, it is called an *external calandria evaporator*.

- The initial separation of vapor from liquid concentrate occurs at the vapor space above the calandria. However, a separate vapor chamber with baffles can be installed to separate the fine droplets of liquid entrained in the vapor.

- These evaporators may be used at atmospheric pressure or reduced pressure (by using a vacuum pump). Steam may also be superheated to have adequate temperature difference for efficient heat transfer.

- The evaporator can be a single pass or a multi-pass type depending on the degree of concentration required and other operating parameters. In a multi-pass system, the product goes one way through some of the tubes and returns through the others.

FIGURE 17.8 Vacuum evaporated kettle (batch type) (Courtesy: M/s Bajaj Processpack Limited, Noida)

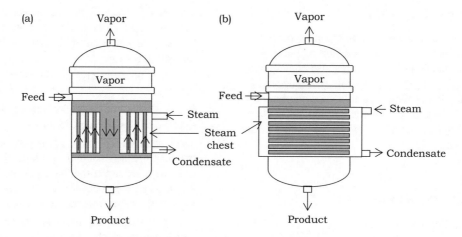

FIGURE 17.9 Schematic of a short tube evaporator (a)vertical tube type, (b) horizontal tube type

Salient Features

- As compared to forced circulation evaporators, these evaporators are low in construction and maintenance costs.

- In comparison to open or closed pans, they offer higher rates of heat transfer and more flexibility for relatively low viscous liquids.

- These are normally not suitable for more viscous foods because of poor circulation of liquors and fouling on heat exchange walls.

- Convection currents and rates of heat transfer are increased in the external calandria evaporators; the external calandria can also be easily cleaned.

- The overall heat transfer coefficient ranges between 4000–8000 kJ.m^{-2}.h^{-1}.K^{-1} for vertical types and 2000–8000 kJ.m^{-2}.h^{-1}.K^{-1} for horizontal types of evaporators.

FIGURE 17.10 Short tube evaporator (Courtesy: M/s Shiva Engineers, Pune)

Applications

- These evaporators can be used for low viscous foods, such as syrups, salt solutions, and fruit juices. Figure 17.10 shows a commercial short tube evaporator.

- In horizontal tube evaporators, there is less natural circulation of liquid, which also reduces the heat transfer coefficient. Thus, they are not suitable for viscous materials. In addition, it is difficult to remove the deposits from outside of the tubes, and, hence, they are not suitable for stone forming (fouling) products. The solid deposits can be easily cleaned in vertical types.

- Most of the short tube evaporators have now been replaced by different types of forced circulation evaporators to obtain the benefit of a shorter processing time. However, in the sugar industry, they are still used to crystallize the refined sugar.

- For heat sensitive foods, including dairy products, external calandria evaporators can be used when operated under a partial vacuum.

17.2.5 Long Tube Vertical Evaporator

Construction and Operation

A vertical shell and tube heat exchanger 3–15 m high (up to 7 m height is very common) is termed a long tube evaporator. The steam flows in the shell and the liquid in the vertical bundle of tubes. The liquid usually enters the tubes at a velocity of 0.3 to 1.2 m.s^{-1}. As vapor is formed in the tubes, the linear velocity increases greatly and the heat transfer rate improves. However, the overall heat transfer coefficients in a natural circulation unit may be low for viscous liquids.

The long tube evaporators can be either rising film or falling film type. Figure 17.11 depicts a comparison of the features of falling film and climbing film type long tube evaporators.

Rising Film Evaporator

- In rising/ climbing film evaporator, the dilute feed enters the evaporator at the bottom and rises up as it evaporates. The tubes are of 25–50 mm diameter and 10–15 m height for low viscous materials.

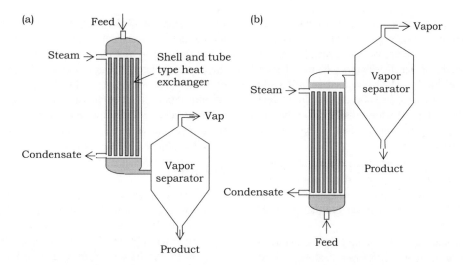

FIGURE 17.11 Comparison of the features of falling film and climbing film type long tube evaporators

- These operate on a "thermo-siphon" mechanism. The food is preheated close to the boiling point, corresponding to the pressure in the heat exchanger before it enters the system. Due to further heating of the food in the heat exchanger, there is formation of vapor. These vapor bubbles tend to move in an upward direction, which causes the food to rise upward inside the tubes. As there is increased production of vapors, the food rises upwards on the walls of the tubes as a thin film.

- The mixture of liquid and vapor is released at the top of the tubes into a separator.

- To prevent the liquid droplets from going out with the vapor, the vapor before leaving the separator is made to impinge on sets of baffle plates. From the separator, a part of liquid is returned to the base of the calandria and mixed with fresh feed before recirculation and the remaining concentrate is withdrawn as product.

Salient Features

- As the vapor carries the liquid with it, it is not necessary to feed the liquid under a head equivalent to the total height of the tube. Instead, the liquid is delivered to the tube under a head of only a few meters. As the liquid rises up, the volume as well as velocity of vapor increases.

- The temperature difference between product and heating medium should be at least 14°C to make a properly developed film in rising film evaporators. Thus, up to four effects of rising film evaporator could be used when the heating steam is supplied to the first effect at 110°C and the product boiling temperature in last effect is 50°C. However, the falling film types can have higher number of effects and the temperature difference in each effect can be 5°C. In the first effect, the boiling point of the food should be limited to 68°–70°C to avoid protein denaturation.

- The fluid, while moving upward against gravity, creates turbulence, which is advantageous during treating viscous products and the products having a tendency to foul on the heating surfaces.

- A part of the concentrated product is recirculated again to the evaporator, which helps in producing sufficient liquid loading inside the heating tubes.

- Larger temperature differences are necessary when using long tubes for the bundle design to function properly. The long tubes allow high concentration ratios for viscous and foaming liquids in a once-through mode.

- The rising film evaporators have high heat transfer coefficients due to the vapor produced in tubes. The overall heat transfer in long tube evaporators is usually 4000–12000 $kJ.m^{-2}.h^{-1}.°C^{-1}$ for low viscous materials and less than 4000 $kJ.m^{-2}.h^{-1}.°C^{-1}$ for highly viscous materials. Forward feed multiple effect systems having a capacity of 45000 $l.h^{-1}$ are reported.

Falling Film Evaporator

Construction and Operation

- This long tube evaporator also has a shell and tube heat exchanger in which the feed liquid flows downward by gravity.

- The heat exchanger has 50–250 mm diameter vertical tubes with a liquid distributor at the top and a liquid-vapor separator at the bottom.

- The product is preheated to a temperature slightly above the temperature of vapor generation in the evaporator. From the preheater, the product flows to the upper section of evaporator. The product flow rates are very high (for a 12 m long tube, it can be up to 200 $m.s^{-1}$ at the end of the tube) due to the collective action of the gravitational force and the push of the steam.

- It is very vital that the product should be uniformly distributed on the inside surface of the tubes in falling film units. Different designs of feeders are available for the purpose. Having, above a carefully levelled tube sheet, a set of perforated metal plates is one such example. Otherwise, there may be inserts in the tube ends. Spider distributors with radial arms are also available by which the feed is uniformly sprayed on the inside surface of each tube. Individual spray nozzles can also be used inside each tube. As the product is slightly superheated, it expands as soon as it leaves the nozzle.

- A moderate amount of liquid is recycled at the top of the tubes for assisting proper distribution of liquid to the tubes. This also helps to increase the flow rate through the tubes as compared to that in once-through system.

- Both the product and the vapors flow downward inside the heating tubes. The vapor is moved to the vapor separation chamber. In some units the volatiles flow upward and are collected at the top of the unit.

- The unit may also be constructed such that the product will first rise in the tubes and then fall, so that it becomes a combination of a rising and falling film evaporator. This gives a better efficiency of evaporation.

Salient Features

- The operation of falling film evaporators does not require a high temperature difference between the heating medium and the feed. Even low temperature differences can result in relatively high heat transfer rates. It is for this reason that short product contact times, typically 5–60 s, are observed in these evaporators.

- The use of small temperature differences makes it possible to use multiple effect systems or vapor compression systems. Automatic control systems can assist in preparing consistent concentrated product.

- The resultant product is of high quality because of the uniform temperature and short product contact times. Falling film evaporators, therefore, are particularly suitable and most frequently used for heat sensitive products.

- The falling film evaporators need a careful design for the operating conditions so that there is uniform wetting of the inside of the tubes, otherwise there may be dry patches and scaling or even clogging of the tubes.

- The system offers the advantage of quick startup and it can be changed to a cleaning mode quickly. The product can also be changed easily. The operating parameters, such as feed rate, concentrations, vacuum, heat supply, etc., can also be easily altered in falling film evaporators.

- Falling film evaporators are the least expensive of evaporators with low residence time and are particularly suitable for large volumes of dilute material. The other advantages include low liquid holding time, less requirement of floor space, and good heat transfer coefficient.

In a combination of the rising and falling film type evaporator, the liquid is fed from below as in a rising film evaporator. When the liquid-vapor mixture reaches the top of the tubes of the heat exchanger, it is again uniformly distributed as a film on the inner surface of the tubes to move downward. Such a system promotes better distribution of the liquid and eliminates the need for long pipes for circulation. The overall headroom requirement is less. The fouling is also less due to high velocities attained in this evaporator.

Applications

- The long tube rising film evaporator is generally not suitable for heat sensitive products. These are used for foods of low viscosity (up to approx. 0.1 N.s.m^{-2}), such as milk and fruit juices. Liquids that tend to foam can be concentrated quite effectively. As the liquid strikes the baffles, the foam is broken up.

- The falling film evaporators are used for more viscous foods and for heat sensitive materials as there is minimum time exposure. These are also used for more concentrated dairy products, fruit juices, etc. A falling film evaporator with short residence times can handle sensitive products that can be concentrated in no other way. Falling film evaporators can easily handle moderately viscous fluids and materials as well as those with mildly fouling characteristics. A number of effects can be used for removing large amounts of vapor; a portion of partly concentrated liquid can also be recirculated at the top.

- Falling film units are well suited for applications requiring a high vacuum and for heat sensitive materials as well as for viscous materials. The temperature difference in the heat exchanger can be kept low. A *thermally accelerated short-time evaporator* (TASTE) is an example of a falling film evaporator commonly used for fruit juice concentration.

17.2.6 Forced Circulation Horizontal or Vertical Tube Type Evaporator

Forced circulation evaporators use pumps to generate high velocity of fluid flow and thus promote high heat transfer coefficient. These are advantageous specifically for foods having a tendency to fouling, for which the boiling needs to be prevented in the tubes as well as for foods having a tendency to crystallize during the evaporation process.

Construction and Operation

- A centrifugal/axial pump is used to move the liquid through the tubes at high velocities. To maintain high circulation rates, axial flow pumps are generally used.

- The heat exchanger may be horizontal (often two pass) or vertical (usually single pass) (Figure 17.12).

- The fluid flow velocity is around 2 to 6 m.s^{-1}. Still higher velocities are usually avoided to reduce the

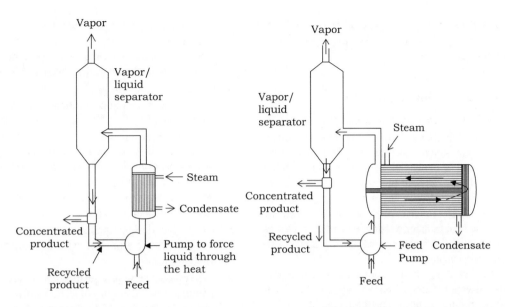

FIGURE 17.12   Schematic diagram of a forced circulation evaporator with (a) vertical heat exchanger, and (b) horizontal heat exchanger

cost of the pumping and the erosion that may occur at high velocities.

- The liquid is fed to the system at a temperature close to its boiling point (under the pressure maintained in the evaporator). The temperature difference between the heating steam and food is usually 3°–5°C. The tubes are kept under hydrostatic pressure so that even though the product is further heated, the boiling in the tubes is prevented. Then this superheated liquid moves to the vapor separator, where the pressure is reduced. This is the place where the liquid flashes into a mixture of vapor and concentrated liquid spray.

- Deflector plates or baffles in the vapor space minimize the entrainment losses. From the separator, the concentrated liquid is collected and a part of it is returned to the pump inlet, where it is mixed with the fresh material and recycled back to the system.

- The vapor removed in the separator moves to a condenser or to the next effect (in case of a multiple effect system).

Salient Features

- Because of the forced circulation, these evaporators have high heat transfer rates.

- In forced circulation evaporators, the temperature difference is maintained at 3°–5°C and thus the liquid product is heated only a few degrees for each pass.

- The overall heat transfer coefficient ranges between 8000–20000 kJ. m^{-2}. h^{-1}. °C^{-1}, and, hence, smaller sized equipment can be used for the same rate of heat transfer as compared to natural circulation systems.

- With less viscous liquids, the extra investment for forced circulation is usually not justified for the relatively lower advantages over natural circulation evaporators. However, they are beneficial for handling viscous materials.

- The recirculation pump is a very important section of the evaporation system and the recirculation rate and the pump capacity are designed using the information on the maximum permitted fluid temperature, velocity in the tubes, vapor pressure of the fluid, and temperature difference between pumping and the heating medium.

- The NPSH (net positive suction head) is also very important as the feed enters the evaporator at or near its boiling point. The pump can handle more volume at a lower head and the required NPSH may be more than what is available when excessive head is developed. Cavitation will be caused in the pump under this situation.

- At higher heads the fluid may boil in the tubes and there may be fouling at low velocities. It will also reduce the heat transfer.

- As the chances of fouling, salting, and scaling are lower, the forced circulation evaporators are easy to clean.

- For forced circulation systems, the high cost of the pumps and associated costs (maintenance and operating costs) are a few disadvantages.

- Some forced circulation systems allow the vaporization of the product in the tubes; it is used only when the head room is small or for non-scaling liquids.

Applications

These evaporators are used for foods that are either heat sensitive or viscous (up to 20 Pa.s) as well as those that tend to foam or foul on the evaporator surfaces and salting liquors. Some examples are pulps and juices, tomato pastes, etc.

This type of evaporator is used in crystallization applications. A special separator is used to collect the crystals from the slurry after the vapor separator.

17.2.7 Agitated Thin Film Evaporators

In the above types of evaporators, during handling of viscous fluids, there may be fouling or "burn on" on heat exchanger surfaces. Hence, a provision for continuously cleaning the heat exchange surfaces can be beneficial. In addition to preventing burn on, it also helps to improve the heat transfer and gives better efficiency of evaporation. Thus, a wiper/scraper is used to continuously clean the inner surface of the tube (i.e., the food side where the heat transfer resistance is much higher as compared to the steam side), and they are known as a mechanical/agitated thin film evaporator.

Two types of agitated thin film evaporators are commonly used. These are the wiped or scraped surface type and the centritherm type having rotating cones to spread the feed liquid.

Wiped Surface and Scraped Surface Evaporator

Construction and Operation

- This consists of a scraped surface heat exchanger, which has been explained in Chapter 6. As discussed, in this device, the steam condenses on the exterior jacket of the tubular heat exchanger and the food moves in the tube. There is a high-speed rotor with short blades along its length that rotates inside the tube (Figure 17.13).
- The product is fed uniformly on to the inner surface of the tube by a feed nozzle. The movement of the rotor, associated with gravity, transports the product downward in a helical path on the inner surface of the tubes.
- The product slurry forms a thin film on the inner surface of the tube. Evaporation takes place as the

liquid moves along the heated tube. The rotor blades continuously sweep the inner surface of the tube to keep the tube surface clean and to prevent formation of any of deposit on it during the evaporation period.

- It is possible to have vertical or horizontal designs of thin-film evaporators and both can have cylindrical or tapered bodies. For the tapered chambers, the rotors are also tapered and the tapered rotors may be further classified as zero clearance, rigid fixed clearance, or an adjustable clearance type. There are many designs available for each of these categories. However, the vertical design is common with cylindrical fixed-clearance rotor.
- The vertical designs can have a control ring to maintain the desired film thickness for the appropriate residence time.
- The scraped or wiped surface evaporators are differentiated based on the food film thickness on the tube walls. The film thickness in wiped film evaporators is about 0.25 mm whereas it is up to 1.25 mm in scraped surface type evaporators.

Salient Features

- The rotor must give about 8 to 10 blade passes per second in any section to prevent fouling and keep the surface clean for deposition of a fresh layer of slurry.
- This creates high turbulence, even in extremely viscous fluids, which helps to achieve high heat transfer coefficients. For similar operating conditions, heat transfer rates are much higher in the case of thin-film evaporator as compared to those in any other type of evaporator.
- The residence time is between 0.5 s to 100 s, which depends on the type of food and the required degree of concentration.
- Temperature differences up to 90°C can be maintained in thin film evaporators. This, along with less residence time, prevents thermal degradation for heat

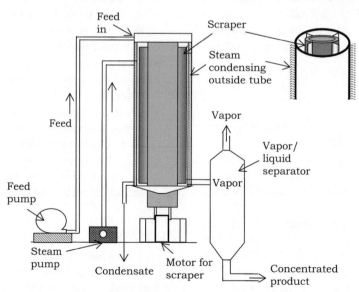

FIGURE 17.13 Schematic diagram of an agitated thin film evaporator (after Singh and Heldman, 2014 with permission)

sensitive foods. For longer residence time, product temperatures should be low.

- In such evaporators, only a single effect is possible, which reduces throughput capacity. Mechanical agitation involves extra capital and running costs. The rotor and the blades should be fitted very precisely so that a uniform thickness of product is scraped and there is no rubbing of the inner walls. This requires a sophisticated manufacturing line. In addition, it needs more power for operation. However, the high heat transfer coefficient means that much less evaporator surface for heat transfer is required than that for other evaporators.

Applications

- These evaporators are highly suitable for viscous foods (in the range of 1 to 50 Pa.s). and foods that may foul or foam on the heat exchanger surface. In addition, these are also suitable for heat sensitive products. Tomato pastes and purees, highly concentrated fruit juices and syrups are some examples of products that are evaporated in agitated film evaporators.
- These can also be used for fruit juices having even more than 25% suspended solids.
- These evaporators need higher capital and operating costs. Therefore, their use is limited to finishing highly viscous products after they have been concentrated in other evaporators/equipment.

Centri-Therm Evaporator

Construction and Operation

- It consists of a number of hollow cones arranged in a stack with a suitable gap in between. The cones rotate along the vertical axis.
- A pipe located at the center feeds the liquid to the underside of the rotating cones. As the cones rotate, the liquor spreads as a thin layer of about 0.1 mm thickness across the heated surface of the cones.

- Steam condenses on the other side of each cone. Thus, as in plate heat exchangers, the steam and liquor flow in alternate layers and there is rapid heat transfer and the liquid evaporates.

Salient Features

- The product residence time is very small at about 0.6–1.6 s.
- The steam, after giving away the heat of vaporization, condenses as droplets on the surface of the cone. As the cones rotate at a high speed, the droplets are flung from the cones almost immediately after formation. A boundary film of condensate is not formed and, therefore, the rate of heat transfer is not reduced. In addition, there is a very thin film of liquid on the other side. Both of these contribute to a high rate of heat transfer.

Applications

Products like fruit juices, coffee, tea extracts, meat extracts, enzymes, etc. can be conveniently processed using these evaporators.

17.2.8 Plate Evaporator

Construction and Operation

- It consists of a series of metal plates stacked onto a frame. The plates serve as the heat exchange surfaces.
- For evaporation systems, two types of heat exchangers, namely, the plate-and-frame type and the spiral-plate type, are used. The first one is alike a plate-and-frame filter press, where embossed plates are mounted on a frame (Figure 17.14).
- The plates may be either flat or corrugated. Corrugated plates provide improved structural rigidity and extended heat transfer surface. There are openings at the corner of the plates for the flow of the fluids. Fluids pass in either a series or a parallel flow through the spaces between the plates. The gaskets

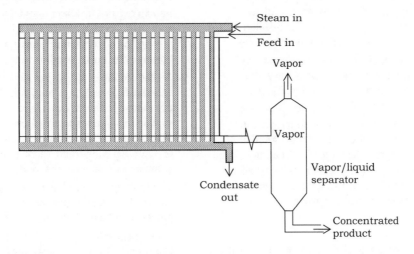

FIGURE 17.14 Schematic diagram of an evaporator with plate type heat exchanger

seal the plates at the peripheries. It is in alternate gaps that the product and heating medium (steam) move.

- The passage is designed such that the product is evenly distributed on the plate surfaces.

- The plate heat exchanger can be either falling film or rising film or rising-falling film. Throughput capacity and the required degree of concentration within a single device determine the number of rising and falling sections. In some applications high vapor velocities create turbulence and aid in removing the rising and falling films from the plate. This also reduces formation of scales on heat transfer surfaces.

- The low-pressure vapor is passed into top openings between the plates.

- A spiral-plate heat exchanger is basically a pair of concentric spiral passages formed by winding two long strips of plates with the center open. To maintain the spacings, spacer studs are welded to the plates. In special cases, the fluid-flow channels are closed at both sides of the spiral plate by welding the alternate channels. In some situations, one of the channels is left completely open and the other is closed at both ends.

- Three flow configurations are possible in the spiral heat exchanger, namely, both fluids in spiral flow, one fluid in spiral flow and the other having a combination of spiral and axial flow, and one fluid in spiral flow and the other having axial flow across the spiral plates.

- In some flat surface plate evaporators, the channels are used alternately for movement of the liquor and the steam. Thus, any scale deposit during movement of liquor on the plates is dissolved subsequently when steam moves through the same channel. However, in such situations, the scale formation in valves used for cycling the fluids does not easily dissolve.

Salient Features

- Plate evaporators are of compact design; they are capable of high throughputs and involve high capital investment.

- The plate package can be opened easily for inspection of the surfaces and for changing any plate, if necessary. The rate of evaporation can also be altered by altering (adding or reducing) the number of plates.

- With a high flow velocity, the heat transfer rate will be very high and the residence time is very short. It also gives less fouling and high energy efficiency.

- There are a number of advantages of spiral-plate evaporators over conventional shell-and-tube units. Its compact design gives a shorter liquid pathway and the centrifugal forces improve heat transfer. In addition, as discussed earlier, there is less fouling and these devices can be easily cleaned. The spiral arrangement accepts the differential thermal expansion.

Applications

Plate evaporators are used for heat sensitive foods and the liquor can be concentrated up to 0.3–0.4 N.s.m^{-2}. For heat

sensitive products they perform better than the long tube evaporators. Curved-flow-in spiral units are useful for handling viscous material or fluids containing solids.

17.2.9 Centrifugal/Conical/Expanding Flow evaporator

Construction and Operation

- The operational aspects of an expanding flow evaporator are almost like the plate evaporator, except that there is a stack of inverted cones instead of the plates.

- As in a plate heat exchanger, the heating medium (steam) moves down in alternate spaces (in between cone) and the product is fed upward in alternate spaces by a central pipe.

- The product moves by vapor pressure difference and the rotation of cones keeps the liquid as a thin film as it flows up. Thus, the evaporation takes place at a rapid rate.

- The vapor concentrate mixture enters a special type of cyclone separator, after it leaves the cone assembly tangentially, for separation of the components.

Salient Features

- As the steam is condensed, its flow area is reduced. This reduction of flow area along each channel helps improve the heat transfer coefficient.

- These evaporators have short residence times.

- The production rate can be changed by changing the number of cones.

Applications

Thermally sensitive products can be suitably processed using these evaporators. However, due to high cost and limited throughput rates, these evaporators are little used in the food industry.

17.2.10 Flash Evaporator

Construction and Operation

- Flash evaporation involves flashing of a heated liquid into an expansion vessel maintained at a lower pressure than the feed (Figure 17.15). The liquid is usually preheated to a temperature much higher than the boiling point corresponding to the expansion vessel.

- As the liquid is fed to the expansion vessel, often called a flash pot, it is adiabatically (enthalpy does not change in the process) expanded to a lower pressure and its temperature reduces to the boiling point of the liquid at the given pressure. Thus, a mixture of saturated liquid and vapor is produced instantly without any change in the total enthalpy.

- The flashing process has three stages of operations, namely, heating, flashing, and heat recovery. A single process combines three stages of operation.

- A multi-stage flash evaporator consists of a series of chambers with successively lower pressures. Thus, the separation is done in stages where vapor flashing

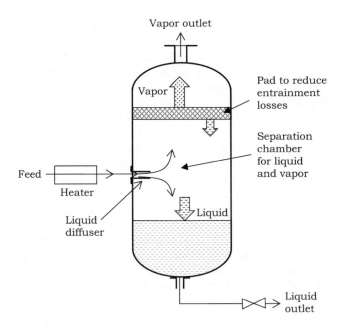

FIGURE 17.15 Schematic of a flash evaporator (after https://www.ou .edu; http://www.oilngasprocess.com)

occurs at each stage. The flashed vapor in each stage is often used for heating the product prior to the heat input zone.

- It is by passing through a throttling valve or other throttling device that the liquid stream undergoes the reduction in pressure. If the throttling valve or device is positioned at the entrance of a pressure vessel, then it is often called a flash drum and the flash evaporation occurs within the vessel.

Applications

- This method where the fruit juice is flashed into a perforated or packed plate column is commonly used for essence recovery and the columns are maintained

at very low pressures. The juice is preheated to 50°– 66°C and the column is maintained at about 3.45 kPa (boiling point about 30°C). As the vapor proceeds up the column, it becomes richer in volatile components. Cooled by a refrigerated system, a surface condenser traps the volatile components. The recovered essence concentrate is blended with the concentrated product.

- Flash evaporation can also be used to purify liquids and, as such, can be used to desalinate water.

17.2.11 Rotary Evaporator

- It is a device, used mostly in laboratories, where solvents are removed from samples efficiently and gently by evaporation.
- The basic parts include an evaporation flask or vial for holding the sample, a hot water (or other fluid) bath to heat the flask, a vacuum system to reduce the pressure inside the flask, a motor to rotate the flask, a vapor duct, a condenser to condense the vapor, and a condensate collecting flask (Figure 17.16).
- The vapor duct is also the axis for sample rotation.
- There is a motorized mechanism to quickly lift the evaporation flask from the heating bath.
- During operation, a thin film of warm solvent is formed on the wall of the rotating flask and, due to the centrifugal and frictional forces, it spreads over the surface. This along with almost no bumping of the liquid in the flask helps in gently evaporating the solvent.

17.3 Evaporator Accessories

17.3.1 Condenser

The condensers are used to liquefy vapors produced from the evaporator. The basic principle involved is heat exchange to

FIGURE 17.16 Rotary evaporator

remove the latent heat from the vapor. Usually, low temperature water is used as the heat sink. The different types of condensers are briefly discussed below.

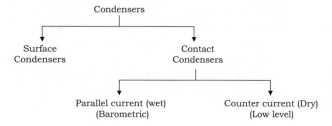

Surface condenser. The cooling liquid and the vapor to be condensed flow on inner and outer sides of a metal tube, respectively. The cooling water usually flows inside the tube and the vapor condenses on the outer surface of the tube. The water removes heat and allows condensation of vapor moving outside the tube. Increasing the water velocity can increase the condensation rate to a certain extent. These have high initial and operational costs. However, they are used when the vapor needs to be recovered, such as in essence recovery systems.

Contact condenser. The cooling liquid and vapor directly mix with each other. It may be a parallel flow type or a counter flow type. There may be some vapor/gases, which remain uncondensed at the end of the process. In a parallel flow system, the non-condensed gases leave at the cooling water and condensed vapor exit point. In the counter current type the non-condensed gases leave at the entry point of the cooling water.

In a *wet condenser*, the same pump removes the cooling water, condensed vapor, and non-condensed gases together; separate pumps remove them in a *dry condenser*. The counter current condensers are dry condensers while parallel current condensers are wet condensers.

A *barometric condenser* (Figure 17.17) is a type of condenser that is kept at a very high level and is connected to a barometric leg such that the water escapes to outside through the barometric leg. The pressure of water in the column ensures air tightness and helps to maintain the vacuum in the system. For a low-level condenser, a pump is used to remove the water.

A *jet condenser* is a direct contact condenser, where cooling water is sprayed onto the vapors to condense them. The rate of water consumption in a jet condenser is considerably higher than that in a barometric condenser and cannot be easily controlled.

The heat that needs to be removed to condense the vapor is termed the condenser duty (q_c) and is given as:

$$q_c = V(h_g - h_{fc}) \qquad (17.20)$$

where, h_g is enthalpy of vapour in the evaporator's vapor chamber and h_{fc} is liquid condensate's enthalpy.

17.3.2 Barometric Leg

As mentioned earlier, most of the time the boiling of liquid is done at reduced pressure. To produce a vacuum at the start

of an operation, auxiliary steam jets can be provided. But the vapor produced from the evaporator under a vacuum has to be discharged into a space under atmospheric pressure while still maintaining the vacuum in the evaporation chamber. This is done by various means. One of them is barometric leg method.

A barometric leg is essentially a vertical column or tube that is higher than 10.4 m (34 ft). If its lower end is kept immersed in a pot of water and the upper end is exposed to a vacuum, the water will rise in the tube depending on the vacuum maximum up to a height of 10m. If the leg is attached to a condenser under vacuum, as the vapor condenses in the leg, the height of the liquid maintains itself. As the condensate in the leg tends to increase, water leaves at the bottom and overflows from the tank (seal pot). The net effect is the removal of liquid from a low pressure to a higher one. The water in the column serves as the barrier between the vacuum and atmospheric pressure (Figure 17.17).

17.3.3 Entrainment Separator

These are devices to prevent the liquid droplets from escaping along with the discharging vapor stream. Some obstacles, like baffle/plates, are placed on the path of the escaping vapor. The high velocity vapor impinges on these stationary plates. The droplets return to the evaporating section after coalescing on these plates. The vapor escapes through the outlets at the top. There can also be deflectors or centrifugal devices. It may be incorporated at the top of evaporator or may be constructed externally connected with the discharge from the evaporator.

17.3.4 Vacuum Pump

The other important accessory is a vacuum pump to reduce pressure in the evaporation chambers.

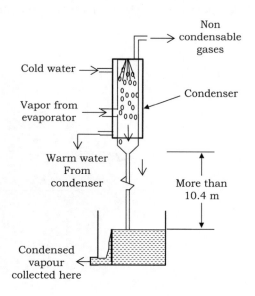

FIGURE 17.17 Schematic of barometric condenser

17.4 Selection of Evaporator

Since many different types of evaporators are available, the selection of an evaporator for any industry broadly depends on the considerations as follows:

- Operating capacity (how much water is needed to be evaporated per hour);
- Required degree of concentration;
- Thermal sensitivity of the product;
- Requirement of volatile recovery facilities;
- Ease of cleaning;
- Reliability and simplicity of operation;
- Capital and operating costs.

Check Your Understanding

1. A liquid food having properties similar to water (density = 1000 kg.m^{-3}, viscosity = 0.001 Pa.s) is flowing through the bottom of a falling evaporator at a rate of 0.006 kg.s^{-1}. The tubes of the evaporator have an inner diameter of 0.023 m and a length of 6 m. Find the

   (i) Reynolds number of flow at the bottom of the tubes;

   (ii) Wetting index (m^3.m^{-1}.h^{-1}) of the liquid at the bottom of the tubes.

   Answer: 332 and 0.30

2. A single effect evaporator is used to concentrate 5000 kg.h^{-1} of a 1.5 wt % sugar solution entering at 50°C to a concentration of 2 wt % at 101.325 kPa. Steam supplied is saturated at 169.06 kPa (115^{0}C). The overall heat transfer coefficient is 1550 Wm^{-2}.K^{-1}. The boiling point of solution is the same as that of water. The specific heat of the feed is 4021 kJ.kg^{-1}. K^{-1}. The latent heat of water at 100°C is 2257.06 kJ.kg^{-1} and the latent heat of steam at 115°C is 2216.52 kJ.kg^{-1}. What will be the required surfaces area for heat transfer?

   Answer: 46 m²

3. A single effect vacuum evaporator has 100 tubes of 25 mm diameter. One thousand kg feed of milk per hour with 15% *TS* is concentrated to 20% *TS* in the evaporator. Film heat transfer coefficients on both sides of the tube are 5000 and 800 w m^{-2}.K^{-1}. Thermal conductivity of 1.5 mm thick SS tubes is 15 W.m^{-1}.K^{-1}. Latent heat of vaporization under vacuum is 2309 kJ.kg^{-1}. What will be height of each tube for 10°C temperature difference across the tube wall?

   Answer: 3.17 m

4. The total solids content in a milk sample is 18%. It is desired to produce 1000 kg of sweetened condensed milk (SCM) having 40% sugar, 25% moisture and rest milk solids. What is the "sugar ratio" (in percentage) in the SCM in terms of sugar and water content in the final product? If the "concentration degree" is 2.5, what is the amount of sugar added in kg in the milk sample?

   Answer: 61.54 and 246.16

5. A single-effect long-tube evaporator has ten tubes each of 2.5 cm diameter and 6 m length. It concentrates pineapple juice from 18^0 Brix to 23^0 Brix. The feed rate into the evaporator is 557 kg.h^{-1} at the boiling point of 70°C (latent heat of vaporization = 2333.82 kJ.kg^{-1}). Neglecting boiling point rise, what is the overall heat transfer coefficient in Wm^{-2}.K^{-1} for 12°C temperature gradient across the tube walls?

   Answer: 1387.164 W.m^{-2}.K^{-1}

6. A fruit juice with a negligible boiling point rise is being evaporated using saturated steam at 121.1°C in a triple effect evaporator having equal area in each effect. The pressure of the vapor in the last effect is 25.6 kPa absolute and the corresponding saturation temperature is 65.7°C. The heat transfer coefficients are U_1 = 2760, U_2 = 1875 and U_3 = 1350 W.m^{-2}.K^{-1}. What is the boiling point (^{0}C) in the first effect?

   Answer: 108.82°C

7. It is desired to concentrate a 20% salt solution (20 kg of salt in 100 kg of solution) to a 30% salt solution in an evaporator. Consider a feed of 300 kg. mi^{n-1} at 30^{0}C. The boiling point of the solution is 110^{0}C, the latent heat of vaporization is 2100 kJ.kg^{-1}, and the specific heat of the solution is 4 kJ.kg^{-1}.K^{-1}). Find the rate of heat that has to be supplied to the evaporator?

   Answer: 3.06 x 10^5 kJ.min^{-1}

8. A single effect evaporator is used to concentrate a fruit juice from 10% solids to 40% solids. The feed enters at a rate of 100 kg.h^{-1} at 15°C and is evaporated under 47.7 kPa pressure (at 80°C). If the steam is supplied at 169 kPa (115°C), determine the quantity of steam used per hour. Assume c$_p$ of the fruit juice and water as 3.96 and 4.186 kJ.kg^{-1}.K^{-1}, respectively. The latent heat of vaporization values of juice and steam are 2309 and 2217 kJ.kg^{-1} at 115°C, respectively. Also assume that the boiling point of juice does not change in the evaporator.

   Answer: 93.24 kg.h^{-1}

9. In the above problem, find the total heat transfer area required if the overall heat transfer coefficient is 2600 W.m^{-2}.K^{-1}. If 2.4 m and 5 mm are the length and diameter of tube, respectively, used in the heat exchanger, determine the heat transfer area and total number of tubes in the heat exchanger.

   Answer: heat transfer area = 0.357 m², number of tubes = 9.

10. In a two-effect evaporator, milk is concentrated from 5% to 50% total soluble solids by weight. Steam is supplied as heating medium in the first effect, where

it condenses at 100°C. The partially concentrated milk from the first effect goes to the second effect, which is maintained so that the boiling point of water is 40°C (in the second effect). The boiling point rise in the first and second effects have been found to be 1°C and 18°C. The heat exchanger surface area in both the effects is 90 m². The heat transfer coefficients are 1100 and 275 $W.m^{-2}.K^{-1}$, respectively. Assume that the total heat transfer is the same for both the effects and the heat required for vaporization of water is 2400 $kJ.kg^{-1}$. Find the amount of water evaporated and the feed rate into the evaporator.

Answer: Feed: 0.752 kg. s^{-1}; vapor 0.676 kg.s^{-1}

11. A triple effect evaporator concentrates a liquid with no appreciable boiling point elevation. The thermal loads and areas in each effect are equal. If the temperature of the steam in the first effect is 395 K and the boiling point in the third effect is 325 K, calculate the approximate boiling points in all the effects. The overall heat transfer coefficients may be taken as 3.1, 2.3 and 1.1 $kW.m^{-2}.K^{-1}$ in the three effects, respectively.

Answer: 381.5 K, 363.2 K and 325 K

12. Cane sugar juice is concentrated from 10 wt% to 35 wt% solids in an evaporator to get 100 kg cane product/h using saturated steam at 140°C, 1 atm pressure (latent heat of vaporization of water = 2145 kJ.kg^{-1}). The liquid boils at 105°C (no boiling point rise) and gives vapor at same temperature (specific enthalpy of water vapor = 2684 kJ.kg^{-1}). The juice enters the evaporator at its boiling temperature. The overall heat transfer coefficient is 3000 $W.m^{-2}.K^{-1}$. The heat capacity of feed and product may be taken as that of water, i.e., 4.187 $kJ.kg^{-1}.K^{-1}$. Find (a) the steam economy of this system, and (b) the area of the heat exchanger.

Answer: (a) 0.96 (b) 1.48 m²

13. A single-effect vacuum evaporator has 100 tubes of 25 mm diameter. One thousand kg of milk per hour with 15% TS is concentrated to 20% TS in the evaporator. Film heat transfer coefficients on either side of the tube are 5000 and 800 $W.m^{-2}.K^{-1}$. Thermal conductivity of 1.5 mm thick SS tubes is 15 $W.m^{-1}.K^{-1}$. Latent heat of vaporization under vacuum is 2309 kJ.kg^{-1}. What will be the height of each tube if the temperature difference across the tube wall is maintained at 10°C?

Answer: 3.164 m

14. A fruit juice with a negligible boiling point rise is being evaporated using saturated steam at 121.1°C in a triple effect evaporator having equal area in each effect. The pressure of the vapor in the last effect is 25.6 kPa absolute and the corresponding saturation temperature is 65.7°C. The heat transfer coefficients are $U_1 = 2760$, $U_2 = 1875$ and $U_3 = 1350$ $W.m^{-2}.K^{-1}$. Find the boiling point of the juice in the first effect.

Answer: 108.82°C

Objective Questions

1. What are the basic parts of an evaporator?
2. Write short notes on
   a. Scraped surface evaporator;
   b. Rotary evaporator;
   c. Barometric leg;
   d. Entrainment separator;
   e. Steam economy;
   f. Vapor recompression;
   g. Boiling point rise during evaporation;
   h. Evaporator accessories.
3. Explain the differences between
   a. Drying and evaporation;
   b. Evaporation and crystallization;
   c. Evaporation and distillation;
   d. Evaporation and condensation;
   e. Short tube evaporator and long tube evaporator;
   f. Falling film evaporator and rising film evaporator.
4. What are the different methods of improving the steam economy in an evaporation process?
5. Explain the principle of multiple effect evaporation. Is it possible to use a very large number of effects in a multiple effect evaporation process?

BIBLIOGRAPHY

Ahmad, T. 1999. *Dairy Plant Engineering and Management* 4th ed.. Kitab Mahal, New Delhi.

Brennan, J.G. (Ed.) 2006. *Food Processing Handbook* 3rd ed. WILEY-VCH Verlag GmbH & Co. KGaA, Weinheim.

Brennan, J.G., Butters, J.R., Cowell, N.D. and Lilly, A.E.V. 1976. *Food Engineering Operations* 2nd ed. Elsevier Applied Science, London.

Charm, S.E. 1978. *Fundamentals of Food Engineering* 3rd ed. AVI Publishing Co., Westport, Connecticut.

Dash, S.K. and Sahoo, N.R. 2012. *Concepts of Food Process Engineering.* Kalyani Publishers, New Delhi.

Earle, R.L. 1983. *Unit Operations in Food Processing* 2nd ed. Pergamon Press, Oxford.

Earle, R.L. and Earle, M.D. 2004. *Unit Operations in Food Processing* (Web Edn.). The New Zealand Institute of Food Science & Technology, Inc., Auckland. https://www.nzifst .org.nz/resources/unitoperations/index.htm

Fellows, P.J. 2000. *Food Processing Technology.* Woodhead Publishing, Cambridge, UK.

Geankoplis, C.J.C. 2004. *Transport Processes and Separation Process Principles* 4th ed. Prentice-Hall of India, New Delhi.

Heldman, D.R. and Hartel, R.W. 1997. *Principles of Food Processing.* Aspen Publishers, Inc., Gaithersburg.

Heldman, D.R. and Singh, R.P. 1981. *Food Process Engineering.* AVI Publishing Co., Westport, Connecticut.

https://www.ou.edu/class/che-design/design%201-2013/Flash %20Design.pdf

Karel, M. 1975. Concentration of foods. In: O.R. Fennema (ed.) *Principles of Food Science*: Part 2- *Physical Principles of Food Preservation*. Marcel Dekker, New York, 266–308.

Kessler, H.G. 2002. *Food and Bio Process Engineering* 5th ed., Verlag A Kessler Publishing House, Munchen, Germany.

Leniger, H.A. and Beverloo, W.A. 1975. *Food Process Engineering*. D. Reidel, Dordrecht.

Lewicki, P.P. and Kowalczyk, R. 1980. Food process engineering. In: P. Linko, Y Malkki, J. Olkku, and J. Larinkari (eds.) *Food Processing Systems*: Vol.1. Applied Science, London, 501–505.

McCabe, W.L., Smith, J.C. and Harriot, P. 1993. *Unit Operations of Chemical Engineering* 5th ed., McGraw-Hill, Inc., New York.

Olsson, B. 1988. Recent advances in evaporation technology. In A. Turner (ed.) *Food Technology International Europe*. Sterling Publications International, London, 55–58.

Rao, D.G. 2009. *Fundamentals of Food Engineering*. PHI Learning Pvt. Ltd., New Delhi.

Rielly, C.D. 1997. Food rheology. In: P.J. Fryer, D.L. Pyle and C.D. Rielly (eds.). *Chemical Engineering for the Food Processing Industry*. Blackie Academic and Professional, London, 195–233.

Sahay, K.M. and Singh, K.K. 1994. *Unit Operation of Agricultural Processing*. Vikas Publ. House, New Delhi.

Singh, R P. and Heldman, D R. 2014. *Introduction to Food Engineering*. Academic Press, San Diego, CA.

Thijssen, H.A.C. 1970 Concentration processes for liquid foods containing volatile flavours and aromas. *Journal of Food Technology* 5: 211–229.

Toledo, R.T. 2007. *Fundamentals of Food Process Engineering* 3rd ed. Springer, New York.

Watson, E.L. and Harper, J.C. 1989. *Elements of Food Engineering*, 2nd ed. Van Nostrand Reinhold, New York.

18

Distillation

Distillation is a process of separating different components of a liquid mixture. The objective is to get both the components in purified or concentrated forms. It consists of a mechanism to energize liquid molecules so that these vaporize and leave the mixture; the vaporized molecules are then condensed back to liquid form. What was vaporized and subsequently condensed is called the *distillate* and the material left behind is known as the *residue*. The liquid mixture may contain only two components (binary mixture) or may have multiple components. The mechanism of energizing the liquid molecules may be heating and/or vacuuming. The liquid that has the maximum volatility vaporizes first. In a multi-component mixture, the vapors are produced in conformity with the volatilization characteristics of the components; the vapor will also be a mixture of different components, higher fraction of the vapors from the component having higher volatility.

Applications of distillation in food industries include manufacturing of potable alcohols like whisky, brandy, wine, and beer. The fermented slurry undergoes several distillation processes depending upon the characteristics of the slurry and the final product using batch and/or continuous distillation. Last traces of solvents from solvent extracted oils are removed using steam distillation or nitrogen stripping. Flavors from fruit juices and extracts are vaporized through vacuum evaporation and then the vapors are passed through a distillation column to concentrate them. Steam distillation, either alone or in combination with vacuum distillation, is used for separating essential oils from herbs, flowers, seeds, etc. Deodorization of certain processed products such as oils, fats, and plant-based milks are also carried out using appropriate methods of distillation.

18.1 General Principles

18.1.1 Pressure of a Mixture of Two Liquids

In the context of a mixture of two liquids, A and B, with a considerable difference in their boiling temperatures, the one with lower boiling point, A, would vaporize faster than the one with higher boiling point, B. The liquid A is said to be more volatile than the liquid B. When the mixture is boiled, the vapors will also be a mixture of A and B and the proportion of A in the vapor would be significantly higher as compared to the component B. For example, if A in the liquid has a mole fraction of 80%, B will have 20% mole fraction. If x_A is the fraction of pure A in the liquid and the vapor pressure of pure A is p_{pA}, then as per Raoult's law,

$$p_A = p_{pA} \, x_A \qquad (18.1)$$

The law states that the partial pressure of one of the components of the ideal solution is equal to the partial pressure of that component in its pure form, at the given temperature, multiplied by its mole fraction in the solution.

According to Dalton's law,

$$P_{total} = p_A + p_B$$

Taking both Raoult's and Dalton's law together,

$$P_{solution} = p_{pA} \, x_A + p_{pB} \, x_B \qquad (18.2)$$

18.1.2 Vapor-Liquid Equilibrium

Vapor-liquid equilibrium (VLE) in a binary mixture can be achieved either under constant pressure or under constant temperature condition. Figure 18.1 presents the VLE for an ideal solution (one that follows Raoult's law) under constant pressure condition. An ideal mixture is one that obeys Raoult's law, i.e., the mixture is a solution of two completely miscible volatile liquids.

Figure 18.2 is the *VLE diagram* for an ideal solution for constant temperature condition. In these figures, BPC is a plot of mole fraction of the more volatile component in the liquid phase versus temperature in a constant pressure vaporization. On the other hand, BPC is a plot of mole fraction of the more volatile component in the liquid phase versus pressure in a constant temperature vaporization. y and x are the respective mole fractions in vapor and liquid phases. Similarly, DPC is a plot of mole fraction of the more volatile component in the vapor phase versus temperature in a constant pressure vaporization or versus pressure in a constant temperature vaporization.

18.1.3 Relative Volatility of Binary Mixtures

Volatility is the ease with which the liquid turns into vapors. At a given temperature and pressure, a liquid with higher volatility is likely to be in vapor form whereas a liquid with lower volatility would still remain in liquid form. Some solids, at specific temperature and pressure, directly convert into vapors and are therefore known as volatile solids.

Relative volatility of the two constituents A and B in a binary mixture is represented as follows:

$$\alpha_{AB} = \frac{y_A / x_A}{y_B / x_B} \qquad (18.3)$$

where, y and x are the respective mole fractions in vapor and liquid phases.

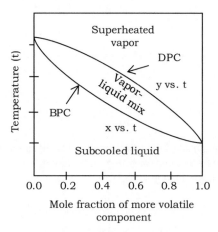

BPC: Boiling point curve

FIGURE 18.1 VLE diagram of a binary liquid at constant pressure,

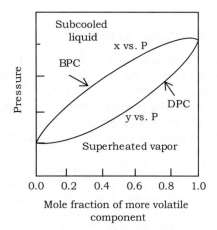

DPC: Dew point curve

FIGURE 18.2 VLE diagram of a binary liquid at constant temperature

The relative volatility of a mixture varies with its composition. For a binary mixture, $x_B = 1 - x_A$. Considering that A is more volatile than B, the relative volatility equation can be rewritten as follows by dropping the subscript A.

$$y = \frac{\alpha_{av} x}{1 - (\alpha_{av} - 1)x} \qquad (18.4)$$

where α_{av} is the average relative volatility.

If Raoult's law is applicable, $p_A = P.y_A$ and $p_B = P.y_B$ and the relative volatility can be expressed as follows:

$$\alpha_{AB} = \frac{p_A / x_A}{p_B / x_B} \qquad (18.5)$$

where p_A and p_B are the partial pressures of the components A and B in the vapor phase. The value of relative volatility should be much higher than 1.0 to permit easy distillation. The separation would be difficult if α_{AB} is close to unity. Examples of relative volatility for a few binary mixtures are given in Table 18.1.

TABLE 18.1

Examples of relative volatility for a few binary mixtures

Components (the first one is more volatile than the second)	Relative volatility
Benzene–toluene	2.34
Benzene–p-xylene	4.82
Hexane–p-xylene	7.0
Chloroform–acetic acid	6.15
Methanol–ethanol	1.56

Boiling points: Benzene, 80.1°C; toluene, 110.6°C; p-xylene, 138.3°C; hexane, 68.7°C; chloroform, 61.2°C; acetic acid, 118.1°C; methanol, 64.6°C; ethanol, 78.4°C.

Example 18.1 At a given temperature, the vapor pressures of pure methanol and pure ethanol are 76 kPa and 43 kPa, respectively. If an ideal solution contains two moles of methanol and one mole of ethanol at that temperature, what would be the total vapor pressure of the solution?

SOLUTION:

Here, the mole fractions of methanol and ethanol are 2/3 and 1/3.

Hence, partial pressure of methanol
= 76 × 2/3 = 50.66 kPa

partial pressure of ethanol = 43 × 1/3 = 14.33 kPa.
Thus, the total partial pressure = 50.66 + 14.33 = 64.99 k Pa.

The total partial pressure of the solution will always lie between the partial pressures of the two pure liquids.

The total partial pressure at a given temperature for different compositions of an ideal solution of two pure liquids can be represented graphically as shown in Figure 18.3.

It may be noted that a liquid having higher vapor pressure at a given temperature indicates its greater volatility as compared to that of the liquid having lower vapor pressure. Also, the liquid with higher vapor pressure will have a lower boiling point.

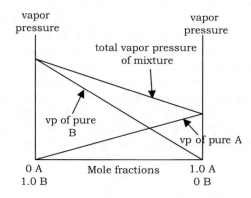

FIGURE 18.3 Total partial pressure at a given temperature for different compositions of an ideal solution of two pure liquids

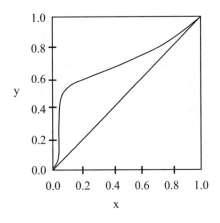
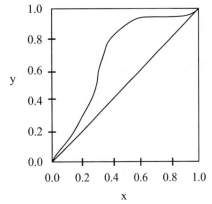

FIGURE 18.4 Vapor-liquid equilibrium diagrams for non-ideal mixtures

The forms of the VLE for non-ideal mixtures would be quite asymmetric, as shown in Figure 18.4.

Azeotrope

An azeotrope is a liquid mixture that produces the vapors of the same composition as that in the liquid when vaporized. An azeotropic mixture, therefore, does not lead to separation when subjected to one of the regular practices except vacuum distillation. Minimum boiling and maximum boiling are the two types of azeotropes. Ethanol-water mixture (at 1 atm, 89.4% mole fraction, 78.4°C) is an example of a minimum boiling type of azeotropic mixture. Hydrochloric acid – water mixture (11.1% mole fraction HCl, 110°C at 1 atm) is an example of maximum boiling type azeotrope. The maximum-boiling type is less common. The VLE curves for azeotropic mixtures are given in Figure 18.5.

18.2 Methods and Equipment

18.2.1 Simple or Double Distillation

A typical distillation process involves boiling the given mixture of volatile liquids in a still (also called a *reboiler*) so that the volatile liquids vaporize. A simple distillation process is shown in Figure 18.6 in which the vapor produced in the first stage is condensed in the second stage and

collected. This can be used, for example, for obtaining pure water from sea water which contains about 3%–4% salt and some other impurities.

In this situation it is assumed that there is nothing volatile in the feed water other than water itself. Even if there is a volatile component in the feed water, it would have a boiling point of say 170°C or more at atmospheric pressure. The amount of distillate over a short interval of time is estimated as follows:

$$m = q / h_v \qquad (18.6)$$

where, q is the total amount of heat transferred to the boiling liquid and h_v is the latent heat of vaporization of water. The above setup could be a continuous process where a supply of sea water is boiled enroute to the condenser and the un-vaporized part of the feed along with some water would need to flow out of distillation plant.

In the case of *double distillation*, the distillate from single-stage distillation, containing a significant amount of the other less vaporized components, is put through the distillation process again for improving the purity of the distillate.

18.2.2 Batch or Differential Distillation

In a batch system, the batch is charged into the flask (the still) and the distillation process is carried out until the complete charge is exhausted. Then, another batch is charged and the

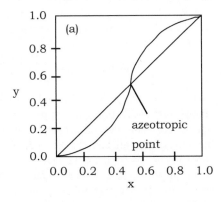
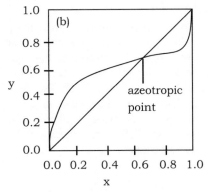

FIGURE 18.5 VLE curves for azeotropic mixtures for (a) maximum boiling azeotrope, (b) minimum boiling azeotrope

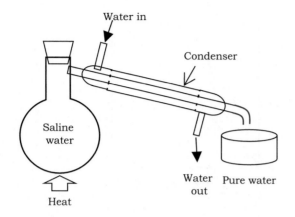

FIGURE 18.6   A simple distillation process

distillation is carried out. Obviously, the batch process is an unsteady process.

Suppose L is the amount of mixture in the still at any given point of time and after the time $d\theta$, a small amount of vapor dL is produced where the fraction of A is y. The concentration of A has changed from x to (x-dx). Thus, the material balance on the more volatile component A can be written as:

Initial amount = amount converted into distillate + amount left in the still

$$x\ L = (x-dx)\ (L-dL) + y\ dL \qquad (18.7)$$

$$\text{i.e. } x\ L = x\ L - dx\ L - x\ dL + dx\ dL + y\ dL \qquad (18.8)$$

Being a very small quantity, dx.dL may be neglected. The equation simplifies as follows:

$$L\ dx = y\ dL - x\ dL = (y-x)\ dL$$

$$\frac{dL}{L} = \frac{dx}{y-x} \qquad (18.9)$$

If the quantity of volatile component A at the start and end of the process can be taken as L_1 and L_2 and the mole fractions as x_1 and x_2, then the above equation can be integrated from L_1 to L_2 and from x_1 to x_2. The result is as follows:

$$\ln\left(\frac{L_1}{L_2}\right) = \int_{x_2}^{x_1} (y-x)^{-1}.dx \qquad (18.10)$$

This equation is known as the *Rayleigh equation* and can be integrated either graphically or numerically by using the data from the equilibrium curve. For ideal mixtures, the Rayleigh equation could be evaluated using relative volatility. If the feed contains n_A moles of A and n_B moles of B, the total feed amount is $n_A + n_B$ moles. The ratio n_A / n_B is equal to x_A/x_B. When a small quantity of feed, dn, undergoes distillation, the change in A is dn_A and the change in B is dn_B. The expression for relative volatility changes as follows:

$$\frac{dn_A/dn}{dn_B/dn} = \frac{dn_A}{dn_B} = \alpha_{AB}\left(\frac{n_A}{n_B}\right) \qquad (18.11)$$

$$\text{Or} \quad \frac{dn_A}{n_A} = \alpha_{AB}\left(\frac{dn_B}{n_B}\right) \qquad (18.12)$$

During distillation, moles of more volatile component A change from no_A to n_A and that of the less volatile component B change from no_B to n_B. Integration of the above equation within limits (the limits of integration for A are from no_A to n_A and the limits of integration for B are from no_B to n_B) yields the following expression.

$$\ln\left(\frac{n_A}{no_A}\right) = \alpha_{AB} \ln\left(\frac{n_B}{no_B}\right) \qquad (18.13)$$

$$\left(\frac{n_A}{no_A}\right) = \left(\frac{n_B}{no_B}\right)^{\alpha_{AB}} \qquad (18.14)$$

18.2.3 Single-Stage Continuous Distillation

The process of distillation could be made continuous by permitting a continuous supply of feed to the distillation apparatus and removal of the undistilled feed from the apparatus. The feed is preheated to a particular temperature and pressure before allowing it to enter the continuous distillation unit; a part of the feed might have vaporized already at the inlet feed temperature. The heated feed then goes to the separator through a pressure reducing valve so that the vapors and the liquid get separated. The vapors are collected at the top and condensed to obtain the distillate. The un-vaporized liquid, also known as bottoms, is collected, and may be recycled. If A is the more volatile material in the liquid mixture, then the amount of vaporization affects the concentration of A in the liquid and the vapor.

Let f be the molal fraction of feed that is vaporized in the separator. Then, (1-f) is the remaining molal fraction of the liquid after vaporization. If all the liquid feed is completely vaporized, f =1 and there is no outflow of the un-vaporized liquid feed.

$$x_F = (f)\ y_D + (1-f)\ x_B \qquad (18.15)$$

where, x_F is the mole fraction of A in the feed; y_D is the mole fraction of A in the vapor; and x_B is the mole fraction of A in the liquid. The above equation can be rearranged as follows:

$$(f)\ y_D = x_F - (1-f)\ x_B$$

$$\text{Therefore,} \quad y_D = -\left(\frac{1-f}{f}\right)x_B + \frac{x_F}{f} \qquad (18.16)$$

Consider an open surface of liquid water in contact with the atmospheric air. At the given temperature and pressure, some liquid water molecules on the surface may gain energy from the atmospheric air and change their phase to vapor. Some vapor molecules from the air may lose energy and fall back on the liquid surface. Under a state of equilibrium, the net change from the liquid surface to the air and from the air to the liquid surface is zero. If the water temperature is raised and/or air pressure is reduced, more water changes its phase.

Conversely, if the water is cooled in relation to the air and/or the air pressure is increased, conversion of liquid water to vapor is reduced. While vaporization requires energy, condensation releases energy. Clearly, under equilibrium, the energy of vaporization is equal to the energy from condensation. If dry air has a pressure p_a and the pressure exerted by the water vapor is p_v, then the total pressure p_t, considering the air and the vapor as ideal gases, is given as follows:

$$p_t = p_a + p_v \qquad (18.17)$$

Vapor pressure can be computed using the Clausius-Clapeyron equation as follows:

$$\ln\left(\frac{p_v}{p_{v1}}\right) = \frac{\lambda}{R}\left(\frac{1}{T_1} - \frac{1}{T}\right) \qquad (18.18)$$

where, p_v and p_{v1} are the partial vapor pressures in pascal at absolute temperatures T and T_1. λ and R are the latent heat of vaporization and the universal gas constant, respectively. Another equation to estimate the vapor pressure is *Antoine equation*, as given below:

$$\ln(p_v) = A + [B/(T + C)] \qquad (18.19)$$

where A, B, and C are the coefficients that depend on the volatile material and the temperature range of interest. For example, if water is the volatile liquid, the coefficients, A, B and C for the temperature range of 284 K to 440 K are 23.1962, 3816.44, and -46.13, respectively.

For a mixture of two liquids in which one of the liquids is more volatile than the other, the boiling point of the more volatile component will depend on the mole fraction of it in the mixture at a constant pressure. If the temperature is held constant, the pressure at which the more volatile fraction boils off depends on its mole fraction in the mixture. For a given mixture composition, the boiling temperature of the more volatile liquid component is called the *bubble point temperature* (BPT), whereas the temperature at which the more volatile vapor component begins to condense is called the *dew point temperature* (DPT). The boiling point temperatures for different mole fractions, when plotted graphically, yield the *boiling point curve* (BPC). Similarly, the *dew point curve* or DPC represents the dew point temperatures of the more volatile component in the mixture for different compositions of the mixture.

If the liquids differ considerably in their volatility, the vapors will contain essentially the more volatile material, which will condense in the condenser. However, if the volatilities of the two liquids are comparable, the vapors will contain both the components and it will be very difficult to obtain pure fractions. The more volatile fraction in the liquid is called the *low boiler*. The liquid remaining in the reboiler is rich in the *high boiler* or *bottoms*. A portion of the distillate is returned back to the distillation column so that the low boiler in the reflux gets vaporized when coming into contact with the rising vapors in the column. The purpose of the reflux is to increase the efficiency of distillation. The purity of the bottoms can be improved through rectification.

18.2.4 Batch Distillation with Reflux

Batch distillation without any reflux does not produce high quality products. Initially, the vapor contains higher fraction of more volatile component but, as time passes, the vapor contains more of the lower volatile component. Thus, the distillate composition varies continuously throughout the distillation process. If it is desired that the distillate composition remains the same, the option is to return a portion of distillate to the still, smaller amount initially and then gradually increasing as the process moves to the completion stage.

McCabe-Thiele Graphical Method

Performance of a batch distillation system with reflux can be analyzed using the *McCabe-Thiele graphical method* with the operating line corresponding to one of the rectifying sections of a continuous distillation process. The reflux ratio required to obtain the top product of the same composition is obtained by trial-and-error method. Alternatively, the value of the reflux ratio could be fixed and the composition of the top product could be allowed to vary until the purity of the top product reaches a certain value.

$$y = \left(\frac{R}{1+R}\right)x + \left(\frac{1}{1+R}\right)x_d \qquad (18.20)$$

where, R is reflux ratio (reflux/distillate), x is mole fraction of more volatile component (MVC) in the liquid phase at any time, and x_d is the mole fraction of MVC in the distillate. As mentioned before, y is the mole fraction of the more volatile liquid in vapor phase.

The process of seeking McCabe–Thiele graphical solution for binary solutions to determine the required number of trays is based on the following assumptions.

1. The liquid mixture is binary;
2. Vaporization heats are equal for the two liquids. It means that the heat required to vaporize one mole of one of the liquids is equal to the heat of condensation of the other liquid;
3. The trays have 100% efficiency. In reality, it is never achieved. However, it permits the determination of the minimum number of trays and then the actual number of trays is determined by taking the tray efficiency into consideration.

The data, which are essentially required, include the top and bottom products' composition, vapor-liquid equilibrium (VLE) data, composition, boiling point, dew point, and actual temperatures of the feed.

Now, the procedure is outlined as follows.

a) Plot the VLE data where x-axis represents the mole fraction of the MVC in liquid phase and y-axis represents the data for vapor phase;
b) Connect the bottom left corner of the VLE plot with the upper right corner, i.e., 45° or diagonal line;

c) Mark the MVC compositions of the top product, bottom product, and feed on the x-axis and extend vertical lines from these points up to the diagonal line;

d) The feed can have the condition somewhere between pure liquid and pure vapor, including the extreme conditions. The slope of the feed line starting from the diagonal and extending until the VLE curve is given by [q / (q-1)], where q is the mole fraction of MVC in the feed;

e) Draw the rectifier operating line starting from the intersection of the top product line and the diagonal to the feed operating line. The slope of this line is equal to the value of the reflux ratio;

f) Draw the stripping operating line starting from the intersection of bottoms product line and the diagonal and then joining this point with the point where the rectifying operating line meets the feed operating line;

g) Starting from the intersection of the top product's operating line and the diagonal, draw a horizontal line until it meets the VLE line. From that point, move vertically down until it meets the rectifying/ stripping operating line;

h) Finally, count the number of steps required to reach the bottoms product line.

The distillation time can be calculated if boil-up rate (mol per unit time) is known.

$$\text{Time} = (n_0 - n_t)\left(\frac{1+R}{V}\right) \qquad (18.21)$$

where, n_0 is the number of mole of MVC in the beginning in still, n_t is the number of moles at the end of the distillation in the still, and V is the boil-up rate. The boil up rate is expressed in mol. s^{-1}.

18.2.5 Continuous Distillation with Rectification

Flash distillation is effective in separating the liquids in a mixture that boils at widely different temperatures. However, the separation becomes difficult when the boiling points of different liquids in a feed have small differences. The distillate and the residual liquid are not pure. The rectification principle is used here to facilitate the separation in such situations.

The basic action in rectification is to recycle a fraction of the distillate into the distillation column so that some of the more volatile liquid in the distillate gets vaporized by the enthalpy of the distillation vapors and the un-vaporized liquid. There are several plates in the distillation chamber; any specific plate would receive un-vaporized liquid from the plate above it and will allow the liquid to flow down to the plate just below it. The vapor from this plate will move upwards (Figure 18.7).

Thus, if we number of plates starting with 1 from top, then for any plate "n," the different components contributing to the mass balance are the mass of vapor rising up from plate n+1 and to plate n-1 and the mass of liquid flowing down from plate n-1 and to plate n+1.

The vapors and the liquid at plate n are in equilibrium, whereas the vapors coming from the lower plate and the liquid coming from the upper plate are not in equilibrium. The temperatures in the column are the highest close to the reboiler and lowest at the top of the column. Moreover, the vapors and the liquid at the top of the column are richer in the more volatile component and possess the lowest concentrations at the bottom of the column. The recycled fraction of the distillate is termed the *reflux*.

Feed is normally supplied directly to the reboiler. However, if the feed is admitted at the plate 'n' somewhere in the central portion of the column above the reboiler, the feed flowing down is subjected to rectification by the vapors rising from the reboiler and the plates below the feed plate and the process is known as *stripping*. The concept is that before reaching the

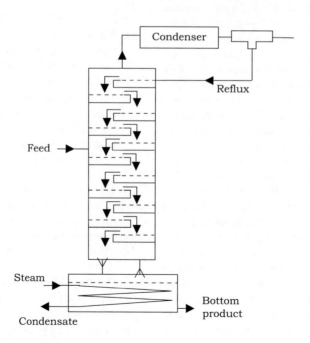

FIGURE 18.7 Schematic of a continuous distillation column

reboiler, the feed is partially stripped off of some of the volatile fraction. As could be appreciated, both rectification and stripping are expected to make the distillation process more efficient. The section of the column above the feed plate is called the *rectification section* and the section consisting of the feed plate and the plates below it is called the *stripping section*.

Material balances in Plate Columns

Considering a binary mixture of liquids, let feed be F, which yields distillate D and bottoms B. The concentration of more volatile liquid in the feed is x_F, in distillate is x_D, and in the bottoms is x_B. Then, the total material balance is expressed as follows:

$$F = D + B$$

The material balance for the more volatile component A is as follows:

$$F. x_F = D. x_D + B. x_B$$

Using the above two equations, we obtain

$$\frac{D}{F} = \frac{x_F - x_B}{x_D - x_B}, \text{ and } \frac{B}{F} = \frac{x_D - x_F}{x_D - x_B} \qquad (18.22)$$

18.2.6 Flash Distillation

It is a simple, continuous, single stage distillation process, as discussed earlier. The liquid feed is pumped through a heating section to raise the temperature and the enthalpy of the feed, and then it is allowed to pass through a pressure reducing valve. The sudden reduction of pressure facilitates instant vaporization of the heated liquid mixture. The vapors and the liquid are in equilibrium. The vapor is collected and condensed. The method of distillation is called flash distillation because the fluid pressure is suddenly reduced by the throttling valve to rapidly vaporize the more volatile fraction of the mixture.

18.2.7 Vacuum Distillation

When the vapors from the reboiler are acted upon by the vacuuming machinery coupled with the condenser, the distillation efficiency increases. Vacuum lowers the boiling point and this phenomenon is utilized to carry out distillation for those fluids that deteriorate in quality at higher temperatures. Conversely, those fluids that boil at higher temperatures could be distilled at lower temperatures. Vacuum distillation along with steam distillation could be utilized for certain class of fluids (Figure 18.8).

18.2.8 Short Path Distillation

Vapors from the still are condensed soon after vaporization to avoid longer distance travel to a condenser (Figure 18.9). Pressure in the system is maintained below atmospheric pressure. This method is specifically suited for high molecular weight organic compounds. Thermally sensitive fluids can be separated using this method. Concentration of vitamin E (tocopherol and tocotrienol) is carried out this way.

18.2.9 Packed Bed Distillation

The distillation column has a packed bed offering the heat and mass transfer opportunities for not only distillation but also for liquid-liquid extraction and gas absorption. The liquid mixture flows down the column and the vapor moves up. Such columns are utilized for large-scale operations. A packed column works best when the liquid and the vapors are distributed uniformly through the column.

There are several advantages of the packed bed columns in comparison to plate-type distillation systems: better handling of toxic liquids, more cost effectiveness for larger capacities, lower pressure drops, more suitability for working with foaming systems, and reduced holding periods for the liquids and vapors. Packing materials could either be structured or be random. Design procedure for a packed bed distillation system includes the selection and the size and shape of the packing material, sizing the column height and diameter, selection and design of packing support, and selection of liquid distributor/ redistributor.

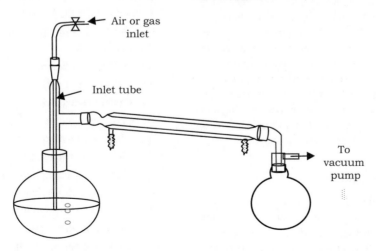

FIGURE 18.8 Schematic of a vacuum distillation process

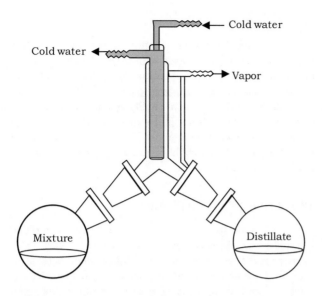

FIGURE 18.9 Schematic diagram of a short path distillation process

18.2.10 Azeotropic Distillation

A liquid mixture is termed *azeotropic* when its constituents cannot be separated because the vapors contain the constituents in almost the same proportions as they exist in the liquid form. A well-known example is a mixture of ethanol and water. Azeotropic distillation is a process to break down the azeotrope with the help of another volatile liquid, called as entrainer or a solvent or *mass separating agent* (MSA), to form lower-boiling heterogenous azeotropes. Common entrainers used in ethanol-water mixture are benzene, toluene, hexane, cyclohexane, pentane, heptane, isooctane, and acetone. The criteria for the selection of entrainers include carcinogenicity, flammability, and other health /safety concerns.

18.2.11 Multi-Component distillation

When a mixture contains several compounds and it is required to separate all the compounds, the process is called multi-component distillation. When the number of components in the mixture are three or more, the analysis of the process of distillation becomes more and more complex. The calculation of equilibrium stages involves enthalpy and mass balances and vapor-liquid equilibria. While the enthalpy balance is one and represents it over the entire column, mass balances are possible for individual stages as well as for the entire column. The issue of phase equilibria for a multi-component mixture is more tedious in comparison to a binary mixture because equilibria depend on temperature, which varies from one stage to another. Analysis of a multi-component mixture, therefore, is generally carried out with the help of digital computers.

18.2.12 Fractional Distillation

A complex mixture containing several components, having closely spaced boiling points, is separated into its components or fractions. A tall fractionating column is placed over the still and fitted with condensers at different heights to collect the different compounds. Petroleum crude oil distillation is done this way. Concentration of flavors and essential oils and deodorization of fats and oils are applications in food industry.

18.2.13 Steam Distillation

As discussed earlier, vacuuming helps the volatile material to boil off at a lower pressure; hence, at a lower temperature. In addition, if steam is added to the mixture, its partial pressure would permit further reduction in the requirement of the partial pressure requirement for the volatile fraction. Thus, reduction in the pressure of the surrounding atmosphere and introduction of steam together permit the distillation process at lower temperatures.

If total pressure required to boil the volatile component off is P and the steam pressure is p_s, the pressure required for distillation is $(P-p_s)$. The ratio of molecules of steam required to vaporize the volatile fraction is equal to the ratio of the partial pressures of the steam and the volatile fraction.

$$\frac{p_A}{p_s} = \frac{P - p_s}{p_s} = \frac{w_A/M_A}{w_s/M_s} \qquad (18.23)$$

Therefore, $\dfrac{w_A}{w_s} = \left(\dfrac{P-p_s}{p_s}\right)\left(\dfrac{M_A}{M_s}\right)$ (18.24)

In the above equation, M_A and M_s are the molecular weights of the volatile material and the steam. The fraction of the volatile liquid in the distillate is w_A, and w_s is the fraction of steam in the distillate. Knowing the concentration of the volatile liquid in the mixture, it is possible to determine the quantity of steam required for the distillation. Steam distillation has high relevance in the distillation of volatile oils and fragrances. It is also used to eliminate unwanted odors and foul smells.

Check Your Understanding

1. Explain the applications of distillation process in commercial food processing.
2. Draw and explain the VLE diagrams of a binary liquid at constant pressure and at constant temperature.
3. Define/ explain the following:
   (a) Dalton's law of partial pressure;
   (b) Raoult's law;
   (c) Double distillation;
   (d) Differential distillation;
   (e) Rayleigh equation;
   (f) Flash distillation;
   (g) Distillation with reflux;
   (h) McCabe-Thiele method;
   (i) Continuous distillation with rectification;
   (j) Short path distillation;
   (k) Packed bed distillation;
   (l) Azeotropic distillation;
   (m) Multi-component distillation;
   (n) Fractional distillation.

BIBLIOGRAPHY

Earle, R. L. and Earle, M. D. 2004. *Unit Operations in Food Processing* (Web Edn.) The New Zealand Institute of Food Science & Technology, Inc., Auckland. https://www.nzifst.org.nz/resources/unitoperations/index.htm

Fellows, P. J. 2000. *Food Processing Technology*. Woodhead Publishing, Cambridge, UK.

Ghosal, S. K., Sanyal, S. K. and Dutta, S. 2004. *Introduction to Chemical Engineering*, Tata McGraw Hill Book Co., New Delhi.

Hines, A. L., Maddox, R. N. 1985. *Mass Transfer: Fundamentals and Applications* 1st ed. Prentice Hall, NJ.

McCabe, W. L., Smith, J. C. and Harriot, P. 1993. *Unit Operations of Chemical Engineering* 5th ed. McGraw-Hill, Inc., New York

Rao, D. G. 2009. *Fundamentals of Food Engineering*. PHI Learning Pvt. Ltd., New Delhi.

Seader, J. D., Henley, E. J. and Roper D. K. 2011. *Separation Process Principles*, John Wiley & Sons, Inc., NJ.

Treybal, R. E. 1981. *Mass Transfer Operations* 3rd ed. McGraw-Hill, Singapore.

19

Crystallization

Crystallization is that technique of material separation where one material is permitted to form crystals in a liquid for subsequently separating it out. It is one of the several processes of material separation, whereby crystals are formed, as opposed to amorphous powdery material, of the solute in a homogenous liquid solution or melt. An example of crystal formation in a vapor is snow formation upon cooling of the water vapor-air mixture. After the formation and/or the growth of the crystals is complete, the remaining solution is called the *mother liquor*. The mother solution together with the crystals, existing in equilibrium, is known as *magma*. The crystals are separated from the mother liquor and dried for further handling, packaging, and storage. In the domain of food processing, crystallization finds its utility in the sugar industry where not only sucrose but also glucose, lactose, and other types of sweeteners are crystalized. Crystallization is also used in the manufacture of salt, citric acid, and ice cream.

19.1 Methods and Principles

19.1.1 Process of Crystallization

Crystallization is a four-step process: (1) creation of a supersaturated state, (2) formation of nuclei (nucleation), (3) growth of nuclei (propagation), and (4) crystal perfection or continued growth (maturation).

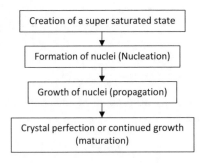

Steps in Crystallization Process

Super-Saturation and Formation of Nuclei

If salt or sugar is dissolved in a given amount of water at a given temperature, it does not dissolve beyond a certain amount and, instead, it starts precipitating at the bottom of the container. The maximum amount of the anhydrous solute in grams that gets completely dissolved in 100 g of the solvent (water in this case) is termed as the solubility of the particular solute in the solvent at that temperature and pressure. However, if the solution is heated to a higher temperature, an additional amount of solute gets dissolved and the solute concentration in the solution increases. It needs to be appreciated that the amount of solute could exceed the amount of the solvent.

> The solute concentration is expressed in many ways, namely, % concentration, i.e., amount of solute per unit amount of solution (also expressed as kg solute per kg of solution), solute mass per unit volume of solvent ($kg.m^{-3}$ or $g.l^{-1}$), molar amount of solute per unit volume of solvent (also known as molarity, $kmol.m^{-3}$ or $mol.l^{-1}$), and mole fractions (ratio of the number of solute moles divided by the number of moles in the solution or the number of solute moles per unit solvent mass (also known as molality, $mol.kg^{-1}$).

Figure 19.1 shows a situation of change in concentration with temperature. The solute concentration can be kept increasing by raising the solution temperature and/or pressure up and adding the solute to the solution to the point of saturation. For a two-component system (one solute and one solvent), the solubility of one component in the other depends on pressure and temperature. At a given temperature and pressure, solubility represents the equilibrium condition. For a solution, the composition no longer changes because the liquid phase is saturated with the solid phase.

The lower curve, in the figure is the *solubility curve* indicating the maximum amount of the solute that could be dissolved at a given temperature and pressure without getting precipitated. Such a curve for sugar in water is given as follows:

$$S = 0.0286\ t^2 - 0.0256\ t + 188.5 \qquad (19.1)$$

where S is the sugar solubility (sugar in g per 100 g of water) and t is the solution temperature in Celsius. The upper curve is the *super-solubility curve*. The zone between the two curves is known as the *metastable zone*. The pressure dependence of solubility is negligible for liquid solutions if the pressure difference is small and is ignored here. It is not that the solubility always increases with a rise in temperature, it can decrease in some cases, e.g., for calcium carbonate, sodium sulfate, and iron sulfate dihydrate.

The potential for the growth of the existing crystals or nucleation, i.e., the production of new crystals, is a non-equilibrium condition called *supersaturation*. Crystallization can occur only if the solute in the solution is more than the solubility limit. The higher extent of supersaturation means more driving force for crystallization. Quantification of the driving

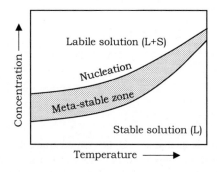

FIGURE 19.1   Solubility curves

force for crystallization and the determination of crystallization yield depend on supersaturation.

If the solute concentration after cooling is above the super-solubility curve, crystals begin to be formed spontaneously and rapidly. However, if the concentration level is in the metastable range, seed crystals are added to obviate the conditions of slow nucleation and solute molecules forming aggregates so as to get dissolved again. Small crystals of the solute, added to the solution, are called *seed crystals* such that the solute keeps getting deposited on the seed crystal surfaces until the stage when the solution concentration falls back to the saturation level. There is no crystal growth below the saturation curve. Instead, crystals may begin to get dissolved. Once crystallization is complete, equilibrium is reached, depending upon the temperature and the solubility conditions, between the crystals and the residual mother liquor. The mother liquor is composed of the liquid with residual dissolved solute and impurities.

As indicated above, while increasing the concentration of solute in the solution, heat is supplied to raise the solution temperature. To set the process of crystallization on, the solution temperature is reduced. The heat removed during the cooling process consists of the latent heat of crystallization plus the sensible heat of the solution. The heat of crystallization is taken as positive because it is opposite to the heat of solution. For practical reasons, the heat of crystallization and the heat of solution, i.e., the heat that was used to form the solution before crystallization, are assumed to be equal.

The constituent particles of crystals are placed orderly in a three-dimensional space lattice. As a result, these crystals appear as polyhedrons having sharp corners and flat surfaces. Although the relative sizes of the faces and surface may be different, the angles that the corners make with the edges and surfaces are the characteristics of the crystalline material. A crystal is called *invariant* if, while growing under ideal conditions, it maintains geometrical similarity. A single dimension of an invariant crystal is sufficient to represent its size. The ratio of surface area to volume of an invariant crystal is represented as follows:

$$\frac{S_p}{V_p} = \frac{6}{\phi_s D_p} \tag{19.2}$$

where, ϕ_s is the shape factor, also known as sphericity. If the characteristic length L of a crystal is defined as $\phi_s \times D_p$, then

$$L = \frac{6V_p}{S_p} \tag{19.3}$$

Crystal Growth

Whether nucleation begins spontaneously or through crystal seeding, the crystal growth continues until the supersaturation conditions prevail. The rate of either the nucleation or the growth depends on the temperature and the level of concentration as well as on the interfacial tension between the solute and the solvent. Maintenance of low level of super-saturation promotes the growth of the existing crystals and new crystals would not form. Conversely, if the level of super-saturation is high, the formation of new crystals is promoted and the growth takes back seat. Large crystals are obtained from slow cooling and low level of super-saturation. Fast cooling produces smaller crystals. Agitation of the solution enhances the rate of nucleation. If the super-saturated solution is cooled and agitated vigorously, nucleation takes place fast and results in a large ensemble of small crystals. *Sonocrystallization* is the process of using power ultrasound to better control the nucleation and the growth of crystals through modulating the crystal size distribution and morphology.

If the crystals are anhydrous, the yield calculation is simple. If the crystals contain some water, the water of crystallization must be accounted for while calculating the crystal yield.

Driving force for crystallization, i.e., the change in Gibb's free energy, is difficult to estimate. It is convenient to work with concentration difference. If C is the concentration of the solution, i.e., the solute mass (g) per 100 g of solvent, at a given instant and S is the concentration at saturation (solute mass in g per 100 g of solvent), concentration difference can be expressed as follows:

$$\Delta C = C - S \tag{19.4}$$

The relative level of supersaturation, σ, is represented as follows:

$$\sigma = \frac{\Delta C}{S} = \frac{C}{S} - 1 \tag{19.5}$$

The driving force for melts is expressed in terms of supercooling as follows:

$$\Delta t = t^* - t \tag{19.6}$$

where, t^* is the equilibrium temperature of the melt and t is the supercooled temperature.

Primary nucleation takes place in a crystal free super-saturated solution. Secondary nucleation occurs due to the influence of larger crystals in magma. Super-saturation levels in the metastable range produce secondary nucleation sites. Crystals smaller than 5 mm are formed by precipitation. These small crystals exhibit size-dependent solubility and it increases for very small crystals. In this context, the process of *ripening* or *aging* is the dissolution of smaller crystals that exist in suspension near the saturation level and subsequent production of larger crystals.

After the nucleation occurs, the growth of crystals is determined by the rate of diffusion of the solute from the solution to the crystal surfaces for deposition. Another factor determining the growth is the reaction rate of the solute molecules at the crystal face to get rearranged into the crystal lattice. These two growth rates are mathematically represented as follows:

$$\frac{dm}{d\theta} = k_d A(C - C_i) \qquad (19.7)$$

$$\frac{dm}{d\theta} = k_s A(C_i - S) \qquad (19.8)$$

where, dm is the incremental mass increase of a crystal during a time $d\theta$, A is the crystal's surface area, C and C_i are the concentrations of the solute in the solution and at the crystal solution interface, respectively, S is the concentration of the saturated solution, k_d is the mass transfer coefficient at the interface, and k_s is the rate constant for the surface reaction.

The above two equations are combined, due to the difficulty of estimating the value of c_i, to give the following:

$$\frac{dm}{d\theta} = kA(C - S) \qquad (19.9)$$

where, $\quad \dfrac{1}{k} = \dfrac{1}{k_d} + \dfrac{1}{k_s}$

Growth of a crystal can be represented as $dm = A.\rho_s.dL$ and, therefore the rate of growth as follows:

$$\frac{dL}{d\theta} = k(C - S) / \rho_s \qquad (19.10)$$

$$= k_2 . (C-S)^g \qquad (19.11)$$

Here $dL/d\theta$ represents the increase in one of the crystal's faces, ρ_s is the crystal's density, k_2 is the coefficient of size change, and g is the system-specific coefficient. Temperature, agitation, and impurities in the solution influence the growth of crystals. If the impurity concentration is high, crystals may not form. When small crystals join together, they form agglomerates, affecting a number of bulk crystal properties, e.g., strength, size, shape, purity, and packing density. Agglomeration is more common in processes producing small (<50 μm) particles.

19.1.2 Multi-Stage Crystallization

The crystallization process could be carried out in a multi-stage fashion to maximize the recovery of solute crystals. Once the crystals have been obtained, the temperature and concentration of the mother liquor solution is brought to a new equilibrium condition to further crystalize the solute. This process of subsequent crystallizations is continued until the impurities in the mother liquor increase to the extent that further crystallization becomes very slow.

However, the purest crystals are obtained in the initial stages.

In the case of the sugar industry, sugar concentration in the solution is increased and crystal seeding is carried out to assist the crystals in attaining the required size. The crystals are then separated using centrifugation. The mother liquor is then further concentrated, seed crystals are added, and then the formed crystals are separated. The process is repeated for 5 to 6 times before discarding the mother liquor as molasses.

Crystals may have cubic, rhombic, tetrahedron, or any other regular shape. The shape is affected by the extent of impurities present in the solution. The crystal shape influences the angle of repose of the crystal bulk and the rate of dissolution. Crystal size uniformity is an important product quality parameter.

If the mass flow rates of feed, crystals, mother liquor, and water (removed during the evaporative crystallization) are denoted as F, CR, ML, and W; concentration factor as CF, quantity of saturated feed as SF; solute concentration as C, concentration at saturation as S, and the water evaporated for bringing the solution to saturation and that evaporated during crystallization are denoted as W_1 and W_2, then the following equations can be written for multi-stage continuous crystallization process.

$$\text{Overall mass balance } F = CR + ML + W \qquad (19.12)$$

$$\text{Crystallizer balance } SF = CR + ML + W_2 \qquad (19.13)$$

Some other equations for the process are as follows:

$$W = W_1 + W_2 \qquad (19.14)$$

$$F = CF \times ML \qquad (19.15)$$

$$F = SF + W_1 \qquad (19.16)$$

$$SF = F \times C/S \qquad (19.17)$$

$$W_2 = CR. (100-S)/S \qquad (19.18)$$

Example 19.1 If salt solution has a concentration of 60% at a given temperature and the solubility at that temperature is 40 g salt per 100 g water, estimate the quantity of the crystallized salt that could be separated in the first stage after the onset of crystallization.

SOLUTION:

Amount of salt in solution = 600 g / kg solution

$$= 1.5 \text{ kg / kg water.}$$

Concentration at saturation = 400 g / kg water
   Amount of salt crystals = 1.5 − 0.40 = 1.1 kg / kg water
Note that 0.4 kg salt per kg of water is still left in the mother liquor which will need to be crystalized in subsequent stages.

Example 19.2 The following parametric values were observed during sugar crystallization operation at a feed rate of 1000 kg. h⁻¹.

Evaporator	Liquor temperature, °C	Concentration at the inlet, %	Concentration in the evaporator, %
1	85	65	82
2	73	82	84
3	60	84	86
4	51	86	89

Calculate the yield of sugar crystals in each evaporator and the concentration of sucrose in the mother liquor leaving the final evaporator.

SOLUTION:

The above tabulated data along with the solubility-temperature relationship (equation 19.1) would give the equilibrium sugar solubility for each of the evaporators.

For the First Evaporator

A. At 85°C, the equilibrium solubility is calculated as:

$$S = 0.0286(85)^2 - 0.0256(85) + 188.5$$

= 392.96, i.e. 393 g per 100 g or 393 kg per 100 kg water

B. The super-saturated concentration of the magma in the first evaporator is 82% and the amount of sugar is 650 kg.h⁻¹ (i.e. 65% of 1000 kg feed), thus the amount of the solution just before crystallization in the first evaporator is 650/0.82 =792.7 kg.h⁻¹.

   i.e. the water available in the solution is 792.7-650 = 142.7 kg.h⁻¹

   and water evaporated = 350-142.7 = 207.3 kg.h⁻¹

C. As the water available in the solution is 142.7 kg.h⁻¹, the maximum sugar that could exist in equilibrium is 142.7 × (393/100) = 560.8 kg.h⁻¹.

D. Thus, sugar crystallized in the first evaporator is 650-560.8 = 89.2 kg.h⁻¹.

E. The composition of the mother liquor exiting the first evaporator is 142.7 kg water.h⁻¹ and 560.8 kg sugar.h⁻¹; the amount of mother liquor is 142.7 + 560.8 = 703.5 kg.h⁻¹.

These quantities are shown below for a better understanding of the mass balances.

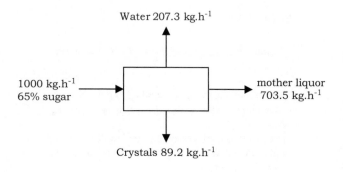

For the Second Evaporator

A) Equilibrium solubility for 73°C is 339 kg/ 100 kg water.

B) Since the super-saturated concentration in the second evaporator is 84%, the amount of the solution just before crystallization is calculated as 560.8/0.84 = 667.6 kg.h⁻¹ and the water available in the solution is 667.6-560.8 = 106.8 kg.h⁻¹.

   Water evaporated = 142.7-106.8 = 35.9 kg.h⁻¹

C) For 106.8 kg water, mother liquor at the outlet of the second evaporator can hold only 106.8 × (339/ 100) = 362.05 kg sugar.h⁻¹.

D) Quantity of crystals from the second evaporator is 560.8-362.05 = 198.8 kg.h⁻¹.

E) The composition of the liquor exiting the second evaporator is 106.8 kg water.h⁻¹ and 362 kg sugar.h⁻¹.

For the Third Evaporator

The outlet conditions of the second evaporator become the inlet conditions of the third evaporator.

A. Equilibrium solubility for 60°C is 289.9 kg/ 100 kg water.

B. Since the super-saturated concentration in the third evaporator is 86%, the amount of the solution just before crystallization is calculated as 362 / 0.86 = 421 kg.h⁻¹ and the water available in the solution is 421-362 = 59 kg.h⁻¹.

   Water evaporated = 106.8–59 = 47.8 kg.h⁻¹

C. For 59 kg water, mother liquor at the outlet of the third evaporator can hold only 59 × (290/100) = 171.1 kg sugar.h⁻¹.

D. Quantity of crystals from the third evaporator is 362–171.1 = 190.9 kg.h⁻¹.

E. Composition of mother liquor exiting the third evaporator is 59 kg.h⁻¹ water and 171.1 kg sugar.h⁻¹. Thus, the amount of solution going out is 230.1 kg.h⁻¹.

For the Fourth Evaporator

A. equilibrium solubility for 51°C is 261.6 kg/ 100 kg water.

B. The super-saturated concentration in the fourth evaporator is 89%, the amount of the solution just before crystallization is calculated as 171.1 / 0.89 = 192.2 kg.h⁻¹ and the water available in the solution is 192.2–171.1 = 21.1 kg.h⁻¹.

   Water evaporated = 59–21.1 = 37.9 kg.h⁻¹

C. For 21.1 kg water, mother liquor at the outlet of the 4ᵗʰ evaporator can hold only 21.1 × (262/100) = 55.28 kg sugar.h⁻¹.

D. Quantity of crystals from the fourth evaporator is 171.1–55.28 = 115.8 kg.h⁻¹.

E. Composition of mother liquor exiting the fourth evaporator is composed of 21.1 kg water.h⁻¹ and 55.28 kg sugar.h⁻¹.

Water 207.3 kg.h⁻¹

1000 kg.h⁻¹
65% sugar

mother liquor
703.5 kg.h⁻¹

Crystals 89.2 kg.h⁻¹

The mass balances in these evaporators can be shown as below:

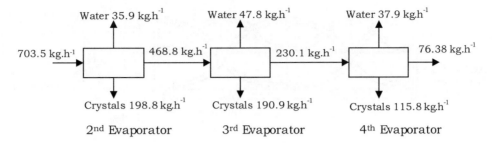

The summary of mass balances of sugar, crystals, and water (in kg. h⁻¹) for the flow rate of 1000 kg. h⁻¹ at the inlet of the first evaporator.

Evaporator no.	Component	Magma temperature, °C	Equilibrium solubility, kg/100 kg	Sugar Concentration in magma, %	Inlet feed composition kg.h⁻¹	Super-saturated magma composition kg.h⁻¹	Outlet Mother liquor Composition kg.h⁻¹	Quantity of crystals, kg kg.h⁻¹
1	Water	85	393	82	350	142.7	142.7	89.2
	Sugar				650	650	560.8	
2	Water	73	339	84	142.7	106.8	106.8	198.8
	Sugar				560.8	560.8	362.05	
3	Water	60	290	86	106.8	59	59	190.9
	Sugar				362.05	362.05	171.1	
4	Water	51	262	89	59	21.1	21.1	115.8
	Sugar				171.1	171.1	55.28	

The above analysis indicates that 594.7 kg.h⁻¹ of sugar is recovered as crystals from the 650 kg.h⁻¹ present in the solution. The recovery percentage, thus, is 91.5%. Obviously, 55.3 kg.h⁻¹ sugar is still there in the 76.38 kg.h⁻¹ mother liquor. The results also indicate that crystal recovery increases in initial evaporators and keeps decreasing after reaching the maximum.

19.2 Equipment

Essential features in a crystallization plant are, (1) mechanism to raise the concentration of the desired solute to a predetermined level, (2) initiation of nucleation and crystal growth, (3) separation of crystals from the residual mother liquor, and (4) drying of crystals, and (5) elutriation, i.e., separation of crystals based on shape and size. Crystallizers can be either batch or continuous type. The crystallizers may be with agitation or without agitation. The mechanisms for achieving super-saturation are raising of solution temperature, evaporation, or vacuum.

An open tank or vat is a simple crystallizer. The super-saturated solution in the crystallizer loses heat to its surroundings and crystals are formed. The rate of nucleation and the size of crystals are difficult to control. Labor costs are high. Steam-heated crystallizers of open pan type are still in use due to their simplicity. Simple solar energy–based evaporators for making crystalline salt from sea water are being used in some countries. The slow evaporation rate results in larger salt crystals.

19.2.1 Cooling Type Crystallizer

In this system, crystallization is achieved continuously, and the rate of cooling is controlled with the help of agitation and a cooling medium. A cylindrical scraped surface heat exchanger, also called a votator is used for crystallization in ice cream preparation. The material to be crystalized flows in the central pipe, crystals are formed on the surface of the pipe, and they get continuously scraped. The cooling medium, flowing through the annular space picks up the heat from the magma formed in the inner pipe.

19.2.2 Swenson-Walker Crystallizer

Swenson-Walker crystallizer, Figure 19.2, is another type of continuous cooling type scraped surface crystallizer. It has about 0.6 m wide open trough with a semicircular cross-section and a water jacket outside. There is a spiral agitator that is rotated slowly to keep the crystals suspended for better growth. The blades of the agitator move close to the crystallizer wall so as to obviate the deposition of crystals on the walls. The resultant crystals have a wide range of sizes.

19.2.3 Evaporative Crystallizer

The solution is heated to let the solvent evaporate, which leads to the super-saturation, nucleation, growth of crystals, and then the separation of crystals. Evaporative crystallization is a popular method for crystallization of salts and sucrose. Steam

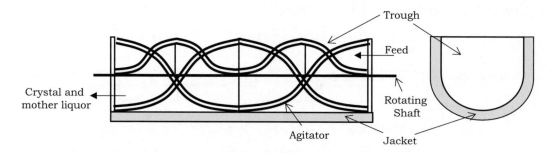

FIGURE 19.2   Swenson-Walker crystallizer (after https://chemico-world.blogspot.com with permission)

is generally the heat source, and forced circulation is an integral component of such a crystallization unit. Since evaporation is the key process involved, the crystallization takes place at an isothermal condition.

In general, forced circulation crystallizers can be either a single effect unit or a multiple effect unit. These units are used when crystal size is not of much importance or if crystal growth rate is generally acceptable. The forced systems employ the vapor recompression concept, either mechanical or thermal. These units usually operate from low vacuum to atmosphere pressure. Depending on the application, the material of construction for the fabrication of these crystallizers could be chosen from a wide range.

19.2.4 Oslo-Krystal Crystallizer

This crystallizer, Figure 19.3, consists of a concentric two-pipe arrangement in which the solution is made to evaporate and crystals are allowed to be formed. The supersaturated liquid falls down through the central down-comer to the lower chamber in which a fluidized bed of crystals is maintained. This lower chamber is also known as the retention or growth chamber. The crystal size can be controlled by removing the crystalline materials from the chamber. The fluidized bed of crystals can also be classified by particle size, which also results in controlled crystal growth. An external pump is used for the recirculation of the mother liquor from the retention

chamber to the upper chamber through the heat exchanger. Thus, it comes under forced circulation type systems.

19.2.5 Vacuum Cooling Crystallization

The device consists of a multi-stage system vacuum crystallizer, into which warm and almost saturated solution is entered. As the pressure is gradually reduced, the solvent evaporates and the solution is simultaneously cooled to the boiling temperature. Vacuum and feed rate are adjusted to obtain the desired crystal size. The solution is first concentrated by evaporation and once the concentration is in appropriate range, seeding is carried out and the vacuum in increased. The salts are crystallized by vacuum-induced lowering of the temperature. Air is also sucked at the bottom of the crystallizer to agitate the crystals and moving them to the outlet. The suspension is subsequently pumped to a cyclone for thickening, the liquid is separated in the centrifuge and the salt dried in a dryer.

Crystallization, as noticed from the foregoing discussion, can give high purity product in a single step. The operating temperature and the energy requirements are lower as compared to distillation. Crystallization plants are simple to construct and operate, resulting in economic production of the crystalline product. However, crystallization is limited in its capability to yield only one pure product from a liquid and the quantity produced is smaller because of phase-equilibria. Process kinetics of crystallization is rather complex to understand.

Check Your Understanding

1. What are the different stages in crystallization?
2. Explain the difference in working principles of a Swanson-Walker crystallizer and an Oslo-Krystal crystallizer.
3. Differentiate between crystallization and evaporation.
4. Define
   a) Magma;
   b) Supersaturation;
   c) Nucleation during crystallization;
   d) Solubility curve;
   e) Sonocrystallization;
   f) Evaporative crystallization.

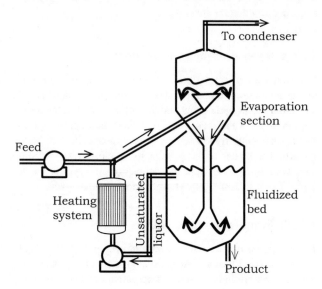

FIGURE 19.3   Oslo-Krystal Crystallizer

BIBLIOGRAPHY

Brennan, J.G., Butters, J.R., Cowell, N.D. and Lilly, A.E.V. 1976. *Food Engineering Operations* 2nd ed. Elsevier Applied Science, London.

Fellows, P.J. 2000. *Food Processing Technology*. Woodhead Publishing, Cambridge, UK.

Geankoplis, C.J.C. 2004. *Transport Processes and Separation Process Principles* 4th ed. Prentice-Hall of India, New Delhi.

https://chemico-world.blogspot.com

https://www.thermopedia.com

Kulkarni, S.J. 2015. A review on studies and research on crystallization. *International Journal of Research and Review* 2(10): 615–618.

McCabe, W.L., Smith, J.C. and Harriot, P. 1993. *Unit Operations of Chemical Engineering* 5th ed., McGraw-Hill, Inc., New York.

Rao, D.G. 2009. *Fundamentals of Food Engineering*. PHI Learning Pvt. Ltd., New Delhi.

Toledo, R.T. 2007. *Fundamentals of Food Process Engineering* 3rd ed. Springer, New York.

20

Extrusion

Extrusion is the process of forcing something out using a device; the device is extruder and the "something" being forced out is extrudate. When it is applied to food materials, moist solids and slurries can be forced out of small holes to impart specific characteristics to the output. Vermicelli and many other savouries have been made in this manner for millennia. Simple devices are used to obtain the extruded products for subsequent drying or frying.

Manual extruders have a long history. The fact that India has had noodle-shaped and related products for well over 4,000 years is indicative of the existence of some sort of extruders operated by human power. As per the available information, the first extruder was developed in 1870 to manufacture sausage in the United States. Since 1930, extrusion technology has been put to use for simple breakfast cereals and dry pasta. The first development of twin screw extruders for food products occurred in the mid-1930s for both co-rotating and counter rotating configurations. Extrusion method found its use in pet food production in 1950.

20.1 Methods and Principles

20.1.1 Hot Extrusion and Cold Extrusion

The extrusion methods that have been mentioned above are fairly low-pressure processes and can be carried out manually. Obviously, the rheology of the material being extruded does not undergo much baro-thermal change in such low-pressure applications. However, under high pressure, a food material could change its phase, i.e., solid to liquid, with concomitant production of heat. Under this situation, if the material is allowed to pass through a small hole of pre-defined shape, the material experiences a sudden drop in pressure allowing the material to expand and cool rapidly. Thus, it is possible to create cooked foods of desirable composition, texture, shape, size, and taste. Extrusion cookers or hot extrusion methods work on this principle.

Food ingredients are heated to the temperature range of 100°C to 200°C in the hot extrusion process. The temperature rises quickly in the extruder due to friction and other supplementary heat sources. The pressure is also very high, and, thus, the conditions in an extrusion cooker are also termed as *baro-thermal conditions* or *hot isostatic pressing conditions*. The food is moved forward to a compression section having small flights to further increase the shear and pressure. Food products, such as weaning foods, snack foods, etc., are manufactured using hot extrusion. Extrusion cooking has also been used widely for the production of designer foods and special feed. Because they are expensive, only large-scale industries have been using hot extruders so far.

Mixing and shaping of food in cold extrusion is accomplished without cooking since heating is limited up to only 100°C. Cold extrusion is used for meat products and pasta where cooking is not needed. Low-pressure extrusion for such products as pet foods is also carried out at temperatures less than 100°C. Colored pasta can be produced using cold extruders by adding tomato purée, spinach paste, or any other natural colorant. Preservation of cold extruded products subsequently is carried out by such practices as drying, chilling, or baking methods followed by packaging. Cold extruders involve lower investment, which make them attractive for small enterprises.

20.1.2 Extruder Performance

Specific Mechanical Energy (SME) is an index of the extruder's performance and defined as the total mechanical energy required to obtain unit weight of extrudate.

The power supplied by the extruder shaft can be written as:

$$P = \frac{2.\pi.N.T}{60} \qquad (20.1)$$

where, P is the power (Watt or $J.s^{-1}$), T is the net torque on the extruder drive (N.m) and N is the screw speed ($rev.min^{-1}$).

Modifying this equation for energy requirement for unit production of extrudate,

$$SME = \frac{2.\pi.N.T}{60.\dot{m}} \qquad (20.2)$$

Here, $\dot{m}$ is the mass flow rate of the materials ($kg.s^{-1}$). Thus, the SME has a unit of $J.kg^{-1}$ (or can be expressed as $kWh.kg^{-1}$).

If the mass flow rate of the product is denoted as m_p ($kg.min^{-1}$), moisture contents (fraction on wet basis) of the raw sample and final product are denoted as M_f and M_p, then the flow rate of feed is determined as follows:

$$\dot{m} = m_p \frac{1 - M_p}{1 - M_f} \qquad (20.3)$$

Increases in moisture and temperature reduce the value of SME. High moisture content of the raw material causes a lubricating effect, which increases the efficiency of energy use, thus reducing SME. A higher screw speed results in higher shear and higher SME.

Measurements in single-screw food extruders have indicated the SME to be in the range of 0.1–0.2 $kWh.kg^{-1}$. This value is much lower as compared to conventional methods. There would be some extra energy required to prepare the raw materials as grinding and conditioning.

DOI: 10.1201/9781003285076-24

20.2 Equipment

Small-scale manufacturing units and food outlets employ manually operated extruders to fully automatic machines. A pasta extruder is a simple device as it has an extruder barrel, mixing chamber, and die for making pasta of specific shapes. Dies are made from plastic, stainless steel, or bronze. The continuous extruders used in industry are described below.

Construction and Operation

- Figure 20.1 shows the main parts and sections of an extruder. As the device is used to prepare a product of desired shape by forcing a product mix through a die or an orifice, the basic parts of an extruder include a barrel, either one or two screws, and a die at the exit end of the barrel. A power unit rotates the screw of the extruder.
- Some specific types of extruders are also available in which the moving member is either a roller or a piston. Similarly, the extruders can also be classified on the basis of their operating temperature, pressure, or on the basis of whether an additional heating device has been integrated or not.

- With the increase in temperature in the barrel, the product mix approaches its melting point or *plasticating point*. Thus, the section after the compression section is known as the liquefying and plasticizing section. The degree of melting varies with the barrel temperature, selected screw pitch, and compression ratio. In principle, mixing improves in the melting section. This also causes a change in the rheological properties of the material.
- Toward the end of the assembly, the temperature may be up to 200°C, and the internal pressure up to 200 bar. During the process of hot extrusion, as the food material gets cooked, such a process is also known as *extrusion cooking*.
- The shear energy, exerted by the rotating screw, provides additional heating of the barrel.
- In this changed rheological state, the food is conveyed under high pressure through a die or a series of dies and the product expands to its final shape. This results in very different physical and chemical properties of the extrudates compared to those of the raw materials used.
- Provisions are also in place for heating and cooling of the barrel. Temperature and pressure sensors also form some essential components of the system.

Extruder classifications

Moving member in the barrel	Number of screws	Operating temperature	Operating pressure	Generation of mechanical/ friction energy
• Screw • Roller • Piston	• Single • Twin	• Cold • Hot	• Low-pressure extruders producing limited shear • High-pressure extruders producing large mechanical energy and shear	• Autogenic (heat is generated by friction of materials in the screw-barrel assembly) • Isothermic (heat provided from external source) • Polytropic

- The designs of screw, barrel, and die vary depending upon the raw food materials and final products.
- Either a single or a mixture of raw food materials is fed to the assembly. The process begins with grinding the raw materials to the correct particle size and then passing the dry mix of raw materials through a pre-conditioner. Here other ingredients, such as water, sugar, fats, steam, etc., are also added depending on the target product.
- In the screw-barrel assembly, the first section is the conveying/transport section, where the food material is forwarded by the screw flight. During this stage, the product mix experiences friction, shear, compaction, increase in pressure, and, consequently, a rise in temperature. Thus heating, mixing, and conditioning take place.

- The semisolid extrudate is finally forced out of a die or a series of dies under high pressure, when there are expansion and changes in surface characteristics due to sudden reduction of pressure. The change in product size in comparison to the size of die is considered as the *expansion ratio*.
- The product achieves its final shape, moisture content, and temperature quickly. Different dies can form the product into different shapes.
- Blades, rotating at uniform speed about the openings of the die, are used for cutting the extrudate in specific lengths.
- The extrudate becomes rigid after cooling and drying while still maintaining porosity.
- The extrudate can either be consumed as it is or processed further for packaging and storage.

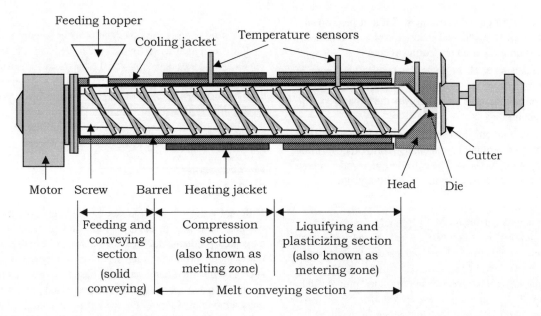

Feeding hopper

Cooling jacket   Temperature sensors

Cutter

Motor   Screw   Barrel   Heating jacket   Head   Die

Feeding and conveying section (solid conveying)

Compression section (also known as melting zone)

Liquifying and plasticizing section (also known as metering zone)

Melt conveying section

FIGURE 20.1 Schematic diagram of a single-screw extruder

Salient Features

- Extrusion methods can produce a large variety of food products by changing the raw materials, die, and operational parameters.

- Extrusion of food materials involves both physical and chemical modifications of raw materials. Moist, starchy, and proteinaceous food materials are mechanically sheared under pressures up to 20 MPa and self-generating heat. The nature of protein, starches, and other constituents undergoes changes leading to changes in chemical structure, physio-chemical properties, and nutritional qualities.

- Extrusion cooking is a HTST process and requires relatively lower amounts of supplementary heat for processing. HTST results in minimum loss of heat-sensitive ingredients.

- The materials get cooked in a hot extruder due to high pressure and friction.

- The extrudate is retained in the extruder for a pre-determined residence time. The final product quality is greatly influenced by such phenomena as transfer of heat, mass, and momentum; residence time; and residence time distribution during extrusion cooking.

- The texture of the extrudates, like puffs, is also determined by the rate of pressure and heat and moisture release after exiting from the die.

- In a normal extrusion application, moisture content of raw material is held between 25% and 30% and the residence time is 30 to 90 s. Normal temperature ranges in the cooking and forming zones of the extruder are 80°–150°C and 65°–90°C, respectively. The change in these parameters can change the physio-chemical properties of the extrudate along with its nutritive value and organoleptic properties.

- The important physical properties of the extrudate include bulk density, expansion ratio, color, water solubility index (WSI), water absorption index (WAI), and textural properties, such as hardness.

- The expansion ratio of the extrudate increases with an increase in barrel temperature and the speed of screw, and it reduces with an increase in the moisture content of raw materials.

- The bulk density of extrudate decreases with increases in barrel temperature, screw speed, and moisture content.

- Co-extrusion is a process where two or more melt streams combine in a die to form an extrudate. The final product is in the form of a crispy tube filled with a fluid. Both the outer crispy layer and the inner components are extruded simultaneously in the form of a continuous filled tube, which is cut subsequently into desired lengths. This is finding applications in the bakery and confectionery industry.

- Certain ingredients, known as *nucleating agents*, are added for cell wall formation and increasing the

FIGURE 20.2 Twin-screw extruder

expansion ratio of extrudates. Sodium bicarbonate, calcium carbonate, and monoglycerides are a few of the commonly used nucleating agents.

- The salt level also affects the texture and color of the extrudate. The addition of salt causes changes in color and airiness and improves the expansion ratio of the product.

- The equipment involves lower handling costs and increased profitability as compared to other forming or cooking methods. The extruders can work for long hours continuously giving a very high yield.

Single-screw extruders are, in general, cheaper with less complicated structure than twin-screw extruders. In some applications, the single screw is sufficient. However, the twin-screw extruder gives a higher output and lower energy consumption per unit product (Figure 20.2). The twin-screw extruder has good feeding characteristics, and the screws help in pushing the materials forward. This also assists in self-cleaning. The

blending is better and the heat transfer is more uniform than the single screw. In the case of the single screw, the shearing takes place only on the barrel surface, whereas in the twin-screw system, there is additional shearing at the meshing positions of screws. Twin-screw extruders are preferred when processing plastics or when dealing with materials having lower thermal stability.

20.3 Applications

Different types of products can be obtained from the extruders based on starch, sugar, cereal, and proteins as directly expanded products, such as corn curls and breakfast cereals; unexpanded products, such as pasta; co-extruded products, such as jelly-filled cores and fruit-based cereals; candies, such as liquorice and chewing gum; and texturized products, such as meat analogues. In addition, the extruders can also be used to prepare half products, such as potato pellets and modified fat mimics and starches. Extrusion technology has also been used to develop new food products.

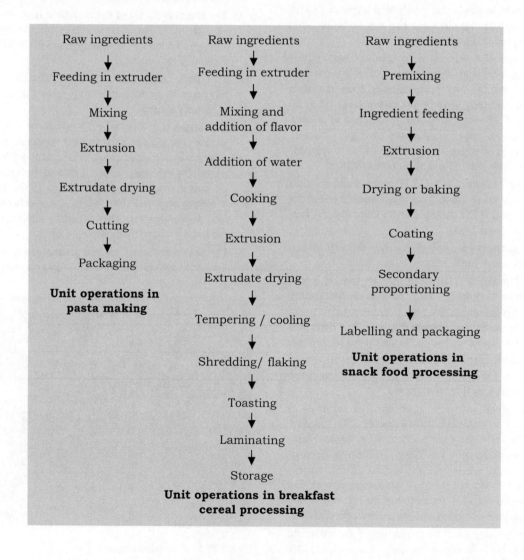

Raw ingredients
↓
Feeding in extruder
↓
Mixing
↓
Extrusion
↓
Extrudate drying
↓
Cutting
↓
Packaging

Unit operations in pasta making

Raw ingredients
↓
Feeding in extruder
↓
Mixing and addition of flavor
↓
Addition of water
↓
Cooking
↓
Extrusion
↓
Extrudate drying
↓
Tempering / cooling
↓
Shredding/ flaking
↓
Toasting
↓
Laminating
↓
Storage

Unit operations in breakfast cereal processing

Raw ingredients
↓
Premixing
↓
Ingredient feeding
↓
Extrusion
↓
Drying or baking
↓
Coating
↓
Secondary proportioning
↓
Labelling and packaging

Unit operations in snack food processing

20.4 Changes in Food Due to Extrusion

In addition to the development of new products, the extrusion process also affects product characteristics in the following ways.

- The porosity of the feed is reduced, thus yielding a product of increased density.

- Hardness increases with a rise in barrel temperature and decreases with an increase in screw speed. Higher moisture content leads to reduced hardness. When the hardness is low, the melt density is lowered and, thus, the screw speed also increases.

- Moisture contents in food extrusion processes are normally less than 40%. The initial moisture content as well as the temperature used for extrusion will affect the final product moisture, which influences the crispiness, degree of cooking, and expansion ratio of the extrudate.

- The moisture content of the extrudate also affects water activity and, thus, the shelf life of the food product. Normally the extruded food products are expected to have a water activity of less than 0.3, i.e., the products are shelf stable.

- The moisture contents of the raw ingredients also influence several other parameters, such as viscosity, torque, product temperature, and bulk density, which, in turn, influence the pressure at the die. An increase in the moisture reduces the melt temperature. At this condition, lower viscosity of the product causes an increase in pressure.

- In the case of puffed products, more moisture content leads to the formation of a thick cell wall and increased hardness.

- There is enhanced bioavailability of nutrients. There is protein denaturation and increased protein absorption. The extruded food products prepared from cereals are rich in lysine, though when legume or cereal legume blends are extruded, there may be loss in lysine at temperatures above 180 °C or shear forces at more than 100 rpm and low moisture (≤15%). There is an increase in the glycaemic index.

- Enzymes and enzyme inhibitors are generally inactivated during extrusion. Inactivation of lipase and lipoxidase caused by extrusion leads to reduced oxidation of the extrudate during storage.

- Some toxic materials present in the raw materials get destroyed. There is inactivation of antinutritional factors in peas when extruded at 145°C and feed moisture of 25%.

- Degradation of ascorbic acid occurs at higher barrel temperatures and low feed moistures. There is also a reduction in vitamin A.

- There is loss of some sugar, which may be due to the sucrose getting converted into fructose and glucose.

- Improvement of the quality of some legume-based extruded food products occurs by obliteration of some oligosaccharides.

- The starch granules get gelatinized during extrusion and give excellent expansion characteristics. Complex starches are converted into simple forms.

- Maillard reaction takes place due to the presence of amino acids and reducing sugars.

- The size, shape, and aspect ratio of dietary fibers may be significantly modified. Total dietary fiber content and its solubility change. Water solubility of dietary fibers increases significantly when specific mechanical energy during extrusion increases.

- Normally there is an increase in low molecular weight soluble fibers, which, thus, tend to decrease the total dietary fiber content. However, the extrusion cooking has been found to increase the amount of dietary fiber in extruded barley.

- Lipids undergo processes of oxidation, hydrogenation, isomerization, or polymerization, leading to some nutritional changes.

- In addition, lipids act as lubricants during the extrusion process, and it has been observed that addition of just 0.5%–1% lipids for low moisture raw materials (less than 25%) help in reducing energy input. However, if the oil content is increased to 2%–3%, extrudate expansion can be adversely affected.

- The water absorption index (WAI) reduces with an increase in screw speed and barrel temperature, which is attributed to higher mechanical shear and higher expansion due to gelatinization. In the case of texturized rice, it has been observed that the increase in feed moisture content and screw speed significantly increased the WAI, but the increase in barrel temperature caused a decrease in WAI.

- An increase in the temperature causes an increase in the water solubility index (WSI), which may be due to the gelatinization that increases the amount of soluble starch. The increase in screw speed also causes higher specific mechanical energy and higher WSI. However, WSI decreases with increase in moisture.

Extrusion technology is still evolving. Although initial developments occurred in the plastics industry, extrusion technology has found enormous applications in food processing. The ability to combine various ingredients into a complete ready-to-eat (RTE) food has endeared the extrusion technology to those seeking functional foods as well as customized designer foods. Smart delivery of nutrients and food bio-actives through extrusion technology has found favor. Earlier, the food industry mostly employed single-screw extruders. Today the use of twin-screw extruders is gradually increasing. Twin-screw extruders with co-rotating, intermeshing, and self-wiping screws can be used as an independent unit or they can be integrated into a system where dryer, coaters, and other peripheral devices are also present.

Extrusion cooking has the potential of preparing edible products from high protein sources. It also appears to be a feasible solution to appropriately neutralize anti-nutritional factors and toxins present in raw materials and, thus, obtain healthier products. Extrusion technology is also showing great promise in adding value to food processing wastes and agricultural residues. This, of course, does not mean that extrusion cooking is ideal for all applications. It is an alternative and, in many cases, a competitive method in relation to other methods of food and feed manufacturing.

Check Your Understanding

1. Differentiate between hot extrusion and cold extrusion and explain their applications in food.

2. Define the specific mechanical energy of an extruder. What are the different factors affecting the specific mechanical energy? Will the SME increase or decrease if the moisture content of the material and/or the temperature increases? Why?

3. What are the different types of extruders? Explain their specific applications.

4. Explain the differences in the working apparatus and applications of a single-screw extruder and twin-screw extruder.

BIBLIOGRAPHY

Berk, Z. 2009. Extrusion. In *Food Process Engineering and Technology* 1st ed. Academic Press, Burlington, 333–350.

Choton, S., Gupta, N., Bandral, J. D., Anjum, N, Chaudhary, A. 2020. Extrusion technology and its application in food processing: A review. *The Pharma Innovation Journal* 9(2): 162–168.

Fellows, P. J. 2009. *Food Processing Technology* 3rd ed. Woodhead Publishing, Cambridge, UK, 456–477.

Harper, J. M. 1981. *Extrusion of Foods*. CRC Press, Boca Raton.

Heldman, D. R. and Hartel R. W. 1997. Food extrusion. In D.R. Heldman and R.W. Hartel (eds.) *Principles of Food Processing*. Aspen Publishers, Inc., Maryland, 253–283.

Offiah, V., Kontogiorgos, V. and Falade, K. O. 2019. Extrusion processing of raw food materials and by-products: A Review. *Critical Reviews in Food Science and Nutrition* 59(18):2979–2998.

Riaz, M. N. (ed.) 2000. *Extruders in Food Applications*. CRC Press, Boca Raton

Singh, S., Gamlath, S. and Wakeling, L. 2007. Nutritional aspects of food extrusion: A review. *International Journal of Food Science & Technology* 42(8): 916–929.

21

Novel Heating Technologies

In this chapter, different novel heating technologies for food applications, such as ohmic heating, infrared heating, microwave, and radio frequency heating, will be discussed. These technologies are reported to yield better quality products as compared to conventional heating methods and also to impart desirable effects on foods.

21.1 Ohmic Heating

21.1.1 Principle

An ohmic heating system, also known as a *joule heater*, is an electrical heater that uses electrical resistance to generate heat, which, in turn, provides energy for food processing. Electrical current is allowed to flow by means of electrodes through the food kept in a vessel, i.e., the product acts as a part of the electrical circuit through which the alternating current flows (Figure 21.1). It generates heat within the product and the quantity of heat generated depends on the electrical resistance of the food and the electric field strength. Domestic water heaters and small-scale space heaters are examples of the ohmic heating effect.

The traditional thermal processing methods require process heat that is generated through either combustion or electrical heating and conveyed to the point of utilization. While generating and transporting the heat, there are losses, reducing the efficiency of heat utilization. Then, the transfer of this process heat from the heat transfer surfaces to the material being processed requires thermal gradients, such as in evaporators and heat exchangers, leading to fouling of the heat transfer surfaces. Also, when the food being processed happens to be highly viscous or contains chunks, the uniformity of heating is difficult to ensure. In such situations, ohmic heating has shown great promise for effective processing. In ohmic heaters, over 95% of the applied electrical power is converted into utilizable heat. Since heat is produced directly within the food, there is greater uniformity of heating.

Rate of Heating and Power Requirement

Ohmic heating involves the dissipation of electrical energy as heat in a conducting substance (in the present context, it is the food). The heat generated is equal to the square of the local electrical field strength multiplied by the electrical conductivity. If the electric field of strength "E" (V.m^{-1}) passes through a food of length, L (m) and cross-section area A (m²) along the direction of the flow of electricity, and R is the ohmic resistance, then the rate of heat energy generated, H (W) is given as follows.

$$H = \frac{\nabla V^2}{R} = \sigma E^2 AL = \sigma E^2 v \qquad (21.1)$$

Here, V is the potential difference applied to the heating system, v is the volume of the food being processed, and σ is the local electrical conductivity (S.m^{-1} or kg^{-1}·m^{-3}·s^3·A^2).

If q (W.m^{-3}) is the rate of internal heat generation per unit volume, then for constant voltage, as per Ohm's law,

$$q = H/v = \sigma E^2 \qquad (21.2)$$

For a situation where the current is constant, the relationship is:

$$q = \frac{|J|^2}{\sigma} \qquad (21.3)$$

where, J is the current density (A.m^{-2}).

However, most commercial ohmic heaters work on the principle of constant voltage.

The ohmic resistance is given as follows:

$$R = \frac{L}{\sigma A} \qquad (21.4)$$

$$E = V/L \equiv \nabla V \qquad (21.5)$$

∇V (=E) is the voltage gradient (V.m^{-1}) or electric field intensity.

Thus, the equation 21.2 can also be written as, $q = \sigma(\nabla V)^2$

Since food is a non-homogeneous material, the potential field distribution in the food material is governed by the Laplace equation.

$$\nabla(\sigma \nabla E) = 0 \qquad (21.6)$$

The local current density, J, is obtained using Ohm's law, as follows:

$$J = \sigma.E \qquad (21.7)$$

The following equation can be used to denote the change in temperature, t, of a section of food.

$$(\rho.c)\frac{\partial t}{\partial \theta} = \nabla(k.\nabla t) + q \qquad (21.8)$$

In this equation, ρ is the density of food, k is the thermal conductivity, and c is the specific heat.

If the material properties are constant and k is reasonably small, then the local heat produced is represented as follows:

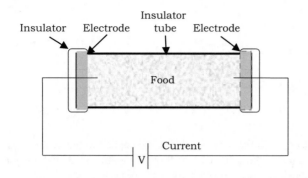

FIGURE 21.1 Schematic of a ohmic heating system (after Sakharam *et al*., 2016; Richa *et al*. 2017; https://moraritsipanificatie.eu)

$$\frac{\Delta t}{\Delta \theta} = \frac{q}{\rho.c} = \frac{\sigma.E^2}{\rho.c} = \frac{\sigma.(\nabla V)^2}{\rho.c} \qquad (21.9)$$

This equation is valid for constant voltage conditions and when there is no other significant heat transfer mechanism, such as convection or/and conduction, and when the heat losses to the surroundings are neglected.

Heating rates of the order of $1°C.s^{-1}$ are normal with 50–60 Hz frequencies and electric field strengths of up to $10 \ V.cm^{-1}$. High heating rates of up to $10°C.s^{-1}$ are also reported.

The temperature increase in each section of a continuous ohmic heater tube can be calculated by (Skudder and Biss, 1987):

$$\Delta t = \frac{V^2 \overline{\sigma} A}{L.m.c_p} \qquad (21.10)$$

Here L (m) is the distance between the electrodes, m is the mass flow rate ($kg.s^{-1}$), A is the surface area, $\overline{\sigma}$ is the average electrical conductivity and c_p is the specific heat.

The required electrical power for the continuous system is obtained as:

$$P = m.c_p(t_i - t_o) \qquad (21.11)$$

In this equation, t_i and t_o represent the temperatures at the inlet and outlet. The power decides the size of the transformer. Normally the maximum power at the transformer is 30% greater than the required power. Any loss of energy should also be considered for calculating the power requirement.

If W is the work done and P is the power supplied to the ohmic heater, then

$$W = P. \ \Delta \theta = VI.\Delta \theta \qquad (21.12)$$

The performance of the ohmic heating system is described by a term, energy efficiency, which is given as follows.

$$\text{Energy efficiency } (\varepsilon) = \frac{\text{Energy utilized to heat the sample}}{\text{Total input energy}}$$

$$(21.13)$$

Factors Affecting the Ohmic Heating Process

The different factors that primarily affect the rate of heating during ohmic heating process are as follows:

Electrical Conductivity of Food

The rate of heat generation during ohmic heating is proportional to the electrical conductivity of the material at a constant voltage gradient. Thus, a knowledge of electrical conductivity values of foods and the factors affecting the variations in these values is very important. Electrical conductivity, σ, depends on temperature, food composition, and structure. Parameters such as material microstructure, free water, and ionic strength also affect the electrical conductivity. In addition, it is temperature dependent for both liquid and solid phases and the dependence on temperature can be expressed as:

$$\sigma (t) = \sigma_i (1 + m (t - t_i)) \qquad (21.14)$$

where, $\sigma (t)$ is the electrical conductivity at the temperature t, σ_i is the electrical conductivity at the initial temperature t_i, and m is a proportionality constant. Electrical conductivity of foods also increases with an increase in water content and temperature due to higher ionic mobility at higher temperature.

The electrical conductivities of pickles, chutneys are approx. $2-3 \ S. \ m^{-1}$ at 25°C. The values for soups and carrot juice are 1.4 to $1.8 \ S.m^{-1}$ (at 25°C) and $1.15 \ S.m^{-1}$ (at 22°C), respectively, as reported by different researchers. Thus, these products experience very rapid electrical heating to the tune of $7–50 \ °C.s^{-1}$. Apple juice, milk, custard, fruits and vegetable pieces, and chicken have electrical conductivity values in the range of approx. 0.05 to $0.5 \ S.m^{-1}$ at room temperature conditions. These can be heated at the rate of $1–5 \ °C.s^{-1}$. The fresh fruits and carrots have the conductivity values of 0.04 to $0.2 \ S.m^{-1}$, which are, thus, electrically slow heating materials. Some materials like syrup, margarine, etc. have still lower electrical conductivity values and are not suitable for ohmic heating.

Frequency and Form of Wave

Ohmic heating of foods normally uses frequencies of 50 Hz and 60 Hz. Values of electrical conductivity, heat and mass transfer properties of foods, as well as the rate of heating of the materials, are affected by the frequency and waveform of applied voltage.

Some studies reveal that the electrical conductivity values are high with low frequency. Similarly, electrical conductivity of turnip tissue has been observed to be considerably higher for sine and saw tooth waves than that for square waves at 4 Hz (Lima et al. 2001). Mass transfer rates increased for frequencies as low as 4 Hz, suggesting further studies on the lowest effective frequencies applicable for ohmic heating (Lima and Sastry, 1999). The AC frequency also affects the extraction efficiency and the heating rate.

If a suspension contains particles of low conductivity, these may be under-heated, which may affect the safety of the food. During ohmic heating, heat transfer through such particles should be taken into consideration along with the monitoring of temperature at the thermal center. In addition, the residence time of the product in a continuous system should be properly supervised.

Product size and Orientation, Heat Capacity, and Viscosity

In the case of a solid-liquid mixture, the conductivities of the different fractions of solids and fluid and their volume affect the rate of heating. If such emulsions and colloids have a lower heat capacity, they would heat faster if the solids and the fluid have similar electrical conductivities. Fluid with higher viscosity would heat faster than fluids with lower viscosity.

Similarly, when there are solids of lower electrical conductivity as compared to the fluid and if they are in low concentrations, these particles will lag thermally. For a high concentration situation, the path of the electrical current through the fluid becomes more tortuous; hence, a higher fraction of the total current passes through the particles.

In the case of emulsions and colloids, the orientation of particles affects the heating, specifically for large particles. It has been observed that for 15–25 mm particle sizes, the orientation of the particles with respect to the electric field affects their electrical properties and, thus, the rate of heating. However, for smaller sizes (say less than 5 mm), there is negligible effect of orientation.

Mass Transfer during Ohmic Heating

There have been some studies on mass transfer during the heating process. A study revealed that, during ohmic heating, the volume of dye from beet root tissue diffusing into solution increased by 40% with respect to lower temperature conventional heating.

21.1.2 Equipment

Construction and Operation

- An ohmic heating system usually consists of electrodes that are separated by a tube or a plane space in which the food is kept. The food is in physical contact with the electrodes to enable the electric current to pass through it. A schematic representation of the ohmic heating device along with the auxiliary equipment is shown in Figure 21.2.

- A power supply system (generator) produces the desired electrical voltage and current combinations, which are then passed through the foods.

- The gap between the electrodes, which determines the electric field strength [V.cm^{-1}], depends on the size of the system.

- The commercial heaters are generally continuous ones. Different configurations of continuous ohmic heaters exist. In the *transverse configuration,* the electrodes are generally plane or coaxial, which are sandwiched between insulating plastic spacers. It is also known as the *plate and frame configuration.* The product flows parallel to the electrodes and the electric field is perpendicular to the mass flow.

- In the *collinear configuration,* the heating column is tubular, and the electrodes are inserted into an electrically insulated tube. The flow of product is parallel to the electric field from one electrode to the other. The operational voltage between the electrodes and the heater casing, which is externally earthed, can reach 3 kV.

- The configurations decide the heating behavior and flow pattern. The transverse systems are prone to a large amount of leakage current to earth through the product material because of the live electrodes close to the inlet and outlet pipe work. Further, phase-to-neutral current density may be non-uniform at the electrode edges in the direction of product flow. Also observed are issues of localized overheating, boiling, and electrode corrosion. Consequently, only fluids containing no particles are treated with transverse mode systems.

- Continuous ohmic heating systems can be classified as either *in-line or cross field* on the basis of

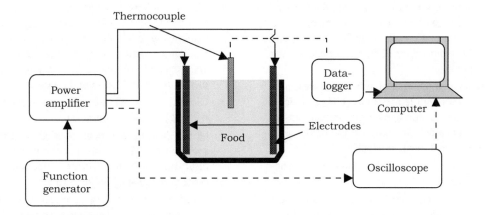

FIGURE 21.2 Schematic representation of ohmic heating equipment used in food processing (after Koubaa et al. 2019; https://www.pmg.engineering /ohmic-heating-of-milk-milk-products; Lee et al., 2013)

the position of the electrodes with respect to product flow. In the first, the electrodes are fixed along the product flow path at various positions. However, due to the voltage drop, the upstream material experiences higher field strength as compared to the downstream material. In the second, the electrodes are placed perpendicular to the flow path and the electric field strength is constant.

- The time and temperature maintained during ohmic heating depends on the product characteristics and need.

- In general, the residence time for food in an ohmic heater is 90 s or less. In commercial heaters, with this residence time the food can be heated by 100°C. The food is then pumped to a holding tank to equilibrate temperature.

Salient Features

- The design of an ohmic heating system basically includes its configuration, the desired rates of flow, the rate of heating, and the temperature rise for the specific food.

- Residence time and rate of heat generation influence the quality of ohmic heat treatment. Research continues for better electrodes, better enhancement in process efficiency, and better controls to impart greater usefulness to ohmic heating technology.

- Uniform heating of the food being processed is the main advantage of ohmic heating in comparison to other thermal processes, including that of microwave heating. There is no overheating of the product, the heat-sensitive constituents are least affected, and there is almost no chance of burning the product. All these together contribute to improved food quality.

- There is significant reduction in cooking time. As compared to conventional heating, there is a faster rise of the center temperature. Thus, it improves the final product sterility and saves power.

- However, as discussed earlier, the liquid with suspended particles will not have uniform electrical conductivity and that creates a shadow region, and there may be temperature differences between the liquid and solids.

- There is rapid attainment of high temperatures, such as those for ultra-high temperature (UHT) processing. Foods processed through ohmic heating possess shelf lives comparable to those of aseptically processed, canned, and sterile products.

- The system can be instantly shut down if the situation demands.

- The capital and energy costs are lower than most other methods with a similar nature of processing. The maintenance cost is also low because of the lack of moving parts.

Fouling on Ohmic Heating Elements

Fouling of electrodes is a common problem during ohmic heating as the electrodes are in direct contact with the material being processed. Application of higher frequency AC supply (about 10 kHz) could greatly mitigate the problem of fouling and, hence, corrosion. Another possible solution envisaged is by using noble materials like titanium, gold, stainless steel, platinized titanium, or platinum for electrodes.

Further, conventional ohmic heating with 50–60 Hz AC has been reported to cause, due to electrolysis of water, evolution of hydrogen and oxygen. Under the specific environment, the generated gas bubbles often lead to the formation of a gas blanket over the electrodes, which affects the heating process.

21.1.3 Applications

Efforts to use direct resistance heating for food processing began in the early 1900s. But these efforts could not be commercialized widely due to the lack of associated technologies, such as packaging and electrodes. The availability of better electrodes in the early 1990s and excellent packaging technologies led to the resumption of research and developments in this area. An economic assessment carried out in the early 1990s indicated that, for premium quality foods, the process would be economically viable. Since then, the capital cost of the systems has gone down and, therefore, even non-premium foods are being processed through this technology.

Some of the major applications (both in commercial scale and in research stages) of ohmic heating are as follows:

- Ohmic heating can be used for blanching, thawing, peeling, evaporation, dehydration, fermentation, and extraction. Common foods that have utilized ohmic heating so far include dairy products, milk-based products, tomato paste, fruit juices, liquid eggs, jams, minced meat, soups, etc.

- Ohmic heating can also be used for heating the food before servicing, which helps in net energy saving.

- This technology especially suits viscous foods, heat-sensitive foods, and foods containing solid particles. Cooking with ohmic heating can be achieved for hamburger patties, chicken, pork cuts, meat patties, and vegetable pieces. It has also been studied for meat slurries and fish-based products.

- Ohmic heating can be used for pasteurization of milk without protein denaturation.

- Ohmic heating has been studied for extraction of sucrose from sugar beets, juice from apples, and soymilk from soybeans. During extraction, the method helps in increasing the juice yield.

- It has been reported to be effective inactivation of microbial spores and enzymes. The time for inactivation of enzymes, such as lipoxygenase and polyphenol oxidase, is lower than normal blanching, and, thus, it helps in inactivation of enzymes without affecting the flavor of the product. It has also proved affective for stabilization of rice bran (inactivation of lipase)

- Ohmic heating can also be used for continuous high temperature short time (HTST) sterilization.

- Some other applications in research as well as in the commercialization stage are fruit peeling, thawing (thawing can be achieved without increasing the product moisture content), and accelerated production of fermented beverages, such as beer and wine, etc.

- In general, the processed food has been observed to retain good quality up to three years of storage. In a commercial ohmic heating system to process liquid eggs, the liquid eggs are heated to 60–65°C and held for 3.5 minutes. The processing capacities of the equipment vary from 3 to 6 t.h⁻¹ and the eggs, processed in this manner, have a shelf life of 12 weeks.

With greater emphasis on efficiency and sustainability, this technology is gaining attention in the food industry. Liquid food pasteurization or sterilization is better through ohmic heating from a food safety angle. For foods with suspended particles, the overall quality is drastically reduced by traditional processing practices due to over-processing of the liquid phase as well as the surface of the suspended particles. Such problems are overcome by ohmic heating methods.

21.2 Infrared Heating

Infrared radiation is the portion of electromagnetic spectrum in the range of 75×10^{-6} cm to $100,000 \times 10^{-6}$ cm (0.000075–0.1 cm). Thus, frequencies of infrared radiation are lower than those of visible light but exceed those of most radio waves. Infrared radiation can be classified into three regions, namely, near-infrared or IR-A (0.75 to 1.4 µm), mid-infrared or IR-B (1.4 to 3 µm), and far-infrared, also termed IR-C (3 to 1000 µm). Temperatures above 1000°C produce near IR waves; temperatures below 400°C generated far infrared waves and medium waves are generated between these temperatures. Direct sunlight includes infrared (47% share of the spectrum), visible (46%), and ultraviolet (7%) light. Infrared radiations have heat penetration directly into the product, high heat transfer capacity, no heating of surrounding air, and fast process control.

21.2.1 Principle

The infrared heating and drying involves absorption of thermal spectrum of electromagnetic radiation (0.75-1000 µm) by the materials.

It was discussed in Chapter 7 that a perfect black body emits the amount of radiation according to the Stefan-Boltzmann law. The *Wein's displacement law* gives the relationship between the wavelength and temperature of a radiating body as follows:

$$\lambda_{max} T = 2897.6 \ \mu m.K \tag{21.15}$$

Higher temperatures produce radiation of shorter wavelengths.

Infrared waves incident on a material are either absorbed, transmitted, or reflected. The degree of heating of the object varies, depending on the amount of absorbed energy. The net rate of heat transfer to a food equals the rate of absorption minus the rate of emission.

$$q = A\varepsilon_1 \sigma(T_1^4 - T_2^4) \tag{21.16}$$

The amount of radiation absorbed by a grey body is the absorptivity, α and it is numerically equal to the emissivity. If there is no transmission of radiation, then reflectivity, $r = 1 - \alpha$.

$$q = \alpha \ \sigma A \ (T_1^4 - T_2^4) \tag{21.17}$$

where T_1 (K) and T_2 (K) are the temperatures of emitter and absorber, respectively. In the case of foods, the absorbed waves are converted into heat, which increases its temperature.

Both regular and body reflections for materials with rough surfaces form the diffuse radiation. The reflection is approximately 50%, at near-infrared ($\lambda < 1.25$ µm) wavelengths and is less than 10% at longer wavelengths. However, the actual reflection for each foodstuff needs to be measured for assessing the actual heat transfer.

The infrared waves can penetrate foods to a certain depth and, thus, can reduce the required heating time as compared to conventional methods. The penetration properties of the waves are also important for final product quality. *Penetration depth* is that at which the incident radiation energy is attenuated to 36.8%. The penetration ability of short waves is ten times higher than that for long waves. The limited penetration depth of infrared heating gives more uniform heating for thin products as compared to thick ones.

21.2.2 Equipment

Construction and Operation

- Radiator is the main component of IR ovens or equipment. The radiators may be classified as gas-heated radiators and electrically heated radiators. They can also be incandescent lamps.

- The electrically heated radiators are of different types, such as long-wave (ceramic, tubular/flat metallic), medium- and short-wave (quartz tube), and ultra-short-wave radiators (halogen). The gas heated radiators are long wave types.

- The short-wave radiators, corresponding to a λ_{max} of 1.3 µm, have a maximum temperature of 2228 K and, for medium infrared radiators, the maximum temperature is 1113 K, corresponding to a λ_{max} of 2.7 µm.

- Various reflectors used in the radiators are either individual gilt twin quartz tube, flat metallic/ceramic cassette reflector, or individual metallic/gold reflectors.

- Thyristors are used to control the degree of heating.

- IR systems may be either a continuous or batch type. A continuous system consists of a conveyor, usually made up of wire mesh. The infrared radiators are usually fixed above the belts, but systems with IR heaters positioned below the conveyor are also available.
- Hot air is often blown through the chamber for convective heating and removing the moisture from the vicinity of the product.

Salient Features

- The infrared source permits rapid evaporation of surface moisture because it heats the product surface. As the surface dries, the moisture from within the commodity is removed as soon as it diffuses to the surface.
- The factors responsible for obtaining an optimal heating result include radiator temperature (that decides the spectrum of wavelength), radiator efficiency, optical and thermo-physical characteristics, and infrared penetration properties of the product. The intensity and uniformity of radiation also affect the radiant heat transfer.
- Products that have better capability of absorbing radiation energy are dried more rapidly in this dryer. Thus, the surface emissivity is a major parameter in deciding the drying rate.
- The temperature at the surface of the commodity can be very high under the effect of radiation and, hence, should be monitored so that denaturation or browning on the surface is prevented.
- With a view to control the surrounding air temperature and humidity, the IR system could be combined with air convection.
- To avoid overheating there may be a need for water or compressed air cooling in some of the high intensity radiators.
- The advantages of the system include reduction in drying time and a better quality product than conventional heating. In addition, it is less expensive compared with dielectric and microwave drying.
- The limitation is that scaling up of this process is difficult, which is not always straightforward.

21.2.3 Applications

- The IR systems are effectively used for different applications, such as drying, frying, baking, roasting, blanching, thawing, and surface pasteurization.
- Fruits and vegetables, fish, etc. can be dried with IR and give better quality than conventional drying methods due to the high heat transfer rate. For drying, the short-wave infrared radiation, combined with convection, has been found more

beneficial. Integration of IR in freeze drying has also been reported to reduce freeze drying time in addition to giving better quality products, as in strawberries.
- Applications of near infrared or short-wave infrared have been studied for deep frying, meat cooking, and drying.
- IR radiation has a better application effect both at the surface and in the inner portions than conventional baking. For continuous baking and grilling, IR heating has been observed to be more useful. IR has been found useful for roasting. High quality dry-roasted almonds can be obtained by the sequential IR and hot air (SIRHA) method.
- IR frying is beneficial to health as the frying time and the residual oil in the finished product are less.
- Some studies have also proved the usefulness of IR for disinfestation and enzyme inactivation of rice. As IR can achieve blanching and dehydration simultaneously, it helps in reducing the processing time and the need for an additional blancher.
- IR dry peeling has been reported to give reduced peeling loss besides having the advantage of non-use of steam and chemicals.
- IR can also be applied for other applications, such as thawing and surface pasteurization of foods and packaging materials.

Safe and value-added consumer-friendly foods, while reducing the consumption of natural resources, are the main merits of the applications of IR in food processing. However, there is a need to optimize the parameters for different applications as baking, drying, etc. for different commodities along with the process control systems, which will aid in greater application of this technology.

21.3 Dielectric Heating

21.3.1 Principle

Dielectric heating, which includes the use of microwaves (MW) and radio frequency (RF) radiation, have enormous potential for heating and processing in food industries. The microwaves are electromagnetic waves with frequencies ranging between 300 MHz and 300 GHz. As the wavelength and frequency are related as, $\lambda = c/f$ ($c = 3 \times 10^8$ m.s^{-1}), the microwave radiation has a wavelength of 1 m to 1 mm. The electromagnetic radiation in the frequency range of 300 kHz to 300 MHz is termed radio frequency (RF) radiation. The wavelength of radiation varies between 1m to 1 km. The applied frequency in RF for heating purpose is between 1 to 200 MHz.

Figure 21.3 shows the frequencies covered under the microwaves and radio waves. Table 21.1 gives the frequencies assigned by the Federal Communications Commission (FCC) for industrial, scientific, and medical (ISM) use to avoid interference with telecommunication. Normally, microwave

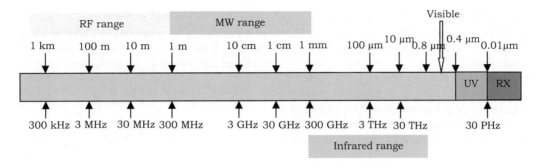

FIGURE 21.3 Electromagnetic spectrum showing the infrared, microwave and RF range

TABLE 21.1

Frequencies assigned by the FCC for industrial, scientific, and medical (ISM) use

Type of wave	Frequency
Microwaves	915 MHz ± 13 MHz
	2450 MHz ± 50 MHz
	5800 MHz ± 75 MHz
	24125 MHz ± 125 MHz
Radio waves	13.56 MHz ± 6.68 kHz
	27.12 MHz ± 160.00 kHz
	40.68 MHz ± 20.00 kHz

heating of foods is carried out at 2450 or 915 MHz and RF heating is done at 13.56 or 27.12 MHz.

The radio waves and microwaves are non-ionizing radiation, unlike X rays and gamma rays. These rays, like other radiations, may get absorbed, reflected, or transmitted through an object while travelling in straight lines. These waves may change direction while travelling from one medium to other.

Water forms a major part of biological materials and forms an electric dipole because of the negatively charged oxygen atom separated from positively charged hydrogen atoms. When a rapidly oscillating electric field is applied to a dielectric material, such as food, the dipoles in the water reorient with each change in the field direction. There is a rise in temperature due to friction among molecules caused by rapidly rotating molecules. This phenomenon, called *dipole* or *orientation polarization*, is strongly dependent on temperature. The following two phenomena explain the conversion of microwave or RF energy into heat.

1. A rapidly changing electric field enables the molecules with permanent dipole moment to change polarity and rotate at a frequency of many millions of times per second. Heat is evolved due to friction among the molecules.

2. Ionic conductance also takes place during MW and RF heating. Two opposite (+ and -) charged particles or ions are produced by the salts dissolved in biological materials. The ions collide with other molecules and drift due to the electric

field in a billiard ball fashion and these collisions give rise to heat.

At frequencies below 1 GHz, it is ionic polarization that predominates, whereas MW heating is more significant at frequencies above 1 GHz.

In studying the interaction between the electromagnetic radiation and the foods, dielectric properties of the food materials have a critical role to play. If an electromagnetic field is applied on a food, it can store electrical energy like capacitors and can dissipate electrical energy as resistors. Macroscopically, *relative complex permittivity* $\varepsilon^*(\omega)$ represents this behavior, which is the square of the complex refractive index $[\varepsilon_1 + i\varepsilon_2 = (n_1 + in_2)^2]$. Many metals at low frequencies have reflectivities close to 100%.

There are two components of the relative complex permittivity (ε^*), indicated as follows:

$$\varepsilon^* = \varepsilon_1 - i\varepsilon_2 \qquad (21.18)$$

where, ε_1 is the dielectric constant, ε_2 is the dielectric loss factor and $i = \sqrt{-1}$. The dielectric constant is defined as the ratio of its capacitance to the capacitance of air or vacuum under the conditions being studied. ε^* is the ability to store energy by a material in response to an applied electric field (the permeability for vacuum is $4\pi \times 10^{-7}$ H.m^{-1} and $\varepsilon^* = 1$) and ε_2 indicates the ability to dissipate electric energy into heat by a material. An increase in the dielectric constant means that the capacitor can store more energy. Therefore, measurement of capacitance can act as a measure of dielectric properties.

The ratio of the magnetic induction to the magnetic intensity is known as *magnetic permeability* and is denoted by the symbol μ. It is a measure a material's resistance to the magnetic field. i.e., how much a magnetic field can penetrate through a material. It is a scalar quantity.

The magnetic permeability (H.m^{-1}) can also be written as:

$$\mu = B/H \qquad (21.19)$$

where, B = magnetic intensity and H = magnetizing field.

The magnetic permeability for most biological materials is almost equal as that of free space ($\mu_o = 4\pi \times 10^{-7}$ H.m^{-1}). It is for this reason that a majority of natural biological materials do not have a strong interaction with the magnetic portion of the electromagnetic field, and only their interaction with the electric field causes heating. Substantial heating occurs in magnetic materials, such as ferrite, however, due to interaction with the magnetic field.

In microwave and radio frequency heating, the dielectric loss factor is primarily responsible in deciding the material's ability to dissipate electric energy into heat. As indicated earlier, the effects of dipole rotation and ionic conduction together leading to energy dissipation into heat can be given as follows:

$$\varepsilon_2 = \varepsilon_{2d} + \varepsilon_{2\sigma} = \varepsilon_{2d} + \frac{\sigma}{2\pi f \varepsilon_0} \qquad (21.20)$$

In the above equation, ε_{2d} is the contribution from dipole rotation and $\varepsilon_{2\sigma}$ is the contribution from ionic conduction (S.m^{-1}), and f is the frequency of electromagnetic waves in Hz (i.e., s^{-1}). Permittivity of free space (ε_0) is 8.854×10^{-12} F.m^{-1}.

Dielectric loss tangent (tan Ω) is another characteristic of dielectric materials, defined as the ratio of dielectric loss factor and dielectric constant.

$$\tan \Omega = \varepsilon_2 / \varepsilon_1 \qquad (21.21)$$

Ω is known as the *loss angle*. The loss tangent expresses the phase shift between the orientation of the molecules and the alternate electrical field. In an alternating current containing an ideal capacitor, the current magnitude will lead the voltage by 90° through its cycle. This angle will be less if there is a dielectric material in the circuit in place of the ideal capacitor and this reduction is known as loss angle Ω. It is commonly expressed as the *loss tangent* (tan Ω). The value of tan Ω for low loss material is ≤ 0.005 and that for medium loss material is 0.005 to 0.01. This is like having a resistor parallel to a capacitor in a circuit.

The dipoles react to the changes in the electric field usually after a fraction of a microsecond; this time gap or delay is termed as *relaxation time*.

Dielectric properties of a material determine the power (P) absorbed by the material in a unit volume. It can be written as:

$$P = 2\pi f \varepsilon_0 \varepsilon_2 E^2 \qquad (21.22)$$

Here, P (W.m^{-3}) is the power conversion per unit volume and E (V.m^{-1}) is the electrical field strength in the material. Considering the value ε_0 as 8.854×10^{-12} F.m^{-1}, the above equation can be modified as:

$$P = 55.63 \times 10^{-12} f E^2 \varepsilon_2 \qquad (21.23)$$

When electromagnetic waves strike a lossy object, a part of these penetrate the object and the other part is reflected back. The depth to which microwaves will penetrate is more for materials with a higher wavelength; however, the strength will reduce with the depth of penetration.

$$\delta_p = \frac{\lambda_0}{2\pi \sqrt{\varepsilon_1 \tan \Omega}} \qquad (21.24)$$

where δ_p is the penetration depth- it is the depth at which power of a penetrating wave reduces to 1/e or 36.8% of its original value. λ_0 is the wavelength of the microwaves in free space. The penetration depth (δ_p) is also expressed as:

$$\delta_p = \frac{c}{2\sqrt{2}\pi f \left[\varepsilon_1 \left(\sqrt{1 + \left(\frac{\varepsilon_2}{\varepsilon_1} \right)^2} - 1 \right) \right]^{1/2}} \qquad (21.25)$$

where, c is speed of light in free space (3×10^8 m.s^{-1}).

Temperature distribution in a microwave irradiated food can be analyzed by solving the time dependent heat conduction equation with a heat source term. The heat source term accounts for the microwave generated heat.

$$\frac{\partial t}{\partial \theta} = \alpha \left(\frac{\partial^2 t}{\partial x^2} + \frac{\partial^2 t}{\partial y^2} + \frac{\partial^2 t}{\partial z^2} \right) + \frac{q_{mw}}{\rho . c_p} \qquad (21.26)$$

Foods absorb energy and heat rapidly because of their high moisture content and high loss factor. Since there is an uneven distribution of energy, heat transfer within the food also takes place by conduction.

Microwave (or RF) energy is not a thermal energy. Heating occurs due to the interactions between waves and the dielectric material. There is conversion of electromagnetic energy to thermal energy and the resulting energy is absorbed throughout the volume of wet material. The increase in internal vapor pressure causes the moisture to move from the inner parts to the surface of the product, where the moisture is removed by evaporation.

As the temperature of food increases, there is also heat transfer by conduction within the product. The conduction heat may further flow to interior parts and heat the portions that are not heated by the microwaves. Heating rates are likely to be low because the materials, which are poor electrical conductors, are also poor heat conductors. Thus, thick materials are often placed between two parallel plates (electrodes) in a dielectric field.

Dielectric Properties of Food Materials

In addition to the frequency of the waves and temperature, the composition of food also influences the dielectric properties of foods. These properties have been extensively studied and are available in the literature. The values of dielectric constant and loss factor for some food materials are given in Table 21.2 for illustrative purposes.

Typically, with minimal variation in temperature and frequency, the dielectric constant for biological materials can be assumed to be constant. The value of the dielectric loss factor in low loss materials is considered to be minimal ($\varepsilon_2 < 0.01$). Low loss materials are those that can absorb more energy and dissipate less. Lossy materials, unlike low loss ones, dissipate rapidly any absorbed electric energy into heat.

Factors Affecting Dielectric Heating

Dielectric materials are better absorbers and transmitters of microwaves. Dielectric properties at the applied frequency determine the ability of a specific product to absorb the

TABLE 21.2

Dielectric properties of some foods

Food	Temp, °C	Frequency (MHz)	Dielectric constant	Loss factor
Beef	25	915	62	27
		2450	61	17
Potato	25	915	65	19
		2450	64	14
Carrot	25	915	73	20
		2450	72	15
Corn (yellow)	30	915	3.58	1.34
		2450	2.71	0.69
Rice	25	2450	3.108	0.491
Wheat	30	915	3.68	1.34
		2450	2.67	0.67

microwave energy. Similarly, the thermal diffusion through the product is controlled by the specific heat and thermal conductivity values.

Heat is generated within the product in the case of dielectric drying, and the moisture evaporated moves to the surface of the material by the vapor pressure difference. However, the moisture is not removed from the surface as efficiently and may re-condense on the surface. Hence, the MW and RF heating process for drying is often supplemented by convective drying.

As discussed in equation 21.24, higher wavelengths imply more penetration depth and, thus, low frequency waves can penetrate more deeply than higher frequency waves. However, even for low frequency waves, the distance between the radiator and the absorber should be less than $1/2\pi$ of the wavelength, which is very less. Thus, heating by the application of a low frequency electric field, as in the case of radio waves, is a near contact process.

The relaxation time is influenced by the viscosity of the material and, thus, is temperature dependent. The dielectric constant of ice is lower than that of water. The dielectric constant also further lowers as the temperature of the ice is reduced. Ice is therefore more transparent to microwaves than water. Thus, during thawing the microwaves are better absorbed by the food.

As per equation 21.22, a dielectric material absorbs electrical power in a way that it is linearly proportional to the square of the electric field, the frequency of the waves, and the relative dielectric loss factor. It has been found experimentally that dielectric heating technology can be used effectively when the loss factor is between 2 and 100. Small penetration depths in MW drying are the result of very high value of the loss factor. A material is transparent to the electromagnetic waves if the loss factor is low. Thus, materials with low loss factor can be added with a small amount of salt to increase the loss factor so that they can be dried with MW and RF.

Some polymer packaging films, glass, and paper do not get heated because they have low loss factors (being transparent to microwaves). Generally, these materials make excellent utensils for cooking in a microwave oven.

A question arises as to why microwave ovens use 915 MHz and 2.45 GHz instead of 20 to 1000 GHz. Energy absorbance

increases and the penetration depth decreases with increasing frequency. The penetration depth is usually a few centimeters for the type of foods cooked/dried in MW ovens. If the frequency applied is 20 GHz, then most of the waves will be absorbed in a thin layer close to the surface of the food, while the interior will still remain cold. Therefore, lower frequency microwaves are used.

In case of an RF or MW heater, if the heat is not dissipated rapidly from the areas heated by the waves, it causes localized heating. The effect is further pronounced when the change in temperature at that region causes changes in the properties of the material and accelerates the dissipation of microwave power in that region. It causes the creation of a hot spot, which is a local *thermal runaway*.

Dielectric applicators vary depending on whether the process is batch or continuous or based on the specific product characteristics. The drying systems are usually combined with other methods, such as hot air, vacuum, infrared, or freeze drying, for more uniform heating and more efficient removal of moisture.

21.3.2 Equipment

As discussed, the microwave (MW) and radio frequency (RF) equipment come under the category of dielectric equipment.

A dielectric heating equipment, in general, is composed of a wave generator and an applicator along with the handling devices and controls. The generator and transportation of electromagnetic waves differentiate between MW and RF dryers.

Microwave Heating Systems

Construction and Operation

The microwave equipment consists of a microwave generator (for example, a magnetron); wave-guides, which are usually aluminum tubes; and a chamber for holding the produce for microwave application. For continuous operation, there is a tunnel in which the material moves on a conveyor belt. Figure 21.4 shows schematically how a batch type microwave heating unit functions.

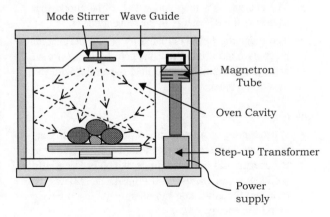

FIGURE 21.4 Schematic representation of a microwave oven (after Lule and Koyuncu, 2017 with permission)

Magnetron

- The magnetron is an oscillator, which converts the supplied power into microwave energy, i.e., a power line electric current at 50 or 60 Hz is converted to electromagnetic radiation at 2450 or 915 MHz. It is the nucleus of the high voltage system, which generates microwave energy.

- Several thousand volts of direct current are needed by the magnetron. Therefore, electrical power is drawn from the line and the power supply system is designed to convert it to the high voltage for the magnetron. (The AC line voltage is stepped up to high voltage, which is then changed to an even higher DC voltage. The high voltage, typically 3000 to 4000 volts, that powers the magnetron tube is produced by a step-up transformer rectifier and filter that converts 120/ 230 V AC to 4000 V DC.)

- With the application of the high voltage system, rapidly oscillating microwave (or RF) energy is liberated by the electrons.

- The basic parts of a magnetron include the anode, the filament/cathode, the antenna, and the magnets.

- The anode is a ring of resonant cavities. The cathode, which produces the free electrons, is a hot metal cylinder located inside the anode ring. The anode (or plate) is a hollow cylinder of iron from which an even number of anode vanes extend inward. The open trapezoidal-shaped areas between each of the vanes are resonant cavities that serve as tuned circuits and determine the output frequency of the tube.

- The magnetron is considered a diode because it does not have a grid like an ordinary electron tube. The magnetic field is created in between the anode and the cathode.

- The anode is so arranged that alternate segments are opposite in polarity, in other words, the cavities are connected in parallel with regard to the output.

- The cathode, also known as the heater, placed at the center of the magnetron, is supported by the large and rigid filament leads.

- The antenna is a loop connected to the anode that transmits the RF energy to the wave guide.

- The magnetic field is created by strong permanent magnets fixed around the magnetron. These are so arranged that the magnetic field is parallel with the axis of the block.

Wave Guide

- When rapidly oscillating microwave radiation is generated, as discussed above, the antenna transmits the RF energy into the wave guide, which is essentially a hollow metal enclosure. It propagates, radiates, or transfers the generated radiation to the oven cavity with little energy loss.

- Sheets of highly conductive metals (e.g., aluminum, copper) are used for wave guides. Wave guides transfer the MW radiation to the heating chamber; the internal reflections also guide the waves to move forward. To reduce energy losses, its inner surfaces should be smooth.

- Wave guides in domestic ovens are just a few centimeters long, whereas these can be a few meters long in industrial units.

Stirrer

- Localized hot and cold spots in food, caused by uneven reflections, are minimized by the use of rotating carousels (turn tables) and mode stirring fans. Both methods reduce shadows and help radiation to reach all parts of the food.

- A fan-shaped device, the stirrer, rotates and distributes the transmitted energy throughout the oven.

- Antennas of different designs are used to direct energy beams over the food in continuous tunnels.

Oven Cavity

- The oven cavity or the MW applicator is a metallic enclosure. MW applicator for batch operations can be a metal chamber whereas it would be a belt-tunnel system for continuous operations. The heating chambers and tunnels are properly sealed to prevent the escape of microwaves. The residence time in continuous industrial microwave dryers is adjusted depending on the commodity and moisture removal requirements.

- Food to be heated is enclosed in the oven cavity within the metallic walls.

- The chamber acts as a *Faraday cage*; a Faraday cage is a container made of a conductor and provides a shield against leakage of electromagnetic radiation from inside to outside and vice versa.

- The food intercepts the distributed MW energy from the stirrer and is reflected by the walls from many directions.

- The metal grids having holes smaller than the wavelength of the MW are used to cover the front door as well as the bulb cavity. These grids act like radiation shield and do not allow the waves to go out.

- Industrial applications use two basic designs, mono-mode and multi-mode, of microwave applicators. Mono-mode reactors have small compact cavities, in which microwave irradiation is directly focused on one single vessel. It gives a high microwave field density and fast heating rates. Multi-mode reactors have larger cavities in which the microwave field is distributed. As shown in Figure 21.4, the waves are reflected from the cavity walls and multiple modes of the waves interact with the load. This is an example of a multi-mode applicator.

- Figure 21.5 shows the schematic depiction of a mono-mode applicator. The selection of a proper design depends on the processing requirements.

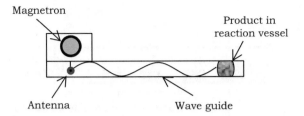

FIGURE 21.5 Schematic depiction of a mono-mode applicator

- In single or mono-mode applicators, the cavities are small, which allow only a single pattern of microwaves. Only a small amount of material, with low effective loss factors, can be processed.

- The multi-mode cavities have at least one dimension of the metal cavities larger than the wavelength of the applied microwave. The domestic MW oven comes under this category. In this case, the wavelength of 2450 MHz microwave is 122 mm, and, hence, the cavity should have at least one dimension more than 122 mm.

- Electric field patterns in a multi-mode applicator are highly non-uniform, thus, mode stirrers are used for improving the uniformity in application. The rotating antenna (shaped as a rotating vaned metallic fan) is the stirring device, which typically rotates at 1–10 rev.s^{-1}.

- Many domestic MW ovens also have turntables for obtaining uniformity of dielectric heating. Specific designs of antenna are used in continuous tunnels.

- Another measure to reduce the area of food not exposed to the MW radiation (*shadowing*) is using multiple MW sources, each with a slightly different frequency to produce different mode patterns. Another method to reduce shadowing is to use multiple MW inputs. As discussed above, the cavity dimension also can be selected so as to allow maximum number of modes.

Salient Features

- The equipment is small and compact and there is easy control of the drying/cooking parameters.

- Microwaves selectively heat moist areas within the food. There is rapid volumetric heating and there is no overheating of the surface. Thus, browning of the surface is avoided. At the same time, the moisture content, volume, shape, and mass of food affect the rate of heating. This leads to uneven heating in some foods, specifically, which have uneven moisture distribution, such as jam-filled donuts.

- Multi-mode cavities are flexible, more popular in industrial applications, and permit heating of products of a wide range of sizes and shapes.

- Complex three-dimensional standing wave patterns are created by the MW radiation that enters a multi-mode cavity and undergoes multiple reflections.

- When the position of food changes or when the dielectric properties change with heating, the wave patterns in multi-mode cavities become highly unpredictable. In fact, the nature of the load, the dimensions of the cavity, as well as the energy spectrum, collectively, determine the type and number of the patterns in a specific cavity.

- For continuous industrial equipment, the power output ranges from 30 to 120 kW. It is important that the size of the heating chamber be matched with the power output from the magnetron.

- The oven walls and cooking utensils are not dielectric materials, and they are not heated by the microwaves. But, usually, they are observed to be heated, which is due to the contact with the hot food.

- The advantages of microwave food processing include short start-up time, volumetric heating, and quick heating, which help in better retention of product nutritional and sensory qualities. It can also heat food selectively depending on the moisture profile. The other benefits include smaller size of equipment, high energy efficiency, and better process control.

- The limitations that microwave food processes pose include lack of color and flavor development (which may be necessary for some processing applications), non-uniform heating, high moisture loss, soggy surface, and firm texture in foods.

Hybrid Type of Systems

Microwave heating systems have been used in conjunction with convective, vacuum, and other drying systems for more efficient removal of moisture from the product. In a microwave-assisted convective dehydration process, microwaves are used to improve the rate of moisture diffusion from inside the produce so as to improve the rate of drying and reduce the drying time. Recent developments have combined vacuum or other forms of moisture removal techniques with microwaves for improving the drying rate and quality of produce.

Radio Frequency Heating Equipment

In the radio frequency heating process, the material is held or moved with the help of a conveyor between two parallel plates or electrodes. As AC supply is passed through the electrodes, it creates an alternating electromagnetic field and the polar water molecules placed in between the electrodes are continuously reoriented to align with the alternating field. This creates heat generation due to friction in water molecules, which eventually rapidly heats the whole mass. It is due to the effect of the RF waves that during drying, the inner portions of the material with more moisture are hot and dry and it is cooler and wetter on the outer surface. Dry materials, such as dry sand, dry paper, and dry food, may not heat at all. The *specific absorption ratio* (SAR) is the RF energy absorbed per mass of tissue expressed as watts per kilogram (W.kg⁻¹). As the material being processed behaves as the dielectric of a capacitor in the applied electric field, RF heating is also known as *capacitive heating*.

In conventional systems, the RF power is generated by a standard oscillator circuit using triode tubes. The electrodes and the material are coupled inductively to the circuit. Another system, known as the 50Ω system, has an automatic tuning device in the applicator circuit that continuously adjusts its overall impedance to 50Ω. This causes the load to receive a stable coupling of RF energy. The 50Ω systems have so far been used only for experimental purposes because such systems are very expensive. In the following, the conventional system is emphasized.

Construction and Operation

- The equipment for RF heating consists of two main components, namely, RF generator (Oscillator) and RF applicator. The generator converts the normal electricity into RF waves. The electromagnetic energy is sent to the applicator, which applies it to the product to be heated.

- The generator constitutes a suitably designed combination of capacitors and inductances connected to a vacuum valve (the triode) and high voltage DC supply. The inductance is an oscillating LC circuit.

- A high voltage alternating electric field is applied to the product placed between the two electrodes.

- The applicators may be of either batch or continuous type. In case of a continuous system, the food material is placed on a conveyor. Often the conveyor belt itself is one of the electrodes (Figure 21.6).

- The whole system is enclosed in a metal chamber to avoid radiation leakage.

- Several designs of RF applicators are available. There are basically three categories, namely, flat or plate electrodes, staggered electrodes, and stray field electrodes. They are illustrated in Figure 21.7. The applicator is selected on the basis of product shape and size.

- In the case of the *plate electrodes*, the material to be heated is placed between two parallel flat plates. The gap between the plates can be varied to adjust the power supply. The material can also be moved as a layer on a conveyor belt in between the plates.

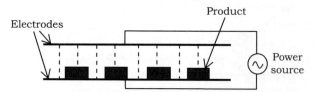

a. Flat electrode or through-field electrode

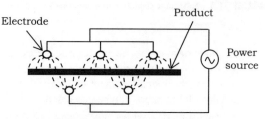

b. Staggered through field electrode

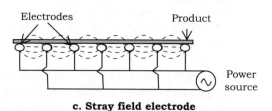

c. Stray field electrode

FIGURE 21.7 Schematic diagram of RF applicators (After Jones and Rowley, 1996 with permission)

Thus, it is possible to dry a relatively thick layer of a material.

- In the case of a *staggered electrode system*, there are two rows of rod or tube electrodes arranged in a staggered manner on both sides of a belt, on which the material moves as a thin layer. As compared to the plate electrodes, this arrangement gives much higher electric field strengths within the thin layer and it can apply 30 to 100 kW.m⁻² to the material in a thin layer. It is also possible to adjust the electric field by changing the gap between the electrodes in lower and upper rows. The electric field can be aligned and concentrated in the layer parallel to the direction of movement of the belt.

- In the case of *stray field electrodes*, rod or tube electrodes are positioned parallel to the layer of material to be heated and the adjacent electrodes are connected to the opposite polarity. The electric field can be aligned in parallel with the material layer by placing the material

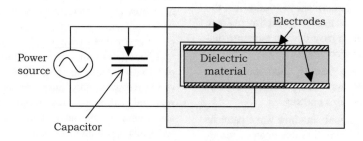

FIGURE 21.6 Schematic of a RF heater

above the banks of the electrodes. These systems are generally used for thin layers of products up to 10 mm, although such systems result in less homogeneous fields as compared to staggered field electrodes.

- Industrial RF heaters have a power rating of 10–300 kW.
- Like MW heating, a combination of RF heating and convective heating offers better results. The combination will heat the water in the interior, move it to the surface, and remove it from the surface, resulting in more uniform drying.

Salient Features

- Because RF heaters use longer wavelengths than MW radiation, RF radiation is capable of deeper penetration into the products, and it provides rapid and uniform heat distribution. In addition, there is neither surface overheating nor hot spots. RF radiation also offers lower energy consumption.
- The microwaves can penetrate to a certain depth into a material and the heating of the remaining parts is by conduction. However, RF heating causes whole volume heating of the material placed in between the plates/electrodes. The thickness of the food in the RF system should conform to the distance between the capacitor plates.
- Due to potential electric hazard, electrode size is restricted, for example, for a 27 MHz equipment, the electrode should not be larger than 1.4 m.
- The effectiveness of RF heating is greatly influenced by the products' dielectric properties. For any specific material, the energy efficiency and heating rate are maximum for the frequency when the dielectric loss factor is maximum. This frequency is known as "Debye resonance."
- The moisture present in the food affects the absorption of energy and, thus, affects the drying rates. Further, the dielectric properties of the food are affected by the material temperature, and, therefore, the penetration depth also varies with the material temperature.
- As the loss factor as well as dielectric constant depend on moisture content and temperature during drying process, the power absorbed by the food material reduces during the process. This effect can be compensated by the application of pulsed field systems or graduated field configurations.
- In addition to the potential benefits as lower drying time, uniform heating and better product quality, the system also involves lower energy, less floor space, and reduced labor costs. This is also a system with no pollution effect.
- Constraints of using RF heating at present include high operational cost and technical problems, such as dielectric breakdown and thermal runaway heating. These could adversely affect both the product and the package.

21.3.3 Applications

Applications of Microwave Heating

- Microwaves are commonly used for cooking food. Microwave cooked rice retains higher levels of protein, fat, and ash contents as compared to conventional boiling and steaming methods. Short cooking times and low temperatures in microwave irradiation, normally, inhibit Maillard reaction. There is substantial reduction in the energy consumption in microwave cooking as compared to normal cooking in the case of both un-soaked and presoaked rice. Reheating of the prepared (ready to eat, RTE) food is also effective with the microwave oven.
- Microwave dryers are increasingly being used for different food materials, including many fruits and vegetables, and several commercial microwave dryers have been developed. Industrial pasta drying using microwave energy is quite prevalent. The overall process uses conventional hot air drying first, then microwave drying, and finally hot air drying. The drying phase using microwaves helps in reducing the drying time and improving the product qualities, through reduced case hardening, in comparison to conventional dehydration methods.
- Microwaves are also used for pasteurization and sterilization. Although it is possible to pasteurize using 2450 MHz microwaves, more uniform food pasteurization is achieved using 915 MHz microwaves. This difference is possibly due to greater penetration depths at 915 MHz microwaves in comparison to those at 2450 MHz. It has been shown that microwave treatments yield superior lethality and higher reduction in D-values as compared to conventional heating. The intense effect of microwave treatment is attributed to the sudden increase in internal pressure generated within the core of pathogens.
- Microwaves have also been used for blanching. As microwave blanching is efficient in heat transfer in foods without requiring much water, the nutrient loss is much lower as compared to conventional methods. It has been demonstrated that the amounts of vitamin C, iron, protein, other nutrients, and ash in microwave blanched fruits and vegetables are much higher as compared to those in hot water blanching.
- Microwaves are commonly used for thawing and defrosting (or tempering) of food. In this the temperature of the frozen food is brought to between -5° to -2°C, i.e., slightly lower than that required for complete thawing, in a short time. Thus, the large space and long duration involved in the conventional thawing process is not required. There is also reduced drip loss and dehydration, microbial growth, and chemical deterioration. In fact, microwave tempering and thawing at 915 MHz has been found to be the most successful microwave heating application in the food industry.
- Microwaves have been successfully used in baking to reduce the time and energy consumption. Cakes

produced by microwave baking are reported to have better textural and sensory qualities as compared to convection methods. Microwaves have also been tested for roasting coffee beans successfully. The product has a better antioxidant activity in comparison to conventionally roasted ones.

With microwaves, thawing does not occur uniformly as the dielectric properties of the material change during the heating process. A material can absorb more microwave energy with an increase in temperature. In other words, the rate of temperature rise will progressively increase as heating progresses. Thus, some parts may be thawed, and some may remain frozen. It is known as *runaway heating*. To mitigate runaway heating effects, heat generated by the microwaves is required to be controlled. The dielectric properties, size, and shape of the biological materials and the wave characteristics affect the thermal non-uniformity in microwave thawing. An effective way to achieve uniform heating is to use lower power levels in a continuous manner. Turning microwaves off intermittently also permits the thermal non-uniformity to equilibrate, leading to more uniform thawing in commercial operations.

When the microwaves enter a food they are transformed to heat energy and when the food is taken out of the microwave application chamber, there is no residual radiation or radioactivity in the food. Food processed using microwaves is completely safe for consumption. However, research efforts are still continuing for obtaining better heating uniformity as well as for studying the feasibility of applying this technology in different types of foods.

Applications of RF Heating

- RF radiation has been found to be advantageous over other methods for several drying, thawing, baking, and sterilization/pasteurization operations. Initial attempts were made for cooking processed meat, heating bread, and dehydrating vegetables.
- A RF heater is more suited for those materials that have low thermal conductivity, high density, and/or high specific heat. RF heating is not suitable for conductive materials and foodstuffs with metals. While comparing RF heating with MW radiation heating, RF radiation is more suitable for larger and thicker food objects.
- RF radiation has been used effectively for finishing drying of baked products to impart desirable texture and mouth feel.
- Applications also include cooking, pasteurization, and roasting of different products and thawing of frozen foods.
- RF radiations have also been used to kill some pests in food after harvest. There are some medical applications also.

In commercial large-scale application of dielectric heating systems, there may be non-uniform heating and the cost of the equipment may be higher compared to other conventional drying methods. However, in comparison to microwave systems, a radio frequency drying system offers more uniform energy distribution and better penetrations in foods, and, hence, it can be expected to be increasingly applied in the food industry.

Check Your Understanding

1. Describe the working principles of
   a. Ohmic heating;
   b. Infrared heating;
   c. Dielectric heating.
2. Differentiate between
   a. Convection heating and microwave heating;
   b. Microwave heating and radio frequency heating.
3. Write short notes on
   a. Factors affecting ohmic heating process;
   b. Ohmic heating equipment;
   c. Applications of ohmic heating;
   d. Infrared heating equipment;
   e. Dielectric heating equipment.

BIBLIOGRAPHY

Ahmed, J., Ramaswamy, H.S. and Raghavan, V.G.S. 2007. Dielectric properties of butter in the MW frequency range as affected by salt and temperature. *Journal of Food Engineering* 82: 351–358.

Altemimi, A., Aziz, S. N., Al-Hilphy, A.R.S., Lakhssassi, N., Watson, D. G. and Ibrahim, S. A. 2019. Critical review of radio-frequency (RF) heating applications in food processing. *Food Quality and Safety* 3: 81–89.

Anderson, A. and Finkelstein, R. 1919. Study of the electro pure process of treating milk. *Journal of Dairy Science* 2: 374–406.

Chandrasekaran, S., Ramanathan, S. and Basak, T. 2013. Microwave food processing—A review. *Food Research International* 52:243–261.

Datta, A.K. and Anantheswaran, R.C. 2001. *The Handbook of Microwave Technology for Food Applications* 1st ed. Marcel Dekker, New York: DOI: 10.1017/CBO9781107415324.004.

de Alwis, A.A.P. and Fryer, P.J. 1990. The use of direct resistance heating in the food industry. *Journal of Food Engineering* 11(1): 3–27.

Hasan, M. U., Malik, A. U., Ali, S., Imtiaz, A., Munir, A., Amjad, W. and Anwar, R. 2019. Modern drying techniques in fruits and vegetables to overcome postharvest losses: A review. *Journal of Food Processing and Preservation* 43(2): e14280. https://doi.org/ 10.1111/jfpp.14280

https://moraritsipanificatie.eu/2019/11/04/coacerea-ohmica-perspective-spectaculoase-in-obtinerea-produselor-de-panificatie-fara-gluten/

https://wiki.anton-paar.com/in-en/which-microwave-synthesis-reactor-is-the-best-one-for-your-research/

https://www.pmg.engineering/ohmic-heating-of-milk-milk-products

Imai, T., Uemura, K., Ishida, N., Yoshizaki, S. and Noguchi, A. 2007. Ohmic heating of Japanese white radish *Raphanussativus L. International Journal of Food Science and Technology* 30(4): 461–472.

Jones, P. L. and Rowley, A. T. 1996. Dielectric drying. *Drying Technology* 14(5): 1063–1098.

Kandasamy, P. and Sarkar, S. 2019. Recent developments in microwave drying technology for food preservation. *Journal of Emerging Technologies and Innovative Research* 6(2): 474–488.

Kar, A., Chandra, P., Parsad, R. and Dash, S. K. 2004. Microwave drying characteristics of button mushroom (Agaricus bisporus). *Journal of Food Science and Technology* 41(6): 636–641.

Koubaa, M., Roohinejad, S., Mungere, T.E., El-Din, B.A., Greiner, R. and Mallikarjunan, K. 201. Effect of Emerging Processing Technologies on Maillard Reactions. In *Encyclopaedia of Food Chemistry. Elsevier*,76–82.

Kudra T. and Mujumdar A. S. 2009. *Advanced Drying Technologies* 2nd ed. Taylor & Francis, Inc., Bosa Roca

Lee, S.Y., Ryu, S. and Kang, D.H. 2013. Effect of frequency and waveform on inactivation of *Escherichia coli* O157:H7 and *Salmonella enterica* serovar Typhimurium in salsa by ohmic heating. *Applied and Environmental Microbiology* 79(1): 10–17.

Lima, M. and Sastry, S.K. 1999. The effects of ohmic heating frequency on hot-air drying and juice yield. *Journal of Food Engineering* 41:115–119.

Lima, M., Heskitt, B.F. and Sastry, S.K. 2001. Diffusion of beet dye during electrical and conventional heating at steady-state temperature. *Journal of Food Process Engineering* 24(5): 331–340.

Lule F. and Koyuncu, T. 2017. Convective and microwave drying characteristics, energy requirement and color retention of dehydrated nettle leaves (*Urtica diocia* L.). *Legume Research- an International Journal* 40(4): https://doi:10.18805/lr.v0i0.8409.

Onwede D. I., Hashim, N., Abdan, K., Janiaus, R. and Chen, G. 2019. The effectiveness of combined infrared and hot-air drying strategies for sweet potato. *Journal of Food Engineering* 241: 75–87.

Onwude, D., Hashim, N. and Chen, G. 2016. Recent advances of novel thermal combined hot air drying of agricultural crops. *Trends in Food Science & Technology* 57:132–145.

Palaniappan, S. and Sastry SK.1991. Electrical conductivities of selected foods during ohmic heating. *Journal of Food Process Engineering* 14: 221–236.

Piyasena, P., Dussault, C., Koutchma, T., Ramaswamy, H. S. and Awuah, G. B. 2003. Radio Frequency Heating of Foods: Principles, Applications and Related Properties—A Review. *Critical Reviews in Food Science and Nutrition* 43(6): 587–606. https://doi.org/10.1080/10408690390251129

Ramaswamy, H.S., Abdelrahim, K.A., Simpson, B.K. and Smith, J.P. 1995. Residence time distribution (RTD) in aseptic processing of particulate foods: A review. *Food Research International* 28(3): 291–310. https://doi:10.1016/0963-9969(95)00005-7.

Richa, R., Shahi, N. C., Singh, A., Lohani, U. C., Omre, P. K. Anil Kumar, G. V. and Bhattacharya, T. K. 2017. *Ohmic heating technology and its application in meaty food: A review. Advances in Research* 10(4): 1–10..

Sadhu, P. K., Pal, N., Bandyopadhyay, A. and Sinha, D. 2010. Review of induction cooking - A health hazards free tool to improve energy efficiency as compared to microwave oven. https://www.researchgate.net/publication/224132587

Sakharam, K.S., Pandey, J.P., Singh, A., Kumar, A., Shukla, A.K., 2016. Development of ohmic heating apparatus for extraction of rapeseed oil. *International Journal for Innovative Research in Science & Technology* 2(11): 211–215.

Sarang, S., Sastry, S.K. and Knipe, L. 2008. Electrical conductivity of fruits and meats during ohmic heating. *Journal of Food Engineering* 87: 351–356.

Sastry, S.K. 2007. A model for heating of liquid-particle mixtures in a continuous flow ohmic heater. *Journal of Food Process Engineering* 15: 263–278. https://doi: 10.1111/j.1745-4530.1992. tb00156.x.

Sastry, S.K. 1997. Measuring the residence time and modeling a multiphase system. *Food Technology* 51(10): 44–48.

Sastry, S.K. and Cornelius, B.D. 2002. *Aseptic Processing of Foods Containing Solid Particulates.* John Wiley and Sons, Inc., New York.

Sharma, G.P. and Prasad, S. 2001. Drying of garlic (Allium sativum) cloves by microwave-hot air combination. *Journal of Food Engineering* 50: 99–105.

Skudder, P.J. and Biss, C. 1987. Aseptic processing of food products using ohmic heating. *Chemical Engineering* 433:26–28.

Sutar, P. P.and Prasad, S. 2007. Modelling microwave vacuum drying kinetics and moisture diffusivity of carrot slices. *Drying Technology* 25(10): 1695–1702.

Sutar, P.P.and Prasad, S. 2011. Optimization of osmotic dehydration of carrots under atmospheric and pulsed microwave vacuum conditions. *Drying Technology* 29(3): 371–380.

Wang, Q. L. 2019. Quality evaluation and drying kinetics of shitake mushrooms dried by hot air, infrared and intermittent microwave-assisted drying methods. *Food Science and Technology* 107, 236–242. https://doi: 10.1016/j.lwt.2019.03.020

Wang, Y., Li, Y., Wang, S., Zhang, L., Gao, M. and Tang, J. 2011. Review of dielectric drying of foods and agricultural products. *International Journal of Agricultural and Biological Engineering* 4(1): doi: 10.25165/IJABE.V4I1.386

Zartha Sossa, J.W., Orozco, G.L., García Murillo, L.M., Peña Osorio, M. and Sánchez Suarez, N. 2021. Infrared drying trends applied to fruit. *Frontiers in Sustainable Food Systems* 5:650690. https://doi: 10.3389/fsufs.2021.650690

B3

Processing Involving Removal of Heat or Moisture

22

Drying

Drying of food refers to the removal of moisture to a recommended level for prolonged storage and for keeping it in good condition until further processing. The food grains, fruits, and vegetables, etc. are harvested at high moisture contents and the recommended levels of moisture to which the commodities are to be dried depend on the nature of the commodities and their use. For example, the paddy grain is harvested at a moisture content of 20%–24% and the recommended moisture content for storage of paddy is 12%–13%. Most of the fruits and vegetables have more than 85% moisture content at harvest. But for storage as a dried product, their moisture contents have to be reduced to less than 10%. For most food products, the activity of microorganisms is greatly reduced when the moisture content is reduced to a value of about 10% or lower.

Artificial drying or drying with a mechanical dryer is generally referred to as *dehydration*. It is also sometimes said that dehydration is almost complete removal of moisture or drying to a very low level, as we do in case of fruits and vegetables. However, the words "drying" and "dehydration" are used interchangeably in the food processing industry. Bone-dry is a condition of no moisture.

The major objectives of drying or dehydration are as follows:

- To preserve the raw food materials against spoilage by lowering moisture content and, consequently, water activity. All biochemical reactions and activities of microorganisms and insects are reduced by reducing the moisture content.
- To lower the cost and difficulty of storage, handling, packaging, and transportation by reducing the weight and, in most cases, volume of the material.

Water activity and its importance in the shelf life of food materials is discussed in Chapter 9.

Evaporation also reduces the moisture content of liquid foods. Removal of water as vapor from a fluid food (or other material) at its boiling point is termed as evaporation, whereas removal of moisture by drying does not require heating the water to its boiling point, rather the removal of water takes place due to the vapor pressure difference between the commodity and the surrounding. The vapor pressure differential for drying is often increased to accelerate the drying process either by heating the commodity (and thus the moisture present within it) or by reducing the pressure of the surrounding medium.

Water or other liquids can also be removed by mechanical means by pressing and centrifuging. But drying essentially involves thermal treatment. Removal of water by mechanical means usually requires lower cost than that by thermal processes. Hence, partial removal of moisture by employing mechanical methods is often practiced before the actual drying to reduce the cost of operation.

22.1 Methods and Principles

First of all, there is a need to know the amount of moisture held by the hygroscopic product (a product that can easily adsorb or desorb moisture) and the level of moisture it is to be dried. The knowledge of this would help to know the amount of moisture removed, which will ultimately guide in designing of the dryer.

22.1.1 Mechanism of Moisture Removal

Drying is a thermo-physical process. There is heating of the commodity to increase the vapor pressure of the moisture within the commodity. When the vapor pressure differential is created in between the commodity moisture and surrounding environment, the moisture is removed. In normal drying processes, as in convective drying (or conduction drying), the hot air (or surface) comes in contact with the surface of the commodity and heats the surface and associated moisture. Thus, removal of moisture takes place at the surface. Initially, there is uniform distribution of moisture in the material to be dried. During drying, the moisture loss from the surface is recompensed by the diffusion of moisture from the inner parts of the product to the surface (Figure 22.1). This involves mass transfer. The drying process is, therefore, a phenomenon of heat and mass transfer. The mass transfer is in the form of diffusion within the commodity and evaporation on the surface. Due to the variations in structures of different types of food products the processes also greatly vary with the food products.

Moisture movement within agricultural and food materials have been explained by many theories by different researchers, namely, liquid movement due to moisture diffusion in the pores (capillary flow), liquid movement due to surface forces (surface diffusion), liquid movement due to moisture concentration differences (liquid diffusion), vapor movement due to differences in vapor pressures (vapor diffusion), vapor movement due to temperature differences (thermal diffusion), and liquid and vapor movement due to total pressure differences (hydrodynamic flow).

22.1.2 Amount of Moisture Removal

The different forms of expressing the moisture content have been discussed in Chapter 3. Unless specifically mentioned, the moisture content is taken to be on a wet basis, i.e., in the previous section, where it is mentioned that the paddy is

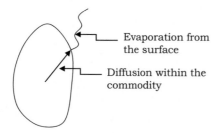

Evaporation from the surface

Diffusion within the commodity

FIGURE 22.1 Moisture removal from a commodity (single grain/ piece of fruit) during the drying process

harvested at 24% moisture level, it is on a wet basis. For all commercial purposes, moisture content is expressed on a wet basis.

The actual amount of moisture to be removed during a drying process is important for designing the dryer and the operational parameters. This can be known by knowing the initial and the final moisture content of the commodity and the amount of commodity.

The following examples explain how to find out the amount of moisture removed in a drying process. Some more examples are given in Chapter 3.

Example 22.1 It is proposed to dry paddy at moisture content of 24% (wet basis) to 14% (wet basis). If four hundred kg is the initial weight of paddy, determine the final weight of the grain and the amount of moisture removed.

SOLUTION:

Weight of paddy = 400 kg

Amount of initial moisture content = 400 × (24/100) = 96 kg

Amount of dry matter in the original sample = 400 – 96 = 304 kg

Final moisture content of paddy = 14% (wet basis)

As we are yet to know the final weight of the dried grain, we cannot use the moisture content (wet basis). Instead, as we know the dry matter in the grain (which remains constant during the drying process), we can use the moisture content (dry basis).

Hence, converting the final moisture content to dry basis,

$$14\%\left(wet\ basis\right)=\frac{14\times100}{100-14}=16.28\%\left(dry\ basis\right)$$

So, the weight of moisture in paddy after drying = 304 × (16.28/100) = 49.49 kg

Amount of moisture removed = 96–49.49 = 46.51 kg

Final weight of the grain = 400–46.51 = 353.49 kg

Otherwise,

$$Initial\ moisture\ content\ on\ dry\ basis = \frac{24\times100}{100-24}$$

$$= 31.58\%$$

Amount of moisture removed = 304 × (0.3158– 0.1628) = 46.51 kg

Final weight of the grain = 400–46.51 = 353.49 kg

Example 22.2 If 500 kg of cabbage at the moisture content of 70% (wet basis) is dried to the moisture content of 8% (wet basis), what would be the weight of the dried material?

SOLUTION:

Weight of dry material in the cabbage = 500 × (1–0.7) = 150 kg

Final moisture content = 150 × (8/(100–8)) = 13.04 kg

Final weight of dried cabbage = 150 + 13.04 = 163.04 kg

Please check that the amount of dry material in the product and the final dried weight are different. It is because the dried material still contains some moisture.

Example 22.3 Find the amount of paddy with 35% moisture content (wet basis) required to produce 3000 kg of dried grain with moisture content of 12% (wet basis).

SOLUTION:

Final weight of paddy is 3000 kg, which has 12% moisture (wet basis).

So dry matter in the grain should be 3000 × (1–0.12) = 2640 kg

Initial moisture content = 35 % (w.b.) = (35/65)×100, i.e. 53.85% dry basis

Hence initial moisture content in paddy = 2640 × 0.5385 = 1421.64 kg

Initial weight of paddy = 2640 + 1421.64 = 4061.64 kg,

i.e. the amount of paddy to be taken is 4061.64 kg.

22.1.3 Equilibrium Moisture Content

It is the final moisture content attained by a *hygroscopic product* (a product that can easily absorb or desorb moisture), when it is kept in a specific atmosphere for a very long time. For example, if a food grain is kept in a go-down, it will either take up or release moisture depending upon the vapor pressure differential between the moisture within it and the surrounding air. The commodity will gain moisture if the surrounding vapor pressure is higher than the vapor pressure exerted by the grain moisture, and it will lose moisture if the surrounding vapor pressure is lower. This process will continue until both the vapor pressures are equal. The moisture content of the grain at that time is known as the *equilibrium moisture content* (EMC). The EMC is usually expressed on a dry basis.

As heat flows between two bodies when there is a difference in temperatures of the two bodies, the moisture transfer takes place between two bodies when there is a difference in vapor pressures. The vapor pressure is the driving force and what flows is the moisture. If there is no difference of vapor pressure, there will not be any flow of moisture.

As vapor pressure is the driving force for moisture removal or gain and the vapor pressure of surrounding air depends on its temperature and relative humidity, the EMC of the commodity will also vary depending upon the temperature and relative humidity of the surrounding environment. Figure 22.2 shows the variations in EMC with respect to the temperature and RH of the surrounding air. A plot of EMC and equilibrium RH (ERH) of a material at a specific temperature (usually 25°C) is known as *equilibrium moisture curve* or *isotherm*. The shapes of the equilibrium moisture curves are sigmoid; if the ERH is plotted on the abscissa and the EMC on the ordinate, the isotherms so obtained will be of inverted S shaped.

The extra moisture present in a commodity than the equilibrium moisture is known as the *free moisture*. In other words, during the process of drying and storage only the free moisture available in the material can be removed.

The *bound moisture* is the moisture held on the adsorbent by intermolecular forces higher than the forces responsible for condensation. These are integral parts of organic molecules such as carbohydrates, proteins, fats, fibres, etc. present in the cells, chemically loosely attached with these materials and are also present in capillaries and crevices within the material. The maximum bound moisture is attained when the commodity is at 100% RH. The bound moisture exhibits a vapor pressure lower than that of liquid water at the same temperature. If the moisture content of the commodity is more than the bound moisture, it is called *unbound moisture*. The unbound moisture is held mainly in the void spaces of the solid; its vapor pressure is the same as that of water at the same temperature.

Further, if the relative humidity of the surroundings is maintained at 0% (vapor pressure is zero), then the product will release all its moisture to the environment and the EMC value will tend to be zero. This situation is rarely observed during drying and storage of agricultural commodities.

There are several factors, such as the type, variety, and maturity of the material, and the temperature and the relative humidity of the surrounding atmosphere that affect the EMC. In addition, the EMC will be different if the product is adsorbing or desorbing moisture. The desorption equilibrium has a higher value than the

adsorption equilibrium and the difference between the EMCs at desorption and adsorption process is due to the *hysteresis effect*, i.e., the history affects the value of an internal state. The EMC for a particular commodity can change depending on its variety and maturity as well as the history of moisture adsorption or desorption. In addition, the EMC determination method and RH measuring technique could also cause variations in the EMC values.

The water contained in the food by capillary forces is known as *absorbed water*, and that contained with polar and valency forces is known as *adsorbed water*.

If the surrounding atmosphere (in which the commodity is stored) or the drying air is heated, the relative humidity of the air goes down, and the vapor pressure of the moisture within the commodity increases (due to heating of the of the commodity), and, hence, and the moisture removal per unit time from the commodity also increases. However, if the temperature is increased without altering the RH, there is still the lowering of the EMC. This is due to the heat of vaporization getting reduced with an increase in temperature, i.e., more moisture can be removed with the same amount of heat supply. In other words, the moisture that is retained by the commodity in response to the corresponding atmosphere is reduced, i.e., the EMC is lowered. This effect is shown in Figure 22.2 and it can be seen that at any particular RH, the EMC value is lowered with increase in temperature.

In general, at low relative humidities the EMC is higher for food materials having more proteins, starch, or other high molecular weight polymers and lower for foods having more soluble solids. Crystalline salts and sugars as well as fats generally adsorb less amounts of water.

A change in the chemical composition or in the history of the product affects the EMC. Oily seeds generally absorb lower amount of moisture from the surrounding air than starch-rich seeds. Hence, oily seeds need to be stored at lower moisture contents than starchy seeds.

The EMC values of a commodity during desorption are higher than the adsorption values at specified RH and temperature conditions, and the difference is due to the hysteresis effect. A number of theories explain the hysteresis effect (Kraemer, 1931; Chung and Pfost, 1967; Young and Nelson, 1967).

Determination of EMC

The EMC values of different types of materials can be determined experimentally. Many theoretical and empirical models are also available for determination of EMC.

The EMC of a commodity can be determined experimentally by keeping a sample in a particular environment and observing its final moisture content. If the air surrounding the grain is still and does not move, the method is known as *static method of determination of EMC*. As the gain and loss of moisture is a slow process it may take even a few weeks to reach the final moisture content (i.e., to attain the EMC) if the atmosphere is static. In a dynamic method, air at specified temperature and RH is continuously moved over the sample in a specified chamber. The method is much faster than the static method. The RH of the atmosphere is maintained at constant level by different concentrations of HCl or H_2SO_4 or salt solutions. Relative humidity obtained at different temperatures with different salt solutions is given in Annexure XIV.

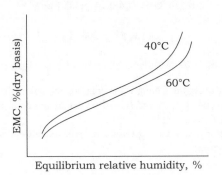

FIGURE 22.2 Equilibrium moisture curves at two temperatures

Another rapid and accurate method, known as *isoteno-scopic method,* measures the vapor pressure exerted by moist grains directly in a previously evacuated jar. A measure of the vapor pressure in the jar, correlated with the temperature and moisture content of the sample, gives the EMC of the sample.

EMC Models

Determination of EMC for specific environmental conditions is carried out with the help of different empirical, semi-theoretical, and theoretical, models. The *Henderson's equation*, based on Gibb's adsorption equation, is given as:

$$1 - \varphi = e^{-C.T.M_e^n} \tag{22.1}$$

In this equation, φ is the relative humidity in decimals, T is the absolute temperature (i.e., temperature in Kelvin) and M_e is the EMC in per cent dry basis. The constants, C and n, depend on the product and temperature.

Chung-Pfost equation for determination of EMC is written as:

$$\ln\left(\frac{p_w}{p_{ws}}\right) = -\frac{A}{R.T}\exp(-B.M) \tag{22.2}$$

Here, T is in Kelvin, R is the Universal gas constant, M is moisture content on dry basis, p_w and p_{ws} are the partial vapor pressure and saturated vapor pressure, respectively. A and B are constants dependent upon grain temperature. The equation is valid between 20% and 90% RH values. There are other equations as Kelvin equation, Langmuir equation, BET equation, Harkins-Jura equation, Smith equation, etc. for the prediction of EMC at any condition.

Bortolotti and Barrozo (2013) have listed the six most frequently cited equations for predicting the equilibrium moisture content, given in Annexure XV.

Example 22.4 If a sample of grain is at 50°C and 50% RH, find its equilibrium moisture content. The constant values of Henderson's equation for RH-vs-EMC are C = 1.48×10⁻⁵ and n=1.89.

SOLUTION:

The Henderson's equation is written as $1 - \varphi = e^{-C.T.M_e^n}$

Hence $\log_e(1-0.5) = -(1.5\times10^{-5})(323.15)(M_e)^{1.89}$

or, $-0.69315 = -(1.48\times10^{-5})(323.15)(M_e)^{1.89}$

or, $(M_e)^{1.89} = 0.69315/((1.48\times10^{-5})(323.15)) = 144.93$

i.e. $M_e = 13.91$ % (dry basis)

Example 22.5 The following observations were taken on the equilibrium moisture content of wheat corresponding to a set of drying conditions.

	Drying air temperature	Drying air RH	EMC
Experiment No. 1	65°C	40%	12% (wet basis)
Experiment No. 2	65°C	65%	17% (wet basis)

Find the values of coefficients, C and n, for Henderson's equation.

SOLUTION:

The given moisture contents expressed on dry basis are,

12% (wet basis) = (12/88) = 13.64% (dry basis)

17% (wet basis) = (17/83) = 20.48% (dry basis)

For the experiment no.1, $1 - 0.4 = e^{-C(338)(13.64)^n}$

or, $\ln(0.6) = -C(338)(13.64)^n$

or, $-0.5108 = -C(338)(13.64)^n$ \qquad (A)

For the experiment no. 2, $1 - 0.65 = e^{-C(338)(20.48)^n}$

or, $\ln(0.35) = -C(338)(20.48)^n$

or, $-1.0498 = -C(338)(20.48)^n$ \qquad (B)

Dividing equation A by equation B,

$0.48657 = (13.64/20.48)^n$

$\ln(0.48657) = n\ln(0.666)$

$n = \ln(0.48657)/\ln(0.666) = 1.77$

Substituting the value in Henderson's equation C = 1.48×10⁻⁵

Example 22.6 In the above example, find the value of EMC if the temperature and RH of the drying air are 70°C and 65% by using (a) Henderson's equation, and (b) Chung and Pfost equation.

SOLUTION:

Using the Henderson's equation

$$1 - 0.65 = e^{-1.48\times10^{-5}(343.15)(M_e)^{1.77}}$$

or, $\log_e(0.35) = -1.48\times10^{-5}\times343.15\times(M_e)^{1.77}$

or, $(M_e)^{1.77} = 206.714$

$M_e = 20.33$ %(dry basis) = 16.89% (wet basis)

For using the Chung-Pfost equation (equation 22.2), the constants A and B will be calculated as in the previous example.

For the two experiments, two equations are obtained.

$$\log_e(0.4) = \frac{A}{R\times338.15}e^{-B\times13.64} \tag{C}$$

$$\log_e(0.65) = -\frac{A}{R \times 338.15} e^{-B \times 20.48} \quad \text{(D)}$$

Dividing equation (C) by equation (D)

$$e^{(20.48-13.64)}B = 2.127$$

or, $6.84 \times B = 0.7574$, and thus, $B = 0.11$

Substituting the value of B in equation (C)

$$-0.916 = -\frac{A}{R \times 338.15} e^{-1.5004}$$

This gives, $A = 1388.74 \times R$
Using the new conditions of drying air as 70°C and 65% RH,

$$\ln(0.65) = -\frac{1388.74 \times R}{343.15 \times R} e^{-0.11 \times M_e}$$

Solving the equation, we obtain

$$-0.11 (M_e) = \log_e (0.1064)$$

$M_e = 20.368$ % (dry basis) = 16.92%(wet basis)

Example 22.7 Experimental results for wheat on relative humidity-vs-equilibrium moisture contents (M_e) have been found to be in a linearized form (y = mx + C) as stated below.

$$\log[-\ln(1-RH)] = n \log (M_e)+\log (CT)$$

During the experiments, the intercept has been found to be –4.025 at T = 303 K. Also it was observed that the EMC of wheat at 50% relative humidity and 30°C is 18.9% (dry basis). Determine the value of EMC for wheat stored at 40°C and 70% relative humidity.

SOLUTION:

Given, $\log[-\ln(1-RH)] = n \log (M_e)+\log (CT)$

Here, T = 303 K

$$\log (CT) = -4.025$$

$$CT = 0.0000944$$

$$C = \frac{0.0000944}{303} = 3.1157 \times 10^{-7} K^{-1}$$

Now, $\log[-\ln(1-RH)] = n \log (M_e) + \log (CT)$

RH = 50%, M_e = 18.9%, T = 303 K

$$\log\left[-\ln\left(1-0.5\right)\right] = n \, log \, (18.9) - 4.025$$

$$-0.1592 = 3.0286 \times \log (18.9) - 4.025$$

or, $n \log (18.9) = 3.866$

Thus, n = 3.0286
For, RH = 70%, and T = 273+40 = 313 K,

$$\log[-\ln(1-RH)] = n \log (M_e) + \log (CT)$$

or, $0.0806 = 3.0286 \times \log (M_e) - 4.025$

Thus, M_e = 22.68%

Check that the EMC decreases with the increase in temperature.

22.1.4 Rate of Drying

The rate of drying of a commodity is usually expressed as the amount of moisture removed per unit time per unit weight of dry material or per m² of drying surface area.

$$\text{Rate of drying} = \frac{\text{amount of moisture removed}}{(\text{amount of dry matter})(\text{time})},$$

$$\text{or Rate of drying} = \frac{\text{amount of moisture removed}}{(\text{drying surface area})(\text{time})}$$

During the drying of high moisture products, initially, a continuous film of water exists on the surface of the solid and the surface is very wet. This unbound water acts as a free water surface. Under this condition the rate of evaporation is almost the same as the rate of evaporation from a free water surface for the specific drying air parameters and it is not affected by the moisture content of the product. The evaporation is also supported by transfer of moisture from within the product. Whatever moisture evaporates from the surface is replaced by the moisture diffusion from the inner parts and, thus, the surface remains completely wet for some time and the drying rate remains practically constant. This period is known as *constant rate period of drying*. During this stage the rate of diffusion of moisture from inner parts of the product to the periphery is the same as the rate of moisture removal from the surface.

> Even though we assume the surface as a flat surface, the surfaces of materials such as grains are in reality rough and the net area of evaporation is much higher than that for a flat surface and the rate of evaporation can be much higher than it would be for a flat water surface.
>
> During the constant rate drying stage, the surface temperature of the product can be considered to be the same as the wet bulb temperature if there is no heat transfer by radiation or conduction.

As drying continues, a time comes when the surface is no longer completely wet, i.e., there is no water film on the surface. It is because the rate of diffusion of moisture from the inner parts of the product reduces as compared to the rate of potential evaporation on the surface. This causes a decrease in the wetted area and so also of the rate of drying. The stage is called *falling rate period of drying*. The moisture content at which the drying rate changes from constant rate stage to falling rate stage is known as the *critical moisture content in drying*.

The falling rate period may be further divided as the *first falling rate period* and *second falling rate period*. The first falling rate period continues until the surface becomes entirely dry. After the surface has become completely dry the evaporation front will proceed from the surface into the interior of the solid. The drying rate is further slowed down and this stage is known as the *second falling rate period*. During this stage, the vaporization of moisture takes place within the commodity and the vapor passes through the capillaries to escape into the surrounding atmosphere. Usually, it is difficult to distinguish the two falling rate periods in many agricultural materials.

If the drying takes place at high temperature, then the surface dries rapidly, but the moisture from the interior portion cannot be conveyed to the surface at the same rate. It causes the surface layers to shrink. The diffusivity also decreases with decreasing moisture concentration. Thus, the surface layers resist the moisture diffusion from interior parts to outside. In extreme cases, the shrunken cells behave like a barrier to moisture migration. The phenomenon is known as case hardening.

In small food pieces the above effect may not be significant. But in large pieces the effect is more distinct. Such effect of case hardening can be reduced by reducing the concentration gradient, which is possible by reducing the rate of drying. Controlling the relative humidity of drying air can control the rate of drying.

22.1.5 Drying Curves

If the amount of free moisture is plotted versus time during a drying process, then Figure 22.3 (a) represents the curve. Similarly, the curve for drying rate vs. free moisture is shown in Figure 22.3 (b).

With reduction in moisture content, the drying rate would normally reduce asymptotically. However, when the drying starts, there is an initial adjustment period (A-B) during which the drying rate may increase along with the increase in product temperature. Usually, this adjustment period is small and is ignored during the analysis of drying. B-C describes the constant rate of drying, and C is the critical moisture content. The first falling rate period C-D has almost a linear rate. At point D, the rate of drying drops further and the second falling rate period starts. If the drying continues for a long time, the product reaches the EMC, denoted by the point E.

The slopes of the tangents drawn to the curve at any point give the rate of drying at that instant of time. Thus, the rate of drying at any time is given as:

$$R = -\frac{(DM)}{A}\frac{dX}{d\theta} \qquad (22.3)$$

where, R (kg water.h^{-1}.m^{-2}) is the drying rate, DM (kg) is the amount of dry solid, and A (m^2) is the exposed surface area of drying. dX/dθ is the change in moisture with time (expressed in fractions per time; here the time unit will be hour as the R has been taken per hour).

Example 22.8 During a drying experiment on garlic, the initial weight of the sample was 125 g and the following observations on the weight of garlic were taken with respect to time at a drying temperature of 80°C. The initial moisture content of the sample was found as 65% (wet basis). Prepare a drying curve and drying rate curve.

Time, min	Weight, g	Time, min	Weight, g	Time, min	Weight, g
0	125	150	75.3	300	51.93
15	118.69	165	72.37	315	50.76
30	113.5	180	69.36	330	50.38
45	107.39	195	66.5	345	49.81
60	101.97	210	63.91	360	49.47
75	98.22	225	61.58	375	49.43
90	94.09	240	58.74	390	49.4
105	88.25	255	57.24	405	49.38
120	83.34	270	55.89		
135	79.45	285	54.43		

SOLUTION:

We can prepare the drying curve by plotting the weight of the sample with respect to time and the drying rate curve by plotting moisture content (dry basis) with respect to time per dry weight of the sample. (The drying rate can also be plotted with respect to surface area of the sample or surface area of the tray instead of weight of the sample.)

Therefore, first we have to find out the dry weight of the sample and the moisture content on dry basis for all observations. Given the weight of the material as 125 g and 65% (wet basis) is the initial moisture content, the dry matter in the sample = 125 × (35/100) = 43.75 kg.

Thus, at any time, the amount of moisture is (total weight -43.75) kg and the value so obtained divided by 43.75 kg is the moisture content on a dry basis. The drying rate for any time interval can be given as amount of moisture removed in that time interval.

For example, for the first 15 minutes the initial moisture was 81.25 g and the final moisture was 74.94 g.

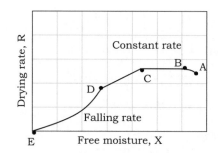

FIGURE 22.3 Typical drying curves showing constant rate and falling rate periods (a) Free moisture-vs-time (b) Drying rate vs. free moisture

Thus, moisture removed = (81.25–74.94)/43.75 kg per kg dry matter

The moisture removal takes place in 15 minutes or 0.25 hours and, therefore, the drying rate for the first 15 minutes

= (81.25–74.94)/(43.75×0.25) = 0.5769 kg/(kg dry matter–h)

Similarly, we find the drying rate for all time intervals. The summary table is as follows. (You can use an Excel worksheet for the purpose).

Time, min	Weight, g	Amount of dry matter (g) = (Initial weight × dry matter fraction)	Amount of moisture, g (= B-C)	Moisture content, (% d.b.) (D/C) ×100	Drying rate, kg/kg dry matter-h (=Moisture removed/ (C × time interval in hours))
A	B	C	D	E	F
0	125	43.75	81.25	185.714	
15	118.69	43.75	74.94	171.291	0.577
30	113.5	43.75	69.75	159.429	0.475
45	107.39	43.75	63.64	145.463	0.559
60	101.97	43.75	58.22	133.074	0.496
75	98.22	43.75	54.47	124.503	0.343
90	94.09	43.75	50.34	115.063	0.378
105	88.25	43.75	44.5	101.714	0.534
120	83.34	43.75	39.59	90.491	0.449
135	79.45	43.75	35.7	81.600	0.356
150	75.3	43.75	31.55	72.114	0.379
165	72.37	43.75	28.62	65.417	0.268
180	69.36	43.75	25.61	58.537	0.275
195	66.5	43.75	22.75	52.000	0.261
210	63.91	43.75	20.16	46.080	0.237
225	61.58	43.75	17.83	40.754	0.213
240	58.74	43.75	14.99	34.263	0.260
255	57.24	43.75	13.49	30.834	0.137
270	55.89	43.75	12.14	27.749	0.123
285	54.43	43.75	10.68	24.411	0.133
300	51.93	43.75	8.18	18.697	0.229
315	50.76	43.75	7.01	16.023	0.107
330	50.38	43.75	6.63	15.154	0.035
345	49.81	43.75	6.06	13.851	0.052
360	49.47	43.75	5.72	13.074	0.031
375	49.34	43.75	5.59	12.777	0.012
390	49.25	43.75	5.5	12.571	0.008
405	49.2	43.75	5.45	12.457	0.005

If we plot the E vs. A then we get the drying curve, and if we plot F vs. A we get the drying rate curve.

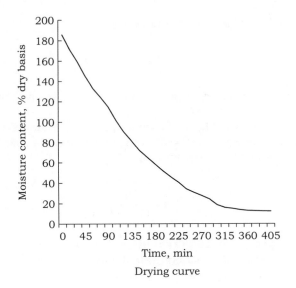

Drying curve

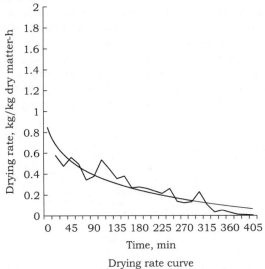

Drying rate curve

Check that

1. The time interval is the same for all the observations, 0.25 h for all calculations. But if the drying interval varies, then we need to take the appropriate values.
2. The drying rate calculated in the above example is the mean drying rate during the time interval. But for our convenience we have placed the values along the specific observation times. A better approach can be to take the mean time to state the drying rate or to take smaller time intervals for taking the observations.

22.1.6 Analysis of Constant Rate Period of Drying

Rate of Drying and Drying Time

The drying curve can be used to determine the time required for the constant-rate period. The equation 22.3 is

$$R = -\frac{(DM)}{A}\frac{dX}{d\theta}$$

$$\theta = \int_0^\theta d\theta = -\frac{(DM)}{A}\int_{X_1}^{X_2}\frac{dX}{R}$$

$$\theta = \int_0^\theta d\theta = \frac{(DM)}{A}\int_{x_2}^{x_1}\frac{dX}{R} \qquad (22.4)$$

If the drying rate is constant, it is denoted as R_c and the above equation is written as:

$$\theta = \frac{(DM)}{AR_C}(X_1 - X_2) \qquad (22.5)$$

Let us consider another alternate way of expressing the rate of drying. The vaporization of moisture during the constant rate period of drying is almost the same as the evaporation from a free water surface. The drying air gives up sensible heat for drying and the vapor again comes back to the air and adds latent heat. At steady state, if there is no heat contribution by radiation and conduction, the process is adiabatic in nature and the heat transfer from the drying air is balanced by the evaporation of moisture. Equations similar to the one used for deriving the wet bulb temperature, t_w, can be written, in this situation.

For analysis, the system is shown in Figure 22.4. In the figure, t (°C) is the temperature, W is the humidity ratio and the

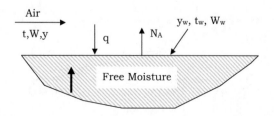

FIGURE 22.4 Heat and mass transfer on a surface during constant rate drying

mole fraction of water vapor is y. The subscript w denotes the water surface. If h $(W.m^{-2}.K^{-1})$ is the heat-transfer coefficient and A (m^2) is the exposed drying area, then the rate of convective heat transfer, q (W) from the air to the surface of the solid is

$$q = h\,A\,(t - t_w) \qquad (22.6)$$

Considering constant drying rate period, i.e. all the heat supplied causes evaporation at the surface, then the rate of drying at this stage, R_c (kg of water.$m^{-2}.s^{-1}$), can be given as:

$$R_C = \frac{q}{A\lambda_w} \qquad (22.7)$$

where, λ_w $(J.kg^{-1})$ is the latent heat at t_w.

Substituting the values from equation 22.6 in equation 22.7, we get

$$R_C = \frac{h(t - t_w)}{\lambda_w} \qquad (22.8)$$

Next, neglecting the small sensible heat changes, the heat of vaporization required for N_W (N_W is the flux of water vapor from the surface, kg mol.$m^{-2}.s^{-1}$) is given as:

$$q = M_W . N_W . \lambda_w . A \qquad (22.9)$$

Equating equations 22.6 and 22.9, we get

$$M_W . N_W . \lambda_w . A = h.A.(t - t_w) \qquad (22.10)$$

$$Thus, \quad \frac{h(t - t_w)}{\lambda_w} = M_W N_W \qquad (22.11)$$

$$i.e. \quad R_C = M_W N_W \qquad (22.12)$$

Now, we will find out the heat loss with the evaporated water vapor.

The value of N_W is given as follows:

$$N_W = k_y . (y_w - y) \qquad (22.13)$$

where, k_y (kg-mol.$m^{-2}.s^{-1}$.(mol frac)$^{-1}$) is the mass transfer coefficient.

If M_A and M_W are the molecular weights of air and water, respectively and W is the humidity ratio, then the mole fraction of water can be given as:

$$y = \frac{W/M_W}{1/M_A + W/M_W} \qquad (22.14)$$

The humidity ratio for the drying air is very small, and, hence, ignoring the second term in the denominator of equation 22.14, it can be simplified as:

$$y \cong \frac{WM_A}{M_W} \qquad (22.15)$$

Thus, equation 22.13 can be written as:

$$N_W = k_y \frac{M_A}{M_W}(W_w - W) \qquad (22.16)$$

$$or, \quad M_W N_W = k_y M_A (W_w - W) \qquad (22.17)$$

Substituting the value of M_W. N_W in equation 22.12, we get the rate of drying as:

$$R_C = k_y M_A (W_w - W) \qquad (22.18)$$

Equation 22.18 indicates that the temperature of the material being dried will be the same as the wet bulb temperature of the drying air and is similar to the equation for the wet bulb temperature in the absence of conduction and radiation during constant rate drying period.

Either the heat-transfer or the mass-transfer coefficient can be used for evaluating the drying rate during constant rate period. However, the error in measuring the temperature t_w at the surface is relatively small, little affecting the value of $(t-t_w)$ as compared to that for (W_w-W). Hence, it is preferable to use the heat-transfer equation.

Time of drying during constant rate period is given as:

$$\theta = \frac{(DM)\lambda_w(X_0 - X_c)}{hA(t - t_w)} = \frac{(DM)(X_0 - X_c)}{k_y M_A A(W_w - W)} \qquad (22.19)$$

In this expression, the X_c is the critical moisture content expressed in fraction, i.e.. at this stage the constant rate period ends. It is assumed that there is no conduction and convection. If these effects are included, the solution becomes more intricate.

The constant rate period during the drying of a solid in bulk can also be simply written as:

$$\theta_c = \frac{X_0 - X_c}{R_c}, \qquad (22.20)$$

Here, X_0 and X_c are the initial moisture content and critical moisture content, respectively in fractions (kg moisture. (kg of dry matter)$^{-1}$). In this case, R_c has the unit of kg water. (kg dry solid)$^{-1}$. time^{-1}. Thus, as discussed earlier, the rate of drying can be expressed per weight basis or per drying area basis.

Factors Affecting Constant Rate Period of Drying

As discussed in the previous equations, during the constant rate period of drying, the rate of drying, R_c, is proportional to the heat transfer coefficient (h). As h depends on the mass flow rate of air or the velocity of air on the surface, R_c is also affected by the velocity of air moving on the surface. However, it has been observed that when radiation and conduction also influence the vaporization of moisture from the surface, the relative effect of air velocity is significantly reduced.

Similarly, when the humidity ratio of air (W) is reduced at a given temperature, the wet bulb temperature, t_w, is reduced and, consequently, R_C increases. It is quite natural that when the air contains less moisture it will increase the rate of drying.

From equations 22.8 and 22.18, we also observe that

$$R_C \alpha \frac{(t - t_w)}{\lambda_w} \quad and \quad R_C \alpha (W_w - W) \qquad (22.21)$$

Thus, the rate of drying depends on the difference between the dry bulb temperature and wet bulb temperature as well as on the difference between the humidity ratios of moisture and air. It must, however, be remembered that t_w increases as the air temperature, t, increases.

The constant rate period of drying continues until a continuous film of water exists on the surface of the product. Neither the rate of diffusion within the solid during this period nor the thickness of the material being dried is important during this period. The rate of diffusion and thickness of material are very important for the analysis of falling rate drying.

22.1.7 Analysis of Falling Rate Period of Drying

Rate of Drying and Drying Time

We start again from equation 22.4.

$$\theta = \frac{(DM)}{A} \int_{X_2}^{X_1} \frac{dX}{R} \qquad (22.22)$$

In the falling rate period, unlike the constant rate period, the drying rate changes. Hence, we plot the value of 1/R versus the moisture content (X) and determine the area under the curve.

The equations for the drying time for some distinct cases are discussed below.

Case I. Drying rate can be approximated as a linear function of X

If the drying rate, R, is assumed to be a linear function of X, then a simple relationship between these two variables can be written as:

$$R = aX + b \qquad (22.23)$$

In the above equation, a is the slope and b is the intercept.

Thus, dR = a.dX,

Substituting this in equation 22.22,

$$\theta = \frac{(DM)}{aA} \int_{R_2}^{R_1} \frac{dR}{R} = \frac{(DM)}{aA} \ln \frac{R_1}{R_2} \qquad (22.24)$$

Since, $R_1 = aX_1 + b$ and $R_2 = aX_2 + b$, the slope is given as:

$$a = \frac{R_1 - R_2}{X_1 - X_2} \qquad (22.25)$$

Substituting in equation 22.24, we get

$$\theta = \frac{(DM)}{A} \cdot \frac{(X_1 - X_2)}{(R_1 - R_2)} \ln\left(\frac{R_1}{R_2}\right) \qquad (22.26)$$

If the drying rate can be approximated as a linear function of X and the R-X curve passes through origin, then the value of b = 0.

$$\text{i.e., } R = aX \qquad (22.27)$$

Differentiating, dX = dR/a The equation 22.24 can be modified for this situation as,

$$\theta = \frac{(DM)}{aA} \int_{R_2}^{R_1} \frac{dR}{R} = \frac{(DM)}{aA} \ln\left(\frac{R_1}{R_2}\right) \qquad (22.28)$$

Considering that the stage starts from the critical moisture content, the slope of the line can be given as, $a = R_c/X_c$, and for initial moisture content (for this stage) $X_1(=X_C)$, the rate of drying, $R_1 = R_C$,

$$or, \quad \theta = \frac{(DM)}{A} \frac{X_C}{R_C} \ln\left(\frac{R_C}{R_2}\right) \qquad (22.29)$$

As $R_C/R_2 = X_C/X_2$, the equation can be modified as:

$$\theta = \frac{(DM)}{A} \frac{X_C}{R_C} \ln\left(\frac{X_C}{X_2}\right) \qquad (22.30)$$

The drying rate is given as, $R = aX = R_C \dfrac{X}{X_C} \qquad (22.31)$

If the drying rate is given in the units of kg water. (kg dry solid)$^{-1}$.min^{-1} (or per hour), the above equation can be simply written as:

$$\theta_f = \frac{X_c}{R_c} \ln\left(\frac{X_c}{X}\right) \qquad (22.32)$$

If there are more than one falling rate stage during the process, the total time to dry up to the final moisture content X in the first falling stage is given as:

$$\theta = \frac{X_0 - X_{c1}}{R_c} + \frac{X_{c1} - X_{r1}}{R_c} \ln \frac{X_{c1} - X_{r1}}{X - X_{r1}} \qquad (22.33)$$

The total drying time to reach the moisture level X in the second falling state is given as

$$\theta = \frac{X_0 - X_{c1}}{R_c} + \frac{X_{c1} - X_{r1}}{R_c} \ln \frac{X_{c1} - X_{r1}}{X_{c2} - X_{r1}}$$
$$+ \left[\frac{X_{c1} - X_{r1}}{R_c}\right]\left[\frac{X_{c2} - X_{r2}}{R_c}\right] \ln\left[\frac{X_{c2} - X_{r2}}{X - X_{r2}}\right] \qquad (22.34)$$

Here, X_{c1} and X_{r1} are the critical moisture content and moisture content after first falling rate and X_{c2} and X_{r2} for the second falling rate drying (obtained by extending the drying rate curves).

Drying in Falling Rate Period by Diffusion and Capillary Flow

If R is assumed to be a linear function of X (i.e. R = aX), we can write the equation 22.3 as:

$$R = -\frac{(DM)}{A} \frac{dX}{d\theta} = aX \qquad (22.35)$$

$$or, \frac{dX}{d\theta} = -\frac{aA}{(DM)} X \qquad (22.36)$$

However, as mentioned before, during the falling rate stage, the main mechanism that control the rate of drying is the rate of liquid diffusion or the capillary movement. We will discuss about the processes below.

Liquid Diffusion of Moisture During Drying

The resistance to mass transfer by evaporation on the surface is very small and, hence, during the falling rate period, the diffusion of moisture controls the rate of drying. Moisture diffusion takes place due to moisture concentration difference, usually in non-porous solids. The equations for diffusion can be used in such situations.

The Fick's second law for unsteady state diffusion is written as:

$$\frac{\partial X}{\partial \theta} = D \frac{\partial^2 X}{\partial x^2} \qquad (22.37)$$

where, D (m².h^{-1}) is the liquid diffusion coefficient, X (kg free moisture. (kg dry solid)$^{-1}$), and x (m) is the distance that the moisture moves in the solid.

The above equation assumes that the diffusion coefficient is constant. But in reality, the coefficient varies with moisture content, temperature, and relative humidity of surrounding air during the process of drying. Similarly, as the constant rate period of drying ends and the falling rate period begins, there is uneven distribution of moisture on the surface. However, for the present analysis, the surface moisture distribution will be assumed to be uniform.

At $\theta = 0$, the above equation can be integrated to obtain

$$\frac{X_\theta - X_e}{X_{\theta1} - X_e} = \frac{X}{X_1} = \frac{8}{\pi^2}\left(e^{-D\theta\left(\frac{\pi}{2a}\right)^2} + \frac{1}{9}e^{-9D\theta\left(\frac{\pi}{2a}\right)^2} + \frac{1}{25}e^{-25D\theta\left(\frac{\pi}{2a}\right)^2} + \cdots\right)$$

$$(22.38)$$

Where, X_e = equilibrium free moisture content in fraction, X = mean free moisture content in fraction at time θ (h), X_1 = free moisture content at $\theta = 0$. In the above equation, a is a shape factor which will be half the thickness of the slab if the moisture is removed from both faces or the total thickness if the drying takes place only on one face, as it happens when a slab is kept on a solid tray for drying with hot air flowing over the slab.

It has been suggested that only the first term in the above equation is significant for long drying times. Therefore, the subsequent terms can be ignored. The equation then reduces to

$$\frac{X}{X_1} = \frac{8}{\pi^2} e^{-D\theta\left(\frac{\pi}{2a}\right)^2} \quad (22.39)$$

Further simplification yields

$$\theta = \frac{4a^2}{\pi^2 D} \ln \frac{8X_1}{\pi^2 X} \quad (22.40)$$

If it is assumed that the diffusion mechanism starts at the critical moisture content, i.e. $X_1 = X_C$, the above equation can be solved as:

$$\frac{dX}{d\theta} = \frac{-\pi^2 DX}{4a^2} \quad (22.41)$$

Multiplying both sides by $-(DM)/A$

$$-\frac{(DM)}{A} \cdot \frac{dX}{d\theta} = R = \frac{\pi^2 (DM).D}{4a^2 A} X \quad (22.42)$$

This equation states that if the controlling factor is the moisture diffusion from the interior to periphery, the drying rate will vary directly with the diffusivity D, the free moisture content, X and will inversely vary with the square of the commodity thickness. It may be noted that with the assumptions taken for the analysis, the falling rate period is not affected by the air velocity or the humidity ratio of drying air. The moisture diffusivity also decreases with decreased moisture content. Thus, in deriving the above equations, average values of diffusivities over the range of concentrations have been used.

Capillary Movement of Moisture during Drying

This analysis considers that the moisture movement through the pores supplies the water for vaporization on the surface. The evaporation mechanism during the process is considered to be the same as that during the constant-rate period. Thus, as in the case of the constant-rate drying period, the process will be affected by the drying air velocity, air temperature, and humidity, among the other factors.

Assuming the moisture flow takes place by capillary movement, the drying rate could be expressed by the Poiseuille's equation for laminar flow, in its modified form and the capillary-force equation.

We start with the equation 22.3,

$$R = -\frac{(DM)}{A} \frac{dX}{d\theta}$$

As the conditions are similar to those in the constant rate period of drying, the drying rate would vary linearly with X.

As we have discussed earlier,

$$R = R_C \frac{X}{X_C} \quad and \quad \theta = \frac{(DM).X_C}{AR_C} \ln \frac{X_C}{X_2} \quad (22.43)$$

If the density and the thickness of the material being dried are given as ρ_s (kg dry solid.m^{-3}) and a (m), then the weight of dry matter is

$$(DM) = A.a.\rho_s \quad (22.44)$$

If θ is the time taken to reduce the moisture content from X_C to X, the above equation can be written as:

$$\theta = \frac{a.\rho_s.X_c}{R_C} \ln \frac{X_c}{X} \quad (22.45)$$

Substituting equation 22.8 for R_C,

$$\theta = \frac{a.\rho_s.\lambda_w.X_c}{h(t - t_w)} \ln \frac{X_c}{X} \quad (22.46)$$

Thus, when the capillary flow controls the falling rate period, the rate of drying will be inversely proportional to the thickness of the material. As the material thickness varies, the drying time to reduce the moisture content within specified limits varies directly. As the rate of drying is also dependent on the heat transfer coefficient, the drying parameters as the air velocity, temperature, and humidity will also affect the drying rate and time.

22.1.8 Moisture Ratio and Drying Constant

A simple drying equation has been proposed based on Newton's law of cooling and heating.

Newton's equation is represented as

$$\frac{dT}{d\theta} = -k(T - T_e) \quad (22.47)$$

$(T-T_e)$ is the differential temperature and $dT/d\theta$ is change in temperature with time. k is a proportionality constant.

A drying equation, analogous to the above equation can be written as follows:

$$\frac{dM}{d\theta} = -k(M - M_e) \quad (22.48)$$

where, M and M_e are initial and equilibrium moisture contents on % (dry basis), θ is drying time, h (or min or s), and k is the drying constant. The drying constant k has a dimension of T^{-1} (given as h^{-1} or min^{-1} or s^{-1}).

From the above equation,

$$\frac{dM}{(M - M_e)} = -k.d\theta \quad (22.49)$$

which can be further integrated within limits; the final equation is obtained as

$$\frac{(M - M_e)}{(M_0 - M_e)} = e^{-k\theta} \qquad (22.50)$$

and the drying time can be obtained as:

$$\theta = \frac{1}{k} \ln\left[\frac{M_0 - M_e}{M - M_e} \right] \qquad (22.51)$$

$\dfrac{(M - M_e)}{(M_0 - M_e)}$ is the moisture ratio (MR).

The drying constant, k, can be determined either by *graphical method* or by *half-life period* method. In the graphical method, a graph is plotted on a semi-log paper between $(M-M_e)/(M_0-M_e)$ as ordinate and θ (h) as abscissa. As per the behavior of the equation, we obtain a straight line, the slope of which is the drying constant, k.

If the time necessary to obtain a moisture content ratio of one-half is denoted as $\theta_{\frac{1}{2}}$, then the equation 22.50 can be written as:

$$\frac{1}{2} = e^{-k\theta_{\frac{1}{2}}} \qquad (22.52)$$

Thus, $\theta_{\frac{1}{2}} = \ln(2)/k$ \qquad (22.53)

Similarly, $\theta_{\frac{1}{4}} = \ln(4)/k$ \qquad (22.54)

Thus, k can be calculated by knowing the values of either $\theta_{\frac{1}{2}}$ or $\theta_{\frac{1}{4}}$.

$\theta_{\frac{1}{2}}$ is also denoted as the *time of one-half response in a drying process.*

Example 22.9 During a drying process, grain at a moisture content of 80% (dry basis) is dried to 55% (dry basis) in three hours. The equilibrium moisture content of the grain corresponding to the drying conditions is 30% dry basis. Determine the drying constant "k."

SOLUTION:

The drying equation is expressed as:

$$\frac{M - M_e}{M_0 - M_e} = e^{-k\theta}$$

$$or, \quad \frac{55 - 30}{80 - 30} = e^{-k\theta_{1/2}}$$

$$\theta_{1/2} = \ln(2)/k$$

$$k = \ln(2)/3 = 0.231 \ h^{-1}$$

Check that k has a dimension of time^{-1}.

Example 22.10 In the previous problem, if the grain's equilibrium moisture content corresponding to the drying conditions is 15% dry basis, what will be the drying constant "k"?

SOLUTION:

The drying equation is expressed as

$$\frac{M - M_e}{M_0 - M_e} = e^{-k\theta}$$

$$or, \quad \frac{55 - 15}{80 - 15} = e^{-k\theta} i.e. \ \ln(1.625) = k\theta$$

$$k = \ln(1.625)/3 = 0.162 \ h^{-1}$$

22.1.9 Drying Methods and Principles

At this stage, we will briefly discuss the working principles of some selected drying methods so as to understand the mass and energy balances related to them. We will discuss the construction and operational features in more details later in this chapter.

Convective Drying

A convective dryer basically constitutes of an air heating unit, an air blowing unit, and a chamber for holding the commodity to be dried (Figure 22.5). After the air blower and heater, a small chamber or space is kept for eliminating the temperature variation of the drying air and allowing the air to be uniformly distributed to all the materials kept in the drying chamber. The space is often known as a *plenum chamber*. In certain large capacity bin dryers, the drying air is distributed to different parts of the bin by conduits connected to the plenum chamber or directly to the air heating unit.

The dryers can be classified as batch or continuous dryers. In continuous systems, a conveyor or trolley system is used for continuous flow of material.

Thin Layer and Deep Bed Drying

In a through flow drying process, if the drying medium (air) crosses the food that is kept in a thin layer, then the drying air's relative humidity and temperature do not change appreciably as it passes through the food, e.g., all the materials are exposed to drying air at similar conditions. For example, if the drying air crosses the grain that is kept in a layer of only 2–3 cm, then the inlet and exit conditions of drying air with respect to the grain layer will not vary significantly. However, if the grain is kept in a depth of 40 cm, then a substantial quantity of heat from the air will be given to the food and a good quantity of moisture will be absorbed by the air. The air may even be completely saturated during the drying process. Thus, the commodity kept at different depths in a deep bed drying is exposed to different drying conditions.

Generally, up to 20 cm thick grain bed with a recommended air-commodity ratio is considered as thin layer for grain

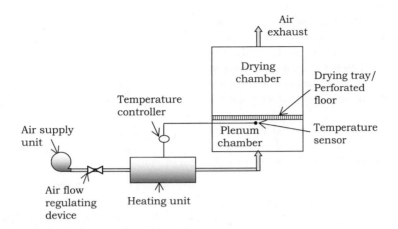

FIGURE 22.5 Schematic view of different components of a simple convective drying system

drying. Fruits and vegetables are normally dried in thin layer. As mentioned above, drying in thin layer involves drying of the entire commodity in similar conditions whereas in the case of drying in a deep bed the rate of air flow per unit mass of commodity is lower as compared to that in drying in a thin layer. The deep bed drying is characterized by different drying conditions at different depths.

The drying process is affected by different commodity parameters as the type and variety of grain, initial moisture content, etc., drying air parameters as the temperature, velocity, relative humidity, etc. and the dryer and operational parameters as the type of dryer, grain depth, exposure time, etc.

For analysis, deep bed drying can be considered as several thin layers taken together. After a considerable period of drying, the material bed can be partitioned in to three zones; the wet zone, the drying zone, and the dried zone. The imaginary layer that separates the drying zone and wet zone can be defined as the *drying front*. The volume of drying zone, in addition to other drying parameters and the grain moisture content, also depends strongly on the air velocity.

The drying of the grain bed occurs at the maximum rate until the drying front reaches the top of the bed. This stage is known as the *maximum drying rate period*. The rate of drying then starts decreasing, and, hence, it is termed as the *decreasing rate period*. Thus, both these periods are taken into account when calculating the drying time in case of deep beds.

The maximum drying rate period is obtained by equating the moisture lost by the commodity to be equal to moisture gained by the drying air.

$$(DM)(X_1 - X_x) = m_a (W_s - W_i) \theta \qquad (22.55)$$

where DM is the amount of dry matter in the commodity, X_1 is the initial fraction of moisture (kg per kg of dry matter), X_x is the fraction of moisture at the end of maximum drying rate period (kg per kg of dry matter), m_a is the mass of dry air going into the system per unit time (kg dry air per hour), W_s is the saturation humidity ratio, W_i the initial humidity ratio, and θ is time (hour).

$$or \quad \theta = \frac{(DM)(X_1 - X_x)}{m_a(W_s - W_i)} \qquad (22.56)$$

Equation 22.56 gives the decreasing drying rate period by taking the moisture content at the end of maximum drying rate period as the initial moisture content.

In deep bed drying, the moisture gradient of the grain bed will increase with increase in the inlet air temperature as well as with increase in bed depth. Thus, very high temperature drying is not used in deep bed. The drying zone will move faster if the initial moisture content of the commodity is less. Higher air velocity will also increase the moisture-holding capacity of the air per unit time and, thus, the drying zone will move faster. Higher air velocities in the turbulent zone do not significantly affect the rate of drying in thin layer drying. Hukill's analysis (1954) has proved very useful for determining drying conditions in deep bed drying.

Mass and Heat Balance

Batch Dryer

In a batch dryer, a specific amount of commodity is kept and the drying air is passed through the grain. The air carries moisture from the commodity and the amount of moisture taken by the air during the drying period is the same as the moisture given off by the commodity. A flow chart of the process can be shown as Figure 22.6.

In this figure DM denotes the dry matter in the commodity, t_c, and X are the temperature and fraction of moisture (kg per kg dry matter) of the commodity. The amount of dry matter does not change during the drying process; it is only the moisture from the commodity that is removed. Similarly, m_a denotes mass of dry air used for drying. When the air takes moisture

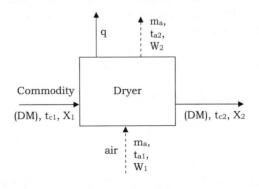

FIGURE 22.6 Mass and energy balance components for a batch dryer

from the commodity, its humidity ratio (W) increases; however, the mass of dry air remains constant. W has the unit of kg moisture per kg of dry air. t_a is the temperature of the drying air. The subscripts 1 and 2 for the air denote the entry to and exit from the dryer and the corresponding subscripts for commodity denote the start and end of drying process. q is the heat loss from dryer.

As the amount of dry matter in the commodity and the amount of dry air remain the same before and after drying (and only the moisture contents/ humidity ratios change), the simple mass balance equation can be written as:

Amount of moisture in the grain before drying + amount of moisture in the air at air inlet = amount of moisture in the grain after drying + amount of moisture in the air at air outlet

$$\text{i.e., } (DM).X_1 + m_a. W_1 = (DM).X_2 + m_a.W_2 \quad (22.57)$$

It can be written as

$$(DM)(X_1 - X_2) = m_a.(W_2 - W_1) \quad (22.58)$$

i.e., loss of moisture by the commodity = moisture gain by the air

The heat balance equation can be written as:

Enthalpy of air going in + enthalpy of commodity before drying

= enthalpy of air going out + enthalpy of grain after drying + heat loss

Here, the enthalpy of commodity and the enthalpy of air will consider both the dry matter (or dry air) and the associated moisture. If the specific heats of the air and the dry matter in the commodity are c_a and c_d, and the specific heats of moisture in the commodity and air are c_{wc} and c_w, then

$$(DM).c_d.t_{c1} + (DM).X_1.c_{wc}.t_{c1} + m_a.c_a.t_{a1} + m_a.W_1.c_w.\, t_{a1}$$

$$+ m_a.W_1.\lambda_w = (DM).c_d.t_{c2} + (DM).X_2.c_{wc}.t_{c2} + m_a.c_a.t_{a2}$$

$$+ m_a.W_2.c_w.\, t_{a2} + m_a.W_2.\lambda_w + q$$

$$(22.59)$$

Unless mentioned otherwise, c_{wc} and c_w values may be considered as equal. Within the ranges of temperature changes for drying air as encountered in normal convective drying processes, the latent heat of moisture can be considered to be constant.

Thus, the above equation can be written as:

$$(DM).c_d.(t_{c1} - t_{c2}) + (DM).c_{wc}.(X_1.t_{c1} - X_2.t_{c2}) =$$

$$m_a.c_a.(t_{a2} - t_{a1}) + m_a.c_w.(W_2.t_{a2} - W_1.t_{a1}) + m_a.\lambda_w(W_2 - W_1) + q$$

$$(22.60)$$

Continuous Counter-Current Dryer
As in the case of any material and energy balance problem, in a continuous dryer the amount of moisture entering with the

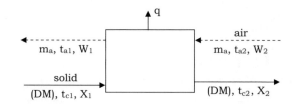

FIGURE 22.7 Mass and energy flow parameters for a counter current continuous dryer

dry air and commodity should be the same as the moisture exiting with the moist air and the dry material. The quantity of dry matter in the commodity remains constant. The energy (sensible and latent heat) going into the system should also be equal to the energy going out.

Let us draw a flow diagram of a simple counter current grain drying system as in Figure 22.7. Let the notations be same as the previous case. The subscripts 1 and 2 stand for the grain inlet and outlet points, respectively. The subscripts a and c stand for air and commodity.

The mass (moisture) balance equation is written as follows:

Amount of moisture in the grain entering the dryer before drying per unit time + amount of moisture in the air at air inlet per unit time = amount of moisture in the grain after drying per unit time + amount of moisture in the air at air inlet per unit time.

Considering the notations as shown in Figure 22.7,

$$(DM).X_1 + m_a.W_2 = (DM).X_2 + m_a.W_1 \quad (22.61)$$

It can be written as

$$(DM)(X_1 - X_2) = m_a.(W_1 - W_2) \quad (22.62)$$

i.e., moisture loss from the commodity
= moisture gain by the air

Here W_1 will be more than the W_2. The dry matter and mass of dry air values will be in kg per unit time as it is a continuous dryer.

The heat balance equation can be written as:

Enthalpy of air going in + enthalpy of commodity before drying = enthalpy of air going out + enthalpy of grain after drying + heat loss $\quad (22.63)$

Here the enthalpy of commodity and the enthalpy of air will consider both the dry matter (or dry air) and the associated moisture entering and leaving per unit time.

$$Thus, (DM).c_d.t_{c1} + (DM).X_1.c_{wc}.t_{c1} + m_a.c_a.t_{a2} + m_a.W_2.c_w.\, t_{a2}$$

$$+ m_a.W_2.\lambda_w = (DM).c_d.t_{c2} + (DM).X_2.c_{wc}.t_{c2} + m_a.c_a.t_{a1}$$

$$+ m_a.W_1.c_w.\, t_{a1} + m_a.W_1.\lambda_w + q$$

$$(22.64)$$

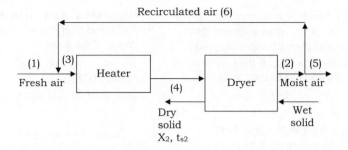

FIGURE 22.8 Flow streams for a drying system with air recirculation

The above equation can be written as:

$$(DM).c_d.(t_{c1} - t_{c2}) + (DM).c_{wc}.(X_1.t_{c1} - X_2.t_{c2}) =$$

$$m_a.c_a.(t_{a1} - t_{a2}) + m_a.c_w.(W_1.t_{a1} - W_2.t_{a2})$$

$$+ m_a.\lambda_w.(W_1 - W_2) + q$$

(22.65)

Dryer with recirculation of the drying air
In many dryers, the air exiting the dryer is not fully saturated and has a substantial amount of heat and thus, has the potential to be used for further drying. Therefore, to improve the economics of the system, a part of the exhaust air is mixed with the fresh air and then recirculated. It also helps in controlling humidity of the drying air. The system is shown in the Figure 22.8.

The parameters of the air at different sections are given below.

Fresh air: t_{a1} and W_1
Recirculated air: t_{a2} and W_2 (This is also the condition of air after drying)
Mixed air: t_{a3} and W_3
Mixed air after heating: t_{a4} with $W_4 (= W_3)$
As can be seen from the flow chart, $W_6 = W_5 = W_2$

In this situation, the mass and heat balances have to be done independently around the heater and dryer and out of the resulting equations, the unknown quantities can be calculated.

Dryer Performance Factors
Under normal situations, ambient air is heated to increase its temperature and thus to reduce its relative humidity after which it is entered into the drying chamber. In the drying chamber, the temperature of the air reduces. On a psychrometric chart, the reduction is along the constant enthalpy line as it is a case of adiabatic cooling. The process is shown in Figure 22.9.

Some dryer performance parameters have been developed that can be obtained from the psychrometric properties of moist air before and after heating and after drying (i.e., the properties of exhaust air). They are as follows.

Heat utilization factor. It is the ratio of temperature reduction due to drying to the temperature increased for drying, i.e.,

$$HUF = \frac{t_2 - t_3}{t_2 - t_1}$$

(22.66)

Note that if $t_3 < t_1$, i.e., if the exhaust air temperature from the dryer is lower than that of the air entering the heating system, then the HUF may be more than 1.

Coefficient of performance for a drying operation is given as:

$$COP = \frac{t_3 - t_1}{t_2 - t_1}$$

(22.67)

Thus, HUF = 1–COP

Effective heat efficiency of the system considers the drying potential of the air that has been actually utilized.

$$EHE = \frac{t_2 - t_3}{t_2 - t_{w2}}$$

(22.68)

where, t_{w2} is the wet bulb temperature of air after heating. Note that the moist air can remove the moisture from the commodity until it reaches saturation or the t_{w2} stage.

Another commonly used parameter in a drying system, the *thermal efficiency*, is given as the ratio of amount of heat utilized to the amount of heat supplied. In a simple form, it can be written as the ratio of heat of vaporization of grain moisture to the heat energy of fuel supplied.

$$Thermal\ efficiency = \frac{(DM)(dX/d\theta)\lambda}{m_a(h_1 - h_0)}$$

(22.69)

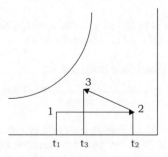

FIGURE 22.9 Temperature change of moist air during drying (1–2 is the process of heating ambient air and 2–3 is the process of drying in which the temperature of air reduces along constant enthalpy line)

where, $(dX/d\theta)$ is the change in moisture content (fraction) on dry basis and $(DM)(dX/d\theta)$ is the amount of moisture removed per unit time, λ is the heat of vaporization of commodity moisture, m_a is the mass of air flow per unit time. If the volumetric flow rate of air is divided by the specific volume of the moist air, the value obtained is m_a.

Example 22.11 Atmospheric air at 25°C and 60% RH is heated to 59°C before allowing it to enter into a dryer. After drying, the air was found to be saturated. Find out the psychrometric condition of air before heating, after heating and after drying. Also find out the heat utilization factor, the coefficient of performance, and the effective heat efficiency.

SOLUTION:

The sensible heating takes place along the constant humidity ratio lines, and the drying proceeds on the constant enthalpy lines.

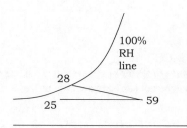

The conditions are shown in the psychrometric chart. The other conditions are given in the table below.

parameter	Condition (1) (Before heating)	Condition (2) (After heating)	Condition (3) (After drying)
t	25°C	59°C	28°C
RH	60%	10%	100%
WBT	19.5°C	28°C	28°C

The other parameters are found out as follows.

(1) The heat utilization factor,

$$H.U.F = \frac{t_2 - t_3}{t_2 - t_1} = \frac{59 - 28}{59 - 25} = 0.912$$

(2) Coefficient of performance, C.O.P = 1 − H.U.F = 1-0.912 = 0.088

(3) Effective heat efficiency,

$$EHE = \frac{t_2 - t_3}{t_2 - t_{w2}} = \frac{59 - 28}{59 - 28} = 1 \; or \; 100\%.$$

Example 22.12 During a drying process, ambient air at 35°C dry bulb temperature and 22°C wet bulb temperature was heated to 60°C and then allowed to enter the dryer. After drying, the exhaust air from the dryer was observed to be at 42°C dry bulb temperature and 28.5°C wet bulb

temperature. **Find out the heat utilization factor, coefficient of performance, and effective heat efficiency of the dryer.**

SOLUTION:

Here, $t_1= 35$°C, $t_2 = 60$°C, $t_{w2}=28.5$°C and $t_3 = 42$°C

$$Heat \; utilization \; factor = \frac{t_2 - t_3}{t_2 - t_1} = \frac{60 - 42}{60 - 35} = 0.72$$

Coefficient of performance = 1−0.72 = 0.28

$$Effective \; heat \; efficiency = \frac{t_2 - t_3}{t_2 - t_{w2}} = \frac{60 - 42}{60 - 28.5} = 0.571$$

Example 22.13 In the above problem if the temperature of the exhaust air is 28.5°C, what will be the effective heat efficiency?

SOLUTION:

The wet bulb temperature remains same during the drying process. Thus, the exhaust air has both the dry bulb and the wet bulb temperatures of 28.5°C, i.e., the air is completely saturated. In other words, the effective heat efficiency will be 1.00.

Example 22.14 Atmospheric air at 30°C and 60% RH enters a grain dryer at a flow rate of 0.8 m³. s⁻¹. The dryer capacity is 1 MT paddy/ hour. The exiting air from the dryer, which is completely saturated, is at 32°C. Find the rate of moisture removed in kg.h⁻¹.

If the air is heated by a fuel having the calorific value of 12500 kJ.kg⁻¹ and the combustion efficiency is 60%, find the rate of fuel consumption for heating the air.

SOLUTION:

The atmospheric air is to be heated to a condition so that it dries the grain and cools down to 32°C after drying. The drying takes place along a constant wet bulb temperature line and heating of air takes place along constant humidity ratio line. So, we can find out the condition of the hot air just before entering the grain as having 60% RH and 32°C wet bulb temperature. Thus, the conditions are as below.

	Condition (1) (Before heating)	Condition (2) (After heating)	Condition (3) (After drying)
t	30°C	66.5°C	32°C
RH	60%	9%	100%
WBT	24°C	32°C	32°C
Humidity ratio	0.016 kg. (kg dry air)⁻¹	0.016 kg. (kg dry air)⁻¹	0.031 kg. (kg dry air)⁻¹
Specific volume	0.88 m³. (kg dry air)⁻¹	0.984 m³. (kg dry air)⁻¹	0.908 m³. (kg dry air)⁻¹
Enthalpy	72.3 kJ. (kg dry air)⁻¹	110.8 kJ. (kg dry air)⁻¹	110.8 kJ. (kg dry air)⁻¹

Air flow rate through the dryer, $Q = 0.8$ m³.s⁻¹

As obtained for condition '1', specific volume of air, $\nu = 0.88$ m³.(kg dry air)⁻¹

Mass flow rate of dry air $= Q / \nu$

$$= 0.8 \left(m^3.s^{-1} \right) / \left(0.88\ m^3.kg^{-1} \right)$$

$$= 0.909\ kg.\ s^{-1}$$

$$= 0.909 \times 3600 = 3272.7\ kg.h^{-1}$$

Humidity ratio change in air is 0.031-0.016 = 0.015 kg. (kg dry air)⁻¹, and this is due to the moisture removed from the grain.

Thus, the amount of moisture removed from the grain

=amount of moisture gained by air

$$= 3272.7 \left(kg\ dry\ air\ /\ h \right) \times 0.015 \left(kg.(kg\ dry\ air)^{-1} \right)$$

$$= 49.09\ kg.h^{-1}$$

Heat added to the air $= 3272.7 \left(kg\ dry\ air.h^{-1} \right)$

$$\times \left(110.8 - 72.3 \right) kJ.\left(kg\ dry\ air \right)^{-1}$$

$$= 125998.95\ kJ.h^{-1}$$

The calorific value of the fuel is given as 12500 kJ.kg⁻¹

Thus, theoretical amount of fuel required = 125998.95/12500 = 10.08 kg.h⁻¹

The combustion efficiency is given as 60%

So, actual fuel required = 10.08 / 0.6 = 16.8 kg.h⁻¹

Example 22.15 In a processing plant, 5000 kg of wheat is to be dried from 22% (wet basis) moisture content to 12% (wet basis) moisture content in a dryer in one hour. What will be the heat requirement for drying? Also find out the rate of air flow required for the drying system. The different conditions/assumptions are dryer efficiency = 50%, grain moisture's latent heat of evaporation = 2390 kJ.kg⁻¹, specific heat of air = 1.008 kJ.kg⁻¹.°C⁻¹. The ambient air is at 35°C, drying air is at 50°C, and ambient air RH is 45% (wet bulb temperature 22°C). The air temperature after drying is 42°C.

SOLUTION:

Dry matter in the grain = 5000 × (1-0.22) = 3900 kg

Amount of moisture removed

$$= 3900 \times \left(\frac{22}{100 - 22} - \frac{12}{100 - 12} \right)$$

$$= 3900 \left(0.282 - 0.1364 \right) = 568.18\ kg.h^{-1}$$

Heat required to remove moisture = 568.18 × 2390 = 1357137.6 kJ.h⁻¹

(The heat required for sensible heat gain by the grain has been neglected here.)

$$\text{Amount of air needed} = \frac{1357137.6}{(50 - 42) \times 1.006}$$

$$= 168396.6\ kg.h^{-1}$$

Dryer efficiency is given as 50%

So the actual air required = 168396.6/0.5 = 336794 kg.h⁻¹

The air flow in m³.h⁻¹ can be found out by finding the specific volume of air at 35°C by using the equation, $\nu = (0.00283 + 0.00456W)\ (t + 273.15)$.

The W (humidity ratio) can be obtained from psychrometric chart.

Example 22.16 Freshly harvested paddy is to be dried from 33% (d.b.) to 15% (w.b.) moisture content. The temperature of the paddy before drying is 25°C and temperature of drying air is 40°C. Determine the energy requirement per kg of water removed. Assume the specific heat of paddy as 1.7 kJ.kg⁻¹.°C⁻¹ and the heat of vaporization of grain moisture at 40°C is 2700 kJ.kg⁻¹.

SOLUTION:

Initial moisture content = 33% (d.b.) = 25% (wet basis)

Final moisture content = 15% (d.b.)

If the initial weight of the grain is assumed as 100 kg,

then initial dry weight = 100 ×(0.75) = 75 kg

Amount of moisture removed = 75 ×(0.33-0.15) = 13.5 kg

Energy required for drying = Sensible heat needed to raise the paddy temperature from 25°C to 40°C + latent heat required to evaporate water from it

Considering that the grain is also heated by the drying air to 40°C at which the vaporization of moisture takes place,

i.e. $q = m.\ c_p.\Delta t + m_w.\lambda$

$$= (100)\ (1.7)\ (40\text{-}25) + (13.5)\ (2700) = 39000\ kJ$$

Energy required per kg of water removed is, 39000/13.5 = 2888.89 kJ. (kg of water)⁻¹.

Example 22.17 It takes five hours for drying a specific amount of vegetables in a tray dryer. The drying air is at 60°C dry bulb temperature and 50°C wet bulb temperature. The heat of

vaporization of moisture at 50°C wbt is 2383 kJ per kg water. Find the drying time if the drying air is at 50°C dry bulb temperature and 45°C wet bulb temperature and the latent heat of vaporization at this condition is 2395 kJ per kg water. The heat transfer coefficients on the surface of the commodity can be assumed to remain the same.

SOLUTION:

The heat supplied during drying is used for increasing the commodity temperature to the drying temperature (sensible heat) and then for vaporization of the moisture (latent heat). As the sensible heat requirement is much lower than the latent heat used for vaporization of moisture, the heat contribution for increasing the temperature of the commodity may be neglected.

Thus,

moisture removed per unit time = heat supplied per unit time/ heat of evaporation

$$\text{i.e. } M_w = \frac{q}{\lambda_w}$$

Considering per unit area basis

$$\frac{M_w}{A} = \frac{q}{A\lambda_w}$$

Where, M_w/A is the drying rate (R); it has a unit of $kg.m^{-2}.s^{-1}$. The unit is $kJ.s^{-1}$ for q and $kJ.kg^{-1}$ for latent heat of vaporization.

The amount of heat supplied can also be written as:

$$q = h.A.\Delta t$$

where, h is the coefficient of heat transfer. If we assume that the air is completely saturated at the exhaust, then $\Delta t = t_{db} - t_{wb}$

Thus,

$$R = \frac{M_w}{A} = \frac{q}{A\lambda_w} = \frac{hA(t - t_w)}{A\lambda_w} = \frac{h(t - t_w)}{\lambda_w}$$

$$\text{Thus, } \frac{R_1}{R_2} = \frac{\dfrac{h(t_1 - t_{w1})}{\lambda_{w1}}}{\dfrac{h(t_2 - t_{w2})}{\lambda_{w2}}}$$

But, as R is the rate of moisture removal with respect to time, $R\alpha \dfrac{1}{\theta}$

$$\text{Thus, } \frac{\theta_2}{\theta_1} = \frac{\dfrac{h(t_1 - t_{w1})}{\lambda_{w1}}}{\dfrac{h(t_2 - t_{w2})}{\lambda_{w2}}}$$

For the present situation,

$t_1 = 60°C$, $t_2 = 50°C$, $t_{w1} = 50°C$, $t_{w2} = 45°C$

$\lambda_{w1} = 2383$ kJ.kg^{-1}, $\lambda_{w2} = 2395$ kJ.kg^{-1}, and $t_1 = 5$ h

$$\text{Hence, } \theta_2 = \theta_1 \frac{(60-50)/2383}{(50-45)/2395} = (5)(2.01) = 10.05 \text{ } h$$

Example 22.18 Drying air at 65°C dry bulb and 50°C wet bulb temperatures is used for drying a vegetable in a tray dryer. Given the surface heat transfer coefficient on the vegetable is 10 W.m⁻².°C⁻¹ and heats of vaporization of moisture are 2346 kJ.kg⁻¹ and 2406 kJ.kg⁻¹ at 65°C and 50°C, respectively, determine the drying rate per square meter of the surface area. Assume that the air temperature reduces to its wet bulb temperature after drying.

SOLUTION:

Assuming that the drying air is completely saturated after removing moisture,

Total heat transferred per unit area, q = h.A.Δt = (10)(1) (65−50) = 150 W

If the heat loss is neglected, then the total amount of heat supplied is used for vaporization of moisture.

As the heat of vaporization at 50°C is given as 2406 kJ.kg^{-1}, amount of moisture removed per m² of surface area

$$= 150(J.s^{-1})/ (2406 \times 10^3) (J.kg^{-1})$$

$$= 6.234 \times 10^{-5} \text{ kg. s}^{-1} = 0.224 \text{ kg. h}^{-1}.$$

Example 22.19 A granular solid is dried in a continuous counter flow dryer at a rate of 400 kg of dry matter per hour from the moisture content of 5% to 0.02% (both on a dry basis). The temperatures of the commodity at the entry and exit of the dryer are 30°C and 60°C. The heat capacity of the commodity is 1.5 kJ. (kg dry matter)⁻¹.°C⁻¹. Hot air enters the dryer at 90°C and an absolute humidity of 0.01 kg. (kg of dry air)⁻¹ and leaves at 40°C. If there is no heat loss in the dryer, calculate the air flow rate and the outlet humidity ratio of the air exiting the dryer. The heat capacity of dry air, water vapor, and water can be taken as 1.005 kJ.kg⁻¹.°C⁻¹, 1.88 kJ.kg⁻¹.°C⁻¹, and 4.187 kJ. kg⁻¹.°C⁻¹, respectively. The latent heat of evaporation of water at 60°C is 2502 kJ.kg⁻¹.

SOLUTION:

Given:

Dry matter weight = 400 kg.h^{-1}
Initial moisture content is 0.05 kg water. (kg dry solid)$^{-1}$

Final moisture content is 0.002 kg water. (kg dry solid)$^{-1}$

The weight of water removed = 400 (0.05 – 0.002) = 19.2 kg.h^{-1}

Assuming no heat loss in the dryer,

Total heat required = heat required for increasing the temperature of dry solid and associated moisture from 30°C to 60°C + heat required to evaporate moisture

$$q = (400)\,(1.5)\,(60-30) + (400)\,(0.05)\,(4.18)\,(60-30) +$$

$$(19.2)\,(2502) = 68546.4 \text{ kJ.h}^{-1} \qquad (A)$$

If m_a is the rate of air flow (kg.h^{-1}), the amount of energy given by the air is

$$q = m_a\,(1.006 + 1.88\,W)\,\Delta t$$

$$= m_a\,(1.006 + 1.88 \times 0.01)\,(90-40) = 51.24\,(m_a) \text{ kJ.h}^{-1} \;(B)$$

From equations (A) and (B), we get m_a= 1337.75 kg.h^{-1}

The absolute humidity of the air coming out of the dryer can be given by

W = 0.01 + (19.2 / 1337.75)

= 0.0243 kg water. (kg dry air)$^{-1}$

Example 22.20 Grains are dried in an experiment in a tray dryer on a solid pan 0.305 m × 0.305 m. The depth of material in the pan is 12.7 mm. The sides and bottom of pan are insulated and hot air flows over the tray at 5 m.s^{-1}. The air has a dry bulb temperature of 60°C and the humidity ratio is 0.01 kg water per kg dry air. The air after drying is completely saturated. The wet bulb temperature of drying air is 27.8°C. The heat of vaporization of grain moisture at the wet bulb temperature is 2430 kJ.kg^{-1}. Calculate the constant rate period. Assume, heat transfer coefficient, h (W.m^{-2}.K^{-1}) = 0.0204(m$_a$)$^{0.8}$, where m$_a$ is the mass flow rate of air in kg.h^{-1}.m^{-2}.

SOLUTION:

Given, Humidity ratio of air, H = 0.01 kg water / (kg dry air)

Air temperature = 60°C

The specific volume of drying air is obtained as

$$\nu = (0.00283 + 0.00456 \times 0.01)\,(60 + 273.15) = 0.958 \text{ m}^3.\text{kg}^{-1}$$

As the air velocity is 5 m.s^{-1}, the mass flow rate of drying air is

$$m_a = v\,(\text{m. s}^{-1})\,/\nu(\text{m}^3.\text{kg}^{-1}) = 5 / 0.958$$
$$= 5.219 \text{ kg.m}^{-2}.\text{s}^{-1} = 18789.144 \text{ kg.m}^{-2}.\text{h}^{-1}$$

The coefficient of heat transfer is given by the relationship,

$$h = 0.0204\left(m_a\right)^{0.8}$$

$$= 0.0204 \times \left(18789.144\right)^{0.8} = 53.55 \; W.m^{-2}.K^{-1}$$

Area of drying bed,

$$A = 0.305 \times 0.305 = 0.093 \text{ m}^2$$

The constant rate of drying will continue until the drying air attains the wet bulb temperature. In this situation, it is 27.8°C.

Thus the rate of drying in the constant rate period is obtained as

$$R_c = q\,/\,\lambda = h\,A\,\Delta t\,/\,\lambda$$

$$= \frac{(53.55)(0.093)(60-27.8)}{(2430 \times 10^3)}(3600) = 0.237 \; kg.h^{-1}$$

Considering per unit area basis, R_c = 0.237/0.093 = 2.55 kg.m^{-2}.h^{-1}

Example 22.21 During a drying experiment with a spherical grain, the moisture content M$_r$ (kg moisture per kg dry material) varies linearly with radius r (m) as M$_r$ = 0.05 + 0.36(r). If a kernel diameter is 0.008 m, calculate its average moisture content.

SOLUTION:

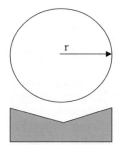

Moisture profile

The moisture variation is shown in the figure.

$$M_r = 0.05 + 0.36r$$

The total area under the curve $= \int_{-r}^{r}\left(0.05 + 0.36r\right)dr$

$$= \left[0.05r + 0.36\frac{r^2}{2}\right]_{-r}^{r}$$

$$= 0.1\,r + 0.72\,r^2\,/\,2$$

Thus, average moisture content,

$$M_r(\text{average}) = \frac{0.1\,r + 0.72\,r^2\,/\,2}{2\,r} = 0.05 + 0.18\,r$$

Here, r = 0.004 m

Thus, M_r (average) = 0.05 + (0.18) (0.004)
= 0.05072 kg. (kg dry matter)$^{-1}$

Example 22.22 A food is being dried in a continuous dryer from 15% (wet basis) to 7% (wet basis). A part of the exhaust air from the dryer is mixed with the fresh hot air for drying. The values of humidity ratio of fresh, recycled and mixed air streams are 0.01, 0.1 and 0.03 kg/kg dry air, respectively. If the feed rate is 100 kg.h^{-1}, find out the required flow rates (kg of dry air per hour) of fresh and recycled air streams. Also find out the dried material obtained per hour.

SOLUTION:

The schematic diagram for the above system is shown below. Let, m_{a1}, m_{a2} and m_a are mass flow rates of fresh air, recycled air and total air flowing into the dryer.

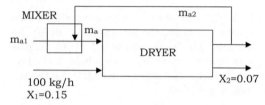

Feed rate = 100 kg.h^{-1}, initial moisture content = 15%

Hence, dry matter = 85 kg.h^{-1}
Final moisture content is 7% (wet basis),
  i.e. (7/93) × 100 % = 7.5268% dry basis

i.e. final moisture with the food is
  85 × (0.075268) = 6.398 kg.h^{-1}
Final weight of the food
  = 85 + 6.398 = 91.398 kg.h^{-1}
Moisture removed
  = 100 – 91.398 = 8.602 kg.h^{-1}

Considering a mass balance around the dryer as shown below

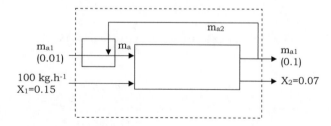

Moisture evaporated into air = moisture removed from grain
  or, m_{a1} (0.03 – 0.01) = 8.602 kg.h^{-1}
  or, m_{a1} = 95.58 kg.h^{-1}
Considering the mass balance around the mixer of air streams,

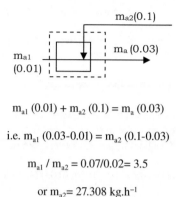

m_{a1} (0.01) + m_{a2} (0.1) = m_a (0.03)

i.e. m_{a1} (0.03-0.01) = m_{a2} (0.1-0.03)

m_{a1} / m_{a2} = 0.07/0.02 = 3.5

or m_{a2} = 27.308 kg.h^{-1}

Example 22.23 Five thousand kg of grain is dried from the initial moisture content of 24% (wet basis) to the moisture content of 13% (wet basis) in a batch dryer. The ambient air at 35°C and 0.018 kg moisture.(kg dry air)$^{-1}$ is heated to 60°C for drying. The air flow rate is 240 m^3.min^{-1}. The humidity ratio of air at the exit of the dryer has been found out to be 0.03 kg.(kg dry air)$^{-1}$. The EMC of the grain under these drying conditions is 8% (wet basis). The grain is expected to be at a mean moisture content of 18% (wet basis) when the drying front just reaches the top layer. Find the drying time both in the maximum and in the decreasing rate drying periods. Take the drying constant as 0.16 h^{-1}.

SOLUTION:

Given, grain's initial moisture content is 24% (w.b.) = 31.58% dry basis

Final moisture content of grain 13% (w.b.)
  = 14.94% dry basis
Moisture content of the grain when the drying front has reached the top or at the end of maximum drying rate period = 18% (w.b.)
  = 21.95% (d.b.)
EMC of the grain = 8% (w.b.) = 8.69% (d.b.)
Initial weight of the grain = 5000 kg
Dry matter in the grain = 5000 × (100-24)/100
  = 3800 kg

The drying time for the maximum drying rate period is given by,

$$\theta_1 = \frac{(DM) \times (M_1 - M_x)}{m_a (W_s - W_i) \times 100}$$

Here, m_a in kg.h^{-1} (or kg.min^{-1}) is the mass flow rate of air through the dryer, and DM is the dry matter in the commodity.

Given the ambient air is at 35°C (dry bulb temperature) and humidity ratio 0.018 kg.(kg dry air)$^{-1}$, specific volume of air at that condition,

$$v = (0.00283 + 0.00456W)(35 + 273.15)$$

$$= (0.00283 + 0.00456 \times 0.018)(308.15)$$

$$= 0.8973 \; m^3 . (kg \; dry \; air)^{-1}$$

Thus, air flow rate, m_a = 240 (m³. min⁻¹) × 60 / (0.8973 m³.kg⁻¹) = 16047.12 kg.h⁻¹

W_i = humidity ratio of air at dryer inlet
= 0.018 kg. (kg dry air)⁻¹

W_s = humidity ratio of air at dryer exit
= 0.03 kg. (kg dry air)⁻¹

Substituting the values in the above equation,

$$\theta_1 = \frac{3800(31.58 - 21.95)}{16047.12 \times (0.03 - 0.018) \times 100} = 1.9\ h$$

The drying time for decreasing rate drying period is given by

$$\theta_2 = \frac{1}{k} \ln \frac{M_x - M_e}{M - M_e} = \frac{1}{0.2} \ln \frac{21.95 - 8.69}{14.94 - 8.69} = \frac{1}{0.2} \ln \frac{13.26}{6.25}$$

$$= 3.76\ hours$$

Thus, the total drying time is
$\theta_1 + \theta_2 = 1.9 + 3.76 = 5.66$ h.

Example 22.24 A freshly harvested vegetable with a moisture content of 84% (wet basis) is dried to 20% (wet basis) level in a tray dryer. It has been earlier observed that the moisture contents after constant rate period of drying, residual moisture content after the first falling rate drying period and residual moisture content after the second falling rate drying period are 60%, 18%, and 4% (all on wet basis), respectively. The constant rate drying continues for 40 minutes. Determine the drying time.

SOLUTION:
First, we calculate all the moisture contents on fractions on dry weight basis.
  Initial moisture content = X_0
= 84 / 16 = 5.25 kg. (kg dry matter)⁻¹
  Final moisture content = X
= 20/80 = 0.25 kg. (kg dry matter)⁻¹
  Critical moisture content = X_c
= 60 / 40 = 1.50 kg. (kg dry matter)⁻¹
  Residual moisture content after the first falling rate period = X_{f1}
= 18/82 = 0.2195 kg. (kg dry matter)⁻¹
  Residual moisture content after the second falling rate period = X_{f2}
= 4/96 = 0.04166 kg. (kg dry matter)⁻¹
  There will not be second falling rate period as the moisture content is above 18% level.
  In this case, the equation 22.33 can be used for calculating the drying time.

i.e. $\theta = \dfrac{X_0 - X_{c1}}{R_c} + \dfrac{X_{c1} - X_{r1}}{R_c} \ln \dfrac{X_{c1} - X_{r1}}{X - X_{r1}}$

The R_c can be found as follows.

$$R_c = \frac{X_0 - X_c}{\theta_c} = \frac{5.25 - 1.5}{40}$$

$$= 0.09375\ kg\ water. \left(kg\ dry\ solid\right)^{-1}.min^{-1}$$

Drying time in falling rate period can be given as

$$\theta_{f1} = \frac{X_{c1} - X_{r1}}{R_c} \ln \frac{X_{c1} - X_{r1}}{X - X_{r1}}$$

$$= \frac{1.5 - 0.2195}{0.09375} \ln \left(\frac{1.5 - 0.2195}{0.25 - 0.2195} \right) = 13.658 \times \ln(45.9836)$$

$$= 52.28\ min$$

Total drying time = 40 + 52.28 = 92.28 min.

Example 22.25 In the above problem, estimate the drying time in the falling rate period if the drying constant (k) is given as 5.1 h⁻¹.

SOLUTION:
In this situation, we can directly use the equation 22.51.

$$\theta = \frac{1}{k} \ln \left(\frac{M_X - M_e}{M - M_e} \right)$$

Here, M_x will be the moisture content when falling rate starts,

thus M_x= 1.5 kg water. (kg dry matter)⁻¹.

M_e = 0.04166 kg water. (kg dry matter)⁻¹

Thus, $\theta = \dfrac{1}{5.1} \ln \left(\dfrac{1.5 - 0.04166}{0.25 - 0.04166} \right) = \dfrac{1}{5.1} \ln(6.999)$

$$= 0.381\ h = 22.9\ min$$

Example 22.26 Air with thermal conductivity 0.03 W.m⁻¹.K⁻¹, density 0.991 kg.m⁻³ and specific heat capacity 1×10³ J.kg⁻¹.K⁻¹ is used for drying. The mass transfer coefficient in equimolar counter-diffusion based on molar concentration gradient is 0.323 m.s⁻¹. Obtain water vapor's mass diffu-sivity in drying operation. The convective heat transfer coefficient is 35 W.m⁻².K⁻¹.

SOLUTION:

Given: ρ_{air} = 0.991 kg.m⁻³, k_{air} = 0.03 W.m⁻¹.K⁻¹

h_m= 0.323 m. s⁻¹, c_p air = 1000 J.kg⁻¹.K⁻¹

h = 35 W.m⁻².K⁻¹

$$\alpha = \frac{k}{\rho.c_p}$$

$$or, \ \alpha = \frac{0.03}{0.991 \times 1000} = 3.03 \times 10^{-5} m^2.s^{-1}$$

$$Now, \ \frac{h}{h_m} = \rho.c_p \left(\frac{\alpha}{D_{AB}} \right)^{2/3}$$

Here, D_{AB} is the mass diffusivity, h and h_m are the heat and mass transfer coefficients.

$$or, \ \frac{35}{0.323} = 0.991 \times 1000 \times \left(\frac{3.03 \times 10^{-5}}{D_{AB}} \right)^{2/3}$$

or, $D_{AB} = 8.38 \times 10^{-4} \ m^2.s^{-1}$

Example 22.27 A vegetable is being dried with air at 65°C. The humidity ratio of the drying air is 0.023 kg water.(kg dry air)⁻¹ and its saturation humidity is 0.035 kg water.(kg dry air)⁻¹. The wet bulb temperature is 36°C. Calculate the heat and mass transfer coefficients during the drying process. The molecular weight of the air can be taken as 28.97 kg.(kg mole)⁻¹. The rate of moisture removal during constant rate period may be taken as 1.5 kg water.m⁻².h⁻¹.

SOLUTION:

Given:

Dry bulb temperature of the air, t = 65°C
Wet bulb temperature of the air, t_w = 36°C
Saturation humidity ratio, W_s
= 0.035 kg water.(kg dry air)⁻¹
Humidity ratio of air, W
= 0.023 kg water.(kg dry air)⁻¹
Molecular weight of air, M_a
= 28.97 kg.(kg mole)⁻¹
Constant rate of moisture evaporation, R_c
= 1.5 kg water.m⁻².h⁻¹.

Now, the rate of water removal can be given as:

$$\frac{dM}{d\theta} = \frac{q}{\lambda_w} = \frac{hA(t - t_w)}{\lambda_w}$$

where, dM is the change in moisture content and q is the heat used for evaporation = h.A. (t-t$_w$). The λ_w is the heat of evaporation of moisture.

The heat of vaporization at 36°C can be given as 2501-2.42(36) = 2413 kJ.kg⁻¹

$$Drying \ rate \ per \ m^2 \ area = \frac{dM/d\theta}{A} = \frac{h(t-t_w)}{\lambda_w}$$

$$or, \ \frac{1.5}{3600} \frac{kg}{m^2 s} = h \frac{(65-36)}{2413 \times 10^3}$$

Thus, h= 34.67 W.m⁻². K⁻¹
The mass transfer coefficient, K_y can be calculated as follows:

$$\frac{dM}{d\theta} = \frac{q}{\lambda_w} = \frac{(K_y M_a \lambda_w) A (W_s - W)}{\lambda_w} = K_y M_a A (W_s - W)$$

or, $R_c = K_y M_a (W_s - W)$
   or, $1.5 = K_y \times (0.035 - 0.023) \times 28.97$
Thus, K_y= 1.5 / ((0.035 - 0.023) × 28.97)
= 4.314 kg mole.m⁻².h⁻¹

Fluidized Bed Drying

The drying of products can be carried out in a natural rest condition or under a fluidized condition. In through flow drying, as the air moves through a granular or particulate bed, resistance is offered by the product. However, when the velocity of air moving in an upward direction is maintained at a point that the weight of the material is just balanced by the force of air, the particle becomes fluidized. Still higher velocity will cause the particles to move away from the bed and this principle is applied in pneumatic drying or *pneumatic conveying*. In other words, when the air velocity is just equal to or marginally more than the terminal velocity of particles being dried, the particle remains in a fluidized state during drying. The drying rate is improved since all the surfaces of the individual particles are exposed to the drying air. A fabric cover or similar device is used to prevent the smaller and dried particles moving out of the drying chamber with the drying air.

The air velocity required to achieve fluidization of spherical particles can be found as:

$$v_f = \frac{(\rho_p - \rho)g}{\mu} . \frac{D_p^2 \varepsilon^3}{180(1-\varepsilon)} \quad (22.70)$$

where, v_f (m.s⁻¹) is the fluidization velocity, ρ_p and ρ are the densities of particle and the fluid (in kg.m⁻³), g (m.s⁻²) is the acceleration due to gravity, μ (kg.m⁻¹.s⁻¹) is viscosity of air, D_p (m) is the diameter of the spherical particles, and ε the porosity of bed (in fraction).

As we will be discussing subsequently in chapter on mechanical separation (under the section settling), the minimum air velocity required for conveying of the particles can be given by

$$v_t = \sqrt{\frac{4}{3} \frac{(\rho_p - \rho)g.D_p}{\rho C_d}} \quad (22.71)$$

where, v_t (m.s⁻¹) is the terminal velocity of the particles under the given situations. Here C_d is the drag coefficient. For Re between 500 and 200000, the value of C_d is 0.44. In that situation,

$$v_t = 1.75 \sqrt{\frac{a D_p (\rho_p - \rho)}{\rho}}$$

In this equation a is the force of acceleration.

Example 22.28 A fluidized bed dryer is used for drying peas at air temperature of 60°C. The peas have an average diameter of 4.2 mm, density of 866 kg.m⁻³ and the porosity of the bed is 38%. What will be the minimum air velocity required to maintain fluidization of the bed? The viscosity and density of air may be taken as 2.18×10⁻⁵ N.s.m⁻² and 1.02 kg.m⁻³.

SOLUTION:

As per the equation 22.70, the air velocity required for fluidization can be obtained as

$$v_f = \frac{(866-1.02)(9.81)}{2.18\times10^{-5}} \cdot \frac{0.0042^2 \times 0.38^3}{180(1-0.38)}$$

$$= 389241000 \times 8.673 \times 10^{-9} = 3.376 \, m. \, s^{-1}$$

Example 22.29 A fluidized bed dryer is used for drying peas using air at 3 atm absolute pressure and 90°C. The average diameter of peas is 5 mm and it has a density and porosity of 1160 kg.m⁻³ and 0.63. The bed contains 5 kg of solids and the bed diameter is 0.2 m. Determine the minimum height of the fluidized bed. If the density of air is 0.964 kg.m⁻³ at the drying temperature and 1 atm absolute pressure, what will be the pressure drop at minimum fluidization condition?

SOLUTION:

The mass of the drying material in the bed can be given as:

$$m = (V) (\rho)(1 - \varepsilon) = (\pi/4) D^2 (h)(\rho) (1-\varepsilon)$$

$$\text{or } 5 = (\pi/4) \, 0.2^2 \, (h)(1160)(1-0.63)$$

$$\text{or } h = 0.371 \, m$$

The ideal gas equation gives, $\dfrac{p_1}{\rho_1} = \dfrac{p_2}{\rho_2}$

Thus, the density of air at 3 atm pressure can be obtained as:

$$\text{or } \rho_2 = (3 \times 0.964)/1 = 2.892 \, kg.m^{-3}$$

The pressure drop in a fluidized bed dryer can be given as:

$$\Delta p = h \, g \, (\rho_p - \rho)(1 - \varepsilon)$$

$$\Delta p = (0.371) \, (20.81) \, (1160 - 2.892) \, (1 - 0.63) = 1558.18 \, Pa$$

Spray Drying

It is a device used for drying liquid foods such as milk. The material is sprayed under high pressure into the drying chamber so that the liquid is atomized into fine droplets. Hot air at a high temperature (say 130°–250°C) is fed into the chamber. Thus, the finely dispersed liquid droplets come in contact with the hot air and there is heat and mass transfer. The tiny droplet sizes (and,

thus, the increased surface area) of the liquid food cause instant removal of moisture and the liquid droplets are converted to powder state. The powder is carried along with the drying air to the exit of the drying chamber. There are devices for collection of powder from the hot air exiting from the system.

In this case the energy and mass going into the system with the liquid food and heating air will be the same as the energy and mass going out of the system with the dried powder and exhaust air.

Example 22.30 One hundred kg of milk is dried in a spray dryer from 80% to 4% moisture content. The drying air is at 160°C and the dryer exhaust air is at 95°C. The temperature of milk powder is 55°C. Determine the amount of air required to dry the milk? Also find out the thermal efficiency of drying process. Assume initial air temperature of milk and air is 30°C, specific heat of milk = 3.89 kJ.kg⁻¹.°C⁻¹, specific heat of air = 1.006 kJ.kg⁻¹.°C⁻¹, and latent heat of vaporization = 2281 kJ.kg⁻¹.

SOLUTION:

Given:

Amount of feed, F = 100 kg
Water in feed = 80%,
or fraction of solids, $X_f = 0.2$
Water in product = 4%,
or fraction of solids, $X_p = 0.96$

The mass balance equation is $F(X_f) = P(X_p)$
or, amount of milk powder or product,

$$P = F(X_f)/(X_p) = 100 \, (0.2)/0.96 = 20.83 \, kg$$

Thus, water evaporated, V = 100 - 20.83 = 79.17 kg
Total heat utilized in the drying system = Sensible heat needed to increase the milk temperature from 30°C (initial milk temperature and ambient temperature are the same) to 55°C and the latent heat required for evaporation of water at 55°C

i.e. $q = F.c_p.\Delta t + V.\lambda$

$$= (100)(3.89)(55-30)+(79.17)(2281) = 190311.77 \, kJ$$

Equating this energy to be same as the energy supplied by the air

$$q = m_a.c_{pa}.\Delta t = (m_a)(1.006)(160-95) = 65.39 \, (m_a)$$

or, the amount of air is

$$m_a = (190311.77/65.39) = 2910.4 \, kg$$

To calculate the thermal efficiency ($\eta_{thermal}$), the denominator is the energy supplied to increase the temperature of air from 30°C to 160°C.

$$Thus, \quad \eta_{thermal} = \frac{190311.77}{(2910.4)(1.006)(160-30)} \times 100 = 50\%$$

Example 22.31 Concentrated milk having 36% solids is dried in a spray dryer. The amount of milk to be dried is 600 kg and the milk initial temperature is 35°C. The final product will have 4% water. The ambient air at 30°C is heated to 140°C before entering the dryer. The temperatures of air and milk particles leaving the dryer are 90°C and 58°C, respectively. Calculate (a) volume of air required to dry the milk, and (b) thermal efficiency of the drying process.

The air is heated by steam in a heat exchanger. If the steam pressure is 0.79 MPa, calculate (c) how much steam is required for the heating the air. The efficiency of the air heater is 80%. (d) Also calculate the amount of steam required per kg of water evaporated.

The specific heat of milk may be taken as 4.00 kJ.kg⁻¹.°C⁻¹. Assume the mass density of air at the mean temperature of drying is 0.9 kg.m⁻³. At 0.79 MPa, the steam temperature is 170°C and the latent heat of condensation of steam is 2049 kJ.kg⁻¹.

SOLUTION:

Given, Initial moisture content = 64% (w.b.)
= 177.78% (d.b.)

Initial dry matter = (600) (0.36) = 216 kg

Final moisture content = 4% (w.b.) = 4.167% (d.b.)

Moisture removed = 216 (1.7778-0.04167) = 375 kg

The amount of dried product = 600-375 = 225 kg

Specific heat of milk = 4.00 kJ.kg⁻¹.°C⁻¹.

Mass density of air = 0.9 kg.m⁻³

Latent heat of vaporization at 58°C can be given as

2501-2.42 (58) = 2360.6 kJ.kg⁻¹.

Thus, heat required to dry 600 kg milk to desired level = sensible heat + heat for vaporization

= 600 (4.00) (58-35) + (375) (2360.6)

= 55200 + 885225 = 940425 kJ

Taking 1.006 kJ.kg⁻¹.°C⁻¹ as the specific heat of air, the heat supplied by air can be given as

$q = m_a.c_p.(t_1 - t_2)$ (m_a = mass of air supplied)

i.e. $m_a.(1.006) (140-90) = 940425$

Thus, m_a = 18696.32 kg

Volume of air, v = m_a/ density = 18696.32/0.9
= 20773.69 m³   (A)

The system will be having a thermal efficiency of 1 if the temperature of hot air exiting the dryer is the same as the powder temperature, i.e., 58°C. But the air exits at 90°C.

Thus, the thermal efficiency of dryer
= (140−90)/(140−58) = 0.6097 = 60.97%   (B)

Heat supplied to the air in the heating section

= (18696.32) (1.006) (140−30) = 2068934.771 kJ

Steam required = 2068934.771 / (2049) (0.8) = 1262.16 kg   (C)

Steam required per kg of water evaporated

= 1262.16/375 = 3.366 kg   (D)

Example 22.32 Concentrated milk (50% total solids) at the flow rate of 1000 kg.h⁻¹ is dried to a moisture content of 5% (dry basis) in a spray dryer. Air used for drying is at 200°C and has a humidity ratio of 0.025 kg water.(kg dry air)⁻¹. The outlet air temperature and humidity are 90°C and 0.06 kg water. (kg dry air)⁻¹. The milk is fed at 70°C and the powder temperature may be assumed to be the same as the air exit temperature. Determine the air requirement per kg of milk supplied to the dryer.

Ambient air temperature is 40°C. Take the specific heat capacities of dry air, powder, and water vapor as 1.005, 3.5, and 1.88 kJ.kg⁻¹.K⁻¹, respectively. The values of heat of vaporization at 40°, 70°, 90°, and 200°C are 2407, 2334, 2283.3, and 1938.5 kJ.kg⁻¹, respectively.

SOLUTION:

Moisture content, 5% (d.b.) = 4.76% wet basis (= 0.0476)

i.e. fraction of solids in the product, X_p = 0.9524

Mass balance: F. X_F=P.X_P

1000 × 0.5 = P × 0.9524

or, P = 525 kg

Evaporated water (V) = F − P = 475 kg

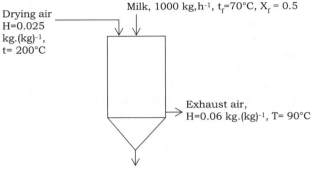

Drying air
H=0.025 kg.(kg)$^{-1}$,
t= 200°C

Milk, 1000 kg,h^{-1}, t_f=70°C, X_f = 0.5

Exhaust air,
H=0.06 kg.(kg)$^{-1}$, T= 90°C

Product, tp = 90°C, c_{pp} = 3.5 kJ.kg^{-1}°C^{-1},
Xp = 0.9524

Now, the enthalpy balance is written as:

$$F.c_{pm}.t_m + m_a.c_{pa}.t_{a1} = m_a. c_{pa}.t_{a2} + V.\lambda + P.c_{pp}.t_p$$

The subscript 'm' denotes milk and 'a' denotes air.

or, (1000) (4.00) (70) + (m_a) (1.005+1.88×0.025) (200)

= (m_a) (1.005+1.88×0.06) (90) +(475) (2283.3)
                 + (525) (3.5) (90)

or, 280000 + 210.4 m_a = 100.602 m_a + 165375

or, 109.798 m_a =114625

m_a = 1043.96 kg.h^{-1}

Thus, the amount of air required per kg of milk = 10.44 kg.

Conduction Drying

In conduction drying, the commodity to be dried is fed on to a heated surface, usually a metallic plate or a rotating drum. In a rotating drum dryer, the material is distributed in a thin layer uniformly on the drum. Steam or hot air is used to heat the drum from the inside. Thus, the product to be dried does not mix with hot air (or steam). The product usually moves on the drum for about 3/4th of a turn, during which it is dried. Thereafter, the dried material is scrapped off the drum with the help of a scrapper blade. The whole assembly is kept within housing. The moisture removed from the commodity is taken away by air or by maintaining vacuum within the housing chamber.

The heat balance in the drum dryer would involve equating the heat supplied by the steam onto the drum surface and the heat taken up by the commodity for drying. The surface area of the drum used for drying is an important parameter here as it affects the total heat exchange.

Example 22.33 In a drum dryer, the diameter and length of drum are 60 cm and 100 cm, respectively. The surface of the drum is under ambient condition. Steam at 140°C is used for heating. The product remains on the surface of drum for 3/4th of its revolution. If the overall heat transfer

coefficient is 5200 W.m^{-2}.K^{-1}, find the evaporation rate on the dryer. Preheated milk at 80°C is fed to the surface of the drum.

SOLUTION:

Given, area of drying surface, A = (¾) (πDL)

= (¾) ×3.14 × 0.6× 1.0

= 1.414 m².

As the milk boils at ambient conditions, the boiling point of milk may be considered as 100°C and the heat required for vaporization may be taken as 2257 kJ.kg^{-1}. (Though the boiling temperature of milk will be little higher than 100°C due to the dissolved solids).

Heat supplied to the milk on the surface of the drum, q = U.A.Δt

Here the temperature difference will be the log mean temperature difference.

$$\Delta t = \frac{(140-80)-(140-100)}{\ln\dfrac{140-80}{140-100}} = 49.33°C$$

$$q = U.A.\Delta t$$

$$= 5200 \times 1.414 \times (49.33)$$

$$= 362684.72 \ W$$

The latent heat of evaporation is 2257 kJ.kg^{-1}

Thus, the rate of evaporation

$$= q \, / \, A$$

$$= 362684.72 \left(J.s^{-1}\right) / \left(2257 \times 10^3\right)\left(J.kg^{-1}\right)$$

$$= 0.16 \ kg.s^{-1}, \ i.e. \ 578.49 \ kg.h^{-1}.$$

Example 22.34 A vacuum drum dryer uses the heating medium (steam) at 140°C and evaporation of moisture takes place at 90°C. Fruit juice having an initial moisture content of 88% and temperature of 60°C is dried to 4% moisture content with the output at a rate of 36 kg.h^{-1}. If the overall heat transfer coefficient is 2400 W.m^{-2}. K^{-1}, find the surface area of the drum. Assume that 70% of the drum surface is used for heating.

SOLUTION:

The dried product is 36 kg.h^{-1}
As per mass balance, F (0.12) = P (0.96)
Thus, F = 36 (0.96)/0.12 = 288 kg.h^{-1}
Vapor removed = 288-36 = 252 kg. h^{-1} = 0.07 kg. s^{-1}
The heat supplied to the system = amount of evaporation × latent heat of evaporation

The latent heat of vaporization at 90°C can be obtained from steam table as 2283 kJ.kg^{-1}.

Thus, q = 0.07 (kg. s⁻¹) × 2283×10³ (J.kg⁻¹) = 159810 J.s⁻¹

The initial juice temperature is 60°C and final juice temperature is 90°C, thus the log-mean temperature difference = [(140-60) - (140-90)]/ ln [(140-60)/ (140-90)] = 63.83°C

Now, q = U.A.Δt

159810 (J.s⁻¹) = 2400 (W.m⁻².K⁻¹)× A (m²)× 63.83 (K)

A = 1.04 m²

As 70% of the drum surface is used for heating, actual area of the drum = 1.04/0.7 = 1.49 m².

Freeze Drying

This process involves reducing the moisture content of foods or other materials by first freezing out the water and then subliming; it is known as *lyophilization* or *cryodesiccation*. The food temperature remains low, and, hence, the product has better nutritional and sensory qualities than conventionally dried products.

Freeze drying basically has three steps, namely, initial freezing, primary moisture removal by sublimation, secondary drying to remove the remaining moisture, and water vapor.

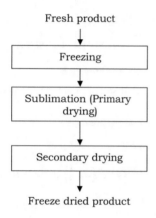

Steps in freeze drying

Initial Freezing

The material is frozen at a low temperature so that no liquid phase is present. The rate of freezing decides the texture of frozen foods. A higher freezing rate implies finer crystal structure, causes less tissue damage and helps in uniform sublimation. It also causes minimum drip and a product with less loss of water-soluble nutrients as amino acids, some proteins, sugars, some vitamins (B and C groups), and minerals. The temperature used for freezing varies with the food, but usually a temperature of -40° to -60°C is used for freezing to assure a quick freezing.

Drying

The frozen products are moved to a drying chamber where heat is supplied to sublime the ice crystals. Considering the triple point of water, and to ensure satisfactory freeze-drying, the recommended vacuum is 0.05–2 mm Hg.

The heat for sublimation is provided by conduction or by radiant heaters or by microwaves. The temperature in the drying (sublimation) chamber should be maintained to be just short of melting of ice.

As the product is heated, the ice on and near the surface of the product sublimes and the vapor moves out forming narrow channels in the product. Gradually the sublimation front (i.e., the layer between the dried zone and ice zone) moves inward. When the heat is further conducted inside, the temperature and water vapor pressure of ice are increased at the sublimation front and ice is changed to vapor form. The vapor then moves through the channels formed by the sublimation of ice in the already dried regions. The sublimation front moves into the food as the drying progresses.

> The triple point of water is the pressure and temperature at which the solid ice, liquid water, and water vapor can coexist in a stable equilibrium; it is 0.006 atm (611.73 Pa or 4.58 Torr or 4.58 mm Hg) and 0.01°C temperature. At pressures below the triple point, liquid water cannot exist and when solid ice is heated at constant pressure, it causes *sublimation* or the phase change from solid to vapor. Thus, the temperature in the sublimation chamber has to be maintained at lower than 0.01°C and the pressure less than 0.006 atm (4.58 mm Hg) for the sublimation to take place. Above the triple point, heating the solid ice at constant pressure causes formation of water, which is not desirable in freeze drying.

At the pressure used in freeze drying, 1 kg water produces about 2000-6000 m³ of vapor. A vacuum pump removes the vapor, which is subsequently condensed by using refrigeration coils.

In laboratory-scale equipment, absorption systems using chemical desiccant can be used. The desiccant can be reactivated by heating.

Heat and Mass Transfer in Freeze Drying

There is formation of a dried layer as sublimation proceeds. When heat is transferred through the dry layer, the following equation relates the partial pressure of water at the sublimation front (p_i) and the partial pressure of water at the surface (p_s).

$$p_i = p_s + \frac{k_d}{\lambda_s b}(t_s - t_i) \qquad (22.72)$$

In this equation, k_d (W.m⁻¹.K⁻¹) and b (kg.m⁻¹s⁻¹) are the thermal conductivity and permeability of the dry layer. λ_s (J.kg⁻¹) is the latent heat of sublimation. t_s (°C) is the surface temperature and t_i (°C) the temperature at the sublimation front. The partial pressures are expressed in Pa.

The drying time, θ_d (s), is given as (Karel, 1974)

$$\theta_d = \frac{x^2 \rho (M_{initial} - M_{final})\lambda_s}{8k_d(t_s - t_i)} \qquad (22.73)$$

where, x (m) and ρ (kg.m⁻³) are the thickness and bulk density of dry material, M is the moisture content of dry layer.

The heat of sublimation also varies with temperature. At 0°C, the value is 2834.5 kJ.kg⁻¹, at -12.2°C, -17.8°C, -20.0°C and at -40°C the heat of sublimation values are 2837, 2838, 2838.2 and 2838.9 kJ.kg⁻¹, respectively.

Remember that the refrigeration of the condensate is the major energy demanding stage in the whole process, and, hence, efficiency of the condenser is used to express economics of freeze drying.

Condenser efficiency

$$= \frac{temperature\ of\ sublimation}{refrigerant\ temperature\ in\ the\ condenser} \quad (22.74)$$

Example 22.35 A food is dried in a freeze dryer from an initial moisture content of 455 % (dry basis) to 5.5% moisture content (dry basis). The pressure inside the ice front can be assumed to remain constant at 72 Pa and the maximum surface temperature of the food is 50°C. Calculate the dryingtime of the food which is dried as a 1.2 cm layer on the tray placed in the freeze dryer, given that the thermal conductivity of dried food = 0.035 W.m⁻².K⁻¹, density of dried food= 450 kg.m⁻³, permeability of dried food = 2.6 × 10⁻⁸ kg.m⁻¹.s⁻¹, and 3.00×10³ kJ.kg⁻¹ is the latent heat of sublimation. The freeze dryer operates at 36 Pa.

SOLUTION:

In the given situation, $p_i = 72$ Pa

$$p_s = 36 \text{ Pa}$$

$$k_d = 0.035 \text{ W.m}^{-2}. \text{K}^{-1},$$

$$b = 2.6 \times 10^{-8} \text{ kg.m}^{-1}. \text{s}^{-1}$$

$$\lambda_s = 3.00 \times 10^3 \text{ kJ.kg}^{-1} = 3.00 \times 10^6 \text{ J.kg}^{-1}$$

$$t_s (°C) = 50°C$$

First, we have to find out the temperature of the sublimation front, t_i (°C) as follows:

$$p_i = p_s + \frac{k_b}{\lambda_s b}(t_s - t_i)$$

$$or, \ 72 = 36 + \frac{0.035}{(2.6 \times 10^{-8})(3.00 \times 10^6)}(50 - t_i)$$

$$or, \ t_i = 50 - \frac{(72 - 36)(2.6 \times 10^{-8})(3.00 \times 10^6)}{(0.035)}$$

or, $t_i = 50 - 80.23 = -30.23°C$

Now, the drying time θ_d (s) can be given as per the equation 22.73. Here, the thickness of food will be taken as 1.2 cm / 2 = 0.6 cm = 0.006 m

Substituting the values in the equation,

$$\theta_d = \frac{(0.006^2)(450)(4.55 - 0.055)(3.0 \times 10^6)}{8\ (0.035)(50 - (-30.23))}$$

$$= 9724.6 \ s = 2.7 \ hours$$

Freeze concentration

Freeze concentration is a method to reduce the moisture content of liquid foods by first freezing the water present in the material and then by separating the ice crystals by a suitable device. As low temperature is used in the process, the loss of volatile aroma compounds is less. This is different from freeze drying in that the final product is a concentrated liquid.

The equipment consists of an indirect (e.g., a scraped surface heat exchanger) or direct (e.g., solid carbon dioxide) freezing system, a mixing vessel to allow the ice crystals to grow, and a separator, usually a centrifugation device to remove the crystals from the concentrated solution.

Unlike freezing, where fast freezing is recommended to get ice crystals as small as possible, in freeze concentration, to minimize the escape of concentrated liquor entrained with the crystals, the size of ice crystals should be as large as possible economically. Thus, a rotating paddle crystallizer is used as a mixing vessel. The separation is carried out by centrifugation, filtration, filter pressing, and wash columns.

The process, as compared to concentration by boiling, has high capital and operating costs and low production rates.

Reversible freeze-dried compression is a modification of freeze-drying method involving freeze drying the food to remove about 90% of the water and then compressing it into bars. The pressure of compression can be as high as 69000 kPa. The product is further dried under vacuum and usually packed in inert gas.

Dielectric Heating and Drying

Dielectric heating is the use of either microwave (MW) or radio frequency (RF) technologies for heating the materials with moisture content. The heating is attributed to combined polarization mechanisms of dipole rotation and ionic conduction effects. The molecular friction in water molecules produces heat, and, thus, the heating is rapid.

In the case of conventional heating methods, such as conduction, convection, and radiation, the heat is applied on the surface of the food, which then is conducted into the inner parts. The surface of the product dries first and then the inner parts diffuse moisture through the dried portions. However, in the case of dielectric heating the heat is generated within the material being dried, i.e., there is volumetric heating.

The applied frequency for RF ranges between 1 to 200 MHz and that for MW is between 300 to 30,000 MHz. For electromagnetic radiation, the frequency and wavelength are related to each other as $\lambda = c/f$, with $c = 3 \times 10^8$ m.s^{-1}. Thus, the microwave radiation has wavelength of 1 mm to 1 m and the RF radiation has wavelength of 1 m to 1 km.

Microwaves or RF radiations and infrared (radiant) energy are transmitted as waves, penetrate food, and are then converted to heat. The main differences between microwave (or RF) and infrared energy are as follows.

- The infrared energy is simply absorbed on the surface of the food and is converted to heat whereas the microwaves (or RF waves) produce heat by inducing molecular friction in water molecules. Heating is limited to the surface of the food in infrared heaters, whereas microwaves or RF radiations heat throughout the food.
- The depth of penetration of the waves in the food is affected by frequency, and, hence, the lower frequency microwaves (or RF) can penetrate more into a food than infrared energy. It also implies that during drying by infrared heating the thermal conductivity of food plays a more vital role than during drying by dielectric heating.
- Moisture content of the material affects the rate of MW and RF heating. The product having low moisture heats slowly. But in case of infrared energy, the rate of heating depends on the material's color and the surface characteristics.

The advantages of the dielectric heating system include shorter times required for heating or drying and more uniform, volumetric heating giving better quality of the dried product. Use of suitable devices make the application of dielectric heating more energy efficient as compared to many other methods. Due to rapid heating and less time exposure to heating conditions, the product characteristics are better retained. Details about the theory of dielectric drying with microwaves and RF waves and factors affecting the dielectric drying process have been given in Chapter 21.

22.2 Dryers

The drying of most food grains/fruits and vegetables is still carried out under the sun in many developing and underdeveloped countries. Except for the cost of labor, sun drying involves almost no capital investment and no operating cost. However, the product is of inferior quality. In particular, sun drying of fruits and vegetables and other heat and light sensitive foods causes degradation in sensory and nutritional qualities and, therefore, it is mostly used for food grains. Some fruits and vegetables, like raisins and chilies, are also dried extensively under the sun. The major problem of sun drying is that it cannot be practiced in adverse weather conditions. Nowadays, dryers are used to dry more materials in less time and in less space and with a better product quality.

22.2.1 Classifications

There are many ways to classify dryers. One way of simple grouping can be made as solar dryers or mechanical dryers. Other types of classification are given below.

On the basis of the type of application of heat, the dryers can be classified as direct contact, indirect contact, or radiation type dryers. In the first type, the heating medium (hot air) comes in direct contact with the commodity during drying. The hot air, in addition to providing the heat for vaporization of moisture, takes away the moisture from the vicinity of the food. These dryers are also known as convection type dryers. As the heat used for vaporization of moisture comes back to the drying air with the vapor, the net enthalpy of the drying air does not change and, thus, these are also known as adiabatic dryers. In the indirect contact types, a metal plate or wall is in between the heating medium and the food material. The metal plate or wall is heated by hot air or steam or any other heating medium, which then heats the material to be dried. These are also known as conduction type dryers or non-adiabatic dryers. There are also direct-indirect dryers, which combine the adiabatic and non-adiabatic drying methods. The radiation type dryers use heating of the commodity by infrared, microwave, or dielectric radiations.

As mentioned earlier, the dryers can also be classified as *batch* or *continuous*. Some examples of batch dryers are bin dryer, kiln dryer, cabinet dryer, and freeze dryer. Some continuous dryers are tunnel dryer, belt or conveyor dryers, belt-trough dryers, etc. There are some dryers, such as the microwave dryer, infrared dryer, fluidized bed dryer, etc., that can be developed either for batch wise or continuous operation.

There are some specific dryers that use combined forms of energy and can be named hybrid type dryers. For example, in a freeze dryer the food is initially frozen (mostly by conduction) and thereafter the frozen water is allowed to sublime under controlled conditions of temperature and pressure.

Dryers can also be classified as *atmospheric dryers* or *vacuum dryers* depending on the operating pressure.

The form and type of food being handled could also be a basis of classifying drying equipment, such as *dryers for solids and pastes* and *dryers for slurry and liquid foods*.

Dryers		
Direct contact dryers	Indirect contact dryers	Radiation dryers
Conveyor dryers	Drum dryer	Solar dryer
Belt-trough dryers	Rotary dryer, etc.	Dielectric
Bin dryers		dryer
Spray dryer,		Microwave
Kiln dryers		dryer
Cabinet dryers		Infrared
Tunnel dryers		
Fluidized bed		
dryers, etc.		

Dryers used for solids and pastes	Dryers for liquid materials
Different types of solar dryers	Spray dryer
Conveyor dryers	Drum dryer
Belt-trough dryers	
Fluidized bed dryers	
Bin dryers	
Cabinet dryers	
Tunnel dryers	
Kiln dryers, etc.	

22.2.2 Dryers for Solids and Pastes

The different types of solar dryers, hot air dryers, such as bin dryer, kiln dryer, tray (or cabinet) dryer, tunnel dryer, belt or conveyor dryer, pneumatic dryer, fluidized bed dryer, are the direct contact type dryers. The rotary dryer, vacuum shelf dryer, and screw conveyor dryer are the indirect contact type dryers. The features are explained in subsequent paragraphs.

Direct Contact Dryers

Different types of hot air dryers and the solar dryers come under the category of direct contact dryers. The heating medium, usually air, is heated by electrical/solar or other forms of energy and the hot air is directed into the dryer.

The solar dryers, i.e., the dryers utilizing solar energy for drying, can be further classified as:

- natural convection or direct dryers;
- forced circulation/ indirect dryers; and
- mixed mode dryers.

Like direct sun drying, solar radiation intensity and other ambient conditions affect the solar drying process. The rate of drying is more than that obtained in open sun drying under similar conditions and lower than that in the hot air dryers.

Solar Cabinet Dryer
Construction and Operation

- It is a direct type solar dryer, which consists of a wooden (or any other material) box inside which the material is kept on trays (Figure 22.10 a). The base and sides of the box are insulated. It has a transparent cover, such as glass or plastic film. The box's inner surfaces are coated with black paint.

- As the solar radiation enters the chamber through the glass cover, the short-wave radiation is absorbed by the black surface(s). The radiation that is reflected back by the surfaces is long-wave radiation and cannot go out of the glass cover. This causes an increase in the inside air temperature. Thus, the commodities kept inside are heated and there is vaporization of moisture.

- The moisture laden air moves in an upward direction by natural convection (since the inside air is heated). Some holes are made on the upper side of the dryer to allow warm moist air to escape. There are also holes

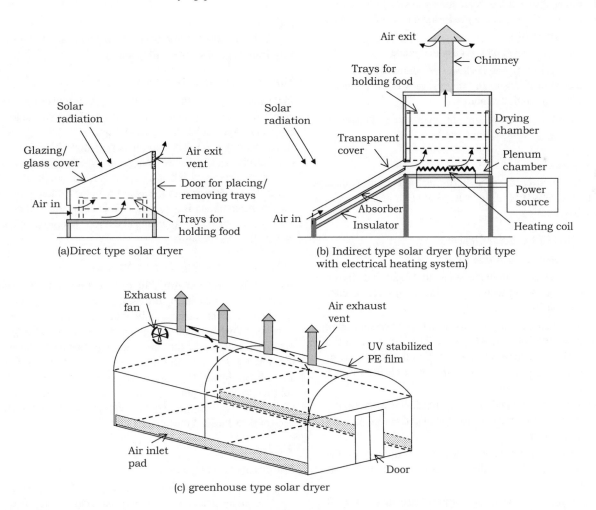

(a)Direct type solar dryer

(b) Indirect type solar dryer (hybrid type with electrical heating system)

(c) greenhouse type solar dryer

FIGURE 22.10 Schematic diagram of (a) a direct type solar dryer, (b) an indirect type solar dryer with auxiliary electrical heating system, and (c) a greenhouse type solar dryer

on the side walls toward the lower part to allow fresh air to enter inside.

- The drying trays are made of perforated base or wire mesh bottom to allow air flow and they are kept at a certain gap from the base of the chamber.
- A door is provided at the rear side of the box for putting the trays inside and for taking them out.
- As the movement of air and moisture is by natural convection, there is no provision of a fan or blower, and, hence, there is no need of electricity.
- Heating and drying of the grains and fruits and vegetables are also done in modified greenhouse type solar dryers, as shown in Figure 22.10 c. The greenhouses are altered by provision of pads/mesh for entry of fresh air from the bottom and chimneys at the top for inducing ventilation and exit of the exhaust air. Exhaust fans are also provided for forced ventilation. The product is kept on trays in layers or on mats kept at a certain height from the base.

Salient Features

- The food product in the enclosure gets heated due to direct absorption of solar energy as well as high temperature. The temperature inside the chamber can go as high as 90°C.
- The cost is low and it is easy to operate.
- The rate of drying is slower as compared to hot air mechanical dryers and there is lack of air temperature and humidity control.
- Direct exposure to the sun may cause changes in color and flavor of some products.

Applications

The solar dryers can be used for drying food grains, fruits, and vegetables. This type of dryer is suitable for drying materials in small quantities.

Mixed Mode Type Solar Dryer

Construction and Operation

- It consists of a simple solar air heater and drying chamber along with a tall chimney to increase convection (Figure 22.10 b).
- The solar air heater has a transparent cover and a box type collector to heat the air. The hot air from the solar air heater enters the bottom of the drying chamber. After passing through the materials the air goes up into the chimney.
- Trays are provided in the drying chamber for holding the food materials, drying the material kept on trays in a thin layer.
- As the solar radiation does not fall directly on the product, it is considered as an indirect type solar dryer.
- In some designs, a layer of burnt rice husk is kept at the base of the air heater to absorb solar radiation, which also adds to heating the air.

- The drying chamber may also have transparent PVC/glass sheet at the top, which allows entry of solar radiation to help in further heating of the commodity.
- The long metal cylindrical chimney is painted black to keep the inside air warm. The warm humid air goes out of the chimney. There is a suitable protecting cover at the top to restrict the rainwater or dust from entering the dryer. In some low-cost devices the chimneys are made of bamboo frame covered with black PVC.
- The heated air in the chimney produces a pressure difference between the top and bottom of the chimney, and thus the air flows in an upward direction. The height of the chimney affects the increase in temperature of air inside it and the drying rate.

Salient Features

- The parameters affecting the drying rate are the design of the dryer, solar insolation, and ambient temperature as well as the initial moisture content of the material, the amount of food kept in the drying chamber, and the depth of food on the tray.
- The material may be stirred to improve the uniformity of drying.

Applications

These dryers are also suitable for food grains and for selected fruits and vegetables in small quantities.

Forced Circulation Type Solar Dryer

In the solar dryers as described above, the air moves by natural convection and no fan or blower is used. However, the forced circulation type solar dryers have blowers for forcing the air through the system. The blower is operated either mechanically or electrically. These dryers are forced circulation solar dryers of direct and indirect types. Hot air obtained in the indirect type dryers from the solar collector is blown through the drying chamber. Bin, tunnel, belt, column, and rotary type are some of the common drying chambers, operated at low as well as high temperatures.

Solar Assisted Hybrid Drying Systems

As the solar radiation availability is often not predictable and the intensity varies during different parts of the day, an auxiliary heating arrangement can be attached to improve the rate of drying. Such systems are known as *solar assisted* or *hybrid drying systems* where the solar energy supplements the auxiliary energy. The indirect type of solar dryer shown in Figure 22.10 b is a hybrid type dryer. The heat storage unit and the solar collectors are put in series. A rock bed system is such a storage device, which stores heat energy during the daytime and releases the heat during the night.

Bin Dryer

Construction and Operation

- It is a large cylindrical or rectangular bin connected to an air heater and a blower. At a certain gap from the bottom of the bin, there is a perforated floor/wire mesh to hold the grain. The height may be up

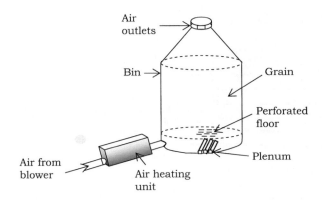

FIGURE 22.11 Schematic diagram of a bin dryer

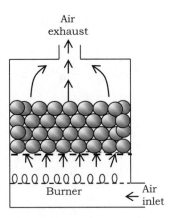

FIGURE 22.12 Schematic diagram of a kiln dryer

to several meters for large ones and even less than a meter for small-size systems (Figure 22.11).

- The drying air enters from the lower part of the bin through the mesh base, and then after passing through the commodity the moist air exits at the top.
- The temperature of air is usually kept below 60°C.
- The volume of the drying air is low, e.g., 0.5 m³ per m² of bin area per second.

Salient Features

- Bin dryers have low capital and running costs.
- These dryers are usually operated as deep bed dryers and can have high capacity.
- If the depth of grain bed is more and hot air is used, there may be considerable variation in the moisture content of the material at different depths of the bed. Use of low temperature air can help to eliminate these variations in moisture content.
- Due to the slow rate of drying, these are generally used for secondary drying (or finishing drying) of a partially dried material. This dryer is also used as a store to smooth out the product temperature fluctuations in a food after drying with other high temperature methods.
- In addition, these are also used to remove small amounts of moisture when the commodity is stored in the system.

Applications

As the structures are tall and the food product is dried in deep beds, the food should have good compressive strength so that the lower layers of food are not damaged. Further, the food should also have sufficient pore space so that the air can move through the product. Hence, these dryers are used for drying food grains, seeds, etc. as well as for finishing drying of some spices, such as ginger and garlic.

Kiln Dryer

Construction and Operation

- It consists of a two-storey building in which the upper one has a slatted (or perforated) floor and is used as the drying chamber. The lower one acts as a furnace (Figure 22.12).

- Natural gas is burnt directly in the dryer. The walls of the lower chamber are provided with some vents to allow the outside air to enter in. The air after being heated in the chamber rises into the drying chamber.
- The food is kept in thin layer, i.e., with a bed of maximum 20 cm depth.
- The heated air comes in contact with the commodity, takes up the moisture from it and then escapes through the top of the kiln.
- In some cases, the furnace is connected to a series of metal pipes, which lead to the stack of food. In some installations the furnace room and radiating pipes have been replaced by steam boilers and steam pipes.
- Some dryers are fitted with fans to increase the drying rate.

Salient Features

- Its construction and operation are easy and involves low maintenance.
- There is limited control over the drying conditions.
- Sensitive products have to be turned regularly, which needs additional labor.
- A major disadvantage associated with the system is that the flavor and aroma of the foods is affected by the combustion products that come in direct contact with them.

Applications

These dryers were earlier used for drying of hops, malt, apple slices, rings, and some other fruits. Most of these dryers have now been replaced by newer methods and designs.

Tower Dryer

Construction and Operation

- It consists of a series of rectangular/circular trays placed one above the other in a tower type drying chamber.
- A simple form of rectangular tower dryer can be an improvement of a kiln dryer, where trays with perforated bottom are stacked in the upper drying chamber. As the hot air enters from the furnace room into

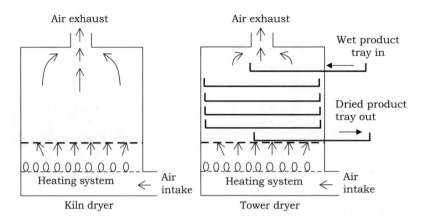

FIGURE 22.13   A comparison between kiln dryer and tower dryer

the drying chamber, it moves in an upward direction through the trays.

- Thus, material in the lowermost tray comes in contact with the hottest air and is dried first. When the material is dried to the desired level, the tray is taken out by a suitable arrangement and the remaining stack of trays comes down by one tray distance. The materials from the dried tray are replaced by wet fresh materials and put back on the stack of trays (Figure 22.13).

- In a circular tray type tower dryer, there is usually a central rotating shaft to rotate the vertical stack. The dryer is usually of 2–7 m in diameter and 2–8 m tall. The stack rotates very slowly (maximum 1 rev.min⁻¹) about the vertical shaft.

- A *turbo dryer*, shown in Figure 22.14, is a tower dryer with internal circulation of the heating gas. The solid material is dropped onto the upper most tray, where it is spread uniformly. There is a wiper on the tray at a particular location and, as the tray revolves, the material is directed by the wiper so as to fall on the immediate lower tray.

- Each tray has a slot to allow falling of the material onto the lower one.

- As the material gradually falls down through the trays it is dried by the hot air. The material is also mixed thoroughly during the travel from one tray to other, and, hence, the drying is uniform. The dried product is discharged from the bottom of the tower.

- Flow of solids and air may be co-current or counter current. In a counter current system, the air is blown from the bottom of the tower. There is also provision for the recirculation of the air.

- The air velocity is about 0.6–2.4 m.s⁻¹, which is maintained by suitable fans installed in the system.

- In the turbo dryer the bottom two trays are often used for cooling purposes.

- In some alternate designs, the material is conveyed by a close pitched screw type spiral conveyor about the vertical axis along the tower during which hot air is circulated on the product.

- The tower dryer can also have provision for recovery of evaporated solvents.

Salient Features

The rate of drying in a turbo dryer is faster than that in a tray or tunnel dryer as there is exposure of all surfaces of the product being dried.

Applications

The tower dryer is used for free-flowing granular materials.

Agitated Pan Dryer

Construction and Operation

- It consists of drying the materials in shallow circular flat bottom pans 1–2 m in diameter with the material depth of 0.3–0.6 m.

- A jacket surrounds the pans in which the hot water or steam flows for supplying of the heat energy.

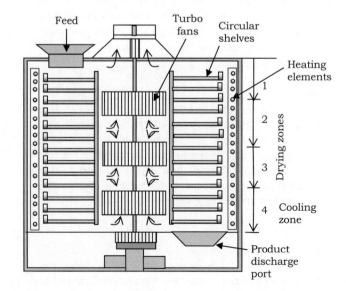

FIGURE 22.14   Turbo dryer (After Pakowski and Mujumdar, 2006 with permission)

Electrical resistance coils may also be used in place of steam or hot water.

- The product kept in the pan receives heat by conduction. Rotary scrapers or plows are provided in the pan to stir and scrap the material and to prevent the material close to the wall from being heated excessively.

- The top of the pan can be open or closed. In closed systems a vacuum pump can be incorporated so as to operate the system under vacuum.

Applications

The dryer is used for pastes and slurries in small batches.

Cabinet Dryer

Construction and Operation

- The cabinet dryer (also known as a *shelf dryer*, *compartment dryer*, or *tray dryer*) consists of an air blower, an air heating unit, and a rectangular drying chamber made of sheet metal in which a number of metal trays are placed one above the other. The walls and door of the dryer are insulated to minimize the heat loss (Figure 22.15).

- A series of electrical resistance coils (or rarely steam coils) are used to heat the air. In the case of electrically operated devices, a thermostatic control device is used to maintain the desired air temperature. In some dryers, auxiliary fans are provided to improve the convection. There may be some heaters above or side of the trays to surge the temperature of drying air and to maintain uniformity of temperature in all parts of the chamber.

- The base of the trays may be solid or perforated and the hot air is allowed to move along/ across the trays. The depth of material on the trays is usually 10–100 mm. There are baffles inside the chamber to distribute the air uniformly to all the trays. The air velocity inside the dryer is kept at 0.5–5 m.s^{-1}.

- The volume of air used is usually high so that the drying air can touch all parts of the dryer. In addition, the drying occurs in a thin layer. Thus, the air at the exit of the dryer is still at a higher temperature than the wet bulb temperature, and, thus, it still has a good

water- holding capacity. Hence, the air is often recirculated after mixing with fresh air; the ratio of the recycled air to fresh air can be up to 80:20 or 90:10. Baffles are fitted at the exhaust for this purpose.

- In some improved devices a stack of trays is placed on the racks on trucks. The trucks (the set of trays) can be easily put into and out of the chamber. Considerable time is saved between drying cycles because the loading and unloading of trucks takes place outside the dryer.

- The drying chamber can be connected to a vacuum, which helps in drying at lower temperature and is particularly helpful for heat sensitive food materials (Figure 22.16).

Salient Features

- It has low capital and maintenance costs. However, being a batch dryer it involves considerable labor for loading and unloading.

- In some devices, the product is often rotated inside the drying chamber to have uniform drying rates for all parts of the product.

- There should be uniform air circulation across all trays for uniform and effective drying. Hence, there should be proper control of temperature as well as a means for proper air circulation to all parts of the dryer. Most tray dryers do not have such provisions, and, hence, there may be variation in the drying rates in different parts of the dryer.

- Occasionally, circulation drying is also used in tray drying, which helps in reducing the drying time.

Applications

The tray dryers are used for fruits and vegetables, pastes, slurries, and varieties of products. These are useful for small production rates and are commonly used in laboratories.

Tunnel Dryer

Construction and Operation

- There is a drying tunnel (instead of a chamber), in which the food products move from one end to the other. Usually, the product is placed on trays that are

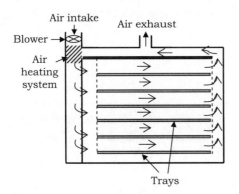
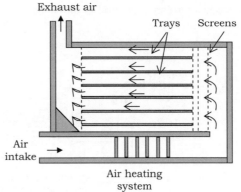

FIGURE 22.15 Schematic diagram of two different arrangements in a tray dryer

FIGURE 22.16   Vacuum tray dryer (Courtesy: M/s Bajaj Processpack Limited)

loaded on trolleys or trucks, which move on a guide or rail. Tunnel dryer is also called a *tray-truck type dryer* (Figure 22.17).

- Drying air is blown into the tunnel by a blower; the flow may be either concurrent, mixed flow, or counter current. In the mixed flow system, air is sometimes concurrent to the product and in other regions, it is counter current.

- The speed of the trolleys/trucks is regulated such that they move in a semi-continuous manner within the drying chamber to the exit end. The semi-continuous manner is adopted to allow some time for loading and unloading of the trolleys at both ends.

- After the trolleys are unloaded at the exit of the dryer, they are returned to the inlet end for further use.

Salient Features

- The direction of air flow is decided on the heat sensitivity of the product at different moisture contents.

Applications

A tunnel dryer is a large capacity equipment for fruits, vegetables, spices, etc. Tunnel dryers, now a day, have generally been replaced by conveyor dryers.

Belt or Conveyor Dryer

Construction and Operation

- The dryer consists of a drying tunnel, but, instead of moving on trucks or trolleys, the product moves on

a mesh or perforated belt type conveyor. The minimum screen size is usually 30 mesh. Belt conveyors without perforations are also used.

- A typical screen conveyor dryer is 2–3 m wide and 4 to 50 m long (Figures 22.18 and 22.19).

- The thickness of the materials on the conveyor is usually 3–5 cm. However, for some types of products it can be as high as 15 cm.

- The drying time of the product can be regulated by the speed of the conveyor and the time for a single pass usually varies between 5 to 120 min. If the product is not dried in a single pass, it can be again fed at the inlet end of the conveyor.

- Air flow can be concurrent, counter-current, mixed flow or through the food product lying on the conveyor.

- In *belt trough dryers*, the belt is in the shape of a trough. The arrangement of the trough may be done to aid in turning and mixing of food, and thus producing more uniform drying.

- To increase the residence time in the dryer, a series of conveyors can be arranged in a chamber one above the other. Such dryers require less floor space but are of more height. The product falls from one conveyor to other by gravity. During the fall from one conveyor to the other the product is mixed.

- The drying chamber may be divided into different sections by providing partitions, with each section maintaining different drying conditions. It can be

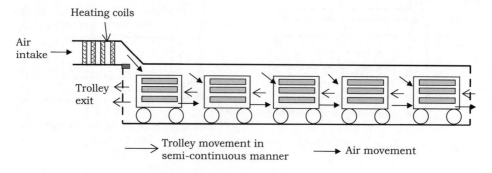

FIGURE 22.17   Schematic of a tunnel dryer (counter-current flow type)

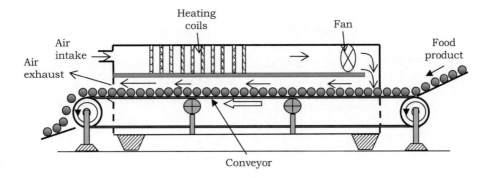

FIGURE 22.18   Schematic diagram of a single stage conveyor dryer (con-current flow type)

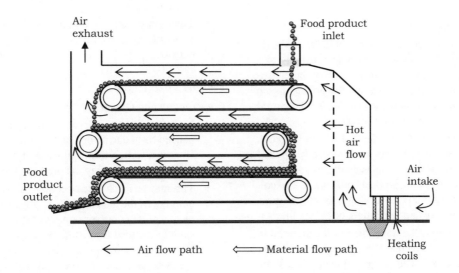

FIGURE 22.19   Schematic of a multi-stage conveyor dryer

helpful for multi-stage drying. There may also be sections for cooling after drying in such systems. Each section can have its own heating and air flow system.

- Usually, the air is blown in an upward direction through the commodity. However, the air flow is in a downward direction toward the end as the product is dry and the dusty materials may be blown away. As the product passes on a screen throughout the drying period, a device is used to collect the fines passing down through the screens.

- In screen conveyor dryers, the pastes and fine particles must be pre-formed. The aggregates usually retain their shape while being dried and do not dust through the screen except in small amounts.

Salient Features

- Such dryers help in more rapid and uniform drying because of better contact with the product and, particularly, when the mesh belt and through circulation are used. Mixing of product further improves the uniformity of drying.

- In the dryers where the materials fall consecutively from one belt to the other, there is mixing of the

material being dried. Further, by reducing the speed of the successive belts the depth of materials over the belts can be increased, i.e., the materials can be dried in deeper beds toward the last part of drying. For example, the material can be dried in a 15-25 mm bed in the first and second stages and then in a 3080 mm bed in the third stage. It helps in better uniformity of drying and in saving floor space.

- Usually, the dryer is used to dry the product in two stages: in the first stage up to about 50%–60% moisture; in the second stage to 15%–20% moisture. The final drying is done very slowly in bin dryers.

- As the drying is mostly in thin layers and the drying air comes in contact with the product for a very less time, much of the heat energy is retained in the drying air, thus the air can be recycled.

- The provision of automatic loading and unloading reduces the cost of labor.

Applications
The dryer is suitable for coarse, granular, and flaky materials, although it is also used for fruits and vegetables. They can also be used for foam mat drying. These are particularly used for large-scale drying.

Foam mat drying is a method in which a liquid food is converted into a stable foam by mixing of a stabilizer and aeration with air or an inert gas such as nitrogen. The foam is then spread as a layer of 2–3 mm thickness on a belt or a tray for drying. The common stabilizer materials are sodium alginate, xanthan gum, sorbitol, agar, and pectin. The foaming of the product increases its volume and helps in rapid drying and gives excellent product rehydration characteristics and sensory qualities.

Pneumatic Dryer

Construction and Operation

- In a pneumatic dryer, the hot air used to dry the product is also used to convey the product, usually to a short distance. Hence, the pneumatic dryer constitutes a metal duct through which the products such as moist powders or particulate foods are conveyed from one end to the other by hot air.

- The movement of the product can be horizontal, slightly inclined upward, or in a vertically upward direction. The air flow is adjusted depending on the terminal velocity of the product being dried. The collection of the material after drying is usually by gravity settling, by cyclone separation, or by bag filters.

- The drying time is the travel time of the material in the duct. Thus, to increase the drying time, longer ducts are required. The conveying duct can also be shaped as a loop to increase the residence time (drying time) in a fixed floor space. Such dryers are also known as *pneumatic ring dryers*. There is also provision of recirculation of the product in such dryers.

- The dryer can also be formed as a *high temperature short time ring dryers* (or *flash dryers*). Drying takes place within 2–10 seconds at very high temperature (even as high as 650°C, though such high temperatures are not used for foods). However, the temperature of solids remains less (about 50°C) as the time of drying is very less and there is evaporative cooling on the surface of foods. This is suitable for heat sensitive materials. A pulverizer is often incorporated in the flash drying system to obtain simultaneous drying and size reduction.

Salient Features

- As the food material remains suspended in drying air, the drying rates are very high. The temperature of air can be closely controlled, and, hence, there is also close control over drying conditions. Heat damage of thermally sensitive products is prevented by evaporative cooling of the particles.

- As the moisture content of the product reduces, its terminal velocity also reduces. Thus, the lighter and smaller particles dry more quickly. The air velocity is monitored so that such particles move rapidly than the large sized and heavier particles to the cyclone separator. This avoids the over-drying of the particles.

- These dryers have relatively low capital and maintenance costs.

- The output can vary over a large range from 10 kg.h^{-1} to 25 t.h^{-1}.

Applications

- The dryers are suitable for heat sensitive foods, moist powders or particulate foods, with particle size ranging from 10–500 μm. It cannot be used for high moisture foods (more than 40% moisture) or for large sized foods.

- The flash dryers can be used to give a rigid, porous structure by expanding the starch in potatoes or carrots.

Fluidized Bed Dryer

Construction and Operation

- In the fluidized bed dryer, the product is kept in suspended form during drying by adjusting the air flow. However, unlike the pneumatic dryer, here there is no intention to convey the product (Figure 22.20).

- This may be batch type or continuous type. In continuous types, the material enters at one end of the dryer on to a mesh conveyor or a set of trays. The trays are slightly inclined toward the exit end. The hot air is blown from below the product at a high speed so that the material moves on the conveyor in a fluid state. The trays are made to vibrate for the movement of the material from one end to the other. The dry product is taken out at the other end of the dryer.

- In the batch type fluidized bed dryer, the material is kept on a drying chamber's perforated floor. The air is blown from below at a velocity higher than the terminal velocity of the commodity to keep the product in a fluidized state. After the material has dried to the desired level, the drying is stopped and the material is collected. Thus, the principle of operation of a batch type fluidized bed drier is almost the same as that of a through flow tray type dryer except that in this case, the air velocity is maintained to keep the product in a fluidized state.

- The air flow should be uniform in all parts of the drying floor. However, since the hot air tries to find the path of least resistance while moving, a distributor is provided in the air flow path below the drying floor to uniformly distribute it. The plenum chamber below the distributor also mixes the air.

- After drying, the high velocity air exits from the dryer top. The high velocity air normally tends to carry the dried and fine particles and dust along with it. Therefore, to allow dis-entrainment of particles moving with the air, there is a disengagement region at the top. Cyclones and bag filters are also used for recovery of the particles from the exhaust air.

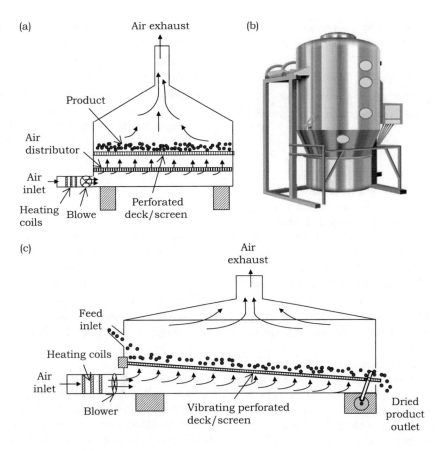

FIGURE 22.20 (a) Schematic diagram of a batch type fluidized bed dryer, (b) outer view of a fluidized bed dryer (after Techno Consultancy Services, Thrissur), (c) schematic diagram of a continuous type fluidized bed dryer

- In another type, the drying chamber has a cone-shaped bottom. The drying air moves upward through the center of the bed rather than uniformly over the cross section in a column and throws the materials to sides at the top like a fountain. The solids circulate downward along the wall of the drying chamber. Thus, the product is continuously mixed in the batch system until it is dried to a desired level.

- *Plug flow dryers* are special rectangular fluidized bed dryers consisting of different sections so that the drying conditions can be varied in different sections when the product is moving in the dryer. The last section may be used for cooling of the dried material.

Salient Features

- Since the product is in a fluidized state during drying, all sides of the product are exposed to the drying air. This improves the rate of drying.

- There is also mixing of the product during drying.

- The minimum air velocity for fluidization depends on the size of the particle, density, moisture content, and amount of food.

Applications

- Fluidization for effective entrainment requires sufficiently small particles. Thus, particulate foods, such as granules, powders, or small pieces of product with low density that will not be mechanically damaged during such drying conditions, can be dried using this type of dryer.

- The food products should not form clumps or become sticky during drying.

- Thus, products commonly dried by such dryers include peas, starch powders, diced vegetables, etc.

Indirect Contact Dryers

Rotary Dryer

Construction and Operation

- The dryer consists of a horizontal metal cylinder or drum that rotates along its axis (Figure 22.21). The drum diameter is usually 1 to 3 m. The length can be 6 to 9 m. The peripheral speed of the drum is kept at about 20–25 m.min^{-1}.

- It can be operated as a direct type or an indirect type dryer. The direct heat systems use hot air for drying. Blown from one end of the dryer, hot air exits at the other end. The flow of hot air may be either concurrent or counter current with the product flow.

- In many designs, the cylinder has an external jacket through which hot air or steam is passed. Here, the drying takes place by conduction or indirect heating. Thus, on the basis of heat supply, the drum dryers can be classified as direct heat counter flow type or direct heat parallel flow type and indirect heat counter flow type or indirect heat parallel flow types.

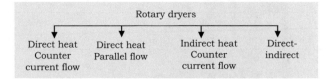

- The drum is slightly inclined (up to 5°) from the feed end to the discharge end. When the material (grain) is fed at the inlet from a hopper, it moves slowly toward the exit end due to gravity and rotation of the cylinder. The rotation of the cylinder causes the product to tumble and to be exposed to the drying air. The drying air contacts all sides of the product, which enhances the rate of moisture removal.

- Some short spiral flights are fixed at the feed end to impart the initial forward motion to the solid. A screw conveyor is placed at the discharge end to carry away the product.

- An exhaust fan may be fixed at the end to regulate the air flow. If the product contains fines or generates fines during drying, suitable devices for the collection of fines from the exiting air, such as a cyclone, filter or washing type solid collection device or a settling chamber, can be connected at the air exhaust.

- In some rotary dryers some longitudinal tubes are provided across the length of the cylinder through which the steam is passed to heat the commodity by conduction. Such a dryer is called a *steam tube rotary dryer*. The heat required for vaporization of moisture is mostly supplied by conduction and the hot air blown in the drum is used for carrying the moisture outside the dryer.

- Another modified form of rotary dryer, named a *rotary louver dryer*, has longitudinal louvers or baffles fixed on the inner surface of the drum along its full length. They lift the materials and make them fall in the drum as a thin layer during which the materials and the hot air come into contact with each other. The materials are moved along the incline by the lifting action. The hot air is admitted below the louvers, which are underneath the bed of solids.

Thus, the drying can be considered to be a combination of through circulation and rotary dryers.

- There are also *rotary vacuum dryers* with the drying chamber connected to a vacuum pump.

- Such dryers can also be used for drying of pastes and slurries in which the heating of the product is by steam jacketed cylindrical shells. There is a central shaft on which some paddles are fitted for stirring the product. The vapor escapes through the opening at the top of the condenser. A vacuum pump is used to remove the non-condensable gases.

Salient Features

- The flow of the hot air in the cylinder can be either concurrent or counter current, which is primarily decided based on the heat sensitivity of the product.

- The dryer may be fed with hot flue gas.

- The allowable mass velocity of the air depends on the dusting characteristics of the material being dried and ranges from 2000–25000 kg.m^{-2}.h^{-1}.

Applications

This dryer is used for granular materials and powders. The common applications are for drying salt, sugar, corn starch, etc. These dryers are specifically advantageous when the particles being dried have a tendency to form lumps, but they should not stick to the surface of the drum. However, the particles which may be damaged by the impact and abrasion, should not be dried in this. The use of rotary louver dryer can overcome this problem.

Vacuum Shelf Dryer

Construction and Operation

- It is a modification of a tray dryer, where the drying chamber is made airtight and is kept under vacuum (usually 1–70 torr).

- The drying chamber consists of a set of hollow shelves arranged within the drying chamber. To raise the surface temperature of the hollow shelves, steam or hot water is passed through them.

- Some flat metal trays are put on the hollow metal shelves and the material is kept on these trays in thin layers for drying. Thus, the heat transfer for drying is by conduction. In some designs the materials may

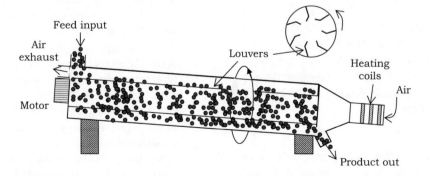

FIGURE 22.21 Schematic of a rotary dryer (counter-current heating type)

be kept directly on the hollow shelves or in others the trays may have spaces for the heating fluid.

- The vacuum pump removes, through a condenser, the vapor from the drying chamber. Rapid removal of moisture is facilitated by the vacuum within the chamber.

Salient Features

- As the vacuum helps in rapid drying, there is reduced heat damage. However, as it involves conduction drying, care should be taken that the dried food does not burn onto the trays.
- The product has a tendency to shrink during drying, which may reduce the contact between the food and heated surfaces. Therefore, additional pressure may be required on the food products to increase the contact between the food and shelves.
- Capital and operating costs are relatively high.
- As such dryers operate in batches, the throughput rate is usually low.

Applications

The vacuum shelf dryer is used for heat sensitive foods and for puff drying.

Screw Conveyor Dryer

Construction and Operation

- It consists of a closed screw conveyor with a jacket for the heating medium. The material fed at one end of the conveyor is heated by the jacket as it moves slowly to the other end (Figure 22.22).
- Steam or hot water is used as the heating medium in the jackets.
- Usually, the shell is 75–600 mm diameter and up to 6 m long. The screw is rotated slowly, usually at 2–30 rev.min^{-1}.
- There may be paddles for the movement and mixing of the product. The screw usually operates 10%–60% full.
- Some pipes are fixed on the roof of the shell for escape of the vapor.

- The chamber can also be connected to a vacuum pump and kept under a moderate vacuum to enhance the rate of drying.
- A number of screws may be used in a series to increase the drying time. The drying conditions may also be kept different in different screws. The last screw can be used as a cooling unit.

Salient Features

- The drying can also be done by convection by blowing air inside the screw.
- As the dryer can be operated under a moderate vacuum and continuous removal of the products is possible, they are used for recovery of volatile solvents from solids. For this reason, they are also known as *desolventizers*.

Applications

These are suitable to handle fine and sticky solids. As mentioned above, such dryers are used for recovery of volatile solvents from spent meal by leaching operations.

In general, the indirect dryers are capital intensive and involve higher operational cost than the direct ones. However, as they can closely monitor the drying conditions and can dry the products at low temperatures and also in the absence of air (so that the damages due to oxidation can be reduced), they are used only for valuable materials with such specific requirements.

Radiation Dryers

The radiation dryers involve heating of the food by radiation sources, such as infrared, microwave, or radio frequency waves. Convective air can also be used for removing the moisture effectively from the vicinity of the food. The infrared source can be resistance elements, incandescent lamps, or gas flames, and it causes heating of the surface of the food. Thus, heating by infrared depends on the food's surface radiation properties. However, the microwaves and radio frequency waves can penetrate to certain extents in the foods and cause volumetric heating of moist foods. RF drying offers deeper penetration of energy in foods and more uniform field distribution as compared to microwave energy. These drying

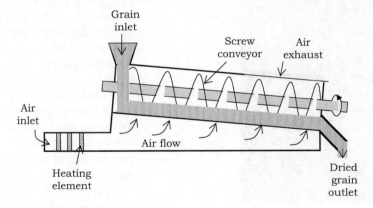

FIGURE 22.22 Schematic of a screw conveyor dryer

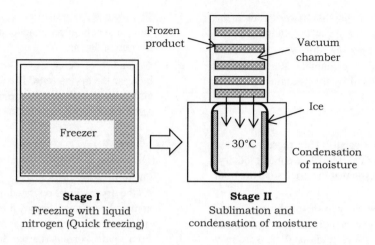

FIGURE 22.23   Stages in freeze drying

technologies and the type of food processing equipment based on these principles have been discussed in Chapter 21.

Freeze Dryer

Construction and Operation

As mentioned earlier in this chapter, freeze drying involves first freezing and then drying (or removing moisture). The food is first frozen and then the moisture is changed from ice form directly to the vapor form by sublimation. Even after the freezing operation, the material may still contain some unfrozen liquid water molecules. Thus, after the sublimation stage, the unfrozen liquid water molecules are vaporized in the next step (Figures 22.23 and 22.24).

- *Freezing device.* The selection of freezing device depends on the type of commodity. For small solids air blast freezers can be used. Immersion in liquid

nitrogen or exposure to CO_2 can be employed for products with uneven shapes. Plate freezers are suitable for uniform brick shaped solids. Drum freezers can also be employed for liquid foods.

- *Drying chamber.* It is a vacuum chamber that holds the food during drying, and heaters to supply latent heat of sublimation. Usually there are trays in the vacuum chamber on which the frozen product is kept.

- Rapid sublimation of ice is required, and it is achieved by maintaining proper vacuum level and heat input. The heat can be supplied either by conduction or by radiation heaters. Microwaves are also used to supply the heat of sublimation.

- For each type of dryer, both batch and continuous versions are available. In continuous types, the drying chamber consists of a long vacuum chamber and trays of food enter and leave the dryer on guide rails. Vacuum locks are provided at both ends. The chamber may be divided into different zones and the heater temperature and product residence times in each zone can be controlled depending on the requirements of food.

- In case of *contact type or conduction freeze dryers*, the frozen food is placed on corrugated trays, which rest on heater plates. The drying rate is slow as the heat is supplied from only one side of food. Besides, the contact between the tray and the product may be uneven. The drying rates are also different for the top and the bottom layers.

- In a modified version, known as *accelerated freeze dryers*, the food is kept under slight pressure between two expanded metal meshes. The metal mesh transfers the heat more effectively. There is an open space within the expanded metal through which the vapor can also easily move out.

- Usually, the velocity of vapor in contact freeze dryers is about 3 m.s^{-1}. It may cause escape of fine particles with the vapor.

- The contact freeze dyers usually have higher capacities than other types.

FIGURE 22.24   A laboratory freeze dryer

- As drying proceeds and the ice front recedes, the porous dried layer acts as an insulator and reduces the drying rate. It also reduces the rate of removal of moisture. Therefore, microwave radiations, which have the capability to penetrate the product up to certain depths, can be used to improve the rate of drying.
- The *radiation freeze dryers* consist of infrared radiation sources to heat thin layers of food on flat trays. The infrared rays offer more uniform heating than conduction types, primarily because the surface irregularities have minimum effect and the drying rate is almost constant. The vapor velocity is approx. 1m.s^{-1} and there is minimum entrainment loss. As there is no need for close contact between the food and heater, flat trays, being cheaper and easier to clean, are used.
- In *microwave heating*, the heat energy is absorbed uniformly by the interface, and sublimation is faster. However, there are certain disadvantages with microwave systems: the absorption of microwave energy by water is several thousand times than that of ice. If by chance a single drop of water is formed in the food, there may be accelerated melting at that point.
- In another not so common method, the vacuum chambers containing the frozen foods are directly dipped into a liquid heating medium, such as polypropylene glycol. This system thus employs both convection and conduction. The heating is uniform but slow because foodstuffs have low thermal conductivity.
- The pressure inside the drying chamber is about 0.05 to 2 mm Hg (Torr), which is normally dependent on the temperature of the vapor trap.
- The vacuum pump is necessary to remove the air from the drying chamber at the start of the process and also to remove non-condensable gases and water escaping the system. As the vapor is removed at a very low pressure, the volume of vaporized water is very large. As mentioned earlier in this chapter, 1 kg water may produce about 2000–6000 m^3 of vapor. Refrigeration coils are used to condense the vapor by contact, where it again forms ice crystals. The collected ice can be reheated with steam to turn it to water and is then drained out from the chamber. The pressure in the drying chamber as well as the temperature of vapor condenser should be as low as possible economically. It has been reported that the lowest economical chamber pressure is 13 Pa. The condenser temperature lower than -35°C is not used.
- The condensers use the major part of the energy supplied to the system, and, hence, the efficiency of freeze drying is determined by the efficiency of the condenser. The condenser efficiency is given as the ratio of temperature of sublimation to the refrigeration temperature in condenser. Thus, for improving the performance of the system, an automatic defrosting mechanism is often fitted with the refrigeration coils.

- During heating the heat has to move through the dried layer as well as the frozen layer and the thickness and thermal conductivity of the ice layer affect the heat transfer through it. With the drying process, the thickness of ice reduces and the rate of heat transfer increases. The temperature of the surface should not in any case increase beyond the level of melting of ice.
- The thickness and area of the food, thermal conductivity of the dry layer, and temperature difference between the surface of food and the ice front also affect the rate of heat transfer. The temperature of the ice front remains constant as a constant pressure is maintained in the drying chamber.
- The dried layer (which is also porous) has low thermal conductivity; as drying proceeds the resistance to heat flow increases. Hence, the thickness of food can be reduced or the temperature difference can be increased for the secondary drying process.
- In the case of microwave heating the heat is generated at the ice front; thus, the parameters such as the thermal conductivities of ice or dry food or the thickness of dry food layer do not affect the drying.
- Care should be taken that the heat is added carefully during drying, so that the temperature is kept at the desired level, or else, the ice will melt.
- During the sublimation, the moisture content of foods is usually brought down to less than about 15%. Then the remaining moisture (usually the unfrozen water) is removed by other desorption methods and the final moisture content can be brought to less than 2% level. In the second stage of drying the temperature of the dryer is raised to near ambient temperature at the low pressure.
- Theoretically, the ice temperature can be just a bit lower than the freezing point. However, to prevent the flow of concentrated solutes in the food, as discussed earlier, the minimum chamber pressure, the maximum ice temperature, and the minimum condenser temperature are used.

Salient Features

- Rapid freezing is used to produce many small ice crystals, so that after sublimation, the final product is very porous. The open porous structure allows rapid and complete dehydration, and the product also has excellent rehydration characteristics. The use of low temperature gives excellent flavor retention, up to 80–100%. Changes occurring to carbohydrates, proteins, vitamins, starches, etc. are only minor.
- The food's texture is well maintained; there is no case hardening and negligible shrinkage.
- The cost of freeze drying is very high due to the high energy cost of refrigeration as well as for the creation of a vacuum.

Packaging of Freeze-Dried Product

- Proper packaging of freeze-dried food is very important. The freeze-dried food is very porous and spongy. Therefore, it has a tendency to reconstitute rapidly, and, hence, quick and proper packaging is required to prevent moisture ingress into the food and absorption of oxygen.

- In addition, improper packaging may also lead to flavor contamination and mechanical damage (the product is fragile due to a highly porous structure). Like any other food, there is also a chance of spoilage due to biological agents and exposure to visible and UV rays.

- The use of moisture proof packages, removal of oxygen from head space (a generally acceptable maximum concentration is about 2%), packing and sealing in nitrogen are the methods used for packing of freeze-dried products. Packaging in nitrogen, in addition to preventing spoilage of the product by pollutants in the air, also prevents absorption of moisture from atmospheric air.

- During freeze drying, the dried food is under a high vacuum. After drying, the product is opened in a nitrogen gas environment and then packing is also done under nitrogen. Opening the packages in normal air will cause filling up of the pores with oxygen, which will lead to degradation.

Applications

As the food is dehydrated at low temperature, the food has excellent sensory and nutritional qualities. However, as higher cost is involved, this method is used in case of high-quality and specialty items, e.g., instant coffee, selected fruits, mushroom, strawberries, tofu and culinary herbs, shrimps, diced chicken meat, etc. Freeze drying is very commonly used for

vaccines and other injectables by the pharmaceutical industries. Microbial cultures for use in food processing are also freeze dried as the enzymes and other quality are better protected.

22.2.3 Dryers for Liquid Foods

In many specific applications, there is a need to dry liquids/ pastes. Drying of liquid or condensed milk to obtain milk powder, drying of gels, and drying of tomato pastes are some such applications. The two common types of dryers used for this purpose are the spray dryer and the drum dryer. The spray dryer is a direct contact type dryer as hot air comes in direct contact with the liquid food. The drum dryer is indirect contact or conduction type dryer. The vacuum band dryer is also used for pastes and liquids. The drum dryer and vacuum band dryer are not very common these days.

Spray Dryer

Construction and Operation

- The basic principle of the spray dryer is to spray the liquid (or paste/slurry) as very fine droplets into a chamber in which it is exposed to drying air. As the product is in the form of fine droplets, the contact area for drying per unit mass of the product is very high, which causes rapid evaporation of moisture. 1m³ of liquid forms approx. 2×10^{12} uniform 100 μm droplets, which gives a surface area over 60,000 m². This high surface area gives rapid vaporization of moisture. The droplets, after releasing the moisture, are converted to fine solids (powders).

- Shown in Figure 22.25 is the typical arrangement of a spray dryer. The product in this system enters at the top into the chamber in atomized form. The air, usually at a high temperature, also enters from the top through

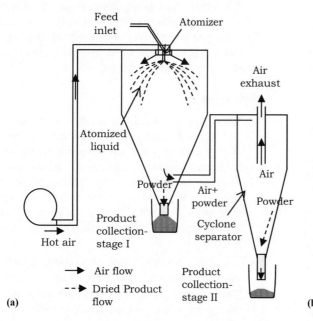

FIGURE 22.25 (a) Schematic diagram of a spray drying system (b) a spray dryer

another inlet. As the hot air moves in a downward direction and as the droplets fall down by gravity, they move together, and the hot air takes away the moisture from the liquid droplets. The height of the chamber is so maintained that the product is converted to dry powder form before it reaches the bottom of the dryer. Some dried particles are collected at the base of the chamber after they settle down. The air also carries a good portion of the product (powder) out of the drying chamber, after which some devices are installed to collect the powder from the air. Thus, the major components of a spray dryer include an atomization system for spraying the liquid as very fine droplets or as a mist, drying chamber, air heater, air handling system to blow the hot air into the chamber, and a separator to separate powder from the air after drying.

- The flow of air and liquid may be counter current, concurrent, or a combination of both (Figure 22.26). In concurrent system both the liquid and the drying air are fed at the top and the product as well as used air are collected at the bottom. The concurrent flow is better suited for heat sensitive materials, as the dried product temperature is low.

- In a counter current system, the air flow is usually from the bottom and majority of the air exits at the top. The liquid product is fed at top and the powder collected at bottom. The air flow rate should not be excessive, or else, it will carry some powders with it at the top. There is also a less common mixed current system where the product is introduced near the center of the drying chamber. The air enters at the bottom portion of the drying chamber from which it moves toward center of the chamber. The product moves upward with the air.

Drying Chamber

- The drying chamber may be either horizontal or vertical. The vertical drying chamber is more common, and the industrial dryers can have a height from a few meters to 25–30 m. The diameter of the chamber can

be 2.5–9 m. The bottom of the chamber is usually conical in shape.

- The liquid is sprayed at the top of the chamber. The residence time is 3–6 seconds in a concurrent system and 25–30 seconds in a counter current system. The chamber size is determined by the air residence time. They should be capable of handling the desired volume of air.

- The dried product moves with the air and the exhaust air is directed to a product separation and recovery unit. The conical base of the drying chamber also guides the large-size powder particles downwards to a collection chamber.

- The conical bottom helps in easy discharge of free-flowing powders. The cone angle is kept higher for better discharge. Hammers or vibrators may be fixed on the conical portion for convenient discharge of slightly sticky materials.

- There are also less common flat bottom chambers, from which the dried material is moved out by a conveying system or auger system. The construction cost is low.

- To prevent heat loss and to protect the people who are working, the drying chamber is insulated.

Air Heating and Handling

- The major components of the air heating and handling system are the air heater, fan, and ducts. As air comes in direct contact with the spray droplets, an air filter is incorporated at the inlet to provide clean air to the system and avoid contamination of the dried product. In addition, a constant amount of air in the dryer can be ensured by using filters where the pressure drop is automatically kept constant.

- Supply of proper quantity of air is very important to provide the heat of vaporization and to take away the evaporated moisture. The ambient air before being delivered into the drying chamber is heated by steam coils or electrical heaters installed in the air supply pathway. Air is usually heated to 130° to 250°C

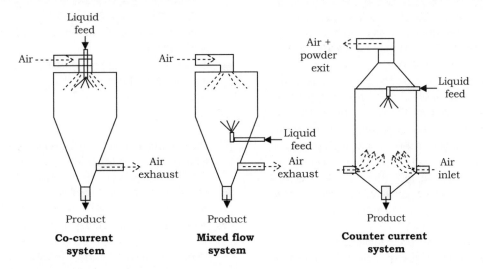

FIGURE 22.26   Air and fluid flow patterns in spray dryers

depending upon the product and other operational characteristics.

- An air disperser disperses the air inside the chamber so that it comes in contact with all parts of the spray.
- The air exhaust system has provisions as cyclone, bag filters, wet scrubber, dampers, and ducts for the recovery of the dried powder.
- After drying, the product may be cooled by ambient air. Very low temperature air (lower than 15°C) should not be used for the cooling.

Atomization

- The atomization system converts the liquid feed to very fine droplets. A feed supply pump carries the liquid from the feed tank to the atomizer. The feed may be pre-treated before being pumped.
- The uniformity of size of droplets is very important for uniform drying. Large sized droplets require a longer time or higher temperature to dry than the smaller droplets.
- The atomizer also acts as a metering device.
- In the food industry, four types of devices are used for atomization, namely, the high-pressure atomization (or one fluid atomization), centrifugal atomization, ring-jet atomization, and pneumatic atomization. However, the first two are more common in industrial spray dryers.

High-Pressure Atomization

- In this system, it is at a high pressure (700 to 2000 kPa) that the liquid food is pumped through a nozzle. The nozzles are made up of hardened stainless steel or tungsten carbide.
- The feed enters tangentially into a swirl chamber preceding the orifice. Thus, the feed has a swirl velocity in addition to axial velocity. The fluid comes out in the shape of a cone, which breaks into droplets.
- A high-pressure pump, such as a three or five piston homogenizer pump, is used for forcing the liquid through the nozzle. The fluid has a very high velocity as it is forced through a very small orifice, e.g., milk is forced at $140–210$ kg.cm^{-2} through an orifice, which may be 0.125 mm dia.
- The droplets may enter the drying chamber at a very high velocity of about 50 m.s^{-1}, but the velocity is quickly reduced to about 0.2 to 2 m.s^{-1} in the chamber.
- Counter current drying is possible in this type of atomizer (this is not possible in centrifugal atomizer).
- The droplet size usually varies from about 100 to 300 microns (average 180). The throughput capacity and droplet size are monitored by controlling the pressure; however, very small droplet sizes (< 30 μm droplets) are not possible.
- The system is sensitive to variations in capacity, wear of nozzles, change in viscosity, and solid content of feed.

Centrifugal Atomization

- In the centrifugal atomization, there is a horizontal disk (atomizer wheel) rotating at a very high velocity on which the liquid falls. The disk with a diameter of about 300 mm rotates at 3000–10000 rev.min^{-1}; thus, the peripheral velocity is about 90–200 m.s^{-1}. The discs may be plane, vaned, or cup shaped.
- A ring-shaped liquid distributor allows the liquid to fall on the disc. As the disk rotates, the centrifugal force throws the liquid as very fine droplets.
- The material is scattered in the shape of an umbrella; it increases the residence time of the material in the dryer significantly. For very fine droplets a mist of cloud is formed.
- As there is no nozzle, such a system does not have a problem in operation even if the liquid has suspended solids.
- The average drop diameter is 20 μm. The parameters, such as the disk diameter, revolution of the disk, and throughput rate, can be varied to regulate the size of the droplets.
- The centrifugal atomizer is used in vertical drying chambers.

Pneumatic Atomization

- It is also known as a two-fluid nozzle atomizer.
- A stream of compressed air is used to break up a jet of liquid. The liquid is fed to a nozzle without pressure and the compressed air is supplied through an annular orifice, which causes the atomization (Figure 22.27a). The spray travels at a very high velocity and a long distance before breaking up.
- The atomizer produces very small particles.
- The method does not require pressure pumping and there is less wear of the orifice jet.
- The system has the advantage that more concentrated liquid can be sprayed as compared with the pressure jet type and is useful for highly viscous materials.
- Pneumatic atomization requires more energy than other two types.
- It operates in two stages, which combine centrifugal force and a blast of air.

Ring-Jet Atomizer

- The liquid is introduced to the center of a centrifugal spray disc that rotates inside an air nozzle, arranged so that a ring space is formed between the nozzle outlet and the periphery of the disc (Figure 22.27 b). The disc rotates at a relatively low speed (300 rev.min^{-1}).
- In the first stage the rotating disc converts a jet of liquid into a fine film and coarse spray with an average droplet size of about 300 μm. In the second stage, the ring-shaped blast of cold air then smashes the drops

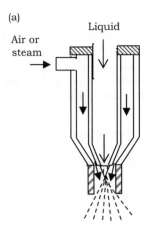 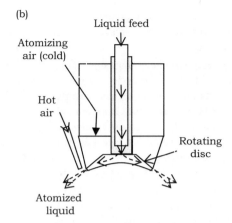

FIGURE 22.27 (a) Pneumatic atomizer, (b) ring jet atomizer

around the periphery of the disc and the size of droplets further reduce to about 50–100 μm.

- Hot air for drying the particle is provided from outside.
- The droplet size can be closely controlled by the intensity of the air blast.

Another type of atomizer, namely, the *ultrasonic nozzle atomizer*, is also used in the industry, in which the liquid is atomized in two stages, the first by a nozzle atomizer, followed by the use of ultrasonic energy, which promotes more cavitation.

Powder Recovery

- In the spray dryer, an air suction system is incorporated for easy movement of the powders to outside the drying chamber. However, as the air and dried powder reach the exit of the drying chamber, a portion of the powder, which usually constitutes larger size particles, is collected at the bottom by gravity settling. The finer particles are carried away by the air. A cyclone separator is connected in a series to the air exhaust to collect the finer particles. Subsequently, for the final separation before expelling the air into the atmosphere, a bag filter is used.
- In conical bottoms, the dried material collected at the base is collected by a rotating discharge valve or suction blower. For flat bottoms, mechanical sweeper, conveyors, or vacuum collectors are used. Any residue that cannot be collected by these devices, e.g., small amounts remaining on the side of walls, is collected at the end of the run.
- Bag filters are used for 100% collection. The fabric used for the bag filter may be cotton, wool, or plastic. The bags are automatically shaken about every 25 min. Some wet scrubbers may be put at the end to arrest the last traces of dust.
- During shut down periods, warm air at about 60°C is circulated to keep the bags dry and to prevent hygroscopic deposits on them.

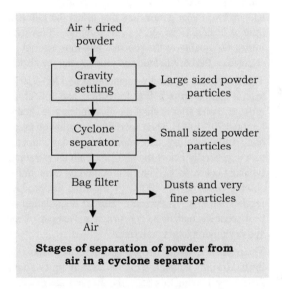

Stages of separation of powder from air in a cyclone separator

Salient Features

- A very short drying time and a good quality product are the main advantages of spray drying. The product is uniform and has good flavor, appearance, and solubility. The spray dryer can have very high throughput capacity and is suited for large-scale continuous production. Further, the operation and maintenance are easy and the labor cost for the operation is low.

In a spray drying process, even for the small particles, the outer layer dries first and the solutes that have solidified as well as the suspended solids form an outer solid layer initially that resists the moisture movement from inside. Subsequently, when the temperature of the particles increases, the liquid within the shell increases in temperature and generates pressure. A stage is reached when the extra pressure breaks the outer casing, and the vapor is released. It gives the products usually as hollow spheres, which offer excellent rehydration quality to the product.

- In the food industry, it is in a pre-concentrated form that the food is usually fed to a spray dryer for increased throughput capacity.

- The air comes in contact with the product for a short time and escapes quickly from the drying chamber. Thus, it has a considerable amount of heat energy. However, as the air coming out of the chamber contains some powder (dust), the drying air cannot be recycled. Complete elimination of dust is usually not possible even by the bag filters.

- The particles should be completely dry before they reach the exit of the dryer, or else moist material will accumulate at the bottom and will adversely affect the process. Therefore, the drying chambers should be large/tall and the drying should be monitored such that wet particles of solid or droplets are not allowed to strike and stick to solid surfaces.

- The temperature of the drying air and diameter of droplets are two major parameters that affect the rate and time of drying. However, very high temperatures may impair the quality of the product, and thus, should be selected on the basis of feed and product characteristics.

- The drying air is usually maintained at 130°–250°C (e.g., it can be up to 250°C for coffee, and 150°C for milk, etc.) and after drying, the air temperature should be about 70°–100°C. The vaporizing water has an evaporative cooling effect, and, hence, the powder temperature is generally below the air's wet bulb temperature, typically below 54°C. The air is cooled by evaporative cooling to about 70°–100°C by the time the product reaches the exit. To decrease thermal degradation of heat sensitive materials, concurrent movement of air and drying droplets is preferred.

- The residence time of droplets is only 5 to 30 seconds in the drying chamber, which is usually up to about 30 m long. During this time, the droplets are dried to less than 10% moisture content from about 40% to 60% moisture content. However, the residence time in some cases can be about 100 sec.

- The spray drying system can have a very low evaporation capacity of 1 l.h^{-1} for laboratory systems up to several tons per hour for large-scale systems.

- The spray dryer involves high capital cost. In addition, the product has to be pumped through an atomizer, which requires a considerable amount of energy. Considerable heat energy is also lost with the outgoing air, and all these add to the cost and inefficiency. Energy requirement in a spray dryer is 16–20 times higher than that in an evaporator for the same amount of water removed.

Applications

Spray dryers are used for producing instant coffee and tea, egg powder, milk powder, soymilk powder, wheat and corn starch products, encapsulated flavors, drying of enzymes, butter, cream, yoghurt, cocoa, potato, ice cream mix, cheese, fruit juices, meat and yeast extracts, etc.

A new application of spray drying is for encapsulation purposes, in which the substance to be encapsulated and the carrier material, such as modified starch, are mixed and homogenized into the form of a slurry and the slurry is then fed to the spray dryer. During the process, as the water is removed, the carrier forms a shell or capsule around the load.

Drum (Roller) Dryer

Construction and Operation

- It is basically a conduction dryer in which there is a single drum or two drums kept side by side, which are heated by steam passing within them. The heated drums are made to slowly revolve, and the liquid food is spread as a thin layer on their surfaces. As the food is heated the moisture gets evaporated.

- The drums are made up of alloy steel, stainless steel, chrome, or nickel plate steel and are usually 0.6 to 4 m in length, 0.6–3 m in diameter, revolving at 1 to 10 rpm.

- The heating medium or steam is at a pressure of 3–6 kg.cm^{-2}. The drum temperature should be kept at less than 130°C to prevent damage to the product.

- The liquid is preheated to about 70°C before it is fed onto the drums, and it is usually in pre-condensed form.

- The drying is so maintained that the slurry is dried to the desired level when it has travelled about 3/4 – 7/8th of a complete rotation. Thereafter, the dried material is scrapped off from the drum surface by a scraper or a doctor's blade. The revolution of the drum is maintained such that the product is on the drum surface for about 30–60 s.

- The dried material in the form of thin flakes is collected in a bin or a conveyor, which is later ground into a powder form.

- There is sufficient air flow above the drum to carry away moisture. The water vapor removal occurs through a vapor hood located above the drums. There is provision at the lower edge of the hood to collect any condensed moisture.

- The condensate (the heating medium after giving away the heat energy) moves from the bottom of drum, which is removed by a pump or siphon.

- The drum dryers can be classified into different groups depending on the number of drums, feeding devices, and whether the drums are kept under atmospheric pressure or a vacuum.

Classification of drum dryers		
Feeding device	Number of drums	Pressure or vacuum operation
Dip feed	Single drum	Atmospheric type
Splash feed	Double drum	Vacuum type
Nip feed		
Roller feed		

- There are different methods of feeding the drums as different types of food products behave differently to adhere to the heated drums. The common feeding

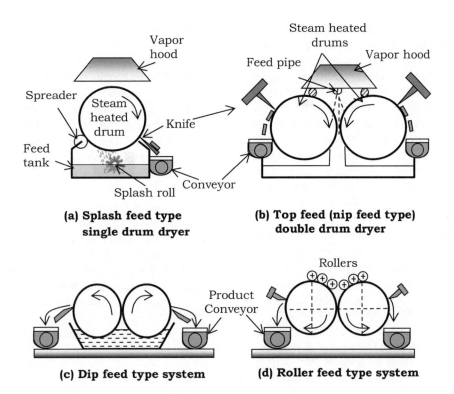

FIGURE 22.28 Schematic diagrams of drum drying system (a) single drum with splash feed, (b) double drum with top feed, (c) dip feed system and (d) roller feed system (After Ahmad, 1997 and Fellows, 2000 with permission)

devices are the dip feed, splash feed, nip feed, and roller feed. In the dip feed type, the liquid is deposited on the surface of the drum by dipping the drum in a container of liquid. In the splash feed type, the liquid is splashed (sprayed) onto the drum surface. The nip feed type feeds the drum from the top. The liquid can also be uniformly distributed on the drum by a suitable spreader or by auxiliary feed rollers (Figure 22.28).

- There may be one or two drums. In the case of double drum types, the drums are kept very close to each other and rotate in opposite directions. The slurry is fed to the gap between two rolls, which is deposited as thin layer on both drums (the gap between the drums controls the slurry thickness). As the drum(s) revolves slowly, the material also moves on the surface of the drum, during which conduction drying takes place. The usual spacing between drums is 0.5–1 mm. Figure 22.29 shows a laboratory double drum dryer.

- With a view to dry foods at lower temperature, the drums may be kept in a vacuum compartment.

Salient Features

- As there is conduction heating and evaporation is also rapid, the dryer gives a faster rate of drying.
- The drying capacity usually ranges between 5 and 50 kg.m^{-2}.h^{-1} of drying surface and is proportional to the active drum area.
- The major factors affecting the capacity of the dryer include the temperature of drum/ steam, feed temperature, thickness of film on drum surface (gap between drums) and speed of rotation. Uniform spreading of material is important for uniform drying. The foods should be in a thin layer because of their low thermal conductivity.
- The drum should also have uniform thickness to have uniform temperature on its surface.
- Preheating of product reduces the residence time required on the drum surface and thus gives better results in addition to giving higher capacity of dried materials per time.
- Pre-concentrating the feed also increases the capacity of dryer. However, thick slurry may lead to an uneven distribution on the drum surface, and, thus, there may be non-uniform drying.

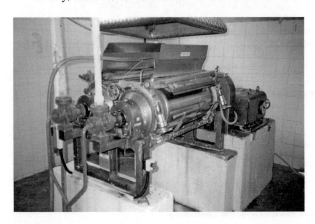

FIGURE 22.29 A laboratory drum dryer

- The drum surface temperature is high (usually more than 100°C), and there will be color and flavor degradation of the product drying in contact with the drum. There may also be some other undesirable reactions, including denaturation of protein. After removal from the dryer, the product is often cooled with chilled air.

- Typical heat consumption in conduction dryers is 2000–3000 kJ.(kg of water evaporated)$^{-1}$ as compared to hot air dryers where it is 4000–10000 kJ. (kg of water evaporated)$^{-1}$. The latent heat of vaporization of water is 2257 kJ.kg^{-1} at atmospheric pressure. The steam economy in a drum dryer varies between 1.2 to 1.6:1.

- The drum dryer has 35%–80% overall thermal efficiency.

- In the single drum design, the product remains on the drum surface for a higher fraction of the drum area. The design also has higher flexibility than double drum designs. In addition, it is easy to maintain.

- A drum dryer occupies less space and, if the product is not impaired by the high temperature of the drum, the drum dryer is more economical than a spray dryer, specifically for a small volume of product.

Applications

Drum dryers can be used for a wide range of products with different solid concentrations. It was earlier used for drying of milk, and now it has been replaced by spray dryers. It is suitable for slurries with large particles that cannot be dried with spray drying. However, the drum dryers are not suitable for salt solutions where salts crystallize during drying or for slurries containing abrasive solids.

Vacuum Band Dryer

Construction and Operation

- It is a continuous form of vacuum shelf dryer. In vacuum band dryer, the feed slurry is dried on a metal belt (or band) conveyor. The belt and its associated parts are kept inside a vacuum chamber where the pressure is maintained at 1–70 Torr. The belt material is usually high-quality steel (Figure 22.30).

- The belt passes over two hollow drums at two ends. One drum is the heating drum, which is kept at a high temperature by passing steam inside. The other drum is the cooling drum and cold water is passed within the drum.

- In the chamber along the path of the band, some radiant heaters are also fixed so that they can heat the food, while it moves on the band.

- The liquid food is sprayed onto the metal conveyor or spread by a feed roller usually at the lower side of the run just after the cooling drum.

- The food is heated combinedly by the heated drum and heaters as it is carried on the belt.

- The travel time of the material is decided by the belt speed and is maintained such that the food is completely dried before it reaches the other drum, which is usually the cooling drum. At the lower side of the cooling drum, a scraper blade removes the dried materials from the belt surface.

- A vacuum pump connected to the chamber takes away the vapor.

Salient Features

During drying, there is shrinkage of the product, reducing the net area of contact between the food and belt. Therefore, good contact between the product and the belt must be ensured. Supplementary heating by the radiant heaters avoids the problem of non-uniform heating due to this reason to some extent.

Applications

The vacuum band heater is used for heat sensitive slurries.

Vertical Thin Film Dryer

Construction and Operation

- It is a contact type continuous dryer that consists of a vertical cylindrical drying chamber with a heating jacket to heat the wall of the chamber.

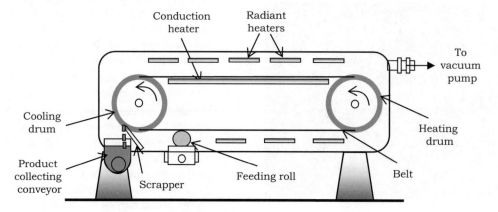

FIGURE 22.30   Schematic of a vacuum band dryer

- There is a rotor along the length of the chamber on which rows of pendulum blades are hinged.
- The feed is allowed to enter the chamber from the top and the blades spread the material as a thin layer over the heated wall. After the product is heated to the boiling point, the moisture evaporates continuously from the feed.
- During the process of evaporation, first solid flakes are formed and then the subsequent actions of the blades convert the flakes to powder form.
- The powdered material falls down and is collected via a suitable air lock.

Salient Features

- The blades have a very narrow clearance with the wall of the chamber to prevent fouling, but at the same time it should be very precise that it never touches the wall.
- The vertical thin film dryer has a small product hold-up, and, thus, there is minimal product disintegration.
- The evaporation rate is high, and the dryer gives high performance even when scaling.

Applications

This dryer is used for heat sensitive liquids/slurries that require drying in short times.

Horizontal Thin Film Dryer

Construction and Operation

- A horizontal thin film dryer consists of a horizontal heated shell on which the product is spread with the help of a horizontal rotor-blades arrangement.
- The wet product is fed continuously through the inlet nozzle and the rotor blades spread the material on the hot wall and simultaneously convey the material toward the other end of the shell.
- As the product is heated and evaporation takes place, the vapor moves in a counter current manner and is discharged through an outlet close to the feed nozzle. Entrained particles from the dry zone are removed in the wet zone.
- The dryer can be designed to operate under a vacuum, atmospheric pressure, and higher pressure.

Salient Features

- The dryer can be used for drying liquids, pastes, as well as solids.
- The dryer has high throughput capacity. The product hold-up time is small, typically between 5 and 15 minutes.

Applications

- These dryers are used for heat sensitive products that require short drying time.

- These dryers can also be used for purposes other than drying, such as cooling solid particles and melting applications.

Check Your Understanding

1. 200 kg of wheat at 23% moisture content (wet basis) is dried to 12% moisture content (dry basis). Calculate (i) the initial moisture content on dry basis and the final weight, (ii) moisture removed, and (iii) moisture content on dry basis when final weight is 185 kg.

   Answer: (i) 29.87% and 172.48 kg, (ii) 27.52 kg (iii) 20.13%

2. Potatoes are dried from 14% to 93% total solids. Considering 8% peeling losses, what is the product yield from 1000 kg of raw potato?

   Answer: 13.85%

3. A wet material is dried inside a tray dryer by using air at 65°C dry bulb temperature and 50°C wet bulb temperature. If the heat transfer coefficient on the surface of the material is 10 W.m^{-2}.K^{-1} and the latent heat of vaporization of water at 65°C and 50°C are 2346 kJ.kg^{-1} and 2406 kJ.kg^{-1}, find the rate of drying in kg water removed per square meter of the exposed surface per hour.

   Answer: 0.2244 kg water.m^{-2}.h^{-1}

4. Three thousand kg of grain is dried to the moisture content of 12% (wet basis) from an initial moisture content of 24% (wet basis) in a batch dryer. The ambient air at 25°C and 0.015 kg moisture.(kg dry air)$^{-1}$ is heated to 55°C for drying. The air flow rate is 200 m^3.min^{-1} and the humidity ratio of air at the exit of the dryer has been found to be 0.026 kg.(kg dry air)$^{-1}$. The drying constant is 0.16 h^{-1}. The grain is expected to be at a mean moisture content of 18% (wet basis) when the drying front just reaches the top layer and the EMC of the grain under these drying conditions is 10% (wb). Find the drying time both in the maximum and the decreasing rate drying periods.

   Answer: 1.437 h, 10.52 h.

5. Fresh onion with a moisture content of 85% (wet basis) is dried in a tray dryer. It has been earlier observed that the moisture contents, on a wet basis, after a constant rate period of drying, residual moisture content after first falling rate drying period and residual moisture content after the second falling rate drying period are 60%, 12%, and 2%, respectively. Constant rate drying continues for 28 minutes. Determine the total drying time to reduce the moisture content of the product to 20% (wet basis) level.

   Answer: 50.9 min.

6. A fluidized bed dryer is used for drying peas. If the peas have an average diameter of 4.5 mm, density of 850 kg.m^{-3}, and the porosity of the bed is 42%, find out the minimum air velocity required to maintain fluidization of the bed. Assume the viscosity

and density of air to be 2.15×10^{-5} N.s.m^{-2} and 0.96 kg.m^{-3}.

Answer: 5.567 m.s^{-1}

7. In a spray dryer 400 kg of milk, containing 60% water and at an initial temperature 25°C, is dried to 3% moisture content. The ambient air is at 30°C, and it is heated to 150°C before entering the dryer. The temperatures of air and milk particles leaving the dryer are 100°C and 55°C, respectively.

   Calculate the volume of air required to dry the milk and the thermal efficiency of the drying process. If steam at 7.2 kg.cm^{-2} pressure is used for heating air, how much steam is required for the heating if the latent heat of condensation of steam at this pressure is 2050 kJ.kg^{-1} and the efficiency of the air heater is 80%? Also calculate the amount of steam required per kg of water evaporated. The specific heat of milk may be taken as 4.00 kJ.kg^{-1}.K^{-1}. Assume the mass density of air at the mean temperature of 125°C as 0.9 kg.m^{-3}.

   Answer: Volume of air = 13355.7 m³; Thermal efficiency of dryer = 52.6%; Steam required per kg of water evaporated = 3.76 kg

8. A drum dryer uses steam at 130°C as the heating medium. The diameter and length of drum are 60 cm and 90 cm. The surface of the drum is under an ambient condition and the milk may be considered to vaporize at 100°C (λ=2257 kJ.kg^{-1}). The product remains on the surface of the drum for 3/4th of revolution of drum. If the overall heat transfer coefficient is 4800 W.m^{-2}.K^{-1}, find the evaporation rate in the dryer. The milk is fed at a temperature very close to its boiling temperature on the drum surface.

   Answer: 292 kg.h^{-1}.

9. A drum dryer uses the heating medium at 150°C and the product dries at 90°C. The liquid food initially at 96% moisture content is dried to 4% moisture content with the output at a rate of 40 kg.h^{-1}. If the overall heat transfer coefficient is 1700 W.m^{-2}.K^{-1}, find the surface area of the roller. Assume that 3/4th of the drum surface is used for heating and there is a negligible increase in temperature of the liquid food on the drum surface.

   Answer: 1.94 m².

10. Milk with 20% dissolved solids is dried at the rate of 60 kg dried milk per hour with 4.5% final moisture content on a drum dryer. The drum dimensions are 0.75 m (D) × 1 m (L). The product remains on the drum up to 70% of a revolution. The drum is heated by steam at 4.7 bar absolute pressure, at which the steam temperature may be taken as 150°C. The milk evaporates at 101°C. Find the overall heat transfer coefficient of the system. The latent heat of vaporization at 101°C (from the steam table) is 2254.3 kJ.kg^{-1}.

    Answer: 1755.34 W.m^{-2}. K^{-1}

11. A continuous dryer is used to dry 12 kg.min^{-1} of a blanched vegetable containing 50% moisture (wet weight basis) to give a product containing 10% moisture. As the dryer could handle feed material with moisture content not more than 25%, a part of dried material was recycled and mixed with the fresh feed. Find the evaporation rate of the dryer and the recycle ratio to achieve the drying requirement.

    Answer: 5.333 kg.min^{-1} and 1.67

12. In a drying experiment, the constant rate of drying is found to be 3.6 kg water m^{-2} h^{-1}. Dry bulb and wet bulb temperatures of the drying air are 75°C and 37°C, respectively. Latent heats of vaporization at the dry bulb and the wet bulb temperatures are 2321 and 2414 kJ.kg^{-1}, respectively. What is the convective heat transfer coefficient for the drying operation?

    Answer: 63.526 W.m^{-2}.K^{-1}

13. A food material having initial moisture content of 400% (dry weight basis) is poured into 10 mm layers in a tray of freeze dryer that operates at 40 Pa. It is to be dried to 8% moisture (dry weight basis) at a maximum surface temperature of 55°C. The dried food has a thermal conductivity of 0.03 W.m^{-1}.K^{-1}, density of 470 kg.m^{-3}, permeability of 2.4×10^{-8} kg.s^{-1}.m^{-1}, and latent heat of sublimation of 2.9×10^3 kJ.kg^{-1}. It is assumed that the pressure at the ice front remains constant at 78 Pa. Find the temperature at the sublimation front and the drying time.

    Answer: 33.16°C, 7 h

Objective Questions

1. Write short notes on
   a. Drying constant;
   b. Cabinet dryer;
   c. Tunnel dryer;
   d. Pneumatic dryer;
   e. Rotary dryer;
   f. Vacuum shelf dryer;
   g. Atomization systems in a spray dryer;
   h. Powder recovery systems in a spray dryer.

2. Differentiate between
   a. Water activity and moisture content;
   b. Bound moisture and unbound moisture;
   c. Free moisture and equilibrium moisture content;
   d. Adsorption EMC and desorption EMC;
   e. Constant rate period of drying and falling rate period of drying;
   f. Conduction drying and convection drying;
   g. Heat utilization factor and effective heat efficiency;
   h. Freeze concentration and freeze drying;
   i. Direct dryers and indirect dryers;
   j. Bin dryer and kiln dryer;
   k. Drum dryer and rotary dryer.

3. Why does the equilibrium moisture content of a food reduce with an increase in temperature?

4. State the situations where the fluidized bed drying is preferred to normal convection dryers.

5. Explain the principle of the spray drying process. What are the advantages of spray drying over other liquid drying methods?

6. What are the different stages in a freeze-drying process? Is fast freezing or slow freezing advised for the freeze-drying process?

7. What is the use of a vacuum during the drying process?

BIBLIOGRAPHY

Ahmad, T. 1999. *Dairy Plant Engineering and Management* 4th ed. Kitab Mahal, New Delhi.

Ashworth, J.C. 1981. Developments in dehydration. *Food Manufacture*.December 25–27, 29.

Bahu, R.E. 1997. Fluidized bed dryers. In C.G.J. Baker (ed.) *Industrial Drying of Foods*. Blackie Academic and Professional, London,65–89.

Barbosa-Canovas, G.V. 1996. *Dehydration of Foods*. Chapman and Hall, New York.

Barr, D.J. and Baker, C.G.J. 1997. Specialized drying systems. In C.G.J. Baker (ed.) *Industrial Drying of Foods*. Blackie Academic and Professional, London, 179–209.

Brennan, J.G. 1992. Developments in drying. In A. Turner (ed.) *Food Technology International Europe*. Sterling Publications International, London, 77–80.

Brennan, J.G., Butters, J.R., Cowell, N.D. and Lilly, A.E.V. 1976. *Food Engineering Operations* 2nd ed. Elsevier Applied Science, London

Brennan, J.G. (ed.) 2006. *Food Processing handbook* 3rd ed. WILEY-VCH Verlag GmbH & Co. KGaA, Weinheim.

Brenndorfer, B., Kennedy, L. Bateman, C.O. , Trim, D.S., Mrema, G.C. and Wereko-Brobby, C. 1985. *Solar Dryers- Their Role in Post harvest Processing*. Commonwealth Science Council, London.

Brooker, D.B., Bakker Arkema, F W, Hall, C W. 1974. *Drying Cereal Grains*. AVI Publishing Co., Westport, Connecticut.

Chakraverty, A. 1999. *Post-Harvest Technology of Cereals, Pulses and Oilseeds*. Oxford & IBH publishing Co. Ltd., New Delhi.

Charm, S.E. 1978. *Fundamentals of Food Engineering* 3rd ed. AVI Publishing Co., Westport, Connecticut.

Chen, C.S., Clayton, J.T. 1971. The effect of temperature on sorption isotherms of biological materials. *Transactions of the ASAE* 14:927–929.

Chung, D.S., Pfost, H.B. 1967. Adsorption and desorption of water vapor by cereal grains and their products. *Transactions of the ASAE* 10:549–551.

Chung, D.S. and Pfost, H.B. 1967. Adsorption and desorption of water vapor by cereal grains and their products. Part II: Development of the general isotherm equation. *Transactions of the ASAE* 10:552–555.

Cohen, J.S. and Yang, T.C.S. 1995. Progress in food dehydration. *Trends in Food Science & Technology* 6:20–25.

Corrêa, P.C., Goneli, A.L.D., Jaren, C., Ribeiro, D.M., Resende, O. 2007. Sorption isotherms and isosteric heat of peanut pods, kernels and hulls. *Food Science and Technology International* 13:231–238.

Day, D.L. and Nelson, G.L. 1965. Desorption isotherms for wheat. *Transactions of the ASAE* 8:293–297.

Earle, R.L. and Earle, M D. 2004. *Unit Operations in Food Processing*. Web Edn., The New Zealand Institute of Food Science & Technology, Inc.,Auckland. https://www.nzifst.org.nz/resources/unitoperations/index.htm

Fellows, P.J. 2000. *Food Processing Technology*. Woodhead Publishing, Cambridge, UK.

Fellows, P.J., Axtell, B.L. and Dillon, M. 1995. *Quality Assurance for Small Scale Rural Food Industries*. FAO Agricultural Services Bulletin 117, FAO, Rome, Italy.

Fennema, O.R. 1996. Water and ice. In: O.R. Fennema (ed.) *Food Chemistry* 3rd ed. Marcel Dekker, New York, 17–94.

Ferrall, A.W. 1976. *Food Engineering Systems Vol. 1: Operations*, AVI Publishing Co., Westport, Connecticut.

Ferrall, A.W. 1979. *Food Engineering Systems Vol. 2: Utilities* , AVI Publishing Co., Westport, Connecticut,

Flink, J.M. 1982. Effect of processing on nutritive value of food: Freeze-drying. In M. Recheigl (ed.) *Handbook of the Nutritive Value of Processed Food* Vol.I, CRC Press, Boca Raton, 45–62.

Garg, H.P. and Prakash, J. 1997. *Solar Energy- Fundamentals and Applications*. Tata McGraw Hill, New Delhi.

Geankoplis, C.J.C. 2004. *Transport Processes and Separation Process Principles*, 4th ed., Prentice-Hall of India, New Delhi.

Goldblith, S.A., Rey, L., and Rothmayr, W.W. 1975. *Freeze Drying and Advanced Food Technology*. Academic Press, New York.

Heldman, D.R. and Hartel, R.W. 1997. *Principles of Food Processing*. Aspen Publishers, Inc., Gaithersburg.

Heldman, D.R. and Singh, R.P. 1981. *Food Process Engineering*. AVI Publishing Co., Westport, Connecticut.

Henderson, S.M. 1952. A basic concept of equilibrium moisture content. *Agricultural Engineering* 33: 29–31.

Henderson, S.M. and Perry, R.L. 1976. *Agricultural Process Engineering*. AVI Publishing Co. , Westport, Connecticut.

Himmelblau, D.M. 1967. *Basic Principles and Calculations in Chemical Engineering* 2nd ed., Prentice-Hall, Englewood Cliffs, NJ.

http://www.wyssmont.com/product_detail.php?section=Dryers&id=1

Hukill, N.V. 1954. Grain drying with unheated air. *Agricultural Engineering* 35(6): 393–405.

Imrie, L. 1997. Solar driers. In C.G.J. Baker (ed.) *Industrial Drying of Foods*. Blackie Academic and Professional, London, 210–241.

Karel, M. 1974. Fundamentals of dehydration processes. In A. Spicer (ed.) *Advances in Preconcentration and Dehydration*. Applied Science, London, 45–94.

Karel, M. 1975. Dehydration of foods. In O.R. Fennema (ed.) *Principles of Food Science, Part 2: Physical Principles of Food Preservation*. Marcel Dekker, New York, 359–395.

Karel, M. 1975. Water activity and food preservation. In O.R. Fennema (ed.) *Principles of Food Science, Part 2: Physical*

Principles of Food Preservation, Marcel Dekker, New York, 237–263.

Kerkhof, P.J.A.M. and Schoeber, W.J.A.H. 1974. Theoretical modeling of the drying behaviour of droplets in spray driers. In A. Spicer (ed.) *Advances in Preconcentration and Dehydration of Foods*. Applied Science, London, 349–397.

Kessler, H.G. 2002. *Food and Bio Process Engineering* 5th ed. Verlag A Kessler Publishing House, Munchen, Germany.

Kjaergaard, O.G. 1974. Effects of latest developments on design and practice of spray drying. In A. Spicer (ed.) *Advances in Preconcentration and Dehydration of Foods*. Applied Science, London, 321–348.

Kraemer E.O. 1931. *A Treatise on Physical Chemistry*, H.S. Taylor (Ed.), MacMillan, New York, 1661.

Kunimitsu, M. and Noriko, T. 2003. Adsorption hysteresis in ink-bottle pore. *The Journal of Chemical Physics* 119, 2301. https://doi.org/10.1063/1.1585014

Leniger, H.A. and Beverloo, W.A. 1987. *Food Process Engineering*, D. Reidel, Dordrecht.

Lewis, M.J. 1990. *Physical Properties of Foods and Food Processing Systems*. Woodhead Publishing, Cambridge, UK.

Lewis, M.J. 1996. Solids separation processes. In A.S. Grandison and M.J. Lewis (eds.) *Separation Processes in the Food and Biotechnology Industries*. Woodhead Publishing, Cambridge, 243–286.

Lorentzen, J. 1981. Freeze drying: The process, equipment and products. In S. Thorne (ed.) *Developments in Food Preservation*, Vol.1. Applied Science, London, 153–175.

Luikov, A.V. 1966. *Heat and Mass Transfer in Capillary Porous Bodies*. Pergamon Press, London.

McCabe, W.L., Smith, J.C. and Harriot, P. 1993. *Unit Operations of Chemical Engineering* 5th ed. McGraw-Hill, Inc., New York.

Mellor, J.D. 1978. *Fundamentals of Freeze-drying*. Academic Press, London.,257–288.

Mohsenin, N.N. 1970. *Physical Properties of Plant and Animal Materials, Vol.1 Structure, Physical Characteristics and Mechanical Properties*, Gordon and Breach, London.

Mohsenin, N.N. 1980. *Thermal Properties of Foods and Agricultural Materials*, Gordon and Breach, London.

Morison, K.R. and Hartel, R.W. 2006. Evaporation and freeze concentration. In D.R. Heldman and D.B. Lund (eds.), *Handbook of Food Engineering* 2nd ed. CRC Press, Boca Raton, 496–550.

Muller, J.G. 1967. Freeze concentration of food liquids: Theory, practice and economics. *Food Technology* 21(49–52):54–56, 58,60,61.

Norrish, R.S. 1966. An equation for the activity coefficient and equilibrium relative humidity of water in confectionery syrups. *Journal of Food and Technology* 1:25–39.

Obert, E., and Young, R.L. 1962. *Elements of Thermodynamics and Heat Transfer*. McGraw-Hill Book Co., New York.

Pakowski, Z. and Mujumdar, A.S. 2006. Drying of pharmaceutical products. In A.S. Mujumdar (ed.) *Handbook of Industrial Drying* 4th ed. CRC Press, Boca Raton, 689–712.

Pandey, P.H. 1994. *Principles of Agricultural Processing*. Kalyani Publishers, New Delhi.

Peleg, M. and Bagley, E.B. 1983. *Physical Properties of Foods*, AVI Publishing Co., Westport, Connecticut.

Pfost, H.B., Maurer, S.G., Chung, D.S., Milliken, G. 1976. Summarizing and reporting equilibrium moisture data for grains. ASAE Meeting. St. Joseph, MI. (Paper, 76–3520).

Rahman, M.S. and Perera, C.O. 1999. Drying and food preservation. In M.S. Rahman (ed.) *Handbook of Food Preservation*. Marcel Dekker, New York, 173–216.

Rai, G.D. 1995. *Solar Energy Utilization*. Khanna Publishers, New Delhi.

Rao, D.G. 2009. *Fundamentals of Food Engineering*. PHI Learning Pvt. Ltd., New Delhi.

Rockland, L.B., and Stewart, G.F. (eds.) 1981. *Water Activity: Influences on Food Quality*. Academic Press, New York.

Ross, K.D. 1975. Estimation of in intermediate moisture foods. *Food Technology*, 29(3): 26–34.

Sahay, K.M. and Singh, K.K. 1994. *Unit Operations in Agricultural Processing*. Vikas Publishing House, New Delhi.

Sanchez, J., Ruiz, Y., Auleda, J.M., Hernandez, E. and Raventos, M. 2009. Freeze concentration in the fruit juices industry. *Food Science and Technology International* 15(4): 303–315. https://doi.org/10.1177/1082013209344267

Schubert, H. 1980. Processing and properties of instant powdered foods. In P. Linko, Y. Malkki, J. Oikku and J. Larinkari (eds.) *Food Process Engineering, Vol.1- Food Processing Systems*. Applied Science, London, 675–684.

Sing, K.S.W. and Williams, R.T. 2004. Physisorption hysteresis loops and the characterization of nanoporous materials. *Adsorption Science & Technology* 22(10): 773–782.

Singh, R.P. and Heldman, D.R. 2014. *Introduction to Food Engineering*. Academic Press, San Diego, CA.

Thijssen, H.A.C. 1975. Current developments in the freeze concentration of liquid foods. In Goldblith, S.A., Rey, L. and Rothmayr, W.W. (eds.), *Freeze Drying and Advanced Food Technology*. Academic Press, London .

Thompson, T.L., Peart, R.M., Foster, G.H. 1968. Mathematical simulation of corn drying – A new model. *Transactions of the ASAE* 11:582–586.

Toledo, R.T. 2007. *Fundamentals of Food Process Engineering* 3rd ed. Springer, New York.

Torreggiani, D. 1993. Osmotic dehydration in fruit and vegetables processing. *Food Research International* 26:59–68.

Watson, E.L. and Harper, J.C. 1989. *Elements of Food Engineering* 2nd ed. Van Nostrand Reinhold Co., New York.

Young, J.H. & Nelson, G.L. 1967. Research of hysteresis between sorption and desorption isotherms of wheat. *Transactions of the ASAE* 10(6): 756–61.

23

Chilling and Freezing

Foods during storage undergo physical (such as moisture loss), physiological (senescence, respiration, and ripening), microbiological (growth of microorganisms), and/or biochemical (lipid oxidation, browning, and pigment degradation) changes affecting their quality and shelf life. Most of these changes can be reduced by reducing the food temperature. Experiments have proved that each 10 K (°C) drop in temperature reduces the rates of most of the chemical and biological reactions by a factor of 1/2 to 1/3. Thus, in the process of refrigeration (cooling), the temperature of a food is reduced and maintained at a desirable level for the required time. The process also includes the maintenance of low temperature during transportation and retailing.

23.1 Methods and Principles

To reach a temperature that most of the reactions discussed above cease, the water must be frozen. In *freezing,* the temperature of the product is reduced to a level at which ice crystals are formed, i.e., there is a phase change of most of the moisture present in the food. Water activity of foods gets reduced on formation of ice and increased concentration of the dissolved solutes in unfrozen water. Thus, improvement in the shelf life of food is achieved on account of both, the low temperature and the reduced water activity. The freezing temperature of the moisture will vary below 0°C depending on the food constituents. Frozen storage for fruits and vegetables is usually at -18°C whereas it is 29°C for meat and fish. It is an interesting historical fact that the use of a temperature at or near -18°C equals 0°F and at or near -29°C equals -20°F. Freezing can increase the shelf life of biological materials for up to a year or so.

In some situations, the formation of ice crystals as well as further concentrations of solutions may not be desirable; hence, food is stored at a low temperature just above the freezing point of water, which is known as chilled storage. Chilling removes sensible heat and the heat of metabolism, but there is no removal of latent heat. The usual storage temperature of foods varies between -1 and 8°C. Thus, chilling involves reducing the temperature of food to a level above the freezing point of water and maintaining that temperature during the storage period. As the temperature is reduced, the rates of biochemical and microbiological changes are reduced, which helps in extending the shelf life of food, both fresh and processed. As we will be discussing later, the recommended chilled storage temperatures and the period up to which the shelf life can be extended depend on the type of food. Commonly available cold stores and household refrigerators are examples of chilled storage.

As discussed in Chapter 9, based on the temperature range of growth, the microorganisms can be categorized as psychrotrophic, psychrophilic, mesophilic, thermophilic, and thermoduric. The low temperature storage primarily aims at reducing the growth and activity of thermophilic and mesophilic organisms.

The main advantage of this method over other food preservation methods is that the frozen and chilled foods most resemble the fresh foods in terms of form, color, flavor, texture, etc. Although freezing of foods aid in longer storage and distribution life, the formation of ice crystals may cause some changes that may diminish the quality.

It is very important to maintain the recommended cold chain for storage, transportation, and retailing of both chilled and frozen food. In fact, it is developments in refrigerated transportation systems that have led to the rapid expansion of the chilled food market.

Although freezing inhibits some enzymes, some specific ones, such as invertase, lipase, lipoxidase, catalase, peroxidase, etc., remain active, and during the long storage, as is intended in frozen storage, these enzymatic reactions, although slow, can cause significant problems. Thus, freezing is often preceded by blanching or scalding to destroy enzymes.

23.1.1 Chilled Storage

Storage Parameters

Chilled storage temperature varies with the category of food. For fresh fish, meats, etc. the storage temperature is -1 to +1°C. Milk, cream, yoghurt, prepared salads, pizzas, pastries, sandwiches, baked products, etc. are stored at a temperature of 0 to 5°C. Cooked rice, soft fruits, fruit juices, butter, hard cheese, margarine, and fully cooked meats are stored at 0 to 8°C. The air is precooled, depending on the material and other conditions, to -1 to -14°C by refrigeration.

Growth of thermophilic and several mesophilic microorganisms is prevented by chilling. However, the microorganisms and enzymes present in food at the time of chilling can cause spoilage, though at a lower rate than that at ambient storage conditions. Some pathogenic microorganisms like *Escherichia coli, Listeria* spp., etc. can survive and grow profusely even at chilled conditions. It is, therefore, required to follow GMP during the production of chilled foods. Besides, chilling is often preceded by other methods of food preservation, such as pasteurization and fermentation, and it is combined with novel packaging technologies, such as modified or controlled atmosphere packaging. It is important to note that the rate of growth of microorganisms is usually high within

DOI: 10.1201/9781003285076-28

the temperature range of 10°C to 50°C and thus when the product is being cooled, the temperature should be brought down to lower than 10°C as quickly as possible.

It is not desirable to chill all foods; some temperate, tropical, and subtropical fruits, when stored below a specific temperature, suffer from some undesirable physiological changes. Such an undesirable change for an individual fruit is a *chilling injury*. This type of injury is normally attributed to a metabolic activity imbalance, resulting in the over-production of toxic metabolites injurious to the tissues. It is observed visually in the forms of internal and external browning, failure to ripen, blemished skin, etc. Apples suffer chilling injury when stored below 2–3°C, lemons at <14°C, bananas at <12–13°C, mangoes at <10–13°C. When stored at less than 7–10°C, pineapples, melons, and tomatoes suffer chilling injury. Further, the storage relative humidity also affects moisture loss from the commodity. Thus, the optimum mean temperature and relative humidity conditions differ for different fruits and vegetables. For some fruits and vegetables, the optimum temperature, RH, and expected storage lives are given in Annexure XVI.

In addition to the temperature and relative humidity of the storage, the shelf life of chilled food is also affected by the type of food and variety or cultivar, the condition of the food and the degree of maturity at harvest and the part of the crop stored. A fruit having bruises and damaged skin will have a shorter shelf life than the healthy fruit. Similarly, the highest metabolic rates and the shortest storage lives correspond to the fastest growing parts of fruits. The temperature during harvest and post-harvest stages such as storage and transportation, etc. also affect the shelf life of the commodity.

Analysis of Cooling Load

Fresh produces continue their respiration process even after harvest. The respiration process also generates heat within the commodity. The following equation gives the heat generation due to respiration at atmospheric pressure and 20°C temperature.

$$C_6H_{12}O_6 + 6O_2 \rightarrow 6CO_2 + 6H_2O + 2.835 \times 10^6 \text{ J.(kmol of}$$
$$C_6H_{12}O_6)^{-1} \qquad (23.1)$$

During chilling and freezing, the heat of respiration has to be removed along with the field heat.

For simplicity, the general heat transfer equations are used for calculating the heat load in cold stores. However, there is heat generation within the commodity, and, hence, unsteady state heat transfer methods are used to determine the refrigeration plant size and any crop's processing time to chill. While the heat removal is by convection from the surface, it is by unsteady state conduction within the material. The overall coefficient h taken after considering all resistances are used in unsteady state charts. The analysis of unsteady state heat transfer often creates problems as it is difficult to get accurate data on the thermal properties of the commodity as well as the convective coefficient. Also if there is evaporation of water during chilling, it affects the accuracy of results.

The following problems on calculation of cooling load are based on use of basic heat transfer equations.

Example 23.1 A cold storage room (8 m×6m×3 m inside dimensions) is constructed of 42 cm thick brick wall with a 12 cm thick cork insulation on the inside of the four walls. The cement plaster on the outside is 1 cm thick. The thermal conductivities of the different layers are: brick wall = 0.52 W.m⁻¹.K⁻¹; cork = 0.03 W.m⁻¹.K⁻¹; cement plaster = 0.3 W.m⁻¹.K⁻¹.

If the heat transfer coefficient of the ceiling is 20% more than that for the walls, what will be the plant capacity needed in tons of refrigeration to maintain the inside temperature at 5°C when the outside temperature is 30°C? Assume that the store is empty.

SOLUTION:

The heat load by conduction through walls can be given as:

$$q = U A \Delta t$$

The overall heat transfer coefficient U can be found out by the following equation,

$$\frac{1}{U} = \frac{x_1}{k_1} + \frac{x_2}{k_2} + \frac{x_3}{k_3}$$

or, $\dfrac{1}{U} = \dfrac{0.12}{0.03} + \dfrac{0.42}{0.52} + \dfrac{0.01}{0.3} = 4.84 \text{ m}^2.\text{K.W}^{-1}$

or, U = 0.2 W.m⁻². K⁻¹.

Total wall area = A = 2 × 6 × 3 + 2 × 8 × 3 = 84 m²

Total wall area also includes the area under doors and we assume that the heat transfer through the doors is same as that through the walls.

Thus, heat transfer through walls =

0.2 × 84 × (30-5) × (24 × 3600)/1000
= 36288 kJ/24 h

For ceiling, U = 1.2 × 0.2 W.m⁻². K⁻¹

Heat transfer through ceiling,

q = U AΔt

$$= (1.2 \times 0.2) \times (8 \times 6) \times (30 - 5) \times (24 \times 3600)/1000$$

$$= 24883.2 \text{ kJ} / 24 \text{ h}$$

Thus, the total heat load through wall and ceiling

= 36288 + 24883.2 = 61171.2 kJ/24 h

The heat load can also be written as 61171.2×1000/ (24×3600) = 708 J.s⁻¹ or 708 W

1 ton of refrigeration = 50.4 kcal.min^{-1} = 3517 W
Thus, the refrigeration capacity needed = 708/3517 = 0.201 ton of refrigeration, TR.

Example 23.2 In the above cold store, 10 tonnes of apples having specific heat of 3.35 kJ.kg^{-1}.K^{-1} are to be cooled from 30 to 12°C in 24 hours. The heat of respiration of apples per 24 hours is 3114 kJ.tonne^{-1}. What will be the extra refrigeration load?

SOLUTION:

Heat load for cooling apples = 10000 × 3.35 × (30-12) = 603000 kJ/24 h
Heat of respiration of apples = 3114 × 10 = 31140 kJ/24 h
Thus, the heat load due to the commodity = 603000 + 31140 = 634140 kJ/24 h
This is same as 634140×1000/(24×3600) = 7339.58 W
Refrigeration load = 7339.58/ 3517 = 2.086 TR.

Example 23.3 In the above cold store loaded with 10 tonnes of apple, it is envisaged that four persons will work for three hours. There is also lighting load of 400 W and air infiltration can be approximated as 3300 kJ/24 hours. What will be the extra refrigeration load due to workers, electrical appliances and the air infiltration, if the heat of respiration for the men is taken as 200 W. What will be the total refrigeration load required for the apple storage?

SOLUTION:

Heat load due to air infiltration = 3300 kJ/24 h
Heat load due to lights = 400 × (24 × 3600)/1000 = 34560 kJ/ 24 h
Heat load because of workers = 4 × 200 × 3 × 3600/1000 = 8640 kJ/ 24 h
Total heat load due to these components

$$= 3300 + 34560 + 8640 \text{ kJ/24 h}$$

$$= 46500 \text{ kJ/24 h}$$

$$= 46500 \times 1000/(24 \times 3600) = 538.19 \text{ W}$$

Refrigeration load = 538.19/3517 = 0.153 ton of refrigeration
Thus, the total refrigeration load required =
= 0.201 + 2.086 + 0.153 = 2.44 tons of refrigeration.

Example 23.4 Find the refrigeration load expressed in tons of refrigeration that is caused by heat loss from four side walls of a small cold room 2.4 m×3.0 m×2.4 m. The walls are made of 20 cm brick, 20 cm cork board and 1.25 cm cement. Outside wall temperature is 21°C. Add a suitable safety factor for losses through joints

etc. **Respective thermal conductivities of brick, cork, and cement plaster are 0.6 W.m^{-1}.K^{-1}, 0.04 W.m^{-1}.K^{-1}, and 0.8 W.m^{-1}.K^{-1}. The cold room is maintained at -30°C.**

SOLUTION:

The equation for the heat load by conduction through walls can be written as:

$$q = U \, A \, \Delta t$$

For the wall with three layers, the overall heat transfer coefficient, U is obtained as:

$$\frac{1}{U} = \frac{x_1}{k_1} + \frac{x_2}{k_2} + \frac{x_3}{k_3}$$

$$\text{or, } \frac{1}{U} = \frac{0.2}{0.6} + \frac{0.2}{0.04} + \frac{0.0125}{0.8}$$

or, U = 0.187 W.m^{-2}.K^{-1}

Area for the four sides contributing to heat transfer is as follows.

$$= 2 \times 2.4 \times 2.4 + 2 \times 3 \times 2.4 = 11.52 + 14.4$$
$$= 25.92 \text{ m}^2$$

Thus, q = 0.187 × 25.92 × (21 − (−30)) = 247.2 W
Considering that a similar amount of heat is lost through the ducts and joints, i.e., taking a safety factor as 2,

$$\text{Total heat load} = 247.2 \times 2 = 494.4 \text{ W}$$

Since 1 ton of refrigeration = 3517 W
Heat load = 494.4 / 3517 = 0.14 ton of refrigeration

Processed foods are also kept in chilled storage for increasing shelf life. Pasteurized milk, processed milk products, and processed juices are good examples. The shelf lives of chilled processed foods, among other factors, also vary with the characteristics of the food, degree of processing, type of package, and temperature during processing, distribution, and storage.

23.1.2 Frozen Storage

Storage Parameters

As mentioned earlier, fruits and vegetables are normally stored in frozen state at -18°C (0°F) and the storage temperature for meat and fish is -29°C (-20°F). Thus, during the process of freezing foods, at the first phase there is removal of sensible heat to bring down the temperature of the product to the freezing point of water, and in the second phase there is removal of latent heat of crystallization so that the water is changed to ice form. In the third and last phase, the sensible heat is further removed to bring the temperature down to the specified storage temperature.

The freezing point of water in the food is usually below 0°C because of the dissolved solutes and the latent heat of crystallization would also be different from pure water. Freezing points of other food constituents as carbohydrates, fat, protein, etc. are lower than water's freezing point and their latent heats may be removed. However, as most of the foods contain a very high fraction of water, the latent heat of other components may contribute only a very small fraction and may be neglected.

At the temperature of -29°C, about 90% of the water present in the food is frozen; the remaining is bound water.

Temperature Profile of the Commodity during Freezing

The change in temperature at the thermal center of the food during freezing can be as shown in Figure 23.1. Considering the figure, the change from point 1 to 2 represents cooling of food to below the freezing point, t_f. This temperature is below 0°C and may be even up to -10°C as the water contains dissolved solutes. At point 2, even though the temperature is lower than the freezing point, the water is still liquid and the ice crystals begin to form. This is known as *supercooling*. During supercooling, the water molecules tend to form groups (which is known as *nucleation*, discussed in the subsequent section) and the type of food and the rate of heat removal determine the length of this supercooling period. Also because the latent heat is released with the formation of ice crystals, the temperature increases to the freezing point (point 3).

After this stage, with the lower freezing temperature, the food continues to lose heat almost at the same rate as that before the formation of ice (3–4). Most of the heat removed at this stage is the latent heat; hence, almost constant temperature is observed. An increase in solute concentration in the unfrozen liquor depresses the freezing point and a small temperature drop is noticed. Most of the ice is formed during this stage.

The change from 4 to 5 indicates that one of the solutes changes phase after becoming supersaturated and, therefore, releases the latent heat of crystallization. This causes the temperature to increase to the eutectic temperature for that solute.

Eutectic temperature is the temperature when a crystal of an individual solute comes in equilibrium with ice and the unfrozen liquor. The knowledge of eutectic temperatures of different food constituents helps in designing the heat load of the freezing equipment. The values of eutectic temperatures for glucose, sucrose, and sodium chloride are -5°C, -14°C, –21.13°C, respectively. As it is difficult to make a distinction between the eutectic temperatures of different solutes in a food, the term *final eutectic temperature* is used, which is defined as the lowest eutectic temperature of the solutes present in a food. The final eutectic temperature for ice cream is about –55°C. During commercial freezing, such low temperatures are not used, and, hence, there is always the presence of unfrozen water in commercially available frozen foods.

Process 5–6 is the continuation of the crystallization process for both water and solutes. Point 6 can be taken as the point where most of the water is frozen. As mentioned before, even at this stage, a portion of the water (7%–15%) is still unfrozen, which is mostly the bound water. It is indicated by stage 6–7 that the temperature would gradually come close to the freezer temperature when kept for a prolonged period.

During freezing, the increased solute concentration causes changes in different properties of the food and its constituents, such as the viscosity, surface tension, pH, and redox potential. There is volume expansion of food because the volume of ice is about 9% more than that of pure water.

Rate of Freezing and Quality of Frozen Food

During the process of freezing, before the formation of ice crystals, a group of water molecules come together to form a nucleus, and, hence, the process is known as *nucleation* of water molecules. When only water molecules reorient and combine to form a group it is known as *homogenous nucleation*, and when the group is formed around some suspended particles or around the cell wall it is known as *heterogenous nucleation*. The second type is more common in foods.

Normally, the water molecules have a propensity to join existing nuclei rather than forming new nuclei. During fast freezing the water molecules do not get enough time to move to nearby existing nuclei before being changed to ice crystal form. Hence, fast freezing produces large number of small ice crystals. But when the rate of freezing is slow, the water molecules are able to join the nearby existing nuclei. Thus, fewer large ice crystals are formed.

Primarily the rate of heat removal from the food controls the rate of growth of ice crystals. In particular, the time taken by the product to pass through the critical zone, say from 0 to -5°C, determines the number and size of ice crystals. The mass transfer rate can also contribute to this effect only when the product becomes more concentrated, i.e., when most of the water has frozen.

Small size ice crystals are desirable in frozen food as it causes less damage to the cells. In the case of slow freezing, the large size of ice crystals in the intercellular spaces deforms and ruptures the cell walls. This does not allow the cells to return to their original shape and turgidity on thawing and results in a softer food. This also causes the cell contents to come out in drops known as drip loss. The small size of ice crystals formed during fast freezing causes much lower damage to the cells. Thus, the fast-freezing method produces a better texture in foods. Of course, very high rates of freezing may cause stress cracks in the food tissues.

A wide variation in the size of the ice crystals is found even with similar freezing rates because of the differences in the nature of the food. In the same food also the differences in

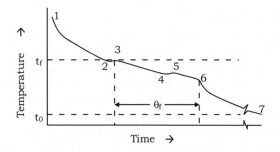

FIGURE 23.1 Temperature change at the thermal center of a commodity during freezing (after Fellows, 2000 with permission)

pre-freezing treatments can cause variability in the size of ice crystals.

Calculation of Freezing Time

Freezing time is usually expressed in two ways.

The *effective freezing time* is the time required to reduce the temperature of the food to the desired temperature at the food's thermal center. Thus, the effective freezing time will be primarily dependent on the initial temperature as well as the rate of freezing per unit amount of produce. It is meant to determine the capacity of a freezing process or equipment.

The *nominal freezing time* is the time taken by the food starting from when its surface reaches 0°C until its thermal center reaches 10°C lower than the temperature of the first ice crystal formation (i.e., 10°C lower than the point 2 in freezing curve). The nominal freezing time does not consider the initial conditions; it indicates the rate of cooling and thus determines the quality of the product.

The calculation of the freezing time is difficult due to many reasons, which include the irregular shape and size of food, changes in the properties of food during freezing, and variation in rates of ice crystal formation within different regions of food. However, with several assumptions, an approximate solution for the freezing time has been derived, which is commonly known as Planck's equation. The following are the assumptions made for the derivation.

1. All of the food is initially at the freezing temperature, but is unfrozen;

2. Temperature and time do not affect the thermal conductivity and specific heat of the frozen portions, i.e. it is a steady state process;

3. The freezing point and the latent heat for all the material do not change;

4. The heat transfer by conduction in the frozen layer occurs very slowly so that steady-state conditions can be approximated.

The derivation is as follows.

Let a slab of thickness a (m) and surface area A (m²) be cooled from both sides by convection, as shown in Figure 23.2. Also consider that a frozen layer of thickness x (m) has been formed on both sides after a given time θ (s). The other parameters are as follows.

Let t_a, t_s, and t_f be the temperatures of the cooling air, the solid, and the temperature at the interface of the frozen and unfrozen layer, respectively. k (W.m⁻¹.K⁻¹) and ρ (kg.m⁻³) are the thermal conductivity and density of the unfrozen material. Let h (W.m⁻².K⁻¹) and λ (J.kg⁻¹) be the surface heat transfer coefficient and latent heat of crystallization.

As it is considered to be a steady state process, the rate of convective heat transfer from the surface of the slab at any time can be given as:

$$q = hA(t_s - t_a) \tag{23.2}$$

The conduction heat transfer through the frozen layer is

$$q = \frac{kA}{x}(t_f - t_s) \tag{23.3}$$

Obtaining the value of t_s from equation 23.2 and substituting in equation 23.3, we get

$$q = \frac{(t_f - t_a)A}{x/k + 1/h} \tag{23.4}$$

Considering dθ to be the time required for freezing a small thickness of the material dx, then

$$q = \frac{(A.dx.\rho)\lambda}{d\theta} = A\rho\lambda \frac{dx}{d\theta} \tag{23.5}$$

Thus, $$\frac{(t_f - t_a)A}{x/k + 1/h} = A\rho\lambda \frac{dx}{d\theta} \tag{23.6}$$

Rearranging and integrating within limits, at $\theta = 0$, x = 0 and at $\theta = \theta$ and x = a/2,

$$(t_f - t_a)\int_0^\theta d\theta = \rho\lambda \int_0^{a/2}\left(\frac{x}{k} + \frac{1}{h}\right)dx \tag{23.7}$$

Integrating and solving for θ,

$$\theta = \frac{\rho\lambda}{t_f - t_a}\left[\frac{a}{2h} + \frac{a^2}{8k}\right] \tag{23.8}$$

In the above equation, 2 and 8 are constants. By denoting the constants as C_1 and C_2, the generalized equation for freezing time is written as

$$\theta = \frac{\rho\lambda}{t_f - t_a}\left[\frac{a}{C_1 h} + \frac{a^2}{C_2 k}\right] \tag{23.9}$$

The values of C_1 and C_2 represent, essentially, the shortest distance between the center to the surface of the food. As we have taken a slab for freezing, the constants C_1 and C_2 have been

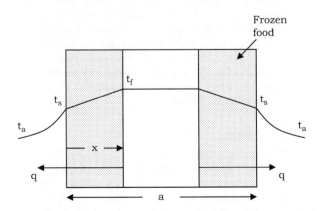

FIGURE 23.2 Temperature profile during freezing of a slab

obtained as 2 and 8. The values of constants for other material shapes have been found as follows:

Material shape	C_1	C_2
Slab	2	8
Cylinder	4	16
Cube / sphere	6	24

If the food is packaged, then the thickness of packaging and its thermal conductivity also need consideration. The modified freezing time for such a situation is given as:

$$\theta = \frac{\rho\lambda}{t_f - t_a}\left[\frac{a}{C_1}\left(\frac{1}{h} + \frac{x_p}{k_p}\right) + \frac{a^2}{C_2 k}\right] \quad (23.10)$$

where, k_p and x_p are the thermal conductivity and thickness of the packaging material, respectively.

Considering $1/C_1 = P$ and $1/C_2 = R$, we may write equation 23.9 as:

$$\theta = \frac{.\rho\lambda}{t_f - t_a}\left[\frac{Pa}{h} + \frac{Ra^2}{k}\right] \quad (23.11)$$

As P and R are also constants, this form of the equation is also used in many freezing time calculations. For a rectangular brick having dimensions $a \times \beta_1 a \times \beta_2 a$, where the shortest side is "a," the P and R values can be taken from the chart prepared by Ede (1952).

Equation 23.9 can also be used for determination of thawing time of a food by replacing the thermal conductivity of the frozen material by that of the thawed material.

Equation 23.9 can be further modified as:

$$\frac{\theta\Delta t}{\rho\lambda} = \left[\frac{Pa}{h} + \frac{Ra^2}{k}\right], \text{ where, } \Delta t = t_f - t_a$$

Substituting ΔH for $\rho.\lambda$,

$$\frac{\theta\Delta t}{\Delta H} = \left[\frac{Pa}{h} + \frac{Ra^2}{k}\right]$$

We can further modify the above equation as:

$$\frac{\theta\Delta t.h}{\Delta H.P.a} = \left[1 + \frac{Rha}{P.k}\right] \quad (23.12)$$

Considering that, $(h.a/k)$ is the Biot number, the above equation can be rewritten as follows:

$$\frac{\theta\Delta t.h}{\Delta H.P.a} = \left[1 + \frac{R}{P}Bi\right] \quad (23.13)$$

Let us define an efficiency term,

$$\eta = \frac{\frac{R}{P}Bi}{1 + \frac{R}{P}Bi} = \frac{Bi}{(P/R) + Bi} \quad (23.14)$$

For a slab, considering that, $P/R = 4$, the efficiency term becomes

$$\eta = \frac{Bi}{4 + Bi} \quad (23.15)$$

The term η, which varies with the Biot number, denotes how closely the freezing medium connects with the food. If we take $\Delta t = 1$ and θ_f' as an intrinsic freezing time, then the relationship can be written as:

$$\theta_f = \frac{\theta_f'}{\eta\Delta t} \quad (23.16)$$

This equation can be used to study the influence of the temperature of the freezing medium and the surface heat transfer coefficient (through Bi) on the actual freezing time.

There are also other theories/equations proposed for freezing time of foods (Lacroix and Castaigne, 1987; Cleland and earle, 2006; Ramaswamy and Tung, 2007). The enthalpy-temperature-composition charts for many types of foods are also available, which help in freezing time calculations.

Factors Influencing Freezing Time

Referring to the equation 23.10, the factors affecting the freezing time can be listed as follows.

- Temperature difference between the freezing medium $(t_f - t_a)$ and the food;
- Density and thermal conductivity of food (ρ and k);
- Thickness of material or the distance of heat transfer within food (a/2);
- Surface area of food for heat transfer (A);
- Surface heat transfer coefficient (h);
- Thickness (x_p) and thermal properties (or, thermal conductivity, k_p) of packaging material, if any.

Example 23.5 The freezing point of a vegetable is -8°C; however, it has to be cooled to –18°C for frozen storage. If 100 kg of the commodity is received at 30°C and then it is cooled in two stages, i.e., in the first stage from 30°C to 4°C (precooling), and in the second stage from 4°C to -18°C, what will be the total heat load? Specific heat of the commodity can be taken as 3.5 kJ. kg⁻¹.K⁻¹ and 2.35 kJ.kg⁻¹.K⁻¹ above and below its freezing point. The vegetable contains 85% moisture. The latent heat of crystallization of water can be taken as 335 kJ.kg⁻¹.

SOLUTION:

The cooling is done in two stages, in the first stage from 30°C to 4°C and in the second stage from 4°C to -18°C. Since the specific heats are different for the commodity above and below the freezing points, the different heat loads will be obtained separately as follows:

Heat load to cool the commodity from 30° to 4°C

$$= 100 \times 3.5 \times (30\text{-}4) = 9100 \text{ kJ}$$

To cool from 4°C to freezing point in freezer

$$= 100 \times 3.5 \times (4\text{-}(\text{-}8)) = 4200 \text{ kJ}$$

For freezing the commodity moisture

$$= 100 \times 0.85 \times 335 = 28475 \text{ kJ}$$

To cool from freezing point to -18°C

$$= 100 \times 2.35 \times (\text{-}8 - (\text{-}18)) = 2350 \text{ kJ}$$

Total heat load = 9100 + 4200 + 28475 + 2350 = 44125 kJ

Actually, the freezing point of dissolved solids in the food is lower than the freezing point of water within the solids. So if we consider the exact temperatures of freezing of the dissolved solids within the food, we can also calculate the sensible heat and latent heats removed for both water and solid constituents separately and approach a more precise answer for such problems.

Example 23.6 An air blast freezer operating at –30°C is used to freeze fish slabs 0.0508 m thick and at a moisture content of 82%. The initial temperature is –2.2°C. If the surface heat transfer coefficient is 20 W.m⁻².K⁻¹, density and thermal conductivity of the frozen portion are 1050 kg.m⁻³ and 0.025 W.m⁻¹.K⁻¹, find the freezing time. Assume the shape factors P = 1/2, R = 1/8 for an infinite slab.

SOLUTION:

Given: t_a = -30°C, h = 20 W.m⁻².K⁻¹, fish thickness (a) = 0.0508 m

Initial temp of fish (t_f) = -2.2°C, moisture content, m = 82 %
Density of unfrozen fish, ρ = 1050 kg.m⁻³
Thermal conductivity of frozen fish, k = 0.025 W.m⁻¹.K⁻¹
shape factors, P = 1/2, R = 1/8

Let us assume the latent heat of crystallization, λ_f = 335 kJ.kg⁻¹,
The freezing time can be given as:

$$\theta_f = \frac{\rho \lambda_f}{t_f - t_a}\left(\frac{Pa}{h} + \frac{Ra^2}{k}\right)$$

i.e. $\theta_f = \dfrac{1050 \times 335 \times 1000 \times 0.82}{-2.2\text{-}(-30)}\left[\dfrac{0.0508}{2 \times 20} + \dfrac{1}{8} \times \dfrac{(0.0508)^2}{0.025}\right]$

$$= (12652877.7 \times 0.82) \times (0.00127 + 0.0129) \text{ s}$$

or, = 147050 s = 40.84 h

Example 23.7 A 8 cm thick slab of food enclosed in a 1.2 mm thick cardboard is frozen in a plate type freezer. The plate temperature is kept at -40°C and the freezing point of the slab is -1°C. If the surface heat transfer coefficient (h) of the freezer is 620 W.m⁻².K⁻¹, latent heat of crystallization of the food is 300 kJ.kg⁻¹, density of the frozen slab of food is 1080 kg.m⁻³, thermal conductivity of the frozen food is 1.8 J.m⁻¹.s⁻¹.K⁻¹, and the thermal conductivity of cardboard is 0.05 J.m⁻¹.s⁻¹.K⁻¹, find

- **a) the freezing time of the food;**
- **b) the freezing time if the plate temperature were changed to -20°C (with other conditions remaining constant);**
- **c) the freezing time if the thickness of the cardboard were 2 mm (other parameters remaining same); and**
- **d) the freezing time if the surface heat transfer coefficient were 380 W.m−2K−1.**

SOLUTION:

(a) The given values are k = 0.05 J.m⁻¹.s⁻¹.K⁻¹, and h_c = 620 W.m⁻².K⁻¹

Thus, the overall heat transfer coefficient =

$$h_s = \frac{1}{\dfrac{x}{k} + \dfrac{1}{h_c}} = \frac{1}{\dfrac{0.0012}{0.05} + \dfrac{1}{620}} = 39.04 \text{ W.m}^{-2}\text{K}^{-1}$$

Bi = (h.a)/k = (39.04)(0.08)/1.8 = 1.73
By using equation 23.15 and taking P/R = 4

$$\eta = 1.73/(4+1.73) = 0.302$$

Thus, the freezing time can be obtained from the Planck's equation as:

$$\theta = \frac{1080 \times 300 \times 10^3}{-1-(-40)}\left[\frac{0.08}{2 \times 39.04} + \frac{0.08^2}{8 \times 1.8}\right]$$

i.e. θ = 12204.287 s = 3.39 h
For performing the analysis for subsequent portions, we calculate the intrinsic freezing time θ_f' using equation 23.16.

$$\theta_f' = \theta_f \times \eta \times \Delta t$$

$$= 3.39 \times 0.302 \times 39 = 39.927 \times h$$

(b) If the freezing temperature is -20°C

$$\theta_f = 39.927 / (0.302 \times 19) = 6.96 \text{ h}$$

(c) If the cardboard thickness is 0.002 m,

$$h_s = \frac{1}{\dfrac{x}{k} + \dfrac{1}{h_c}} = \frac{1}{\dfrac{0.002}{0.05} + \dfrac{1}{620}} = 24.03 \text{ W.m}^{-2}\text{K}^{-1}$$

Bi = (24.03×0.08)/1.8 = 1.07

$$\eta = 1.07/(4+1.07) = 0.21$$

so, $\theta_f = 39.927/(0.21 \times 39) = 4.85$ h

(d) If the surface heat transfer coefficient is 380 $W.m^{-2}.K^{-1}$

$$h_s = \cfrac{1}{\cfrac{x}{k} + \cfrac{1}{h_c}} = \cfrac{1}{\cfrac{0.0012}{0.05} + \cfrac{1}{380}} = 37.54 \ W.m^{-2}.K^{-1}$$

$$Bi = (37.54 \times 0.08)/1.8 = 1.668$$

Therefore, $\eta = 1.668/(4+1.668) = 0.2943$

so, $\theta_f = 39.927/(0.294 \times 39) = 3.48$ h

In the above problem, check how the freezing time changes due to changes in the thickness of the packaging material (resistance to heat transfer), overall heat transfer coefficient, as well as the temperature of the freezing medium.

23.1.3 Evaporative Cooling

Temperature in an enclosure could be reduced, with a concomitant increase in the relative humidity, in a cost-effective manner by evaporative cooling. Therefore, evaporative cooling has been used for short-term storage of perishable horticultural commodities in many developing countries. As we have discussed in chapters 5 and 8, the evaporative cooling is a constant enthalpy process represented by a line of constant wet bulb temperature on the psychrometric chart. Relative humidity of the surrounding air strongly influences the efficiency of an evaporative cooler. Depending on the ambient conditions, the use of evaporative cooling brings the air temperature down in an enclosure by about $10°–15°C$ below ambient and increases the relative humidity significantly.

23.1.4 Vacuum Cooling

In this method, the commodity is placed in a vacuum chamber, maintained at 530–670 Pa, where its heat is removed by evaporative cooling. As a thumb rule, for every 1% of water evaporated, the product temperature reduces by 5°C. Since there is removal of moisture from the commodity, the product is pre-wetted before vacuum cooling to prevent desiccation of the product tissues.

Vacuum cooling (or chilling) works well for commodities like leafy vegetables, where the commodity has a large surface area to volume ratio and can release the moisture quickly. It is commonly used to remove field heat from vegetables in on-farm precooling devices.

23.1.5 Cryogenic Freezing

As we have discussed in Chapter 8, temperatures below 223 K (-50°C) or lower are considered cryogenic for food applications. The two common cryogens used in food processing are nitrogen and carbon dioxide.

Liquid nitrogen is colorless, odorless, and inert. Nitrogen liquefies at a temperature of -196°C (77 K) and possesses 378 kJ per kg of liquid nitrogen as the refrigerating capacity at normal pressure. It is stored and supplied at a pressure of 36 bar in general. Liquid nitrogen has 88 K to 96 K boiling points at 3–6 bar, respectively. As a rule of thumb, mechanical refrigeration equivalent to 100 kW is produced by 1 ton.h⁻¹ of liquid nitrogen. Solid carbon dioxide (dry ice) has a cooling capacity of 620 $kJ.kg^{-1}$. Liquid CO_2 at 25°C and 65 bar gives a cooling effect of 199 $kJ.kg^{-1}$ and at -16°C and 22 bar, the cooling capacity is approx. 310 $kJ.kg^{-1}$. These high rates of heat extraction are possible due to the high temperature difference between the commodity and refrigerant and also due to high surface heat transfer coefficients.

During analysis of cryogenic freezing, it can be assumed that all the heat is removed from the product, which is at the freezing point and that the product is homogenous and isotropic. It is also assumed that the surrounding fluid is at a uniform temperature. The product is assumed to be of spherical shape. Under these conditions, the freezing period (θ) can be estimated using Planck's formula.

$$\theta = \frac{\rho\lambda}{t_f - t_a}\left[\frac{r_0}{3h} + \frac{r_0^2}{6k}\right] \qquad (23.17)$$

Where, h is convection coefficient, t_f is frozen product temperature, t_a is the cryogen temperature, r_0 is equivalent radius of the product, λ is freezing latent heat, and k is thermal conductivity.

A typical cryogenic system for foods is designed to freeze the product down to -18°C or -29°C over a freezing time of a few minutes. We have discussed earlier why the freezing temperatures have been so determined.

23.2 Equipment

23.2.1 Classifications of Freezing/ Chilling Equipment

The systems used for chilling/freezing foods may be categorized as follows.

A. *On the basis of type of equipment used*

The equipment used for chilling/freezing can be classified based on the method used to remove heat as mechanical refrigeration systems and the cryogenic systems. The latter is also known as total loss refrigeration system.

In the cryogenic system (cryogenic means very low temperatures), there is direct contact of food with the refrigerant, which changes from a liquid to a gaseous state as it receives heat from the food. Further heat removal increases the temperature of the gaseous refrigerant. Earlier, the refrigerant was not recovered and was usually released to the atmosphere/ environment. This is why it is known as a total loss refrigeration system. Solid carbon dioxide (known as dry ice) and liquid nitrogen are common "cryogens." Recent systems permit recycling of the cryogens back to the food.

Ice is a common total loss refrigerant used for cooling and transporting fish. The ice receives 333 kJ.kg⁻¹ to melt.

Cryogens can also be used for chilling of foods through indirect methods in which the cryogen can be used to reduce the temperature of air and then the air can be used for chilling the foods.

In a *mechanical refrigerator*, a low-pressure, cold, liquid refrigerant passing through an evaporator receives heat from the surrounding air to change to gaseous form (Figure 23.3). This low-pressure hot gas then moves to a compressor, where it is compressed to form a high-pressure hot gas. The gas then moves through a condenser coil, where the heat is released to the surrounding and the gas changes to form a high-pressure cold liquid. The refrigerant then moves through the expansion valve, where the pressure is reduced before it is recycled back to the evaporator. Thus, heat is continuously removed from the commodity by the refrigerant and released to the atmosphere.

The evaporator coil may be in direct contact with the food or the air (or other medium) surrounding the food. In indirect types, a liquid is cooled by the evaporator coil, which is then used to cool the cooling medium that is used to extract heat from the food.

Examples of freezers/chillers using cooled air as refrigerant are a chest freezer, air blast freezer, spiral freezer, fluidized bed freezer, and cabinet freezer. The immersion freezer uses cooled liquid system. The plate freezer and the scraped surface freezer are examples of cooled surface freezers. Air-cooled systems have lower heat transfer rates than the water cooled or other systems.

The basic equipment may be the same both for chilling and freezing, but the temperature of the cooling medium used in chilling is comparatively higher than that used in the freezing.

The surface heat transfer coefficient (h) depends on several factors, such as type and velocity of the freezing medium, shape and surface roughness of food, etc. The air at about 1–2 m.s⁻¹ has the heat transfer coefficient value of approximately 20 W.m⁻².K⁻¹. If the velocity increases to 6 m.s⁻¹, the heat transfer coefficient is about 60 W.m⁻².K⁻¹. Some typical values of heat transfer coefficients in cooling systems are given in Annexure XVII.

B. On the basis of type of contact between the food and chilling/freezing medium

On the basis of type of contact between the food and chilling/freezing medium, the freezers can be classified as direct contact and indirect contact types.

The air blast freezer, fluidized bed freezer, and immersion freezer are examples of direct contact type freezers, in which the chilling medium (air or other fluids) comes in direct contact with the food. The cabinet freezer, plate freezers, and scraped-surface continuous freezers are indirect contact freezers. In these, the chilling medium lowers the temperature of the plate or shelves or surface of the tube, which remove heat from the food.

C. On the basis of rate of movement of ice front

As discussed earlier, the rate of freezing affects the quality of the frozen products, and, hence, the freezers are also classified by the rate at which the ice front moves (Fellows, 2000).

* Slow freezer and sharp freezers (rate of movement of ice front: 0.2 cm.h⁻¹);
* Quick freezers (rate of movement of ice front: 0.5–3 cm.h⁻¹);
* Rapid freezers (rate of movement of ice front: 5–10 cm.h⁻¹);
* Ultra-rapid freezers (rate of movement of ice front: 10–100 cm.h⁻¹).

The first category includes the cold stores and still-air freezers. The air blast and plate freezers come under the quick-freezing category. The fluidized bed freezing is a rapid freezing method. The cryogenic freezer comes under the category of ultra-rapid freezers.

The different freezers used in the food processing industry are as follows. As mentioned before, the same type of equipment having higher temperatures of the freezing medium can be used for chilling.

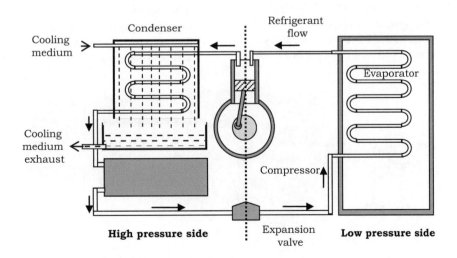

FIGURE 23.3 Components of a mechanical refrigeration system

23.2.2 Direct Contact Freezers

Chest Freezer

Construction and Operation

- In consists of a big chamber in which the air temperature is maintained between -20°C and -30°C by a refrigeration system.
- The commodities are kept in the chamber usually on trays in racks.
- The air is usually still; however, large chambers may have fans for the circulation of air.
- Large chest freezers are regarded as commercial cold stores.

Salient Features

- Heat transfer coefficient of this freezer is low as the air used for cooling is almost still. Fans are often used to induce forced convection and thus the heat transfer coefficient.
- As moist air enters into cold stores (through loading doors) and there may be dehydration of unpacked fruits and vegetables, there may be condensation of moisture on cold surfaces. In such a situation a dehumidifier is essential.

Applications

Due to low freezing rates, these are not used for commercial freezing. However, they find use as hardening rooms for ice cream and for storing foods frozen by other methods.

Air blast freezer

Construction and Operation

- These are big chambers in which cooled air between –30°C and –40°C usually at a velocity of 1.5-6.0 m.s⁻¹ is circulated over the food. Some systems even use air velocities up to 30 m. s⁻¹.
- The system can either be continuous or batch type.
- Food in batch type systems is stacked on trays in the freezer chamber, whereas in the continuous system, the material is moved on a conveyor belt or by

trolleys through the chamber/ tunnel during which it comes in contact with the cold air.

- In continuous systems, as shown in Figure 23.4, the air flow may be either parallel or perpendicular to the food.
- The air flow ducts are arranged such that the air passes evenly over all food pieces. The trolleys or belts (or trays) should be loaded evenly so that the air does not bypass through spaces between the trays or trolleys.
- Multi-pass tunnels can have a number of belts and it is from one belt to another that the materials fall. It helps in controlling the material depth on the belts as well as in breaking up any food clumps.
- The food may be packed and kept in the chamber, which is a case of *indirect contact air-blast freezing system*.

Salient Features

- The thickness of boundary layer surrounding the commodity is reduced by the high air velocity increasing, thereby, the surface heat transfer coefficient. The impingement systems with their high velocity can have much higher surface heat transfer rates.
- For thick products, higher air speeds are not suited because the heat transfer within the product cannot match that at the surface. Even in many situations, for cooling of large items, the air velocity greater than 1 m.s⁻¹ may not give the desired economical advantage as the power required by the fans increases 8 times if the air velocity is doubled. It has been observed that an increase in air velocity from 0.5 to 2 m.s⁻¹ may cause about 20% reduction in chilling time, while the fan power needs a 64-fold increase.
- The continuous equipment can have high throughput capacity (200–1500 kg.h⁻¹).
- As compared to most of the other freezing methods, the equipment is relatively economical because it involves a relatively low capital investment and is compact.
- Usually, the continuous systems have concurrent air flow system.
- Multi-pass tunnel type freezers offer a smaller surface area-to-volume ratio and can allow even 30% saving in energy and 20% less floor space.

FIGURE 23.4 A continuous air blast chiller (Courtesy: M/s Bajaj Processpack Limited)

- Usually, the depth of material on the belt is increased in successive passes by reducing the belt speed successively, e.g., the material depth may be kept at 25–50 mm initially (to have higher freezing rates by more exposure of surface area, say for 5–10 min) after which the thickness of material is kept about 80–100 mm (say) on the second belt.
- Oxidative changes and *freezer burn* could occur in un-packaged or individually quick frozen (IQF) foods because of a large volume of recycled air.
- Frequent defrosting is necessary since moisture from unpacked foods, carried by the air, is deposited as ice on the refrigeration coils.
- The unwrapped products tend to dehydrate, which can be avoided by saturating the air with water, which can be done by recirculating air over ice cold water. This is also called wet air cooling. For wet air cooling, the evaporator (plate or coil) may be immersed in a tank of 0°C water.

Applications

The air blast freezer can freeze foods having different sizes and shapes, and, hence, it can be used for most fruits and vegetables and other products. Impingement systems are more suitable for products with high surface area to weight ratios, such as slabs, pizzas, etc.

Spiral Freezer

Construction and Operation

- The spiral freezer is a modified continuous form of air-blast freezer, in which the material to be frozen is moved on a continuous vertical-axis spiral mesh belt.

- Figure 23.5 is a schematic diagram of material flow in a spiral freezer. The material enters the chamber at the lower side of one end of the chamber and moves in a spiral path to the exit at the top at the other end.
- The freezing time is dependent on the speed of the conveyor.
- Forced in a downward direction through the belt stack is either cold air or liquid nitrogen sprays, creating a counter current effect.
- In some devices, a self-stacking belt is used in which during the movement of the belt, after each loop, the belt is positioned to rest on the tier below it, i.e., the belt is self-stacking. There is no need of supporting rails. It helps to increase the number of loops that can be made for a given height of the chamber, and thus, increases the capacity.

The *tunnel freezer* is also a continuous freezer, which consists of a straight mesh belt on which the product moves. A larger floor area is required as compared to the spiral freezer.

Salient Features

- The loading and unloading of the food can be made automatic.
- These freezers can have high capacity while requiring relatively small floor space (for example, a capacity of up to 3000 kg.h⁻¹ can be obtained by a 32-tier spiral freezer having 50–75 cm belt width).
- The downward movement of air helps to reduce weight losses, which would be high otherwise due to evaporation of moisture.

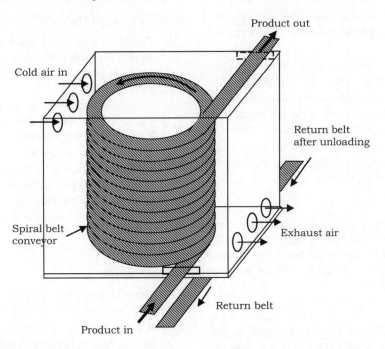

FIGURE 23.5 Schematic diagram of material flow in a spiral freezer

• The freezers require low maintenance costs.

Applications

Spiral freezers (and tunnel freezers) are used for a variety of foods, including fruits and vegetables, meat products, ice cream, pizza, etc.

Fluidized Bed Freezer

Construction and Operation

• In this system, particles are in fluidized state during freezing, exposing more surface area of the product to the chilling/freezing air. It can be continuous type of batch type.

• In a continuous system, the cold chamber is like a sloping tunnel and the commodity travels on a perforated belt through it. In batch systems the materials are kept on a perforated tray in the chamber. The cold air, directed upward at high velocities through the conveyor causes vibrations and the product moves on the conveyor.

• The temperature of cold air is usually between -25°C and -35°C.

• The air velocity required for fluidization depends on the shape and size of the food pieces. Usually this ranges between 2 and 5 m.s^{-1} and the depth of bed ranges from 2 to 13 cm depending on the commodity.

• In some applications, freezing is done in two stages: the first stage aims at creating an ice glaze on the food surface by quick freezing with the help of fluidization. The second stage is the finishing freezing at a depth of 10–15 cm. This is helpful for fruit pieces and similar products to prevent clumping.

Salient Features

• As there is almost complete exposure of surface area to the freezing air, the rate and uniformity of freezing is very high.

• Fluidization helps better exposure of the materials to the freezing air, and the time of freezing is reduced.

• This produces higher heat transfer coefficient and, thus, reduces freezing time as compared to the air blast systems, thus giving higher production rates.

• The equipment needs less frequent defrosting because there is less dehydration of unpacked food as compared to air blast freezing.

Applications

It is for free-flowing products such as peas, green beans, sweet corn kernels and sliced carrots, that the freezer is used. They are also used for cooked shrimps, diced meats, etc. Fluidized bed freezing and air blast freezing are used for individual quick frozen (IQF) foods.

In a *through-flow freezer*, another similar type of equipment, the air moves through the bed of food, but not at

a high enough speed to cause fluidization. This type is suitable for larger pieces of food. Suited for the production of IQF foods, it is compact and has a high capacity.

Cabinet Freezer

Construction and Operation

• It is a batch type freezer consisting of cooling chamber or cabinet in which the product to be frozen is kept on trays.

• Loading can be done through a door on one side or from the top.

• Some fans are provided in the chamber for movement of freezing air.

Salient Features

• Being a batch system, the capacity is usually less and the freezing time is the length of time that the product remains in the cabinet.

• The heat transfer coefficient is low and almost the same as still air freezers.

• The products may be packaged prior to freezing for reducing loss of moisture.

Applications

The cabinet freezer is used for freezing fruits and vegetables, packaged goods like ice cream, etc. This is also used for home, laboratories, and retail storage applications. Common laboratory deep freezers are good examples of cabinet freezers.

Immersion Freezer

Construction and Operation

• This is a cooled liquid freezing system in which the commodity is dipped in the liquid refrigerant (Spray types are also used in which a cold liquid is sprayed on to the food).

• The batch immersion freezers can have 2–3 m^3 size tanks. In a continuous system, as shown in Figure 23.6, the commodity is carried on a conveyor through a bath of liquid refrigerant.

• Refrigerants for use include refrigerated propylene glycol brine, liquid nitrogen or CO_2, glycerol, or a calcium chloride solution. In immersion freezing, in contrast to cryogenic freezing, the refrigerant liquid remains liquid without any change of state throughout the freezing operation.

Salient Features

• There is direct and intimate contact between the material being frozen and the refrigerant, and, hence, it gives a high rates of heat transfer.

• There is reduced product dehydration and coil frosting problems.

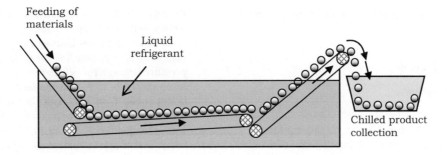

FIGURE 23.6 Schematic diagram of a continuous immersion freezing system (after Singh and Heldman, 2014 with permission)

- Even for unwrapped commodities, there is no loss of moisture or weight.

- Very low temperature refrigerant (such as liquid nitrogen) may cause thermal shock to the product. In many foods, such shock may cause cracking and splitting of the tissues, which should be considered before using ammonia for freezing any specific food.

- In continuous systems, the freezing time can be controlled by the speed of the conveyor.

- The capacity can be very high for continuous systems. The IQF method freezes foods even faster and liquid nitrogen bath, 1.5 m long, may have a throughput capacity of 1 ton of small particulate foods per hour.

- If unwrapped food is being frozen, the liquid used in the process must be safe to ingest. Use of either cryogens, sugar salt, or alcohol solutions could be made for the purpose. The solution should not contain any unsafe or toxic material.

- It involves relatively low capital costs.

Applications

It is commercially used to freeze products with uneven shapes, such as film-wrapped poultry and many times it is taken as a pre-freezing device before blast freezing.

23.2.3 Indirect Contact Freezers

Indirect contact refrigeration methods involve putting uniform flat pieces of food in between two cold metal surfaces, which are cooled by a refrigerant. Plate coolers, belt coolers, jacketed heat exchangers, and falling film systems are such cooling systems. The heat transfer is mostly by conduction and these offer more efficient heat transfer and considerable energy savings over air cooling methods.

Plate Freezer

Construction and Operation

- The plate type freezer can be of batch type, semi-continuous, or continuous types. A batch system constitutes a stack of vertical or horizontal hollow plates, through which the refrigerant is pumped at -40°C.

- Foods, in the form of a thin slab, are placed in between the plates and a slight pressure is applied by

moving the plates closer (Figure 23.7). There is no direct contact between the food and the refrigerant.

- In semi-continuous or continuous systems, the product moves in between two belts, which are cooled by the refrigerant.

Salient Features

- Direct and close contact between the product and the plate surfaces is permitted by the product configuration. Thus, the heat removal is better.

- As the freezing medium does not come in direct contact with the food, the moisture loss from the commodity is almost insignificant and thus, the requirement of defrosting of refrigeration coils is minimized.

- In case of packaged foods, the pressure from the plates prevents bulging of the packages.

- It gives relatively low operating costs, better economy and space utilization as compared to other methods. Production rates can be up to 2500–3000 kg.h^{-1}.

- The main limitations are restrictions on the shape of foods, flat and relatively thin, and relatively high capital costs.

- If the package contains air spaces, it may cause uneven cooling.

Applications

Plate freezers are used for freezing of thin pieces of foods which are in the form of slabs, such as filleted fish, fish fingers.

FIGURE 23.7 Arrangement of plates in a plate freezer

The maximum thickness of product in such freezers is limited to 50–70 mm.

Scraped-Surface Freezer

Construction and Operation

- A scraped-surface freezer constitutes a scrapped-surface heat exchanger, similar to the equipment used for evaporation, in which liquid food is frozen on either the outer or the inner surface of a cooled cylinder. It is also known as a cylindrical freezer.

- It is in ice cream manufacture where the scraped-surface freezers are specifically used. Frozen food is scraped by the rotor from the wall of the cylinder. As air is incorporated, the product becomes light. During this stage, reduction in temperature is between -4°C and -7°C. The product is then transferred to containers and then placed in a hardening room to finish the freezing process. (Expressed as *overrun*, it is the increase in the product volume in relation to the original volume due to the incorporated air.)

Salient Features

- The scraped-surface freezers are specifically used when there is a need to incorporate air into the frozen food. The food is frozen as a thin layer on the surface of the heat exchanger, which gives a rapid rate of freezing. The scrapping keeps the surface of the heat exchanger clean and thus increases the heat transfer rate.

- Freezing is very fast, so very minute crystals are obtained.

- The product consistency is smooth and creamy.

Applications

The freezer is specifically used for making ice cream besides freezing of liquid or semi-solid foods.

> **Example 23.8 In a scraped-surface freezer there is one scraper blade to scrap the solid from the heat exchanger body. It rotates at 150 rev.min⁻¹. Find the maximum time that a food remains on the surface of the freezer. If there are four blades, what will be the corresponding time?**

SOLUTION:

Given: $N = 150$ rev.min⁻¹

Thus, time required for 1 revolution = $(1/150)$ min = $(60/150)$ s = 0.4 s

As the material will be scraped with each revolution of the blade at any point, it will remain on the surface for maximum of 0.4 s.

In the second case, if there are 4 blades, all the blades will go on scraping the surface, i.e., the material will remain on the surface for 1/4 of the time of revolution

i.e., the material will remain for 0.1 s.

> **Example 23.9 A continuous scrapped-surface freezer with two scraper blades is used to cool ice cream mix. The rotor holding the scraper blades rotates at 150 rev.min⁻¹. If the difference in temperature between the refrigerant and the product being cooled is 25°C and the heat transfer coefficient is 600 W.m⁻².K⁻¹, calculate the maximum thickness of the ice layer formed on the cylinder. Take the density of ice as 917 kg.m⁻³ and the latent heat of crystallization as 334 kJ.kg⁻¹.**

SOLUTION:

Given:

Overall heat transfer coefficient, $U = 600$ W.m⁻².K⁻¹

Temperature difference, $\Delta t = 25$ K

No of scraper blades = 2

Speed of the scraper blades = 150 rev.min⁻¹ = $(150/60)$ rev.s⁻¹ = 2.5 rev.s⁻¹

Latent heat of crystallization, $\lambda = 334$ kJ.kg⁻¹

Density of ice, $\rho = 917$ kg.m⁻³

Therefore, the time interval between two scraping operations is

$$\theta = 1/(2 \times 2.5) = 0.2 \text{ s}$$

The thickness of the ice layer (x) frozen within this time period is found by making heat balance.

If A is the area of material deposited on the freezer surface,

Heat extracted by refrigerant, $q = (U. A. \Delta t) \theta$

Latent heat removed, $q = m \lambda = (A. x. \rho) \lambda$

Thus, $U .A. \Delta t. \theta = A. x. \rho .\lambda$

$$\text{or, } x = \frac{U .A. \Delta t .\theta}{A . \rho. \lambda} = \frac{(600)(25)(0.2)}{(917)(334000)}$$

$$= 9.79 \times 10^{-6} \text{ m}$$

23.2.4 Chilling with Ice

Commonly used for some fruits, vegetables, and fish is chilling with crushed ice or an ice/water mixture. It is a common practice to pack individual fish between layers of crushed ice. It is mostly due to the direct contact and cold melt water percolating through commodity that the cooling takes place.

Ice can maintain a very constant temperature. When sea water is used or salt is added to ice, temperature of slightly lower than 0°C can be maintained. Ice is capable of providing a large amount of refrigeration rapidly.

23.2.5 Evaporative Coolers

In general, an evaporative cooler has one or more porous wall (s) that are kept moist with water. Water evaporates from the moist porous walls when outside warm and dry air is pulled through them, reducing the air temperature while raising its humidity.

There are direct, indirect, and mixed type evaporative cooling systems in vogue. In the first type, water evaporates directly into the air, cooling and humidifying the air simultaneously. This method is commonly used with residential systems where air is passed through a moist material and then forced into the room. The examples are dessert coolers. Effectiveness of the evaporative cooler, expressed in terms of the actual wet bulb depression achieved in relation to the theoretical value of the wet-bulb depression, is in the range of 55% to 70%.

High level of humidity produced by direct evaporation of water in the air, approaching saturation, can be undesirable for some applications. In the case of indirect cooling methods, the cool moist air obtained through evaporative cooling is used through an appropriate heat exchanger to lower the air temperature without raising its humidity. The typical effectiveness is about 75%. Note that power is required to operate both the water pump and fans.

If it is required that cooling of air should accompany only a limited humidity increase, it is possible to combine both direct and indirect methods. Indirect-direct evaporative (IDEC) cooling devises are, thus, the mixed type evaporative cooling systems. The first stage cools the air without increasing humidity; the second stage cools and adds moisture. Effectiveness of these mixed type systems can be increased to 100% to 115%, thereby, bringing the air temperature down to even slightly below the outdoor wet-bulb temperature.

Different types of evaporative coolers have been developed and recommended for different situations. The simplest one consists of a wet porous bed through which air is drawn and cooled and humidified. Another device comprises of two pots, one kept inside the other. The inner pot stores food and the gap between two pots is filled with water or wet sand.

An evaporative cooler developed by the Indian Agricultural Research Institute, New Delhi, is named a *Zero energy cool chamber*; its side walls are made of two layers of brick, with a gap of 75 mm filled with sand. There is a cover made of cane or other plant materials, porous sacks, or cloth. Larger capacity evaporatively cooled storage structures with the two wall construction with extra provisions of exhaust fans, perforated floor, perforated side walls and/or solar and wind assisted ventilators have been developed and demonstrated extensively. Installation of desert coolers with large storage chambers is also a form of evaporative cooling.

However, as discussed, the potential cooling depends on ambient conditions and this system is not suitable when the outside relative humidity is very high. The system permits removal of field heat of produce and short duration storage of horticultural produce on a farm.

23.2.6 Cryogenic Freezers

Some of the commercially available cryogenic freezers are described below.

Continuous Belt Type Cryogenic Freezer

Construction and Operation

- Continuous cryogenic freezing consists of a tunnel in which the material to be frozen moves on a perforated belt and the cryogen (liquid refrigerant at a very low temperature) is sprayed on the material.

- Figure 23.8 shows the flow of material in a liquid nitrogen freezer. The food (either packed or unpacked) moves on a perforated belt through the tunnel, in which gaseous nitrogen first cools the food and then liquid nitrogen sprays freezes it. In other words, the gas removes the sensible heat and then the liquid removes the latent heat, while evaporating itself. The typical design uses the vapor formed by the evaporation of the refrigerant during the freezing process for precooling before freezing.

- The freezing time is controlled by the rate of movement of the conveyor and microprocessors provided in the system control the temperature and belt speed

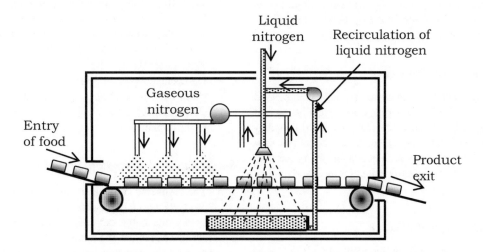

FIGURE 23.8 Schematic diagram of a cryogenic (liquid nitrogen) freezer (after Fellows, 2000 with permission)

so that the product attains the desired temperature at the exit of the freezer.

- To complete the freezing process, the food is then passed to a mechanical freezer.
- Liquid nitrogen and solid or liquid carbon dioxide are the two most common refrigerants used in the equipment.

Salient Features

- The basic characteristic of cryogenic freezing is that the refrigerant (or cryogen) changes phase as it absorbs heat from the food being frozen.
- The choice of refrigerant depends on cost, availability, environmental impact, and performance of the product. The boiling point of liquid nitrogen is -196°C and that of CO_2 is -78.5°C. Thus, the temperature difference between the refrigerant and commodity can be kept very high. The refrigerant also remains in close contact with the food. Thus, there is very high rate of heat removal.
- As discussed before, rapid freezing does not cause many changes in food's sensory and nutritional qualities. The power consumption is also reduced.
- The advantages of cryogenic systems over mechanical freezing systems are the flexibility to process a number of products and lower capital cost without major changes in the system.
- Copper, brass, bronze, monel, and aluminum are the metals suitable for cryogenic temperatures.

Cryogenic Spiral Freezer

The spiral freezer equipment used for cryogenic freezing is the same as discussed in the previous section. The traditional designs use only one cryogen. However, advanced methods involve spraying of liquid oxygen with liquid nitrogen on the food. The mixture is advantageous in terms of greater operators' safety. This is achieved by avoiding build-up of gaseous nitrogen in the vicinity of the freezer, which also causes lack of oxygen. Tunnels have a warm and a cold end. Liquid N_2 is sprayed at the exit end and the cold gas is drawn toward the entrance. Gas is exhausted as warmly as possible.

Cryogenic Immersion Freezer

In cryogenic immersion freezing, the commodity is immersed in cryogen bath for 5 to 50 s to freeze it superficially. A crust is formed in the product, leading to a reduction in the dehydration rate and product clumping. Delicate food items that may crack under the thermal stress are not suitable for processing by this method, whereas it is better suited for meat and poultry products.

Cryogenic Impingement Freezer

These systems involve applying high-velocity air jets and atomized nitrogen in a vertically downward direction onto the product surface. The use of air jets increases the freezing rate by reducing resistance to heat transfer. The products that have high surface-to-weight ratios are better suited to be frozen by the impingement freezers.

Cryogenic Freezer for Liquid Products

A free-flowing freezer is a cryogenic freezer where liquid foods, dairy, and other fluid products are dripped into a freezing chamber through a number of orifices. In the freezing chamber, the liquid product comes in contact with a mixture of gaseous and liquid cryogen, and it is frozen to small beads.

General Considerations Regarding Cryogenic Freezers

Some other general points for consideration regarding cryogenic freezers are as follows.

- Heat transfer from a product is accelerated by using cryogen at high velocity. In addition to giving higher heat transfer rates, the units are also more compact than mechanical systems.
- The high rate of freezing better preserves the texture when products are thawed, retains the initial color and taste. It also gives improved yield resulting from lower moisture loss.
- Well-insulated containers, called Dewars, are used to store cryogenic fluids to minimize loss due to boil off. Dewar flasks and cans are non-pressurized, vacuum jacketed vessels.
- Cryogenic processing involves a high operating cost at present.
- Working with extremely low temperatures requires precautions for human and food safety. Note that the cryogens displace oxygen in non-ventilated confined spaces. If the oxygen level in a work space falls below 19.5%, it is unsafe for people. Therefore, well-ventilated areas are required to handle cryogenic liquids.
- Cryogens, when in contact with skin, can cause frostbite due to extreme low temperature. Splashing of cryogens can result in permanent eye damage. A person working with cryogenics must use the complete set of protective gears.
- Un-insulated pipes or vessels containing liquefied gases must be well segregated to obviate any possibility of accidental touch with any unprotected part of the human body.

Applications of Cryogenic Freezing

- The cryogenic systems are very suitable for cooling products with a high surface area to weight ratio. In such products the extraction of heat from within the surface can match the removal of heat at the surface, or, else, the cryogenic systems cannot be of significant advantage. Some examples are bread, cake, pizza, selected fruits and vegetables, sliced/diced meats, and seafoods.

- The cryogen used for freezing delicate products like muffins and cakes is nitrogen. Reduction in the cooling time and space requirements is possible by using liquid nitrogen vapor to cool baked foods. Rapid reduction in temperatures can be achieved by spraying liquid nitrogen, which helps to minimize the deterioration during storage of commodities.
- Beverage industry uses both carbon dioxide and nitrogen as cryogens. For sparging to reduce the negative effects of dissolved oxygen, nitrogen is used in brewery products, beverages, oils and fats, and milk products.

23.3 Storage and Transportation of Chilled/Frozen Food

To achieve the actual advantage of freezing/ chilling to increase the shelf life of the foods, the commodity should also be stored and transported in chilled conditions.

The following points should be considered.

- Since the refrigeration systems are intended to just maintain the temperature and not to remove heat used in most storage/transport containers, the food should be precooled and kept in the storage/transportation chambers at the correct temperature.
- The extra heat generated by the respiring products has to be removed during the transportation to maintain the temperature.
- During storage and transportation, along with the maintained temperature, the fluctuations in temperature also greatly affect the storage life. Poor temperature control can also lead to degradation in color of the materials.
- The type of packaging also affects the shelf life of commodities. The unpacked/ unwrapped foods may show color changes due to loss of moisture.
- One major problem in chilled storage and transport vans is the inadequate and uneven air flow to all parts of storage. A system that circulates air uniformly around the load is needed. Natural circulation of the air is acceptable for small chambers.
- Sufficient air flow below the bottom layer of the commodity is required. Care has to be taken during loading to ensure that no product should touch the inner surfaces of the vehicle, which are usually at higher temperatures than the inner temperature of the chamber and may cause heat gain to the product by conduction.

The refrigerated containers (reefers) have inbuilt refrigeration systems. They are operated electrically by generators fitted for the purpose.

Refrigerated display equipment and food retail outlets can have integral refrigeration system or remote refrigeration system. In the latter, the evaporator (or cooling coils) is fixed within the storage chamber and the other parts are located separately.

23.4 Selection of a Freezing Equipment

The selection of freezing equipment depends mostly on the characteristics of the product to be stored as follows.

- Amount of the commodity;
- Size and shape of the product;
- Packaging material, if any, on the food;
- Required rate of freezing.

If the freezer is to be used for more than one type of product, then the characteristics of each product need to be considered. In addition, the selection of freezing equipment will also depend on whether a batch type or continuous type operation is desired.

23.5 Thawing

23.5.1 Principle

Thawing is the opposite to freezing and is the process of bringing back a frozen food to normal temperature. The surface ice melts resulting in the formation of a moisture layer on the food surface when it is thawed. As the thermal conductivity and thermal diffusivity of water are lower than those of ice, this moisture layer on the surface reduces heat flow into the interior portion of the commodity. This insulating effect increases as the thickness of thawed layer increases. It is in contrast to the freezing process, in which the rate of heat transfer is enhanced as the thickness of frozen layer increases. Therefore, thawing is a slower process than freezing under similar conditions and the same temperature difference.

As discussed earlier, slow freezing causes large ice crystals and cellular damage. During thawing, the damaged cells cause release of the cell constituents and water-soluble nutrients, which come out of the food as drip loss. Thus, the texture, sensory, and nutritional qualities of the food are changed. It has been reported that fruits can lose up 30% of vitamin C through drip loss.

The drip losses and moisture deposit on the surfaces of the food encourage enzyme and microbial activity. We must remember that freezing does not kill most of the bacteria, but keeps them dormant. Thus, if the food is not properly cleaned and blanched prior to freezing, then the microbial and enzyme activity may cause spoilage of the thawed food, particularly when the product temperature is more than 5°C. If the thawed food is heated quickly to a temperature, which is sufficient to kill the microorganisms (as in cooking), then such spoilage can be arrested. If the food is not intended to be heated, it should be consumed within a short time of thawing.

If the thawing is carried out at a small temperature difference (as normally practiced at home, where we keep the frozen food at ambient condition), the thawing period is longer and the risk of spoilage is more. However, as per FDA, the temperature of food should not be increased to more than 5°C during the thawing process. Commercially, to retain a firm texture, foods are normally thawed to a level just below the freezing point. Thus, the process of cooking must be planned ahead so that an appropriate method for thawing can be used.

23.5.2 Equipment/Methods

Four equipment/ methods are used for thawing food, as follows.

- Refrigerator;
- Microwave;
- keeping the product under cold running water;
- partial cooking.

Some important points for consideration during thawing are as follows.

- Thawing time should be as far as possible low and the drip loss should be minimum.
- If the frozen food is thawed in the refrigerator, the temperature of the product should remain below 5°C.
- The microwave should be used for thawing frozen food only if it is to be cooked immediately.
- If water is used for thawing, the food should be completely submerged in water; the water should be clean, hygienic, and a maximum temperature at 20°C. Moving water at a sufficient velocity can be helpful to increase the heat transfer rate, to cause agitation of the food, and to take away the loose particles, if required.
- A part of the cooking process can also be used as thawing for foods like frozen soup, vegetables, pizza, etc.
- The temperature of ready-to-eat foods during thawing should not increase to more than 5°C.
- The time gap between thawing and cooking should be, in general, minimum.
- Commercially, foods are also thawed in a vacuum chamber by condensing steam, or by circulating moist air over the food.

Check Your Understanding

1. Ice cream is frozen to -8°C in an ice cream freezer operating at a temperature of -20°C. The ice cream mix with a water content of 65% is fed into the freezer at 5°C. The initial freezing temperature of the mix is -2.5°C. The specific heat of the mix is 3.35 kJ $.kg^{-1}.K^{-1}$ and that of the mix at 100% over run is 2.5 $kJ.kg^{-1}.K^{-1}$. It is found that at -8°C about 70% of the water is frozen. Find the refrigeration load of the ice cream freezer in tons for 450 kg.h^{-1} feed load.

Answer: 23.9125 kW (6.8 TR)

2. Find the refrigeration load expressed in tons of refrigeration that is caused by heat loss from the four side walls of a small cold room 2.4 m × 3.0 m × 2.4 m. The walls are made of 20 cm brick, 20 cm cork board, and 1.25 cm cement. The inside wall temperature is 21°C and outside temperature is -30°C. Add a suitable safety factor for losses through joints. The thermal conductivity of brick, cork, and cement plaster area 0.6 W.m^{-1}.K^{-1}, 0.04 W.m^{-1}.K^{-1}, and 0.8 W.m^{-1}.K^{-1}, respectively.

Answer: 0.14 tons of refrigeration

3. Ice cream at a temperature of -18°C is being transported through a refrigerated truck having outside dimensions of 6 m length, 3 m width, and 2 m height. The truck is travelling at a speed of 90 km.h^{-1} on a highway where the air temperature is 45°C. The truck is insulated in a way such that the outside surface temperature of the truck is maintained at 15°C. Assume that there is no heat transfer from the front and back of the truck. Properties of air at 30°C are: $\rho = 1.1514$ kg.m^{-3}, $\mu = 1.86 \times 10^{-5}$ Pa.s, $c_p = 1.007$ kJ.kg^{-1}.K^{-1}, k = 0.0265 W.m^{-1}.K^{-1}. Use the relation; Nu = 0.036 Re$^{0.8}$ Pr$^{0.33}$. Find the average heat transfer coefficient of the system and rate of heat transfer at the four surfaces.

Answer: h = 53.18 W.m^{-2}.K^{-1}, Rate of heat transfer = 95.7 kW

4. A long cylindrical piece of meat having a diameter of 0.02 m containing 80% moisture is being frozen with air at -30°C. Initial temperature of the meat is -2.5°C (freezing point). The heat transfer coefficient of the freezer unit is 20 W.m^{-2}.K^{-1}. If density of the unfrozen meat is 1050 kg m^{-3} and the thermal conductivity of the frozen meat is 1.025 W.m^{-1}.K^{-1}, the latent heat of fusion for water is 335 kJ.kg^{-1}, shape factors P and R are (1/4) and (1/16) respectively, what will be the freezing time?

Answer: 0.373 h

5. A cold storage plant is required to store 25 MT apples. The initial temperature of the apples is 30°C, the refrigerated storage temperature is 2°C, and the specific heat of the apples before the freezing point is 0.874 kCal.kg^{-1}°C^{-1}. If the cooling is achieved within 8 h, determine (i) capacity of refrigeration plant, (ii) COP of reverse Carnot cycle between the temperature range, and (iii) hp required to run the plant if the actual COP is 25% of the Carnot COP.

Answer: 15.955 HP

6. The higher and lower temperature in a refrigerator working on reversed Carnot cycle are 35°C and -15°C. The capacity of the machine is 35.16 kW. Calculate (i) COP, and (ii) heat rejected from the system per hour, and (iii) power required.

Answer: COP = 5.16, heat rejected = 41.97 kW and power input = 6.81 kW

7. In a continuous belt freezer, fish fillets at a feed rate of 1000 kg.h^{-1} is frozen. The unfrozen fish having

moisture content of 85% (wet basis) enters the freezer at 25°C and complete frozen fish exits at -20°C. The properties of the fish are latent heat of crystallization = 330 kJ.kg^{-1}, fixed freezing point = -25 °C, density = 1100 kg.m^{-3}, specific heat capacity above freezing point = 3.60 kJ.kg^{-1}.K^{-1} and specific heat capacity below freezing point = 1.97 kJ.kg^{-1}.K^{-1}. Neglecting other heat losses in the freezer, the power requirement of the compressor (in kW) having a coefficient of performance of 2.50 is

Answer: 46 kW

Objective Questions

1. Name the factors that affect the shelf life of fresh fruits and vegetables.

2. Name the factors that affect the shelf life of processed fruits and vegetables.

3. Name some climacteric fruits and vegetables. How are their considerations different from the non-climacteric fruits for freezing purposes.

4. Write down the equation of heat generation due to respiration of agricultural commodity.

5. Categorize the different microorganisms on the basis of their temperature sensitivity.

6. Differentiate between mechanical freezing and cryogenic freezing.

7. Write down the important properties of a refrigerant.

8. What are the commonly used refrigerants in cryogenic freezing? What are their boiling points?

9. What is chilling injury? Write down the temperature limits for chilling injury of a banana, mango, tomato, and apple.

10. Write short notes on
    a. Eutectic temperature;
    b. Nucleation;
    c. Factors affecting freezing time;
    d. Planck's equation;
    e. Cryogenic freezing;
    f. Spiral freezer;
    g. Fluidized bed freezer;
    h. Immersion freezing;
    i. Vacuum cooling;
    j. Thawing equipment.

11. Differentiate between
    a. Chilled storage and frozen storage;
    b. Direct type freezing and indirect type freezing;
    c. Slow freezing and fast freezing;
    d. Homogenous nucleation and heterogeneous nucleation;
    e. Chest freezer and air blast freezer.

12. What are the relative advantages and disadvantages of immersion freezing over normal air freezing methods?

13. Explain the typical situations where a scrapped-surface freezer will be preferred to other methods.

BIBLIOGRAPHY

Alvarez, J.S. and Thorne, S. 1981. The effect of temperature on the deterioration of stored agricultural produce. In S. Thorne (ed.) *Developments in Food Preservation*, Vol.1, Applied Science Publ, London, pp. 215–237.

ASHRAE. 1965. *ASHRAE Guide and Data Book. Fundamentals and Equipment for 1965 and 1966*. American Society of Heating. Refrigerating and Air Conditioning Engineers, New York.

Betts, G.D. 1998. Critical factors affecting safety of minimally processed chilled foods. In S. Ghazala (ed.) *Sous Vide and Cook chill Processing for the Food Industry*. Aspen Publications, Gaithersburg, 131–164.

Brennan, J.G. (ed.) 2006. *Food Processing Handbook* 3rd ed. WILEY-VCH Verlag GmbH & Co. KGaA, Weinheim.

Brennan, J.G., Butters, J.R., Cowell, N.D. and Lilly, A.E.V 1976. *Food Engineering Operations* 2nd ed. Elsevier Applied Science, London.

Campbell-Platt, G. 1987. Recent developments in chilling and freezing. In: A Turner (ed.) *Food Technology International Europe*. Sterling, London, 63–66.

Charm, S.E. 1978. *Fundamentals of Food Engineering*, 3rd Edn., AVI Publishing Co., Westport, Connecticut.

Cleland, A.C. and Earle, R.L. 1984. Freezing time predictions for different final product temperatures. *Journal of Food Science* 49(4):1230–1232.

Cleland, A.C. and Earle, R. 2006. Assessment of freezing time prediction methods. *Journal of Food Science* 49(4): 1034–1042.

Creed, P.G. and Reeve, W. 1998. Principles and applications of sous vide processed foods. In: S. Ghazala (ed) *Sous Vide and Cook-chill Processing for the Food industry*. Aspen Publications, Gaithersburg, 25–56.

Dash, S. K. and Chandra, P. 2001. Economic analysis of evaporatively cooled storage of horticultural produce. *Agricultural Engineering Today (ISAE)*. 3-4: 1–9.

Desrosier, W. and Desrosier, N. 1978. *Technology of Food Preservation*, 4th ed., AVI Publishing Co., Westport, Connecticut .

Earle, R.L. and Earle, M.D. 2004. *Unit Operations in Food Processing*, Web Edn., The New Zealand Institute of Food Science & Technology,Inc., Auckland. https://www.nzifst .org.nz/resources/unitoperations/index.htm

Evans, J. and James, S. 1993. Freezing and meat quality. In A. Turner (ed.) *Food Technology International Europe*. Sterling Publications International, London, 53–56.

Farrall, A.W. 1976. Cooling and refrigeration. In A.W. Farrall (ed.) *Food Engineering Systems*, AVI Publishing Co., Westport, Connecticut, 91–117.

Fellows, P.J. 2000. *Food Processing Technology*. Woodhead Publishing, Cambridge, UK.

Fennema, O.R. 1975a. Freezing preservation. In O.R. Fennema (ed.) *Principles of Food Science, Part 2: Physical Principles of Food Preservation*, Marcel Dekker, New York, 173–215.

Fennema, O.R. 1975b. Effects of freeze-preservation on nutrients. In R.S. Harris and E. Karmas (eds.) *Nutritional Evaluation of Food Processing*, AVI Publishing Co., Westport, Connecticut, 244–288.

Fennema, O.R. 1982. Effect of processing on nutritive value of food: Freezing. In M. Rechcigl (ed.) *Handbook of the Nutritive Value of Processed Food* Vol.1, CRC Press, Boca Raton, 31–44.

Fennema, O.R. 1996. Water and ice. In O.R. Fennema (ed.) *Food Chemistry*, 3rd ed. Marcel Dekker, New York, 17–94.

Geankoplis, C.J.C.. 2004. *Transport Processes and Separation Process Principles*, 4th ed. Prentice-Hall of India, New Delhi.

Heap, R.D. 2000. Refrigeration of chilled foods. In C. Dennis and M. Stringer (eds.) *Chilled Foods* 2nd ed., Ellis Horwood Ltd, Chickester.

Heldman, D.R. and Hartel, R.W. 1997. *Principles of Food Processing*. Aspen Publishers, Inc. Gaithersburg.

Heldman, D.R. and Singh, R.P. 1981. *Food Process Engineering*. AVI Publishing Co., Westport, Connecticut.

Holdsworth, S.D. 1987. Physical and engineering aspects of food freezing. In S. Thorne (ed.) *Developments in Food Preservation* Vol.4. Elsevier Applied Science, London, 153–204.

Hougen, O.A. and Watson, K.M. 1946. *Chemical Process Principles Part-II. Thermodynamics*. John Wiley & Sons, New York.

Jacob, M. and Hawkins, G.A. 1957. *Heat Transfer* Vol. II. John Wiley & Sons, New York.

Jul, M. 1984. *The Quality of Frozen Foods*. Academic Press, London, 44–80, 156–251.

Kalia, M. and Sood, S. 2019. *Food Preservation and Processing*. Kalyani Publishers, New Delhi.

Kessler, H.G. 2002. *Food and Bio Process Engineering* 5th ed. Verlag A Kessler Publishing House, Munchen, Germany.

Lacroix, C. and Castaigne, F. 1987. Simple method for freezing time calculations for infinite flat slabs, infinite cylinders and spheres. *Canadian Institute of Food Science and Technology Journal* 20(4): 252–259.

Leniger, H.A. and Beverloo, W.A. 1975. *Food Process Engineering*. D. Reidel Publishing Co., Boston.

McAdams, W.H. 1954. *Heat Transmission*, 3rd ed. McGraw-Hill Book Co., New York.

Olsson, P. and Bengtsson, N. 1972. *Time Temperature Conditions in the Freezer Chain*. Report, No.30 SIK, Swedish Food Institute, Gothenburg.

Rahman, M.S. 1999. Food preservation by freezing. In M.S. Rahman (ed.) *Handbook of Food Preservation*. Marcel Dekker, New York, 259–284.

Ramaswamy, H.S. and Tung, M. A. 2007. A review on predicting freezing times of foods. *Journal of Food Process Engineering* 7(3): 169–203.

Rao, D.G. 2009. *Fundamentals of Food Engineering*. PHI Learning Pvt. Ltd., New Delhi.

Rose, D. 2000. Total quality management. In C. Dennis and M. Stringer (eds.) *Chilled Foods* 2nd ed. Ellis Horwood, Chichester.

Sebok, A., Csepregi, I. and Baar, C. 1994. Causes of freeze cracking in fruits and vegetables. In A Turner (ed.) *Food Technology International Europe*. Sterling Publications International, London, 66–68.

Siebel, J.E. 1918. *Compend of Mechanical Refrigeration and Engineering* 9th ed. Nickerson and Collins, Chicago.

Singh, R.P. and Heldman, D.R. 2014. *Introduction to Food Engineering*. Academic Press, San Diego, CA.

Toledo, R.T. 2007. *Fundamentals of Food Process Engineering* 3rd ed. Springer, New York.

Van Beek, G. and Mefert, H.G.Th. 1981. Cooling of horticultural produce with heat and mass transfer by diffusion. In: S.Thorne (ed.) *Developments in Food Preservation* Vol.1. Applied Science, London, 39–92.

Watson, E.L., and Harper, J.C. 1989. *Elements of Food Engineering* 2nd ed. Van Nostrand Reinhold, New York.

B4

Processing at/ near Ambient Temperature

24

Non-Thermal Processing of Foods

Many foods are heat sensitive due to the presence of minerals, vitamins, nutrients with functional properties as antioxidants, and bioactive compounds. The quality of such foods may deteriorate when conventional heat treatments are used for controlling the pathogens. Further, the demand for minimally processed foods is increasing and methods to reduce the microbial load without applying heat have been a focused area of research. Such methods, which can be broadly considered as non-thermal methods, aim at destroying the pathogens with minimum effect on the sensory attributes and nutrient content of the products.

The non-thermal processing methods that are being explored include high-pressure processing and high-pressure homogenization, low pressure, pulsed electric field, high-voltage arc discharge, oscillating magnetic fields, different forms of ionizing radiations (gamma irradiation, electron beam), gases (ozone, chlorine dioxide, cold plasma), light (ultraviolet, pulsed light, high-intensity laser), ultrasound and chemical sanitizers (chlorine, surfactants). These methods yield products that are safer to consume than the untreated products. However, there is a need to validate the methods for specific food products. Usually, low temperature storage will be required for non-thermal processed foods.

24.1 Vacuum Processing

24.1.1 Principle

When pressure around a food material is either increased or reduced, the food changes in its physical and chemical states. These changes, due to a change in pressure, are exploited to impart desirable characteristics to the food.

When the surrounding fluid pressure is reduced for the purpose of food processing, it is called vacuum processing. Recall that vacuum is the value obtained after subtracting the absolute pressure around the food from the atmospheric pressure. Creation of a vacuum means removing the surrounding fluid from the container. The removal of fluid from the container reduces the number of fluid molecules that are available to interact with materials in the container.

Pressure strongly affects many of the physical properties of the fluids. A few of these properties are electrical conductivity, thermal conductivity, optical transmission, optical absorption, and propagation of sound. The properties of solids and liquids under vacuum also show significant changes; if a solid is kept under vacuum, solid food material atoms will spontaneously leave its surface. The rate of material vaporization under a vacuum depends on the system pressure and the material's vapor pressure.

Hygroscopic food materials are stored under a vacuum. The materials those tend to oxidize readily are stored under high vacuum or in an inert atmosphere (nitrogen or argon gas). The vacuum range is 1–600 mbar absolute in the food industry. The different vacuum ranges can be given as in Table 24.1 (Berman, 1992). Packaging of food in vacuumized plastic film bags is very common.

24.1.2 Equipment

Liquids, such as water, boils in an adequately vacuumed chamber without any application of heat. At normal atmospheric pressure, which is around 1.01 bar, water boils at 100°C. If the pressure is reduced to 0.042 bar, the vaporization takes place at 30°C. The vaporization occurs soon after the water vapor pressure exceeds that of the vacuum environment. Due to this rapid evaporation of product moisture, the cooling takes only 2 to 6 minutes to reach 30°C. The equipment is also known as vacuum cooling equipment.

Construction and Operation

- A vacuum cooling system consists of a cooler chamber, refrigeration system, and vacuum pump working together to achieve optimum and efficient cooling.
- The entire cooler chamber is evacuated by the vacuum pump. This evacuation provides the opportunity for the product inside the chamber to lose thermal energy and cool down by means of evaporative cooling as a result of the pressure drop in the chamber. Evaporative cooling is the result of the product moisture getting boiled off under vacuum in the chamber.
- The refrigeration system helps in condensing the warm water vapor given off by the product being cooled. The vapor is condensed and is not allowed to enter the vacuum pumps.
- The vacuum system continues to lower the pressure within the chamber, resulting in cooling the fresh product.

Salient Features

- This is a process of rapid evaporative cooling to reduce the temperature of high moisture foods in short time duration.

DOI: 10.1201/9781003285076-30

TABLE 24.1

Ranges of vacuum

Pressure Description	Range (k Pa)
Low vacuum	3.3 to 101.325
Medium vacuum	1.0×10^{-4} to 3.3
High vacuum	1.0×10^{-7} to 1.0×10^{-4}
Very high vacuum	1.0×10^{-10} to 1.0×10^{-7}
Ultrahigh vacuum	1.0×10^{-13} to 1.0×10^{-10}
Extreme ultrahigh vacuum	below 1.0×10^{-13}

- The advantages include improved product life and better retention of quality, specifically for fruits and vegetables. The short processing time also gives energy savings.

24.1.3 Applications

Vacuum cooling can be applied to different fruits and vegetables, lettuce and mushrooms, etc. There is the possibility of vacuum cooling for meat and bakery products as well.

There are several other applications of vacuum in food processing, such as vacuum evisceration of poultry, vacuum pumping of fish, vacuum flash cooling of milk, vacuum evaporation, de-aeration of process water and liquid foodstuffs, deodorization of vegetable oil, fractional distillation, vacuum filtration, vacuum drying, freeze drying, bottling of beer and soft drinks, dry and wet vacuum cleaning, opening of bags and lifting of loads, vacuum pumping of liquids, etc. The types of equipment differ for different types of applications.

24.2 High-Pressure Processing

When the surrounding pressure is increased, it is high hydrostatic pressure processing. The term "hydrostatic" is used because if there is no fluid surrounding the food, the intended benefits would not be achieved.

24.2.1 Principle

The process involves exposure of food to high pressure (100–800 MPa) for a short time (few seconds to over 20 minutes). Typical processing conditions are 600 MPa and 3 to 5 minutes with the temperature ranging from less than 0°C to above 100°C. It was in 1899 that the West Virginia Extension Service, USA, reported the use of pressures of 650 MPa over an extended period of 60 hours for storage of milk without refrigeration (Hite, 1899). This piece of work could be called the precursor of today's high-pressure processing (HPP).

The following two principles govern high-pressure processing.

1. **Isostatic Pressure Principle**. When pressure is applied uniformly on a food material from all directions simultaneously, it is instantaneously and uniformly transmitted throughout the sample. As the pressure is uniform from all sides, the food is not crushed. The transmitted pressure also remains uniform irrespective of the size and shape of the food.

2. **Le Chatelier's Principle**. Le Chatelier's principle refers to states of thermodynamic equilibrium. It states that a change in pressure or volume will result in an attempt to restore equilibrium by creating more or less mols of gas. For example, if the pressure in a system increases, or the volume decreases, the equilibrium will shift to favor the side of the reaction that involves fewer mols of gas. Similarly, if the volume of a system increases, or the pressure decreases, the production of additional mols of gas will be favored.

The HPP treatment process maintains the original freshness, taste, color, and flavor of food. The nutritional content is better as compared to traditionally processed foods and the foods are safer. Both pathogens and viruses are effectively destroyed and the shelf life is extended. The processing time is less; the HPP can achieve the same level of reduction of bacterial spores within seconds, which would normally take several minutes at the same temperature and atmospheric pressure. In addition to destruction of microorganisms, there may be activation or inactivation of enzymes, denaturation of proteins, changes in properties of carbohydrates, and fats. HPP may be considered as a finishing process for some food processing applications.

24.2.2 Equipment

Construction and Operation

- A typical HPP system has a pressure vessel, a pressure generation mechanism, hydraulic compressors, a material handling device, and a heating/ cooling unit. The fluid that is used to transmit the pressure to the food is kept inside. The systems also have different controls for the pressure, temperature, and time (Figure 24.1).

- Food products can be processed in either semi-continuous or batch process. Solid foods or bulk products can be processed in batch systems only. Semi-continuous high-pressure processing is possible in liquid foods, such as juices, without any packaging requirements.

- The basic steps involved in a batch process involve loading of the food in perforated baskets and then, after keeping the baskets in the chamber, the pressure is increased. After holding the product for the desired time under pressure, the pressure is reduced quickly and the vessel is opened. The time required to attain the desired pressure after closing the chamber is known as *come-up time*.

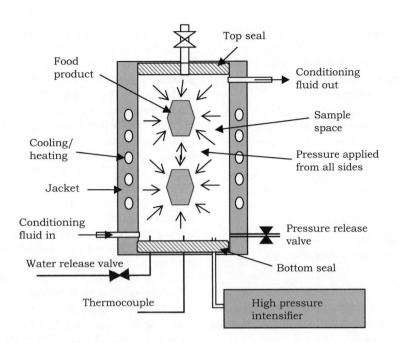

FIGURE 24.1 Schematic diagram of a HPP system for treating food products

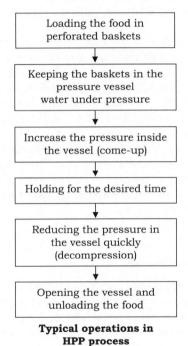

Typical operations in HPP process

- As mentioned earlier, typically the products are held at 600 MPa for 3–5 min. Thus, about 5–6 batches can be achieved per hour considering the times for loading, come-up, decompression, and unloading the product.
- After the processing, the treated product is removed from the application chamber and then forwarded to be stored in a conventional manner.
- Packaging materials such as polypropylene (PP), polyethylene (PE), polyester, and nylon-cast polypropylene pouches are compatible with HPP of food products. Glass, paper, and metallic containers are not suitable for HPP.

Salient Features

- It is an instantaneous process and acts uniformly throughout the food mass, independent of the size and shape of food.
- There is an adiabatic heating due to compression and a consequent rise in temperature of about 3°C per 100 MPa in food moisture whereas the temperature increase is about 8°–9°C per 100 MPa for fats and oils.
- The vessel, where the pressure is increased for HPP, stays relatively cold. The heat that is generated during HPP is lost through the walls of the vessel. Thus, there may be temperature variations in the system.
- Deprotonation of charged groups and disruption of salt bridges and hydrophobic bonds cause the inactivation of microorganisms. Cell morphology also changes due to collapse of intercellular gas vacuoles, anomalous cell elongation, and interruption of movement of microorganisms.
- Both the temperature of the food before processing as well as the processing temperature are critical for the effectiveness of the system. The foods stored at ambient are processed at 70° to 90°C at up to 700 MPa, whereas the refrigerated commodities can be processed at ambient temperature. Inactivation of spore-forming bacteria such as Clostridium botulinum needs higher pressure than 700 MPa (with temperature ranging from 90°–110°C).

- High pressures produce compression and adiabatic heating; the increase in temperature depends on the applied pressure and composition of the food. If a lipid at 70°C is compressed to 700 MPa, the increase in temperature will be about 40°–60°C.

- Foods become more acidic during the process by a pH change of 2 or 3 and the pH returns to its original value after the pressure is released. This relates to Le Chatelier's Principle.

- The critical factors that control quality of food include the applied pressure range, temperature, composition of food, water activity, pH, protein structure, composition of solvent, and microbial load.

- The process may also help access to an enzyme to its substrate, which may accelerate deterioration.

24.2.3 Applications

Refrigerated foods like fresh juices, ready-to-eat meals and meats, and oysters and guacamole dips are processed using high pressures, up to 600 MPa.

High-pressure processing has proved more effective when combined with other non-thermal technologies, such as gamma irradiation, ultrasound, and anti-microbial agents, such as lacticin, nisin, etc.

24.3 Pulsed Electric Field Processing

Short pulses of electricity have proved effective in food preservation. Instead of continuous supply, electricity is converted into electric pulses such that voltage over the pulse duration is magnified up to 1,000 times. Energy received from a high-voltage power supply is stored in a capacitor and, then, discharged through foods in the PEF process. These electric pulses, when applied to foods, affect the vegetative and pathogenic cells in the food in a way that the output has desirable characteristics.

24.3.1 Principle

Microbial Load Reduction

Electric pulses of about 5–100 kV.cm^{-1}(commonly 20-80 kV .cm^{-1}) are used at or near ambient temperature with periods of micro-seconds to cause the microbial inactivation. Cell

membranes of most of the pathogenic bacteria rupture in liquid media during the process. The food may be static or may flow through the treatment chamber. The process has shown to achieve 5 log cycle reduction in microbial population.

The PEF has been reported to cause electroporation leading to increased permeability of microbial cell walls causing microbial inactivation. When PEF is applied, the cell membrane behaves as a capacitor and the cell constituents as a dielectric material of low conductance. This creates a transmembrane potential and in the range of applied pulses, perpendicular transmittance of about 10 mV is generated. The continuous increase in transmembrane potential creates pores in the membrane. Electro-compressive forces, produced by attraction of opposite charges on both sides of the membrane, increase the permeability. When the electric field is continuously applied, the pores increase in size and ultimately cause the rupture of the cell wall and loss of cell content (Figure 24.2).

The vegetative cells of bacteria and yeasts are effectively inactivated by PEF, but the spores are not affected. The main focus, therefore, in using PEF is to control the food-borne pathogens, especially, in acidic foods. Short bursts of electricity (microseconds to milliseconds) are used in PEF, which does not cause any detrimental effects on quality in pumpable foods.

Below a particular strength of the electric field, the changes are reversible and the area of pores is lower than the membrane area. It can be called the critical strength. If the applied electric field is beyond this, then the pores will be large and the changes will be irreversible leading to destruction of the cell membrane and cell death. Critical electric field strength is usually about 1–2 kV.cm^{-1} for plant cells and 10–14 kV.cm^{-1} for microbial cells such as *E. coli*. The most common equipment generates short square waves and the polarity is reversed to avoid erosion of electrodes. The sinusoidal and exponential decay type wave forms can also be produced.

Energy Requirement

In a capacitor, the applied low power electrical energy is stored for some period and then discharged almost instantaneously at a very high power level. Thus, an electrical impulse is created. This impulse is made to pass through a small gap between two

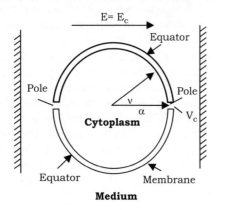

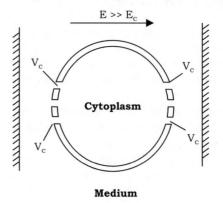

FIGURE 24.2   Schematic representation of the changes in a cell exposed to electric field

electrodes, of the order of millimeters, to cause electric field strength of up to 100 kV.cm⁻¹.

The cell membrane in the foods can be considered as a capacitor filled with a dielectric material of low permittivity, $\varepsilon_m = 2$, which is lower than that of the materials outside the cell.

The treatment time, θ, is given as:

$$\theta = n.\tau_p \tag{24.1}$$

where, n is the number of pulses required for processing and τ_p is the pulse duration.

$$\text{and } \tau = RC = R_f C_o \tag{24.2}$$

where, R is electrical resistance and C is capacitance. For the present analysis the subscript f is for food and C_o is the capacitance of the capacitor bank in the PEF system.

In a RC circuit, i.e. for an exponential pulse, pulse duration, $\tau_p \approx 5\tau$

$$\text{Energy/pulse} = 0.5\, CV^2 \tag{24.3}$$

where, V is the measured potential across the treatment chamber.

$$\text{Total energy} = n\, (\text{Energy/pulse}) \tag{24.4}$$

If d is the gap between the electrodes, the average electric field strength, E, is given as:

$$E = V/d \tag{24.5}$$

$$\text{Thus, } V(\theta) = V_o\, e^{-(\theta/\tau)} \tag{24.6}$$

$$W = n.W_p = n\int^{\tau p} V(\theta).I(\theta).d\theta \tag{24.7}$$

where, W is the total energy required for the PEF treatment, W_p is the specific energy required for the pulse period, and I

is the current. If v is the volume of the treatment chamber, the specific energy per unit volume is given by

$$q = W_p / v \tag{24.8}$$

Energy per unit mass of food, q_m, is defined as:

$$q_m = W_p / m_f = W_p / (\rho_f v_f) \tag{24.9}$$

When pulse duration is large, the treatment chamber needs a cooling system in order to keep the food temperature below the acceptable level.

$$W_p = m_f c_p \Delta t \tag{24.10}$$

$$\text{or } \Delta t = n\, q / c_p \tag{24.11}$$

24.3.2 Equipment

Construction and Operation

Major components of a PEF electrical circuit are shown in Figure 24.3.

The basic parts of a PEF processing system for foods include a treatment chamber in which the food is kept, a high-voltage power source, a capacitor bank to store the electrical energy, a current limiting resistor, and a switch for discharging energy from the capacitor through the food (Figure 24.4).

- The normal supply of 220 V AC is converted into high-voltage AC, and then rectified to high-voltage DC power by a DC generator.

- The capacitor bank stores the energy released by the power source and then discharges it through the food in the form of a strong electric field for micro-seconds.

- The treatment chamber has two electrodes. In continuous systems the electrodes can be of parallel plate or any other shape such as disk-shaped and

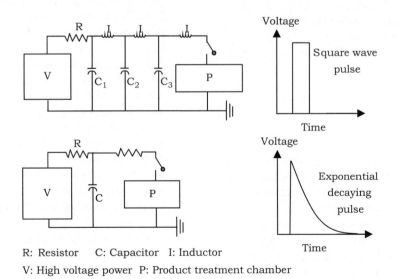

R: Resistor   C: Capacitor   I: Inductor
V: High voltage power P: Product treatment chamber

FIGURE 24.3   A pulsed electric field circuit with (a) square wave pulse (b) with exponential decaying pulse (After Toepfl *et al*. 2014 with permission)

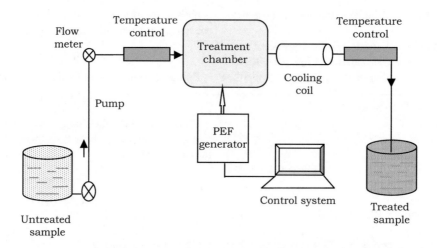

FIGURE 24.4   Schematic diagram of a PEF food processing system (after Nowosad et al. 2021 with permission)

round edged, concentric cylinder, concentric cone, converged electric field, and co-field types.

- The parallel plate electrodes having a small gap between the electrodes are the most preferred as the electric field generated is uniform.

- In the co-field method, the electric field is applied through various treatment chambers at a cycle of 1000 times a second.

- Treatment chamber in a bench top model can have about one mm electrode gap and a 100 µl volume, which may give 25 kV.cm⁻¹ electric field intensity.

- The electrodes are usually made up of stainless steel, though gold, platinum, metal oxides, carbon and carbon-brass have also been used as electrodes or electrode surfaces.

- Polythene, polypropylene, nylon, polysulfone, PVC, and plexiglass are some of the insulator materials used in the treatment chamber.

- Need-based additional devices and controls may be necessary, such as a pump for a continuous system, cooling system for the treatment chamber, etc. Since high-pressure develops rapidly in the chamber due to the electrical sparks through the liquid food, a pressure release device is essential to ensure the safety of the operation.

Salient Features

- The critical factors that affect the PEF include electric field strength, pulse characteristics, treatment temperature, food properties as the electrical conductivity, density, viscosity, pH, and water activity as well as the type of microorganisms.

- An oscilloscope is used to measure the voltage across the treatment chamber. To minimize electromagnetic interference, the oscilloscope should be placed in a shielded area. A current probe also measures the current flowing through the food. The temperature probe is also located in the chamber.

- Though PEF is considered to be a non-thermal and the process typically operates at 35–50°C, the processing chamber often experiences an increase in temperature. Previous experiments have mentioned a temperature increase of up to 30°C for orange juice.

- Design of the chamber and whether the food is flowing or non-flowing are two important considerations, among several factors, to decide the time of exposure.

- Cell membrane pores enlarge during the process in reversible or irreversible manner, which may cause the loss of cell constituents and / or entry of the external matter, ultimately causing the cell death. This is helpful in improving the extraction of juice from fruits and vegetables.

- PEF is more lethal under low ionic strength, low conductivity, and high resistivity. However, the enzyme inactivation is not satisfactory as it affects only a few enzymes. Also there is very less effect on spores.

- The operating conditions in a continuous treatment chamber with parallel plate electrodes are as follows: chamber volume 20–80 cm³, electrical gap 0.95 cm, PEF intensity 15–70 kV.cm⁻¹, pulse width 2–15 µs, pulse rate 1 Hz and food flow rate 600–1200 ml.m in⁻¹.

- Foods use about 15 kV.cm⁻¹ of critical field strengths, whereas for using PEF as a disinfectant, critical field strength of 35 kV.cm⁻¹ is used.

- Average power needs are typically in the range of 30–400 kW for PEF units. However, design of the pulse generator is the deciding factor for the actual demand of power. The requirements are voltages of 40–100 kV and currents of 100A–5kA for industrial applications. Semiconductors used in the equipment at optimal operating conditions have a life time of 10^{12} pulses.

- Polished electrode surfaces reduce the dielectric breakdown of the food to be processed.

- Materials for the treatment chamber need to be washable (cleaning in place) or autoclavable and chemically inert with respect to foods.

- In addition to retaining the nutritional value, color, and other properties of the products in a better way as compared to conventional methods, PEF improves energy use, making the process efficient and economical. It is also a waste free method.
- It involves high cost of generation of pulses. The initial investment is also high.

24.3.3 Applications

- It is a continuous processing method, used for pumpable foods like milk, yoghurt, juices, etc. Further the solids have low electrical conductivity, which make them unfit to be processed by this.
- PEFs of 25–45 kV.cm^{-1} intensity are used for pasteurization of apple juice, peach juice, skim milk, etc. It is more effective in pasteurization for killing vegetative bacteria, yeast, and molds in comparison to conventional methods. As the product temperature is not very high during the process, the food retains better sensory and nutritional qualities.
- PEF has been found to be useful for the extraction of sugars and starches from root vegetables. Studies on juice processing indicate almost 70% more yield and 15% more dry matter in the case of carrots when processed at 520 V.cm^{-1} for a total duration of 100 microseconds. The method can also be used for extraction of polysaccharides and peptides.
- PEF can help in drying at low temperatures. Product quality deterioration occurs in conventional methods of drying, due to either slow drying rates or high processing temperatures. PEF enhances the mass transfer during drying; a 30% improvement in mass transfer rates while drying meat and fish has been reported. In the case of red peppers, also drying time reduction from 360 to 220 minutes was observed. There is reduced effect on volatile compounds.
- With the use of PEF, color extraction at lower temperature can be achieved with improvement in the preservation of the extracted color.
- The method has certain limitations that it cannot be used for products with low electrical conductivity and those containing or forming air bubbles. Electric arcs produce dielectric breakdown and produce unwanted products due to gas bubbles. Hence, to prevent the formation of air bubbles, pressure is applied and at fields above 20 kV.cm^{-1}, this would lead to electrical arcing.

PEF, in general, results in energy savings due to the enhanced heat and mass transfer rates. The process is highly promising from an energy consumption point of view. Moreover, since a PEF process requires much less than a second of time to accomplish the job as compared to minutes and hours required by other technologies; the product quality deterioration is minimized. Availability of commercial units has been a limitation.

24.4 Ultrasonication

Ultrasonic waves are the sound waves of 20 kHz or higher frequencies. These waves generate gas bubbles in liquid media, and, when these bubbles burst, there is very high increase in local temperature and pressure killing the microorganisms. Ultrasound is applied in food processing for different applications, such as food preservation, improving the effects of thermal treatments, modification of texture, and analysis of food. As it is a non-thermal processing method, it is very useful for heat-sensitive foods.

24.4.1 Principle

Ultrasound is the sound wave having frequency more than 20 kHz, i.e., exceeding our hearing limit. In food processing applications, these waves are applied to agitate particles in a food. Ultrasound waves can be divided into three frequency ranges as follows below and is shown in Table 24.2.

The commonly applied frequency is between 20 kHz and 500 MHz. Frequencies higher than 100 kHz at intensities below 1 W.cm^{-2} come under the low-energy (low power, low intensity) category, whereas the frequencies between 20 and 500 kHz at intensities higher than 1 W.cm^{-2} are categorized as high-energy (high power, high-intensity). High-frequency ultrasound is used for analytical equipment to study the physicochemical properties of food.

Ultrasound in liquid systems causes acoustic cavitation. After the formation of bubbles, the bubbles expand until a certain critical size and then burst in the liquid (Figure 24.5). The collapsing bubbles act as hotspots; it generates energy to increase the temperature and pressure up to 5000 K and 500 atm, respectively. This creates the desirable processing effect. In a pure liquid, which possesses high-tensile strengths, the common ultrasonic generators cannot produce high enough negative pressures to cause cavitation. However, most of the liquids are usually impure and allow formation of cavities, which, in turn, cause the thermal, mechanical, and chemical effects.

The expansion or compression of the materials due to the effects as mentioned above results in rupture of cells. The hydrolysis of water inside the oscillating bubbles lead to formation of H$^+$ and OH$^-$ free radicals that lead to subsequent chemical reactions and microbial destruction. The method effectively inactivates microorganisms; the degree of inactivation depends on the characteristics of the microorganisms.

TABLE 24.2

Frequency ranges of ultrasound waves

Frequency range	Nomenclature
20–100 kHz	Power ultrasound
100 kHz–1 MHz	High frequency ultrasound
1–10 MHz	Diagnostic ultrasound

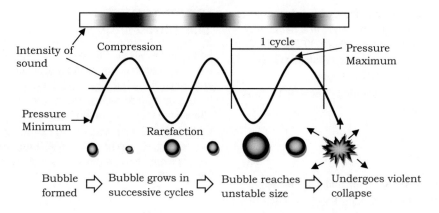

FIGURE 24.5 Cavitation caused by ultrasonication (after www.altrasonic.com; www. hielscher.com)

24.4.2 Equipment

Construction and Operation

- The main component of an ultrasonicator is the ultrasonic generator. This device produces a signal to power a transducer, which, in turn, converts the electrical signal to mechanical vibrations.

- The ultrasonication probe transmits the vibrations to the solution being treated.

- This probe moves up and down at a very high speed, where the amplitude could be controlled and chosen based on the requirements of the product being treated.

- As mentioned earlier, after the formation of the bubbles, they increase in size, which leads to a situation where the applied ultrasonic energy cannot retain the vapor phase in the bubble and thus, rapid condensation takes place. The condensed molecules collide violently, causing the temperature and pressure to increase. The temperature can reach up to 5500°C and pressure up to 50,000 kPa. This high pressure kills the bacteria. The hot zones can also kill some bacteria, but they are very localized.

Salient Features

- Different factors that affect the minimum oscillation of pressure to produce cavitation are the temperature, hydrostatic pressure, dissolved gas, specific heat of the liquid, tensile strength of the liquid, etc. The temperature is inversely proportional to the cavitation threshold. The frequency above 2 .5 MHz does not cause cavitation.

- The characteristics of microorganisms affect the bacterial destruction. Larger cells are more sensitive than smaller ones. The younger cells are more sensitive than older ones. Coccal forms are more resilient than rod-shaped bacteria. Similarly, the Gram-positive bacteria are more resistant than Gram-negative and aerobic bacteria are more resistant than anaerobic bacteria. Besides, spores are also much more resistant than vegetative cells.

- A major limitation of ultrasound is that the released products are inactivated due to the shock waves. It may also cause degradation of food constituents, development of off-flavors, and changes in physical properties due to the formation of radicals.

- Ultrasound application needs more energy input.

24.4.3 Applications

Ultrasonic waves have found many varied types of applications in food industries. Some important applications are as follows.

- Ultrasound together with other treatments (chlorination, ozonation) has been found to be efficient and economically feasible for water treatment.

- In the milk processing industry, in addition to waste water treatment, this technology is also used for removal of fat from dairy wastewater using enzymes, whey ultrafiltration, crystallization of ice and lactose, etc. It has also found applications in changing the functionality of milk proteins.

- It helps in extraction of intracellular constituents such as soybean oil or lycopene.

- It can be used for assisting in production of protein conjugates and enhancing the enzymatic hydrolysis of proteins.

- It assists in improving the separation of protein-starch in the wet-milling.

- It helps in increasing the rate of trans-esterification of fat as well as in emulsification.

- The functional characteristics of food proteins as well as textural properties of fat products can be changed by application of ultrasonic waves.

- It has also found applications in the tenderization of meat.

- The ultrasonic waves accelerate anaerobic digestion process and enhances the cellulolytic activity in cellulose preparation.

- Manosonication is a method of incorporating high static pressure in an ultrasound treatment chamber, which helps in inactivation of enzyme in vegetables.

- A very important application of ultrasonication is cleaning fruits and vegetables (ultrasonic cleaners operate at 20–400 kHz). Low-frequency ultrasonic waves are also used for enhancing pretreatment processes like degassing, crystallization, precipitation, and leaching.

- It has important applications in analysis of food constituents as well as contamination.

Ultrasound is a non-toxic method and there is considerable energy saving. Though many different applications have been suggested and are practiced, it remains an important area of research. The potential still exists to develop ultrasound-based cost-effective food processing systems at commercial scale, promising savings in labor and energy.

24.5 Gamma Irradiation/ Ionizing Radiation

Food irradiation involves exposing foods to γ-rays, X-rays, and electrons. This damages the DNA of microorganisms, which inactivates them and prevents their multiplication. The irradiation used within regulatory parameters has been found to control pathogens in food products; it can reduce or eliminate microorganisms as *E. coli* O157:H7, *Salmonella*, *L. monocytogenes*, and *Campylobacter jejuni*.

24.5.1 Principle

The microbial inactivation occurs primarily due to the direct effect of the ionizing radiations on the cell constituents and transformation of the cytoplasmic membrane through chromosomal DNA in the cell nuclei. Besides the formation of radiolytic products, such as water radicals H+, OH– also contribute to the inactivation of microorganisms.

Photons or electromagnetic waves emitted from an atom's nucleus are called Gamma rays. These rays penetrate to a considerable depth in the food and dislodge the electrons from food molecules, producing ions. The ionizing gamma irradiation is common for killing the pathogens and the primary source of gamma radiation is ^{60}Co. The mean energy of ^{60}Co radiation is 1.25 MeV (mega-electron-volt). Electrons and X-rays have limits of 10 and 5 MeV, respectively.^{137}Cs is also used for generating gamma radiation.

An electron-volt is the amount of kinetic energy gained by a single electron accelerating from rest through an electric potential difference of one volt in vacuum. 1 eV = $1.602176634 \times 10^{-19}$ J.

The radiation doses required for different types of food products depend on the type of product and the end products' quality. A high dose (10–74 kGy) of irradiation is used for sterilization, whereas pasteurization uses a lower dose (0.1 kGy). Usually, low dose radiation is employed for high moisture foods such as spinach, mushrooms, lettuce, fresh fish, etc. Dried spices and herbs are irradiated with slightly higher doses.

Radiation at low doses can affect some vitamins, such as thiamin, but the overall organoleptic and nutritional qualities of food do not change. Both gamma and UV radiations affect the anthocyanin content of foods. The critical factors affecting the effectiveness of the process include the composition of foods, molecular weight of organic compounds, and efficiency of repair mechanisms for DNA of microorganisms.

Although the technology is in use worldwide for a variety of food products for more than 50 years, it is still misunderstood in terms of induced radioactivity. The process does not make the food radioactive; the irradiation energies used in the process are much lower than what could induce radioactivity. Irradiation also does not leave any chemical residue in the product as chemical or fumigation methods do.

24.5.2 Equipment

Usually, foods to be irradiated are packaged and then moved into a chamber behind a labyrinth. The radionuclide sources are kept inside permanent concrete shields. Neutron bombardment of ^{59}Co and ^{136}Cs produces the radionuclides ^{60}Co and ^{137}Cs as fission fragments of a nuclear power reactor operation.

The radiation source is raised from a water pool, which causes the product to absorb the gamma radiation. After the treatment, the product moves out of the chamber and the radiation source returns to the pool for shielding. Electrically driven radiation sources are also available, which can be switched on and off as per requirements.

24.5.3 Applications

Irradiation has been used for many types of foods for extending shelf life. It is used to prevent sprout formation in potatoes, disinfection of grains, pulses, spices and some fruits and vegetables such as melons, strawberries, beans, lettuces, gourds, vegetables, and tubers.

A low dose 1kGy is applied to green onions for increasing shelf life.

Irradiation has proved useful in color retention in fresh meats and to reduce the microbiological load in meat, poultry, eggs, etc.

Irradiation can also be used on frozen foods.

Irradiation, at a dose of 1.5 kGy, increased the shelf life of ground beef, in combination with reduced oxygen packaging and refrigeration, to more than 15 days as compared to 4 days for non-irradiated food kept in a refrigerator.

As declared by the World Health Organization, irradiation up to 10 kGy is not a toxicological hazard in case of any food commodity. Food irradiation is categorized as a food additive in the USA, thus, all radiation processes need the approval of the Food and Drug Administration (FDA). The doses have also been specified for different types of food. It is mandatory that the foods treated by radiation should have a "radura" symbol or the text "treated with radiation" on the package (Figure 24.6). However, if the food has not been irradiated, but with some irradiated ingredients added, there is no need to put the symbol. Food irradiation has been approved for over 100 food items in more than 40 countries.

FIGURE 24.6   Radura symbol

All foods are not suitable to be irradiated. Specifically, milk and similar protein rich foods can degrade in color, odor, and flavor, specifically at higher dose levels.

Some studies are also available for application of gamma radiation to pomegranate juice, carrot juice, etc. and UV radiation to orange, guava, and pineapple juice for reducing the microbial load.

24.6 Ultraviolet Radiation Processing

24.6.1 Principle

Ultraviolet rays are a part of the electromagnetic spectrum and lie between the wavelength range of 400 nm–100 nm. A comparison of different rays is given in Annexure XVIII. Although it is high energy radiation, UV rays are non-ionizing. About 9% to 10% of solar radiation outside the Earth's atmosphere is in UV range. While passing through the atmosphere, UV radiation diminishes and the amount available at sea level is less than one-third the value outside the atmosphere. Humans cannot see ultraviolet rays; however, some creatures, such as insects, birds, and some mammals, can see near-UV. UV light when applied at high levels acts on the DNA, which results in blockage of transcription and replication. Thus, the cells are made dysfunctional, ultimately pushing them to death.

Classification of UV radiations

UV radiations can be grouped into UV-A, UV-B, and UV-C as given in Table 24.3.

In some classifications, the distinction between UV-A and UV-B is at 315 nm. The shorter the wavelength, the more the energy of the radiation.

UV-C makes up only 0.5% of all solar radiation on the Earth's surface because the atmospheric ozone layer absorbs most of it. UV-B band causes skin cancer in humans. It also can impair photosynthesis in many plants. It is partially absorbed in the ozone layer. UV-A (320–400 nm) is not absorbed by ozone in the atmosphere, but it can be blocked by clouds. This is responsible for sunburn in humans.

Sources of UV radiation

The natural source of UV radiation on the Earth is solar radiation. Arc-welding machines, fluorescent lamps, mercury vapor, metal halide, and quartz halogen lamps are artificial UV radiation sources. UV radiation can be produced either by heating a body to an incandescent temperature or by passing an electric current through vaporized mercury gas. The latter is commercially used. UV LEDs are now available as a substitute for conventional UV sources, which can be used for small housing.

Two different modes of UV treatments can be applied to food products such as continuous and pulsed mode. Low- and medium-pressure mercury lamps and excimer lamp technologies are available. Conventional low-pressure mercury UV lamps used for surface decontamination emit UV rays at 254 nm. UV-excimer lamps have the advantage of functioning at different wavelengths, which are produced by combining with different gases. UV rays are energetic enough to destroy chemical bonds or to modify the existing bonds. The UV radiation can control the intended chemical and biological processes without using any chemical.

The pulsed UV light is reported to be effective in reducing microbial contamination; however, their low penetration and temperature rise in the product have been a concern.

24.6.2 Equipment

Construction and Operation

- Different types of continuous systems have been developed for processing liquid products such as milk-based beverages and fruit juices.
- The reactors used for the processing include annular, thin film, dynamic mixers, and coiled tube devices.
- In the case of annular type reactors for treatment of beverages, the gap between two concentric cylinders and the length of the annular reactor vary depending on the composition of the flowing liquid.
- In the case of thin film reactor, the fluid is passed on a fixed surface and the gap is usually kept less than 1 mm.
- The static mixers provide for continuously changing the flow pattern in the system so that the microbial cells are exposed to the ultraviolet radiation to a greater extent.
- In dynamic mixers, the reactor has a stationary outer cylinder and a rotating inner cylinder, which brings fluid from the dark outer cylinder to a high-intensity inner region.

Measurement of UV radiation

A beam of radiation passing through space may be characterized by its radiant energy (J) and radiant flux

TABLE 24.3

Wavelengths of UV radiations

Radiation	Wavelength, nm (nanometers)
UV-A	320–400
UV-B	280–320
UV-C	100–280

(W.m⁻²). Radiation intensity and radiance are terms relating to a source of radiation. Irradiance relates to the object struck by the radiation. Spectroradiometers and radiometers are meant to measure UV radiation. Spectro radiometry relates to spectral measurements of radiometric quantities. Radiometers could be either thermal or photonic. In bio-dosimetry a sample is inoculated with microorganisms and log reductions are counted after the radiation exposure. Two main types of UV sensors, namely, broadband UV sensor and spectro-photometer are commonly used in industry.

Salient Features

- Different applications have specific requirements for UV radiation since both the wavelength of the radiation and certain minimum irradiation (dose) are needed to achieve the desired effect. In addition, the efficacy of UV treatment is dependent on transmission of UV light through the liquid food, which, in turn, is influenced by the food's constituents like organic solutes, pigments, and suspended particles.

- Vitamin C is light sensitive and degrades in the presence of UV light. The UV treatment parameters, such as wavelength and intensity, need to be optimized for each of the products independently.

- Different protocols for validation of UV decontamination system are available, such as the NSF International standard and NWRI guidelines in the USA.

- The energy requirement of UV technology is also lower as compared to thermal processing.

24.6.3 Applications

- Thus, UV irradiation has been a preferred method for purification of drinking water and are commonly employed, at present, for municipal water, water treatment systems, and shellfish wet storage.

- UV is being applied for minimal processing of food products; liquids, emulsions and liquid with particles or suspensions can be treated with UV. The USFDA has approved UV-light application for pasteurization treatment of fruit juice products in the year 2000 and the basic performance criteria set for UV- application is 5 log reductions in number of pathogens present in fruit juices. Absorbance of UV light in fruit juices has been found to be significantly more as compared to water.

- The UV-C range is reported to have germicidal properties, which inactivates the bacteria and viruses. The maximum germicidal effect is achieved in the range of 250 to 270 nm. Ultra-violet radiation (100 to 400 nm) is used for disinfection of solid surfaces and liquids. To accomplish microbial inactivation, the radiant exposure from UV should be more than

400 J.m⁻² in all parts of the product. The bacteria are more vulnerable to UV in air than in liquid medium. Thus, the UV can also be used for disinfecting the process air.

- UV processing for juices and cider has been helpful to reduce *E. coli* O157:H7 and *Cryptosporidium parvum*. The UV application is also reported to reduce patulin mycotoxin in apple cider and allergenicity compounds in peanut- and soybean-based products.

- UV light is used in disinfection of microflora of the food contact surfaces in equipment, packaging materials, conveyors, etc. The packaging materials should not degrade when exposed to the UV treatment.

- In addition, UV radiation can also be used to identify and quantify pathogenic germs and other hazardous materials as well as polluting gases in air.

The UV application has greater consumer acceptance due to the use of non-ionized radiation compared to gamma radiation. Although encouraging results have been obtained from lab scale studies, the validation of the process on a large scale is important to ensure the process effectiveness for food applications. Convergence of UV radiation with other technologies such as pulsed electric field and *mano-thermo-sonication* (treating with a combination of temperature, pressure and ultrasound on a product simultaneously) would become more common to obtain the safe and nutritious food products in future in comparison to the solo application of UV radiation.

24.7 Pulsed X-ray Processing

X-ray pulses of high intensity are generated by radionuclide sources with a solid state opening switch. X-ray irradiation has a kinetic energy limit of 5MeV.

In a Linear Induction Electron Accelerator (LIEA), the accelerated electron beam is directed to hit a heavy metal converter plate, which converts the electron beam to X-rays. This is in the form of a broad-band photon-energy spectrum, which is then filtered to obtain the resultant high-energy and highly penetrating radiation.

These have very good depth of penetration (60–400 cm) as compared to electrons, which can penetrate only about 5 cm in food. The direction of the electrically produced radiation and shape of the radiation field geometry can be controlled to cater to different package sizes. X-ray treatment has low energy requirements, among other food preservation methods.

Pulsed X-rays have proved to be effective against *Salmonella serovars* in poultry. It can also reduce sprout development in potatoes and mold growth on strawberries.

24.8 Pulsed Light Processing

Pulsed light (PL) is composed of visible light (45%), ultraviolet (about 25%), and infrared (30%) radiation in the form of short and intense pulses. This is also known as pulsed ultraviolet light or pulsed white light. The intensity of light is 20,000 times brighter than sunlight; when this light is applied on a food for a very short time, it kills the microorganisms. There are 1–20 flashes per second; the duration of each flash is 1 μs–0.1s. Energy density ranges about 0.01 to 50 J.cm^{-1}.

The microbial killing is mostly due to photochemical action of the shorter ultraviolet wavelengths, though photo-thermal effects may also be present. UV absorption by DNA causes DNA mutation. and changes the DNA and RNA structures. The theories proposing photo-thermal effect suggest that absorption of UV light from a flash lamp by bacteria causes its temporary heating and rupture. FDA (21 CFR179.41) approves the pulsed light treated foods.

The basic equipment comprises a power unit for a pulse generator and the chamber in which the light is transformed to high-power pulses (Figure 24.7). The xenon-flash lamps are commonly used in such systems.

As discussed, the light is very intense, but as that lasts only for a fraction of time, there is no thermal effect. It does not affect the quality and nutrient content of food.

Pulsed light is used commercially for disinfecting food contact surfaces and packaging material and equipment, but commercial application for foods is limited. Studies with food products such as cooked meat, fresh cut mushrooms, fruit juices, and milk have produced positive results. However, some studies note the effect of pulsed light treatment on the color and texture of food. The critical factors include the wavelength of power and number of pulses, distance of food from the PL source, transmissivity of the light, and thickness of food.

It has a limitation in that the foods with uneven shapes or fissures may not receive the treatment uniformly at all surfaces. Thus, it may be necessary to combine other methods for complete protection of food. In addition, pulsed light treatment has minimal effect on some strains of micro-organisms as *Listeria monocytogenes*.

24.9 Cold Plasma Processing

Plasma is considered as the fourth state of matter after the solid, liquid, and gaseous states. The state of matter can be changed, releasing free electrons and ions by breaking intra-molecular and

intra-atomic structures, when a matter gains energy. Plasma is an ionized gas containing positive and negative ions, neutral molecules, and electrons, which transfer energy by collision in between the gas molecules. The collisions give rise to the development of different substances, such as hydrogen peroxide, ozone, hydroxyl radicals, nitrogen oxide, and UV radiation, which interact with the food. The reactive oxygen free radicals cause changes in the DNA and proteins of the microbial cells. There is also accretion of charged particles on the microbial cells, oxidation of cell components, and breakdown of the membrane. All these effects in combination result in microbial inactivation.

Different methods such as heating, electricity, and the use of lasers are used for ionizing gases into plasma. The carrier gases can be air, oxygen, helium, nitrogen, and argon. The basic parts of a cold plasma system include a device that generates the cold plasma, a treatment chamber, and controls for gas and pressure (Figure 24.8). The plasma generator can use either radiowave or microwave or plasma jet or dielectric discharges. The type of carrier gas and the type of plasma generator affect the plasma composition along with other factors such as the operating pressure and temperature.

Cold plasma has been found to be an effective surface sterilizer for food packaging materials, and foods as almonds, cherry tomatoes, strawberries, apples, etc. It has been found to have inactivation effect on *Salmonella enterica serovar Typhimurium*, *Escherichia coli* O157:H7, *Salmonella* Stanley, etc. However, the effects would be different depending on food properties such as pH, cell structure, water activity, etc.

The process has the specific advantage that the process does not need water and antimicrobial products are formed in the air. Thus, it can be combined with an air cooling process. There is no heating of the food as the cold plasma is produced at or near room temperature.

24.10 Ozone Treatment

Ozone (O$_3$) is a powerful disinfectant that kills pathogens and, thus, the shelf life of foods is extended. Ozone, and the by-products of ozone decomposition contribute in killing microorganisms. Ozone is used commercially for extending the shelf life of specific foods. To use ozone for washing, the gas is bubbled into water.

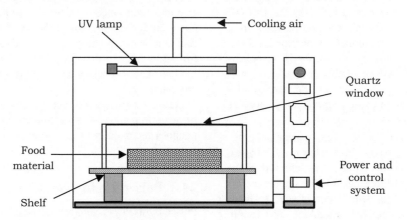

FIGURE 24.7   Schematic of a pulsed UV light treatment chamber

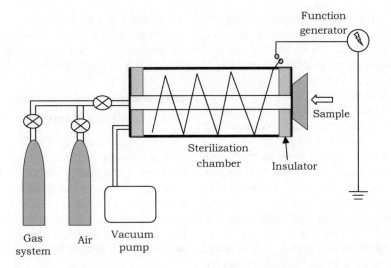

FIGURE 24.8 Schematic diagram of a low-pressure cold plasma sterilization system (after Basaran et al., 2008 with permission)

Ozone is toxic to humans at very high doses, but it decomposes quickly into oxygen after the treatment, and there is no residual effect in the food. It is more advantageous than other common sanitizers like chlorine or common bleach in that ozone leaves no residue in the food or on utensils, thereby avoiding the development of cancer-causing by-products such as organochloride.

Ozone causes oxidation of fat and development of rancid off-flavors; therefore, it is not recommended for treating high fat foods.

24.11 Electron Beam Processing

Electron beam irradiation (E-beam) application involves solid or liquid foods targeted by high energy accelerated electrons (moving at a speed close to that of light) to eliminate/reduce the number of insects, pests, and pathogens. No radioactive isotopes, unlike Gamma irradiation, are used in this technology. The heart of this technology is an electron beam generator that can be turned on and off. Essential parts of the organisms like viruses and bacteria are disrupted by the E-beam by severing the linkages in DNA or RNA. It does not heat the product. The application of E-beam can be precisely controlled, however, it has a limited penetration depth. The cost of the technology is also very high.

E-beam is used effectively for killing *E. coli* O157:H7 and such similar pathogens in packaged ground meat.

24.12 Static and Oscillating Magnetic Fields

Static and oscillating magnetic fields (SMF and OMF) can be used for reducing the microbial load. The SMF has a constant magnetic field intensity. The OMF involves sinusoidal waves of constant or decaying amplitude.

The magnetic fields may be heterogeneous, in which the intensity of magnetic field reduces with the distance from the coil, or homogeneous, in which the magnetic field intensity is uniform throughout.

For treatment, the food is put in a plastic bag and subjected to OMF with 1 to 100 pulses at 5 to 500 kHz frequency and at 0° to 50°C temperatures. The treatment time ranges from 25 to 100 milli-seconds. Due to the treatment the food temperature may increase by 2–5°C.

Two theories have been proposed for the reduction of microorganisms. The first theory suggests that the enzymes and hormones, vital for the cell processes, are damaged due to the weakening of the bonds between ions and proteins by OMF. The other theory suggests SMF and OMF loosen the bond between the calcium ion and the calmodulin (a calcium-binding protein). Cell death is ultimately caused by the metabolic disorder.

24.13 Dense Phase Carbon Dioxide

Dense phase carbon dioxide (DPCD) treatment has the capability to reduce the microorganisms. Considering that it is a non-thermal treatment, there is minimum loss of nutrients or quality.

The mechanism of microbial killing involves cell membrane modification, decrease of intracellular pH, inhibition of cellular metabolism due to lowered pH, inactivation of key enzymes, disturbance in the intracellular electrolyte balance, breakage of cell membrane, and release of vital cell constituents of microorganisms. Another theory explains the CO_2 reacts with water and increases cell permeability and then the pressurized CO_2 enters through the bacterial cell membrane, which disturbs the order in the lipid chain and thus structurally and functionally disrupts the cell membrane. As the dissolved CO_2 move into the cytoplasm, there may be an increase of pH. This may impair the cell viability, may reduce cell metabolism, and may cause denaturation of certain proteins and enzymes. All these lead to microbial inactivation and internal damage of the metabolic processes.

The basic parts of a batch type DPCD application system include a CO_2 gas cylinder, a pressure regulator, a chamber or vessel to hold the material, a water bath or heater, and a CO_2 release valve. After placing the material in the vessel, CO_2 is released into the vessel and appropriate pressure and temperature levels are maintained. After the desired treatment time, the CO_2 gas is released through the release valve. An agitator may be equipped in the system to reduce the time required to saturate the sample with the gas. The batch or semi-continuous processes can have a treatment time of 120 to 140 min.

A continuous high-pressure CO_2 system can also be used for higher capacity. In such a system, developed for a capacity of 1 liter of liquid food per hour there are two 5 l HDPE vessels for holding the liquid. These are connected to the pump through which CO_2 gas is pumped at 40.0 MPa. The mixture then moves through a temperature-controlled tube. The treated liquid is degassed in separate containers after depressurizing it in a capillary tube maintained at constant temperature. The treatment times are usually 3 to 9 minutes.

The method can be applied for liquid foods, such as juices, which can be thoroughly mixed with CO_2. The method has shown to give better retention of β-carotene, antioxidants, phytochemicals, and organoleptic attributes.

The DPCD has been found to lower the vitamin C concentration; however, the loss is lower than the untreated samples. There is unavailability of commercial scale equipment. Main bottlenecks for extensive use of this preservation method are stringent regulations related to the release of CO_2 into the atmosphere and relatively higher cost of the operation.

24.14 High-voltage Arc Discharge

An arc in the liquid food (media) is created by high-voltage arc discharge (HVAD) method of food processing by the application of an electric pulse. While passing through an electrode gap, the liquid food experiences intense electric waves and electrolysis, killing the microorganisms.

The process does not increase the temperature of food and, thus, the nutritional and sensory qualities are better retained. The high-voltage pulses are useful for non-thermal sterilization and pasteurization as 90% reduction of microorganisms can be attained with 10 discharges. Energy from an electric field can be transformed to plasma in case of indirect arc discharge, and then to shock waves, leading to the production of oxidizing agents and free radicals within the product. The free radicals and oxidation reactions can also inactivate enzymes.

The electrical charges can also be reversed for foods kept in between two electrodes, and in such situations the voltage is increased to a peak while the pulse width is about 1 to 5 s. Voltage is then reduced gradually until the voltage reaches its peak at the opposite polarity. The vertical pulses have energy of 0.1–25 J/pulse with field strengths of 15–120 kV.cm^{-1}.

The method has been tested on grapefruit juice; the shelf life of the juice and flavor could be retained for more than 100 days.

The pulsed high-voltage arc discharge is also being studied for reducing surface contamination of foods and beverages.

Chances of recontamination of the food by disintegration of food particles by shockwaves and chemical products of electrolysis put limitations on the utility of this method. The effectiveness of chemical action depends, in addition to the applied voltage, on the initial load and type of microorganism, the amount of the product, and the distribution of chemical radicals and electrode material.

24.15 Sanitizer Washing

Chlorine is commonly used in the food industry to eliminate pathogens on equipment/utensil surfaces and in food. There are legal regulations for the use of chlorine and other sanitizers; the concentrations must be carefully monitored. The effectiveness of a sanitizer is reduced in removing pathogens because of the ability of the microbes to attach to food. Enhancement in the effectiveness of sanitizers by the use of surfactants is, therefore, being studied.

24.16 Combination of Non-Thermal Processing Technologies

It is possible to use combinations of two or more non-thermal processing technologies to augment the effects of individual non-thermal processing methods. Reports are available on the combination of HPP with nisin or lacticin and PEF with ultrasonication, PEF with UV, ultrasonication and UV treatment for food preservation. These combination methods often give a synergistic effect and improve the inactivation of microorganisms in addition to better retainment of the product quality and shelf life as compared to the individual methods.

Check Your Understanding

1. Write short notes on
   a. Vacuum processing applications in food industries;
   b. High-pressure processing;
   c. Different parts of a high-pressure processing equipment;
   d. Pulsed electric field heating;
   e. Ultrasonication in food processing;
   f. Gamma irradiation applications in food;
   g. UV radiation processing;
   h. Pulsed X-ray processing;
   i. Pulsed light processing;
   j. Cold plasma processing.

BIBLIOGRAPHY

Abida, J., Rayees, B. and Masoodi, F.A. 2014. Pulsed light technology: A novel method for food preservation. *International Food Research Journal* 21: 839–848.

Balny, C. Hayashi, R., Heremans, K. and Masson, P. 1992. *High Pressure and Biotechnology*. John Libby & Co. Ltd., London.

Basaran, P., Basaran-Akgul, N. and Oksuz, L. 2008. Elimination of Aspergillus parasiticus from nut surface with low pressure cold plasma (LPCP) treatment. *Food Microbiology* 25(4): 626–32.

Berman, A. 1992. *Vacuum Engineering Calculations, Formulas and Solved Exercises.* Academic Press, Inc., San Diego.

Bialka, K.L. and Demirci, A. 2008. Efficacy of pulsed UV-Light for the decontamination of *Escherichia coli* O157:H7 and *Salmonella* spp. on raspberries and strawberries. *Journal of Food Science.* 73(5): M 201-7. https://doi.org/10.1111/j.1750-3841.2008.00743.x

Brennan, J.G. (ed.) 2006. *Food Processing Handbook* 3rd ed., WILEY-VCH Verlag GmbH & Co. KGaA, Weinheim.

Brennan, J.G., Butters, J.R., Cowell, N.D. and Lilly, A.E.V. 1976. *Food Engineering Operations* 2nd ed. Elsevier Applied Science, London.

Chandrapala, J. and Leong, T. 2014. Ultrasonic processing for dairy applications: Recent advances. *Food Engineering Reviews* 7: 143–158.

Chakraborty, S. and Rao, P.S. 2015. Impact of HPP on antioxidant capacity of fruit beverages. FnBnews.com/Beverage/impact-of-hpp-on-antioxidant-capacity-of-fruit-beverages-37935

Charm, S.E. 1978. *Fundamentals of Food Engineering* 3rd ed. AVI Publishing Co., Westport, Connecticut.

Crawford, L.M. and Ruff, E.H. 1996. A review of the safety of cold pasteurization through irradiation. *Food Control* 7(2): 87–97.

DeVito, F. 2006. *Application of Pulsed Electric Field (PEF) Techniques in Food Processing.* Ph.D. thesis. Department of Chemical and Food Engineering, Universita Delgi Studi Di Salerno, Italy. 147 pages.

Dunn, J., Ott, T. and Clark, W. 1995. Pulsed light treatment of food and packaging. *Food Technology* Sept, 95–98.

Earnshaw, R.G. 1992. High pressure technology and its potential use. In A. Turner (ed.) *Food Technology International Europe.* Sterling Publications International, London, 85–88.

Earnshaw, R.G. 1998. Ultrasound: A new opportunity for food preservation. In M. J. W. Povey and T. J. Mason (eds.). *Ultrasound in Food Processing.* Blackie Academic and Professional, London, 183–192.

Eugen, H., Astanei, D., Ursache, M. and Hnatiuc, B. 2012. A review over the cold plasma reactors and their applications. International Conference and Exposition on Electrical and Power Engineering (EPE). DOI:10.1109/ICEPE.2012.6463884

Farrall, A.W. 1976. *Food Engineering Systems.* AVI Publishing Co., Westport, Connecticut.

Fellows, P.J. 2000. *Food Processing Technology.* Woodhead Publishing, Cambridge, UK.

Galazka, V.B. and Ledward, D.A. 1995. Developments in high pressure food processing. In: A. Turner (ed.) *Food Technological International Europe.* Sterling Publications International, London, 123–125.

Gould, G.W. and Jones, M.V. 1989. Combination and synergistic effects, In: G.W. Gould (ed.) *Mechanisms of Action of Food Preservation Procedures,* Elsevier, London, 401.

Guneser, O. and Yuceer, Y.K. 2012. Effect of ultraviolet light on water- and fat soluble vitamins in cow and goat milk. *Journal of Dairy Science* 95(11): 6230–6241.

Hendrickx, M. Ludikhuyze, L., Van Den Broeck, I. and Weemaes, C. 1998. Effects of high pressure on enzymes related to food quality. *Trends in Food Science and Technology* 9: 197–203.

Hite, B.H. 1899. *The Effect of Pressure in the Preservation of Milk: A Preliminary Report. W. V. A. E. Station.* West Virginia University, Morgantown, WV.

Hoover, D.G., Metrick, C., Papineau, A.M., Farkas, D.F. and Knorr, D. 1989. Application of high hydrostatic pressure on foods to inactivate pathogenic and spoilage organisms for extension of shelf life. *Food Technology* 43(3): 99.

https://www.altrasonic.com/ultrasonic-acoustic-cavitation-in-sonochemistry_n34

Jan, A., Sood, M., Sofi, S.A. and Norzom, T. 2017. Non-thermal processing in food applications: A review. *International Journal of Food Science and Nutrition* 2(6): 171–180.

Kessler, H.G. 2002. *Food and Bio Process Engineering* 5th ed. Verlag A Kessler Publishing House, Munchen, Germany.

Knorr, D. 1993. Effects of high hydrostatic pressure processes on food safety and quality. *Food Technology* 47(6): 156.

Knorr, D. 1995a. High pressure effects on plant derived foods. In D.A. Ledward, D.E. Johnson, R.G. Earnshaw and A.P.M. Hasting (eds.) *High Pressure Processing of Foods.* Nottingham University Press, Nottingham, 123–136.

Knorr, D. 1995b. Hydrostatic pressure treatment of food: Microbiology. In: G.W. Gould (ed.). *New Methods of Food Preservation.* Blackie Academic and Professional, London, 159–175.

Ledward, D.A., Johnston, D.E., Earnshaw, R.G. and Hasting, A.P.M. (eds.).1995. *High Pressure Processing of Foods.* Nottingham University Press, Nottingham.

Leighton, T.G. 1998. The principle of cavitation. In: M.J.W. Povey and T.J. Mason (eds.) *Ultrasound in Food Processing.* Blackie Academic and Professional, London, 151–182.

Leistner, L. and Gorris, L.G.M. 1995. Food Preservation by Hurdle Technology. *Trends in Food Science and Technology* 6: 41–46.

Lewis, M.J. 1990. *Physical Properties of Foods and Food Processing Systems.* Woodhead Publishing Ltd., Cambridge, 287–290.

Loaharanu, P. 1995. Food irradiation: current status and future prospects. In: G.W. Gould (ed.). New Methods of Food Preservation. Springer, Boston, MA. https://doi.org/10.1007/978-1-4615-2105-1_5

Mertens, B. 1995. Hydrostatic pressure treatment of food: Equipment and processing. In: G.W. Gould (ed.) *New Methods of Food Preservation.* Blackie Academic and Professional, London, 135–158.

Messens, W., Vancamp, J. and Huyghbaert, A. 1997. The use of high pressure to modify the functionality of food proteins. *Trends in Food Science and Technology* 8: 107–112.

Miller, B.M., Sauer, A. and Moraru, C.I. 2012. Inactivation of Escherichia coli in milk and concentrated milk using pulsed-light treatment. *Journal of Dairy Science* 95(10): 5597–5603.

Mir, S.A., Shah, M. and Mir, M.M. 2016. Understanding the role of plasma technology in food industry. *Food Bioprocess Technology* 9: 734–750.

Nowosad, K., Sujka, M., Pankiewicz, U. and Kowalski, R. 2021. The application of PEF technology in food processing and human nutrition. *Journal of Food Science and Technology* 58: 397–411.

Odriozola-Serrano, I., Aguilo-Aguayo, I., SolivaFortuny, I. and Martín-Belloso, O. 2013. Pulsed electric fields processing effects on quality and health-related constituents of plant-based foods. *Trends in Food Science and Technology* 29: 98–107.

Palou, E., Lopez-Malo, A., Barbosa-Canovas, G.V. and Swanson, B.G. 1999. High pressure treatment in food preservation. In: M.S. Rahman (ed.) *Handbook of Food Preservation.* Marcel Dekker, New York, 533–576.

Qin, B., Zhang, Q., Barbosa-Canovas, G.V. 1994. Inactivation of microorganisms by pulsed electric fields of different voltage waveforms. *IEEE Transactions Dielectrics and Electrical Insulation* 1: 1047–1057.

Rahman, M.S. 1999. Light and sound in food preservation. In: M.S. Rahman (ed.) *Handbook of Food Preservation.* Marcel Dekker, New York, 669–686.

Toepfl, S., Heinz, V. and Knorr, D. 2005. Overview of pulsed electric field processing for food In Da Wen Sun (ed). *Emerging Technologies for Food Processing.* Academic Press, Cambridge, 69–97. https://doi.org/10.1016/B978-012676757-5/50006-2.

Toepfl, S., Siemer, C., Saldana-Navarro, G., Heinz, V. 2014. Overview of pulsed electric fields processing for food. In Da-Wen Sun (ed.) *Emerging Technologies for Food Processing*, 2nd ed.. Academic Press, Cambridge, MA, 93–114. 10.1016/B978-0-12-411479-1.00006-1

Vega-Mercado, H., Gongora-Nieto, M.M., Barbosa-Canovas, G.V. and Swanson, B.G. 2007. Pulsed electric fields in food preservation. In M. S. Rahman (ed.) *Handbook of Food Preservation*, 2nd ed. CRC Press, Boca Raton.

Watson, E.L. and Harper, J.C. 1989. *Elements of Food Engineering* 2nd ed. Van Nostrand Reinhold, New York.

Yang, B., Shi, Y., Xia, X., Xi, M., Wang, X., Ji, B. 2012. Inactivation of food borne pathogens in raw milk using high hydrostatic press. *Food Control* 28: 273–278.

Zhang, Z.H., Wang, L.H., Zeng, X.A., Han, Z. and Wang, M.S. 2017. Effect of pulsed electric fields (PEFs) on the pigments extracted from spinach (Spinacia oleracea L.). *Innovative Food Science & Emerging Technologies* 43: 26–34.

Zhang, Z.H., Wang, L.H., Zeng, X.A., Han, Z., Brennan, C.S. 2018. Non thermal technologies and its current and future applications in the food industry: A review. *International Journal of Food Science & Technology* 54(9): https://doi.org/10.1111/ ijfs.13903.

25

Food Fermentation and Biotechnology Applications

When any technology is applied to biological systems including living organisms or derivatives thereof for the purpose of making or modifying processes or products, it is called "biotechnology." Modern biotechnology is capable of examining at the molecular level and is also known as genetic engineering. Food biotechnology is the branch of biotechnology that aims at enhancing food qualities or yield or to develop new and improved food or food-related products.

The primary goal of food biotechnology today is to provide more abundant, less expensive, more nutritive, and safer food to address the needs of the growing population globally in a sustainable way. For seeking the desired results, it involves a large variety of processes on plants, animals, microbes, or any part of these organisms.

25.1 Food Fermentation

Food fermentation is an important area of application of biotechnology in food processing. The word "ferment" comes from the Latin verb *fervere,* which means "to boil." Fermented dairy products, pickles, vinegar, sauerkraut, and other fermented products are essentially obtained through the process of controlled fermentation. For thousands of years people have been practising the conversion of fruit juices into wine, milk into cheese or yoghurt, or barley into beer through a process that we know today as fermentation. The products of fermentation find their applications in chemical, pharmaceutical, and food industries.

25.1.1 Principle

The process involves microorganisms that act on the raw materials in such ways as to result in the products of interest. Fermentation is, thus, a process, generally anaerobic, that consumes sugar in the raw material and produces organic acids, gases, or alcohol. The most common groups of microbes involved in food fermentation are bacteria, yeasts, and molds. Industrial fermentation is carrying out the process intentionally on a large scale to manufacture products useful to humanity.

The science of fermentation is also known as *zymology.* German chemist Eduard Buechner, 1907 chemistry Nobel laureate, demonstrated that the enzymes produced by the microbes were responsible for fermentation. Enzymes are complex proteins produced by microbes to catalyze the initiation and/or control bio-reactions. Being proteinaceous in nature, enzymes are sensitive to fluctuations in concentrations of substrate, temperature, pH, moisture content, and ionic strength. Extremes of temperature and pH are responsible for denaturing the proteins and destroying enzyme activity.

Acetic acid producing *Acetobacter* and *Lactobacillaceae* species are important, useful bacteria in food fermentations. The desirable yeasts for food fermentation are derived from the *Saccharomyces* genus, especially *S. cerevisiae.* Yeasts in general are larger than most bacteria. Yeasts produce enzymes that favor chemical reactions, such as production of alcohol, leavening of bread, and inversion of sugar. Ripening and flavoring of cheeses are achieved through the genus *Penicillium.* Molds, being aerobic, require oxygen for growth; have the largest array of enzymes; can colonize; and grow on most types of food.

Microbes during fermentation produce by-products, which help in changing the texture and flavor of the food substrate and act as preservatives. Lactic and acetic acids in larger proportions, propionic, fumaric, and malic acids in smaller proportions, are produced during fermentation. Acetic acid, subsequently, affects the flavor and has a preservative effect on the food. Fermentation may also change the food texture. In milk, the acid causes the precipitation of milk protein to form solid curd.

25.1.2 Types of Fermentation Processes

Fermentation process can be carried out as a *batch process* in which after the preparation and inoculation of the vessel, it is left to run for a few days or up to a week, leading to densities of 2×10^6 cells/ml. The process stops when a vital nutrient ends up or there is accumulation of metabolic waste products that inhibit the microbial growth. Another process, the *Fed batch fermentation process* is not left to run until a key nutrient is exhausted. Instead, concentrated nutrients are added over time and the fermentation process is extended. *Chemostat* permits the removal of a part of the cells produced as part of the process and their retention in the fermenter. This process is better suited for fast-growing microbes. *Perfusion,* the fourth fermentation process, requires the removal of the medium and is suited for slow growing microbes. The loss of cells going away with the medium is prevented by providing a perfusion device. This device prevents the loss of cells but permits the removal of media such that fresh media can be added.

25.1.3 Fermentation Equipment

- Fermentation is carried out in a bioreactor or fermenter; it is a vessel in which sterile nutrient media and pure culture of microorganism are mixed for fermentation under aseptic and optimum conditions.

DOI: 10.1201/9781003285076-31

- Provisions for aeration, agitation, control of temperature, pH, foaming control, and sterilization are made in the fermenter (Figure 25.1).

- There are inoculation points for microorganisms, inlets to provide sterile compressed air, and hot and cold water outlets for the fermented media and sampling ports.

- The material of the fermenter construction should be non-corrosive under conditions of steam sterilization, high pressure, and pH changes or due to the presence of any toxic substance in the fermentation media. Wooden fermenters are common for beer, wine, and lactic acid fermentation. Stainless steel (SS 304 and SS 316), coated with epoxy or glass lining, is the other common material for fermenter construction.

- Impellers facilitate mixing of media, oxygen, and microorganisms uniformly.

- Foam formed in the head space during the fermentation process can cause contamination and impellers help in breaking the foam bubbles.

- Baffles fixed on the walls of the fermenter prevent vortex formation and thereby prevent the media from spilling out of the fermenter.

- Fermentation media, inoculum, and substrate are added in the fermentation tank through the inoculation port in an aseptic manner.

- Spargers provide aeration in the fermentation tank through pipes containing 5–10 mm holes. The sterile air, released in the form of tiny air bubbles in the fermenter, also helps in mixing of media in addition to meeting physiological needs.

- A sampling point provides aseptic withdrawal of samples with the goal of monitoring the fermentation process and quality control.

- The pH of media is checked at specific intervals by the pH control device, which is then adjusted to its optimum level by addition of acids or alkalis.

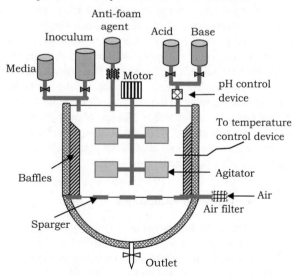

FIGURE 25.1 Schematic diagram of an industrial fermenter

- A temperature control device generally contains a thermometer and cooling coils or jackets around the fermenter and is aimed at removing the excess heat generated by exothermic reactions in the fermenter.

- A foam controlling device, placed on top of the fermenter, has a small tank containing an anti-foaming agent and an inlet connected to the fermenter. The foam generated during fermentation is either destroyed mechanically or neutralized.

- An aseptic outlet at the bottom of the fermenter removes the fermented media and products.

- *Stirred tank reactor* (STR) is the most common fermenter; other types include hollow fiber chamber, tower, and airlift types. The fermenters may be autoclavable, if small scale, or stainless steel steam-in-place type. Single use disposable fermenters are also available.

- Temperature, nutrients, pH, moisture, oxygen concentration, and inhibitors are the six parameters that influence microbial growth. Therefore, microbial growth is controlled through manipulation of these six parameters.

25.1.4 Applications

- The fermenters are used for preparation of foods such as cheese, wine, beer, and bread.

- Lactic acid fermentation is used for products as yoghurt, sauerkraut, pickles, and kimchi. Acetic acid fermentation converts starches and sugars from grains and fruits into sour tasting vinegar and condiments, including apple cider, vinegar, and kombucha.

- It is also used for preparation of functional foods and nutraceuticals as well as bio preservatives for food applications. Production of single-cell protein is another important application.

25.2 Other Applications of Biotechnology in Food Processing

Biotechnology during the past six decades has undergone a transformation enabled by such discoveries as DNA structure, gene mapping, genetic decoding, and gene splicing. In view of the scientific capabilities at gene level, the present-day biotechnology has become highly potent in manipulating the downstream processes. Therefore, the present-day biotechnology is distinguished from the traditional biotechnology and is better known as modern biotechnology. There are three important techniques of modern biotechnology. These are (1) recombinant DNA technology or genetic engineering, (2) plant tissue culture, and (3) transgenic or genetically modified organism (GMO).

Genetic engineering is either the synthesis of artificial gene or repair of gene or combining of DNA from two organisms and manipulating the artificial gene together with the recombinant DNA (modified DNA). The objective is the improvement of microbes, plants, animals, and humans. Tissue culture is the process of regenerating plants in a laboratory or a controlled environment from organ explants, tissue explants, cells, and protoplasts. A genetically modified organism or transgenic is created from the process of genetic engineering and can be a plant, an animal, or a microbe such as fungi, bacteria, and virus. The process of transfer, integration and expression of transgene (foreign or modified gene) in an organism is called genetic transformation. Transgenes may be from the same species in their improved form, related wild species, unrelated species, or microbes.

Modern food biotechnology involves working at DNA level; joining of two pieces of DNA from different organisms leading to a single DNA of desired characteristics, thereby, transferring individual specific genes from parent organisms to improve the resultant food. When the modified or recombinant DNA expresses, it manifests into the desired product. These techniques are faster and more precise. In the case of the food processing sector, it is intended to improve molds, bacteria, and yeasts with the objective of improving the efficiency, control, and yields of the process and the safety, consistency, and quality of the product.

A genetically modified (GM) food is prepared from the produce of genetically modified crops, animals, and microbes. A GM food is different from the food prepared from the conventionally developed varieties in the sense that it contains the proteins produced by the transgene in question, the enzyme produced by the antibiotic resistance gene that was used during gene transfer by genetic engineering, and the antibiotic resistance gene itself. Protein engineering has facilitated the development of novel enzymes with modified structures and desirable properties, such as improved activity, thermal stability, etc.

Some such important applications are listed below.

Development of crops and other foods. There have been significant achievements in development of crop plants, meat, poultry, and fish, etc. with desired functionalities. For example, tomatoes, melons and papaya have been developed that ripen at the right time, canola, soybean, and sunflower have been developed that do not produce trans fats and provide the specific omega 3 fatty acids, pigs and cows have been bred to produce higher omega-3 fat content in the meat.

Production of processing aids. GM micro-organisms have been engineered to extract high yields of products such as vitamins, organic acids, enzymes, amino acids, flavoring agents, certain complex carbohydrates, and polyunsaturated fatty acids for healthy food formulations. Biotechnological approaches have led to the development of xanthan gum for stabilizing and thickening, vitamins, organic acids, vanillin for flavoring, etc. Amino acids have been produced through recombinant DNA technology to form building blocks for active ingredients used in many industrial processes. Aspartame and thaumatin are non-nutritive sweeteners and represent a breakthrough for diabetic people.

The genetically modified enzymes used in food industry today are chymosin widely used in cheese manufacture, catalase used in mayonnaise production, glucose oxidase used in baking, α-amylase to convert starch into maltose and protease used for meat tenderization, baking, and dairy products.

In addition to the above, applications of modern biotechnology have been made successfully in the areas of fermentation, including brewing and wine making and for food ingredient production through plant cell bioreactors.

Bio-preservation of food. Bio-preservation involves using the antimicrobial metabolites or microorganisms that induce bactericidal effects to prevent undesirable changes in the food. Fermentation bacteria (such as lactic acid bacteria) are of utmost importance as the lactic acid bacteria can produce bio-preservatives such as nisin, flavor compounds such as di-acetyl, and products such as enzymes and probiotics. The bio-preservatives can be made available in the form of powder, granules, pressed cake, or concentrated liquid. The dried protective cultures can be sprayed on the food. The concentrated mixture is diluted and sprayed directly at recommended levels to the food surface or package prior to packing.

Microbial metabolism also generates antimicrobial compounds (e.g., bacteriocins and lysozyme). Certain microorganisms, such as the *Lactobacillus* species, associated with fermented foods are probiotics. The active microbial inoculants deliberately added in such foods (called a starter culture) contribute to the pre- and pro-probiotics compositions that offer digestion abetting properties.

Another commercial achievement in this field is "defined starter cultures," a mixture of microorganisms that facilitates the fermentation and preserves the food simultaneously. Similarly, ready-made inoculants, called "fitting cultures," produced by an advanced process called back-slopping, consist of the previous batch of a fermented product usable in successive fermentations, such as fermented fish sauces and vegetables. The dairy industry uses such cultures in the production of yoghurt, kefir, and cheeses.

Many cost-effective and cost-competitive bio-preservatives are available on the market. Currently, efforts are being made to develop genetically designed milk, food grade bio-preservatives, functional foods, bioactive peptides, oral vaccines, recombinant enzymes, cultures, flavors, colors, and bacteriocins for bio-preservation. Fermentation is industrially demanding; the process automation and sustenance of microbial purity are very critical.

Monitoring of food quality and safety. Biotechnology has facilitated the development of rapid and precise tests for ascertaining food quality and safety, which can help in real-time monitoring of food quality. Characterization and monitoring the presence of normal flora, spoilage flora, and microflora in foods can be carried out effectively through such molecular methodologies as pulsed-field gel electrophoresis

(PFGE), polymerase chain reaction (PCR), and ribotyping.

Bio-waste utilization. Biotechnological solutions involve treatment of food industry waste with enzymes or microorganisms for reducing this organic substance into products like protein mass or bio-fuels. Biogas plants, advanced bio-based techniques for treating wastewater such as oxidation, biofilters and reactors, are some means for treating food wastes. Enzymolysis technology converts solid food substrate into bio-ethanol using an enzyme, one of the ever-demanding industrial bio-fuels of the time.

Use of GM microorganisms involving enzymes gained importance majorly due to the ease of growing and accessing of the microorganisms in food systems. The domain of food biotechnology is highly fertile and, therefore, immense possibilities exist for tomorrow. In deference to environmental concerns and economy, manufactured meat products are on the horizon. The area of nutri-genomics is being pursued to prescribe individual's diets for health and wellness. Integration of biotechnology with nanotechnology would unleash tremendous opportunities.

Check Your Understanding

1. Explain the principle and different applications of food fermentation.
2. What are the essential features of a food fermentation equipment?
3. State different applications of biotechnology in food processing.

BIBLIOGRAPHY

Admassie, M. 2018. A review on food fermentation and the biotechnology of lactic acid bacteria. *World Journal of Food Science and Technology* 2(1):19–24.

Bauer, W. 1986. The use of enzymes in food processing. *Food Europe* 5: 21–24.

Dimidi, E., Cox, S. R., Rossi, M. and Whelan, K. 2019. Fermented foods: Definitions and characteristics, impact on the gut microbiota and effects on gastrointestinal health and disease. *Nutrients* 11(8): 1806, https://doi.org/10.3390/nu11081806

Falk, M. C., Chassy, B. M. , Harlander, S. K., Hoban, T. J. IV, McGloughlin, M. N. and Akhlaghi, A. R. 2002. Food biotechnology: Benefits and concerns I. The Journal of Nutrition 132(6):1384–1390.

Frazier, W. C. and Westhoff, D. C. 1987. *Food Microbiology* 3rd ed. McGraw-Hill, New York.

Ghosal, G. G. 2018. Biotechnology in food processing and preservation: an overview. In M. Holban and A.M. Grumezescu (eds.) *Advances in Biotechnology for Food Industry* 1st ed. Academic Press, UK, 27–54.

https://www.generalmicroscience.com/industrial-microbiology/fermentor-design

Hui, Y. H., Meunier-Goddik, L., Josephsen, J. Nip, W-K., Stanfield, P. S., and Toldra, F. (eds). 2006. *Handbook of Food and Beverage Fermentation Technology (Food Science and Technology, Vol. 134).* Marcel Dekker, New York.

Kessler, H. G. 2002. *Food and Bio Process Engineering* 5th ed. Verlag A Kessler Publishing House, Munchen, Germany.

Lee, B. H. 2014. *Fundamentals of Food Biotechnology* 2nd ed. John Wiley & Sons, Ltd., New York.

Maryam, B. M., Datsugwai, M. S. S. and Shehu, I. 2017. The role of biotechnology in food production and processing. *Engineering and Applied Sciences* 2(6): 113–124.

Matz, S. A. 1972. *Bakery Technology and Engineering.* The AVI Publishing Co., Inc., Westport, Connecticut, 165–236.

Patra, J. K., Das, G., Paramithiotis, S.and Shin, H.-S. 2016. Kimchi and other widely consumed traditional fermented foods of Korea: A review. *Frontiers in Microbiology* 7: 1493. https://doi.org/10.3389/ fmicb.2016.01493.

Stanbury, P. F. and Whitaker, A. 1984. *Principles of Fermentation Technology.* Pergamon Press, Oxford.

Whitaker, J. R. 1972. *Principles of Enzymology for the Food Sciences.* Marcel Dekker, New York, 151–253, 287–348.

26

Nanomaterial Applications in Food Processing

Nanotechnology involves working with matter at nanoscale for design, characterization, production, and use of new products. The word "nano" means "dwarf" in Greek. Nanotechnology includes the domains of nanoscience, engineering, and technology. Imaging, measuring, modeling, and material handling operations at nano scale are the important activities. At nano scale, materials exhibit properties that are very different from those at micro and macro scales. For example, as particles get smaller, the surface area and its reactivity with the environment increases many times. The optical properties such as color may also change as the reflection and absorption of light varies with the particle size. The physical, magnetic and electric properties of materials also change dramatically at such small sizes, making them novel.

26.1 Nanomaterials

Nanomaterials may include nanoparticles, nanotubes, fullerenes, nano-fibers, nano-whiskers, and nanosheets. Nanoparticles, ubiquitous naturally in the form of smoke and dust, are defined as discrete entities having each of the three dimensions of 100 nm or less (1 nm = 10^{-9} m). In industry, the production of a nanomaterial involves preparing the material with one or more dimension in the range of 1 to 100 nm. For example, one kg of copper with 50 micron particle sizes will have a total surface area of 15 m², this will be 75000 m² if the particle size is 5 nm. Table 26.1 provides a comparison of dimensions of some common objects.

Nanomaterials are known to have been used for millennia. Indian families have been producing nanocarbon through vapor deposition on metallic or earthen surfaces for subsequent use as "kajal" for eye health. The Chinese used gold nanoparticles to color ceramic porcelains. Use in Roman culture has also been reported. However, in was in 1974 that the word "nanotechnology" was used for the first time and since then considerable work has been done in the application of this technology. The 1996 Nobel Prize in Chemistry went to Harold Kroto, Robert Curl, and Richard Smalley for their discovery of Carbon 60, also known as Buckyball or fullerene. Two-dimensional graphene characterization earned Andre Geim and Konstantin Novoselov the 2010 Nobel Prize in Physics.

There are three types of nano materials/devices. The first category consists of isolated, substrate-supported or embedded nano-particles, nano-wires, or nano-films with reduced dimensions and/ or dimensionality. The second category includes a bulk material with nanometer-sized surface. The third category is of bulk solids with a nanometer-scale internal structure.

26.2 Manufacture of Nanomaterials

There are two approaches for nanomaterial manufacturing: top-down and bottom-up. Smaller components arrange themselves into more complex assemblies in the bottom-up approach while the top-down approach seeks to create nanoscale materials and devices by using larger, externally controlled components.

As the name suggests, in the first method, physical or chemical means are used to break down the material to nanometer-size particles. Attrition is a common method. Some examples are dry milling of wheat and homogenization of milk to reduce the size of fat globules, which change the reactivity and characteristics of the product. Lasers are also used for nanomaterial manufacture.

Bottom-up methods of nanomaterial manufacture include crystallization, solvent extraction/ evaporation, layer-by-layer deposition, microbial synthesis, self-assembly, and biomass reactions. In this approach, the molecules can be arranged in a phased manner for obtaining specific desirable characteristics and more complex molecular structures can be obtained. An example of nanomaterial caused by self-assembly is casein micelle. Colloidal dispersion is another bottom-up approach. Nanolithography is a hybrid approach in which the growth of thin films and etching are both carried out.

The methods of nanoparticle manufacturing can also be classified on the basis of states of the materials during manufacture.

- Solid phase processes such as grinding, milling, and alloying;
- Liquid phase methods involving preparing colloids and aerosols from solvents;
- Gas phase processes, such as high temperature evaporation, vapor deposition, flame pyrolysis, and plasma synthesis;
- Sol-gel technique.

26.3 Applications

Potential applications of nanotechnology are immense and pervade all domains. Some of the commercial products already on the market include coatings (self-cleaning windows and stain-proof clothing), catalysts (Envirox™ cerium oxide), nano-remediation (SAMMS technology to remove mercury), photographic paper, filters (nano-fibers), toothpaste (to remineralize teeth), paint (improved adhesion and anti-fungal qualities/anti-graffiti), clothes (non-staining and anti-radiation),

TABLE 26.1

Dimensions of some common materials

Material	Dimension (nm)
Table salt	300,000
Human hair	60,000
Red blood cell	7,000
Virus	50–100
DNA helix	2
Buckyball	1

Source: As per Buzea et al. (2007).

batteries (phosphate nanocrystal technology), and several cleaning products. Some food applications are as below.

Food Ingredients and Controlled Release

- Nano-foods use nanomaterials and nanotechnological techniques to cultivate, produce, process, and package them. The impact of nanotechnologies in the food industry is already visible from primary production at the farm level to processing. Engineered nanomaterials (ENMs) could be inorganic, surface functionalized, and organic. Nanofoods are aimed at obtaining health-promoting additives, longer shelf life, or new flavors.

- Nanostructured food ingredients and food nano-sensing are two main categories of the applications in food sector. Nanostructured food ingredients encompass a wide range of interventions in food processing, including packaging. These have been found to improve the shelf life of different foods and to improve the taste and texture of foods.

- The nanostructured ingredients have also many important applications as antimicrobial agents. Nano-silver (Ag) particles are antimicrobial in nature and are used in many products/proposed products/research topics. Silver nanoparticles easily penetrate cells of microorganisms and have been observed to sterilize more than 650 types of bacteria. Zinc oxide (ZnO) is another popular nanomaterial used for light activated sterilization of surfaces.

- Nanotechnology has also been applied for emulsions, biopolymer matrices, and association colloids. Nano-emulsions have been developed for applications in the food industry, which help in giving excellent rheological and textural properties to food. Nano-emulsions in food products is a case in point where reduction in fat content can be achieved without compromising creaminess. Size reduction of fat globules in an emulsion stabilizes it against break down and separation. Thus, nano-emulsification will reduce the need for stabilizers.

- Nano-carriers are used for controlled and targeted release of a substance in food products without disturbing their basic morphology. Nanoencapsulation provides protection of encapsulated compounds, such as vitamins, antioxidants, proteins, and lipids, until their use.

- Reduction in the particle size of bio-actives can help in improving their solubility and bioavailability and for efficient delivery of active ingredients. Thus, nanotechnology has found good acceptance in developing nutraceuticals and functional foods.

- Application of nanotechnology has led the development of 5 nm edible coatings. The edible nanocoatings also act as a carrier for imparting specific colors, flavors, or specific antioxidants, enzymes, etc.

- Nanomaterials have also been used as anti-caking agents.

Food Packaging

- Nano polymers are now replacing conventional food packaging materials.

- Nanocomposites have found excellent applications in food packaging. Nano-particulates (100 nm or less) are incorporated into plastics for improving their properties such as mechanical strength, barrier properties, heat resistance, durability, etc.

Sensing

- Nano-sensors are being used to detect the presence of contaminants, mycotoxins, and microorganisms in food.

- Incorporation of nano-sensors in packaging materials has permitted the development of smart/intelligent packaging systems that enable the consumers in monitoring the packed food in real time.

- Biosensors and lab-on-a-chip systems have been developed through bio-nanotechnology research, which have found wide acceptability.

It can be appreciated that nanotechnology has enormous potential for its use in food sector. When combined with biotechnology, the potential grows beyond comprehension. There are, however, huge gaps in knowledge to offer better and safe products. The bio-nano convergence is presently seen with great suspicion. The constraints relate to the acts of detection and elimination. The problem is further compounded by the absence of robust regulatory framework. Consumers need to be assured of health benefits on a sustainable basis without any environmental hazards.

Check Your Understanding

1. Explain the different methods of preparation of nanomaterials.

2. What are the applications on nanomaterials in food processing and packaging?

BIBLIOGRAPHY

Berekaa, M. M. 2015. Nanotechnology in food industry: Advances in food processing, packaging and food safety: A review. *International Journal of Current Microbiology and App Sciences* 4(5):345–357.

Bradley, E. L., Castle, L. and Chaudhry Q. 2011. Applications of nanomaterials in food packaging with a consideration of opportunities for developing countries. *Trends Food Science and Technology* 22:603–610. https://doi.org/10.1016/j.tifs.2011.01.002.

Buzea, C., Pacheco, I. I.and Robbie, K. 2007. Nanomaterials and nanoparticles: Sources and toxicity. *Biointerphases* 2(4):MR17 –71. doi: 10.1116/1.2815690

Cushen, M., Kerry, J. P., Morris, M. A., Cruz-Romero, M. C.and Cummins, E. 2012. Nanotechnologies in the food industry: Recent developments, risks and regulation. *Trends in Food Science and Technology* 24:30–46. https://doi.org/10.1016/j.tifs.2011.10.006.

Hayes, A. W. and Sahu, S. C. 2017. Nanotechnology in the food industry: A short review. *Food Safety Magazine,* Feb-March, 2017.

Llorens, A., Lloret, E., Picouet, P.A., Trbojevich, R. and Fernandez, A. 2012. Metallic-based micro- and nanocomposites in food contact materials and active food packaging. *Trends in Food Science and Technology* 24:19–29.

Moraru, C. I., Panchapakesan, C P., Huang, Q., Takhistov, P. and Liu S. 2003. Nanotechnology: A new frontier in food science. *Food Technology* 57(12):24–29.

Singh, T., Shukla, S., Kumar, P., Wahla, V., Bajpai, V. K. and Rather, I. A. 2017. Application of nanotechnology in food science: Perception and overview. *Frontiers in Microbiology* 8. https://doi.org/10.3389/fmicb.2017.01501.

B5

Food Packaging and Material Handling

27

Filling and Packaging

The shelf life of a food product is extended or shortened by the way it is packaged. Proper packaging helps to reduce/eliminate the spoilage of food during storage caused by the mechanical forces (impact, vibration, compression, or abrasion); physical and chemical parameters (moisture vapor, oxygen, UV light, temperature changes, etc.); contamination on account of soil, insects, and microorganisms; as well as other factors such as pilferage, tampering, and adulteration. In addition, good packaging helps in safe and efficient transportation and in enticing the customer to purchase the product (specifically true for retail packages). The packaging also helps in giving the customer information about the product (sometimes a legal requirement). Basically, it ensures safe delivery to the end user in a sound condition.

27.1 Basics of Food Packaging

27.1.1 Basic Requirement of a Package

Faulty packaging can completely spoil the efforts and investments in production of the food. Thus, the packaging material and device should be appropriate. The basic requirements of the packaging material are as follows.

- It should retain the food in appropriate form and it should have functional size and shape for retailing and distribution.
- It should protect the product, and at the same time it should not contaminate it.
- The packaging material should not fail (break) during the packaging and distribution.
- It should be cost effective and should have aesthetic value.
- It should be suitable for easy disposal or reuse.

With the significant growth of the plastic industry and use in food and other packaging, it has also become important to have a quality waste management plan. Thus, the five Rs of waste management have become significantly important. These are renew, reduce, reuse, recycle, and recover. Often, these five Rs are also mentioned as refuse, reduce, reuse, repurpose, and finally, recycle.

Further, the selection of proper packaging system is also important. The major parameters for selecting a packaging system are given in Table 27.1.

27.1.2 Classification of Packages

The different levels of packaging can be classified as follows.

- Primary package. The food or beverage is in direct contact with the packaging material, e.g., bottle and cap, flexible milk pouches, chocolate wrappers, metallic cans, etc.
- Secondary or transit package. This is used to contain and assemble primary packs, e.g., a shrink-wrapped assembly of water or beverage bottles, corrugated fiber board case carrying retail oil packages
- Distribution package. These are aimed for efficient handling and protecting the packages and product during distribution, e.g., pallet, roll cage. These are also known as shipper or tertiary packages.
- Unit load: Often a group of distribution packages are collated into a single unit for convenience in handling, storage, and transportation.

Packages can alternatively be classified as consumer packages and industrial packages. The consumer units contain and protect preferred quantities for retail sale, permit home storage, and advertise the food. The industrial packages are intended for delivering goods from manufacturer to manufacturer.

The packaging can also be classified as primary, secondary, and tertiary packaging depending on the level and functions of the packaging material. Packaging can also be standard or custom made depending on the product and other requirements.

27.1.3 Packaging Materials

Different types of materials such as glass, metal, plastics, films, textile, paper and carton, wood, etc. are used for food packaging (Table 27.2). The different packaging materials used and their important features in relation to food packaging are discussed below and summarized in Table 27.2.

Plastics, the most common packaging material, are gradually replacing glass and metals. The major plastics used for food packaging are the polyolefins, styrene polymers (polystyrene or expanded polystyrene) and carbonate group (polycarbonate). Polyolefins include polyethylene, polypropylene, polyvinyl group (PVC), and condensation polymers (PET, Nylon-6 or polyamide).

Each plastics group has multiple subgroups. For example, the polythene can be low-density polythene (LDPE), linear low-density polyethylene (LLDPE), very linear low-density polyethylene (VLLDPE), high-density polythene (HDPE), high molecular high-density polyethylene (HMHDPE), and copolymers like primacor (EAA) or surlyn. Polypropylene includes cast polypropylene (CPP), biaxially oriented polypropylene (BOPP), and tubular quenched poly propylene (TQPP).

DOI: 10.1201/9781003285076-34

TABLE 27.1

Major parameters for selecting a packaging system

Product parameters
- Composition and quality of product
- Required shelf life
- Method of preparation

Packaging requirements
- Pack size
- Packaging system, automation requirements
- Display type, printing requirements, image of product and packing, aesthetic value, etc.
- Amenability to fit into the existing and/or future systems
- Legislation requirements (integrated with the product)
- Cost effectiveness

Distribution conditions
- Distribution type
- Protection of the product during storage, distribution, and retail sale

Selection of packaging materials depends on the nature of the commodity, the storage environment, and the duration of storage. In addition, transportation and conveying methods also decide the type of package material.

27.1.4 Manufacture of Packaging Enclosures

Metal Cans

Steel and aluminum are used for metal containers. It is the low carbon mild steel, produced as black plate initially, that is used for packaging. The black plate is converted into tin plate or tin-free steel (TFS) and then the containers as well as closures are manufactured. The black plate is electrolytically coated with a thin layer of tin (about 0.38 μm thick). TFS, also known as electrolytic chromium/chrome oxide coated steel (ECCS), is obtained by electrolytically coating both sides of the black plate with a thin layer of chrome/chrome oxide. For food packaging applications, an organic coating is applied to

TABLE 27.2

Different food packaging materials and their salient features

Materials	Features
Glass	• Rigid • Inert with respect to foods • Impermeable to gases and vapors • Transparent to light and may be colored • Can be easily returned and reused • Surface coating helps to increase abrasion resistance • Mouth can be made up of different sizes for easy filling, such as containers with wide mouth fitted with easy-open-caps • Containers are fragile and, therefore, breakable • Needs a separate closure.
Metals as tin plate (tin coated cans using 99.75% mild steel with low carbon) and aluminum	• Rigid • Good tensile strength • Lacquered internally to prevent corrosion • Barrier to light, moisture and gases • Strong enough to withstand the processing and handling stresses as well as external environmental conditions • Usually resist chemical action of the product; however, the metal may react with product if not properly lacquered or coated, causing metal contamination. • Require closures, seams and crimps to form packs; however, offers convenient opening and safe product removal • The packaging material can be recycled, thus, obviating disposal problem
Plastic containers (rigid and flexible films made up of PE, PP, polyolefins and polyvinyl chloride, and different combinations/laminations, etc.)	• Low density • Wide range of tensile and tear strength, barrier, and optical properties (as obtained from different types of plastics) • Can also be transparent • Permit use over a wide range of temperatures (with proper selection of plastics) • Flexible, crease formation can be assured if required • Usually of low stiffness • Selective gas and vapor permeabilities
Paper and paper board	• Low-density • Good stiffness and can be creased and folded • Can be glued • Grease resistance is possible • Low absorption resistance to moisture vapor and liquids • Printing is not expensive. • Low tear resistance • This material is not brittle but it has low tensile strength as compared to metal. • Needs coatings or laminations for resistance to light penetration • Wrappings, coatings or laminations impart adequate barriers to liquids, gases and vapors.

the inside surfaces to prevent any reaction between the product and the metal.

The metal cans can be shallow to tall and the cross section may be circular and round. The food and drink cans are either two-piece or three-piece containers. A two-piece can is obtained by making a cylinder from a disc of metal and there is no seam. Three-piece cans are obtained when a piece of flat metal is rolled into a cylindrical body and a longitudinal seam is usually formed by welding with the two ends. The open end is used to fill the product after which the can is closed, usually by double seaming method.

Other shapes are also available, though not very common for the food industries. Closure systems are also different as per different requirements of mode of operation. Some common examples are full aperture easy-open ends (FAEOs) and screw top closures along with wads or sealant material.

Plastic Materials

The plastics used in food packaging applications are in the form of either rigid plastic containers, such as tubs, trays, bottles, jars, pots, flexible film pouches, bags, sachets, and heat-sealable flexible lidding materials. Plastic films are also used, such as cling, stretch and shrink wrapping, plastic laminations for paperboard, expanded or foamed plastic, plastic bands to provide external tamper evidence, plastic lids and caps, diaphragms on plastic and glass jars, pouring and dispensing devices, etc. Fresh produce trays, drums, crates, tote bins, and sacks are some other forms of plastics used in food industries. By definition, films are less than 100 μm thick. Semi-rigid packages are produced from plastic sheets of up to 200 μm thickness.

Plastic films and sheets are manufactured by forcing the molten plastic through a narrow slot or die in the extruder. The rigid packages, such as bottles and closures, are manufactured by forcing the molten plastic into shape using a precisely machined mold. Profile extrusion involves the forcing of the molten polymer coming out of the extruder through a shaped orifice in the die; this method is used for manufacture of plastic tubes.

The films are made either by sheet extrusion or blown film extrusion. Sheet extrusion uses "T-shape" or "coat-hanger" dies with a slot orifice. The film obtained by this process is referred to as "cast film." In the blown film/ sheet method, the polymer is extruded through a circular die into a closed circular bubble and then the bubble is expanded with air. The latter is more common for the preparation of PE and other films. This method can produce very wide sheets. A 2 m diameter bubble, when slit, will make a 6 m sheet. Cast films cannot be prepared of such width.

Plastic containers such as bottles, jars, etc. are prepared by *blow molding* (blow forming) or *injection molding* process. The blow molding can be of three types, namely, extrusion blow molding, injection blow molding, and stretch blow molding.

The different methods to improve the properties and performance of plastics are as follows.

Coextrusion. It is the process of combining two or more layers of different plastics together at the point of extrusion.

Lamination. It is the process of joining two or more layers of plastics together using adhesives.

Blending. It involves mixing different plastic granules before extrusion.

Coating. Coating can be done by several methods, such as vacuum deposition, deposition from either solvent or aqueous mixtures, and extrusion.

Glass Containers

The molten glass is gravity fed through spouts to the bottle forming machines. The glass containers are made by two processes, namely, "blow-and-blow" and "press-and-blow." Both processes need two molds: a blank mold that forms an initial shape, or "parison," and a blow mold in which the final shape is produced. These processes differ slightly from each other only in the way the parison is produced and the selection of the process depends on the container to be prepared.

Paper and Paper Board

Pulping is the process to derive virgin or primary fiber directly from wood. The lignin, the natural adhesive that binds cellulose fiber together, are not water soluble and so cannot be easily removed by simple water washing; hence, grinding wood or chemical treatment is used. These are known as mechanical pulping and chemical pulping. In mechanical pulping, the extraction of cellulose from wood is done by mechanically abrading or cutting wood.

In the chemical method, chemicals are used to dissolve away the natural lignin binders (noncellulose components) in wood, leaving the fiber bundles intact and undamaged. Several chemical processes can be used; the two most common are based on alkali sulphate and acid sulphites. The strongest chemical pulps are made by alkali sulphate extraction. This is also commonly known as the kraft ("Kraft" is the German word for strength) process. It is preferred for softwoods because it is able to emulsify and remove resinous components.

Subsequent stages are refining and bleaching of pulp. Cellulose fiber bundles separated from the wood mass are refined (beaten) to release small fibre strands, or fibrils. The amount of refining has to be adjusted depending on the required strength and other characteristics of the final product. The chemically separated fiber, usually brown, can be bleached and all traces of non-cellulosic material could be removed. Pulps are whitened by bleaching with chlorine containing compounds or hydrogen peroxide, though the chemical bleaching reduces the strength of final paper.

The paper and paper boards are then prepared from the pulps by different unit operations, such as stock preparation, sheet formation, pressing, drying, coating, reel up, and finishing. Paper and paperboards can be treated using one or more of the following processes for additional barrier and functional performance.

1. Treatments during manufacture;
2. Lamination;
3. Plastic extrusion coating and laminating;
4. Printing and varnishing;
5. Post printing roller varnishing/coating/laminating.

Subsequently, the different packages, such as paper bags, sachets, pouches, multiwall paper sacks, folding cartons, rigid boxes, drums, tubes, composite containers, corrugated fiber board packaging, are prepared with the help of different machines. Paper-based flexible packages, involving paper with plastics, are associated with FFS machinery.

27.2 Equipment for Filling and Metering Food in Packages

In food industries, it is important that proper quantity of the prepared food is filled in the packages, which may be made up of plastic, glass, metal, or other materials. Different types of metering and filling devices are available for the purpose.

27.2.1 Filling of Liquids and Pastes

The liquid and pasty food products can be filled either by gravity (Figure 27.1) or by a metering system (Figure 27.2). The first one is used mostly for glass bottles, which are not distorted during the filling. The rigid bottles are filled up to a preset filling height. In the second type, the metering mechanism helps filling a definite volume; hence, this is also called volumetric filling. As the pre-metered amount of liquid is filled, the shape and the distortion of the package do not matter. The commercially available equipment are described below.

Gravity Filler

Construction and Operation

- It basically consists of a circular filler bowl under vacuum.
- At the bottom of the bowl, a number of filler valves are fitted, which are held by a spring. Vacuum pipes connected to the filler valves rise above the level of the milk.
- A pedestal raises the empty bottles such that the rim of the neck of the bottle is pressed against the rubber of the filler valve.
- When the bottles are further raised, it is connected to vacuum. The air and foam are sucked out from the bottles through the hollow central vacuum tube. The valve rubber is then lifted compressing the spring, which creates an annular orifice in the valve that makes the liquid to flow down to the bottles.
- As the bottle drops away from the valve, the spring returns the rubber seal and the milk flow is stopped.

Salient Features

- The filler valves do not open unless a bottle is in proper position. Even if the bottle is put in an inclined position or if the rim of the bottle is broken, then a vacuum will not be created and the liquid will not flow. Even cracked bottles will not be filled.
- The use of the vacuum speeds the rate of filling and also prevents milk from dripping from filler valves when a bottle is not under the valve.

Applications

This device is used for filling rigid bottles made up of glass or plastic for milk and other liquid foods.

Sachet Form-Fill-Seal (FFS) System

Construction and Operation

- The important features of the machine are the constant level tank with liquid feeding device, film feeding

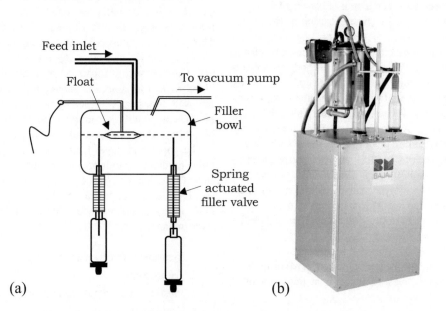

(a)   (b)

FIGURE 27.1   (a) Schematic diagram of a gravity filling system (after Ahmad, 1999 with permission), (b) A gravity filler (Courtesy: M/s Bajaj Processpack Limited, Noida)

FIGURE 27.2 An automatic can filling machine (Courtesy: M/s Bajaj Processpack Limited, Noida)

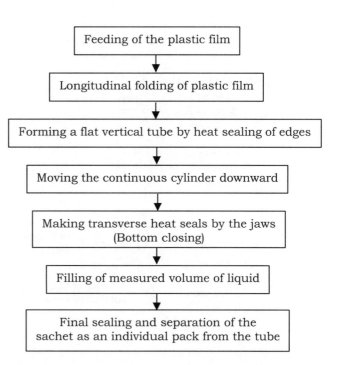

Basic operations in FFS packaging system

device, horizontal seal, vertical seal, paper looping, and forwarding and filling systems (Figure 27.3).

- This system uses filling by metering, in which the filling machine consists of a constant level tank. A float valve maintains the constant head of liquid.

- In the equipment, the machine feeds the film over a cone-shaped tube called a forming tube and the plastic films are formed into flat vertical tubes by vertical seaming (by two vertical sealing jaws), where the two edges of the film come together. This vertical tube or continuous cylinder moves downward.

- There are two transverse jaws, which move to make a seal that forms the bottom of the package. (Actually the transverse seal of the previous pouch closes the bottom of the tube.) After the bottom is formed, the specified amount of liquid is allowed to flow into the system.

- The liquid milk flow is controlled by a valve operated by an electronically controlled sensor device. The valve opens for a pre-set time after getting the signal from the sensor.

- After the liquid has been poured into the package, the package moves down by the length of a pouch. The transverse jaws again heat seal the top edge and cut it from the main cylinder. The top edge of this package also forms the bottom edge of the next package, which is filled in the succeeding step.

- Often the filler consists of two filling heads with a single chamber and each can operate independently and can fill different sizes of pouch at the same time.

- The packing film drawing mechanism is by clutch and brake. The packing film unwinding is done by dancing roller, proximity sensors, and unwinding motor. There is no tension on the packing film while unwinding.

Salient Features

- As the level of the liquid in the tank is kept constant, the flow rate also remains constant. Thus, the flow of liquid is directly proportional to the time for which the valve is kept open.

- The width of the bag is set by the forming tube's design. The distance that the tube passes in one step determines the length. To change the bag width, the forming tube has to be changed along with using the reel with required film width.

- Sealing is either a lap seal or a fin seal. In the lap seal, the two overlapping edges of the film are sealed together. In the fin seal, the inside surfaces of both the edges are sealed together. The type of film being sealed influences the selection of the type of seal. The compatibility of sealing of the outer and inner layers in a multi-layered film must be checked.

- The total weight that can be dispensed in a calibrated time is a function of specific gravity of liquid. There may be variation in the bulk density and flow characteristics in different batches, and, hence, the operating parameters have to be adjusted depending on the density.

- The liquid filling range for a machine is specified. For example, it can usually be from 1 ml to 10 ml or 10 ml to 50 ml or from 50 ml to 100 ml or so. Milk or oil are filled in volumes even up to 2 liters per unit.

- The production speed for small sachets can be approx. 50–60 per minute per track.

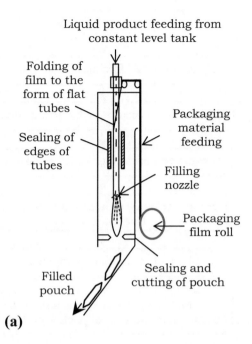

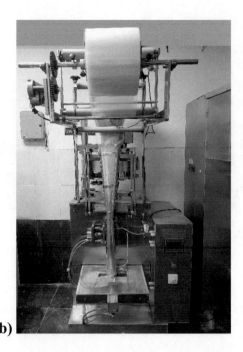

(a)   **(b)**

FIGURE 27.3 (a) Schematic diagram of a form-fill-seal (FFS) system of packaging (after Ahmad, 1999 with permission), (b) a FFS packaging machine

Applications

This is commonly applied for filling sachets and widely used in packaging of pasteurized milk, butter milk, RTS, etc.

Piston Type Filler

Construction and Operation

- This also operates on filling by metering.
- The device consists of a filler bowl, on the lower side of which there are a number of small buffer chambers for the liquid (Figure 27.4). Each chamber is connected through an operating valve (say valve 1) with the filler bowl. One side of the buffer chamber is connected to a piston cylinder device through another valve (say valve 2). A third valve (valve 3) is located at the bottom of the buffer chamber that helps filling of the bottles placed below.
- The steps of operation are as follows.
  - When the piston moves outward, it opens the valves 1 and 2 (with valve 3 closed) and the liquid flows down from the filler bowl to the buffer chamber.
  - When the piston moves inward, the compression opens the valve 2 and 3 and closes valve 1. Thus, the liquid that is in the buffer chamber flows down to the bottle placed below.

Salient Features

The displacement of the piston determines the amount of liquid to be filled in the bottle.

Applications

This filling system can be used for filling thin liquids as well as more viscous products. Figure 27.5 shows a commercial piston type filler.

Metering Cup Type Filling System

Construction and Operation

- The system consists of some filler tubes, which have openings at a certain height. If the tube is immersed in a liquid, it is filled in the tube through this opening. The opening also allows overflow of the excess liquid, and, thus, when the tube is taken out of liquid, it maintains a specified height.
- There is a rod (called steering rod) at the axis of the tube. The lower part of the rod forms a valve/stopper at the bottom of the tube and the top portion is connected with a spring system.
- When the top end is pressed, the valve at the bottom opens such that any liquid present in the tube flows down.
- The steps of operation are as follows (Figure 27.6).
  - The calibrated tube is lowered in the filler bowl until the filler tube dips into the liquid. The tube is, thus, filled.
  - Then the bottle is placed tightly below the tube and the bottle moves on an incline so that the tube is also raised upward by the bottle. The bottle seals the cup and excess liquid in the tube flows back into the filler bowl. At this stage the liquid in the tube maintains the specified height.

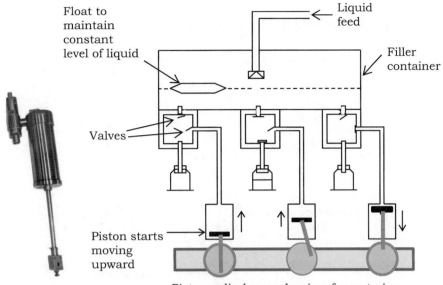

Piston-cylinder mechanism for metering

FIGURE 27.4 Operation of a piston type filler (after Ahmad, 1999 with permission)

FIGURE 27.5 Piston filler (Courtesy: M/s Bajaj Processpack Limited, Noida)

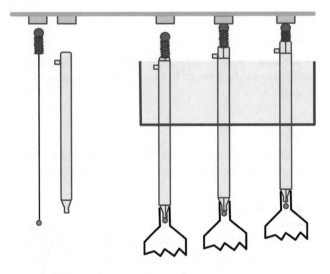

FIGURE 27.6 Schematic of filling process in a metering cup filling system (Check that when the bottle presses the filler tube, the valves open at the bottom of filler tube)

- Further raising of the tube presses the rod to a stopper at the top and it opens the discharge valve at the lower end of the rod. It causes the metered amount of liquid to flow into the bottle.
- Then, as the metering cup and the bottle are lowered, the discharge valve is closed.
- The calibrated cup is again filled with liquid while the filled bottle is removed from the filler.

Salient Features

- The system gives a high accuracy of filling.

Applications

The system is used for low viscous liquid filling in rigid bottles.

Paste Filler

Construction and Operation

- A conical bowl holds the viscous material or paste, and there is a rotor (screw operated by a motor) at the bottom to push the product from the bowl into the package (Figure 27.7).
- An electronic device is used to control the revolution time or metering time of the rotor.
- The filler bowl level is controlled so as to maintain uniformity in filling. A pump is used to fill the materials in the filler bowl. Also there may be an auxiliary mixing device to properly mix the contents in the bowl so that the density remains the same at all levels in the bowl.

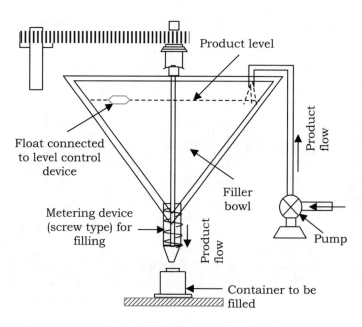

FIGURE 27.7   Schematic diagram of a paste filler (after Ahmad, 1999 with permission)

Salient Features

In the case there are some solid particles or the paste is highly viscous, care should be taken so that the product is not damaged.

Applications

This filling system can be used for liquids of low or high viscosity and even for pastes, which may contain solid particles.

FFS System for Pastes

The pastes can also be filled by FFS systems in which volumetric injection type filling systems are used. The motion of the roll is mechanical intermittent type as for the liquids.

27.2.2 Filling of Dry Products

Dry products, such as powders and granules, need to be metered and packed. Normally these are metered for a specific volume and filled. A dry product that is sticky, because of high moisture content or because it is thermoplastic at too high temperature, is not suitable for continuous metering.

The common metering and filling devices used for powders and grains are discussed below and some schematic diagrams are given in Figure 27.8.

Screw Metering Device

- A screw, placed below the hopper containing the product, moves the material and fills the desired quantity in the containers placed below.
- The screw can be either horizontal or vertical axis type.

- The screw should be properly filled for which the pitch and the diameter of the thread should be more near the hopper and should gradually reduce towards the outlet. Revolving spiral bands can also be installed to loosen the dry product before moving into the screw threads.
- The revolution of the screw controls the product flow rate. Normally, the mass flow rate is more accurate and uniform, if the diameter of the screw is less and the number of revolutions is more than vice versa.
- It is also important that the hopper should be always kept filled to a specified level to have the product exit at a constant pressure.

Bucket Wheels

- The device has a rotating wheel, which meters the material from the hopper to the container placed below.
- The system is relatively inaccurate and, hence, is suitable only for metering dry products into large packs.
- As the content of a whole bucket of the rotating wheel is emptied at once, the metering cannot be said to be strictly a continuous one.
- In this case also the hopper should be always kept filled to a specified level to have a constant pressure.

Vibrating Conveyor Chutes and Conveyor Belts

- Vibrating conveyors can be used for continuous and accurate metering.

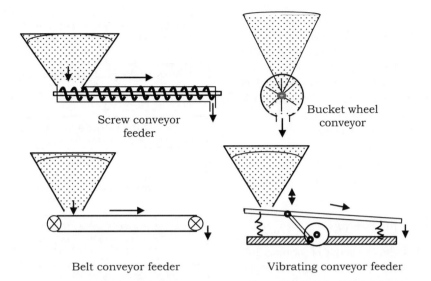

FIGURE 27.8 Different filling mechanisms for powders and granules (after Ahmad, 1999 with permission)

- The flow rates can be easily adjusted by regulating the frequency of the vibrations or the speed of the conveyor belt. In addition, adjusting the dimensions of the orifice outlet and installing side walls can also help adjust the product flow.

FFS System for Granules and Powders

- The form-fill-seal machines for powders and granules help filling the desired volume of the product in pouches, after which the pouch is sealed in the manner it is done for liquids.
- A multi-head scale type filling machine can be used for filling the pouches. Alternatively, any other device, such as an auger or volumetric filler,, can be used. The formation of the tube and seals is as discussed earlier for liquid FFS systems. As soon as the bag is ready, an electronic synchronized system is used to drop the product automatically.
- After the product has been filled, the bag is sealed at the top and cut away. The top seal of the bag being cut away becomes the bottom of the next bag.
- There may be additional features and devices to print a date/batch code and/or make a tear notch for easy opening and/or make a hole to let it hang on a wall display.
- The FFS machine can also be horizontal type, which normally are intended to meet the requirement for heavier packages, even up to 40–50 kg. These should not be confused with other horizontal systems, also known as "flow wrappers."
- The vertical FFS machine takes the flat film from the reel and creates the bag by first vertical sealing to form the film tube and then horizontal top

and bottom sealing. However, in horizontal FFS machines, a preformed tubular film (with or without gussets) is wound onto a reel. The machine cuts and forms the bag's lower seal. The bag then moves horizontally through the machine, where it is filled and then sealed.

Some general points to be remembered for filling of dry and granular products are as follows.

- The products are filled by metering the volume. However, there may be deviations in weight as the bulk densities may differ.
- Such continuous metering devices cannot be useful for sticky products.
- High mechanical pressure should be avoided to prevent damage of product and also to prevent formation of lumps.

27.2.3 Aseptic Filling

The basic difference between the traditional filling system and aseptic filling is that in case of conventional in-bottle system, the unsterilized liquid food product is first filled in the bottle and then the filled bottles are put to sterilization. In *aseptic filling*, the sterilized product is filled in pre-sterilized packages. After filling, there is no heat treatment of the product. Thus, the filling and packaging of high-water activity products in FFS systems should be done in aseptic conditions. Aseptic systems are also available for rigid bottles and containers.

Some basis requirements in an aseptic filling system are as follows.

- The container and the method of closure need to be suitable for aseptic filling. The sealed containers must not let microorganisms enter into them during distribution and storage.
- The container surfaces that come in contact with the product must be sterilized after the container is formed and before it is being filled.
- The filling must be without contamination.
- The closure must be sterilized before applying it to the container and the environment where the sealing of the closure with the container is carried out should also be sterilized.

Aseptic Filling System for Pouches

- In case of pouches, the operations of forming, filling, and sealing need to be carried out in a sealed chamber where sterilized air is continuously supplied (Figure 27.9 a).
- The air in the system is also sterilized using a bacterial filter capable of removing 99.9% of particles of 0.3 μm diameter or more.
- The film is made to pass through a bath of H_2O_2 solution kept in a chamber. The usual time is about 20 s for 1-liter size pouches. The film feed rate for smaller pouches is lower and the sterilizing time is more.
- Mechanical scrapers placed at the outlet of the bath remove the surplus liquid from the surface of films.
- The film then moves through a chamber in which sterile air is kept at 45°C to further remove the peroxide. The final H_2O_2 concentration in the filled pouch should be lower than 1 ppm.
- Thereafter the film is moved forward and there is formation of the tube by vertical sealing followed by the formation of bottom of the pouch by transverse sealing, as discussed before.
- The filler for low viscosity products is supplied from a balance tank by gravity. For thick products, piston type filling device is connected to a nozzle. A pump may also be used for more viscous products.
- The filled pouch goes down as discussed earlier.

Aseptic Filling for Plastic Bottles

The process of filling of blow-molded bottles aseptically is carried out as follows (Figure 27.9 b).

1. First sterilizing a bottle and then filling and sealing aseptically.
2. Blowing the bottle in sterilized condition and then filling and sealing aseptically.
3. Blowing the bottle aseptically and filling and sealing consecutively (i.e., blowing, filling, and sealing operations are integrated).

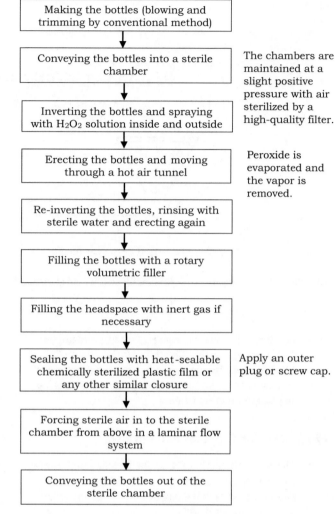

Sequence of operations in aseptic filling of plastic bottles

27.2.4 Hot Filling of Liquids

Fruit juices are packed when hot in PET bottles. The liquid cools in the bottles after capping, which causes a vacuum in the containers. It may cause deformation of the bottles. Therefore, the side wall of the containers are provided with vacuum panels. The circumference and base of the bottle are also designed with reinforcing ribs and grooves. The materials used for such applications have a faster rate of crystallization and/or higher glass transition temperature. The package is also heat-set so that it has a good temperature resistance.

Laminated flat type or stand-up pouches, used for hot fill packaging method of acidic fruit juices can be made from metallized polyester or polyethylene. The laminations provide barrier to oxygen transmission.

Combination of thermoplastic films with paperboard and aluminum foil are normally used for making the aseptic rigid packages. The aluminum foil provides barrier properties

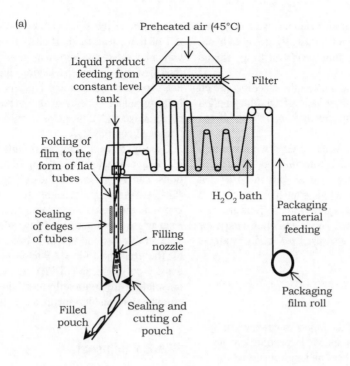

(a)

Preheated air (45°C)

Liquid product feeding from constant level tank

Filter

Folding of film to the form of flat tubes

H₂O₂ bath

Sealing of edges of tubes

Filling nozzle

Packaging material feeding

Filled pouch

Sealing and cutting of pouch

Packaging film roll

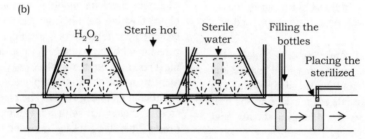

(b)

H₂O₂   Sterile hot   Sterile water   Filling the bottles

Placing the sterilized

FIGURE 27.9 Schematic diagram of (a) sterilization of packaging material in an aseptic filling system for flexible pouches, (b) sterilization and filling of plastic bottles in aseptic packaging system (After Ahmad, 1999 with permission)

against oxygen and light. The inner polyethylene helps sealing of the package and the outer paper gives the stiffness and forms the brick shape. It also helps with printing on it.

27.3 Shrink Packaging

27.3.1 Principle

This is a method in which a thin plastic film is wrapped around an object and heat is applied. The heat makes the loosely surrounding plastic film to shrink on the object and causes the film to conform to the shape of the object. The film forms a protective layer around the object against transmission of moisture, odor, etc.

The shrinking temperature is usually kept as 120–130°C; the film usually melts at about 175°C. The shrinkage temperature for fresh food is kept lower. Recent developments in the resins have allowed shrink wrapping machines to package products at as low as 65°C for just a few seconds, the tunnel temperature may be up to 90°C. After shrink wrapping, the food products need to be immediately cooled.

Earlier, PVC was the most commonly used shrink film material; however, low-density polyethylene and polyolefin (POF)

have almost replaced the PVC for many reasons, including the sustainability. Polypropylenes, vinylidene chloride, polyester, polystyrene, and coextruded films and laminations are also used for the purpose. Different varieties of shrink wrap films are available with different thicknesses, clarities, strengths, and shrink ratios.

The major characteristics on which the shrink film is rated are the shrink energy or the shrink force (the amount of force that the shrink film will exert on the product), shrink orientation, shrink percentages, resistance to puncture/tear, moisture vapor transmission rate, anti-fog properties, barrier to odor, etc.

In the shrink-packaging machine, three kinds of film wrapping methods can be used depending on the items, namely, open at both ends, open at one end, and fully enclosed. The first one is suitable for cylindrical items, such as wine bottles. The wrapping with one end open method is generally to pile the goods on the pallet and wrap them together with the pallet, which is mostly used as a transportation package. The fully enclosed systems are suitable for individual packaging of fruits and vegetables as well as for specific secondary packaging requirements.

Often the shrink film and stretch film are confused. Both can be made of polyethylene. However, when PE is used in stretch applications, it is prepared differently so that it can stretch around items. The stretch film usually stretches about 100%–300%, whereas shrink film can stretch to around 50%–75%.

Stretch films can also be classified as the machine stretch film (which is applied by a machine), hand stretch film (applied by hand), pre-stretched type of stretch film, UV stabilized stretch film, etc.

Cling wrap, cling film, Saran wrap, or Glad wrap is a thin film used primarily for sealing and protecting food in containers. The cling wrap is commonly sold on rolls in boxes that have a cutting edge. It maintains a tight stretch over some time without the use of adhesive. Though the stretch films are usually intended for pallets or boxed goods and cling wraps for food containers, the former is also used for food containers and the latter for cartoons.

27.3.2 Equipment

- As the method suggests, the object is wrapped in a plastic film and heat is applied. Depending on the mode of application of heat and movement of the product during heating, the machines can be classified as tunnel type, oven type, frame type, and gun type.
- In a tunnel device, the conveying device moves the materials into the packaging machine, where the material is wrapped and sealed with shrinkable film and then the materials are moved into the heat shrinkable channel, so that the film shrinks and tightly wraps the goods.
- The tunnels are usually electrically heated. An automatic temperature adjustment device and a forced air circulation system are incorporated to ensure even flow of hot air in the packaging tunnel (and thus to obtain a uniform shrink). Often there is a speed-regulating device for the conveyor for controlling the heat shrinkage time.
- The automatic machines have a wrapping machine and heat shrinkage channel in sequence. If the packaging machine is not an integrated one, then the product is first wrapped and then placed on the conveyor moving through the heat tunnel. It can be single packaging or multi-packaging.
- For batch systems, an electric or gas heat gun can also be used.

27.4 Modified Atmosphere Packaging

27.4.1 Principle

The internal atmosphere of a package is modified in case of modified atmosphere packaging for commodities like foods, drugs, etc., to improve the product shelf life. The normal atmosphere is 21% oxygen and 0.03% carbon dioxide. The oxygen helps the respiration of the (respiring) foods and enhances biochemical reactions such as lipid oxidation. In addition, oxygen also promotes the growth of aerobic spoilage microorganisms. All of these lead to short shelf lives of many commodities, which are otherwise known as perishable commodities. In the modified atmosphere packaging, the oxygen in the package head space is reduced and replaced with other gases, which helps to reduce or delay oxidation reactions and microbiological spoilage and, thus, helps in enhancing the shelf life of the packaged commodity.

Usually, oxygen is replaced with nitrogen or carbon dioxide. Argon or helium are also used for some specialty applications. The oxygen level in the package is usually maintained at 1%–3%. As the products are often respiring, the level of oxygen and other gases are maintained by using active techniques, such as gas flushing or passively by using "breathable" films.

Desirable gas and moisture barrier properties are essential for the films used for MAP applications. The films like low-density polyethylene (LDPE), polyvinyl chloride (PVC), and polyolefins are commonly used, though other materials and co-extrusions and laminations are also used.

27.4.2 Equipment

The equipment for modified atmosphere packaging is capable of flushing out air from the package, replacing it with a designed gas mixture or a different gas, and then sealing the package with the product. For packaging food with MAP, the most common gas that replaces oxygen is nitrogen. The equipment in which the gas is replaced and the package is sealed is called a *chamber sealer*. The chamber sealer draws the air from the chamber with a vacuum pump and puts nitrogen inside. Then the package is sealed while still in the chamber by using a heating element.

The different types of equipment used for modified atmosphere packaging are as follows.

MAP tray packaging equipment. The tray contains the food on which the film lid is applied in the packaging chamber. This can be used for fruits and vegetables, fish, meat, etc.

MAP vertical form fill and seal equipment. The product is automatically weighed and packaged with the help of a high-speed vertical form fill seal bagging system. The packages can be of different forms, such as fin seal, lap seal, side-gusset, or pillow bags. This is well suited for snack food packaging, where oxygen is replaced by nitrogen. The equipment is extremely flexible, can run at high speeds (can pack even 100 bags per minute), and can have rapid changeover (i.e., less down time). The metering of the product is done usually by weigh scales, volumetric cup fillers, or other special devices.

MAP fin seal flow wrap equipment. This is specifically a fin seal equipment, which is also called a flow wrapper. The machine provides fast horizontally wrapped packages with or without printed film. This is useful for single-item food products. A wide range of options for customizing the conveying and packaging needs of the product are available.

MAP band sealing equipment. This gives gas flushed bags and pouches that are hermetically sealed. As discussed earlier, nitrogen gas flush systems are common, though gas mixtures can be used.

Hand-held oxygen analyzers are available for monitoring the oxygen content in the packages. If the oxygen content in the package is not appropriate, then adjustments have to be made with the packaging system, i.e., process, machinery, or materials. Recent innovations have led to the use of interactive and digital components that alert the consumer in case there is a gas leak, a change in temperature, or the package is tampered with.

Other Forms of MAP

Desiccant packets are placed inside the food package, which absorb moisture from certain types of packaged foods. The desiccants do not have direct contact with the food and are kept in a permeable package. The common desiccant is silica gel and the desiccant package may be made up of materials like plastic, clay, etc. The desiccants are non-edible and usually toxic; hence, a label on the package warns the consumer not to ingest the contents of the packet.

Sometimes small package valves (usually made up of plastic) are provided on the package that allow the gases and smells to escape, but they do not allow the oxygen to enter inside. These are better suited for stand-up pouches and for products like coffee.

Vacuum packaging in which the air is taken out of the package with the help of a vacuum pump or the packaging done in a vacuum chamber can also be considered MAP.

27.5 Controlled Atmosphere Packaging

27.5.1 Principle

Controlled atmosphere packaging (CAP) or controlled atmosphere storage (CAS) involves flushing the desirable gas mixture into an impermeable package to replace the air for extending the shelf life of the commodity. The respiration and physiological changes in perishable foods can be retarded by modifying the storage environment as reducing but not eliminating the oxygen, increasing the carbon dioxide, and reducing the temperature. The optimum temperature, gas composition as well as the relative humidity differ for different fruits and vegetables. For most fruits and vegetables, the oxygen level is kept at about 2%–3% and carbon dioxide is kept at 3%–5%. The water vapor, and other trace gases are also often controlled. While maintaining the oxygen and carbon dioxide at recommended levels, the nitrogen gas is used as the make-up gas.

In the case of normal atmosphere storage of food products like meat, poultry, and fish, in addition to the growth in the number of microorganisms, several of the food properties as color and texture change due to the reactions with oxygen. One such example is the pigment changes and discoloration of red meat. However, under CAP, the carbon dioxide reacts with the animal tissue and produces acid carbonate. It reduces the pH of the meat and thus retards or inhibits the growth of spoilage microorganisms. Eggs are coated with oil or wax to minimize the loss of water and carbon dioxide; the use of CAP further increases the period of retention of freshness. The combination of oxygen and ethylene (C_2H_4) absorbers, together with CO_2 release agents, is adopted in the initial storage period of some commodities.

Hypobaric storage is another type of CA storage in which the storage chamber is kept under reduced pressure and high humidity. As the amount of air is decreased, the availability of oxygen is also reduced and the deteriorative reactions are retarded. The high humidity is intended to prevent the dehydration of the commodity. Such storage is used in warehouses as well as in trucks used for transportation of the commodities.

27.5.2 Equipment

- The equipment used for CAP has different devices for the generation and maintaining of CA, including oxygen removal, excess carbon dioxide removal, and addition of air to replace the oxygen consumed by respiration, removal of ethylene and addition of carbon dioxide. The storage chamber is made air impermeable.

- The most common oxygen scavenger is ferrous oxide; ascorbic acid, sulphites, some nylons, unsaturated hydrocarbons, enzymes, etc. are also other agents. $Ca(OH)_2$ reacts at sufficiently high humidity with the CO_2 to produce $CaCO_3$. The ethylene scavengers are mostly based on $KMnO_4$. Other ethylene scavengers are titanium dioxide catalyst, activated carbon, activated clays, and charcoal containing PdCl as a metal catalyst. CO_2 can be produced by using dry ice or the reaction of the exudates from the product with a mixture of sodium carbonate and citric acid inside the drip pad.

- The humidity for dry foods can be controlled by using silica gel, calcium oxide, or activated clay or minerals. For high water activity foods, moisture-absorbent pads, sheets and blankets and superabsorbent polymers, such as carboxymethyl cellulose (CMC), polyacrylate salts, and starch copolymers, zeolite, cellulose and their derivatives, are available.

- As the recommended storage conditions vary with the commodity, the selection of these devices for generating and maintaining CA depends on the type of commodity and the storage condition required.

- Recent innovations have led to faster establishment of desired atmosphere, less fluctuation in gas compositions, reduction of the ethylene in the chamber, and ability to change the gas composition as required during storage.

It is very important that the gas compositions and the temperature should be precisely maintained. Any lapse causes detrimental effects on the quality of food. Further, the CA stored produce spoils quickly when exposed to normal atmospheres in the supply chain. CA storage equipment is relatively sophisticated and bulky and thus, it limits its use in transportation and retail. The packages are also more sophisticated and costly.

27.6 Smart Packaging

There are several functions attributing to the smartness of a food packaging practice. These functions include

- retaining the package integrity;
- actively preventing food spoilage;
- enhancing product attributes (e.g., look, taste, flavor, aroma, etc.);
- responding actively to changes in product or package environment;
- communicating about the product, its history, or condition to the user;
- assisting with opening and indicating seal integrity; and
- confirming product authenticity.

Different types of *active packaging* methods that help in modifying the package environment and *intelligent packaging* methods, enabling sensing, monitoring and indicating about the food come under the category of smart packaging.

The active packaging methods can again be grouped under active scavenging systems or absorbers (e.g., oxygen absorber), active releasing systems or emitters (e.g., for CO_2 release), or controlled release packaging (e.g., for antimicrobials). These agents are usually put in a small sachet or label in the package itself. The active agent may also form a part of the package material or is integrated during the preparation of the package material. In the latter, the packaging material as a whole is the active agent. The different methods/devices used in active and intelligent packaging technologies are given in Table 27.3. A list of common active packaging agents used to modify the packaging environment is given in Annexure XIX.

Food packaging preserves and protects the food, extends the shelf life, and, thus, creates the possibility of extending the food distribution area. This is very important when developing and underdeveloped countries are struggling to establish a proper value chain for different commodities, especially for

perishables. Different packaging techniques and materials are available to suit the different types of food products based on their form, shape and size, bulk density/specific gravity, perishability etc. Improper packaging can cause more harm than benefit. Thus the food manufacturer should have proper knowledge of the behavior of the food being packaged, the packaging system, and the packaging material.

Check Your Understanding

1. What are the basic objectives of food packaging?
2. What are the basic requirements of a good food package?
3. Name important packaging materials used in the food industry. For each individual packaging material, give some examples of uses from your own experience.
4. Differentiate between
   (a) Primary packaging, secondary packaging, and tertiary packaging;
   (b) Shrink film, stretch film, and cling film;
   (c) Controlled atmosphere storage and modified atmosphere packaging;
   (d) Active MAP and passive MAP;
   (e) Active packaging and smart packaging.
5. Write short notes on
   - Filling milk by gravity;
   - FFS system;
   - Piston type filling system;
   - Metering cup filling system;
   - Filling of pasty products;
   - Filling of dry products;
   - Aseptic filling of plastic pouches;
   - Aseptic filling of blow molded bottles.
6. What are the different types of flexible films used in food packaging? State the important parameters of such films along with their merits, demerits, and applications.
7. Explain the merits and demerits of different types of packaging materials used for milk.

TABLE 27.3

Different methods/devices used in intelligent and active packaging technologies

Active packaging technologies	Intelligent packaging technologies
Oxygen scavenging	Time-temperature history
Anti-microbial	Microbial growth indicators
CO_2 scavenging	Light protection (photochromic)
CO_2 emitting	Physical shock indicators
Ethylene scavenging	Leakage, microbial spoilage
Odor and flavor absorbing/releasing	indicating
Moisture absorbing	
Heating/cooling	

BIBLIOGRAPHY

Ahmad, T. 1999. *Dairy Plant Engineering and Management*. 4th ed. Kitab Mahal, New Delhi.

Coles, R., McDowell, D. and Kirwan, M. J. (eds.) 2003. *Food Packaging Technology*. Wiley-Blackwell, NJ.

Crosby, N. T. 1981. *Food Packaging Materials: Aspects of Analysis and Immigration of Contaminants*. Elsevier Applied Science, London.

Dash, S. K. 2015. Modified atmosphere packaging of food. In: Alavi, S., Thomas, S., Sandeep, K. P., Kalarikkal, N., Varghese, J., Yaragalla, S. (eds.) *Polymers for Packaging Applications*. CRC Press, pp. 337–378.

Dash, S. K., Kar, A. and Gorrepati, K. 2013. Modified atmosphere packaging of minimally processed fruits and vegetables. *Trends in Post-harvest Technology* 1: 1–19.

John, P. J. 2008. *A Handbook on Food Packaging.* Daya Publishing House, New Delhi.

Mahadevia, M. and Gowramma, R. V. 2007. *Food Packaging Materials.* Tata McGraw Hill, New Delhi.

Mangaraj, S., Yadav, A., Bal, L. M., Dash, S. K., Naveen Kumar, M. and Arora, S. 2019. Application of Biodegradable Polymers in Food Packaging Industry: A Comprehensive Review. *Journal of Packaging Technology and Research* 3(2): 77-96. https://doi.org/10.1007/s41783-018-0049-y.

Natarajan, S., Govindarajan, M. and Kumar, B. 2014. *Fundamentals of Packaging Technology.* 2nd ed. PHI Learning, Delhi.

Paine, F. A. and Paine H. Y. (eds.) 1992. *A Handbook of Food Packaging.* 2nd ed. Springer Science and Business Media, Dordrecht .

Palling, S.J. (ed). 1980. *Developments in Food Packaging.* Elsevier Applied Science, London.

Rao, D. G. 2009. *Fundamentals of Food Engineering.* PHI Learning Pvt. Ltd., New Delhi

Robertson, G. L. (ed.) 2009. *Food Packaging and Shelf life: A Practical Guide.* CRC Press, Boca Raton.

Robertson, G. L. 2005. *Food Packaging Principles and Practice.* 3rd ed. CRC Press, Boca Raton.

Sacharow, S. and Griffin, R. C. 1980. *Principles of Food Packaging* 2nd ed. AVI Publishing Co., Westport, Connecticut.

Yadav, A., Mangaraj, S., Singh, R., Dash, S.K., Naveen Kumar, M. and Arora, S. 2018. Biopolymers as packaging material in food and allied industry. *International Journal of Chemical Studies* 6(2): 2411–2418.

28

Material Handling

Raw materials, partially processed products, end products, packaging materials, fuel and other utilities, wastes, containers, etc. need to be conveyed from one section to another in a food processing plant. Although people are engaged for such operations in most small and medium enterprises, mechanical handling is preferable as it involves less labor, reduces drudgery, and makes the operations more efficient. The activity must be carried out with utmost safety and minimum time, wastage, and expenditure. Proper material handling helps to achieve correct control of materials and processing conditions, to maintain product quality, and to minimize production costs. The performance of a processing plant is measurably affected by the efficiency of the movement of materials from one unit operation to another. Modern plant layouts give prominence to material conveyance in the plant for better economics and product quality. On the whole, mechanical handling results in reduced total processing time, storage space, cost, risk of accidents, and wastage of materials; improved working conditions and product quality; and better inventory control.

28.1 Material Handling Equipment

28.1.1 Classifications

The types of devices used for the shifting of materials can be considered under two broad categories, namely, the conveying and elevating equipment for short movements and the transport devices on road, rail, air, or water for long-distance transportation. The short movements are usually within the plant or in a godown. Devices such as forklifts are also used for loading and unloading pallets and square or rectangular packed materials as well as removing and transporting items from delivery vehicles; these can be considered as intermittent conveyors.

Common conveying equipment used in the food industry include the belt conveyor, roller conveyor, screw (auger) conveyor, drag/chain/tow conveyor, bucket elevator, vibrating conveyor, and pneumatic conveyor. In addition, other conveyors, such as chute, overhead, vertical, walking beam, and wheel conveyors, are also used. These conveyors use gravity and mechanical power, either alone or in combination, to move the material from one point to another, and they consist of frames that hold belts, buckets, rollers, or other moving components.

28.1.2 Belt Conveyor

Construction and Operation

- This consists of a continuous belt to convey the materials. The looped belt of metal or rubber linked sections is held between two rollers (or end-pulleys) tightly.

- One of the pulleys, known as the drive pulley, is driven by a motor and the other pulley is called the return pulley. Power is transmitted to the drive roller using either a variable or constant speed reduction device. The transmission system helps to operate the belt at different speeds depending on the throughput required.

- The other major parts of the conveyor are the idlers, loading and discharging devices, and the supporting framework. The idlers are antifriction bearings supporting the load of the belt and conveyed materials.

- The belt may either be flat or V (trough) shaped. Flat belts permit the conveyance of packed foods and trough-shaped belts are good for bulk materials. The conveyor belt may be supported by a metal slider pan for light loads.

- The conveyor can be operated either in horizontal or in inclined planes (Figure 28.1).

- The belts for transportation of food materials are usually made up of different types of materials depending on the material being conveyed and operating conditions. The common materials are rubber (synthetic), leather, plastic, fabric, and metal. Thick and strong belts are used for heavier loads. Often a rubber coating or cover on the belt protects it from wear and tear. Composite materials made up of steel and polyurethane, canvass, or polyester are also used. For very high temperature operations, the carbon steel wire belt is used. Very simple installations as a light canvas belt sliding over a long table is also used for fruit sorting applications.

- The unloading from the belt conveyor is done either by (i) unloading funnel, (ii) plough type unloader (scrapper), or (iii) pulley type unloader (tripper) (Figure 28.2). The unloading funnel is placed at end of the belt below the pulley to collect the materials. However, if the material needs to be collected at an intermediate point along the path of the belt, any one of the other two are used. The scrappers or ploughs can divert the materials from the normal flow path, which are then collected in a container placed below. There may be one or more than one scrapper depending on the requirement of the operation. A belt tripper is made of two pulleys arranged one above the other. One of the pulleys raises the belt to a height more than the other one in such a manner that a discharge chute or funnel can be placed below that. Thus, the material falls down form the upper pulley straight to the discharge chute. The tripper can be supported on a fixed or a movable frame so that the location of discharge of the materials along the belt can be changed.

DOI: 10.1201/9781003285076-35

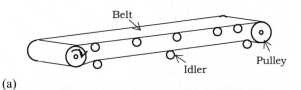

(a)

(b)

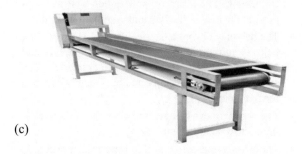

(c)

FIGURE 28.1 (a) Schematic of a belt conveyor (b) an inclined belt conveyor with slats (c) a flat belt conveyor used for food sorting applications (Courtesy: M/s Bajaj Processpack Limited, Noida)

Salient Features

- The carrying capacity is high. It has a long service life.
- Belt conveyors have a high mechanical efficiency as in larger installations all the load is carried on antifriction bearings.
- Since there is little or no relative motion between the product being carried and the belt, there is no damage to the product.
- The belt conveyors are suitable for moving materials to long distances.
- The angle of operation is limited, which is primarily decided by the coefficient of friction of the material against the belt in motion (so on the material and state of the surface of the belt), the angle of repose of the material, the method of loading and the speed

of the belt. The angle of inclination of belt conveyor should be 10°–15° less than the angle of friction of the material at rest on the belt (because the angle of friction at a roller support may be larger due to sagging). If the belt is fitted with cross slats to prevent the slippage of the product being conveyed, that may be inclined up to 45°.

- The initial cost of belt conveyor for short distances is usually high compared to other types of horizontal conveying systems. However, the initial cost of a belt conveying system for longer distances is low or competitive.

Conveying capacity of a belt conveyor can be given as,

$$Q = A. v \qquad (28.1)$$

$$W = A .v. \rho \qquad (28.2)$$

Where, Q (m³ .h⁻¹) is the volumetric capacity, A (m²) is the area of cross section of load, v (m.h⁻¹) is the belt speed, W (tonne. h⁻¹) is the weight carrying capacity and ρ (tonne.m⁻³) is the loading density of material being conveyed.

The cross-sectional area of load depends on the width of belt, the trough angle, and the surcharge angle of the material kept on the belt (Figure 28.3). A trough angle of 20° is commonly used for paddy and other food grains. The angle of inclination of the conveyor also affects the carrying capacity of the belt. The surcharge angle of the material on the belt is less than its angle of repose as the belt moves at a considerable speed. The material is kept on the belt with at least 4–5 cm clearance on both sides.

Applications

- The belt conveyor is used for conveying both unit loads and bulk loads of fruits and vegetables, food grains, etc. The trough-shaped belt is commonly used for moving granular bulk materials like food grains.
- It can be used for conveying materials in a horizontal or a slightly inclined plane.
- The belt conveyor is also used for sorting fruits and vegetables; the material is moved on the belt on a table, while the operators standing on both sides of the belt to carry out the sorting operation.
- Belt conveyors are found in applications such as airport luggage carousels.

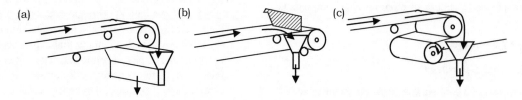

FIGURE 28.2 Different types of discharging devices used in the belt conveyor (a) by gravity at the end a pulley, (b) by a scraper, and (c) by arranging two pulleys at an intermediate location along the path at different elevations to allow gravity discharge

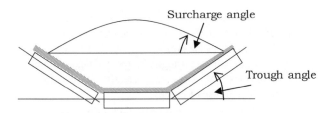

FIGURE 28.3 Trough angle and surcharge angle of load on a belt conveyor

28.1.3 Roller Conveyor

Construction and Operation

- It consists of parallel evenly spaced free running rollers mounted in frames, on which loads such as packed foods or crates are conveyed (Figure 28.4). Many overlapping types of roller conveyors are available.

- The rollers are made up of steel, aluminum, and plastic. There may be a coating of plastic or rubber on the aluminum or steel cylinders to increase friction.

- Different accessories, such as hanging brackets, slide rails, pop up stops, curves, and supports, are used on roller conveyors as per requirements. The path may be straight or curved.

- The rollers can be powered or unpowered. The rollers, which are not powered and are on a horizontal plane, the product is moved manually. If there is a slope, the product moves by gravity. This is also known as a *gravity roller conveyor*.

- The powered rollers can have different drive types such as motorized rollers, rollers driven by belts or chains/ sprockets.

- The belt in a *belt-driven live roller conveyor* is motorized to power each roller; it helps the control of the motion. The movement of the belt is opposite to the required material movement and the belt can be V-shaped for curved roller conveyors. *A line shaft roller conveyor,* powered by a rotating shaft, has each of its rollers belted via urethane belts and drive

FIGURE 28.4 Rolls in a roller conveyor

spools. As a result, this type of conveyor is easy to maintain and quiet.

- The chain driven roller is used for heavy-duty applications and when the working environment may cause damage to the belt below the rollers. The chain in a *roll-to-roll chain conveyor* is wrapped around each roller and holds on to two sprockets on each roller. A single non-looped continuous chain drives the rollers in the *single strand chain driven roller conveyor.*

- In one advanced type, known as *zero pressure roller conveyors* (or pressure-less accumulating conveyors), sensors and motors are used to create a buffer zone between moving products and there is no accumulation or contact of individual products. Certain "zero pressure" zones are created on the track and, therefore, the conveyors are called as *zero pressure roller conveyors.*

- When each roller is connected to a motor with the help of a belt/chain/shaft, the conveyor is known as a *drive or live roller conveyor*. It helps to control the rate of movement of the material even when the material moves down on an inclined plane on the rollers. It can also be used to move the materials on a slope, usually up to a maximum of 10°–12°. It can also be used in bi-directional applications.

Salient Features

- Key specifications of a roller conveyor include the axle center dimensions and the roller diameter. The roller material and the drive type in the case of powered rollers are two other important parameters. The frame, which is usually made up of aluminum or steel, must be capable of supporting the load.

- The surface of the material to be conveyed must be rigid and supported by a minimum of three of the rollers to allow smooth skating of the material on the surface of the rollers. Thus, the size of the material primarily determines the size of the roller. In addition, heavier and high-impact loads require larger rollers. A high set roller conveyor is recommended if the material is likely to experience overhang from the conveyor.

- A fall of about 10 cm in 3 m for most purposes is sufficient in the case of gravity conveyors. Steeper inclines would produce greater acceleration of packages, causing problems.

Applications

- Different types of roller conveyors have different types of applications. For flat bottom, light to medium weight products in food industries, gravity roller conveyors are generally used. Also, gravity roller conveyors are often used as alternative for skate wheel conveyors that do not require any specialized timing. They are also used for milk crates or even milk cans in the receiving section of a dairy plant.

- These are primarily used for baggage handling on loading docks or on assembly lines.
- Roller conveyors have been found ideal for accumulating the products after high-speed sorting machinery. If there is a need to stop the material movement for some time at control points or there is a need to turn the material, roller conveyors are preferred over typical belt conveyors.

28.1.4 Chain Conveyor

Construction and Operation

- This consists of a large chain fashioned in a loop around two pulley wheels, by which goods are moved (Figure 28.5). A suitable motor drives one of the wheels and thus the chain. The material is either directly moved by the chain or pushed by an attachment on the chain.
- Chain conveyors and their minor variations are classified as plain chain, drag, slat, trolley, or overhead conveyor and other types.
- The *drag conveyor* (also known as scarper conveyor and flight conveyor) consists of a single- or double-stranded chain of flights to pull or scrap the material kept on a stationary surface or trough (which is usually the bottom part of the conveyor). A motor powers the chain. About 50% of the available space within the conveyor is used for material movement whereas the space utilization is only about 20%–45% in other conveyors. Multiple inlets and outlets, in addition to top loading of materials, are provided since the gaps between consecutive flights are large enough. They can be horizontal or inclined. Inclined or vertical movement may require special designs of the paddles.
- Besides the standard type of drag conveyor, as mentioned above, there is an *en-masse* type of drag

FIGURE 28.5   Chain conveyor (Courtesy: M/s Shiva Engineers, Pune)

conveyor having a single-stranded chain with paddles attached to its both sides. It is also known as "skeletal chains" type of drag conveyor. There may be a single trough or two troughs for the paddles attached to both sides. This gives more effective space use and more efficient use of energy.

- A *tubular drag conveyor* has round (or disk shaped) flights on a single chain and is placed in a tubular housing. It helps feeding and exiting of the material at any angle relative to the tube. It also allows for airtight, vertical conveyance.
- The *slat conveyor* uses a chain-driven loop of slats. The slats are typically made of plastic or metal and can be given shapes as per requirements. A motor drives the chain. The slats give a rigid, flat surface for the movement of the product. The path can be horizontal or slightly inclined. Optional side railings provide further support. Though the slat conveyors usually have a straight path, there can be a curved path also to meet specific requirements in food and bottle manufacturing processes.
- The slat conveyors can be further grouped as the *standard slat conveyor* and *apron conveyor*. The standard slat conveyor has a straight pathway and often has side frames. The apron conveyors have interlinked slats with raised edges.
- An *overhead conveyor* has an endless loop of chain that revolves on a rigid track or a straight run of trolleys (also called as trolley conveyors). The carriers connected to the chain/trollies hold the material to be conveyed. A conveyor is called as unpowered or powered overhead conveyor depending on whether the material is pushed manually or by using motor power. A monorail type track can be either completely enclosed or open. Designed into any desired shape, the track and the carriers are very simple and versatile in form and application. They can also be classified as *synchronous* (or non-accumulating type) and *asynchronous* types. Each item is equally spaced in the synchronous type and moves at the exact same time. In the latter, each item moves independent of the others. The asynchronous types are typically unpowered.

Salient Features

- The chain conveyors are not as expensive as belt conveyors. They are operated at lower speeds than the belt conveyor. Unlike the belt conveyors, the chain conveyors can have sharp turns.
- The flights of standard drag conveyors use up to 50% of the space within the conveyor for material movement. Therefore, the drag conveyors are more efficient in comparison to the other conveyors. The conveying capacity of drag conveyors is decided by the surface area of the paddle and the speed of the chain.

- The conveyors with stainless steel slats/flights are not damaged even if they get wet or oily. Plastic slats/flights are common, but also should be avoided for high-temperature products.

- Chain speeds are maintained low (6–10 mm/min) for flight conveyors and the inclination is limited to 30° because the material would slip back if the recommended limits are exceeded. Such a conveyor may be permanent or portable, the portable farm grain elevator being the most widely used adaptation.

- The trolley conveyor with I-beam track is advantageous in that its direction of motion is extremely flexible. It can have steep elevations and sharp turns up to 180°. They have a high strength and small elongation. Load pushing and other elements can be fastened to them firmly and conveniently. Even if in some situations the overhead conveyors require power, the power requirement is very less.

- One problem associated with the chain conveyors is their wearing and stretching over time, which necessitates constant maintenance and lubrication of joints. In addition, the material may come in contact with the chains.

Applications

- The slat conveyor can be used for products that would normally damage most conveyors, for irregular shaped materials, as well as for hot items. Heavy-duty slat conveyors are used in the automobile industry. In The food industry, the standard slat conveyor is used to convey small to medium-sized objects. Apron conveyors can be used for both bulk and unit loads for both horizontal and inclined paths. It can be made leak proof for movement of slurries.

- Scraper conveyors are used for dry, granular, free flowing, and nonabrasive materials. These are used extensively for moving raw materials into processing plants. The standard drag conveyor, or simply drag chain conveyor, has been found ideal for grains and is often used as a feeder for other machines. Wet and sticky material can move on a tubular drag conveyor with low losses because it operates at low speed.

- The flexibility of the overhead conveyor makes it suitable for many applications. The trolley conveyor can be used for products of large unit size or for those that are handled in boxes or baskets, barrels, crates, churns, etc. Meat products, fruits, and vegetables are handled in this manner. As the direction of motion of the trolley conveyor is very flexible, it can be advantageously used for operations, such as immersing the product in a bath for operations as blanching, cooking, or cooling. Meat carcasses can be moved using a monorail conveyor on an overhead track.

28.1.5 Screw Conveyor

Construction and Operation

- The screw conveyor, also known as a spiral, worm, or auger conveyor, has a large helical screw that moves in a trough, casing, or compartment. The material is fed through one inlet provided on the conveyor, which passes to the other end of the equipment due to the turning auger (Figure 28.6).

- The auger is typically in coupled sections. Bearings on both ends hold it and the motor torque is transmitted to the conveyor through the coupling bolts. Either a direct drive, gears, or a chain and sprocket can drive the screw. Long screws are also supported by brackets at required spacings. The diameter of the

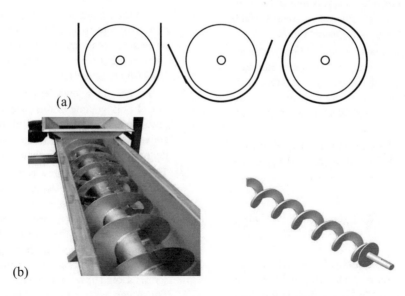

(a)

(b)

FIGURE 28.6 (a) Different types of troughs, (b) a screw conveyor with U-trough (Courtesy: M/s Shiva Engineers, Pune), and (c) a shaftless screw

screw in the conveyors used in the food industry may be 100–600 mm.

- For food industry applications, the screw flights and trough should be made from food- grade materials such as high-grade stainless steel.

- On the basis of the travel path, the screw conveyors can be horizontal, vertical, or inclined screw conveyors. The horizontal screws usually operate at 15%–45% trough loading, the lower value corresponds to more viscous and sluggish material and the higher value corresponds to lighter free-flowing material. The troughs for horizontal conveyors are mostly U-shaped with or without cover. The inclined ones are used up to an inclination of 45°. The *vertical screw conveyor* uses inclination angles greater than 45°. For steep inclines, the screws move the material in a complete cylindrical housing. In such vertical screw conveyor systems, which usually operate at full loads there is no bracket between the screw ends, as this would interfere with the material movement.

- The feeding device in the vertical screw conveyor has to be chosen carefully as a simple gravity hopper may not allow proper feeding due to blowing out of the material by the rapidly turning auger. This often necessitates the use of a screw type feeder. The screw feeder also facilitates maintaining a constant speed of the vertical auger, increased efficiency, and decreasing power consumption.

- Different types of screw conveyor designs are also available. A *ribbon* or *shaftless screw* has no central shaft and, instead, has an empty helix. It is helpful for high-moisture or non-flowing (sticky) products.

- A live bottom screw conveyor has multiple screws on a trough, typically placed under large bins. When the material falls down the bin, it is moved by many augers. They come as the mass flow cone, variable pitch, tapered flight screw, and other types.

- Special cut flight, paddle type, and ribbon screws are used for mixing, both singly and in connection with conveying. Double and triple-flight, variable-pitch, and stepped diameter screws are available for moving non-free flowing materials and controlling the feed rate.

- During rotation of the screw, the product moves one pitch for every revolution. The theoretical capacity of a standard screw conveyor is given as

$$Q = \frac{\pi}{4}\left(D^2 - d^2\right)P.N \qquad (28.3)$$

Q (m³.min⁻¹) is the capacity, D (m) and d (m) are the inner diameter of the housing and the diameter of the screw, P (m) is the pitch of the screw and N is the revolutions per minute. The actual capacity is about 30% to 50% of the theoretical capacity, because of the trough loading percentage and other factors as fluid characteristics of the material, screw length, head of material, and elevation or lift.

Salient Features

- The pitch of a standard pitch screw is approximately equal to the diameter. Horizontal installations and those with inclines of up to 20° can use this screw. For inclines greater than 20°, half standard pitch screws may be used.

- The trough should be made of food-grade frictionless material. The characteristics of the material being moved and the service conditions influence the selection of the type trough.

- The trough may have optional covers for a cleaner and more contained process.

- Although simple and relatively inexpensive, power requirements of screw conveyors are high and single sections have length limitations. The length of horizontal screws is usually limited to 6 m, beyond which high friction forces result in excessive power consumption.

- For inclined conveyors, there is minimal effect on conveyor efficiency if the inclination angles are lower than 10° and the loss can be compensated by increasing the auger speed. However, more inclination reduces the efficiency of conveyance and requires more torque for the movement, and, hence, the inclination should be decided accordingly.

- For a higher inclination of screws, there is a greater possibility of the material falling back to the lower end of the conveyor. Therefore, the design should consider the simulated 100% trough loading at its lowest point.

- Efficiency obtained for ribbon screws is better in comparison to that obtained for any other conveyor or even for other screw conveyors and the conventional designs with no internal bearings. It increases flexibility of operation.

- A live bottom screw conveyor can control the flow rate of bulk volume of the materials that pack under pressure. Even if materials of different sizes enter the conveyor at different densities/speeds from the inlet, the material is evenly discharged.

- The selection of the screw conveyor, screw diameter, pitch, revolution, etc. depends on the type of material to be conveyed, the load-carrying requirement, and the trough loading percentage. The information will help in finding the power requirement. For specific applications, such as mixing, the auger has to be accordingly chosen. For inclined conveyers, the screws should have short pitch to maintain efficiency. The trough has to be selected based on the load type; if the load is light and dry, a round trough will be better for duct protection. Sticky materials will require a shaftless screw and a lower trough loading factor. The distance of travel, change in elevation, and material collection device are some other factors in selecting the screw.

- The horsepower required by a screw conveyor for a horizontal movement is given as:

$$hp = \frac{Q.L.F}{4500} \qquad (28.4)$$

Here, Q (kg.min^{-1}) is the capacity of the bucket elevator, L (m) is the length of the conveyor, F is a factor that depends on the conveyed material. For conveying paddy, the value of F has been found as 1.5. Another empirical correction factor w needs to be multiplied to find out the actual horsepower. The value of w is 2 if the calculated hp is less than 1. The values of w will be 1.5, 1.25, and 1.1 if the calculated horsepower is between 1 and 2, between 2 and 4, and between 4 and 5, respectively.

Applications

- The screw conveyors are used to move granular materials, fine powders, as well as damp, sticky, and heavy viscous materials. Some screw conveyors have specialty applications for moving wet and caking products also. It can work in an open or closed housing, depending upon the environment.
- Screw conveyors are also used for batch or continuous mixing operations or to maintain solutions. Ribbon screws are mostly used for the purpose.
- These are also used as feeders or metering devices under bins or hoppers. For such uses, the portion of the screw just below the bin has one-half or one-third pitch and the balance of the screw has full pitch. In the case of a LSU dryer, a screw conveyor fitted at the discharge end of the dryer helps to control the grain flow rate.
- Screw conveyors are used widely in different farm machineries, such as a thresher, a baler, and a grain and crop combine.

28.1.6 Vibrating Conveyors

Construction and Operation

- The conveyor is made of a slightly inclined trough or deck, which is vibrated or shaken lengthwise. The particulate materials kept on the deck move along the incline due to the vibration due to hopping.
- A drive system imparts an oscillatory motion to the trough at the specific frequency and amplitude. Usually the stroke (the movement of the trough) is kept as twice the amplitude of the vibration. The

minimum amplitude and maximum frequency are desirable.

- They can also be modular and portable. Standard vibrating conveyors can have capacities even up to 40 tons.h^{-1} at speeds up to 20 m.min^{-1}.
- Natural frequency vibrating conveyors are those that have springs and other components to make the trough vibrate; this reduces the use of energy and uniformly distributes vibration across the trough. Most vibrating conveyors use some natural frequency techniques.
- The main drive mechanisms used for vibration are the cranks/springs, rotating weights, and electromagnets. In the first type, a four-bar mechanism is rotated by a motor. In the other systems, the vibrations are carried out either by using rotating eccentric weights or by a rotating magnetic field.

Salient Features

- The function of a vibrating conveyor is highly dependent upon its motor and its construction.
- The speed of movement is controlled by the amplitude of vibration. The vibrating conveyors are also known as "controlled" conveyors, as the frequency and amplitude are exactly set and controlled.
- Different types of materials behave differently to the chosen amplitude and frequency of the conveyor; hence, the design parameters should be decided accordingly.
- Oscillating conveyors are specific type of vibrating conveyors in which lower frequency and a larger amplitude of motion than those typically used in vibrating conveyors are used. However, often they are not specifically differentiated.
- The major design parameters include the size, shape, and duty rating of the conveyor, those also decide the load capacity, trough length, and vibrational frequency.
- They have long service life and require less maintenance.

Applications

- Vibrating conveyors are used for horizontal or shallow incline conveying of bulk products as powders and grains and for irregularly shaped materials. They can convey fragile materials, such as potato chips, without crushing them.
- The vibrating conveyor is used extensively where sanitation and low maintenance are vital.
- Vibratory feeders are also commonly used for continuous weighing applications or as feeding units for processing equipment. Vibratory feeders can move sticky foods quite well.

- Very heavy-duty vibrating conveyors are also used in the rock and metal industries. Capacities can be more than 500 tons.h^{-1}.

28.1.7 Pneumatic Conveyors/Elevators

Construction and Operation

- Powders or small particulate foods are conveyed in a pipe by using pressure differentials (Figure 28.7). The medium of flow can be compressed air or nitrogen. This pressure differential can be created by using a blower.

- The transfer of material can be achieved by a suction system (or negative pressure system or pull system, where a blower is provided at the exit end) or a push system or positive pressure systems, where the blower is near the feed end. The negative pressure systems are also often referred as *vacuum conveyors*.

- In the case of push systems, materials from one inlet can be pushed to many outlets, whereas the material from many inlets can be pulled to one outlet using the pull systems. A combination of a push-pull system is also possible.

- The material has to be fed uniformly at a specified rate in a dispersed manner into the air stream so that it can be lifted and conveyed by the gas. A rotary feeder is a commonly used device having a bladed rotor with pockets at its inlet port. The products are dropped at the outlet port and to the conveying pipeline.

- After the materials are conveyed to the desired point, they are recovered from the air stream by gravity or by using devices such as cyclone separators or bag filters.

- Thus, the essential parts of a pneumatic conveyor include a source of power for blowing air and/or suction, feeding device, ducts, and a cyclone or receiving hopper for collection of the airborne materials.

- The pneumatic conveyors can be pressure or vacuum-driven and classified as *dense phase, semi-dense phase, and dilute phase* conveyors. A dilute phase pneumatic conveyor uses a high-velocity air stream at low pressure to fluidize fine particles. It can be either a dilute phase pressure conveyor or a dilute phase vacuum conveyor. Vacuum conveyors are meant for short distances. Pressure conveyors are good for conveying powders with bulk densities lower than 1000 kg.m^{-3}. A dense phase pneumatic conveyor utilizes a high-pressure stream at low velocity to move those products that are prone to breakage. The material does not get fluidized. Instead, the material is "pulsed" though the system. Some air injectors along the conduit supplement the force required for the flow. In the case of semi-dense phase systems, only a portion of the material gets fluidized. These provide some increased speed, but at the same time does not cause breakage of fragile materials. These are essentially used when either the dense or dilute conveyor does not work.

Salient Features

- Pneumatic conveyors have the advantage of low initial cost and simple construction. Since there are fewer moving parts as compared to the other conventional conveyors, maintenance is easier.

- The investment cost is lower for pneumatic conveyors in comparison to other conveyor types. However, pneumatic conveyors exhibit lower efficiency because more energy is needed to keep the blowers/vacuums on. The other disadvantage is the possible damage to conveyed materials.

- The conveying pathway can be changed easily, a wide variety of materials can be conveyed, and the system is self-cleaning.

- The gas velocity is critical; the solids settle down if it is too low, there is abrasion damage to the pipe surfaces if it is too high. Thus, the minimum velocity required to move the product should be used.

- The negative pressure system reduces the dust. In addition, if there is any leakage in the duct, the air is pulled inward so that material loss and release of dust to the surrounding area are prevented.

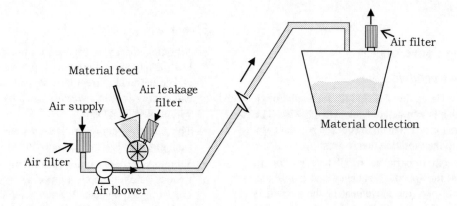

FIGURE 28.7   Schematic diagram of a pneumatic conveying system

- Technical specifications of a pneumatic conveyor system include product density, gas flow rate, gas pressure, pipe size, length of travel, bends, etc. If the product is very dusty, proper filtration may be required to protect the workers.

Applications

- The pneumatic conveyors are well suited for low bulk density, small size, and dry products, such as powders, granules, flakes, wheat, sugar, etc.
- The dilute phase pneumatic conveyors find their applications for conveying light, non-abrasive, and not easily breakable materials, such as flour. They are also useful for chemical and toxic materials because the risk of exposure to them is reduced. Heavy bulk density, blended, abrasives (sugar, salt, etc.), and other fragile materials are conveyed using dense phase pneumatic conveyors.
- These are also commonly used as burner feeders.

28.1.8 Bucket Conveyor/Elevator

Construction and Operation

- It consists of metal or plastic buckets (Figure 28.8) fixed at equal distances on two endless chains or on an endless belt. There is an engine connected with reducing gears to drive the belt. There are accessories for loading the buckets, receiving the materials from the buckets, and for maintaining the belt tension.
- The lower end near the feeding section is known as the boot section and the upper section is known as the head hood. The engine, constituting the motor and the gear box, is fixed near the head hood. The discharge spout is fitted with the head hood.
- The buckets may be enclosed in a casing, made up of steel or wood, which is also called a leg. There may be either one leg for both the runs (elevation and return) or two legs (Figure 28.9 a). In case of the latter, the return leg is at a gap from the elevator leg.

FIGURE 28.8 Types of elevator buckets

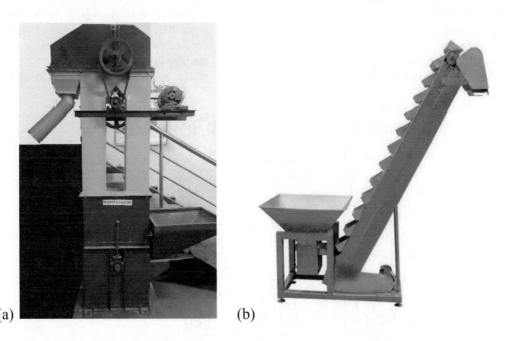

(a) (b)

FIGURE 28.9 (a) A vertical bucket elevator with two legs (Courtesy: OUAT, Bhubaneswar) (b) an inclined bucket elevator (Courtesy: M/s Shiva Engineers, Pune)

Buckets are also used to elevate the material in an inclined plane (Figure 28.9 b). Z-shape bucket elevators carry the load to a vertical distance and then convey the material horizontally before discharging. Figure 28.10 shows schematic diagrams of (a) centrifugal discharge type bucket conveyor (b) continuous bucket elevator.

Salient Features

- The discharge may be centrifugal type, where evenly spaced buckets on a chain or belt dig the material from the feeding section and throw out at the top, typically just after the top turn of the chain. The material goes through the discharge spout. A small portion of the material may fall into the casing and get collected at the bottom feeding section for lifting again. It can be operated at both low and high speeds and can be operated even at 120 m.min⁻¹. Some centrifugal elevators used for lighter material can even move at 300 m.min⁻¹.

- The continuous bucket conveyors have no gap between buckets to prevent any spillage and are moved at low speeds of 60–100 m.min⁻¹. The feeding to the buckets is done individually at the inlet by gravity and there is no digging as in the previous case. They can be in S, Z, or C shape and different ratings, depending upon need.

- Positive discharge conveyors are a special kind of centrifugal conveyors where the buckets pass over an extra sprocket allowing the discharge of the material to the discharge spout even at low speeds. The speed is usually kept at 30–75 m.min⁻¹; this prevents undue spilling and breakage of materials. The buckets are usually kept large to compensate for their slower speeds. Positive discharge conveyors are, typically, more expensive than the other two bucket conveyors.

- Gravitational and centrifugal forces together influence a product mass that turns around the pulley (Figure 28.11). The magnitude of the centrifugal force oriented downward is given as:

$$C_f = Wv^2\!\big/_{g.r} \qquad (28.5)$$

where, W is the weight of the material, v is the velocity of the material, r is the radius of rotation, and g the acceleration due to gravity. The gravitational force is the same as the weight of the material. For an optimum centrifugal discharge, the resultant of the product weight and the centrifugal force should preferably be directed toward the lip of the bucket. Thus, the best possible condition occurs when both these forces are equal in magnitude.

$$C_f = Wv^2\!\big/_{g.r} = W \qquad (28.6)$$

$$\text{or, } v^2 = g.r, \text{ i.e., } v = \sqrt{g.r}$$

Since, $v = 2\pi rn / 60$, where n is the rev / min,

$$n = 29.9\!\big/_{\sqrt{r}} \qquad (28.7)$$

The above relationship indicates that if the diameter of the wheel (pulley) is less, the number of revolutions will be more for the same discharge angle.

- Bucket elevators can have a high capacity and are very efficient due to the absence of frictional loss from sliding the material on the housing.

- The selection of the bucket elevator type and parameters, such as the bucket shape and size, are primarily based on the type and amount of material that must be conveyed and the conveying distance. The design and number of buckets are also decided by the height and orientation of the elevator. The drive is selected on the basis of the speed of the conveyor. The protection of the material

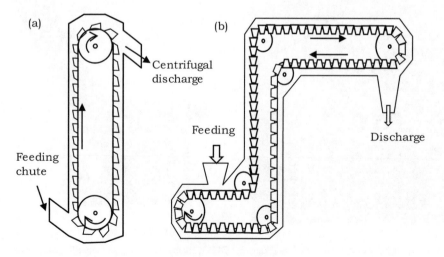

FIGURE 28.10   Schematic diagram of (a) centrifugal discharge type bucket conveyor (b) continuous bucket elevator

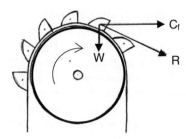

FIGURE 28.11 Forces acting on the conveyed material at the point of discharge

from aeration, sticking, or the work environment also influences the selection of the elevator. For example, if the material is very fine and has a tendency to generate dust in the environment, then a slower speed of the elevator is adopted. The selection of the elevator is also different for chemically active or hazardous materials.

- The capacity of a bucket elevator is given as:

$$\text{Capacity (m}^3 \cdot \text{min}^{-1}) = C \times v \times N \qquad (28.8)$$

where, v= bucket elevator speed (m.min^{-1}), C = capacity of each bucket (m^3), N = number of buckets per meter length of conveyor (m^{-1})

Applications

- Centrifugal elevators are used for grains, sugar, and chemicals. As the products are filled by gravity, continuous buckets can be used to handle light, fragile, or friable products. Such elevators are used in those situations that require minimal agitation, e. g., the plastics and mining industries, severe duty uses, etc. Positive discharge conveyors are used where the priority is to preserve product integrity, such as for grains, nuts, dried fruits, etc.
- The bucket elevators are commonly used as feeding devices for continuous dryers (where the material falls down during drying) or rice mills.

28.1.9 Magnetic Conveyor/Elevator

Construction and Operation

- Magnetic conveyors use magnets to hold the materials. These can be either a magnetic belt conveyor or a beltless conveyor. In the belt type, equally spaced electromagnets are placed against a conveyor belt (on the opposite side of the surface where the load is held). Electromagnets are preferred for the flexibility to change their strength.
- In the belt less conveyor, which is also known as slider bed conveyors, an internal chain of permanent magnets is rotated, which imparts magnetic force through some stationary bed.

- The magnetic materials can be moved by the conveyor in inclined, vertical, or even upside-down situations.

Salient Features

- The strength of the magnets is high enough to resist gravity effects.
- They can operate at high speeds with minimal noise.

Applications

These are used for conveying of cans within canneries. They are capable of holding the cans in place even during vertical movement and are also able to invert empty cans for cleaning.

28.1.10 Vertical Conveyors

Construction and Operation

- These are special systems used to lift or lower unit loads. Often they are placed between two horizontal conveyors.
- The load can be carried by buckets, belts, chutes, screws, and other holding devices.
- These can be operated by gravity or by powered devices.
- The vertical conveyors can be of reciprocating type, which are also known as reciprocating mezzanine lifts (it can move up and down multiple levels to serve many floors), scissor lift conveyor (which has a strong platform to move heavy loads up to about 5 to 6 feet height, continuous type (which has different configurations like C, Z, or S type to move the materials continuously through vertical sections) or spiral types. The continuous types often connect two horizontal conveyors with one vertical conveyor. The vertical screw conveyor is another form, which has been discussed earlier.

Applications

- They are used for low frequency, intermittent vertical transfers. These are specifically used to transfer materials like boxes and cartoons, crates, etc. in between two floors or for vertical accumulation in processing, packaging, and distribution systems.
- The spiral conveyor gives more residence time for the same lift and are often used in spiral freezers or blanchers.

28.1.11 Other Systems

There are also many other special conveying systems, such as skate wheel conveyor, vacuum belt conveyor, etc. In the skate wheel conveyor, wheels are used in place of rollers. The vacuum belt conveyor has perforations on the belt, which are connected to a vacuum. Thus, any material kept on the belt will be sucked in and it will have a better grip to enable the belt to move at higher inclinations.

Water is used to move the fruits and vegetables upon receipt in a food processing plant. Usually, washing and conveying are integrated in a single operation.

28.2 Selection of a Conveyor System

The major characteristics of a conveyor system are the carrying capacity per unit length, maximum load capacity or the rate of flow, throughput (volume conveyed per unit time), rated speed, and frame configuration. The location of drive is another important specification of the system. The drive is found on the discharge side of the conveyor in case of the head or end drive system. A center drive system has its drive mounted underneath along its length. This arrangement is helpful in those situations where the direction of the conveyor needs to be reversed.

For achieving higher efficiency in materials handling, the following three important aspects need to be observed.

(1) Plan the handling scheme using systems approach;

(2) Application of bulk handling; and

(3) Automation wherever possible.

In addition, it is important that the materials should be grouped together for easier handling using all layers of a building's height. Use of gravity is preferred, where possible, for improving the efficiency and reducing the power requirement of the system.

Material handling is an important operation in any food processing plant or food supply chain. In the food processing plant, the raw and partially processed products, fuel, containers, packaging materials, etc. are transferred into the plant and shifted from one place to the other within the plant. The processed products, wastes, etc. need to be transferred out of the plant. The materials must be moved in the correct quantities to and from the correct place. It should be done with a minimum of time, labor, wastage, and expenditure and with maximum safety. Efficient material handling is very important to optimize the product quality and minimize costs of production. In addition, it helps in reduced processing time, in better stock control, and in saving storage space and cost, improved working conditions, improved product quality, lower risk of accidents, and less wastage of materials. Different types of material handling equipment are available, which have to be selected on the basis of the type of load, distance of transfer, investment required, and many other parameters.

Check Your Understanding

1. If a material is moved on a flat belt of 1 m width with 5 cm gap on both sides, find the maximum capacity of the conveyor for paddy at a belt speed of 2 m.s^{-1}.

The material may be taken to have an angle of repose as 30° and density as 550 kg.m^{-3}.

2. A screw conveyor is 5 m long. The diameter of the screw and shaft are 30 cm and 15 cm. The pitch is 28 cm. What is the conveying capacity of the screw if the actual capacity is 40% of the theoretical capacity? If the conveyor is used for wheat (bulk density = 750 kg.m^{-3}), the material factor is 0.4, and the power transmission efficiency is 50%, what will be the power required?

Answer: Actual capacity = 2.227 m^3. min^{-1}, Power = 1.5 hp = 1.137 kW

Objective Questions

1. What is efficient material handling?

2. Write down the important parts and working principles of a belt conveyor.

3. What are the different unloading devices of a belt conveyor?

4. State the different applications of (a) belt conveyor, (b) roller conveyor, (c) chain conveyor, (d) overhead conveyor, (e) shaftless screw conveyor, (f) vibrating conveyor, (g) magnetic conveyors.

5. In which case do we prefer the powered roller conveyors to the unpowered ones.

6. Compare the construction features of different types of chain conveyors. Explain the applications of different types of chain conveyors in the food industry.

7. State the different applications of screw conveyors. Where is the cylindrical housing / tube type housing used in a screw conveyor?

8. What is the difference between a centrifugal discharge and a continuous discharge in the case of a bucket elevator?

9. Write short notes on

   a. Pneumatic conveyor;

   b. Vibrating conveyor;

   c. Drag conveyor;

   d. Slat conveyor.

10. State the advantages of a belt conveyor over a screw conveyor.

11. Explain the difference between a roller conveyor and a skate wheel conveyor.

12. State the differences between different types of buckets for their application point of view.

13. What are the different factors for selecting a conveying system?

BIBLIOGRAPHY

Ahmed, T. 1999. *Dairy Plant Engineering and Management* 4th ed. Kitab Mahal, New Delhi.

Chakraverty, A. 1999. *Post-Harvest Technology of Cereals, Pulses and Oilseeds*. Oxford & IBH Publishing Co. Ltd., New Delhi

Fellows, P. J. 2000. *Food Processing Technology.* Woodhead Publishing, Cambridge, UK.

Henderson, S. M. and Perry, R. L. 1976. *Agricultural Process Engineering.* The AVI Publishing Co., Westport, Connecticut.

https://www.thomasnet.com/articles/materials-handling/all-about-bucket-conveyors/

Pandey, P. H. 1994. *Principles of Agricultural Processing.* Kalyani Publishers, New Delhi.

Rao, D. G. 2009. *Fundamentals of Food Engineering.* PHI Learning Pvt. Ltd., New Delhi.

Sahay, K. M. and Singh, K. K. 1994. *Unit Operation of Agricultural Processing.* Vikas Publishing House, New Delhi.

Spivakovsky, A. O. and Dyachkov, V. K. 1985. *Conveying Machines. Vol I and II.* Mir Publications, Moscow.

Section C

Food Quality, Safety and Waste Disposal

29

Shelf-Life Estimation of Food Products

A long shelf life for a food product, processed or fresh, is expected to ensure its effective utilization. Shelf life is defined as the time duration for which it retains its quality up to an acceptable level for consumption. Clearly any food product is expected to have a long shelf life if it could remain worthy of consumption, and without deterioration, for a long time. The purpose of this chapter on shelf-life estimation has been to provide a basic knowledge about various considerations needed for determining the product shelf life.

29.1 Importance of Shelf-Life Determination

The shelf-life declaration on the package products is mandatory for selling the product while ensuring the safety of consumer health. In food industry, the shelf life of a food product is the duration between the time when it is processed and packaged until the time it ceases to be worthy of consumption (expiry date). The product *expiry* or *use-by date* relates to its perishability, indicating that the product beyond the indicated date could develop some level of food poisoning. Another phrase, *best before*, is used for durable food products and indicates that deterioration in the food quality would begin after the indicated date, but it may remain edible for a more period beyond the specified date. For example, an apple after plucking it from a tree could remain edible for several weeks under cool ambient conditions, whereas a tomato would begin losing its lustre and turgidity only after a few days: the apple has a longer shelf life than the tomato. Table 29.1 gives the requirement of the use-by or best-before dates of different categories of foods.

Expiration or best-before date is a guide to a consumer while making the purchase and these declarations assure food safety. Shelf life is not only affected by the producer or manufacturer but also by the distributor, retailer, and consumer in the supply chain.

29.2 Factors Influencing Shelf Life

Several variables influence the product shelf life, i.e., biochemical composition, extent of processing, packaging method, handling method, and storage condition. It is possible to predict the individual as well as cumulative effects of these variables on the product shelf life. Actually, each parameter has a specific relationship with the shelf life. However, ultimately, it is the shelf life that matters. The factors that influence the shelf life of a food product can be grouped under extrinsic and intrinsic factors.

The extrinsic factors include exposure to light, ambient air temperature, humidity, packaging defects, and handling during transport and distribution. Intrinsic factors include raw materials, biochemical composition of the product, water activity, available oxygen, pH, and redox potential. The water activity and temperature together are the two most important factor that affect the shelf life of food products. Food grains could remain in good edible condition for several years because of lower water activity. Cold storages prolong the shelf life of perishables.

29.2.1 Intrinsic Factors

Raw material. The quality of the raw material determines the quality of a finished food product; if the raw material is of poor quality, the product quality will suffer and the shelf life will be lower. Raw materials arriving from farmers' fields need to be checked for damage, bruises, infections, and contaminants before initiating the processing operations. The quality of the raw materials can be enhanced by implementing Good Agriculture Practices. Raw materials should have minimum microbial loads, have attained correct maturity stages, and be devoid of physical contaminants.

Biochemical composition. The biochemical composition of the food product is another intrinsic factor that affects the shelf life. A higher oil content of the product renders it prone to rancidity. Lower pH values and higher salt/sugar contents lead to increase in shelf life. Certain ingredients are intentionally added at the formulation stage precisely for enhancing the product shelf life.

Texture of the raw material. It becomes important during processing. The raw materials should have desirable texture. In addition, non-uniformity of the products leads to uneven biochemical reactions and microbial growth. As a result, the final form of the processed product could be considerably different within batches with changed shelf life. Obviously, under such non-uniformity conditions, the predictive models achieve only limited success. Therefore, proper mixing during the processing may result into better shelf life and greater success of predictive microbiological models.

Water activity. It indicates the moisture availability for biochemical reactions; it is expressed as the relative humidity of the surrounding atmosphere at which the product is in equilibrium and denoted as a_w. We have discussed about water activity in Chapter 9. The relationship between water activity and the moisture content of the product is non-linear. It is water

TABLE 29.1

Types of foods and the requirement of use-by or best-by date

Type of Food	Use-by date	Best-before date
Shelf-stable foods[a]		✓
Frozen food		✓
Raw food that requires processing to reduce or eliminate food poisoning bacteria to make the food safe		✓
Chilled ready-to-eat food	✓	
Food may contain one or more food-poisoning microbes, such as *Listeria monocytogenes* and *Clostridium botulinum*	✓	

[a] Shelf-stable foods are those foods that, due to their composition (low moisture content, high salt or sugar content), do not need to be refrigerated during storage. In the case of foods that are normally refrigerated, the processing and packaging are such that the product can be stored under ambient conditions for a usefully long period.

activity that determines the product stability as well as potential for microbial growth, chemical and biochemical reactions, and moisture migration. While food products such as fresh juices, fruits, vegetables, meat, and fish have water activity values greater than 0.98, cooking and concentration bring down the water activity to the 0.98–0.93 range. Drying and fermentation further reduce the a_w to the 0.93–0.86 range. Most dried fruits, confectionary products, heavily salted fish, and aged cheese will have water activities in the 0.8–50.6 range. Water activities lower than 0.6 are possible for dried vegetables, biscuits, cookies, noodles, sugar, and cornflakes.

Redox potential (E_h). The redox potential expresses the ease with which a food product could gain or lose electrons and thus is prone to oxidation. Loss of an electron signals oxidation; therefore, when a compound loses an electron, the compound is oxidized. The redox potential is measured in terms of millivolts (mV) because the movement of electrons causes electric current to flow. It is dependent on the pH of the food. Oxidation also occurs when a compound reacts with oxygen. Obviously, the presence of oxygen during processing operations is responsible for the oxidation-reduction (redox) potential the system. The redox potential affects the biochemical reactions as well as microbial survival and growth. Different microbes respond to the redox potential differently. Aerobic microbes require the highest amount of oxygen and the redox potential for aerobes has been observed to be in +300 to +500 mV. Anaerobic microbes require no oxygen and microaerophilic microbes require limited amount of oxygen; both could function in +100 to ≤ 250 mV. Facultative anaerobic microbes can grow at redox potential in +300 to -100 mV range. Thus, redox potential becomes a very important parameter for shelf-life determination of a food product.

pH. The pH of a product depends on its constituents and represents the concentration of hydrogen ions present in the product on account of the acid ingredients. These ions get dissociated when brought in contact with water and can now be measured. Growth and survival of microorganisms is strongly related to the pH values; lower values not favoring microbial growth. Because the food product is continuously undergrowing chemical and biochemical changes, the resultant pH also keeps varying with time. In addition to its effect on microbial growth, pH also influences the chemical and biochemical reactions, e.g., enzymatic and non-enzymatic browning and degradation of some food components and colors.

The available oxygen in the food is also an important parameter that supports biochemical reactions and affects the shelf life.

29.2.2 Extrinsic Factors

Extrinsic factors include processing conditions, exposure to light, ambient air temperature, humidity, packaging materials and defects, and handling during transport and distribution.

Processing operations. The different processing activities, such as cleaning, sorting, and grading; size reduction, separation; thermal and non-thermal treatments; mixing; enrobing; filling; and packaging, all affect the product shelf life in one way or the other. All these operations influence one or more of the intrinsic factors affecting, in turn, the shelf life.

Hygiene. Hygiene is a practice that influences the microbial load on the food product. Good hygiene leads to reduced microbial load and, in turn, to higher shelf life. Hygiene levels acceptable for food safety assurances can be achieved through enforcement of Good Hygiene Practices (GHP) during the processing activities.

Packaging materials and practices. The packaging materials and practices provide convenient handling and isolation of the food products from surroundings and contaminants to assure the desired shelf life. Packaging provides protection against light, loss/gain of moisture, gas exchange, and several other deteriorative effects affecting the shelf life.

Storage facilities are required for providing a congenial environment for the stored product to minimize the rate of product deterioration with the aim to control such factors as light, temperature, humidity, and the gaseous composition of the surrounding atmosphere. Successful storage along with packaging goes a long way in ensuring long shelf lives of the products. Storage environmental parameters have been determined and are available in the literature for practically all categories of raw and processed food products. It is expected that the storage environments are closely regulated and monitored for facilitating the attainment of the maximum shelf lives.

These factors affect the product shelf life individually as well as in combination; there are interactive effects. As an example, consider a canned product destined to a temperate climatic location. If the same product is sent to a location in a tropical climate, it would be unacceptable. In fact, the product for a location in a tropical climate would need more severe thermal treatment to keep the microbial population under a permissible limit during the shelf life.

29.3 Kinetics of Shelf-Life Deterioration

Shelf-life deterioration of a food product may be a function of one or more of the intrinsic and extrinsic factors. Let F represent the shelf-life deterioration as a function of time as a zeroth order function.

$$F = F_0 - k\theta \tag{29.1}$$

The parameter F_0 is the initial value of the chosen shelf-life variable. The coefficient k is an apparent deterioration rate and θ is time. The minus sign indicates that F is decreasing with time. The coefficient k would depend upon the factors influencing the shelf life; temperature T is one of these factors. If it is assumed that all other factors could be controlled through proper storage and packaging, k would be a function of T alone.

Following an analogy with the Arrhenius equation, k is represented by an Arrhenius type equation.

$$k = k_0 e^{E_a/RT} \tag{29.2}$$

One could easily recognize that k_0 is the Arrhenius's equation constant at 0 K; E_a is activation energy; R is universal gas constant, and T is absolute temperature. Clearly, k_0 is impractical, it would be necessary to determine the Arrhenius equation constant at some easily achievable temperature, say T_{ref}. Then k_0 can be replaced with k_{ref} and the equation is modified as follows:

$$k = k_{ref} e^{-\frac{E_a}{R}\left(\frac{1}{T} - \frac{1}{T_{ref}}\right)} \tag{29.3}$$

Reference temperatures used are normally 273 K for frozen and chilled foods and 293 K for ambient temperature stored foods. It is now possible to represent the shelf-life of a food product analytically as follows:

$$F = F_0 - k_{ref} e^{-\frac{E_a}{R}\left(\frac{1}{T} - \frac{1}{T_{ref}}\right)} . \theta \tag{29.4}$$

Although the above-mentioned approach is generally used, one must realize that there could be conditions such as non-linearity in the above equation. If instead of a fixed temperature, there are temperature variations over the storage period, the shelf-life function can be evaluated as follows:

$$F = F_0 - k_{ref} \int_0^\theta e^{-\frac{E_a}{R}\left(\frac{1}{T(\theta)} - \frac{1}{T_{ref}}\right)} . d\theta \tag{29.5}$$

Temperature $T(\theta)$ represents the variable temperature. If an effective temperature can be found somewhere in the range of $T(\theta)$ such that the shelf-life deterioration evaluated at the effective temperature is equivalent to that found for the actual temperature variation, the function F can be evaluated as follows:

$$F = F_0 - k_{ref} e^{-\frac{E_a}{R}\left(\frac{1}{T_{eff}} - \frac{1}{T_{ref}}\right)} . \theta \tag{29.6}$$

The above expression is very simple and can be used for the shelf-life estimation. Given the upper or lower limit of acceptable F, the resultant shelf life can be calculated. There can be several such functions representing the trend of the time-deterioration variations.

If the shelf-life reducing factor is microbial spoilage, the microbial population from the initial count would need to be estimated as affected by time and the duration after which the microbial population crossed the threshold level would indicate the maximum shelf life. If N_0 is the initial microbial count, the population N after time θ can be represented as follows:

$$dN/d\theta = k'N \tag{29.7}$$

where, k' is the specific growth rate of the microbial population. On integration the result is as follows:

$$\ln(N) = k'\theta + C \tag{29.8}$$

Since $N = N_0$ at $\theta = 0$, the coefficient of integration, C, can be evaluated and the final expression for the microbial population is as follows:

$$N = N_0 e^{k'\theta} \tag{29.9}$$

The specific growth rate k' is a function of different extrinsic and intrinsic factors influencing the shelf life. Under the most favorable conditions, k' can be taken as k_{max}, accelerating the population growth and reducing the resultant shelf life. Assuming the dependence of k_{max} on temperature, one of the most important independent parameters, as Arrhenius equation type, the value of k_{max} can be estimated as follows:

$$k_{max} = k_{ref}e^{-\frac{E_a}{R}\left(\frac{1}{T_{eff}} - \frac{1}{T_{ref}}\right)} \qquad (29.10)$$

The reference specific growth rate, k_{ref}, is the value obtained for the reference temperature, T_{ref}. The minimum shelf life predicted by any of these functions would be the predicted shelf life. This theoretically determined shelf life should be adjusted allowing a safety factor to arrive at a more realistic shelf life of the food product.

Pulungan, et al. (2018) used the above-mentioned analysis to determine the shelf life of an apple brownie. They conducted the experiments at 25°, 35°, and 45°C for FFA, a_w, and organoleptic quality. The shelf life of an apple brownie based on FFA was determined to be 110, 54, and 28 days at 25°, 35°, and 45°C.

29.4 Determination of Shelf Life

The following methods are generally employed to estimate the shelf life of a food product.

29.4.1 Direct Method

Experiments are conducted using a food product and storing it under the most likely conditions before consumption. The factor (s) that affect the shelf life and tests that should be conducted to determine the effect of a particular factor on shelf life are first of all determined. Samples are drawn at predetermined storage periods and the tests are conducted. The data are analyzed and the shelf-life is evaluated. Prospective consumers are asked for their opinions about the suitability of the food product. This method is quite reliable only if the conditions of storage and handling during the experiment are truly representative of the actual consumption conditions. The tes-consumers should be representative of the actual consumer population.

This method takes a long time to determine the shelf life and is expensive. This method is also used to estimate the shelf life of ambient-stable acidified foods such as salad dressings and sauces. Estimation of shelf life of chilled cooked meat products are carried out using the direct method. The study needs to be repeated if there is any change in the product formulation and/or processing.

29.4.2 Indirect Methods

There are several indirect methods of shelf-life determination; each method has its merits and demerits.

Challenge Test

This method entails inoculating the food product with the target microbe and observing the time duration until the microbial population in the food product increases above the safe threshold level. The test is based on the real determination of the duration for the target microbe only. However, the product may get contaminated with other microbes for which the tests are not performed. If N number of microbes are involved, a minimum N number of such trials would be required. Still, the result may not be reliable as the issue of microbial interactions cannot be addressed in this manner. Another limitation is the true representation of the storage conditions.

Predictive Microbiological Method

As the name suggests, this method is based on using mathematical and statistical tools for determining the safe storage/shelf-life periods based on the data available in the literature or collected under laboratory conditions. Depending upon the accuracy of the analytical tools and the data precision, the predicted shelf-life estimates can be highly reliable or unreliable. Some of the predictive models in vogue are Growth Predictor, ComBase, FORECAST (by Campden and Chorleywood Food Research Association (CCFRA)), and Pathogen Modelling Program from USDA.

The predictive microbiology methods are fast and reasonably reliable. However, it should be remembered that a microbial population under an actual situation may be a lot different in comparison to the population for which the predictions are available. Realistic predictions require that the initial microbial load, variations in food matrix, likely competition among different microbes, and the physiological state of microbial growth are all considered in the prediction models.

Survival Method

This method consists of asking the potential consumers whether they would accept the food products that were manufactured at different dates. It is a sort of opinion poll regarding the acceptability or otherwise of these product samples and, thereby, deriving a shelf life and the perceived product quality relationship. The survival method is necessary but not sufficient to estimate the shelf life. In reality, the survival method along with another test would meet the requirements of shelf-life determination.

Accelerated Test

A food industry would like to launch its food product as expeditiously as possible to capture the market before competitive products are introduced by other industries. This requires shelf-life determination using accelerated testing. Sets of extrinsic factors are selected such that the shelf-life estimation can be made in a reasonably short period. The results of the accelerated test are then suitably interpreted for realistic sets of parameters.

Sensor Technologies

Sensors can be used as dynamic and simple methods to evaluate shelf life. Sensors, in the form of stickers on the food product packages, aim to signal the shelf-life deterioration and to indicate the time when the product needs to be pulled out of the supply chain. Time-temperature indicators, which could be either mechanical, electrical, electro-chemical, enzymatic, or microbiological sensors, have been found to be effective visual indicators of shelf-life deterioration. There are hundreds of patented indicators available today. One such indicator is FRESHCHECK based on polymerization reaction from TEMPTIME (www.freshcheck.com).

Check Your Understanding

1. What is the importance of shelf-life determination for food processors?
2. What are the different factors affecting the shelf life of a food?
3. What is redox potential and how does it affect the shelf life of a food?
4. How does the texture of raw material affect the shelf life of a food?
5. What are the different indirect methods for determining shelf life of a food product?

6. Write short notes on
   a) Effect of pH on shelf life;
   b) Challenge test for determination of shelf life;
   c) Use of sensors for shelf-life assessment.

BIBLIOGRAPHY

Floros, J. D. dan, Gnanasekharan, V. 1993. *Shelf Life Prediction of Packaged Foods: Chemical, Biological, Physical, and Nutritional Aspects.* Elsevier, London

Pulungan, M., Sukmana, A. D. and Dewi, I. A. 2018. Shelf life prediction of apple brownies using accelerated method. *IOP Conference Series: Earth and Environmental Science* 131, 012019. doi:10.1088/1755-1315/131/1/012019

Robertson, G. L. 1999. *Shelf Life of Packaged Foods, Its Measurement and Prediction in Developing New Food Products for a Changing Marketplace.* CRC Press, Florida

30

Food Quality, Safety, and Hygiene

Food quality and safety are essential features of food and its trade today. The declaration of June 7 as World Food Safety Day (WFSD) by the United Nations Organization (UNO) since 2018 underscores the increasing importance of food safety today and tomorrow. Adherence to predetermined quality attributes of nutritional composition, processing, packaging, and marketing determine the price that a food product commands. However, the food quality becomes relevant only when the food product is certified to be safe for human consumption. Today, one out of every ten persons globally falls sick attributable to the consumption of unsafe food. Food safety consists of all the important practices that any food business must use to ensure that the food is acceptable for its consumption in the recommended way. Thus, food safety is a necessary condition for any food business. Additional levels of food quality would create differentiated food products suiting different strata and categories of consumers. While safety is essential and a non-negotiable feature of a food product, quality can be a compromise between price and the intended consumption.

30.1 Food Hygiene and Sanitation

Hygiene owes its origin to *hygeia*, the presiding Greek goddess of cleanliness, sanitation, and health. It is the process of cleaning an environment of all factors, including microorganisms, that may cause health problems. In the context of food, hygiene constitutes all actions required to control food hazards so that the food remains safe for intended human consumption. Food hygiene is the guiding principle for preservation and processing of food such that safety of the food for its intended consumption is ensured.

While food hygiene and food safety are related, they are not the same. Food safety consists of all the important practices that any business must carry out to render the food fit for its intended consumption. In fact, food hygiene is a component of food safety. Clearly, food safety cannot be ensured without assurance of food hygiene.

When food is prepared at the individual or the family level, food hygiene is not a serious concern because issues pointed out earlier are easier to take care on a small scale as compared to an industrial level. While food hygiene refers to practices at individual and family levels, food sanitation refers to the creation of a clean and hygienic environment at the food industry level and making the environment sustainable.

30.1.1 Sources of Contamination

Contamination is the unintentional presence of objectionable, unwanted, or harmful material, e.g., microorganisms,

chemicals, and dilutants before, during, or after processing. When the objectionable, unwanted, or harmful materials are intentionally added, it is termed "adulteration." The contaminants and adulterants make the food products unsafe for human consumption.

The contamination may originate from polluted air, water and soil, or the intentional use of such chemicals as pesticides, animal drugs, and other agrochemicals. Contamination is the main concern for food hygiene, leading, subsequently, to public health concern and trade disputes.

30.1.2 Reduction of Contamination

Food contamination probability can be minimized by undertaking the following steps.

a) Adequate personal hygiene: Foods can get infected with pathogens by not following strict hand washing and, thereby, transferring trace amounts of fecal matter from hands to the food.

b) Avoidance of cross-contamination: Cross-contamination is the term denoting the physical transfer of harmful microbes from one person, object, or place to another.

c) Selection of correct storage and cooking temperatures: There are microbes, including pathogens, present in a raw or cooked food. These microbes grow, especially during storage to levels that become harmful to consumers. The temperatures used for cooking as well storage need to keep the growth of microbes under check.

d) Avoidance of food contamination by animal waste: Even healthy animals raised for food may contain many foodborne microbes, which can contaminate other foods by not following proper hygiene. One such situation is that of meat and poultry becoming contaminated during slaughter by small amounts of the contents in the animals' intestines. There is also the possibility of fresh fruits and vegetables becoming contaminated by using the water contaminated by animal manure or human sewage.

30.1.3 Cleaning in Place

The objective of practicing cleaning in place (CIP) protocols is to prepare equipment hygienically for food processing without dismantling the system. The CIP protocols include all the mechanical and chemical systems necessary for the purpose. The underlying principle is to pump the required cleaning agents through the path that the product traversed in order to remove the product soil from internal surfaces by cleaning, rinsing, and sanitizing the processing equipment. The CIP protocols for different commodities are different.

Since CIP systems are designed for and integrated into the processing equipment, product and employee safety are enhanced. The main economic benefits include sustainable product quality and reliability, reduced product recalls, greater employee efficiency and productivity, reduced loss of production time, water and energy savings, and lower cost of effluent treatment. Five steps of CIP protocols for dairy and beverage processing plants are as follows.

1) Pre-rinse cycle: It removes most of the remaining residues, wets the inner surfaces, dissolves sugars, partially melts fat, and carries out the non-chemical pressure test.
2) Caustic wash (60°–80°C): It is meant to soften the fats for easy removal. Alkali used in the wash has high pH and the concentration ranges from 0.5% to 4.0%.
3) Intermediate rinse: Freshwater is used to flush out the residual detergent traces left from the caustic wash.
4) Final rinse: Deionized, reverse osmosis treated, or water from a city supply is used to wash out the cleaning agent residues.
5) Sanitizing rinse: It is usually carried out before starting the next production batch with a view to minimizing the probability of the presence of microorganisms in the processing line.

Today food technologists prefer use of peracetic acid (solution of hydrogen peroxide + acetic acid) in place of bleaches for better performance of CIP.

In addition, one or more of the following steps may also be followed the general five-step process.

a) Push-out: carried out before the pre-rinse step and is meant for better product recovery;
b) Acid-wash (55°–65°C): dissolves mineral scales on account of hard water or protein residues and neutralizes system pH;
c) Air blow: gets residual moisture out of the processing line.

30.2 Food Safety and Quality

The main purpose of a food safety assurance plan is to make sure that, when the food is prepared and consumed as per its intended use, the consumer will not feel any discomfort or fall sick. Thus, the health risks that are likely to occur from the food are reduced. Food quality, on the other hand, takes into consideration the food characteristics that are acceptable to consumers. ISO defines food quality as the sum total of product features and properties needed to satisfy the needs of the customer.

30.2.1 Food Safety Attributes

As indicated above, any food safety system needs to protect public health through the minimization of risks associated with unsafe food. This expectation is fulfilled by ascertaining that the preparation of the foods has meticulously followed the safety standards formulated for the purpose. Food safety components include hazards (physical, chemical, and microbial), allergens, radiation hazards, commercial frauds, legal risks, and pests.

Food safety is enforced through an education, research, and communication-based interaction among the food industry, government, and consumers. The food industry produces, processes, imports, prepares, and distributes food. The government, on its part, undertakes the duties of education, research, formulation of laws, enforcement, surveillance, and regulation. Consumers contribute through feedback, education, and market demand.

30.2.2 Food Quality Attributes

Ten main attributes that characterize the quality of a food product are nutrition, size, thickness, viscosity, texture, succulence, color, flavor, consistency, and turbidity. Some food technologists consider consistency, viscosity, turbidity, and succulency as components of texture itself. The sensory quality of a food is based on the human senses and consists of four components, i.e., to be sensed by the eyes (appearance), to be sensed by the nose and tongue (flavor), to be sensed by the nervous system (texture/kinesthesis), and to be sensed by the ears (sound). Kinesthesis refers to the group of attributes that a consumer evaluates through a sense of feel, especially by means of the mouth.

30.2.3 Food Safety Hazards

The aim of a food safety plan is to ensure that the food consumed according to its intended use will not cause any disease or health hazard. The probability that the food, if not consumed in an intended manner and quantity, will cause harm or injury is defined as a hazard. The hazards are classified as physical, chemical, and biological.

Physical hazards, such as hair, wood, plastics, soil, stones, metal, and even parts of pests, can cause disease or injury to the person when consumed as a component of food.

Chemical hazards include harmful substances found in food. This hazard category consists of a large range of substances, such as chemical residues, pesticides, preservatives, food colors, toxic metals, polychlorinated biphenyls, and many other additives.

Biological hazards include the presence of all unintended life-forms, e.g., bacteria, viruses, and parasites. Food-borne pathogens are the disease-causing category of microorganisms present in food. Food-borne pathogens cause infections and poisoning.

30.2.4 Essential Requirements for Food Safety Assurance

Food safety assurance includes the following three critical components.

a) Pre-harvest food safety plan (on-farm food safety)

It aims to assist in implementing good on-farm practices at the producers' level and in building quality consciousness at the grass-roots level. The first category of stakeholders of on-farm food safety plan includes the agricultural inputs suppliers and farmers involved in food production, Farmers receive pesticides, veterinary drugs, and other agricultural chemicals for food production. This category poses food safety risks and requires specific attention. Animal feed poses risks because it may contain pathogens or toxic chemicals. Unsafe levels of pesticides, fertilizers, and veterinary drugs have been found to be present in foods such as red meat, poultry products, farmed fish, and agricultural crops.

Certain foods serve as vehicles of health hazards. For example, animals raised on the farm for meat and milk require special attention be given to animal hygiene. Similarly, the animals slaughtered at abattoirs must receive proper hygiene treatment.

b) Post-harvest food safety plan

The objective is to help in implementing good management practices at the processing plant level, including storage, handling, packaging, and distribution systems, that will bring quality improvement and establishment at all stages of post-harvest processes.

i) Note that there is a wide range of processed foods that use complex and highly technical methods of manufacture for a variety of purposes, such as food safety assurance, shelf-life extension, spoilage reduction, and trade facilitation. There is a high risk of food safety hazards in all such cases. There is a need to strictly apply food safety management based on HACCP principles in order to minimize the risks.

ii) Food retailing is another food safety hot spot whether it is the sale of foods in supermarkets, shops, market stalls, or by street-food vendors. Food is prepared, handled, stored, transported, and sold by this class of stakeholders in food value chains, having considerable influence on food safety. Therefore, whether in formal or informal sectors, these food service providers need to be sensitized and trained with regard to hygiene, safety, and pest and sanitation control. It is common knowledge that incorrect food handling, storage, and preparation increases the intensity of foodborne diseases when such food is consumed.

c) Education and training for human resource development

Foodborne illness is a large and growing public health problem. There is a significant increase over the past few decades in the incidence of diseases caused by pathogens and parasites in food. Therefore, human resource development is an important aspect of a food safety assurance plan. Such a team of trained people will also help in bringing about sustained improvement in products and in the supply of safe and hygienic food. It is because of a lack of technical and trained manpower that an effective institutional framework cannot be put in place. Insufficient knowledge about the underlying hazards and risks, combined with poor financial resources, will certainly lead to poor food safety assurance.

30.2.5 Food Quality and Safety Management

Food quality and safety management is required to be robust with adequate science inputs and a background of risk analysis and prevention. In addition, the safety management system should be based on a national food law that provides for comprehensive surveillance and monitoring activities for risk analysis. It is important that because conflicting views at times of crisis can create chaos, there must be one national voice responsible for food safety. This individual must be given the authority and resources to formulate and implement science-based food policies for all national activities. The national food policy requires allocation of adequate resources for active involvement of stakeholders from state and local governments, industry, and consumers, and for the discharge of their roles and responsibilities in the context of effective food safety system.

The two classical approaches for ensuring food quality are quality control and quality assurance. The process of quality control begins after the food has been produced or processed. Random samples are drawn, the quality is assessed, and the conclusion is drawn whether the samples are acceptable or not. If the samples are not acceptable, the sampled food lot is destroyed, and a huge economic loss is incurred. Alternatively, quality assurance requires that the probable occurrences of quality compromises are foreseen, measured solutions applied, and results are verified and documented. The complete process of quality assurance, in contrast to quality control, is a dynamic approach to ensure that the problems, if they occur, are nipped in the bud. Obviously, the quality assurance approach is sustainable and customer satisfaction is guaranteed. Moreover, the approach, better known as Total Quality Management (TQM), encourages the participation of all employees of the manufacturing unit, including suppliers, in this endeavor.

Agricultural produce from production catchments is taken to processing industries, storages, and wholesale markets located largely in urban areas since the required skills and infrastructure are not available in the production catchments. During the entire food production to consumption value chain, the processed food's safety for human consumption must be ensured in a way that product quality remains commensurate with consumer expectations.

In addition to local food production and processing, food products in any country are sourced from all corners of the globe, generating increased vulnerability of its citizens to food quality and safety related issues including contaminants, pathogens, diseases, and adulterants. The way to improve the availability of safe food requires real time functional linkages among the food protection agencies at the local, state, and national levels. Every country needs a food safety system in place that is efficient, integrated, and seamless. The system defines the roles and responsibilities of various functionaries in clear terms, permits as much information flow as required among government agencies, and recognizes the contributions of all parties. Resultantly, the mechanism yields food safety results that have a higher degree of uniformity across the nation.

Good Agricultural Practices

Raw materials for food industries come from agriculture. If agricultural practices are not appropriate, the produce could become a major source of hazards, creating a formidable challenge for food safety assurance. Good Agricultural Practices (GAP) are a collection of science-based rules that apply to on-farm production and post-production activities. The implementation of GAP leads to cleaner and safer produce with due respect for workers' health, environment friendliness, economics, and sustainability. These are voluntary guidelines for farmers to reduce the risks associated with chemical and biological contaminations in their production. Complete elimination of risks is impractical; however, GAPs can guide farmers to reduce the contamination risks where possible. GAP for farmers is a set of modules indicated as follows.

1. Record keeping and internal self-assessment;
2. Site history and site management;
3. Worker health, safety and welfare;
4. Wastes and pollution management;
5. Environmental management;
6. Complaints;
7. Traceability;
8. Visitors' safety.

Each of the above-mentioned modules enumerates criteria/requirements for controls required for implementation on a single farm or a group of farms. Each of the modules gives a checklist in tabular form containing verifiable indicators for the criteria. This checklist along with checkpoints can be used by farmers for self-assessment before auditors begin their certification process. GAP allows structured feedback against each criterion. The feedback from clients and continued scientific developments keep the GAP dynamic. Compliance criteria/requirements can be critical, major, or minor. Critical criteria compulsorily require 100% compliance. In case of major criteria, 90% compliance is compulsory. However, in case of minor criteria, the mandatory compliance level is 75%.

Good Hygiene Practices

According to the Codex Alimentarius Commission, Good Hygiene Practices (GHP) are those that prevent food contamination at all stages from primary production to processing and handling of the final product so that consumers could have access to safe food. Contamination arising out of improper practices is the prime source of foodborne diseases. The following are the eight broad categories of operations requiring GHP.

- Personal hygiene is not limited to hand washing only. It also includes practices and instruments related to personal care. Therefore, personal hygiene extends to such activities as sterilization of instruments used for tattoo marking, hairdressing, and body piercing.

- The design of facilities and equipment for food storage, processing, and handling requires special considerations while constructing the premises and preparing the operating plans from the point of view of minimizing contamination. Special considerations also include proper maintenance, cleaning, disinfection, and protection against pests.

- Primary production (e.g., farming and animal husbandry) considers implementing hygienic practices to minimize the probability of attracting those hazards that may become unwieldy and impossible to control later. Examples of hazardous materials are microorganisms, pesticides, mycotoxins, and antibiotics in foods eaten raw or fresh.

- Establishment maintenance, cleaning and disinfection, and pest control can be effectively carried out when consideration is given in locating the premises, equipment, surfaces, and facilities. This convergence of design and operation efforts is effective in protection against pests and minimization of contamination.

- Control of operations is based on HACCP to keep food hazards away by following hygiene control, control of materials inflow, and smart management practices.

- Enough details about food hygiene are available for product and consumer information so that informed choices could be made toward prevention of contamination and checking the growth of food-borne pathogens by storing, preparing, and using the food correctly.

- Food, during transportation, needs protection from contamination sources and damage so that it does not become unsuitable for consumption. The enclosed environment during food transportation must not be congenial for the growth of pathogenic microorganisms and the production of toxins.

- Training and competence include the creation of awareness and the establishment of training programs, including refresher trainings and supervision.

Good Manufacturing Practices

Good manufacturing practices (GMP) are the concern of manufacturers/processors and are meant to encourage them to take proactive steps, as a part of quality assurance, to make sure that the manufactured products can be consumed. GMP enable the manufacturers to put in place a system whereby contamination is either minimized or eliminated altogether while completely avoiding false labeling. GMP help the manufacturers/producers to become food safety reliant.

Good handling practices present an inclusive approach from production to consumption to identify potential risks and their sources and to suggest the steps to be undertaken to minimize contamination risks. Essentially, all persons involved in handling food need to have good personal hygiene and keep the relevant equipment sanitized.

Hazard Analysis Critical Control Point System

Effective management of food safety hazards is enhanced through the use of Hazard Analysis and Critical Control Point (HACCP) systems. On the basis of their in-house research, Pillsbury Company in the late 1950s developed HACCP and subsequently used it to provide safe food for American astronauts. HACCP is a mechanism, a protocol, for process control to be able to produce, process, and provide safe and wholesome food for consumers.

In food production, food safety is of greater significance than food quality and the HACCP system is the method to achieve such safety. The basis of HACCP is common sense, and it aims to prevent the food from becoming unsafe. HACCP protocols require a critical assessment of the entire process so as to identify the points along the food production chain where the probability of hazard occurrence can be high. The risks associated with the hazard are evaluated. Then appropriate measures are taken to contain the hazard and the results are monitored and verified.

Ever since HACCP was developed in the late 1950s, people have become aware of its usefulness and its implementation in food sector. It is now accepted globally for food safety assurance. Implementation of HACCP is a preventive approach for ensuring food safety. Unlike the traditional practice of end product inspection, HACCP saves time, detects the hazards at an early stage of processing, initiates appropriate remedial action right away, ensures a good quality product, resulting in higher economic returns, sustainability, and brand value. Resources are efficiently utilized by all functionaries in the value chain for assuring food safety. Although GAP, GMP, GHP, and HACCP are tools to assist and facilitate the goal of making safe food available, the primary responsibility lies with the producers, processors, suppliers, food handlers, and traders.

A meticulous exercise is carried out to determine those points along the food production and processing chain where contamination can occur. Such points are called critical points and methods are then developed to apply appropriate controls on the critical points to obviate the hazards. HACCP protocols rely very heavily on bookkeeping for quickly locating the origin of a specific problem. The safety program effectiveness requires a reliable verification system when carrying out the HACCP process.

At the present state of scientific inquiry, it is not possible to mandate HACCP programs for all products and processes. Even then, the incidence of foodborne contamination can be minimized through adopting suitable science-based strategies. The seven HACCP principles are stated as follows.

1. Carry out the hazard analysis of the process;
2. Identify the critical control points (CCPs) in the process;
3. Specify the critical limits for each CCP in the process;
4. Develop a monitoring system for each CCP;
5. Specify the corrective actions for process deviations;
6. Record keeping;
7. Establish verification procedures.

Establish standard operating procedures (SOPs) and good manufacturing practices (GMPs) before undertaking the development of a HACCP plan for a product. The seven principles of HACCP provide confidence about the correctness and verification of the food safety assurance. When the verification is performed by independent nationally or internationally accredited agencies, it becomes a certification procedure. ISO 22000 is the food safety management system that constitutes an internationally recognized certification procedure.

Let us consider a potential hazard as the presence of Enteric pathogens, e.g., *Salmonella* and *E. coli*. The hazardousness has been justified since enteric pathogens have been observed to be associated with outbreaks of foodborne illness when undercooked ground beef is consumed. Therefore, it is a critical point for which cooking is the control measure.

30.3 Food Regulations

Food regulations address the food safety and quality issues for the following domains.

- Design of food processing plant and subsequent construction, cleanliness, and worker hygiene;
- Food raw materials, their physical and chemical compositions (including contaminants and adulterants), microbiological quality, and processing characteristics;
- Correct and unambiguous product labeling.

A food processor or food handler will incur severe penalties in case of failure to comply with the regulations. Defaulters may, in extreme cases, be forced to close the processing plant and pay heavy fines. There is a greater awareness among consumers about food safety and quality and consequences in terms of falling sick and medical costs. Therefore, it is not uncommon these days for consumers to hold processors/food handlers responsible for deficiencies, if any, in products and services and approach designated authorities for a redress of their complaints. In such cases, there is a greater likelihood of inspection of production premises and products leading to the drafting of charges and the awarding of punishment.

Food is a matter of life and death. Therefore, laws are stringent against those who do not abide by food laws. It is essential for food processors to know the legal requirements applicable to their food operations. All those concerned with the management of the food plant must keep themselves well informed about the applicable latest laws. Food processors, handlers, and traders must ensure that food safety is never compromised during the production to consumption value chain.

Food exporters need to understand the legal positions in the country of food manufacture and the destination country to avoid incurring penalties. In case of exports, for example, the product being exported may need reformulation and/or special labeling to meet the legal requirements of the importing country. It is important that appropriate protocols, in line with the prevalent regulations and standards, are formulated and practiced for successful food processing enterprises.

30.4 Food Standards

Standards form an essential component of any food safety and quality framework. Both safety and quality components for all aspects of food production and services are embedded in food standards with a view to achieving safety and wholesomeness throughout the food supply and trade. The standards are formulated at international, national, regional, and company levels. International standards are published by the ISO and the Codex Alimentarius Commission (CAC).

The government of India enacted the Food Safety and Standards Act (FSSA) in 2006, consolidating a multitude of food safety–related laws under one umbrella. The implementing agency is the Food Safety and Standards Authority of India (FSSAI), established under the FSSA in 2006. The FSSAI formulates science-based food standards and regulates the manufacture, storage, distribution, sale, and import of food.

The Bureau of Indian Standards (BIS), set up under the Ministry of Consumer Affairs, formulates standards for foods and various other consumer goods. BIS also operates a voluntary certification scheme.

30.5 International Initiatives

Several international organizations and agreements have come into existence, especially since World War II, to address the safe food trade as well as research and training in safety, quality, and security of food. The Codex Alimentarius Commission (CAC), International Standards Organization (ISO), World Trade Organization, World Health Organization (WHO), and Food and Agriculture Organization (FAO) are the main international bodies. CAC seeks to protect the health of consumers through the formulation of international food standards. The ISO and CAC work jointly to publish international food standards. The CAC has published the Codex Alimentarius which means "Food Code." The Food Code is a collection of food standards that have been internationally adopted. The Food Code includes not only standards and codes of practice but also guidelines and other recommendations relevant in the context of the health of consumers and of fair trade. Codex standards serve as the basis for national food standards established by different countries.

While CAC undertakes to facilitate the production, processing, and trade of safe food, the ISO promotes standardization and related activities to facilitate the exchange of goods and services globally. The ISO is a non-governmental federation of national standards bodies (ISO member bodies). The ISO seeks to promote international cooperation in the areas of intellectual, scientific, technological, and economic activities for food and all other commodities. International standards published by the ISO stem from cooperative efforts by the countries of the world and, therefore, are acceptable to all.

Check Your Understanding

1. Write short notes on
   (a) Food safety hazards;
   (b) CIP;
   (c) GAP;
   (d) GHP;
   (e) GMP.
2. Differentiate between
   (a) Food safety, food hygiene, and food sanitation;
   (b) Food regulations and food standards.
3. Name the different food safety and food quality attributes from the point of view of the producer and the consumer.
4. What are the different steps in HACCP? Explain with an example from a food processing industry.
5. What are the different sources for contamination of food in a food production and processing chain? Explain the steps that can be taken for reducing contamination.
6. What are the essential requirements for food safety assurance?
7. Name the international agencies related to food standards and trade and the type of involvement of these agencies.

BIBLIOGRAPHY

Jood, S. and Khetarpaul, N. 2009. *Food Preservation*. Agrotech Publishing Academy, Jaipur.

Khetarpaul, N., Jood, S. and Punia, D. 2011. *Food Analysis*. Daya Publishing House, New Delhi.

Manay, N. S. and Shadaksharawamy, M. 2008. *Foods, Facts and Principles* 3rd ed., New Age International (P).Ltd., New Delhi.

Mudambi, S. R. and Rajgopal, M. V. 2007. *Fundamentals of Foods, Nutrition and Diet Therapy* 5th ed. New Age International (P) Ltd, New Delhi.

Sethi, M. 2008. *Institutional Food Management*. New Age International (P) Ltd., New Delhi.

Subbulakshmi, G. and Udipi, S. A. 2006. *Food Processing and Preservation*. New Age International (P) Ltd., New Delhi.

Swain, S., Ali, N., Mangaraj, S. and Dash, S. K. 2017. *Agricultural Process Engineering Vol IV (Process Machinery, Standards and Research Management)*. Kalyani Publishers, New Delhi.

31

Effluent Treatment

Practically, all food processing operations use raw materials along with water, energy, and some other resources and result in a) desirable and b) undesirable products. Desirable products are utilized for food purposes and the undesirable category of products are termed as wastes and effluents and need to be suitably treated before their disposal. These effluents can be in solid, liquid, or gas form. The solid wastes and effluents require treatments to render them harmless to human life and the environment. Effluent is derived from 'effluere', a Latin verb which means 'to flow out'. In the present-day context, effluent is any non-useful material in fluid form that is discharged in to natural environment, be it a landfill, water stream or air. Solid wastes in case of fruit and vegetable processing would include leaves, stem, pomace, seeds, peel, rind, spoiled raw material, etc. Solid wastes are also treated for their safe disposal to avoid legal action and help minimize environmental impact.

No matter what the extent of treatment before discharging the wastes and effluences, the environment does get adversely affected, leading to poor air quality, contaminated water and diseased soils. The world is already seized of the seriousness of the problem and both mitigation and adaptation measures are being developed to deal with the waste and effluent disposal issues. ISO 14000 is a set of standards that can help the food processors, food handlers, and food operators in minimizing the impact of their operations on the environment. Lately, there is increasing interest in circular food economy because it aims at eliminating the wastes and effluents altogether by cascading the food production and processing operations in such a way that the by-products of one production /processing operation becomes the raw material of the next production/processing operation and so on until the by-products of the previous operation become the raw material of the specific operation.

The purpose of this chapter is to characterize the food processing wastes and effluents and understand the different available methods of treatment/equipment so as to be in a position to apply the technologies under field conditions.

31.1 General Principles

The concept of waste is akin to that of weed. It cannot be claimed that the material being wasted has no value, except that we in society have no practical means of utilizing it. Take, for example, the case of a mango pulp processing unit. After squeezing out and extracting the pulp, the remaining fruit skin and stone become a waste product or, at most, composting material. If practicable technologies are available, these fruit skins and stones can become the raw materials for manufacturing various valuable food and non-food products. There are innumerable cases where by-products and effluents need to be discarded for want of suitable technologies.

Every waste and effluent disposal proposition needs to be considered in depth to determine if it could be minimized through some processing breakthrough. There is an effective strategy of five "R's" (Refuse, Reduce, Reuse, Repurpose, and Recycle) for waste and effluent management. Refuse is the first step toward efficient waste and effluent management strategy. It involves selecting such materials whose utilization will indeed reduce the production of wastes and effluents. An example is to select a packaging material that can be reused in place of single-use plastic films. The next step is to reduce the use of harmful, non-recyclable, and wasteful materials to minimize the amount of their disposal. Reuse entails determining if the material being intended for disposal could be reused by adopting some minimum treatments. For example, there is no justification to discard a PET bottle as long as its structural integrity is assured. If it cannot be reused, it can be altered for use in another application, i.e., repurposed. Finally, if it is not possible to reuse or repurpose, then consideration should be given to recycling the material meant to be discarded.

Reduce wastes and effluent to the greatest extent possible and then determine if the remaining wastes and effluent can be reused as a raw material or if they can be recycled to make another valuable product. If nothing else, the waste material or the effluent should be processed to recover energy. Whey in the dairy industry is an effluent that can be converted into a beverage or it can be digested microbiologically to recover energy in the form of biogas. Thus, Reuse, Recycle, and Recover energy is an excellent management strategy for food industry wastes and effluents.

31.1.1 Wastes and Effluent Characteristics

Oxygen in the water stream needs to be balanced with the oxygen demand of the wastes. The oxygen demand consists of two components; chemical oxygen demand (COD) and biochemical oxygen demand (BOD). Thus, the total oxygen demand for degradation of the wastes is the sum of the COD and the BOD.

Chemical oxygen demand (COD) represents the quantity of oxygen needed to chemically oxidize the organic matter and the inorganic nutrients at a given temperature. It is expressed as $mg\ L^{-1}$, and varies in the range of 5–50 $mg\ L^{-1}$ for rivers, 25–250 $mg\ L^{-1}$ for polluted rivers, 500–1200 $mg\ L^{-1}$ for raw municipal sewage, and 1000–50000 $mg\ L^{-1}$ for contaminated industrial effluents.

Biochemical oxygen demand (BOD) is the amount of oxygen required by microbes to decompose the organic constituents at a given temperature under aerobic conditions. Like

DOI: 10.1201/9781003285076-39

COD, BOD is also expressed in terms of mg of oxygen consumed per liter (mg L^{-1}) of the sample. Clean river water will have a BOD value of about 1.0 mg L^{-1}, whereas the BOD value for an untreated sewage can be 600 mg L^{-1}. BOD in kg for an amount of waste can be obtained from laboratory determined BOD as follows:

$$BOD \ (kg) = (C \times V) / 1{,}000{,}000$$

where, C is the BOD value in mg L^{-1} or ppm and V is the volume of wastewater in liter.

Total oxygen demand (TOD) is the amount of oxygen required for complete combustion of a total organic carbon (TOC) present in the sample, measured by a TOD analyzer and expressed as mg L^{-1}. Correlations between TOD or TOC and COD / BOD for a given sample can be determined.

Acidity/alkalinity, pH, suspended solids, flow (rate of effluent flow, say, Ls^{-1}), oils and greases, chloride, toxic compounds, etc. are other parameters that can characterize the effluents.

As the food processing wastes and effluent contain organic materials with high BOD and COD values, which produce a foul smell and have the potential to contaminate soil, water, and air, landfill is not a preferred disposal method. Instead, it is better to either convert the solid wastes from food processing industries into value added products or, if nothing else, make slurry, treat it to reduce BOD and COD values, and then discharge the treated slurry into a water body.

The major food processing sectors contributing to wastes are a) fruits and vegetable processing, b) dairy processing, c) meat processing, and d) oil processing.

Fruits and vegetables are high moisture commodities and the majority of the processed products also contain high moisture. These processed products invite rejection if the product quality goes bad during processing or post-processing handling and storage. In addition, non-edible portions of fruits and vegetables, such as peel, pomace, seeds, rind, and leaves, become part of rejections.

The dairy industry produces large volumes of mainly liquid wastes containing whey, oils, lubricants, suspended particles, traces of cleaning agents, organic compounds, etc. The composition of the liquid effluent will vary according to the processes and products manufactured from milk.

Meat industry wastes originate at slaughterhouses and processing plants. Slaughterhouse wastes may include animal/bird body hair, blood, fats, manure, bones, and feathers. The waste stream from the meat industries has been found to contain a fairly large fraction of solids (25%–30%).

The oil processing industry's major output is edible oil. The wastewater, organic solids, and inorganic residues make up the waste stream.

Substantial wastewater volumes and solid wastes are generated from fish processing, sugar industries, and food grain industries.

31.1.2 Natural Purification Ability of a Water Stream

When some quantities of liquid and solid wastes are discharged into a water stream, the concentrations of the organic and inorganic ingredients of the wastes keep decreasing downstream mainly due to dilution, oxidation, sedimentation, and microbial degradation. Water has some dissolved oxygen depending on its temperature and this dissolved oxygen is responsible for organic matter degradation through oxidation. The temperature dependence of dissolved oxygen (DO) in water is depicted in Figure 31.1. In the range of 0°–50°C temperature, the DO values vary between 14 and 5 ppm. The biodegradation process of the organic waste begins soon after it enters the water stream and continues until the waste is completely exhausted. Since the stream water is always in the process of mixing, the waste gets mixed with the large amount of water in the stream, reducing the waste concentration. Some of the non-biodegradable wastes either settle down or remain floating and undergo some mechanical fragmentation due to hydraulic stresses. There is microbial population in the water stream and these microbes undertake further digestion of the waste. The waste matter keeps decreasing due to the combined action of physical, chemical, and biological forces as it moves along the water stream. At sufficient distance downstream, the water in the stream is almost as pure as it was before the waste was discharged into the stream. Depending upon the flow and its microbial and chemical compositions, each water stream has its waste-carrying capacity, which can be determined from the waste composition.

31.1.3 Water Usage and Wastewater Generation

Food processing industries require large amounts of water for washing, peeling, blanching, steam generation, heating/cooling, etc. The freshwater requirements vary from 7 to 30 liter

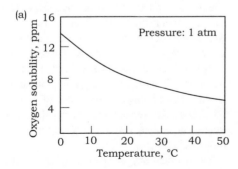
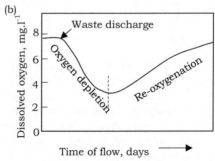

FIGURE 31.1 Conceptual presentation of (a) capability of water to dissolve oxygen at different temperatures, and (b) variation of dissolved oxygen with time in a water stream

TABLE 31.1

Characteristics of a few selected commodities[a]

Commodity	Hydraulic load L/ kg	BOD, mg L^{-1}	SS, mg L^{-1}	Organic load, mg L^{-1}	pH	Temperature, °C
Apple	12	10	3	13	5.6	54
Citrus	16	7	3	18	6.5	79
Potato	16	24	20	44	–	–
Tuna fish	26	22	–	–	–	–
Fish processing	20	3000	2500	–	–	–
Whey	10	20,000	56,782	76,782	4.1	–

[a] Katsuyama, 1979

per kg of raw fruits or vegetables. Some of the freshwater is reused in a bid to reduce the net water requirements. The reuse percentage can be as high as 70% in some cases, such as concentrated tomato juice products.

BOD and suspended solids together constitute the *organic load*. The amount of water in the wastewater is called the *hydraulic load*. The hydraulic and organic loads together make up the *raw waste load* (RWL). Effluent characteristics relevant for effluent treatment include the hydraulic and organic loads. Representative values for a few commodities are listed below (Table 31.1).

Waste generation is influenced by the type of commodity, raw material quality, product that is being manufactured, plant size, age of the plant, processes and equipment, processing rate, and plant hygiene.

31.2 Methods of Treatment

As explained in the previous sections, effluents from food processing industries may contain one or more a) suspended solids, b) dissolved gases, c) microbes, d) organic matter, and e) minerals. Selection of treatments depends on the effluent composition and the permissible quality parameters of the treated effluent. The four major categories for treatment include a) physical, b) chemical, c) biological, and d) incineration methods.

31.2.1 Physical Methods

The physical methods include 1) sedimentation, 2) filtration, 3) dissolved air flotation, 4) flocculation, and 5) centrifugation. The sedimentation, filtration and centrifugation methods have earlier been presented in the book; the other physical methods are explained as follows.

Dissolved Air Flotation System

Dissolved air flotation systems (DAF) are intended to cause the suspended solids in the wastewater to separate out along with the contaminants and move up to the water surface. Polymers may be introduced into the wastewater prior to DAF to facilitate flocculation. The raw untreated water is admitted to the treatment tank from the waste tank with the help of a pump automatically with flow control. Air is forced into the

DAF tank using an aeration pump to create turbulence and a circulatory motion from top to bottom. Air gets dissolved into the wastewater in the process and air pressurizes the vessel. The wastewater flow is so controlled as to leave half of the vessel to be occupied with pressurized air. The dissolved air under pressure in the vessel causes the oils and suspended material to move up, float on the surface, and get removed by skimmers. The water underneath the scum is treated clean water and can be drained off from the bottom.

Coagulation/Flocculation

Solutions of some salts are used to destabilize and increase the particle size of dissolved organic compounds in the wastewater. Alum and FeCl$_3$ are common coagulants used in the process. Lignosulfonate (LSA), bentonite (BEN), and carboxymethyl cellulose (CMC) have also been used as coagulants. Coagulation-flocculation is followed by decantation or filtration to separate the scum from the clean effluent.

Incineration

Incineration involves combustion of the organic components of the solid waste resulting in production of heat, ash, and flue gases. Heat can be captured for further use. The flue gases may also contain some toxic gas, which may require scrubbing. The biomass through incineration is reduced to only 15%–20% in weight and 5%–10% in volume.

31.2.2 Chemical Methods

Chemical methods include the use of chlorine-based sanitizing chemicals and precipitation of the suspended particles. Besides chlorine-based sanitizing chemicals, other sanitizing chemicals frequently used are quaternary ammonium compounds, amphoteric bactericides, and iodophores.

31.2.3 Biological Methods

Microbes degrade the organic compounds and, in the process, reduce the BOD levels in the effluents. Untreated effluents may have BOD levels of about 150 ppm and the permissible BOD levels before discharging the effluent in the natural environment may not be more than 20 ppm. The two common methods of biological treatment are mechanical agitation (activated sludge method) and non-mechanical methods of mixing.

Activated Sludge Method

There are two tanks; air is pumped from the bottom in one of the tanks using some sparger unit and the air bubbles rise in the tank, creating good agitation of the contents. The oxygen dissolved into the water meets the respiration needs of the microbes present in the tank, leading to greater production of a microbial biomass. The microbial biomass, called sludge, is transferred to the second tank where no agitation is provided, and the sludge settles down at the bottom. Clean water is decanted from the top of the tank, a portion of the sludge is fed back to the first tank to serve as microbial culture, and the remaining sludge is removed from the bottom of the second tank.

Trickling Bed Filters

This type of filter is a sort of packed bed containing such inert materials as pebbles and crushed rocks. The raw effluent is trickled over the bed and air is pumped through the bottom of the packed bed. When the effluent comes in contact with the pebbles/crushed rocks, there is slime (fungi, algae) formation on their surface. When the effluent and the pumped air come in contact in the presence of the slime, more biomass is produced. When sufficient biomass has been produced, it slips down in the packed bed, requiring its collection. The packed bed ensures close contact between the effluent and the air.

Landfill

A landfill is designed to be a place to deposit the non-useful biomass in such a way that it does not pose a health hazard or cause environmental pollution. The deposited biomass will eventually degrade and disintegrate over a fairly long period of time. A well-designed landfill has a lined bottom over which the solid wastes are deposited in layers. Provisions are made to collect leachates and gases and for monitoring and final capping. Landfills also become filled, requiring either another landfill or some other method of solid waste treatment.

Composting

Any organic biomass, finding no other application, can be composted for its subsequent application to farms as a soil amendment. Compost is produced by microbes present in the organic biomass or added separately. The microbes metabolize the biomass in a controlled manner, consuming the organic portion of the refuse and converting the biomass into a nutritionally rich organic matter, also called black gold. Composting is environment-friendly and reduces the bulk of the biomass to half. Vermicomposting is a process of organic matter disintegration carried out by earthworms.

31.2.4 Advanced Treatment Methods

Technological development is a dynamic process that takes contemporary science and engineering advances into account. Great strides have been made in the fields of biotechnology, nanoscience, and communication that promise better technologies for application in all fields. Several new technologies have been developed in the recent past to effectively and economically treat effluents from different manufacturing and service sectors. A few of them are indicated below.

Anaerobic Treatment

Specific anaerobic microbes clean up the organic loads. It requires setting up large reactors where anaerobic microbes are added to digest the organic waste material from the food processing industries. The resultant output from the anaerobic treatment is harmless and is in the form of useful slurries and gaseous fuels. For example, dairy plant effluent in the form of whey, when subjected to mesophilic anaerobic fermentation, produces biogas for fuel use and water suitable for irrigation.

Ozonation

Ozone (O_3) is a strong oxidizing agent that leaves no toxic residues in the treated wastewater. Ozone is 50% stronger than chlorine as a disinfectant. There are more efficient ozone generation and application methods now to make ozonation attractive for wastewater treatment. Ozonation has been found to be quite effective in removing colors from the effluents.

Photocatalysis

Electromagnetic radiation is used to produce semiconductor excitation and, thereby, generating electrons and holes to remove pollutants. TiO_2, ZnO, and SnO are some catalytic materials that have been used.

Adsorption

The process of adsorption, occurring naturally, can be utilized for the treatment of wastewater. A solid surface acts on the adsorbing surface and the molecules of dissolved substances in the wastewater adhere to the surface.

31.3 Equipment

This section discusses the equipment and machines that have been utilized for effluent treatment. Much equipment is the same as that used in food processing operations. The capacities and construction materials for treatment of effluent may differ from those for food processing. In addition, experience has enabled differentiation among different technological solutions, identifying some as more appropriate than others.

31.3.1 Settling Tank

Settling or sedimentation tanks can be either batch or continuous types. Continuous type settling tanks are also known as *thickeners* or *clarifiers*. Except for small volumes of effluents, the majority of sedimentation operations are of the continuous type. In continuous settling tanks, raw effluent is fed horizontally into the tank. Depending on the concentration and nature of suspended solids in the raw effluent, it may sometimes be useful to pretreat it for flocculation. The settled sludge at the bottom of the tank is pushed out to a discharge trench. A variant of a settling tank is a sludge blanket clarifier, where the raw effluent enters the tank at the bottom under a blanket of sludge. The blanket acts as a filter. The blanket also reduces the inlet flow velocity. The clean water is discharged as overflow. There can be other variations to facilitate the settling of the suspended solids and the removal of the sludge.

31.3.2 Filter

There is a great deal of filter variations. The two major types of cake filters are constant pressure filters and constant volume filters. In industrial applications, the pump sizes are fixed and, therefore, the preference is for constant pressure types of filtration systems. Either positive static pressure gradient or vacuum drive the filtration. The three common types of filters in widespread use are plate-and-frame filter presses, leaf filters, and rotary vacuum filters. As discussed in Chapter 12, rotary vacuum filters are of the continuous type. Filter aids are the materials added to the raw effluents to improve filtration efficiencies.

31.3.3 Centrifuge

The process of separation of suspended solids is essentially sedimentation; however, the process of sedimentation is accelerated by the use of centrifugal force. Commercial centrifuges develop as high as 20,000 times the gravitational acceleration (g) to enhance the separation efficiency as well as throughput. Immiscible and insoluble materials can very conveniently be separated out. Three configurations of centrifuges, i.e., tubular bowl, disc bowl, and basket, are mostly used.

A tubular bowl centrifuge is a long and narrow centrifuge rotating at rotational speeds of around 15000 rpm. Raw effluent is fed at the bottom of the narrow rotating bowl at the center of the centrifuge. The heavier fraction moves away from the center and the lighter fraction is removed from the center.

A disc bowl centrifuge is so named because of several concentric conical discs enclosed in a bowl. The discs have two holes all in similar locations to form a continuous path for the exit of solid particles. Disc bowl centrifuges are more prevalent in the dairy processing industry.

A basket centrifuge has a perforated drum that rotates at a predetermined speed. The feed enters the rotating perforated drum from either the top or the bottom and water is expelled through the perforations. More details of these centrifuges have been given in Chapter 12.

31.3.4 Treatment Plant

Actual effluent treatment plants will use more than one mechanism to achieve the acceptable effluent quality. A case study of poultry processing (Rao, et al, 2013) is given here. The main components of poultry processing wastes include fats, biological solids, and feathers. The objective of the waste treatment is to reduce BOD and COD of the effluent to the legally permissible levels. The processing unit produced 700,000 LPD of effluent with 2000 ppm BOD and 500 ppm suspended solids. The effluent was first pumped through a rotary screen and into the DAF system. The pH was raised to 7–8. It was then followed by coagulation-flocculation treatment. Floccin™ 1105 was the flocculant used in the treatment. This combination of treatments brought the effluent within the legal limits of the quality parameters acceptable to the municipality. In a given situation, the quantity and quality of the effluent in terms of hydraulic and organic loads need to be determined along with any other specific constituents. A suitable treatment plant is then designed meeting the local waste disposal regulations and with due regard for economics and sustainability.

31.4 Value Addition to the Waste

As indicated earlier, food processing industries produce nutritionally rich residues and by-products that should be considered for producing value-added products and extracts instead of discarding them as worthless. A 2019 report from the FAO notes that India, China, the USA, and the Philippines generate huge quantities of food wastes. Extraction of bio-active compounds for cosmetic and pharmaceutical industries is a viable proposition. These residues from food processing industries are an excellent substrate for fermentation, offering immense possibilities and challenges.

31.5 Zero Solid Discharge

The approach is to treat the solid wastes such that the wastes are largely recovered. The solid wastes from food processing units are rich in nutrients and organic matters, making them eligible as raw materials for the production of biofertilizers, biofuels, and many other value-added products. These solid wastes can be pretreated enzymatically using commercial enzymes. Subsequent downstream processing of the pretreated biowastes substantially raises the recoveries and the residual solid quantities are reduced. The remaining solid wastes can then be subjected to processes like incineration to completely eliminate the solid wastes. Incineration permits recovery of thermal energy by combusting the organic solid wastes.

Check Your Understanding

1. What are the different types of wastes generated from food processing industries?
2. Define COD and BOD and state their importance in relation to waste management.
3. Define raw waste load and hydraulic load in reference to wastewater.
4. State the different physical, chemical, and biological methods of treatment of food processing wastes.

5. What are the different types of equipment used for waste management in food processing plants? State their operation principles.

6. What is zero solid discharge? Why it is important? How it can be achieved?

7. State some methods of value addition of food processing wastes with suitable examples.

BIBLIOGRAPHY

Arvanitoyannis, I. S. 2007. *Waste Management for the Food Industries*. Academic Press, New York.

Brennan, J. G., Butters, J. R., Cowell, N. D. and Lilly, A. E. V. 1976. *Food Engineering Operations* 2nd ed. Elsevier Applied Science, London.

Green, J. H. and Kramer, A. 1979. *Food Processing Waste Management*. AVI Publishing Co., Westport, Connecticut.

Jha, S. N. 2004. *Dairy and Food Processing Plant Maintenance: Theory and Practice*. International Book Distributing Co., Lucknow.

Katsuyama, A. M. (ed.). 1979. *A Guide for Waste Management in the Food Processing Industry*. The Food Processors Institute, Washington, DC.

Kelleher, B. P., Leahy, J. J., Henihan, A. M., O'Dwyer, T. F., Sutton, D. and Leahy, M. J. 2002. Advances in poultry disposal technology: A review. *Bioresource Technology* 83:27–36

Nand, K. 1999. Biomethanation of agro-industrial and food processing wastes. In V.K. Joshi and A. Pandey (eds.) *Biotechnology: Food Fermentation Vol.II*. Educational Publishers and Ditributors, New Delhi, 1349-72.

Ockerman, H. W. and Hansen, C. L. 2000. *Animal by-Product Processing and Utilization*. CRC Press, Boca Raton.

Rao, D. G. 2005. *Introduction to Biochemical Engineering*. Tata McGraw-Hill, New Delhi.

Rao, D. G. 2010. *Fundamentals of Food Engineering*. PHI Learning Pvt. Ltd., New Delhi.

Rao, D. G., Meyyappan, N., and Feroz, S. 2013. Treatment of effluent waters in food processing. In D.G. Rao, R. Senthil Kumar, J.A. Byrne and S. Feroz (eds.) *Wastewater Treatment: Advanced Processes and Technologies*. CRC Press, Boca Raton.

Zall, R. R. 2004. *Managing Food Industry Waste: Common Sense Methods for Food Processors*. 1st ed. Wiley-Blackwell, NJ, USA.

Annexures

Annexure – I

Conversion of Units

Quantity	Conversion factors
Length	1 m = 3.2808 ft
	1 ft = 0.3048 m
	1 km = 0.621 mile
	1 A° = 10^{-10} m
	1 micron = 1μm = 10^{-6} m
Mass	1 lbm = 0.453 kg
Area	1 m^2 = 10.769 ft^2
	1 acre = 0.4047 ha
Volume	1 ft^3 = 0.0283 m^3
	1 liter = 61.024 in^3
Density	1 $g.cc^{-1}$ = 62.43 $lbm.cft^{-1}$
	1 $lbm.ft^{-3}$ = 16 $kg.m^{-3}$
Temperature	X°C = (X–273.16) K
	°C = (°F–32)/1.8
	°R = 459.67 + °F
Force	1 N = 10^5 $g.cm.s^{-2}$(dyn)
	= 1 $kg.m.s^{-2}$ = 0.12 kgf = 0.2481 lbf
	1 kgf = 9.806 N = 2.205 lbf
	1 lbf = 0.453 kgf = 4.4482 N
	1 dyn = 10^{-5} N
Pressure	1 atm = 1.01325 × 10^5 Pa (N/m²)
	= 760 mm Hg
	= 1.03 $kgf.cm^{-2}$
	= 1.01325 bar
	1 torr = 1 mm Hg = 133.3224 Pa
	1 $kp.m^{-2}$ = 1 mm WS (water column) = 9.80665 Pa
	1 bar = 750.06 mm Hg
	= 401.85 mm water
	= 0.987 atm
	= 10^5 $N.m^{-2}$
	1 $N.m^{-2}$ = 1 pa
	= 10^{-5} bar
	= 10^{-2} $kg.m^{-1}.s^{-2}$
Work or energy	1 erg = 10^{-7} J
	= 2.39 × 10^{-8} cal
	1 Joule = 10^7 erg
	= 0.239 cal
	= 3.7251 × 10^{-7} hp.h = 2.778 × 10^{-7} kWh
	1 btu = 251.16 cal
	= 1.055 × 10^3 Joule
	= 2.9307 × 10^{-4} kWh
	1 kcal = 4.1868 kJ
	= 1.1622 × 10^{-3} kWh
Absolute viscosity	1 poise ($g.cm^{-1}.s^{-1}$) = 0.1 $kg.m^{-1}.s^{-1}$ (Pa.s)
	1 poise = 1 $dyn.s.cm^{-2}$
	1 $kg.m^{-1}.s^{-1}$ = 1 $N.s.m^{-2}$ =1 Pa.s
	1 cP = 0.001 Pa.s

Quantity	Conversion factors
Kinematic viscosity	1 Stoke = 10^{-4} $m^2.s^{-1}$
Power	1 hp = 746 W
	= 596.8 $kcal.h^{-1}$
	= 0.7068 $btu.s^{-1}$
	1 kW = 0.8 $kcal.h^{-1}$ = 14.34 $cal.min^{-1}$
	1 $btu.h^{-1}$ = 0.293 W
Energy flow rate	1 W = 0.86 $kcal.h^{-1}$
	1 $btu.h^{-1}$ = 0.252 $kcal.h^{-1}$
	1 ton of refrigeration = 200 $btu.min^{-1}$
	= 50 $kcal.min^{-1}$
	= 3.517 kW
Enthalpy	1 $btu.lb^{-1}$ = 0.5556 $kcal.kg^{-1}$
Specific heat	1 $kcal.kg^{-1}°C^{-1}$ = 4.1868 $kJ.kg^{-1}.K^{-1}$
	1 $btu.lbm^{-1}.°F^{-1}$ = 4.186 $J^{-1}.g^{-1}.°C^{-1}$
	= 1 $cal.g^{-1}K^{-1}$
Heat flow rate	1 $kcal.h^{-1}$ = 1.163 W (1 W = 1 $J.s^{-1}$)
Thermal conductivity	1 $kcal.m^{-1}.h^{-1}.°C^{-1}$ = 1.163 $W.m^{-1}.K^{-1}$
	1 $W.m^{-1}.K^{-1}$ = 0.8598 $kcal.m^{-1}.h^{-1}.°C^{-1}$
	1 $kcal.m^{-1}.h^{-1}.°C^{-1}$ = 0.672 $btu.ft^{-1}.h^{-1}.°F^{-1}$
	= 1.1623 $W.m^{-1}.K^{-1}$
Heat transfer coefficient	1 $kcal.m^{-2}.h^{-1}.°C^{-1}$ = 1.163 $W.m^{-2}.K^{-1}$

Annexures II

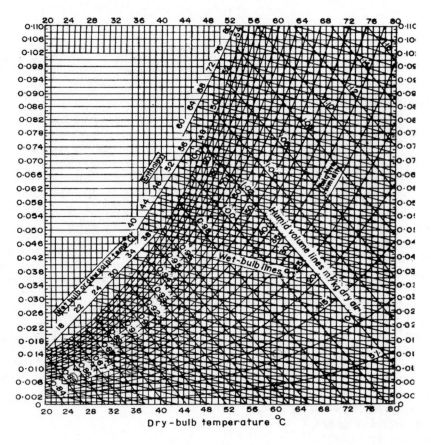

Dry-bulb temperature °C

Psychrometric chart (Adapted from: Sahay, K.M. and Singh, K.K. 1994. *Unit Operations in Agricultural Processing.* Vikas Publishing House, New Delhi.)

Annexure III

Viscosities of Some Common Food Materials

	Temperature, °C	Specific gravity at 16°C	Absolute viscosity, Pa.s
Apple juice, 20 °Brix	27	–	0.0021
Apple juice, 60 °Brix	27	–	0.03
Apple sauce	80	1.1	0.50
Butter fat	43	–	0.042
Chocolate	49	–	0.280
Coconut oil	24	0.93	0.055
Condensed milk, 75% solids	20	1.3	2.16
Corn oil	25	0.92	0.0565
Corn oil	57	0.92	0.028
Cottage cheese	18	–	30

	Temperature, °C	Specific gravity at 16°C	Absolute viscosity, Pa.s
Cottonseed oil	20	0.88	0.0704
Cream 10% fat	40	–	0.00148
Cream 30% fat	60	1.0	0.00289
Cream 30% fat	16	–	0.014
Cream 50% fat	16	0.98	0.112
Cream 50% fat	32	–	0.055
Custard	85–90	1.6	1.50
Fruit juice	18	–	0.055–0.075
Milk, homogenized	20		0.0020
Olive oil	38	0.91	0.04
Orange juice, 30 °B	20	–	0.63
Palm oil	38	0.92	0.043
Peanut oil	25	0.92	0.066
Safflower oil	25	0.925	0.052
Soybean oil	30	0.93	0.04
Tomato paste, 30% solids	18	–	0.195
Water	20.2	1	0.001

	Temperature, °C	Specific gravity at 16°C	Absolute viscosity, Pa.s
Whole egg	4.5	–	0.15
Yoghurt	40	1.15	0.152

(References: Charm, S.E. 1978. *Fundamentals of Food Engineering*, 3rd ed. AVI Publishing Co., Westport, Connecticut; Dash, S.K. and Sahoo, N.R. 2012. *Concepts of Food Process Engineering*. Kalyani Publishers, New Delhi; 15bookFellows, P.J. 2000. *Food Processing Technology*. Woodhead Publishing, Cambridge, UK; Toledo, R.T. 2007. *Fundamentals of Food Process Engineering* 3rd ed. Springer, USA; https://www.engineeringtoolbox.com; www.michael-smith-engineers.co.uk)

Annexure IV

Thermal Conductivities of Different Materials used in Food Processing Equipment, Storage Structures, and Other Applications

Material	Thermal conductivity, W.m⁻¹.K⁻¹	Material	Thermal conductivity, W.m⁻¹.K⁻¹
Air at 0°C	0.024	Glass, ordinary	0.8
Aluminum	205.0	Ice	1.6
Asbestos	0.08	Iron	79.5
Brick, red	0.6	Polystyrene (Styrofoam)	0.033
Brick, insulating	0.15	Polyurethane	0.02
Concrete	0.8	Steel	50.2
Cork board	0.04	Water at 20°C	0.6
Fiberglass	0.04	Wood	0.12–0.04

(Sources: https://www.engineeringtoolbox.com; www.michael-smith-engineers.co.uk)

Annexure V

Thermal Conductivities of Different Foods

Material	Thermal conductivity, W. m⁻¹. K⁻¹
Apple juice, 87% water (at 20°C)	0.559
Apple juice, 87% water (at 80°C)	0.631
Apple, dried, 41% moisture content (at 23°C)	0.219
Apple, freeze dried 21 Pa (at 35°C)	0.0185
Apple, whole fruit (at 8°C)	0.418
Beef ground, 53% moisture 22% fat (at 3°C)	0.364
Broccoli (at –6°C)	0.385
Butter, 0.56 g.cm⁻³ (at 4°C)	0.197
Butter milk, 0.35% fat (at 20°C)	0.569
Carrot, 0.6 g.cm⁻³ (at –16°C)	0.669
Chicken breast, 69–75% moisture, 0.6% fat (at 20°C)	0.412
Corn (yellow), 14.7% moisture (at 32°C)	0.16
Corn syrup, 1.16 g.cm⁻³ (at 25°C)	0.562
Egg albumin gel, freeze dried, 4.4 Pa (at 41°C)	0.0129
Egg yolk, 50% moisture, 33% fat (at 31°C)	0.420
Egg, white, 88% moisture (at 36°C)	0.558
Egg, whole 0.98 g.cm⁻³ (at –8°C)	0.960
Fish cod, 83% moisture, 0.1% fat (at 3°C)	0.534
Glucose solution, 70% water juice (at 20°C)	0.504
Grape juice, 89% moisture (at 20°C)	0.567
Grape juice, 89% moisture (at 80°C)	0.639
Honey, 80% moisture (at 70°C)	0.415
Meat sausage, 68% moisture (at 25°C)	0.427
Milk, evaporated, 72% water 5% fat (at 20°C)	0.504
Milk, whole, 90% water, 3% fat (at 28°C)	0.580
Oats 12.7% moisture (at 27°C)	0.130
Oil, mustard (at 25°C)	0.170
Oil, olive (at 32°C)	0.168
Oil, peanut (at 25°C)	0.169
Onion (at 8.6°C)	0.575
Peas, 0.7 g.cm⁻³ (at 7°C)	0.315
Potato starch gel, freeze dried 4.3 Pa, room temperature	0.0091
Potatoes, mashed 0.97 g.cm⁻³ (at –13°C)	1.09
Raisins, 32.2% moisture (at 23°C)	0.336
Sorghum, 13% moisture content (at 5°C)	0.131
Squash (at 8°C)	0.502
Strawberries (at –14°C)	1.10
Strawberry jam, 41% moisture (at 20°C)	0.338
Sucrose solution, 90% water (at 20°C)	0.566
Sucrose solution, 90% water (at 80°C)	0.636
Wheat, 10% moisture	0.155
Whey, 90% water (at 20°C)	0.567

(Source: www.cae.tntech.edu/~jbiernacki/CHE 4410 2016/Thermal Properties of Foods.pdf; ASHRAE Handbook 2006)

Annexure VI

Approximate Specific Heat of Selected Foods

Food	Moisture content	Specific heat above freezing, kJ. kg^{-1}.K^{-1}	Specific below freezing, kJ. kg^{-1}.K^{-1}	Freezing point
Apples, fresh	84	3.81	1.98	−1.1
Banana	74	3.56	2.03	−0.8
Broccoli	90.7	4.01	1.82	−0.6
Cabbage	92	4.02	1.85	−0.9
Candy, marshmallows	16.4	2.02	−	−
Candy, milk chocolate	1.30	1.83	−	−
Carrot	88	3.92	2.00	−1.4
Cauliflower	92	4.02	1.85	−0.9
Chicken	66	4.34	3.32	−2.8
Corn, sweet, yellow	76	3.62	1.98	−0.6
Egg, dried	3.1	2.04	2.00	−
Egg, white	87.8	3.91	1.81	−0.6
Egg, white dried	15	2.29	2.10	−
Egg, whole	75	3.63	1.95	−0.6
Fish, whole	70–82	3.2–3.78	2.14–2.27	−2.2
Ice-cream	60	3.19	2.74	−5.6
Juice, apple	88	3.87	1.78	−
Juice, grape	87.4	3.85	1.78	−
Juice, lemon	92	3.99	1.73	−
Juice, orange	89	3.9	1.76	−
Juice, tomato	94	4.03	1.71	−
Milk, butter	18	2.40	2.65	−
Milk, cream	74	3.59	2.21	−2.2
Milk, dried	2.5	1.85	−	−
Milk, evaporated	74	3.56	2.08	−1.4
Milk, whole	88	3.89	1.81	−0.6
Mushroom	92	3.99	1.84	−0.9
Peas, green	79	3.75	1.98	−0.6
Pineapples	86.5	3.85	1.91	−1.0
Sausage	40–54%	2.82–3.15	2.3–2.4	−1.7
Strawberries	92	4.00	1.84	−0.8

(Source: www.cae.tntech.edu/~jbiernacki/CHE 4410 2016/Thermal
Properties of Foods.pdf; ASHRAE Handbook 2006)

Annexure VII

Approximate Specific Heat of Materials used in Food Engineering Calculations

Material	Specific heat, c$_p$, kJ.kg^{-1}.K^{-1}
Air, 0°C	1.0035
Air, room conditions	1.012
Aluminum	0.897
Ammonia	4.70
Asbestos cement board	0.84
Brick	0.840
Concrete	0.880
Copper	0.385
Fat	1.93
Fiber hard board	2.1
Fiberboard, light	2.5
Glass	0.84
Ice (-10°C)	2.05
Sand	0.835
Silica aerogel	0.84
Soil, dry	0.80
Soil, wet	1.48
Steel	0.466
Tin	0.227
Water	4.1813
Wood	1.7

(Source: https://www.engineeringtoolbox.com/
specific-heat-solids-d_154.html;
wikipaedia.org)

Annexure VIII

Approximate Thermal Diffusivity Values of Selected Foods

Food	Moisture content	Thermal diffusivity, mm^2.s^{-1}	Reference temperature
Apple, whole	85	0.14	0–30
Apple, dried	42	0.096	23
Applesauce	80	0.14	65
	80	0.12	5
Bananas, flesh	76	0.12	5
Jam, strawberry	41	0.12	20
Jelly, grape	42	0.12	20
Potato, whole		0.13	0–70
Potato, mashed and cooked	78	0.12	5
Raisins	32	0.11	23

Food	Moisture content	Thermal diffusivity, mm².s⁻¹	Reference temperature
Strawberries, flesh	92	0.13	5
Codfish	81	0.12	5
Beef stick	37	0.11	20
Cake, chocolate	32	0.12	23

(References: www.cae.tntech.edu/~jbiernacki/CHE 4410 2016/ Thermal Properties of Foods.pdf; ASHRAE Handbook 2006)

Annexure IX

M and n Values for Use in Equation 7.46 for Natural Convection

Geometry	$Gr_f \times Pr_f$	M	n
Vertical plate/cylinders	$10^4 - 10^9$	0.59	0.25
	$10^9 - 10^{13}$	0.10	0.33
Horizontal cylinders	$10^4 - 10^9$	0.53	0.25
	$10^{9} - 10^{12}$	0.13	0.33
Horizontal plates			
Upper surface of heated plates or lower surface of cooled plates	$2 \times 10^4 - 8 \times 10^6$	0.54	0.25
Upper surface of heated plates or lower surface of cooled plates	$8 \times 10^6 - 10^{11}$	0.15	0.33
Lower surface of heated plates or upper surface of cooled plates	$10^5 - 10^{11}$	0.27	0.25
Vertical cylinder, height=diameter	$10^4 - 10^6$	0.775	0.21

(Reference: Holman, J. P. 1997. *Heat transfer* 8th ed. Mc-Graw Hill, Inc., NY)

Annexure X

Simple Equations for Natural Convection Heat Transfer Coefficients

Vertical plates
  Laminar range $h = 1.42(\Delta t/L)^{0.25}$
  Turbulent range $h = 1.31(\Delta t)^{0.33}$
Horizontal plates facing upward when cooled or downward when heated
Always laminar $h = 0.59(\Delta t/L)^{0.25}$
Horizontal hot plate facing upward or cold plate facing downward
  Laminar range $h = 1.32(\Delta t/L)^{0.25}$
  Turbulent range $h = 1.52(\Delta t)^{0.33}$
Horizontal cylinders
  Laminar range $h = 1.32(\Delta t/d)^{0.25}$
  Turbulent range $h = 1.24(\Delta t)^{0.33}$
Vertical cylinders can be considered as vertical plates.

Δt is the difference between solid temperature and surrounding temperature in K,
  L is the characteristic dimension, vertical or horizontal, in m,
  h is in W.m⁻². K⁻¹ and d is diameter in m

For laminar flow: $10^4 < GrPr < 10^9$ and for turbulent flow: $GrPr > 10^9$
(References: Albright, L.D. 1990. *Environment Control for Animals and Plants*. ASAE, St. Joseph, Michigan; Holman, J. P. 1997. *Heat transfer* 8th Edn. Mc-Graw Hill, Inc., NY; Rao, D.G. 2009. *Fundamentals of Food Engineering*. PHI Learning Pvt. Ltd., New Delhi; ASHRAE Handbook of fundamentals)

Annexure XI

D and z Values of Some Spoilage Microorganisms and Nutrients in Different Types of Foods

Component	Source	z (°C)	D_{121} (min)
Cl. botulinum 213-B	Peas (canned)	8	0.22
Cl. botulinum 62A	Corn (canned)	10	0.30
Cl. botulinum spores	Various sources	5.5–10	0.1–0.3
Bacillus stearothermophilus	Peas	11	6.16
	Spinach	12	4.94
	Various sources (pH more than 4.5) and at temperatures above 110°C	7–12	4.0–5.0
Thiamin	Carrot puree (pH 5.9) at 109°–149°C	25	158
Lysine	Soyabean meal at 100°–127°C	21	786
Chlorophyll a	Spinach at 100°–130°C	45	34.1
Chlorophyll b	Spinach at 100°–130°C	59	48
Anthocyanin	Grape juice (natural) at 20°–121°C	23.2	17.8
Peroxidase	Peas at 110°–138°C	37.2	3.0

(References: Fellows, P. J. 2000. *Food Processing Technology*. Woodhead Publishing, Cambridge, UK; Toledo, R.T. 2007. *Fundamentals of Food Process Engineering*, 3rd ed. Springer, USA; Reed, J.M. Bohrer, C.W. and Cameron, E.J. 1951. Spore destruction rate studies on organisms of significance in the processing of canned foods. Food Research 16: 338-408

Annexure XII

Test Sieves and the Size of Openings

US Sieve Size	Tyler Equivalent	Indian Standard sieves	Opening	
			mm	in
No. 2½	2½ Mesh		8.00	0.312
No. 3	3 Mesh		6.73	0.265
No. 3½	3½ Mesh		5.66	0.233
No. 4	4 Mesh	480	4.76	0.187
No. 5	5 Mesh		4.00	0.157
No. 6	6 Mesh	340	3.36	0.132
No. 7	7 Mesh	280	2.83	0.111
No. 8	8 Mesh	240	2.38	0.0937
No.10	9 Mesh	200	2.00	0.0787
No. 12	10 Mesh	170	1.68	0.0661
No. 14	12 Mesh	140	1.41	0.0555
No. 16	14 Mesh	120	1.19	0.0469
No. 18	16 Mesh	100	1.00	0.0394
No. 20	20 Mesh	85	0.841	0.0331
No. 25	24 Mesh	70	0.707	0.0278
No. 30	28 Mesh	60	0.595	0.0234
No. 35	32 Mesh	50	0.500	0.0197
No. 40	35 Mesh	40	0.420	0.0165
No. 45	42 Mesh	35	0.354	0.0139
No. 50	48 Mesh	30	0.297	0.0117
No. 60	60 Mesh	25	0.250	0.0098
No. 70	65 Mesh	20	0.210	0.0083
No. 80	80 Mesh	18	0.177	0.0070
No.100	100 Mesh	15	0.149	0.0059
No. 120	115 Mesh	12	0.125	0.0049
No. 140	150 Mesh	10	0.105	0.0041
No. 170	170 Mesh	9	0.088	0.0035
No. 200	200 Mesh	8	0.074	0.0029
No. 230	250 Mesh	7	0.063	0.0025
No. 270	270 Mesh	6	0.053	0.0021
No. 325	325 Mesh	5	0.044	0.0017
No. 400	400 Mesh	4	0.037	0.0015

(Reference: Sahay K.M. and Singh KK. 1994. *Unit Operation of Agricultural Processing*. Vikas Publishing House, New Delhi)

Annexure XIII

F_1 Values for Selected z Values at Retort Temperatures Below 121°C for Analysis of Thermal Processing

121-T_f(°C)	z value					
	4.4°C	**6.7°C**	**8.9°C**	**10°C**	**11.1°C**	**12°C**
5.6	17.78	6.813	4.217	3.594	3.162	2.848
6.1	23.71	8.254	4.870	4.084	3.548	3.162
6.7	31.62	10.00	5.623	4.642	3.981	3.511
7.2	42.17	12.12	6.494	5.275	4.467	3.899
7.8	56.23	14.68	7.499	5.995	5.012	4.329
8.9	100	21.54	10.00	7.743	6.31	5.337
9.4	133.4	26.10	11.55	8.799	7.079	5.926
10.0	177.8	31.62	13.34	10.00	7.943	6.579
10.6	237.1	38.31	15.40	11.36	8.913	7.305

Selected f_h/U and g Values When z = 10 and j_c = 0.4–2.0 for Analysis of Thermal Processing

f_h/U	Values of g for the following j_c values					
	0.40	**0.80**	**1.00**	**1.40**	**1.80**	**2.00**
0.50	0.0411	0.0474	0.0506	0.0570	0.0602	0.0665
0.60	0.0870	0.102	0.109	0.123	0.138	0.145
0.70	0.15	0.176	0.189	0.215	0.241	0.255
0.80	0.26	0.267	0.287	0.328	0.369	0.390
0.90	0.313	0.371	0.400	0.458	0.516	0.545
1.00	0.408	0.485	0.523	0.600	0.676	0.715
2.00	1.53	1.80	1.93	2.21	2.48	2.61
3.00	2.63	3.05	3.26	3.68	4.10	4.31
4.00	3.61	4.14	4.41	4.94	5.48	5.75
5.0	4.44	5.08	5.40	6.03	6.67	6.99
10.0	7.17	8.24	8.78	9.86	10.93	11.47
20.0	9.83	11.55	12.40	14.11	14.97	16.68
30.0	11.5	13.6	14.6	16.8	18.9	19.9
40.0	12.8	15.1	16.3	18.7	21.1	22.3
50.0	13.8	16.4	17.7	20.3	22.8	24.1
100.0	17.6	20.8	22.3	25.4	28.5	30.1
500.0	26.0	30.6	32.9	37.5	42.1	44.4

Adapted from: Stombo, C. R. 1973. *Thermobacteriology in Food Processing 2nd ed.* Academic Press, New York

Annexure XIV

Relative Humidity Obtained at Different Temperatures with Different Salt Solutions

Salt solution	Temperature, °C			
	30	**40**	**50**	**60**
Lithium chloride	0.113	0.112	0.111	0.110
Potassium acetate	0.216	0.204	0.192	0.160
Magnesium chloride	0.324	0.318	0.312	0.306
Potassium carbonate	0.432	0.432	0.433	0.432
Sodium nitrite	0.635	0.616	0.597	0.578
Sodium chloride	0.750	0.748	0.746	0.745
Potassium chloride	0.950	0.950	0.950	0.950

References: Sahay, K.M. and Singh, KK. 1994. *Unit Operation of Agricultural Processing*. Vikas Publishing House, New Delhi.; http://ecoursesonline.iasri.res.in/mod/ page/view.php? id=863; Chakravorty, A. 1981. *Post-Harvest Technology of Cereals, Pulses and Oilseeds*. Oxford and IBH, New Delhi.

Annexure XV

Some Equations for Predicting the Equilibrium Moisture Content

Name	Equation
Bradley (1936)	$\ln(a_w) = A[B]^M$
BET (1938) (BET stands for Brunauer, Emmet and Teller)	$M = \dfrac{M_0 A.a_w}{1 - a_w} + \dfrac{M_0 A}{(A-1)(1-a_w)}$
Smith (1947)	$M = A - \ln(1 - a_w)^B$
Henderson (1952)	$M = \dfrac{1}{AT}\left[\ln(1-a_w)\right]^{1/B}$
GAB (1966) (GAB stands for Guggenheim, Anderson and de Boer)	$M = \dfrac{M_0 A.B.a_w}{(1 - A.a_w)(1 - A.a_w + ABa_w)}$
Henderson-Thomson	$M_e = \left[\dfrac{\ln(1 - RH)}{-a(T_s + c)}\right]^{1/b}$
Chung-Pfost (1967)	$M_e = -\dfrac{1}{b}\ln\left[\dfrac{(T_s + c)\ln(RH)}{-a}\right]$

In the above equations, M is the equilibrium moisture content, decimal dry basis, M_0 = monolayer moisture content, decimal dry basis, a_w is the water activity in decimals, A and B are constants specific to the equations, and T is the absolute temperature in Kelvin.

References: Sahay, K.M. and Singh, K.K. 1994. *Unit Operation of Agricultural Processing*. Vikas Publishing House, New Delhi; http://ecoursesonline.iasri.res.in/mod/ page/view.php ?id=863

Annexure - XVI

Recommended Cold Storage Conditions for Some Common Fruits and Vegetables

Food	Temperature, °C	Relative humidity (%)	Storage life (days)*
Storage temperature at around 0°C			
Cherry	−1–0°C	85–90	2–3 weeks
Pear	−1–0°C	85–90	2–6 months
Table grape	−1–0°C	80–85	3–6 months
Strawberry	0–2°C	80–85	14–21 days
Garlic	−1 to 0°C	65	6–9 months
Broccoli	0°C	95	2–3 weeks
Cabbage	0°C	92–95	2–6 months
Carrot	0°C	95–98	4–7 weeks
Cauliflower	0–2°C	85–95	3–4 weeks
Lettuce	0–1°C	95–100	3–4 weeks
Peas	0°C	88–92	1–3 weeks
Spinach	0°C	95	1–2 weeks
Sweet corn	0–1.5°C	90–95	Less than 1 week
Storage temperature at 2° to 6°C			
Apple	−1–4°C	85–90	1–12 months
Orange	3–8°C	85–90	Up to 3 months
Pomegranate	5–7°C	85–90	2 months
Potato seed tubers	4–5°C	90–95	5–8 months
Potato for fresh consumption	7–10°C	90–95	5–8 months
Potato for frying and processing	10–15°C	90–95	1–2 months
Beans	5–7.5°C	90–95	7–10 days
Capsicum	7.5°C	90–95	3–5 weeks
Chilli	7.5°C	90–95	3–5 weeks
Cucumber	10–12.5°C	90–95	10–14 days
Brinjal	10–12°C	90–95	1–15 days
Pumpkin	10–13°C		2–3 months
Storage temperature at around 12°–13°C			
Watermelon	10–15°C	80–90	2–3 weeks
Mango	10°C	85–90	4–7 weeks
Banana	13–14°C	80–85	2–3 weeks
Lemon	14–15°C	85–90	4–6 months
Ginger	12–14°C	75	4–6 months
Tomato Mature Green	12.5–15°C	85–90	2–3 weeks
Tomato Firm-ripe	7–10°C	85–90	2 weeks

* The actual storage conditions and storage life may vary greatly due to variations in variety, maturity and other factors.

References: UC Davis Postharvest Center 2017, 'Produce fact sheet', University of California; Fellows, P. J. 2000. *Food Processing Technology*. Woodhead Publishing, Cambridge, UK; Kalia, M. and Sood, S. 2012. *Food Preservation and Processing*, Kalyani Publishers, New Delhi; Lal, G., Siddappa, G. S., Tondon, G. L. 2009. *Preservation of Fruits and Vegetables*, ICAR, New Delhi; https://www.publications.qld.gov.au; https://www.crscoldstorage.co.uk/news/cold-storage-fruit-and-veg.html

Annexure XVII

Typical Heat Transfer Coefficients for Different Cooling Systems

Freezing system and temperature of freezing medium	Heat transfer coefficient, $W.m^{-2}.K^{-1}$
Still air freezer (air at 0.1–0.5 m.s⁻¹ and –35 to –37°C)	5 to 10
Still air freezer (air at 1–2 m.s⁻¹ and –35 to –37°C)	20
Air blast (air at 6 m.s⁻¹ and –20 to –40°C)	10-200
Spiral belt (air at –40°C)	25–50
Fluidized bed (air at –40°C)	90–140
Plate (plate temperature –40°C)	100–500
Immersion freezer (still liquid at –50 to –70°C)	100
Immersion freezer (mildly agitated liquid at –50 to –70°C)	500
Cryogenic freezer (cryogen temperature –50 to –196°C)	1250–1500

References: https://eng.libretexts.org; Fellows, P. J. 2000. *Food Processing Technology*. Woodhead Publishing, Cambridge, UK

Overall Heat Transfer Coefficients for Fluids (practically still)

Fluids	Material of heat transfer	Overall heat transfer coefficient, $W.m^{-2}.K^{-1}$
Water–air (gas)	Mild steel	11.3
	Copper	13.1
Water–water	Mild steel	340–400
	Copper	340–455
Steam–air	Mild steel	14.2
	Copper	17
Steam–water	Mild steel	1050
Steam–water	Copper	1160
	Stainless steel	680

Reference: ww.engineeringtoolbox.com/overall-heat-transfer-coefficients-d_284.html

Annexure - XVIII

Comparison of Different Rays

Name	Wavelength	Frequency (Hz)	Photon energy (eV)
Gamma ray	less than 0.02 nm	more than 15 EHz	more than 62.1 keV
X-ray	0.01 nm–10 nm	30 EHz–30 PHz	124 keV–124 eV
Ultraviolet	10 nm–400 nm	30 PHz–750 THz	124 eV–3 eV
Visible	390 nm–750 nm	750 THz–430 THz	3.2 eV–1.7 eV
Infrared	750 nm–1 mm	430 THz–300 GHz	1.7 eV–1.24 meV
Microwave	1 mm–1 meter	300 GHz–300 MHz	1.24 meV–1.24 μeV
Radio	1 mm–100,000 km	300 GHz–3 Hz	1.24 meV–12.4 feV

Annexure XIX

Common Active MAP Agents and Food Applications

Active packaging system	Mechanisms	Food applications
Oxygen scavengers	1. Iron based 2. Metal/acid 3. Metal (e.g., Platinum) catalyst 4. Ascorbate/ metallic salts 5. Enzyme based	Bread, cakes, cooked rice, biscuits, pizza, pasta, cheese, cured meats and fish, coffee, snack foods, dried foods, and beverages
Carbon dioxide scavengers/ emitters	1. Iron oxide/calcium hydroxide 2. Ferrous carbonate/metal halide 3. Calcium oxide/activated charcoal 4. Ascorbate/sodium bicarbonate 5. Dry ice 6. Pressurized gas cylinder	Coffee, fresh meats and fish, nuts and other snack food products, and sponge cakes
Ethylene scavengers	1. Potassium permanganate 2. Activated carbon 3. Activated clays/zeolites 4. Catalytic oxidation	Fruit, vegetables, and other horticultural products
Preservative releasers	1. Organic acids 2. Silver zeolite 3. Spice and herb extracts 4. Bha/bht antioxidants 5. Vitamin e antioxidant 6. Volatile chlorine dioxide/ sulphur dioxide	Cereals, meats, fish, bread, cheese, snack foods, fruit, and vegetables
Ethanol emitters	1. Alcohol spray 2. Encapsulated ethanol	Pizza crusts, cakes, bread, biscuits, fish, and bakery products
Moisture absorbers	1. PVA blanket 2. Activated clays and minerals 3. Silica gel	Fish, meats, poultry, snack foods, cereals, dried foods, sandwiches, fruit, and vegetables
Flavor/odor adsorbers	1. Cellulose triacetate 2. Acetylated paper 3. Citric acid 4. Ferrous salt/ascorbate 5. Activated carbon/clays/zeolites	Fruit juices, fried snack foods, fish, cereals, poultry, dairy products, and fruit
Temperature control packaging	1. Non-woven plastics 2. Double walled containers 3. Hydrofluorocarbon gas 4. Lime/water 5. Ammonium nitrate/water	Ready meals, meats, fish, poultry, and beverages

Index

For Product Safety Concerns and Information please contact our
EU representative GPSR@taylorandfrancis.com Taylor & Francis
Verlag GmbH, Kaufingerstraße 24, 80331 München, Germany